T0135292

Emergence, Complexity and Computation

Volume 24

Series editors

Ivan Zelinka, Technical University of Ostrava, Ostrava, Czech Republic
e-mail: ivan.zelinka@vsb.cz

Andrew Adamatzky, University of the West of England, Bristol, UK
e-mail: adamatzky@gmail.com

Guanrong Chen, City University of Hong Kong, Hong Kong, China
e-mail: eegchen@cityu.edu.hk

About this Series

The Emergence, Complexity and Computation (ECC) series publishes new developments, advancements and selected topics in the fields of complexity, computation and emergence. The series focuses on all aspects of reality-based computation approaches from an interdisciplinary point of view especially from applied sciences, biology, physics, or chemistry. It presents new ideas and inter-disciplinary insight on the mutual intersection of subareas of computation, complexity and emergence and its impact and limits to any computing based on physical limits (thermodynamic and quantum limits, Bremermann's limit, Seth Lloyd limits...) as well as algorithmic limits (Gödel's proof and its impact on calculation, algorithmic complexity, the Chaitin's Omega number and Kolmogorov complexity, non-traditional calculations like Turing machine process and its consequences,...) and limitations arising in artificial intelligence field. The topics are (but not limited to) membrane computing, DNA computing, immune computing, quantum computing, swarm computing, analogic computing, chaos computing and computing on the edge of chaos, computational aspects of dynamics of complex systems (systems with self-organization, multiagent systems, cellular automata, artificial life,...), emergence of complex systems and its computational aspects, and agent based computation. The main aim of this series it to discuss the above mentioned topics from an interdisciplinary point of view and present new ideas coming from mutual intersection of classical as well as modern methods of computation. Within the scope of the series are monographs, lecture notes, selected contributions from specialized conferences and workshops, special contribution from international experts.

More information about this series at http://www.springer.com/series/10624

Andrew Adamatzky

Editor

Emergent Computation

A Festschrift for Selim G. Akl

 Springer

Editor
Andrew Adamatzky
Unconventional Computing Centre
University of the West of England
Bristol
UK

ISSN 2194-7287 ISSN 2194-7295 (electronic)
Emergence, Complexity and Computation
ISBN 978-3-319-83505-1 ISBN 978-3-319-46376-6 (eBook)
DOI 10.1007/978-3-319-46376-6

Printed on acid-free paper

This Springer imprint is published by Springer Nature
The registered company is Springer International Publishing AG
The registered company address is: Gewerbestrasse 11, 6330 Cham, Switzerland

Foreword

Selim G. Akl

This book is dedicated to Prof. Selim G. Akl to honor his major research achievements in computer science over four decades. In this way his colleagues, students, and friends wish to express their gratitude to this great scientist.

Dr. Akl completed his Ph.D. at McGill University in 1978. He has been a faculty member in the Queen's School of Computing (formerly Department of Computing and Information Science) since 1978. Dr. Akl serves as Director of the Queen's School of Computing since July 2007. He has held visiting positions at the University of California Berkeley, Simon Fraser University, the University of Puerto Rico, Clarkson University, and Kent State University. He was an SRI International Fellow at the Stanford Research Institute, in Menlo Park, California, and an NSERC Senior Industrial Research Fellow at MacDonald Dettwiler and Associates, in Richmond British Columbia. In 1990, he held the Louis Néel Chair at the École Normale Supérieure de Lyon, France.

Throughout his career, Dr. Akl has made significant contributions in multiple areas of computer science. He has published 170 journal articles and presented 180 conference papers. Dr. Akl has written four influential monographs on parallel algorithms and parallel computation that are widely considered as landmarks in parallel computation research and have been translated to many languages. As will be discussed below, Dr. Akl's research has made a strong impact on the development of each area of computer science he has worked on. Dr. Akl has collaborated with a large number of scientists and has more than 100 co-authors in publications in a wide range of areas.

Dr. Akl possesses the enviable skill of making complex ideas and visionary concepts accessible and engaging, whether they appear in written form or in a presentation. He is a brilliant scientific communicator. Dr. Akl has given plenary lectures at many international conferences and invited talks at universities world-wide. He has supervised 24 Ph.D. theses and numerous M.Sc. theses. Dr. Akl has always been very dedicated and supportive to his students and many of his former students have leading positions in academia and industry.

Dr. Akl has made substantial contributions to the computer science community through his selfless service. He is the Editor in Chief of Parallel Processing Letters and has held editorial positions in 10 professional journals. He has served as a program committee or organizing committee member of 70 computer science conferences. In 2007, Dr. Akl served as General Chair of the Sixth International Conference on Unconventional Computation held in Kingston, Ontario. Dr. Akl has received a number of academic honors for research and teaching, including the Queen's University Prize for Excellence in Research in 2008 and the Queen's University Award for Excellence in Graduate Supervision in 2012.

Dr. Akl has made important contributions to many areas of theoretical computer science. The following sections briefly summarize some of these achievements.

Parallel Computation and Parallel Algorithms

The goal of parallel computation is to reduce the time needed to solve a computational problem by using several processors that are working simultaneously. Designing parallel algorithms, i.e., methods for solving problems efficiently on parallel computers is a nontrivial task, requiring creativity and a way of thinking completely different from the one used for sequential algorithms.

Dr. Akl was one of the pioneering researchers in this field and, since the early 1980s, he has contributed many efficient parallel algorithms to the literature. These include algorithms for problems such as selection, sorting, computing convex hulls, or enumerating combinatorial objects. The algorithms were designed to run on a wide spectrum of computational models, such as shared memory machines, combinational circuits, and interconnection networks. Dr. Akl was among the first to demonstrate the links between parallel computation and optical computing, and the importance of parallel algorithms in real-time applications.

In work dealing with the foundations of parallel computing, Dr. Akl was the first to demonstrate superlinear speedup in the number of processors, i.e., he established that there exist computational problems that can be solved with n processors *more than n times faster* than on a sequential computer. Dr. Akl uncovered three general computational paradigms within which parallel computation leads to a superlinear improvement in performance. The results were groundbreaking and could be viewed even as counter-intuitive.

In 2004, Dr. Akl demonstrated the existence of classes of *inherently parallel problems,* that is, classes of problems that can be solved on a parallel computer with an appropriate number of processors, but not on a sequential computer or even a parallel computer with fewer processors. Furthermore, Dr. Akl has established the impossibility of constructing a *universal computer* that could simulate any other computation. This is a significant foundational result with philosophical implications and, at least initially, it went against much of the commonly held beliefs in computer science. By questioning common wisdom in parallel computation—computations with uncertain time restrictions, computations under the influence of the laws of nature, and computations subject to mathematical constraints—Dr. Akl has achieved surprising results, with implications that continue to be timely and important.

Dr. Akl has written four influential monographs that are considered landmarks of parallel computation research, education, and practice. Each of these books was in some sense a 'first'. Before *Parallel Sorting Algorithms* (Academic Press, 1985) there existed no book that covered parallel algorithms exclusively. This monograph became a model for many other books that were published on parallel algorithms in the years since. Dr. Akl's second book *The Design and Analysis of Parallel Algorithms* (Prentice Hall, 1989) was broader in scope and offered a unified and rigorous treatment of the different techniques for designing and analyzing algorithms for parallel computers. The book is considered a classic and has been used extensively worldwide. To this day, the only book offering an in-depth treatment of parallel algorithms for computational geometry problems is Dr. Akl's *Parallel Computational Geometry* (Prentice Hall, 1993). Finally, *Parallel Computation: Models and Methods* (Prentice Hall, 1997) is an encyclopedic volume that presents the major approaches to designing efficient parallel algorithms as well as the different parallel computer models on which the algorithms would be executed. The book is listed on Amazon among the 21 best books on *algorithmics,* not just parallel algorithms. The creator of the list, Prof. Christoph Kögl from the University of Kaiserslautern, described the book as "Possibly the best book on parallel algorithms".

Unconventional Computation and Natural Computation

Unconventional computation investigates the possibility of building computers unlike any that are currently in use. The models investigated include biomolecular, chemical, optical and quantum computers, and analog neural networks or cellular automata, among many others. Often parallelism captures the essence of unconventional computing and, consequently, Dr. Akl's transition into unconventional computation can be viewed as a logical continuation to his extensive work in parallel computation.

Dr. Akl's work in unconventional computation has various motivations and goals. One goal is to understand the natural processes by modeling them as algorithms, e.g., plant respiration as a cellular automaton algorithm or photosynthesis as a quantum process. Dr. Akl used two-dimensional cellular automata to provide efficient solutions for computational problems in plant respiration that had remained open for several years. Another goal is to seek inspiration from nature for computational models such as genetic algorithms, neural networks, or swarm intelligence. A further goal is to use natural laws to perform more efficient computations, for example, in biomolecular computing, chemical computing, or quantum computing. Significantly Dr. Akl's work has uncovered computational problems that can be performed on a quantum computer but *not even in principle* on any classical computer. Furthermore, in considering physical systems as computational models the work tackles philosophical questions about the definition of computation, that is, what it means 'to compute'.

Work in Other Research Areas

In addition to the areas of Parallel Computation and Unconventional Computation, Dr. Akl has made important contributions in at least four other areas which we briefly mention here. Selim Akl is a pioneer in the field of *Computational Geometry* working in the nascent field while still a Ph.D. student at McGill. Dr. Akl co-authored several seminal papers among them a very effective and often-cited heuristic to compute the convex hull of a planar point set. Dr. Akl continued his exploration of Computational Geometry and published several parallel algorithms to solve geometric problems. In the area of *Design and Analysis of Combinatorial Algorithms* he was the co-editor of the book Algorithms and Data Structures (Springer 1995) and served on the Editorial Board of the journal *Information Processing Letters* (North Holland). Dr. Akl was a pioneer of modern *Cryptography and Data Security,* presenting a paper on digital signatures as early as 1981. Dr. Akl's elegant and ingenious solution to the problem of controlling access to information in a hierarchical organization opened up an entire subfield of research. Over the years many researchers have attempted, without success, to improve on his solution which remains the state of the art for access control in a hierarchy. Generalizations to this solution are described in Dr. Akl's book *Adaptive Cryptographic Access Control* (Springer, 2010) and applied to a number of computer security issues, most notably the protection of information in data warehouses.

Since 2009, Dr. Akl has collaborated with a cardiologist in the area of *Biomedical Computing and Computer Assisted Medicine.* The work uses unconventional algorithmic techniques to analyze electrocardiograms, towards a better diagnosis and treatment of cardiac arrhythmias. While the clinical work is still in early stages, the research has produced a substantial number of publications and two completed Ph.D. degrees supervised by Dr. Akl.

Personal Recollections

The undersigned have had the good fortune to be Selim's long time colleagues at the Queen's School of Computing and we conclude with a few more personal remarks.

Selim is truly a Renaissance man. Along with his eclectic research interests Selim is also actively involved in the creative arts. He has a background in theater and was a prominent fixture of the Kingston theater scene. Selim founded the French speaking theater company Les Tréteaux de Kingston, shortly after he arrived to Kingston, and the company has been continuously active to this day. Currently Selim's creative outlet is photography. The expert care and attention to detail found in his research articles can be seen in his beautiful photographs. Since 2005 his pictures grace the cover of the monthly magazine Vista.

As a colleague Selim has always been supportive and encouraging. His unbounded energy and creativity is stimulating to colleagues and students alike. Selim leads a weekly seminar group that has continued without interruption for decades. The group includes professors, researchers, clinicians, and students, all from diverse backgrounds that, in addition to computing, include biology, philosophy, and surgery. It is Selim's affable nature that is the force of attraction that brings this diverse group of individuals together.

Selim has natural leadership qualities that impressively enhance his productivity. Perhaps his most significant human characteristic is his generosity and kindness. He works tirelessly to assist his colleagues and students, always with encouragement, and never with derision. There are many researchers today, those who were supervised by Selim, or worked with him as colleagues, whose careers are marked indelibly by his strong influence. We are all richer for knowing and working with him.

We, together with all contributors of this volume, wish Selim continued success in the years to come.

Kingston, Canada   David Rappaport
July 2016   Kai Salomaa

Contents

Simple Deterministic Algorithms
for Generating "Good" Musical Rhythms

Godfried T. Toussaint

Abstract The most economical representation of a musical rhythm is as a binary sequence of symbols that represent sounds and silences, each of which have a duration of one unit of time. Such a representation is eminently suited to objective mathematical and computational analyses, while at the same time, and perhaps surprisingly, provides a rich enough structure to inform both music theory and music practice. A musical rhythm is considered to be "good" if it belongs to the repertoire of the musical tradition of some culture in the world, is used frequently as an *ostinato* or *timeline*, and has withstood the test of time. Here several simple deterministic algorithms for generating musical rhythms are reviewed and compared in terms of their computational complexity, applicability, and capability to capture "goodness."

1   Introduction

The Oxford dictionary defines aesthetics as a set of principles concerned with the nature and appreciation of beauty, especially in art. Traditionally it is also a branch of philosophy concerned with the nature, expression, and perception of beauty and artistic taste [1]. The design of algorithms that generate "good" musical rhythms thus falls in the domain of *computational aesthetics* which is concerned with questions such as: How can the computer generate aesthetic objects without human intervention? [2]. A related question also asked is: How can the arts influence computer generated aesthetic objects? This question is sometimes attributed to another emergent field called *aesthetic computing* [3]. These two fields are, not surprisingly, inextricably intertwined. Not only do artistic principles provide artificial intelligence researchers with new ideas, but the results of computer programs influence artistic practices. Closely related to these two emerging computational fields is the area concerned with constructing mathematical measures of aesthetics, which goes back most notably to at least the work of Birkhoff [4–6] and should not be confused with the field that studies the aesthetics of mathematics [7]. The former attempts to develop mathematical

G.T. Toussaint (✉)
New York University Abu Dhabi, Abu Dhabi, United Arab Emirates
e-mail: gt42@nyu.edu

© Springer International Publishing Switzerland 2017
A. Adamatzky (ed.), *Emergent Computation*, Emergence, Complexity
and Computation 24, DOI 10.1007/978-3-319-46376-6_1

1

measures of aesthetics that predict human aesthetic judgments, whereas the latter is concerned with studying the role of aesthetics in the mathematical research carried out by mathematicians. Not surprisingly these two fields are also inextricably intertwined. Indeed, the philosopher Oswald Spengler [8] wrote: "The mathematics of beauty and the beauty of mathematics are ... inseparable."

Most research in these fields has been limited to the visual arts, such as painting [9], and less attention has been paid to music in general [10, 11] and musical rhythm in particular [12–14]. Mathematical measures of aesthetics explore a variety of features such as symmetry [15], the Golden section [16–18], and complexity [19] to determine how well they correlate with human judgments [20].

It is useful to distinguish between *algorithmic* generation of music, and generation of music using electronic digital computers. The word 'algorithmic' specifies the use of well defined rules, without necessarily implying that these rules must be implemented on an electronic digital computer. Indeed the algorithmic approach to music composition may use any other method such as rolling dice with human hands to generate rhythms and melodies, as was popular in 18th century Europe [21], when more than twenty algorithmic processes were devised, inspired by the new developments in mathematics and probability theory that were receiving public attention at the time. In this sense algorithmic composition predates the advent of the electronic digital computer by centuries if not millennia. Following the introduction of the simple dice-rolling methods employed in 18th century Europe, much work has been done using more advanced approaches for incorporating randomness to compose music. Such methods usually involve the application of Markov processes [22]. Markov processes work well in general for applications where short-term dependencies are sufficient to capture relevant information, such as in text recognition [23, 24]. To exploit long term dependencies in musical rhythm, probabilistic methods that incorporate the distributions of distances between subsequences have been shown to be superior to more traditional Markov methods [25]. Other approaches that incorporate randomness to generate musical rhythms include genetic algorithms that use probabilistic rules to mutate rhythms to obtain new better rhythms [13, 26]. Some systems, such as *The Continuator*, interact with a musician during a performance, and either modify the music that the performer plays, or generate music that complements what is being played by the performer [27].

The methods described above generate rhythms using complex probabilistic algorithms that involve parameters that must be tuned in order to yield "good" rhythms, with sufficiently high probability. Furthermore, the "goodness" or quality of the rhythms produced by these methods is evaluated either by mathematical measures of aesthetics such as fitness functions in simulated annealing and genetic algorithms [28, 29], or by human beings who are usually the designers of the algorithms.

In contrast to the complex and probabilistic methods outlined above, this chapter provides a description of some simple and deterministic rules that are guaranteed (within specified limits of rhythm length) to generate "good" musical rhythms. Furthermore, in contrast to the above methods that measure "goodness" by either human evaluations, or with mathematical measures of "goodness," the methods described in the following consider a rhythm to be "good" if it belongs to the repertoire of the musical tradition of some culture in the world, is used frequently as an *ostinato* or *timeline*, and has withstood the test of time. Typical examples of "good" rhythms that satisfy this definition are the *timelines* in Sub-Saharan music [12, 30–33], the *talas* in Indian music [34], the *compas* in the Flamenco music of southern Spain [35], and the *wazn* in Arabic music [36–38].

The most convenient and economical representation of a musical rhythm is in *box notation* as a sequence of binary symbols that represent sounds and silences, each of which has a duration of one unit of time. In text the simplest visualization uses the symbols "x" and "." to denote the onset of a sound and a silent pulse (unit rest), respectively. Thus the sixteen-pulse, five-onset clave son rhythm would be represented by the sequence [x . . x . . x . . . x . x . . .]. Such a skeletal representation is eminently suited to objective mathematical and computational analyses, while at the same time, and perhaps surprisingly, encapsulates a rich enough structure to provide considerable musical insight into both the theory and practice of musical rhythm. Here several simple deterministic algorithms for generating musical rhythms are reviewed and compared in terms of their ability to capture "goodness" as defined above.

2 Maximally Even Rhythms and the Euclidean Algorithm

The most salient simple deterministic rule that generates "good" rhythms yields rhythms that have the property that their onsets are distributed in the rhythmic cycle as evenly as possible. In 2004 the author discovered that the ancient Greek algorithm for determining the greatest common divisor of two numbers, known as the *Euclidean Algorithm* [39], generates scores of traditional musical rhythms from cultures all over the world. For this reason they are called *Euclidean rhythms*. This discovery was first published in the Bridges-2005 conference held in Banff (Canada) [40], and most recently re-published in *Interalia Magazine* [41]. Furthermore, it turns out that the Euclidean algorithm generates rhythms that have their onsets distributed as evenly as possible in the rhythmic cycle. Sets that have this property are termed *maximally even sets* in the music theory literature, where it was originally introduced in the context of pitch-class sets (chords and scales) by Clough and Douthett [42]. The first appearance of the Euclidean algorithm is in Propositions 1 and 2 of Book VII of Euclid's *Elements* written circa 300 BC [43]. Given two positive integers, n and k, the Euclidean algorithm repeatedly subtracts the smaller number from the larger until either 1 or 0 is obtained. The greatest common divisor of the two numbers is 1 if the algorithm terminates with 1, and the number just preceding 0 if 0 is obtained. However, for the purpose of rhythm generation we are in fact not interested in calculating the greatest common divisor of the two numbers, but rather in the process by which the answer is obtained. In this setting n denotes the number of pulses (onsets and silent rests) in the rhythm, and k denotes the number of onsets (sounded pulses). The repeated subtraction process in the Euclidean algorithm is illustrated in Fig. 1 with $n = 16$ and $k = 5$. A similar implementation of the Euclidean algorithm

Fig. 1 The rhythm obtained
by the Euclidean algorithm
with $n = 16$ and $k = 5$

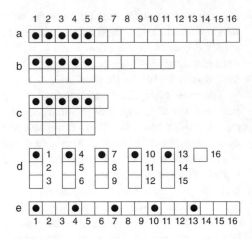

was used by Bjorklund [44, 45] to design timing systems for spallation neutron source accelerators in the Los Alamos National Laboratory, that evenly distribute a specified number of electrical pulses within a given interval. In row (a) the 16 pulses are first organized so that the sounded pulses (here denoted by squares filled with black disks) fill the first 5 positions going from left to right, and the remaining 11 silent pulses (denoted by empty squares) fill the remaining positions. Since there are more empty boxes than filled boxes, 5 of them are "subtracted" and placed flush to the left under the remaining boxes, as in row (b). At this stage the "remainder" of 6 empty boxes is still larger than 5, so a second subtraction is performed to yield the pattern in row (c). This process terminates when the remainder consists of a single column of boxes shorter than the others, such as one box in row (c), or an empty column. The generated rhythm is then obtained by reading row (c) in a top-to-bottom and left-to-right fashion, as illustrated in rows (d) and (e). The rhythm obtained in row (e) with $n = 16$ and $k = 5$ has inter-onset intervals (IOI) 33334, and is the signature rhythm of electronic dance music (EDM) [46], and one of the ways the shamans on the east coast of South Korea subdivide a 16-pulse cycle in their ritual drumming music [47]. Since the type of rhythm considered here is cyclic, and thus repeats throughout a piece of music, it is also useful to consider rhythm *necklaces* consisting of all rotations of a given rhythm. Note that in the EDM rhythm the long interval occurs at the end. On the other hand, the rhythm heard on the piano of Radiohead's recent song 'Codex' places the long IOI at the start of the pattern to obtain 43333 [48]. Furthermore, there are traditional rhythms that situate this interval at other locations in the cycle. For example the *bossa-nova* rhythm from Brazil has IOI = 33433 [49].

Given the two positive integers $n = 16$ and $k = 5$, the Euclidean algorithm terminates with a remainder of 1, establishing that the numbers 16 and 5 are *relatively prime*. Two integers are relatively prime if there exists no integer greater than 1 that divides both. On the other hand, with $n = 16$ and $k = 4$ the remainder is 0, the arrangement of boxes forms a 4×4 rectangle yielding the regular rhythm [x . . . x

. . . x . . . x . . .], a house kick drum (four-on-the-floor) pattern [46]. Therefore the Euclidean algorithm generates regular rhythms as well. However, the most interesting rhythms are obtained when n and k are relatively prime numbers [14, 31, 32, 40, 50]. In addition, if the starting point of the cyclic rhythm is not important and all rotations are included in the set, then these rotations are known as Euclidean *necklaces*. If mirror reflections are also included in the set then the set is referred to as a Euclidean *bracelet* [14].

By varying the values of n and k one may generate scores of Euclidean rhythms that are used in traditional music all over the world. The number n is generally smaller than 24 [51]. Usually the value of k is between one fourth and one half that of n. The most frequent values of n the world over are 4, 6, 8, 12, 16, and 24. When n is 8, 12, or 16, a popular value of k is 5. Figure 2 depicts the Euclidean algorithm at work with $n = 12$ and $k = 5$. The resulting rhythm with pattern [x . . x . x . . x . x .] is the Venda clapping pattern of a South African children's song [31]. As a final example consider the case when $n = 8$ and $k = 5$ pictured in Fig. 3. The resulting rhythm with pattern [x . x x . x x .] is a rhythm found in the music of many cultures around the world, known in Cuba as the *cinquillo* pattern [52]. When it is started on the second onset it is the Spanish *tango* [53] and a thirteenth century Persian rhythm, the *al-saghil-al-sani* [37].

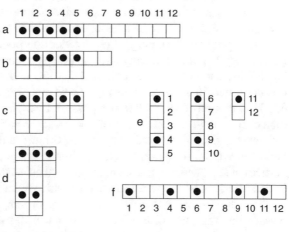

Fig. 2 The rhythm obtained by the Euclidean algorithm with $n = 12$ and $k = 5$

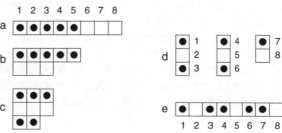

Fig. 3 The rhythm obtained by the Euclidean algorithm with $n = 8$ and $k = 5$

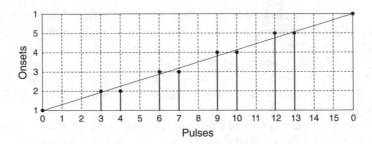

Fig. 4 Generating a maximally even rhythm by the snapping algorithm with $n = 16$ and $k = 5$

The Euclidean algorithm described above, based on repeated subtraction, generates rhythms that are maximally even sets [42], in the sense that the IOIs of the rhythms obtained are distributed as evenly as possible in the necklace cycle. Maximally even rhythms may also be generated by means of a simple geometric process that consists of snapping real numbers to integers in a $d \times n$ grid of squares, as illustrated in Fig. 4 for the case $n = 16$ and $k = 5$. The vertical y-axis denotes the number of onsets desired in the rhythm, whereas the horizontal x-axis denotes the units of time at which the onsets should occur. First connect the lower left corner of the $d \times n$ grid to the upper right corner with a straight line. This diagonal line intersects the horizontal dashed lines at equally spaced x-coordinates. The first intersection is at $x = 16/5 = 3.2$, the second intersection at $x = 2(16/5) = 6.4$, and so forth. The final step involves "snapping" these intersection points to their next lower integer (unless they happen to already have an integer x-coordinate). The resulting rhythm has IOI pattern 33334, the electronic dance music rhythm (EDM) [46]. Alternately one can "snap" the intersections to the next higher integer to obtain the IOI pattern 43333, a rotation of 33334. It is also possible to implement the "snapping" algorithm on a circular lattice. For the case of $n = 16$ and $k = 5$ the circle is first divided into a circular lattice of 16 equidistant points. On this lattice place an inscribed regular pentagon with one of its vertices on the first lattice point. Finally the remaining four vertices of the pentagon are snapped to the their nearest counter-clockwise integer lattice point, unless they are located on a lattice point.

Since the discovery that the Euclidean algorithm generates almost all the most popular rhythms that occur in traditional music all over the world in such a simple fully automatic manner, and can in addition generate "good" new rhythms that seem not to have appeared before in traditional music, by specifying unusual numbers for n and k, it has been frequently implemented electronically, and is now available in a variety of commercial open-source hardware sequencers such as Ableton Live [54]. Sequencers for generating Euclidean rhythms have also been applied to distributed multi-robot systems (swarm robotics) in which the motions of the robots control a group of Euclidean rhythms played concurrently [55].

3 Almost Maximally Even Rhythms and the Snapping Algorithm

For given values of n and k the Euclidean algorithm generates only a single "good" rhythm, which for $n = 16$ and $k = 5$ is the EDM rhythm [x . . x . . x . . x . . x . . .]. However, there exist other "good" rhythms with $n = 16$ and $k = 5$ used in traditional world music, such as the distinguished clave son: [x . . x . . x . . . x . x . . .] [12], which although not maximally even, are close to being maximally even. This motivates the generalization of the concept of maximally even, in order to obtain a simple deterministic algorithm that captures these additional "good" rhythms found in practice. There exists a plethora of mathematical possibilities for defining rhythms that are *approximately* maximally even. For example, one can define a measure of the distance between any rhythm with say $n = 16$ and $k = 5$, such as the *edit* distance [56], and consider a rhythm to be approximately even if the edit distance between the rhythm in question and a maximally even rhythm with $n = 16$ and $k = 5$ is below a specified threshold.

The algorithm for generating maximally even rhythms with the snapping algorithm on the grid illustrated in Fig. 4 suggests a natural generalization of maximally even rhythms by permitting each intersection point to be "snapped" to either its left (floor function) or right (ceiling function) nearest integer (pulse). This generalized "snapping" algorithm is conveniently described as a traversal in a *nearest pulse directed acyclic graph* (NP-DAG) constructed as follows (refer to Fig. 5). The source vertex of the NP-DAG is the lower left corner of the grid that corresponds to the occurrence of the first onset at time zero. Directed edges are connected from the source vertex to both the left and right nearest integer pulse locations (vertices) corresponding to the first intersection point of the diagonal line with the dashed line of

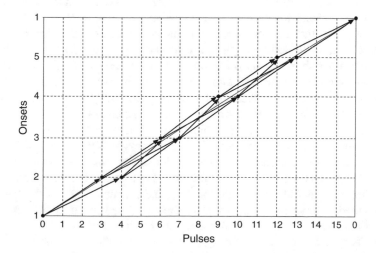

Fig. 5 The nearest pulse directed acyclic graph (NP-DAG) obtained with $n = 16$ and $k = 5$

Fig. 6 Almost maximally even rhythms with $k = 5$ and $n = 16$

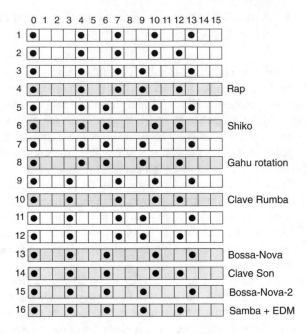

the second onset. This process is continued from the two vertices created, connecting directed edges to the two vertices determined by the succeeding intersection point. Finally the last two vertices are connected to the upper right target vertex, which corresponds to the starting onset at time zero. In this NP-DAG every path from the source vertex to the target vertex corresponds to an IOI pattern along the x-axis and thus a generated rhythm. The rhythms generated with this algorithm are termed *almost maximally even*. Since the vertices of this DAG other than those corresponding to the last intersection point of the diagonal at level 5, have degree 2, the number of distinct paths from the source vertex to the target vertex is $2 \times 2 \times 2 \times 2 = 16$. Therefore for $n = 16$ and $k = 5$ there are sixteen almost maximally even rhythms. These sixteen rhythms are shown in box-notation in Fig. 6. Note that among this collection are present eight well known traditional rhythms (shaded) including the clave son, and a rotation of the gahu rhythm which has IOI pattern 33442 [57]. Note also that rhythms No. 1 and 9 are a rotations of the samba or EDM rhythm as well as the bossa-nova and its variant, and rhythm No. 11 is a rotation of the clave son. Furthermore, rhythm No. 5 is a rotation of the mirror image of the clave son. Therefore the notion of almost maximally even is a much more encompassing characterization of "good" rhythms than the stricter definition of Euclidean maximally even rhythms, and includes some, but not all, the rotations and mirror images of the traditional rhythms used in practice, suggesting that some of these transformations of "good" rhythms also produce "good" rhythms. Recall that if a rhythm is maximally even, then all its rotations and mirror images are also maximally even. However, not all the rotations or mirror images of an almost maximally even rhythm are almost maximally even. For example the clave son has an IOI of length 2, but none of the sixteen almost maximally even rhythms start with an IOI of length 2.

Although the unshaded rhythms in Fig. 6 do not appear to be used in traditional music, and are thus not "good" according to the definition used in this study, this does not imply that they would not be considered "good" rhythms by present-day musicians. Indeed, as has already been pointed out in the preceding, rhythm No. 1 with IOI pattern 43333, which is a rotation of the EDM rhythm, is used by Radiohead. Also, a rotation of the clave son by 180 degrees when viewed on a circle, or equivalently, starting the rhythm on the silent pulse No. 8, yields the rhythm [. . x . x . . . x . . x . . x .], which is a popular way to play the rhythm in salsa music [58]. Aesthetic judgments in general, and of the "goodness" of a musical rhythm are of course partly dependent on cultural upbringing and musical experience [59]. Hannon et al. provide evidence that supports the hypothesis that culture-dependent familiarity of musical meter has a significant influence on rhythmic pattern perception [60]. There is also explicit evidence that language has an influence on the rhythmic aspects of music composition, and implicitly on the perception of musical rhythm [61]. To the author, all 16 almost maximally even rhythms in Fig. 6 sound good, although some are less familiar than others.

The 16 almost maximally even rhythms with $k = 5$ and $n = 12$ are shown in Fig. 7. Note that as with $k = 5$ and $n = 16$, half of the rhythms generated by the NP-DAG algorithm (shown shaded) are well established traditional rhythms used in practice in Sub-Saharan Africa, Andalusia in Southern Spain, and Cuba [14]. A noteworthy feature that distinguishes the rhythms used in practice from the other eight (unshaded) is the absence of two onsets located in adjacent pulses. None of the former have an IOI = 1, and all but one (No. 13) of the latter contain an IOI = 1. Due

Fig. 7 Almost maximally even rhythms with $k = 5$ and $n = 12$

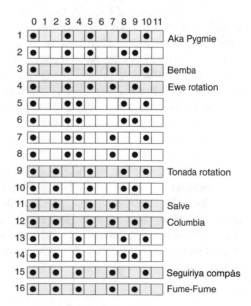

Fig. 8 Almost maximally
even rhythms obtained with
$k = 5$ and $n = 8$

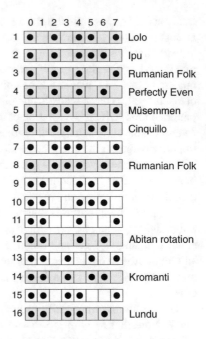

to the decrease in available temporal space for 5 onsets to be distributed among 12 rather than 16 pulses, the NP-DAG algorithm creates these short IOIs, which appear to be an undesirable feature in the rhythms used in practice.

The sixteen almost maximally even rhythms obtained when the values of k and n are set to 5 and 8, respectively, are shown in Fig. 8. As with $n = 12$ and $n = 16$, more than half (ten) of the rhythms generated by the NP-DAG algorithm (shown shaded) are well established traditional rhythms used in practice in Rumanian folk music, vodou rhythms, Sub-Saharan Africa, Cuba, and the Arab world [14, 36–38]. A noteworthy feature that distinguishes the rhythms used in practice from the other seven (unshaded) is the absence of two groups of contiguous onsets. None of the former contain one group of two contiguous onsets and one group of three continuous onsets. Due to the further decrease in available temporal space for 5 onsets to be distributed among 8 rather than 12 pulses, the NP-DAG algorithm tends to create fewer groups of onsets, whereas three groups appear to be preferred in practice. Another feature present in these rhythms is that some of them (Nos. 4, 11 and 12) contain only four onsets. Due to the fact that 5 is more than one half of 8, the snapping rule used in the NP-DAG algorithm sometimes creates "collisions" whereby the rightward-snapped onset and the leftward-snapped onset of two consecutive input onsets coincide, resulting in the loss of one onset. Nevertheless, the regular 4-onset rhythm No. 4 is used all over the world, and the irregular 4-onset rhythm No. 12 when started on the last onset has IOI = 2132, which is the Abitan vodou rhythm [62].

4 Mutating "Good" Rhythms

The NP-DAG algorithm for generating almost maximally even rhythms described in the preceding section is limited to generating, from one maximally even rhythm made up of n pulses, fifteen offspring rhythms with the same number n of pulses. However, with a slight modification the snapping algorithm may transform a "good" rhythm with n pulses into one with m pulses where $n \neq m$. Such a modification, besides serving as a model of the trans-cultural evolution of musical rhythms, and as a fully automatic algorithm for generating additional "good" rhythms, also provides a tool for changing the meter or introducing metrical ambiguity during performances on the fly [63–66]. This version of the snapping algorithm is most conveniently illustrated using concentric circular notation of cyclic rhythms [14, 67]. Figures 9 and 10 depict the algorithm for the most ubiquitous values of the number of pulses $n = 8$, 12, 16 and the number of onsets $k = 5$. The input rhythms are displayed as polygons composed of solid lines, and the output rhythms as polygons with dashed lines. The snapping

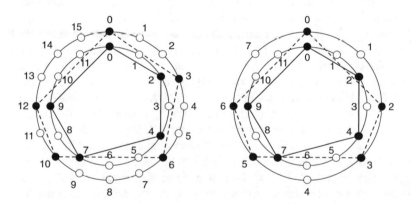

Fig. 9 Binarization from $n = 12$ to $n = 16$ (*left*), and from $n = 12$ to $n = 8$ (*right*)

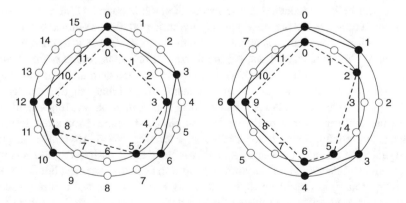

Fig. 10 Ternarization from $n = 16$ to $n = 12$ (*left*), and from $n = 8$ to $n = 12$ (*right*)

algorithm is similar to the algorithm used to generate Euclidean rhythms, except that here the onsets on the input circle are snapped to selected pulses on the output circle. If an onset on the input circle is flush with a pulse on the output circle, then it does not move. Otherwise several possibilities exist: (1) the onsets may be snapped to their nearest clockwise neighboring pulse, (2) their nearest counter-clockwise neighboring pulse, or (3) simply to their closest neighboring pulse in either direction. In Figs. 9 and 10 the nearest clockwise rule is used. Rhythms that are made up of 8 or 16 pulses (numbers divisible by 2 and not by 3) are here called *binary* rhythms, whereas rhythms with 12 pulses (divisible by 3) are here called *ternary* rhythms. The process of snapping a non-binary rhythm to a binary rhythm is called *binarization* [64–66], whereas snapping a non-ternary rhythm to a ternary rhythm is called *ternarization* [63]. Figure 9 (left) shows the binarization of the ternary 12-pulse, 5-onset *fume-fume* rhythm (on interior circle) to a 16-pulse binary rhythm, the clave son (on exterior circle). The diagram on the right shows the binarization of the fume-fume to an 8-pulse binary rhythm, in this case the cinquillo. Note that binarizing a ternary Euclidean rhythm does not necessarily yield a binary Euclidean rhythm. The fume-fume rhythm is Euclidean, and so is the cinquillo, but the clave son is not, although it is almost maximally even.

Figure 10 (left) shows the ternarization of the binary 16-pulse, 5-onset clave son rhythm (on outer circle) to a 12-pulse ternary rhythm with IOI = 32313 (on inner circle). The diagram on the right shows the ternarization of the binary 8-pulse lundu rhythm (on outer circle) to an 12-pulse ternary rhythm with IOI = 23133 (on inner circle). Note that in this case both output ternary rhythms are rotations of each other.

The algorithm for generating almost maximally even rhythms described in the preceding may be viewed as a method for transforming a single maximally even rhythm that is established as being "good" according to our definition of "good," to a larger family of rhythms that are expected to be "good," by means of small local changes to the maximally even rhythm, in the form of minimal shifts of onsets, while maintaining the even distribution of the onsets in the rhythmic cycle as much as possible. These small changes fall into the much broader category of rhythm mutations. Mutations are typically defined in a biological context involving a modification of a DNA molecule that is modeled as a sequence of symbols each of which may take on one of four values. In the present context a rhythmic mutation is defined broadly as a transformation of one binary sequence to another. It is useful to distinguish between *local* and *global* transformations. A global transformation is guided or constrained by one or more properties of the rhythm as a whole, such as maintaining maximal evenness or almost maximal evenness, or transforming a binary rhythm to a ternary rhythm (or vice-versa). On the other hand, local transformations are implemented by local rules that may disregard their effect on global structural properties. Intuition suggests that a natural simple local rule for generating "good" rhythms is to make small judicious changes to existing "good" rhythms. One possible method is simply to take an established "good" rhythm such as a maximally even Euclidean rhythm or a ubiquitous non-Euclidean rhythm that has withstood the test of time, such as the clave son, and shift one or more of its onsets (other than the first) in either direction by one or more pulse positions. Application of this rule to the maximally even

Fig. 11 Mutations of the maximally even rhythm obtained by shifting a single onset by one pulse

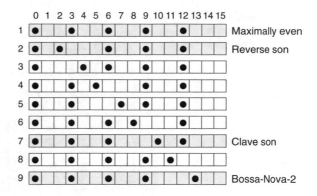

(Euclidean) EDM rhythm with the minimal restrictions that only a single onset may be shifted by only one pulse position *minimal onset shifting* (MOS) rule, yields the eight mutations shown in Fig. 11, four of which (shown shaded) are rhythms used in practice. Rhythm No. 2 may be viewed as the clave son run backwards starting at the last onset, or as the clave son run forwards starting at the third onset. The unshaded rhythms all sound good and it would not be surprising to find them used in practice somewhere, and thereby satisfy our definition of "good." Rhythm No. 5 is a more syncopated version of a popular rap rhythm given by [x . . . x . . x . x . . x . . .] by virtue that the second onset in the rap rhythm is anticipated by one pulse. Rhythm No. 8 is also a more syncopated variant of the clave son obtained by anticipating the two last onsets.

Recall that the sixteen almost maximally even rhythms in Fig. 1 were generated by snapping each intersection point to its nearest left and right pulse positions. Note that all sixteen rhythms have the property that not one of their onsets is more than one pulse away from its nearest onset in the maximally even rhythm (No. 16). However, this does not imply that a rhythm obtained with a single shift of one of the onsets of a maximally even rhythm, by one pulse position, implies that the resulting rhythm is almost maximally even. Indeed, some rhythms in Fig. 11 are not almost maximally even, such as rhythm No. 8 which ends with an IOI of length 5, whereas no almost maximally even rhythm in Fig. 6 has such a long IOI.

Application of the MOS rule to the distinguished "good" rhythm, the clave son, yields the eight mutations shown in Fig. 12, six of which (shown shaded) are used in practice. However, both rhythms numbered 4 and 7 are "good" rhythms as well. Rhythm No. 4 anticipates the second and third onsets of the shiko by one pulse each, making it more syncopated than the shiko. Rhythm No. 7 introduces hesitation on the last pair of adjacent onsets of the soukous by starting one pulse later, thus placing greater emphasis on the closing response portion of the rhythm. The two examples of the MOS rule applied to the EDM and clave son rhythms suggest that this method may be a viable alternative to the Euclidean and NP-DAG algorithms. In terms of computational complexity the MOS rule is certainly efficient once a "good" rhythm is given as input. However, compared to the Euclidean algorithm it requires too much memory (and concomitant search time) in terms of a table of existing

Fig. 12 Mutations of the clave son obtained by shifting a single onset by one pulse

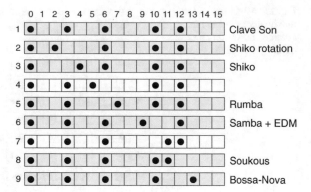

rhythms, whereas the Euclidean algorithm requires no knowledge of any existing rhythms, generating rhythms automatically by merely varying n and k. Furthermore, comparing the MOS rule with the NP-DAG algorithm, the former yields fewer "good" rhythms than the latter. Of course one could relax the MOS constraint that only one onset may be shifted by only one pulse position, thereby generating many more rhythms. However, then the property of maximal evenness will be grossly violated and the chance of generating good rhythms will dwindle.

A variety of other local mutation algorithms are possible that sometimes yield a "good" rhythm. However, they are rather ad hoc and thus lack generality and applicability. For example, an extremely simple rule is to just delete one onset from a "good" rhythm in the hope that the remaining rhythm is still "good." Here deletion means replacement of an onset with a silent pulse. If the last onset of the clave son [x . . x . . x . . . x . x . . .] is deleted one obtains the rhythm [x . . x . . x . . . x . . . . .] which is often heard in practice and is therefore "good." However, deleting the third onset of the clave son yields [x . . x . . . . . . x . x . . .], which is not a successful mutation. So an algorithm that uses this rule requires the solution of the difficult problem of finding a general rule to determine which onsets of any given rhythm may be deleted without losing "goodness." Another approach is to change rhythms by some rule, and pass the resulting rhythms through similarity filters in the hope that admitting a rhythm that is similar to a "good" rhythm must also be "good." Such methods depend on measures of similarity or distance between rhythms [56, 68]. However, the relationship between "goodness" and similarity (or distance) is not yet well defined, making it difficult to select an appropriate similarity measure that will guarantee good results. The edit distance, often used in music applications, is known to correlate well with human judgements of rhythm similarity [56], but this does not imply it also correlates with rhythm "goodness." Assume for instance that a mutation of the clave son is accepted by a filter that uses the edit distance, if the distance is at most 1. Both of the above mutations obtained by deleting either the third or last onsets have edit distance 1 from the clave son, and yet one is "good" and the other is "bad." Furthermore, this approach may incur a heavy computational burden, if the distance between a candidate rhythm and all the "good" rhythms stored in some table must be computed and compared to some acceptability threshold.

5 Conclusion

In contrast to the computationally complex randomized and probabilistic methods, outlined in the introduction, that are used to generate musical rhythms without any guarantees that the resulting rhythms are "good," and with the requirement that parameters must be tuned by their designers in order to yield rhythms that are good enough, this chapter focused on two computationally efficient and conceptually simple deterministic algorithms that are guaranteed to generate "good" musical rhythms: (1) the Euclidean algorithm, which for specific numbers of pulses n and onsets k yields a single maximally even (Euclidean) rhythm, and (2) the NP-DAG (Nearest Pulse Directed Acyclic Graph) algorithm that generates a family of almost maximally even rhythms. It is argued that although other simple deterministic algorithms for mutating "good" rhythms to obtain new "good" offspring rhythms are easy to concoct, they fall short of the Euclidean and NP-DAG algorithms on several counts. They not only lack generality and applicability, but are less efficient in terms of memory requirements and computational complexity, and are not guaranteed to yield "good" rhythms without "human intervention," the latter being one of the hallmarks of the field of *computational aesthetics* [2]. A word of clarification is in order here concerning the words "without human intervention," regarding the selection of the values of the number of pulses n and the number of onsets k in either the Euclidean or the NP-DAG algorithms. Clearly, selecting n and k arbitrarily does not guarantee that these algorithms will always yield "good" rhythms. For instance, if $n = 128$ (as happens for some Indian talas) and k is too large ($k = 50$) or too small ($k = 5$) relative to n, then the resulting rhythms are guaranteed to be terrible. Are not n and k then, parameters that must be tuned in order to obtain good results, thus implying that the algorithms depend on human intervention in order to perform well? To clarify this seeming contradiction it helps to distinguish between *parameters* that must be *tuned*, and *constraints* that must be *satisfied*. The parameters that must be tuned in typical approaches to rhythm generation, such as genetic algorithms, use complicated fitness functions that depend on statistics compiled from music corpora, and that encapsulate parameters including frequencies of notes, saliency weights attached to notes, and relations between note duration intervals [29]. These parameters (including the weights) must be tuned by trial and error to yield good results. On the other hand, the Euclidean and NP-DAG algorithms assume that the values of n and k are selected so as to lie in the range of values found in existing styles of music practice, and therefore are musical constraints that must be satisfied, rather than parameters that must be tuned. Once these values are fixed, the rhythm generation is automatic and completely free of human intervention. In this sense these algorithms fall also in the area of *aesthetic computing* [3], which asks how the arts can influence computer generated aesthetic objects. The values of n and k that are used in musical traditions all over the world have evolved over many years, even millennia, and have been adopted as part of the artistic practices of different cultures, providing the artistic influence on the computational generation of "good" musical rhythms.

Acknowledgments This research was supported by a grant from the Provost's Office of New York University Abu Dhabi, through the Faculty of Science, in Abu Dhabi, United Arab Emirates.

References

1. Hamilton, A.: Aesthetics and Music. Continuum International Publishing Group, London (2007)
2. Hoenig, F.: Defining computational aesthetics. In: Neumann, L., Sbert, M., Gooch, B., Purgathofer, W. (eds.) Computational Aesthetics in Graphics, Visualization and Imaging, pp. 13–18 (2005)
3. Fishwick, P.: Aesthetic Computing. MIT Press (2006)
4. Birkhoff, G.D.: Aesthetic Measure. Harvard University Press, Cambridge (1933)
5. Boselie, F., Leeuwenberg, E.: Birkhoff revisited: beauty as a function of effect and means. Am. J. Psychol. **98**(1), 1–39 (1985)
6. Garabedian, C.A.: Birkhoff on aesthetic measure. Bull. Am. Math. Soc. **40**, 7–10 (1934)
7. Montano, U.: Explaining Beauty in Mathematics: An Aesthetic Theory of Mathematics. Springer, Switzerland (2014)
8. Spengler, O.: The Decline of the West. I. Knopf, New York (1926)
9. Zhang, K., Harrell, S., Ji, X.: Computational aesthetics: on the complexity of computer-generated paintings. Leonardo **45**(3), 243–248 (2012)
10. Edwards, M.: Algorithmic composition: computational thinking in music. Commun. ACM **54**(7), 58–67 (2011)
11. Pachet, F., Roy, P.: Musical harmonization with constraints: a survey. Constraints J. **6**(1), 7–19 (2011)
12. Toussaint, G.T.: The rhythm that conquered the world: what makes a "good" rhythm good? Percussive Notes. November Issue, pp. 52–59 (2011)
13. Toussaint, G.T.: Generating "good" musical rhythms algorithmically. In: Proceedings of the 8th International Conference on Arts and Humanities, Honolulu, Hawaii, USA (2010)
14. Toussaint, G.T.: The Geometry of Musical Rhythm. Chapman-Hall-CRC Press (2013)
15. Harary, F.: Aesthetic tree patterns in graph theory. Leonardo **4**(3), 227–231 (1971)
16. Ahmed, Y., Haider, M.: Beauty measuring system based on the Divine Ratio. In: Proceedings of the International Conference on User Science and Engineering, pp. 207–210. IEEE (2010)
17. Davis, S.T., Jahnke, J.C.: Unity and the golden section: rules for aesthetic choice? Am. J. Psychol. **104**(2), 257–277 (1991)
18. Pallet, P.M., Link, S., Lee, K.: New "golden" ratios for facial beauty. Vision Res. **50**(2), 149–154 (2010)
19. Rigau, J., Feixas, M., Sbert, M.: Conceptualizing Birkhoff? Aesthetic measure using Shannon entropy and Kolmogorov complexity. In: Cunningham, D.W., Meyer, G., Neumann, L. (eds.) Computational Aesthetics in Graphics, Visualization, and Imaging. The Eurographics Association (2007)
20. Sinha, P., and Russell, R.: A perceptually-based comparison of image-similarity metrics. Perception **40** (2011)
21. Hedges, S.A.: Dice music in the eighteenth century. Music Lett. **59**, 180–187
22. Xenakis, I., Kanach, S.: Formalized Music: Mathematics and Thought in Composition. Pendragon Press (1992)
23. Shinghal, R., Toussaint, G.T.: Experiments in text recognition with the modified Viterbi algorithm. IEEE Trans. Pattern Anal. Mach. Intell. **PAMI-1**, 184–193 (1979)
24. Shinghal, R., Toussaint, G.T.: The sensitivity of the modified Viterbi algorithm to the source statistics. IEEE Trans. Pattern Anal. Mach. Intell. **PAMI-2**, 181–185 (1980)
25. Paiement, J.-F., Grandvalet, Y., Bengio, S., Eck, D.: A distance model for rhythms. In: International Conference on Machine Learning, New York, USA, pp. 736–743 (2008)

26. Burton, A.R., Vladimirova, T.: Generation of musical sequences with genetic techniques. Comput. Music J. **23**(4), 59–73 (1999)
27. Pachet, F.: Interacting with a musical learning system: The Continuator. In: Proceedings of the 2nd International Conference on Music and Artificial Intelligence, Edinburgh, Scotland, UK, September 12–14, pp. 119–132 (2002)
28. Horowitz, D.: Generating rhythms with genetic algorithms. In: Proceedings of the 12th National Conference of the American Association of Artificial Intelligence, Washington, USA, Seattle, p. 1459 (1994)
29. Maeda, Y., Kajihara, Y.: Rhythm generation method for automatic musical composition using genetic algorithm. In: IEEE International Conference on Fuzzy Systems, Barcelona, Spain, pp. 1–7 (2010)
30. Agawu, K.: Structural analysis or cultural analysis? Competing perspectives on the standard pattern of West African rhythm. J. Am. Musicol. Soc. **59**(1), 1–46 (2006)
31. Pressing, J.: Cognitive isomorphisms in World Music: West Africa, the Balkans. Thailand and western tonality. Stud. Music **17**, 38–61 (1983)
32. Rahn, J.: Asymmetrical ostinatos in Sub-Saharan music: time, pitch, and cycles reconsidered. In Theory Only **9**(7), 23–37 (1987)
33. Toussaint, G.T.: Mathematical features for recognizing preference in Sub-Saharan African traditional rhythm timelines. In: Proceedings of 3rd Conference on Advances in Pattern Recognition, Bath, United Kingdom, pp. 18–27 (2005)
34. Thul, E., Toussaint, G.T.: A comparative phylogenetic analysis of African timelines and North Indian talas. In: Proceedings of 11th BRIDGES: Mathematics, Music, Art, Architecture, and Culture, pp. 187–194 (2008)
35. Guastavino, C., Toussaint, G.T., Gómez, F., Marandola, F., Absar, R.: Rhythmic similarity in flamenco music: comparing psychological and mathematical measures. In: Proceedings of 4th Conference on Interdisciplinary Musicology, Thessaloniki, Greece, pp. 76–77 (2008)
36. Hagoel, K.: The Art of Middle Eastern Rhythm. OR-TAV, Kfar Sava, Israel (2003)
37. Wright, O.: The Modal System of Arab and Persian Music AD 1250–1300. Oxford University Press, Oxford (1978)
38. Touma, H.H.: The Music of the Arabs. Amadeus Press, Portland, Oregon (1996)
39. Franklin, P.: The Euclidean algorithm. Am. Math. Mon. **63**(9), 663–664 (1956)
40. Toussaint, G.T.: The Euclidean algorithm generates traditional musical rhythms. In: Proceedings of BRIDGES: Mathematical Connections in Art, Music, and Science, Banff, Canada, pp. 47–56 (2005)
41. Toussaint, G.T.: The Euclidean algorithm generates traditional musical rhythms. Interalia Mag. **16** (2015) (Electronic publication: http://www.interaliamag.org)
42. Clough, J., Douthett, J.: Maximally even sets. J. Music Theory **35**, 93–173 (1991)
43. Heath, T.L.: The Thirteen Books of Euclid's Elements (2nd ed. [Facsimile. Original publication: Cambridge University Press, 1925] ed). Dover Publications, New York (1956)
44. Bjorklund, E.: A metric for measuring the evenness of timing system rep-rate patterns. Technical Note SNS-NOTE-CNTRL-100, Los Alamos National Laboratory, U.S.A. (2003)
45. Bjorklund, E.: The theory of rep-rate pattern generation in the SNS timing system. Technical Note SNS-NOTE-CNTRL-99, Los Alamos National Laboratory, U.S.A. (2003)
46. Butler, M.J.: Unlocking the Groove: Rhythm, Meter, and Musical Design in Electronic Dance Music. Indiana University Press, Bloomington and Indianapolis (2006)
47. Mills, S.: Healing Rhythms: The World of South Korea's East Coast Hereditary Shamans. Ashgate, Aldershot, U.K. (2007)
48. Osborn, B.: Kid Algebra: Radiohead's Euclidean and maximally even rhythms. Perspect. New Music **52**(1), 81–105 (2014)
49. Morales, E.: The Latin Beat-The Rhythms and Roots of Latin Music from Bossa Nova to Salsa and Beyond. Da Capo Press, Cambridge, MA (2003)
50. Kubik, G.: Africa and the Blues. University of Mississippi Press, Jackson (1999)
51. Arom, S.: African Polyphony and Polyrhythm. Cambridge University Press, Cambridge, UK (1991)

52. Floyd Jr., S.A.: Black music in the circum-Caribbean. Am. Music **17**(1), 1–38 (1999)
53. Evans, B.: Authentic Conga Rhythms. Belwin Mills Publishing Corporation, Miami (1966)
54. Sasso, L.: Drum Mechanics: Ableton Live Tips and Techniques. In: Sound on Sound (2014). http://www.soundonsound.com/sos/dec14/articles/live-tech-1214.htm. Accessed 5 April 2016
55. Albin, A., Weinberg, G., Egerstedt, M.: Musical abstractions in distributed multi-robot systems. In: Proceedings of the IEEE/RSJ International Conference on Intelligent Robots and Systems, Vilamoura, Algarve, Portugal, pp. 451–458 (2012)
56. Post, O., Toussaint, G.T.: The edit distance as a measure of perceived rhythmic similarity. Empirical Musicol. Rev. **6** (2011)
57. Locke, D.: Drum Gahu: An Introduction to African Rhythm. White Cliffs Media, Tempe, AZ (1998)
58. Peñalosa, D.: The Clave Matrix; Afro-Cuban Rhythm: Its Principles and African Origins. Bembe Inc., Redway, CA (2009)
59. Masuda, T., Gonzales, R., Kwan, L., Nisbet, R.E.: Culture and aesthetic preference: comparing the attention to context of East Asians and Americans. Pers. Soc. Psychol. Bull. **34**(9), 1260–1275 (2008)
60. Hannon, E.E., Soley, Ullal, S.: Rhythm perception: a cross-cultural comparison of American and Turkish listeners. J. Exp. Psychol.: Hum. Percept. Perform. Advance online publication (2012). doi:10.1037/a0027225
61. Patel, A.D.: Music, Language, and the Brain. Oxford University Press, Oxford (2008)
62. Wilcken, L.: The Drums of Vodou. White Cliffs Media, Tempe, AZ (1992)
63. Gómez, F., Khoury, I., Kienzle, J., McLeish, E., Melvin, A., Pérez-Fernández, R., Rappaport, D., Toussaint, G.T.: Mathematical models for binarization and ternarization of musical rhythms. In: BRIDGES: Mathematical Connections in Art, Music, and Science, San Sebastian, Spain, pp. 99–108 (2007)
64. Toussaint, G.T.: Modeling musical rhythm mutations with geometric quantization. In: Melnik, R. (ed.) Mathematical and Computational Modeling: With Applications in Natural and Social Sciences, Engineering, and the Arts, pp. 299–308. Wiley (2015)
65. Pérez-Fernández, R.: La Binarización de los Ritmos Ternarios Africanos en América Latina. Casa de las Américas, Havana (1986)
66. Pérez-Fernández, R.: El mito del carácter invariable de las lineas temporales. Transcult. Music Rev. **11** (2007)
67. Liu, Y., Toussaint, G.T.: Mathematical notation, representation, and visualization of musical rhythm: a comparative perspective. Int. J. Mach. Learn. Comput. **2** (2012)
68. Toussaint, G.T.: A comparison of rhythmic dissimilarity measures. FORMA **21** (2006)

Is Universal Computation a Myth?

Selmer Bringsjord

Abstract Akl has claimed that universal computation is a myth, and has offered a number of ingenious arguments in support of this claim, one of which features the challenge of tracking the locations of multiple, ever-moving robots on Mars. I provide what I see as a refutation of this argument; my counter-argument is based on a thesis that is less informal and more plausible than the Church-Turing Thesis, and on my own generalized variant of Kolmogorov-Uspensky machines. While I concede that it doesn't deductively follow from the success of my refutation that universal computation is, or can be, real, I conclude by pointing toward a route that I believe can vindicate the counter-claim that universal computation is specifiable, and instantiable.

1 Introduction

Selim Akl's remarkable oeuvre provides innumerable opportunities for one to write about the foundations, both formal and philosophical, of computation. For the present volume, I've seized upon a single opportunity: his ingenious and provocative "The Myth of Universal Computation" [1]. My analysis, in a further narrowing, is specifically targeted at a key argument of Akl's within this paper, a fascinating one involving the tracking of multiple robots (assumed to be) on Mars. I denote this argument as '$\mathcal{A}_{\bar{u}_{TM}}$.' Because I shall use '$\bar{u}_{TM}$' to denote the statement that no Turing machine can be a universal computer, the subscript in '$\mathcal{A}_{\bar{u}_{TM}}$' is just a convenient reminder that $\bar{u}_{TM}$ is the conclusion of this argument.

Akl's overall goal in Akl [1] is in fact much more ambitious than establishing $\bar{u}_{TM}$, for he doesn't think *any* rigorous, fixed, abstract model of computation can be

---

I'm indebted to Selim Akl for bringing to my attention countless stimulating ideas, only one of which I explore herein.

S. Bringsjord (✉)
Rensselaer AI and Reasoning (RAIR) Lab, Department of Computer Science, Rensselaer
Polytechnic Institute (RPI), Troy, NY 12180, USA
e-mail: selmerbringsjord@gmail.com

© Springer International Publishing Switzerland 2017
A. Adamatzky (ed.), *Emergent Computation*, Emergence, Complexity
and Computation 24, DOI 10.1007/978-3-319-46376-6_2

universal. This is made clear by Akl at the very outset of his paper, in fact in his paper's abstract. There he says this about what the paper by his lights accomplishes:

> It is shown that the concept of a Universal Computer cannot be realized. ... This result applies not only to idealized models of computation, such as the Turing Machine and the like, but also to all general-purpose computers, including existing conventional computers, as well as contemplated ones such as quantum computers [1].

It obviously follows from this quote that Akl takes himself to have shown not only that the Turing machine (TM) isn't a universal computer, but that any other candidates for the title of 'universal computer' will likewise fail to reach universality. Leveraging the notation that we have already allowed ourselves, we can hence observe that Akl sees his paper as providing a sound argument $\mathcal{A}_{\overline{u}_{QC}}$ for the conclusion $\overline{u}_{QC}$; here, 'QC' is an acronym referring to quantum computers. Indeed, letting 'C' be a variable ranging over any established class of idealized computing machines, we can safely say that Akl's ultimate goal (which he believes he has reached in the paper in question) is to establish

$$\overline{u} := \neg \exists \, C \, \overline{u}_C;$$

and we can denote his overarching argument by '$\mathcal{A}_{\overline{u}_C}$.' However, again, my objective is the narrow, focused one of showing that Akl's multiple-robot argument $\mathcal{A}_{\overline{u}_{TM}}$ for $\overline{u}_{TM}$ is unsuccessful. While it doesn't follow deductively from my refutation that Akl's overarching argument $\mathcal{A}_{\overline{u}_C}$ is overthrown (because he gives additional arguments for $\overline{u}_{TM}$ beyond the one I target), if his other arguments for $\overline{u}_{TM}$ fail, his overarching case $\mathcal{A}_{\overline{u}_C}$ would fall, and hence despite his clever analysis and argumentation there may well be a form of *bona fide* universal computation. I contend, but do not prove, that my counter-argument against $\mathcal{A}_{\overline{u}_{TM}}$ can in fact be generalized into a recipe that overthrows the other arguments Akl gives against universal computation. At the end of the present chapter I suggest a logic-based route toward formalizing a form of universal computation.

My selection of Akl's paper and the specific $\mathcal{A}_{\overline{u}_{TM}}$ within it, I confess, is not without an element of selfishness, since the topics with which Akl deals in this important work are ones I too have thought a bit about. Nonetheless, as will soon be seen, our respective points of view are fundamentally different. Put with brutal brevity, I come to computation after reflecting upon the cognition of animals and persons, and from there move to the relevant logico-mathematics for modeling and computationally simulating that cognition; Akl, on the other hand, draws morals about the nature of computation after considering "de-agentized" information flowing at the mercy of time and change, in the real, physical world. (His multiple-robots-on-Mars scenario is a perfect case of his orientation in action.) We both move on from our respective starting points to consider the limits of computation, but our respective conclusions turn out to be quite different: Akl (obviously) regards $\mathcal{A}_{\overline{u}_{TM}}$ (and $\mathcal{A}_{\overline{u}_{QC}}$, and indeed $\mathcal{A}_{\overline{u}_C}$) to be sound; I don't. Moreover, as I've already indicated, I think that the concept of universal computation can in principle be formally defined via increasingly powerful logics, and that the concept can in fact be instantiated in our universe (in some mind sufficiently powerful to reason in these logics).

The plan for the remainder of the chapter is straightforward: The first step (Sect. 2) is to explain that Akl's understanding of the Church-Turing Thesis (CTT) is inaccurate. Next, in Sect. 3, using my analysis of CTT as a springboard, I present a different and much "safer" thesis: "Selmer's Safer Thesis," or just 'SST' for short. More accurately, the safer thesis is actually a thesis *schema*. Whereas CTT (as I shall point out) relies on the concept of *effective* computation, my safer thesis schema relies instead upon what I call *reflective-C* computation. Here, where C is again (recall above) a variable ranging over any of the established idealized frameworks for computing at the Turing level, reflective-C computation is a semi-formal description of the fully formal computation in C. With SST in hand, I next (Sect. 4) recapitulate and analyze Akl's multiple-robot argument $\mathcal{A}_{\overline{u}_{TM}}$, drawing directly from his paper to do so. Then in the next Sect. 5, I refute Akl's argument. Some concluding remarks that gesture toward a universal computer wrap up the paper (Sect. 6).

2 The Church-Turing Thesis (CTT), for Real

Our first step is to isolate and analyze what Akl takes to be the "Church-Turing Thesis" (CTT). Doing so is easy, for here is a verbatim quote from Akl [1, p. 172]:

> While fairly simple conceptually, the Turing Machine is a truly powerful model of computation. So powerful in fact, that it was believed until recently that no model more powerful than the Turing Machine can possibly exist (in other words, a model that would be able to perform computations that the Turing Machine cannot perform). This belief is captured in the following statement, known as
>
> **Church-Turing Thesis**: Any computable function can be computed on a Turing Machine [73, 54].

Unfortunately, this is not CTT. The reason is perfectly simple and uncontroversial: It must be a particular *kind* of function that is said in the thesis to be a Turing-computable one. Church [14, p. 356] originally used the informal phrase 'effectively calculable' to label the kind of function in question. The phrase 'effectively computable' is the syntactic variant of Church's phrase that is currently used. Now, notice that Akl, in the quote immediately above, gives two citations immediately after typographically setting out his version of the thesis in question. Could it be that Akl has been led astray by the authors in question? I investigated; sure enough, this appears to be exactly what happened. For example, here is how [28, p. 209] puts it: "The Turing machine (TM) is believed to be the most general computational model that can be devised (the **Church-Turing thesis**)." This is what Akl is referring to when he offers the citation '[54].'[1]

---

[1]Unfortunately for Savage, the super-recursive computational models explored by Turing in his doctoral dissertation under Church (i.e. [30]) refute Savage's claim that (at least at the time of his writing) it is believed that the Turing machine is the "most general computational model." A wonderful discussion of these matters in relation to Turing's dissertation is provided in Feferman [18]; cf. Bringsjord [5].

But why do I say that Akl has been led astray? The reason, again, is simple, and quite decisive: The Church-Turing Thesis, CTT as we abbreviate it, is not what Akl says it is, because the core idea in CTT is the equivalence of what is "effectively" or "mechanically" or "algorithmically" computable, with what is Turing-computable. Hence we must be more precise and accurate.[2]

Again, the heart of CTT is the informal notion of an *algorithm*,[3] which has been nicely characterized (in traditional fashion) by Mendelson as

> an effective and completely specified procedure for solving a whole class of problems. ....
> An algorithm does not require ingenuity; its application is prescribed in advance and does
> not depend upon any empirical or random factors [24, p. 225].

An *effectively computable* function is thus the computing of a function by an idealized "worker" or "computist" following an algorithm.[4] (Without loss of generality, we can for present purposes view all functions as taking natural numbers into natural numbers; that is, for some arbitrary f, $f : \mathbb{N} \mapsto \mathbb{N}$).

CTT also involves a more formal notion, that of a so-called *Turing-computable* function. If the formal notion is wed to a different paradigm, then we would no longer have the Church-*Turing* Thesis. For example, we could refer instead to a *recursive* function, or a register machine-computable function, etc. Mendelson employs Turing's approach, and Turing machines are what Akl focuses upon in the paper we're analyzing. A function $f : \mathbb{N} \mapsto \mathbb{N}$ is *Turing-computable* iff there exists a TM m which, starting with n on its tape (perhaps represented by n |s), leaves $f(n)$ on its tape after processing. (The details of the processing are harmlessly left aside.) Given this definition, CTT amounts to

> CTT A function f is effectively computable if and only if it's Turing-computable.

Most scholars, as the reader herself is likely to know, regard CTT to be true. However, I'm not one of them. So while I have on hand a counter-argument against $\mathcal{A}_{\bar{u}_{TM}}$ that employs CTT, I certainly can't use it here. Not only that, but in a rather interesting twist, even if I was inclined to affirm CTT, I still couldn't use it as a premise in a counter-argument against $\mathcal{A}_{\bar{u}_{TM}}$. The reason is that a careful reading of Akl [1] reveals that Akl himself is quite prepared to give up CTT. In fact, he appears to hold that $\bar{u}_{TM}$ entails the falsity of CTT.

---

[2] And here I follow my own prior work, and the work of others, including those who have instructively sought to *prove* CTT. In my own case, devoted in part to arguments *against* CTT, see e.g. Bringsjord and Arkoudas [6], Bringsjord and Govindarajulu [7]; for an attempt to prove CTT, see the chapter on CTT in Smith [29]; and for a wonderful exposition of CTT and its history, including coverage of the trap of stating CTT erroneously as in the case of Savage [28], see Copeland [15].

[3] Interestingly enough, Lewis and Papadimitriou [23], the pair of authors Akl [1] draws from in order to formally characterize Turing machines, well understand that CTT asserts an equivalence between an intuitive notion of algorithm and Turing-computability, for—in a quote isolated by Akl himself—we read that CTT consists in the proposition that "the idea of a 'computational procedure' or an 'algorithm' is equivalent to the idea of a Turing Machine."

[4] Turing [31] spoke of "computists" and Post [27] of "workers," humans whose sole job was to slavishly follow explicit, excruciatingly simple instructions.

3 Selmer's "Safer" Thesis (SST)

The received view is that CTT is not only true, but unprovable.[5] The main rationale in support of this view is the claim that the concept of effective computation is too informal to allow a proof of CTT [4]. Whether or not this rationale is correct, the fact certainly remains that the "left side" of the biconditional that constitutes CTT is not formal, while the right side, which refers to the Turing-computability of a function, can be rendered formal. Turing-computability (and of course also therefore Turing-decidability, etc.) can be formalized in various ways, as Akl [1] points out. He draws heavily on Lewis and Papadimitriou [23] to recount some of the formalism in question. And we could even be more formal and rigorous about Turing-computability, since we could move from the naïve set theory of Lewis and Papadimitriou [23] to axiomatic set theory (e.g., ZFC), and laboriously build up to CTT from there. But a move to greater rigor in defining the right side of CTT would still leave the left side vague. I introduce now the promised thesis schema SST, which includes still on its left side a somewhat informal concept, but not one as intuitive and informal as effective computation. As I've said, while I can't for the reasons given above use CTT in my counter-argument against $\mathcal{A}_{\overline{u}_{TM}}$, I can, and do, use SST.

The first step toward SST is to recall that above we quantified over idealized computational schemes C to introduce $\overline{u}$. We can leverage this simple idea in order to formulate a thesis schema that is at once both much more plausible and much less informal than CTT. Instead of employing the concept of *effective computation* as in the case of CTT, SST employs the concept of *reflective-C* computation, where C here is once again functioning as a variable ranging over the space of established idealized computational schemes that are provably equivalent to that of the Turing machine. This space includes not just Turing machines (T), but also for example Post machines (P) [26], register machines (R) [16], the μ-recursive functions (M) [23], unrestricted (= Type 0) grammars (G) [25], the λ-calculus (Λ) [12, 13], and my favorite formal model of Turing-level computation that doesn't explicitly use logic and deduction, and one that is clearly the most cognitively realistic category under C, Kolmogorov-Uspensky (KU) machines (K) [21]. Each of these idealized frameworks is an acceptable instance of the general variable C that ranges over all established idealized frameworks equivalent to Turing machines; and all of these frameworks are equivalent. Hence, for instance, a function f is G-computable if and only if (iff) f is T-computable iff f is Λ-computable iff f is R-computable, and so on.

In this context, I now introduce the new concept of *reflective-C* computation. This concept is not fully formal, but it's much more formal than the very vague and intuitive concept of effective computation. To see the basic idea is this, start by bringing to mind some formal description of one of the idealized frameworks listed in the previous paragraph. For focus and to ease exposition, but without loss

---

[5] A few have held that CTT is provable, but they are in the extreme minority, and, joined by others, I have shown that defenders of the provability of CTT are incorrect. E.g., while Smith [29] has tried to prove CTT, see Bringsjord and Govindarajulu [7].

of generality, let's first choose T, Turing machines. Many examples of reflective-T computation can be given, and in fact many are given in the literature; here's one:

Imagine an old-fashioned railroad track that starts at a certain point and extends infinitely in one direction. Imagine as well that this track is laid upon railroad ties spaced at a regular interval, and that on the ground, bounded by two ties and two stretches of track, is a blackboard. In addition, suppose that there is a small boxcar that can roll on the track, powered by a simple mechanical lever, and a switch onboard the boxcar that controls the direction of movement (left or right on the track). The boxcar is occupied by a well-trained chimpanzee who can power the car by pushing the lever up and down, and who can through toggling of the switch move left or right. In addition, the bottom of the boxcar is hollow, so when the boxcar is positioned over a blackboard, the chimpanzee can reach down to erase symbols appearing on the blackboard, and to write symbols on the blackboard. He does so in accordance with very simple instructions. Finally, the combined ensemble of the chimp, the boxcar, and his simple tools, at any one moment, are assumed to be in any one of a particular number of finite, pre-defined configurations. Whatever a—as we shall call them—*chimp machine* can (reflective-T-compute) a Turing-machine can compute, and *vice versa*. The reason for this, in a word, is that chimp-machine computation, while intuitive, is directly reflective of T computation. And of course we didn't need to refer to chimps and boxcars. Instead, we could have referred to any number of an infinite number of other props, and we could still be depicting computation that is directly reflective of the formal Turing-machine model. The general truth in play can be elevated to the following statement:

SST$_T$ A function f is reflective-T-computable if and only f is T-computable (= Turing-computable).

To make sure there is no misunderstanding or resistance, let me explain that we can do the same kind of trick for register machines, formal, idealized machines which are reflected by the less formal concept of *raven machines*, as I now explain.[6]

Raven machines include, first and foremost, a raven: Roger. Roger is a thoroughly obedient bird whose range of activity is highly restricted. Roger is shown in Fig. 1. You will note that he is holding something in his beak. What is it? It's a little round stone. Roger doesn't fly (at least when he is working); when we tell him to start a work session, he simply moves little round stones around, in accordance with programs that we provide to him, and he halts when we tell him to conclude a work session. More specifically, his movement of the stones is confined to moving them into and out of numbered boxes. For any given work session, we provide Roger with n boxes to start, and if his program makes reference to the number m of a box beyond the ones he intially has, Roger calls out "More," and instantly a new box numbered m appears for him to employ.[7] Raven machines consist of the combination of: programs to

---

[6]Here I draw upon my Bringsjord and Taylor [9], which I use to teach introductory formal logic and computability theory. Most readers will be familiar with register machines, which are elegantly and economically defined in Ebbinghaus et al. [17].

[7]Alternatively, we could imagine that Roger's call for another box results in a box from an infinite supply provided at the outset of his efforts, moving into his work area.

Fig. 1 Roger the raven

Fig. 2 Snapshot of a raven machine in operation

instruct Roger, Roger himself with his perception and action powers, and the stones and boxes. Figure 2 shows a snapshot of a raven machine during some computation.

We are interested in having raven machines compute number-theoretic functions, that is functions from $\mathbb{N}^n$ to $\mathbb{N}$. In order to enable this, we shall understand a given natural number n to correspond to n stones located in a given box. The natural number 0 will correspond to the absence of any stones; so an empty box is assumed to be holding 0. Hence Fig. 2 shows a configuration in which box B_1 holds stones representing the numeral '2,' and box B_2 holds stones representing the numeral '3.' Ordinary addition $+$ of natural numbers is of course such that

$$+ : \mathbb{N}^2 \mapsto \mathbb{N}.$$

Can Roger compute it? Yes, easily. But in order to see how, we need to specify the format of the instructions that we provide him with.

Each instruction to Roger is one of five possible types. We now define this quintet, by giving the schema for each one, and in addition an intuitive explanation of what the instruction communicates to our bird. Note that every instruction begins with a natural number l that serves as its label.

1. l SET $B_i = B_i - \bullet$

    This instruction tells Roger to take away one stone from box B_i. If this box happens to be empty, Roger doesn't do anything, and simply moves on to the next instruction.

2. l SET $B_i = B_i + \bullet$

    This instruction tells Roger to add one stone to box B_i.

3. l IF $B_i = e$ THEN j ELSE k

    This instruction tells Roger that if box B_i is empty, he should shift his attention to, and follow, instruction with label j; otherwise he should move to instruction with label k.

4. l ROGER, POINT TO B_i

    This instruction tells Roger to point to box B_i, in order to inform us that this is output he wishes us to have.

5. l ROGER, HALT

    This instruction simply tells Roger to halt. In any set of instructions given to Roger (i.e., in any *raven program* given to him), there can only be one instruction of this type.

Let's now put these schemas into action, in the form of a raven program for Roger that carries out addition. In order to do that, we shall assume that box B_1's contents denotes the first of the two numbers to be added, and that box B_2's contents denotes the second. Here then is a program for addition:

$$
\begin{aligned}
&0 \ \text{IF} \ B_1 = e \ \text{THEN} \ 5 \ \text{ELSE} \ 1 \\
&1 \ \text{IF} \ B_2 = e \ \text{THEN} \ 6 \ \text{ELSE} \ 2 \\
&2 \ \text{SET} \ B_2 = B_2 - \bullet \\
&3 \ \text{SET} \ B_1 = B_1 + \bullet \\
&4 \ \text{IF} \ B_2 = e \ \text{THEN} \ 6 \ \text{ELSE} \ 2 \\
&5 \ \text{ROGER, POINT TO} \ B_2 \\
&6 \ \text{ROGER, POINT TO} \ B_1 \\
&7 \ \text{ROGER, HALT}
\end{aligned}
$$

With an initial input of $\bullet\bullet$ in box B_1, and $\bullet\ \bullet\ \bullet$ in box B_2 (i.e., an initial configuration that constitutes a request to Roger that he tells us what $2 + 3$ is), the program given here will cause Roger to point to B_1 when it contains $\bullet\ \bullet\ \bullet\ \bullet\ \bullet$, at which point he will stop.

The overall point of this account of raven machines is to flesh out and make rather obvious the proposition that raven-machine computation, while somewhat informal, is equivalent to register-machine computation ($= R$ computation). In short, raven computation, albeit informal, is nonetheless directly reflective of register-machine computation. In addition, raven machines could be replaced by an indefinite number of schemes of the same intuitive sort, but ones featuring different animals, and/or objects other than stones to be manipulated, etc. All these schemes would preserve reflection of R computation. We hence have the general statement:

SST$_R$ A function f is reflective-R-computable if and only f is R-computable.

Now let's spend just a bit of time on the idealized computing framework that I said above was my favorite: KU machines (or the space of idealized machines denoted by just 'K'). It's my favorite because it's the most cognitively robust model of information-processing to emerge from the mid 20th century, as far as I know. I have my own variant of the original conception that is set out in Kolmogorov and Uspenskii [21]. I have neither the time nor the space to set out either the original conception or my own formal generalization (*workbook machines*) in full detail, but I can certainly give a sufficiently detailed explanation of the latter, and while my formalization is more general that KU machines, workbook machines can serve as an adequate stand-in in the present paper for the more primitive KU-machine framework. (While KU machines are equivalent to Turing machines, workbook machines have settings that can be configured in such a way as to allow these machines to compute functions beyond the Turing Limit in the Arithmetic Hierarchy, e.g. the halting problem.) Also, because workbook machines are built from scratch to be reflected by the ordinary notebooks used by systematic human thinkers through their careers, the account that I now give of workbook machines will serve both to introduce such machines, and to do so by providing an informal correlate of workbook machines. The informal correlate is what I call *notebook machines*. Obviously, the situation is thus such that notebook machines are reflective-K machines.

A workbook machine has an associated formal language L of the type customarily used to specify the syntax of the formulae allowed in a logic, and to specify the inferential machinery by which formulae can be linked to each other. The language is composed of formulae that can be constructed according to a formal grammar of a familiar type (e.g. a BNF grammar) from a list of syntactic ingredients: variables, constants (= names), function symbols, relation symbols, quantifiers, operators (e.g., modal operators), connectives, and punctuation symbols. In addition, and this is unusual, L includes the elements necessary to precisely write expressions that are typically in meta-theory; but we can leave this aside in the present context.[8] The language may also include things necessary to tap into abstract algebra, and thereby move beyond what readers are used to seeing in standard logics to regiment typical symbolic formulae. For example, the language of a given workbook might need to be extended to allow for the precise specification of diagrams; such a language is used in the Vivid family of logics for reasoning over symbolic formulae and

---

[8]The formal language associated with a workbook machine for information-processing in line with logic programming would thus allow formulae in the language of horn-clause logic, and would also allow for meta-logical expressions like $\forall \mathcal{I} : \mathcal{I} \models \mathrm{Ra} \leftrightarrow \mathrm{Ra}$, where $\mathcal{I}$ is an interpretation from model theory.

diagrams [2]. Finally, another part of L is the machinery needed to specify proof theories and argument theories; for example, rules of inference.

We have already implicitly made use of such languages above; for example, when we defined the statement $\bar{u}$ above we made implicit use of the formal language that underlies first-order logic, which has only the two quantifiers $\exists$ and $\forall$, and no operators. But workbooks allow formal languages that are simpler or more complex than the language of first-order logic. In fact, the language for a given workbook may allow formulae that are infinitely long. Most readers will have seen grammars used to define formal languages for logics, so I spend no more time on formulae.

Workbooks are composed of *pages* that come in sequences like a conventional book of the type that people read. Pages can come in any size, as long as the size is finite. Workbooks can have arbitrarily many pages, though here let's confine our attention to books that have only a finite number of pages. What can be written on the pages that are in workbooks? The answer is that formulae in L can be written inside labeled nodes (e.g. inside ovals), the nodes can then be connected by directed arcs, and the arcs can be labeled by such things as the names for inference schemas. For convenience and clarity, the labels can be put inside their own shapes (e.g. boxes).

How does computation happen in a workbook machine? It happens when a *scribe* is given instructions for what modifications to make on a page, within a proper subset of space on the page that is the focus of attention of the scribe. As a profitable example for the reader to consider, imagine that a scribe is given instructions for how to carry out long division. The only differences between how people in the real world carry out long division (on a piece of paper or a blackboard) versus how such an algorithm gets mechanized in a workbook machine is that in the latter case each number must be encased within a labeled node, the arrangement of nodes relative to each other is enforced by arcs connecting them, and a scribe can make what would for some humans be a number of sequential actions on page in one step.

I now provide an example of computation by a scribe in an implemented workbook machine, the Slate environment for producing proofs.[9] The example is shown through two snapshots of a page in Slate. In the first snapshot (see Fig. 3), the scribe's attention is focused on the simple theorem that $0 \neq 5$ (notice that it is highlighted), to be proved from the axioms of Peano Arithmetic (PA) (shown on the left), plus some helpful definitions (shown on the right). In the next snapshot, the theorem has in one step been proved. In order to do this, the scribe has moved AXIOM 1, and cited this axiom along with the definitions for support of a provability claim (viz., that the theorem can be proved). It's very important to realize that this progress has been made in a single step, because when below I model the tracking of Akl's Mars robots it will be a key fact that in a single step the distances of multiple robots from a landmark can be computed.

I have described workbook machines, which are formal, idealized machines that subsume KU machines, by way of the less formal class of notebook machines. Notebook machines, like the chimp machines and raven machines also characterized

---

[9] Slate is provided with Bringsjord and Taylor [9]; an early version is described in Bringsjord et al. [10].

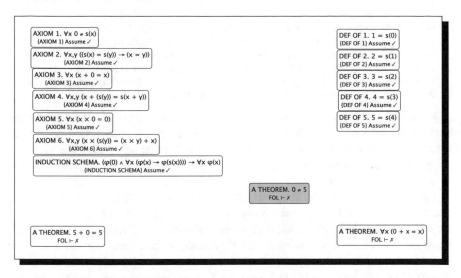

Fig. 3 Snapshot of page built in slate system as workbook machine

above, are not fully formal; however, notebook machines are clearly directly reflective of workbook machines/KU machines (although, again, the former can be set to allow super-Turing computation). In addition, instead of my own conception of notebook machines, which is based on the concept of a scribe, any number of other quasi-formal description of KU machines could be created[10]—and in all these other variants, computational equivalence between them and KU machines would be preserved. Summing up, we have:

SST_K A function f is reflective-K-computable if and only f is K-computable.

While my refutation of Akl's $\mathcal{A}_{\overline{u}_{TM}}$ can be articulated with only SST_K (see Sect. 5), there is no reason, in general, to stay at the level of only instances of SST, and a better version of my counter-argument uses SST itself, as the schema that it is. In order to move to SST itself, in its fully general schematic form, we have only to invoke again quantification over the entire space of Turing-level idealized frameworks for computation, via C. Doing so yields the following general proposition:

SST   For all established idealized computational frameworks C, a function f is reflective-C-computable if and only f is C-computable.

4   Akl's Robot Argument ($\mathcal{A}_{\overline{u}_{TM}}$)

To start his argument $\mathcal{A}_{\overline{u}_{TM}}$, Akl presents a scenario involving multiple robots:

---

[10]Smith [29] provides an alternative.

On the surface of Mars n robots, $R_0, R_1, \ldots, R_{n-1}$, are roaming the landscape. The itinerary of each robot is unpredictable; it depends on the prevailing conditions in the robot's environment, such as wind, temperature, visibility, terrain, obstacles, and so on. At regular intervals, each robot R_i relays its current coordinates $x_i(t) = (a_i(t), b_i(t))$ to mission control on Earth. Given the coordinates of the n robots at time t, mission control determines the distance of $R_i, 0 \leq i \leq n - 1$, to a selected landmark $L(t)$ using a function F_i [1, p. 181].

A key additional aspect of the scenario, which Akl has introduced before giving the scenario described in the quote immediately above (he introduces it on p. 180), is that there is a composite n-ary function $\mathcal{F}$ that takes as its input tuple that which is returned by the F_i. Akl says (p. 180) that $\mathcal{F}$ might return for instance the sum or the minimum of the values of the F_i.

Leaving aside the exotic imaginary setting of Mars, this type of scenario is one that is quite relevant for my own laboratory, which has more than its share of robots, and which specifically often investigates the coördination of multiple robots acting simultaneously in various environments. Of course, we don't use any such low-level formalism as that used to specify Turing machines to track and reason about the diachronic attributes of robots through time. For that matter, *no one* writes sophisticated software to control and coördinate multiple robots in the language of Turing machines, certainly if the robots in question do anything cognitive.[11] Instead, my approach, and derivatively that of the lab I direct, is a logicist one; specifically, we draw from a space $\mathscr{CC}$ of *cognitive calculi* to model both cognitive and physical states of artificial agents (including robots) through time [8, 11, 19, 20]. A cognitive calculus is a highly expressive computational logic, one that in some instances subsumes so-called "BDI" logics, and in some cases includes provision for natural-language understanding and/or generation, uncertainty, and non-monotonic reasoning. In the present paper, it would be inappropriate to review in detail any of the calculi in $\mathscr{CC}$; instead, in the Sect. 5, I will make informal and rapid use of a particular cognitive calculus [2] that allows for the representation of, and reasoning over, pictorial information in human-level fashion. Importantly, the reasoning is such that single steps can comprise what humans working with paper and pencil would need to do in a number of sequential steps; recall the discussion above centering around Figs. 3 and 4. With information expressed pictorially, it turns out that the kind of rapid and real-time processing of Akl's function F_i can be accomplished in the manner he says is beyond the reach of Turing machines. But before we get to this, we of course need to have before us the remainder of Akl's $\mathcal{A}_{\bar{u}_{\mathrm{TM}}}$, which, in his own words, runs as follows:

A Turing Machine fails to compute all the F_i as desired. Indeed, suppose that $x_0(t)$ is read initially and placed onto the tape. It follows that $F_0(x_0(t))$ can then be computed correctly (perhaps at a later time). However, when the next variable x_1, for example, is to be read, the time variable would have changed from t to $t + 1$, and we obtain $x_1(t + 1)$, not $x_1(t)$. Continuing in this fashion, $x_2(t + 2), x_3(t + 3), \ldots, x_{n-1}(t + n - 1)$ are read from the input. In [my example], by the time $x_0(t)$ is read, robots $R_1, R_2, \ldots, R_{n-1}$ would have moved away from $x_1(t), x_2(t), \ldots, x_{n-1}(t)$.

---

[11]Cognitive robotics is, at least in its original form, a logic-based affair; see e.g. Levesque and Lakemeyer [22].

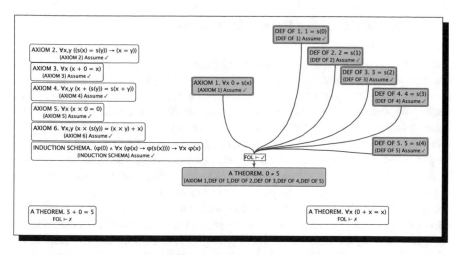

Fig. 4 Snapshot of successor page built in slate system as workbook machine

Since the function according to which each x_i changes with time is not known, it is impossible to recover $x_i(t)$ from $x_i(t+i)$, for $i = 1, 2, \ldots, n-1$. Consequently, this approach cannot produce $F_1(x_1(t)), F_2(x_2(t)), \ldots, F_{n-1}(x_{n-1}(t))$ as required [1, p. 181].

There are some important aspects of this argument that can and should be revealed by some analysis. First, when Akl says "A Turing Machine fails to compute all the F_i as desired." he is not to be interpreted as saying only that there exists a Turing Machine that fails to compute all the F_i. The kernel of the logical shape of this trivial proposition would be

$$\exists m\, (TM(m) \wedge FailsCompute(m, F_i)], \tag{1}$$

whereas what Akl is claiming is (as must be the case if he is to succeed) in line with the claim $\bar{u}_{\text{TM}}$, and the logical shape of his claim is the much more ambitious and much more interesting

$$\neg \exists m\, (TM(m) \wedge Compute(m, F_i)]. \tag{2}$$

This really is a very ambitious claim indeed. For there are a lot of Turing machines; there are overall a countably infinite number of them, of course. How does Akl know that no Turing machine in this vast space can compute F_i? His reasoning appears to be that while the configuration of the robots and the landmark at a given time t can be placed into memory (= onto a tape of a given Turing machine), there isn't enough time to compute all the distances, because by the next timepoint $t+1$ the robots have moved and the distances will be different—and to further complicate matters the environmental forces that partially determine the itineraries (I would say *plans*, and I imagine that the robots are running AI planners) of the robots aren't retrievable. But is this reasoning sound? I don't think so, and now I explain why.

5   A Refutation of the Robot Argument

Since Akl concedes F_i to be a Turing-computable function, there exists a Turing machine able to compute the distance of each robot R_j from the landmark at every timepoint t, *before* another such computation needs to occur. By definition, we are dealing with number-theoretic functions, so we are dealing with a digitization of the entire scenario. In 'the 'real world" the robots must take some time to travel to new locations, and when they have arrived at those new locations, a new configuration can be perceived, and processed afresh. So yes, I certainly agree that the solution for a Turing machine is not to be found in computation carried out, as Akl says, "later." But the solution is to be found in the fact that there exists some Turing machine $m^\star$ that computes F_i, for all relevant i, very quickly and at once, in parallel, in an *intervening* moment, *before* the robots arrive at their new locations. (Even if robot movement is staggered in time, it remains possible, in principle, to *any* configuration of any subset of the entire collection of robots to be perceived and distances to be computed at a single intervening timestep.) And having computed that, m^* can compute $\mathcal{F}(F_i)$, for all relevant i, in another intervening moment. As Akl says, the composite function $\mathcal{F}$ might return something like the sum of all the distances of each robot from the landmark, at a given timepoint.

Of course, Akl's claim is that the "intervening" activity I have described is excluded. But what excludes it? It is true that Akl can stipulate a constraint according to which there is no intervening time available to be used. But such a stipulation merely establishes Eq. 1; it doesn't establish what he needs: Eq. 2.

An even more severe problem for Akl is that such back-and-forth dialectic is entirely irrelevant, because there is a non-constructive way of establishing that there does exist the Turing machine $m^\star$ that can compute F_i. In fact, I now give such a non-constructive way: a formally valid argument whose conclusion is that the problem in this case, *contra* Akl, can be solved by some TM (i.e., I establish the negation of Eq. 2). (Since my argument is formally valid on any standard proof theory, I could classify my argument as an outright proof, save for the fact that "Selmer's Safer Thesis," SST, isn't itself proved in the present paper—though certainly it can be proved.) My refutation is in defense of the proposition that, relative to standard idealized computation — that is computation characterized by P, R, M, G, Λ, K, and the other frameworks that can be readily found in the literature—Turing machines are universal.

As mentioned above (Sect. 2), I can't employ CTT in any refutation of Akl's argument (since, again, I believe that, and indeed believe that I've shown that, CTT is in fact false). However, I'm able to use SST and a the particular instance of it for KU machines. Here's my argument:

Refutation of $\mathcal{A}_{\overline{u}_{TM}}$

SST tells us that any number-theoretic function f that is reflective-C-computable is C-computable. It follows directly by universal instantiation on C that any f that is reflective-K-computable is K-computable. (Or we could of course use SST_K directly.) Where F_i is the function (informally) defined by Akl that maps location information regarding n robots to the distance of each robot from a given landmark (all indexed, of course, to a particular time t), F_i is reflective-K-computable; hence F_i is K-computable. But it's a theorem that K-computability and T-computability (= Turing-computability) are equivalent. Therefore, *contra* Akl, F_i is Turing-computable (and with it also $\mathcal{F}$). **QED**

In order to undergird this argument, I introduce a simple pictorial framework that enables us to represent snapshots of the locations of robots and the landmark on Mars. In this framework, which is grid-based, $\bullet_j$ indicates a robot at the relevant location, and $+$ indicates the location of the landmark. Each configuration of the grid corresponds to a page in a workbook machine. (I leave out messy ovals to define nodes, and explicit arcs, in order to increase readability.) Consider the following configuration:

$$
\begin{array}{cccccc}
\circ & \circ & \circ & \circ & \circ & \circ \\
\circ & \circ & \circ & \circ & \circ & \circ \\
\circ & \circ & \circ & \circ & \circ & \circ \\
\circ & \circ & \circ & \circ & \circ & \circ \\
\circ & \circ & \circ & \bullet_1 & + & \circ \\
\circ & \circ & \circ & \circ & \bullet_2 & \circ
\end{array}
$$

Here there are only two robots (but there is no loss of generality). Please try to make a quick ruling as to how far each robot is from the landmark. . . . Correct, both robots are the same distance (1 unit) from the landmark. Notice that you rendered this verdict by taking in the workspace as a whole. Now here is a second configuration:

$$
\begin{array}{cccccc}
\circ & \circ & \circ & \circ & \circ & \circ \\
\circ & \circ & \circ & \circ & \bullet_1 & \circ \\
\circ & \circ & \circ & \circ & \circ & \circ \\
\circ & \circ & \circ & \circ & \circ & \circ \\
\circ & \circ & \circ & \circ & + & \circ \\
\circ & \circ & \circ & \circ & \bullet_2 & \circ
\end{array}
$$

To ensure that you understand what is being depicted in this second configuration, I ask again: How far is robot R_2 from the landmark, and how far is robot R_1 from the landmark? Once again, I'm quite sure that you can see what the answer is: R_1 is 3 units away, and R_2 is 1 away. This shows that a scribe in a workbook machine could, presented with any such configuration, write down in one step, the distance for each robot R_i. This in turn shows that F_i is reflective-K-computable, since in principle there is no reason why what you have instantly seen in the two configurations just given can't be seen in *any* such configuration, by a scribe perceiving a page in a workbook. The area of focus in any given configuration will never be too large to in principle be taken in, since it is always going to be finite, and since it will never increase from one page to the next in an expansion that grows non-recursively fast. (There are only a finite number of robots in Akl's scenario, separated from each other and the landmark by only finite distances.) This suffices to undergird the non-constructive deductive counter-argument given above.

6 Concluding Remarks on Another Path

Alert-and-astute readers know that the failure of Akl's robot argument $\mathcal{A}_{\bar{u}_{TM}}$, as revealed above, is formally consistent with the statement $\bar{u}$ that there doesn't exist a universal computer. (This is a fact that Akl himself, in rebuttal, might well convey.) Hence, in the current state of the inquiry into whether or not a universal computer is a myth, we are unable to resolve the central question. After all, perhaps one of the *other* arguments given by Akl for $\bar{u}_{TM}$ succeeds. However, for what it's worth, I've analyzed each of these arguments and find each to be at best inconclusive. I encourage those interested in getting to "ground truth" on $\bar{u}$ not to accept on faith this report on my analysis, but to study Akl's inventive arguments for themselves. That said, I do want to end, as promised, by introducing the reader to what I see as a better route for settling the central question. My intuition, based on initial reflection upon this other route, is that universal computation does in fact exist, and that therefore u (notice that the overline is gone) holds.

So what is the route I recommend? To see its general shape and direction, suppose, first, that *universal computation*, which we'll symbolize by the predicate U, is stipulated to be a *disjunctive* concept, one with so much in-built latitude that it ranges across all forms of information-processing, not just computation as it's systematized in standard Turing-level-and-below information processing, and indeed not just information-processing as it's formalized in the entire Arithmetic Hierarchy (of which the Turing-computable portion is only a small part). (Akl's writings never allow under U forms of information-processing beyond AH, as far as I can tell.) We therefore admit information-processing over uncountable sets. So far this is quite imprecise, of course. But the disjunction can be made precise by appeal to formal logics — as long as we countenance formal logics of more and more power, including those that exceed information-processing in AH. Let me explain, at least to a degree.

We know that to capture the behavior and power of standard Turing machines, and any rigorous form of information-processing at and below this level in AH (e.g., to harken back to the categories deployed above, register machines, the λ-calculus, KU machines, etc.), we can use standard first-order logic $\mathscr{L}_1$. This is shown in excruciating detail in traditional proofs of the undecidability of theoremhood in $\mathscr{L}_1$ (= the undecidability of the *Entscheidungsproblem*), wherein deciding theoremhood in $\mathscr{L}_1$ is (frequently) reduced to the halting problem [4]. Encapsulating, we know that a number-theoretic function f is Turing-computable if and only if, from a suitable theory Γ, representing the operation of a standard Turing machine and initial information (including a given element a of the domain of f),

$$\Gamma \vdash \phi_{f(a)},$$

where $\phi_{f(a)}$ is the formulae expressing the value that f returns on a as an argument, and where $\vdash$ is interpreted in standard form, that is to indicate provability in some typical proof theory for $\mathscr{L}_I$. (Such a proof theory underlies the single-step inference shown in Fig. 4.)

For distinctions within the sub-space of Turing-computable functions that pertain to how much time it takes for Turing machines and automata below them to compute a relevant function (the sub-space that covers such categories as NP-complete), that too can be captured by $\mathscr{L}_I$ with suitable function symbols and relation symbols to capture time and change, number of steps in a computation, and size of input. But we also know that once we for instance move from finitary logic (and $\mathscr{L}_I$ is of course certainly finitary: all its formulae are of finite length, as are all its formal, object-level proofs) to infinitary logic, we can quickly move to information-processing that is beyond Turing machines.[12] (In parallel, we know that such a move allows us to surmount Gödelian incompleteness, since such results are based on Turing-level axiom systems in $\mathscr{L}_I$, such as Peano Arithmetic.) For a quick example, note that the "small" infinitary logic $\mathscr{L}_{\omega_1\omega}$ allows countably infinite disjunctions and conjunctions, and countably infinite proofs. Using infinitary logics, we can build up coverage of increasingly challenging functions to compute, where we express the computing of a function g in terms of what is expressible and deducible in the relevant logic, following the general recipe sketched above in the case of $\mathscr{L}_I$, where what is to be proved is that from a declarative representation of a given argument a in g's domain, $g(a)$ is what is returned. If we take this route, we can say that a universal computing framework, that is a framework to which can be accurately ascribed the relation U for universal computation, is one which, given any well-defined function g and input a, can prove $g(a)$, in either logic $\mathscr{L}_1$, or $\mathscr{L}_2$, or $\mathscr{L}_3$, or $\ldots$, where this is a progression of increasingly powerful logics. I wonder what Akl would say about this route. One thing certainly seems clear: This disjunctive, logicist route, without all that much work, would yield a precise framework on which all the challenges in the remarkably fertile and suggestive [1] can be modeled and thereby met. In fact, the more rigorous and accurate is Akl's reasoning in setting out a challenge, the easier such modeling, for some logic $\mathscr{L}_k$, becomes.

---

[12]Readers interested in learning more, can consult as a starting point the excellent [3].

References

1. Akl, S.: The myth of universal computation. In: Vajteršic, M., Trobec, R., Zinterhof, P., Uhl, A. (eds.) Parallel Numerics '05: Theory and Applications, University of Saltzburg, pp. 167–192. Saltzburg, Austria. https://www.cosy.sbg.ac.at/events/parnum05/book/prolog.pdf (2005)
2. Arkoudas, K., Bringsjord, S.: Vivid: an AI framework for heterogeneous problem solving. Artif. Intell. **173**(15), 1367–1405. http://kryten.mm.rpi.edu/vivid_030205.pdf (2009). (The URL http://kryten.mm.rpi.edu/vivid/vivid.pdf provides a preprint of the penultimate draft only. If for some reason it is not working, please contact either author directly by email)
3. Barwise, J.: Infinitary logics. In: Agazzi, E. (ed.) Modern Logic: A Survey, pp. 93–112. Reidel, Dordrecht, The Netherlands (1980)
4. Boolos, G.S., Burgess, J.P., Jeffrey, R.C.: Computability and Logic, 4th edn. Cambridge University Press, Cambridge, UK (2003)
5. Bringsjord, S.: Theorem: general intelligence entails creativity, assuming .... In: Besold, T., Schorlemmer, M., Smaill. A. (eds.) Computational Creativity Research: Towards Creative Machines, pp. 51–64. Atlantis/Springer, Paris, France. http://kryten.mm.rpi.edu/SB_gi_ implies_creativity_061014.pdf (2015). (This is Volume 7 in *Atlantis Thinking Machines*, edited by Kühnbergwer, Kai-Uwe of the University of Osnabrück, Germany)
6. Bringsjord, S., Arkoudas, K.: On the provability, veracity, and AI-relevance of the Church-Turing thesis. In: Olszewski, A., Wolenski, J., Janusz, R. (eds.) Church's Thesis After 70 Years, pp. 66–118. Ontos Verlag, Frankfurt, Germany. http://kryten.mm.rpi.edu/ct_bringsjord_ arkoudas_final.pdf (2006). (This book is in the series *Mathematical Logic*, edited by W. Pohlers, T. Scanlon, E. Schimmerling, R. Schindler, and H. Schwichtenberg)
7. Bringsjord, S., Govindarajulu, N.S.: In defense of the unprovability of the Church-Turing thesis. J. Unconv. Comput. **6**, 353–373. http://kryten.mm.rpi.edu/SB_NSG_CTTnotprovable_ 091510.pdf (2011). (Preprint available at the url given here)
8. Bringsjord, S., Govindarajulu, N.S.: Given the web, what is intelligence, really? Metaphilosophy **43**(4), 361–532. http://kryten.mm.rpi.edu/SB_NSG_Real_Intelligence_040912.pdf (2012). (This URL is to a preprint of the paper)
9. Bringsjord, S., Taylor, J.: Logic: A Modern Approach: Beginning Deductive Logic. Motalen, Troy, NY (2015). (This is an e-book edition of January 25 2016. The book is accompanied by the Slate software system, ISBN of 78-0-692-60734-3, and version of January 25 2016)
10. Bringsjord, S., Taylor, J., Shilliday, A., Clark, M., Arkoudas, K.: Slate: an argument-centered intelligent assistant to human reasoners. In: Grasso, F., Green, N., Kibble, R., Reed, C. (eds.) Proceedings of the 8th International Workshop on Computational Models of Natural Argument (CMNA 8), pp 1–10. University of Patras, Patras, Greece. http://kryten.mm.rpi.edu/Bringsjord_ etal_Slate_cmna_crc_061708.pdf (2008)
11. Bringsjord, S., Licato, J., Govindarajulu, N., Ghosh, R., Sen, A.: Real robots that pass tests of self-consciousness. In: Proccedings of the 24th IEEE International Symposium on Robot and Human Interactive Communication (RO-MAN 2015), pp 498–504. IEEE, New York, NY. http://kryten.mm.rpi.edu/SBringsjord_etal_self-con_robots_kg4_0601151615NY.pdf (2015). (This URL goes to a preprint of the paper)
12. Church, A.: A set of postulates for the foundation of logic (Part I). Ann. Math. **33**(2), 346–366 (1932)
13. Church, A.: A set of postulates for the foundation of logic (Part II). Ann. Math. **34**, 839–864 (1933)
14. Church, A.: An unsolvable problem of elementary number theory. Am. J. Math. **58**(2), 345–363 (1936)
15. Copeland, J.: The Church-Turing thesis. In: Zalta, E. (ed.) The Stanford Encyclopedia of Philosophy. http://plato.stanford.edu/archives/sum2015/entries/church-turing (2002)
16. Ebbinghaus, H.D., Flum, J., Thomas, W.: Mathematical Logic. Springer, New York, NY (1984)
17. Ebbinghaus, H.D., Flum, J., Thomas, W.: Mathematical Logic, 2nd edn. Springer, New York, NY (1994)

18. Feferman, S.: Turing in the land of O(Z). In: Herken, R. (ed.) The Universal Turing Machine, 2nd edn, pp. 103–134. Springer, Secaucus, NJ (1995)
19. Govindarajalulu, N.S., Bringsjord, S., Taylor, J.: Proof verification and proof discovery for relativity. Synthese **192**(7), 2077–2094 (2015)
20. Govindarajulu, N.S., Bringsjord, S.: Ethical regulation of robots must be embedded in their operating systems. In: Trappl, R. (ed.) A Construction Manual for Robots' Ethical Systems: Requirements, Methods, Implementations, pp. 85–100. Springer, Basel, Switzerland. http://kryten.mm.rpi.edu/NSG_SB_Ethical_Robots_Op_Sys_0120141500.pdf (2015)
21. Kolmogorov, A., Uspenskii, V.: On the definition of an algorithm. Uspekhi Matematicheskikh Nauk **13**(4), 3–28 (1958)
22. Levesque, H., Lakemeyer, G.: Chapter 24: cognitive robotics. In: Handbook of Knowledge Representation. Elsevier, Amsterdam, The Netherlands. http://www.cs.toronto.edu/~hector/Papers/cogrob.pdf (2007)
23. Lewis, H., Papadimitriou, C.: Elements of the Theory of Computation. Prentice Hall, Englewood Cliffs, NJ (1981)
24. Mendelson, E.: Second thoughts about Church's thesis and mathematical proofs. J. Philos. **87**(5), 225–233 (1986)
25. Partee, B., Meulen, A., Wall, R.: Mathematical Methods in Linguistics. Kluwer, Dordrecht, The Netherlands (1990)
26. Post, E.: Finite combinatory processes-formulation 1. J. Symb. Log. **1**(3), 103–105 (1936)
27. Post, E.: Recursively enumerable sets of positive integers and their decision problems. Bull. Am. Math. Soc. **50**, 284–316 (1944)
28. Savage, J.: Models of Computation: Exploring the Power of Computing. Addison-Wesley, Reading, MA. http://cs.brown.edu/~jes/book/pdfs/ModelsOfComputation.pdf (1998). (The URL given here goes to a free version of the book available online)
29. Smith, P.: An Introduction to Gödel's Theorems. Cambridge University Press, Cambridge, UK (2013). (This is the second edition of the book)
30. Turing, A.: Dissertation for the PhD: "Systems of Logic Based on Ordinals". Princeton University, Princeton, NJ (1938)
31. Turing, A.M.: On computable numbers with applications to the entscheidungsproblem. Proc. Lond. Math. Soc. **42**, 230–265 (1937)

A Hierarchy for $BPP//\log\star$ Based on Counting Calls to an Oracle

Edwin Beggs, Pedro Cortez, José Félix Costa and John V Tucker

Abstract Algorithms whose computations involve making physical measurements can be modelled by Turing machines with oracles that are physical systems and oracle queries that obtain data from observation and measurement. The computational power of many of these physical oracles has been established using non-uniform complexity classes; in particular, for large classes of deterministic physical oracles, with fixed error margins constraining the exchange of data between algorithm and oracle, the computational power has been shown to be the non-uniform class $BPP//\log\star$. In this paper, we consider non-deterministic oracles that can be modelled by random walks on the line. We show how to classify computations within $BPP//\log\star$ by making an infinite non-collapsing hierarchy between $BPP//\log\star$ and BPP. The hierarchy rests on the theorem that the number of calls to the physical oracle correlates with the size of the responses to queries.

1 Introduction

Consider algorithms that request and receive data from an external source in the course of their computations. These algorithms abound and can be found in all sorts of monitoring and control systems. We suppose these algorithms are modelled by Turing machines with oracles that are physical systems, and whose oracle queries ask

E. Beggs · J.V. Tucker
College of Science, Swansea University, Singleton Park, Swansea,
SA2 8PP Wales, UK
e-mail: e.j.beggs@swansea.ac.uk

J.V. Tucker
e-mail: j.v.tucker@swansea.ac.uk

P. Cortez · J.F. Costa (✉)
Department of Mathematics, Instituto Superior Técnico and Centro de Filosofia
das Ciências da Universidade de Lisboa, Lisboa, Portugal
e-mail: fgc@math.ist.utl.pt

P. Cortez
e-mail: pedro.cortez.91@gmail.com

© Springer International Publishing Switzerland 2017
A. Adamatzky (ed.), *Emergent Computation*, Emergence, Complexity
and Computation 24, DOI 10.1007/978-3-319-46376-6_3

and obtain data by means of some process to measure a physical quantity. Essentially, through a measurement procedure, the Turing machine will access a sequence of approximations to a real number.

Starting in [5, 6], we began a theoretical investigation of such physical oracles, focussing on classic deterministic physical experiments. To guide our thinking we conceived an abstract experimenter using some physical equipment to undertake an abstract experiment to measure a physical quantity. The Turing machine modelled the experimental procedure and the data from the oracle modelled observations of the equipment: see [5, 6, 10, 12, 13, 16] inter alia.

Technically, we examined what was involved in an algorithm requesting and receiving data from a physical process, and especially interface properties to do with

(a) the error margins involved in the data: the queries could have *infinite precision*, being exact or having finite but vanishingly small errors; or have a *finite precision* that is a fixed error margin;
(b) the time taken by the algorithm to acquire the data: the queries need not take one computational step or unit time, but may take time depending on the size of the query.

We also placed complexity constraints on the computations, especially *polynomial time*.

The computational power of many of these physical oracles has been established using non-uniform complexity classes. These have the general form $\mathcal{B}/\mathcal{F}$ consisting of a complexity class $\mathcal{B}$ equipped with class $\mathcal{F}$ of special oracles called advice functions. An advice function is a map $f : \mathbb{N} \to \Sigma^*$ that provides extra data $f(n)$ to the Turing machine when computing with inputs of size $n \in \mathbb{N}$. Advice functions are suitable for representing real numbers (in binary, say). Typically, we take $\mathcal{B}$ to be the class P, defined by polynomial time deterministic Turing machines; or to be the class BPP, defined by polynomial time Turing machines governed by fair probability distributions. We take $\mathcal{F}$ to be based on logarithms.

Through a detailed investigation of protocols between analogue and digital components of many types of system (see [8, 13]), we established the computational power of these oracles as follows.

For infinite precision measurements, in deterministic polynomial time, the computational power was shown to be $P/\log\star$. However, in the more realistic case of finite fixed precision measurements in deterministic polynomial time, the computational power was shown to be $BPP//\log\star$. This was done for a wide variety of physical oracles and led to a thesis proposing $BPP//\log\star$ as a limit to computation [15]. The probabilistic form of $BPP//\log\star$ is due to the use of probabilities to handle fair choices of data from within the fixed-size error intervals of the deterministic physical oracle. Probabilistic oracles are the subject of [4].

Our attempts to model measurement algorithmically addressed a longstanding question, first formulated by Geroch and Hartle in their intriguing paper [20]: What are the physically measurable numbers? Are the measurable numbers computable numbers? Measurement is a scientific activity supported by a full theory developed throughout the last century as a chapter of mathematical logic (see [21]). Our computational theory of measurement started in [9, 10] and focussed on the time needed to make a measurement; here we consider the amount of data involved in making a measurement.

The data provided by the oracle is constrained by

(i) the size of responses to queries, and
(ii) the frequency of calls to the oracle.

The size of the data can be controlled by the size of the values of the advice functions $|f(n)|$. We will show that for $BPP//\log\star$, for inputs of size n, the amount of bits translates into a modest number of calls to the oracle, which is poly-logarithmic in n.

In this paper, we also introduce the possibility of using physical oracles whose behaviour is modelled stochastically, as one finds in statistical mechanics. Imagine a physical experiment modelled by a random walk on the line, as discussed in [19]. The oracle is non-deterministic and can be connected to a Turing machine that can be deterministic or non-deterministic: we will need both. Specifically, we will use Turing machines and fair probabilistic Turing machines.

Let $\log^{(k)}$ be the class of advice functions $f : \mathbb{N} \to \Sigma^*$ such that $|f(n)| \in O(\log^{(k)}(n))$. Let poly($\log^{(k)}$) be the class of polynomial functions in $\log^{(k)}$. We prove the following:

Theorem 1 *The class of sets decidable in polynomial time by RW fair probabilistic Turing machines that can make up to* poly($\log^{(k)}(n)$) *calls to the RW oracle, for inputs of size n, is exactly $BPP//\log^{(k+1)}\star$.*

The hierarchy of complexity classes within $BPP//\log\star$ we establish starts with $BPP//\log\star$ and approaches arbitrarily close to BPP.

We show strict boundedness, i.e., $k \geq 0$, $\log^{(k+1)} \prec \log^{(k)}$. In particular, this is true for $k \geq 1$ and we have the following infinite descending chain

$$\cdots \prec \log^{(4)} \prec \log^{(3)} \prec \log^{(2)} \prec \log,$$

which can generate a hierarchy as in the figure.

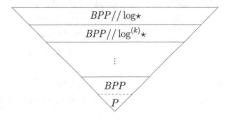

Theorem 2 *The classes of sets decided by RW fair probabilistic Turing machines that can make up to*

$$\cdots \subsetneqq \text{poly}(\log^{(3)}(n)) \subsetneqq \text{poly}(\log^{(2)}(n)) \subsetneqq \text{poly}(\log(n)) \subsetneqq \text{poly}(n)$$

calls to the RW oracle coincides with the descending chain of sets

$$\cdots \subsetneq BPP//\log^{(4)}\star \subsetneq BPP//\log^{(3)}\star \subsetneq BPP//\log^{(2)}\star \subsetneq BPP//\log\star ,$$

respectively.

While measuring a physical magnitude, a slight amount of bits of the binary representation of a real number, relative to the size of the input, can originate hyper-computation.

It is striking the extent to which the class $BPP//\log\star$ arises naturally in exploring physical systems and in physically inspired computational models. However other non-uniform classes have been found useful. The computational power of determin-istic neural networks having access to real numbers in polynomial time was proposed to be P/poly in [24]. These results contrast with our many results involving $P/\log\star$: our reduction of power in deterministic time is due to the fact that measurement takes time in non-linear systems, while in [24] the systems considered are piecewise linear. However, inspired by the work in [24], the authors of [26] specify hardware presum-ably designed to be capable of computing a non-decidable fragment $BPP//\log\star$. In our view such systems will not support programming, since programming in such a context will the introduction of a real number into the system with unbounded precision. Eventually, such systems will be capable of emergent computation due to arbitrary unknown reals (if real numbers exist in Nature) specifying their compo-nents. Emergent computational activities might well be relevant in learning tasks.

2 Random Walk Oracles

2.1 Random Walk

Consider the random walk experiment (RWE) of having a particle moving along an axis. The particle is sent from position $x = 0$ to position $x = 1$. Then, at each positive integer coordinate, the particle moves right, with probability σ, or left, with probability $1 - \sigma$, as outlined in Fig. 1. If the particle ever returns to its initial position $x = 0$, then it is absorbed. In this process, the particle takes steps of one unit, at time intervals also of one unit, postulated to be the time step of a Turing machine transition (see [25]).

We are interested in the probability that the particle is absorbed (see [22]). Let p_i be the probability of absorption when the particle is at $x = i$. In our model, the

Fig. 1 Random walk on the line with absorption at $x = 0$

particle is launched from $x = 0$ but it only starts its random walk at $x = 1$. It is easy to see that $p_1 = (1 - \sigma) + \sigma p_2$. From $x = 2$, to be absorbed, the particle must initially move from $x = 2$ to $x = 1$ (not necessarily in one step), and then from $x = 1$ to $x = 0$ (again, not necessarily in one step). Both movements are made, independently, with probability p_1, thus, p_2 is just p_1^2. More generally, we have $p_k = p_1^k$. Therefore, the equation for the unidimensional random walk with absorption at $x = 0$ is given by the equation

$$p_1 = (1 - \sigma) + \sigma p_1^2,$$

with solutions $p_1 = 1$ and $p_1 = \frac{1-\sigma}{\sigma}$. For $\sigma = \frac{1}{2}$, the solutions coincide and $p_1 = 1$. For $\sigma < \frac{1}{2}$, the second solution is impossible, because $\frac{1-\sigma}{\sigma} > 1$, so, we must have $p_1 = 1$. For $\sigma = 1$, the particle always moves to the right, so $p_1 = 0$. Thus, for the sake of continuity of p_1, for $\sigma > \frac{1}{2}$, we must choose $p_1 = \frac{1-\sigma}{\sigma}$. Consequently, we get

$$p_1 = \begin{cases} 1 & \text{if } \sigma \leq \frac{1}{2} \\ \frac{1-\sigma}{\sigma} & \text{if } \sigma > \frac{1}{2} \end{cases}.$$

So, if $\sigma \leq \frac{1}{2}$, with probability 1 the particle always returns, but the number of steps is unbounded. In Fig. 2, we illustrate this situation, for the case $\sigma = 1/4$, giving the possible locations of the particle, and the respective probabilities, after the first steps.

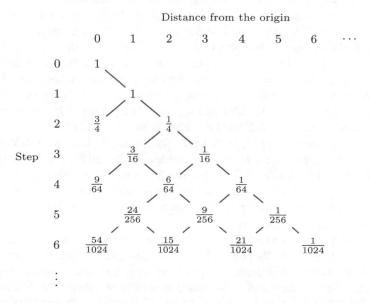

Fig. 2 Diagram showing probabilities of the particle being at various distances from the origin, for the case of $\sigma = 1/4$

2.2 Machines with Random Walk Oracles

We will combine the RWE with both Turing machines and fair probabilistic Turing machines. Probabilistic Turing machines have been around since the 1950s and have a number of equivalent formulations. For example, the machine may randomly choose between the available transitions at each step with probability $\frac{1}{2}$. Perhaps the most elegant and easiest way to describe them is to say that they have access to a fair independent coin toss oracle, returning values 'heads' or 'tails' with probability $\frac{1}{2}$. Whilst the definition of the machines can be shown to converge, the different criteria in use for recognising strings do not.

Definition 1 Consider any form of Turing machine that gives probabilistic results, e.g. a Turing machine with any form of random oracle. A set $A \subset \{0, 1\}^*$ is accepted by such a Turing machine $\mathcal{M}$ in polynomial time if there is a $\gamma < 1/2$ so that for for every input w, $\mathcal{M}$ halts in polynomial time and

- If $w \in A$, $\mathcal{M}$ accepts w with error probability bounded by γ;
- If $w \notin A$, $\mathcal{M}$ rejects w with error probability bounded by γ.

For example, fair probabilistic Turing machines are used to define the class *BPP* with the criterion that any given run of the algorithm, it has a probability of (say) at most $\frac{1}{3}$ of giving the wrong answer, whether the answer is accept or reject. Fair probabilistic Turing machines are required for our main theorems.

Now, let us consider a Turing machine coupled with a random walk experiment, as introduced in [19]. To use the RWE as an oracle, we admit that the probability σ that the particle moves forward, encodes some advice. Unlike scatter machine experiments in [1, 6, 12], the RWE does not need any parameters to be initialized, i.e., the Turing machine does not provide the oracle with any dyadic rational, it just "pulls the trigger" to start the experiment. We consider both a Turing machine with added RWE oracle, a *RW Turing machine*, and a fair probabilistic Turing machine with added RWE oracle, a *RW fair probabilistic Turing machine*.

For every unknown $\sigma \in (0, 1)$, the time that a particle takes to be absorbed is unbounded. We introduce a constant time schedule to bound the oracle consultation time. If the particle is absorbed during that time, the finite control of the Turing machine changes to the 'yes' state, otherwise, the finite control changes to the 'no' state. The experiment has two possible outcomes and a constant time schedule.

We analyse the probability of 'yes'.

A path of the random walk is a possible sequence of moves that the particle makes until it is absorbed. Note that all such paths are made of an even number of steps. Paths of the random walk along the positive x-axis with absorption at $x = 0$ are isomorphic to a specific set of well-formed sequences of parentheses. For instance, in a random walk of length 6, the particle could behave as $((()))$ or $(()())$, where a movement to the right is represented by "(" and a movement to the left is represented by ")". The first opening parenthesis corresponds to the first move of the particle from $x = 0$ to $x = 1$. The probability of answer in 6 steps is the sum of two probabilities corresponding to the two possible paths. All paths of a certain length have the same probability; namely, for every even number n, the probability of each path of length n is

$$\sigma^{\frac{n}{2}-1}(1-\sigma)^{\frac{n}{2}}.$$

Therefore, we only need to know the number of possible paths for each length, i.e., the number of well-formed sequences of parentheses satisfying some properties. In [17], the authors generalize the Catalan numbers and prove the following interesting result:

Proposition 1 *(Blass and Braun [17]) For every $\ell, w \in \mathbb{Z}$, $\ell \geq w \geq 0$, let X be the number of strings consisting of ℓ left and ℓ right parentheses, starting with w consecutive left parentheses, and having the property that every nonempty, proper, initial segment has strictly more left than right parentheses. Then*

$$X = \frac{w}{2\ell - w} \binom{2\ell - w}{\ell}$$

Note that when $w = \ell = 0$, the undefined fraction $w/(2\ell - w)$ is to be interpreted as 1, since this gives the correct value $X = 1$, corresponding to the empty string of parentheses. From this proposition, we derive the probability $q(t)$ that the particle is absorbed in even time $t + 1$, for $t \geq 1$. It suffices to take $\ell = (t+1)/2$ and $w = 1$:

$$q(t) = \frac{1}{t}\binom{t}{\frac{t+1}{2}}(1-\sigma)^{\frac{t+1}{2}}\sigma^{\frac{t+1}{2}-1}.$$

Therefore, the probability that the particle is absorbed during the time schedule T is given by

$$F(\sigma, T) = \sum_{\substack{t=1 \\ t \text{ odd}}}^{T-1} \frac{1}{t}\binom{t}{\frac{t+1}{2}}(1-\sigma)^{\frac{t+1}{2}}\sigma^{\frac{t+1}{2}-1}.$$

This is the probability of getting the outcome 'yes' from the oracle. Figure 3 allows us to understand the behaviour of the probability $F(\sigma, T)$ as a function of σ. We see that, as T increases, $F(\sigma, T)$ increases as well, since the longer the machine waits, the more likely it is that a particle is absorbed. We can also see that as T approaches infinity, $F(\sigma, T)$ approaches the probability p_1 that the particle is absorbed, which makes sense, since p_1 represents a probability of absorption with unbounded time. For analytical reasons, we will consider only $\sigma \in [\frac{1}{2}, 1]$, corresponding to a variation of p_1 from 1 to 0. Note that we could consider any interval contained in $[0, 1]$. For every T, this probability is a function of σ that satisfies the following conditions:

(a)　$F(\bullet, T) \in C^1([\frac{1}{2}, 1])$,
(b)　for every $\sigma \in [\frac{1}{2}, 1]$, $F'(\sigma, T) \neq 0$ and
(c)　n bits of $F(\bullet, T)$ are computable in time $O(2^n)$.

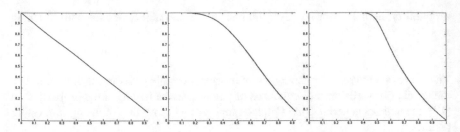

Fig. 3 Graphs of $F(\sigma, T)$ for $T = 2$, $T = 10$ and $T = 100$

These conditions are the basis of an axiomatisation *SPO* of stochastic physical oracles in the forthcoming paper [4], and from which take the following theorem:

Theorem 3 *For every set A, $A \in BPP//\log\star$ if, and only if, it is decidable by a RW Turing machine in polynomial time.*

3 Computational Resources

Consider that we have a limiting number of particles that the RW Turing machine can launch, i.e., a bound in the number of oracle calls that the machine can make. We study now how the precision in the measurement of σ depends on the number of oracle calls.

Theorem 4 *A RW Turing machine, or a RW fair probabilistic Turing machine, that can make up to $\xi(n)$ calls to the RW oracle, on input w of size $|w| = n$, can read $\frac{1}{2}\log(\xi(n)) + c$ bits of the unknown parameter σ, where c is a fixed constant, in polynomial time.*

Proof The proof is common to both types of Turing machine. We know that each particle has probability of absorption $F(\sigma, T)$ in time T. Thus, if we make $\xi(n)$ oracle calls on an input of size n, the number of times α that the experiment returns 'yes' is a random variable with binomial distribution. Let us consider $X = \alpha/\xi(n)$, the random variable that represents the relative frequency of absorption ('yes'). We have the expected value $\mathbb{E}[X] = \mathbb{E}[\alpha]/\xi(n) = \xi(n)F(\sigma, T)/\xi(n) = F(\sigma, T)$ and the variance $\mathbb{V}[X] = \mathbb{V}[\alpha]/\xi(n)^2 = \xi(n)F(\sigma, T)(1 - F(\sigma, T))/\xi(n)^2 = F(\sigma, T)$ $(1 - F(\sigma, T))/\xi(n)$. Chebyshev's inequality states that, for every $\delta > 0$,

$$P(|X - \mathbb{E}[X]| > \delta) \leq \frac{\mathbb{V}[X]}{\delta^2} \leq \frac{F(\sigma, T)(1 - F(\sigma, T))}{\xi(n)\delta^2} \leq \frac{F(\sigma, T)}{\xi(n)\delta^2}.$$

Let k be the number of bits of σ to be read.[1] This means that we have to find σ up to an error of 2^{-k-5}. To do this, we first estimate the probability $F(\sigma)$ up to an error δ, and then run a bisection algorithm to find the value of σ (this may require polynomial time). The value of δ needed to ensure the required accuracy of σ depends on the lower bound of the derivative of F. To allow for this we set $\delta = C\,2^{-k}$ for some $C > 0$, and then

$$P(|X - F(\sigma, T)| > C\,2^{-k}) \le \frac{2^{2k}C^{-2}F(\sigma, T)}{\xi(n)} \le \frac{2^{2k}C^{-2}}{\xi(n)},$$

and if we want an error probability of at most γ, we set

$$\frac{2^{2k}C^{-2}}{\xi(n)} \le \gamma.$$

Applying logarithms, we get

$$2k - 2\,\log(C) - \log(\xi(n)) \le \log(\gamma)\,,$$

therefore,

$$k \le \frac{\log(\xi(n)) + \overbrace{\log(\gamma) + 2\,\log(C)}^{\text{constant value}}}{2}.$$

For the RW Turing machine, for every σ, $F(\sigma, T)$ increases with T and the term $\log(1/F(\sigma, T))$ decreases; contrary to what one might expect, for every input word w of size n, the longer we wait for the particles to return, the less precision we can obtain for σ.[2] We take the particular case that in every oracle call the machine will wait exactly two time steps for the particle to return ($T = 2$). Therefore, $F(\sigma, 2) = (1 - \sigma)$. Now, with $k \in O(\log(\xi(n)))$, we have

$$P(|(1 - X) - \sigma| = P(|X - (1 - \sigma)| > 2^{-k-5}) \le \gamma.$$

With value $1 - X$ we can estimate σ.   □

This result suggests a non-collapsing hierarchy of classes can be defined by the magnitude of the number of queries to the oracle. As we want this to be a hierarchy built on *BPP* and within *BPP*// log⋆, we must ensure that all of the machines we consider can compute *BPP*. Thus we consider a RW oracle added to a probabilisitic Turing machine, to give an RW fair probabilistic Turing machine.

---

[1] It is proved in [1, 13] that, for every $\sigma \in C_3$ and for every dyadic rational z, if $|\sigma - z| \le 2^{-k-5}$, then the binary expansions of x and z coincide on the first k bits.

[2] This statement makes sense, since, if we wait too long, then we will lose information about the absorption time of the particle.

4   Lower and Upper Bounds

We encode advice functions in order to compare RW Turing machines with Turing machines with advice. We define the iterated logarithmic functions $\log^{(k)}(n)$:

- $\log^{(0)}(n) = n$;
- $\log^{(k+1)}(n) = \log(\log^{(k)}(n))$.

Similarly, we define the iterated exponential $\exp^{(k)}(n)$:

- $\exp^{(0)}(n) = n$;
- $\exp^{(k+1)}(n) = 2^{\exp^{(k)}(n)}$.

The iterated exponential is a well known bound on the number of computation steps of elementary functions (e.g. see [23]). For every $k \in \mathbb{N}$, the functions $\log^{(k)}$ and $\exp^{(k)}$ are inverse of each other. Let $\log^{(k)}$ also denote the class of advice functions f such that $|f(n)| \in O(\log^{(k)}(n))$.

Let $c(w)$ be the encoding of a single word w. We define the encoding $y(f) = \lim y(f)(n)$ for an advice function $f \in \log^{(k)}\star$ in the following way:

- $y(f)(0) = 0.c(f(0))$;
- if $f(n+1) = f(n)s$, then

$$y(f)(n+1) = \begin{cases} y(f)(n)c(s) & \text{if } n+1 \text{ is not of the form } \exp^{(k)}(m) \\ y(f)(n)c(s)001 & \text{if } n+1 \text{ is of the form } \exp^{(k)}(m) \end{cases}$$

So, for example, if we want to encode a function $f \in \log\log\star$, we just have to place the separator 001 when $n+1$ is of the form 2^{2^m}, for some $m \in \mathbb{N}$.

For every k and for every $f \in \log^{(k)}\star$, we have that $y(f) \in \mathcal{C}_3$. Also, for every n, in order to extract the value of $f(n)$, we only need to find the number $m \in \mathbb{N}$ such that $\exp^{(k)}(m-1) < n \leq \exp^{(k)}(m)$ and then read $y(f)$ in triplets, until we find the $(m+1)$-th separator. Then, it is only needed to ignore the separators and replace each 100 triplet by 0 and each 010 triplet by 1. Since $f \in \log^{(k)}\star$, we know that $|f(\exp^{(k)}(m))| = O(\log^{(k)}(\exp^{(k)}(m))) = O(m)$. We conclude that $3O(m) + 3(m+1) = O(m)$ bits are enough to get the value of $f(\exp^{(k)}(m))$ and, consequently, $O(\log^{(k)}(n))$ bits to get the value of $f(n)$.

Definition 2 Denote by $\text{poly}(g(n))$ the class of functions $f : \mathbb{N} \to \mathbb{N}$ for which there is a polynomial $p(x)$ so that $f(n) \leq p(g(n))$ for all $n \in \mathbb{N}$.

We can use this to prove the following result:

Theorem 5 (Lower bounds) *For every k, every set in $BPP//\log^{(k+1)}\star$ is decidable in polynomial time by a RW fair probabilistic Turing machine that can make up to $\xi(n) \in \text{poly}(\log^{(k)}(n))$ RW oracle calls on inputs of size n.*

Proof Let A be an arbitrary set in $BPP//\log^{(k+1)}\star$ and $\mathcal{M}$ a probabilistic Turing machine with advice $f \in \log^{(k+1)}\star$, which decides A in polynomial time with error probability bounded by $\gamma_1 \in (0, 1/2)$.

Let $\mathcal{M}'$ be a RW fair probabilistic Turing machine with unknown parameter $y(f)$, the encoding of f, and let $\gamma_2 \in \mathbb{R}$ be such that $\gamma_1 + \gamma_2 < 1/2$. Let w be a word such that $|w| \le n$. Theorem 4 assures that $\mathcal{M}'$ can estimate, up to adding constants, $\frac{1}{2} \log(\xi(n))) = \frac{1}{2} \log\left((\log^{(k)}(n))^m\right)$ (which for m large gives an arbitrary constant multiple of $\log^{(k+1)}(n)$) bits of $y(f)$, and, thus, $\mathcal{M}'$ can read $f(n)$ in scheduled protocol time $T = 2$ and in machine polynomial time, with an error probability bounded by γ_2. We have that $P('yes') = 1 - \sigma$ and $P('no') = \sigma$. By definition, the machine can also make a sequence of fair coin tosses of polynomial length. Therefore, $\mathcal{M}'$ can decide A in polynomial time, with error probability bounded by $\gamma_1 + \gamma_2 < 1/2$. □

Taking the special case $k = 0$, we have the following complementary result to Theorem 3:

Corollary 1 *Every set in BPP//log⋆ is decidable in polynomial time by a RW fair probabilistic Turing machine that can make up to $\xi(n) \in \mathrm{poly}(n)$ RW oracle calls on inputs of size n.*

In order to state and prove upper bounds, we need the following auxiliary result. This uses the query tree $\mathcal{T}$, a tree with two branches—'yes' and 'no'—every time a query is made. The probability of taking a path down the tree is just the product of the probabilities of the edges taken at every vertex.

Theorem 6 *Let A be the set decided by a RW Turing machine, or RW fair probabilistic Turing machine, $\mathcal{M}$ with unknown parameter σ that can make up to $\xi(n)$ calls to the RW oracle, for inputs of size n, with error probability bounded by $\gamma < 1/4$. If $\mathcal{M}'$ is an identical RW machine, except with unknown parameter $\tilde{\sigma}$ and the probability of absorption $\tilde{F}$, such that*

$$|F(\sigma, T) - \tilde{F}(\tilde{\sigma}, T)| < \frac{1}{8\xi(n)},$$

then, for any word of size $\le n$, the probability of $\mathcal{M}'$ making an error when deciding A is $\le 3/8$.

Proof We know that $\mathcal{M}$ and $\mathcal{M}'$ make at most $\xi(n)$ calls to the oracle, in such a way that the query tree $\mathcal{T}$ associated to both, has maximum depth $\xi(n)$. Let w be of size not greater than n. Let D be the assignment of probabilities to the edges of $\mathcal{T}$ corresponding to the unknown parameter σ and 'yes' probability $F(\sigma, T)$ and D' be the assignment of probabilities given by the unknown parameter $\tilde{\sigma}$ and 'yes' probability $\tilde{F}(\tilde{\sigma}, T)$. Since $|F(\sigma, T) - \tilde{F}(\tilde{\sigma}, T)| < 1/8\xi(n)$, the difference between any particular probability is at most

$$\kappa = \frac{1}{8\xi(n)}.$$

Invoking Proposition 11 of [1], we have two different cases:

- $w \notin A$: In this case, an incorrect result corresponds to $\mathcal{M}'$ accepting w. The probability of acceptance $P_A(\mathcal{T}, D')$ for $\mathcal{M}'$ is

$$P_A(\mathcal{T}, D') \leq P_A(\mathcal{T}, D) + |P_A(\mathcal{T}, D') - P_A(\mathcal{T}, D)|$$
$$\leq \gamma + \xi(n)\kappa$$
$$\leq \gamma + \xi(n)\frac{1}{8\,\xi(n)} = \frac{1}{4} + \frac{1}{8} = \frac{3}{8}$$

- $w \in A$: In this case, an incorrect result corresponds to $\mathcal{M}'$ rejecting w. The probability of rejection $P_R(\mathcal{T}, D')$ for $\mathcal{M}'$ is

$$P_R(\mathcal{T}, D') \leq P_R(\mathcal{T}, D) + |P_R(\mathcal{T}, D') - P_R(\mathcal{T}, D)|$$
$$\leq \gamma + \xi(n)\kappa$$
$$\leq \gamma + \xi(n)\frac{1}{8\,\xi(n)} = \frac{1}{4} + \frac{1}{8} = \frac{3}{8}$$

In both cases, the error probability is bounded by $3/8$. □

Let $F(\sigma, T)|_m$ denote the first m bits of the probability $F(\sigma, T)$. The next theorem is a corollary of the previous:

Theorem 7 *Let A be the set decided by RW fair probabilistic Turing machine $\mathcal{M}$ with unknown parameter σ that can make up to $\xi(n)$ calls to the RW oracle, for inputs of size n, with error probability bounded by $\gamma < 1/4$. If $\mathcal{M}_n$ is an identical fair probabilistic Turing machine, with unknown parameter $\tilde{\sigma}$, but with the exception that the probability that the oracle returns 'yes' is given by $F(\sigma, T)|_{\log \xi(n)+3}$, then $\mathcal{M}_n$ decides the same set as $\mathcal{M}$ in the same time, but with error probability bounded by $3/8$.*

Now we state and prove upper bounds.

Theorem 8 *(Upper bounds) For every k, every set decided in polynomial time by a RW Turing machine, or RW fair probabilistic Turing machine, that can make up to $\xi(n) = \mathrm{poly}(\log^{(k)}(n))$ calls to the RW oracle, where n is the size of the input, is in $BPP//\log^{(k+1)}\star$.*

Proof Let A be a set decided in polynomial time $p(n)$ and with error probability bounded by $1/4$ by a RW Turing machine $\mathcal{M}$ with unknown parameter σ that can make up to $\xi(n) \in \mathrm{poly}(\log^{(k)}(n))$ calls to the oracle. We specify a probabilistic Turing machine $\mathcal{M}'$ with advice $f(n) = F(\sigma, T)|_{\log \xi(n)+3}$ to decide A. We have $f \in \log^{(k+1)}\star$.

By Theorem 7, we know that an RW Turing machine with 'yes' probability $f(n)$ decides the same as $\mathcal{M}$ for words of size $\leq n$, but with error probability $\leq 3/8$. The value $f(n) = F(\sigma)|_{\log \xi(n)+3}$ is a dyadic rational with denominator $2^{\log \xi(n)+3}$. Thus, $m = 2^{\log \xi(n)+3}f(n) \in [0, 2^{\log \xi(n)+3})]$ is an integer. Consider $\kappa = \log \xi(n) + 3$

fair coin tosses, interpreted as a sequence of bits. The machine $\mathcal{M}'$ then tests if $\tau_1 \tau_2 \ldots \tau_k < m$, where $\tau_1 \tau_2 \ldots \tau_k$ is now interpreted as an integer. If the test is true, the machine returns 'yes', otherwise it returns 'no'. The probability of returning 'yes' is $m/2^k = f(n)$, as required. The time taken is polynomial in n. □

From Theorems 4 and 8, we get the following corollary:

Theorem 9 *The class of sets decidable in polynomial time by RW fair probabilistic Turing machines that can make up to* $\text{poly}(\log^{(k)}(n))$ *calls to the RW oracle, for inputs of size n, is exactly* $BPP//\log^{(k+1)}\star$.

As we want the RW Turing machines to run in polynomial time, the maximum number of oracle calls that we can allow is polynomial. For that bound, the corresponding class is $BPP//\log\star$. Thus, if we restrict more and more the number of queries to the oracle, we can obtain a fine structure of $BPP//\log\star$. Observe that if k is a very large number, the machine is allowed to make only few calls to the oracle, but the advice is smaller, so the number of bits that the machine needs to read is also smaller.

5   The Hierarchy

We explore some properties of advice classes (see [3, 7, 24]).

If $f : \mathbb{N} \to \Sigma^*$ is an advice function, then we use $|f|$ to denote its size, i.e., the function $|f| : \mathbb{N} \to \mathbb{N}$ such that $|f|(n) = |f(n)|$, for every $n \in \mathbb{N}$. For a class of functions, $\mathcal{F}$, $|\mathcal{F}| = \{|f| : f \in \mathcal{F}\}$.

Definition 3 A class of advice functions is said to be a class of reasonable advice functions if:

1. for every $f \in \mathcal{F}$, $|f|$ is computable in polynomial time;
2. for every $f \in \mathcal{F}$, $|f|$ is bounded by a polynomial;
3. for every $f \in \mathcal{F}$, $|f|$ is increasing;
4. $|\mathcal{F}|$ is closed under addition and multiplication by positive integers;
5. for every polynomial p of positive integer coefficients and every $f \in \mathcal{F}$, there exists $g \in \mathcal{F}$ such that $|f| \circ p \leq |g|$.

Definition 4 Let r and s be two total functions. We say that $r \prec s$ if $r \in o(s)$. Let $\mathcal{F}$ and $\mathcal{G}$ be classes of advice functions. We say that $\mathcal{F} \prec \mathcal{G}$ if there exists a function $g \in \mathcal{G}$ such that, for every $f \in \mathcal{F}$, $|f| \prec |g|$.

We have $\log^{(k+1)} \prec \log^{(k)}$, for all $k \geq 0$. Now, we just need to know the relation between the non-uniform complexity classes of *BPP*, induced by the relation $\prec$ in the advice classes. Remember that a set is said to be tally if it is a language over an alphabet of a single symbol (e.g. $\{0\}$). Now, consider the set of finite sequences over the alphabet Σ ordered first by size and then alphabetically. The characteristic function of a set $A \subseteq \Sigma^*$ is the unique infinite sequence $\chi_A : \mathbb{N} \to \{0, 1\}$ such that, for every n, $\chi_A(n)$ is 1 if, and only if, the n-th word in that order is in A. The characteristic function of a tally set A is a sequence where the i-th bit is 1 if, and only if, the word 0^i is in A. The following theorem generalizes the related theorem of [3, 7, 24], where it is proved for the deterministic case.

Theorem 10 *If $\mathcal{F}$ and $\mathcal{G}$ are two classes of reasonable sublinear advice functions[3] such that $\mathcal{F} \prec \mathcal{G}$, then $BPP//\mathcal{F} \subsetneq BPP//\mathcal{G}$.*

Proof Trivially, $BPP//\mathcal{F} \subseteq BPP//\mathcal{G}$. Let *linear* be the set of advice functions of size linear in the size of the input and $\eta.linear$ be the class of advice functions of size ηn, where n is the size of the input and η is a number such that $0 < \eta < 1$. There is an infinite sequence γ whose set of prefixes is in $BPP//linear$ but not in $BPP//\eta.linear$ for some η sufficiently small.[4] Let $g \in \mathcal{G}$ be a function such that, for every $f \in \mathcal{F}$, $|f| \prec |g|$. We prove that there is a set in $BPP//g$ that does not belong to $BPP//f$, for any $f \in \mathcal{F}$.

A tally set T is defined in the following way: for each $n \geq 1$,

$$\beta_n = \begin{cases} \gamma|_{|g|(n)} \, 0^{n-|g|(n)} & \text{if } |g|(n) \leq n \\ 0^n & \text{otherwise} \end{cases}.$$

T is the tally set with characteristic string $\beta_1 \beta_2 \beta_3 \ldots$. With advice $\gamma|_{|g|(n)}$, it is easy to decide T, since we can reconstruct the sequence $\beta_1 \beta_2 \ldots \beta_n$, with $(n^2 + n)/2$ bits, and then we just have to check if its n-th bit is 1 or 0. We conclude that $T \in P/g \subseteq BPP//g$.

We prove that the same set does not belong to $BPP//f$. Suppose that some probabilistic Turing machine $\mathcal{M}$ with advice f, running in polynomial time, decides T with probability of error bounded by[5]

$$2^{-\log(4|g|(n))} = \frac{1}{4|g|(n)}$$

Since $|f| \in o(|g|)$, then, for all but finitely many n, $|f|(n) < \eta|g|(n)$, for arbitrarily small η, meaning that we can compute, for all but finitely many n, $|g|(n)$ bits of γ using an advice of length $\eta.|g|(n)$, contradicting the fact that the set of prefixes of γ is not in $BPP//\eta.linear$. The reconstruction of the binary sequence $\gamma|_{|g|(n)}$ is provided by the following procedure:

---

[3]$\mathcal{F}$ is a class of reasonable sublinear advice functions if it is a class of reasonable advice functions such that, for every $f \in \mathcal{F}$, $|f| \in o(n)$..

[4]We can take for γ the Chaitin Omega number, Ω.

[5]E.g. see Proposition 6.17 in [2]. The probability of error of a given probabilistic machine that decides T in polynomial time can be reduced below any fixed value just by iteration.

```
procedure
  begin
    input n;
    x := λ;
    Compute |g|(n);
    for  i := n²-n/2 to n²-n/2 + |g|(n) do begin
        Query 0ⁱ to T by running machine M with advice f(i);
        if "YES" then x := x1 else x := x0;
    end for;
    output x
  end.
```

The queries are made simulating machine $\mathcal{M}$ which is a probabilistic Turing machine with error probability bounded by $2^{-\log(4|g|(n))} = \frac{1}{4|g|(n)}$. Thus, the probability of error of $\mathcal{M}'$ is bounded by

$$\frac{1}{4|g|(\frac{n^2-n}{2})} + \cdots + \frac{1}{4|g|(\frac{n^2-n}{2} + |g|(n))}.$$

As $|g|$ is increasing, the error probability is bounded by

$$\frac{1}{4|g|(\frac{n^2-n}{2})} \times |g|(n),$$

which, for $n \geq 3$, is bounded by

$$\frac{1}{4|g|(n)} \times |g|(n) = \frac{1}{4}.$$

$\square$

As we are considering prefix advice classes, it is useful to derive the following corollary:

Theorem 11 *If $\mathcal{F}$ and $\mathcal{G}$ are two classes of reasonable sublinear advice functions such that $\mathcal{F} \prec \mathcal{G}$, then $BPP//\mathcal{F}\star \subsetneq BPP//\mathcal{G}\star$.*

Proof The proof of Theorem 10 is also a proof that $BPP//\mathcal{F} \subsetneq BPP//\mathcal{G}\star$, because the advice function used is $\gamma\!\downarrow_{|g|(n)}$, which is a prefix advice function. Since $BPP//\mathcal{F}\star \subseteq BPP//\mathcal{F}$, the statement follows. $\square$

We have already seen that, for all $k \geq 0$, $\log^{(k+1)} \prec \log^{(k)}$. In particular, this is true for $k \geq 1$ and we have the following infinite descending chain

$$\cdots \prec \log^{(4)} \prec \log^{(3)} \prec \log^{(2)} \prec \log.$$

Therefore, by Theorem 11, we have also the descending chain of sets

$$\cdots \subsetneq BPP//\log^{(4)}\star \subsetneq BPP//\log^{(3)}\star \subsetneq BPP//\log^{(2)}\star \subsetneq BPP//\log\star,$$

that, according with Theorem 9, coincide with the classes of sets decided by RW fair probabilistic Turing machines that can make up to

$$\cdots \subsetneq \text{poly}(\log^{(3)}(n)) \subsetneq \text{poly}(\log^{(2)}(n)) \subsetneq \text{poly}(\log(n)) \subsetneq \text{poly}(n)$$

calls to the RW oracle, respectively.

6 Conclusion

Summary. We introduced RW fair probabilistic Turing machine specified as fair probabilistic Turing machines having access to a random walk experiment on a line. We then proved that the class of sets decidable in polynomial time by RW fair probabilistic Turing machines that can make up to $\text{poly}(\log^{(k)}(n))$ calls to the oracle is exactly $BPP//\log^{(k+1)}\star$, where $\log^{(k)}$ is the class of advice functions f such that $|f(n)| \in O(\log^{(k)}(n))$.

We proved that, if $\mathcal{F}$ and $\mathcal{G}$ are two classes of reasonable sublinear advice functions such that $\mathcal{F} \prec \mathcal{G}$, then $BPP//\mathcal{F} \subsetneq BPP//\mathcal{G}$. Although this result was already discussed for the deterministic case in [3, 7, 24], the probabilistic case seems not to have been considered.

Then, we presented a fine structure of $BPP//\log\star$ based on counting oracle calls:

$$\cdots \subsetneq BPP//\log^{(4)}\star \subsetneq BPP//\log^{(3)}\star \subsetneq BPP//\log^{(2)}\star \subsetneq BPP//\log\star,$$

that coincide with the structure of classes of sets decided by RW fair probabilistic Turing machine that can make up to

$$\cdots \subsetneq \text{poly}(\log^{(3)}(n)) \subsetneq \text{poly}(\log^{(2)}(n)) \subsetneq \text{poly}(\log(n)) \subsetneq \text{poly}(n)$$

calls to the RW oracle, respectively.

Open Problem. Together with the transfinite chain of advice classes presented in [7, 18], we also have a transfinite chain of non-uniform probabilistic classes:

$$\cdots \subsetneq BPP//\log^{(2\omega)}\star \subsetneq \cdots \subsetneq BPP//\log^{(\omega)}\star \subsetneq \cdots \subsetneq BPP//\log^{(2)}\star \subsetneq BPP//\log\star.$$

In fact, the chain of non-uniform classes can be continued, where $\log^{(\omega)} = \bigcap_{k\in\mathbb{N}} \log^{(k)}$ is a non-empty class (as shown in [7, 18] for diverse transfinite classes). However, we do not know if there is a correspondence between these complexity classes and the classes decided by RW fair probabilistic Turing machines with bounded number of oracle calls, since we only proved such a correspondence for advice classes of the form $\log^{(k)}$, with $k \in \mathbb{N}$. At present, we do not know how to encode a function $f \in \log^{(\omega)}\star$ into a real number.

Acknowledgments The research of José Félix Costa is supported by Fundação para a Ciência e Tecnologia, projeto FCT I.P.:UID/FIL/00678/2013.

References

1. Ambaram, T., Beggs, E., Costa, J.F., Poças, D., Tucker, J.V.: An analogue-digital model of computation: turing machines with physical oracles. In: Adamatzky, A. (ed.) Advances in Unconventional Computing, vol. 1(theory), p. 38. Springer (Sept 2016, to appear)
2. Balcázar, J.L., Días, J., Gabarró, J.: Structural Complexity I, 2nd edn. Springer, 1988 (1995)
3. Balcázar, J.L., Gavaldà, R., Siegelmann, H.T.: Computational power of neural networks: a characterization in terms of Kolmogorov complexity. IEEE Trans. Inf. Theor. **43**(4), 1175–1183 (1997)
4. Beggs, E., Cortez, P., Costa, J.F., Tucker, J.V.: Classifying the computational power of stochastic physical oracles (2016)
5. Beggs, E., Costa, J.F., Loff, B., Tucker, J.V.: Computational complexity with experiments as oracles II. Upper bounds. Proc. R. Soc., Ser. A (Math., Phys. Eng. Sci.) **465**(2105), 1453–1465 (2009)
6. Beggs, E., Costa, J.F., Loff, B., Tucker, J.V.: Computational complexity with experiments as oracles. Proc. R. Soc., Ser. A (Math., Phys. Eng. Sci.), **464**(2098), 2777–2801 (2008)
7. Beggs, E., Costa, J.F., Loff, B., Tucker, J.V.: Oracles and advice as measurements. In: Calude, C.S., Costa, J.F., Freund, R., Oswald, M., Rozenberg, G. (eds.) Unconventional Computation (UC 2008). Lecture Notes in Computer Science, vol. 5204, pp. 33–50. Springer (2008)
8. Beggs, E., Costa, J.F., Poças, D., Tucker, J.V.: Computations with oracles that measure vanishing quantities. Math. Struct. Comput. Sci. (in print)
9. Beggs, E., Costa, J.F., Tucker, J.V.: Computational models of measurement and Hempel's axiomatization. In: Carsetti, A. (ed.) Causality, Meaningful Complexity and Embodied Cognition. Theory and Decision Library A, vol. 46, pp. 155–183. Springer (2010)
10. Beggs, E., Costa, J.F., Tucker, J.V.: Limits to measurement in experiments governed by algorithms. Math. Struct. Comput. Sci. **20**(06), 1019–1050 (2010)
11. Beggs, E., Costa, J.F., Tucker, J.V.: Physical oracles: the turing machine and the Wheatstone bridge. Studia Log. **95**(1–2), 279–300 (2010)
12. Beggs, E., Costa, J.F., Tucker, J.V.: The impact of models of a physical oracle on computational power. Math. Struct. Comput. Sci. **22**(5), 853–879 (2012)
13. Beggs, E., Costa, J.F., Poças, D., Tucker, J.V.: Oracles that measure thresholds: the turing machine and the broken balance. J. Log. Comput. **23**(6), 1155–1181 (2013)
14. Beggs, E., Costa, J.F., Tucker, J.V.: A natural computation model of positive relativisation. Int. J. Unconv. Comput. **10**(1–2), 111–141 (2013)
15. Beggs, E., Costa, J.F., Poças, D., Tucker, J.V.: An analogue-digital Church-Turing thesis. Int. J. Found. Comput. Sci. **25**(4), 373–390 (2014)
16. Beggs, E., Costa, J.F., Tucker, J.V.: Three forms of physical measurement and their computability. Rev. Symb. Log. **7**(12), 618–646 (2014)
17. Blass, A., Braun, G.: Random orders and gambler's ruin. Electr. J. Comb. **12**, R23 (2005)
18. Costa, J.F.: Incomputability at the foundations of physics (A study in the philosophy of science). J. Log. Comput. **23**(6), 1225–1248 (2013)
19. Costa, J.F.: Uncertainty in time. Parallel Proc. Lett. **25**, 1540007, 13 (2015)
20. Geroch, R., Hartle, J.B.: Computability and physical theories. Found. Phys. **16**(6), 533–550 (1986)
21. Krantz, D.H., Suppes, P., Luce, R.D., Tversky, A.: Foundations of Measurement. Academic Press, vol. 1 (1971), vol. 2 (1989) and vol. 3 (1990)
22. Mosteller, F.: Fifty Challenging Problems in Probability with Solutions. Dover Publications (1987)

23. Odifreddi, P.: Classical Recursion Theory II. North Holland, Studies in Logic and the Foundations of Mathematics (1999)
24. Siegelmann, H.T.: Neural Networks and Analog Computation: Beyond the Turing Limit. Birkhäuser (1999)
25. Venegas-Andraca, S.E.: Quantum Walks for Computer Scientists. Morgan and Claypool Publishers (2008)
26. Younger, A.S., Redd, E., Siegelmann, H.T.: Development of physical super-turing analog hardware. In: Obara, O.H., et. al. (eds.) Unconventional Computation and Natural Computation—13th International Conference (UCNC 2014). Lecture Notes in Computer Science, vol. 8553, pp. 379–391 (2014)

On Computable Numbers, Nonuniversality, and the Genuine Power of Parallelism

Selim G. Akl and Nancy Salay

Abstract We present a simple example that disproves the universality principle. Unlike previous counter-examples to computational universality, it does not rely on extraneous phenomena, such as the availability of input variables that are time varying, computational complexity that changes with time or order of execution, physical variables that interact with each other, uncertain deadlines, or mathematical conditions among the variables that must be obeyed throughout the computation. In the most basic case of the new example, all that is used is a single pre-existing global variable whose value is modified by the computation itself. In addition, our example offers a new dimension for separating the computable from the uncomputable, while illustrating the power of parallelism in computation.

1 Introduction

The universality principle is the cornerstone of computing and the reason for the rapid ascendancy of the discipline as the most influential science of our time. According to this principle, any general-purpose computer A can execute, through simulation, and more or less efficiently, any computation that is possible on any other general-purpose computer B [42]. In essence, the principle expresses a deep and important insight into the relationship between computability and universality. Perspicuously stated, it says that a function is computable if and only if its value can be obtained by simulation on any general-purpose computer [1, 26, 29, 36–38, 42, 44, 45, 56]. Here, general-purpose computers, our domain of discourse, are to be understood as ones that are defined and fixed once and for all; the capabilities of a general-

S.G. Akl (✉)
School of Computing and Department of Mathematics and Statistics,
Queen's University, Kingston, ON K7L 3N6, Canada
e-mail: akl@cs.queensu.ca

N. Salay
Department of Philosophy and School of Computing, Queen's University, Kingston,
ON K7L 3N6, Canada
e-mail: salay@queensu.ca

© Springer International Publishing Switzerland 2017
A. Adamatzky (ed.), *Emergent Computation*, Emergence, Complexity
and Computation 24, DOI 10.1007/978-3-319-46376-6_4

purpose computer are never modified in order to fit the computational problem to be solved. In theoretical computing, general-purpose computers are represented using computational models such as the Turing Machine, the Random Access Machine (RAM), the Cellular Automaton, and the like [54]. In practice, they are the processors in our tablets, our mobile phones, our cars, and so on. It follows from the universality principle that any function that can be evaluated on any general-purpose computer is a computable function. In other words, being universally simulatable is a sufficient condition of computability. It also follows from this principle that a function that is not universally simulatable must not be computable. In other words, being universally simulatable is a necessary condition of computability. A function that *could not* be evaluated on some general-purpose computer, but *could*, nevertheless, be computed by another, would be a counter-example to the necessity clause of the universality principle and would show us that the connection between computability and universal simulatability is weaker than is generally assumed: simulatability is sufficient for computability, but it is not necessary; that is, a function can be computable in some contexts, but not in all contexts.

The universality principle does in fact hold for *conventional* computations, such as, for example, sorting into non-decreasing order a list of numbers that are given in arbitrary order, searching a list for a given datum, numerical computation, text processing, and so on. To illustrate, consider the following parallel algorithm for sorting a sequence $S = g_0, g_1, \ldots, g_{n-1}$ of n distinct integers on a linear array of processors $p_0, p_1, \ldots, p_{n-1}$. Processor p_i contains g_i and can communicate with its two neighbours p_{i-1} and p_{i+1} (except for p_0 and p_{n-1} which have only one neighbour each, namely, p_1 and p_{n-2}, respectively). In the "compare and swap if needed" operation of the algorithm, processors p_i and p_{i+1} compare their integers, placing the smaller in p_i and the larger in p_{i+1}.

Parallel Sort
 for $k = 0$ **to** $n - 1$ **do**
  for $i = 0$ **to** $n - 2$ **do in parallel**
   if $i \bmod 2 = k \bmod 2$
   then p_i and p_{i+1} compare and swap if needed
   end if
  end for
 end for. ∎

Algorithm Parallel Sort completes the sort in $O(n)$ time [3]. If it so happens that only one processor, namely p_0, is available, then the parallel algorithm can be easily simulated by having the single processor methodically imitate the operations of the n processors. The sequential solution uses an array S to store the sequence to be sorted. Initially, $S[i]$ contains g_i, for $i = 0, 1, \ldots, n - 1$. The algorithm is given in what follows. In it, the operation "compare $S[i]$ and $S[i + 1]$ and swap if needed" compares the two integers currently in $S[i]$ and $S[i + 1]$, placing, as a result, the smaller in $S[i]$ and the larger in $S[i + 1]$.

Sequential Sort
  for $k = 0$ **to** $n - 1$ **do**
   for $i = 0$ **to** $n - 2$ **do**
    if $i \bmod 2 = k \bmod 2$
    then compare $S[i]$ and $S[i + 1]$ and swap if needed
    end if
   end for
  end for. ■

Algorithm Sequential Sort completes the sort in $O(n^2)$ time (clearly not the best sorting algorithm sequentially, but a sufficient illustration of the idea of simulation for our purposes).

But here's the rub: simulation is always feasible *only for conventional computations*. Several classes of *unconventional computations* have been uncovered recently for which simulation is not always possible and, consequently, for which universality does not hold. These classes include computations that involve time-varying variables, time-varying computational complexity, rank-varying computational complexity, interacting variables, uncertain time constraints, mathematical constraints, and so on [11–18]. While these unconventional computations can be executed successfully on certain computers, they cannot be simulated on a unique fixed computer. Because simulation is not always possible, the universality principle as currently understood is false. This conclusion is referred to as nonuniversality in computation [6–10].

Let time be divided into discrete time units. The nonuniversality result is usually stated as follows: no computer U can be universal if it is capable of only a finite and fixed number of basic arithmetic and logical operations, such as addition, comparison, exclusive-or, and so on, per time unit.

Nonuniversality Proof: Assume that computer U can perform $D(i)$ operations during time unit i of a computation, for $i = 0, 1, \ldots$ For any computation C requiring $E(i)$ operations during time unit i, where $E(i) > D(i)$ for at least one i, U will fail to successfully complete C. Therefore, U cannot be universal. Note that C is computable on another computer U' capable of $E(i)$ operations during time unit i. However, U', in turn, will be defeated by another computation C' requiring $F(i)$ operations during time unit i, where $F(i) > E(i)$, for at least one i, and consequently U' cannot be universal either.

One example of such a computation C calls for sorting an input sequence of elements, while imposing an extra condition to be satisfied by any candidate algorithm. Thus, in this unconventional version of the standard sorting problem presented earlier in this section, a sequence $S = g_0, g_1, \ldots, g_{n-1}$ of n distinct integers is given. It is required to transform the sequence S in situ into a sequence $S' = g_0', g_1', \ldots, g_{n-1}'$

whose elements are the same as those of S, with the difference that, at the end of the computation, $g_0' < g_1' < g_2' < \cdots < g_{n-2}' < g_{n-1}'$. So far, this is the classic sorting problem. The unconventional variant adds a new requirement: at no time, once the sorting process has begun, should there be three consecutive elements of an intermediate sequence $S'' = g_0'', g_1'', \ldots, g_{n-1}''$, such that $g_i'' > g_{i+1}'' > g_{i+2}''$, for $i = 0, 1, \ldots, n - 3$. A complete description of this example, and its implications can be found in [14]. It suffices to note here that algorithm Parallel Sort succeeds in carrying out this computation for all input sequences S of size n, while algorithm Sequential Sort fails when presented with a sequence $S = g_0, g_1, \ldots, g_{n-1}$ in which $g_0 > g_1 > g_2 > \cdots > g_{n-2} > g_{n-1}$. A parallel computer with fewer than $n - 2$ processors also fails to solve this problem.

Every reasonable model of computation is, by definition, capable of only a finite and fixed number of operations per time unit. The same is obviously true for any *practical* computer which is built once and for all; it too can only perform a finite (and fixed) number of operations per time unit. Given the nonuniversality proof, it follows that unless the unreasonable assumption is made that a computer is capable at the outset of an infinite number of operations per time unit, universality cannot be achieved by any computer [6–18, 46–51].

In this paper we present an even stronger result: there are computable functions that are not computable universally, even on systems capable of an infinite number of operations per time unit, so long as the general purpose computers in question are constrained to perform operations *sequentially*. This is supported by the simplest counter-example to the universality principle of which we are aware. As a bonus, the counter-example that we propose illustrates the true, often unappreciated, power of the idea of parallelism in computation: parallelism does not just speed up sequential computations; it makes certain computations *possible*. An example of the lack of appreciation of what parallelism brings to computing is the *Speedup Theorem* and specifically its 'proof' [2, 35, 39–41]. This theorem states that the best sequential (that is, single-processor computer) solution to a given problem P can be sped up, at most, by a factor of n if an n-processor parallel computer is used instead. The proof goes as follows.

> Let t_1 be the running time of the best sequential algorithm for P, and let t_n be the running time of a parallel algorithm. Assume that the speedup t_1/t_n is larger than n. In that case we can simply simulate the parallel algorithm on a single processor, resulting in a running time of $n \times t_n < t_1$, which is impossible, since t_1 is already, by definition, the best possible sequential running time. The assumption is therefore false and t_1/t_n cannot be larger than n.

A demonstration of the fallacy of this 'theorem', through several counter-examples that achieve speedups *exponential* in n, is provided in [3, 4], where a number of additional references can also be found. Even in popular science writing, claims can be found to the effect that parallel computing can do no more, in principle, than sequential computing [52].

The remainder of this paper is organized as follows. Our counter-example is described in Sect. 2. Some consequences of our result are derived in Sect. 3. In Sect. 4 we generalize our counter-example using two different models of parallel computation. Conclusions are offered in Sect. 5.

2   The Global Variable Paradigm

Our computation, call it C_0, consists of two distinct and separate processes P_0 and P_1 operating on a global variable x. The variable x is *time-critical* in the sense that its value throughout the computation is intrinsically related to real (external or physical) time. Actions taken throughout the computation, based on the value of x, depend on x having that particular value at that particular time. Here, time is kept internally by a global clock. Specifically, the computer performing C_0 has a clock that is synchronized with real time. Henceforth, real time is synonymous with internal time. In this framework, therefore, resetting x artificially, through simulation, to a value it had at an earlier time is entirely insignificant, as it fails to meet the true timing requirements of C_0. At the beginning of the computation, $x = 0$.

Let the processes of the computation C_0, namely, P_0 and P_1, be as follows:

P_0 : **if** $x = 0$ **then** $x \leftarrow x + 1$ **else** loop forever **end if**.

P_1 : **if** $x = 0$ **then** read y; $x \leftarrow x + y$; return x **else** loop forever **end if**.

In order to better appreciate this simple example, it is helpful perhaps to put it in some familiar context. Think of x as the altitude of an airplane and think of P_0 and P_1 as software controllers actuating safety procedures that must be performed at this altitude. The local nonzero variable y is an integral part of the computation; it helps to distinguish between the two processes and to separate their actions.

The question now is this: on the assumption that C_0 succeeds, that is, that both P_0 and P_1 execute the "**then**" part of their respective "**if**" statements (not the "**else**" part), what is the value of the global variable x at the end of the computation, that is, when both P_0 and P_1 have halted?

We examine two approaches to executing P_0 and P_1:

1. **Using a single processor:** Consider a sequential computer, based, for example, on the RAM model of computation [24], equipped, by definition, with a single processor p_0. The processor executes one of the two processes first. Suppose it starts with P_0: p_0 computes $x = 1$ and terminates. It then proceeds to execute P_1. Because now $x \neq 0$, p_0 executes the nonterminating computation in the "**else**" part of the "**if**" statement. The process is uncomputable and the computation fails. Note that starting with P_1 and then executing P_0 would lead to a similar outcome, with the difference being that P_1 will return an incorrect value of x, namely y, before switching to P_0, whereby it executes a nonterminating computation, given that now $x \neq 0$.

2. **Using two processors:** The two processors, namely, p_0 and p_1, are part of a shared memory parallel computer, based, for example, on the Concurrent-Read Exclusive-Write Parallel Random Access Machine (CREW PRAM) model of

computation [3]. In this model, two or more processors can read from, but not write to, the same memory location simultaneously. In parallel, p_0 executes P_0 and p_1 executes P_1. Both terminate successfully and return the correct value of x, that is, $x = y + 1$.

Two observations are in order:

1. The first concerns the sequential (that is, single-processor) solution. Here, no ex post facto simulation is possible or even meaningful. This includes legitimate simulations, such as executing one of the processes and then the other, or interleaving their executions, and so on. It also includes illegitimate simulations, such as resetting the value of x to 0 after executing one of the two processes, or (assuming this is feasible) an ad hoc rewriting of the code, as for example,

if $x = 0$ **then** $x \leftarrow x + 1$; read y; $x \leftarrow x + y$; return x **else** loop forever **end if**.

   and so on. To see this, note that for either P_0 or P_1 to terminate, the **then** operations of its **if** statement must be executed *as soon as* the global variable x is found to be equal to 0, and not one time unit later. It is clear that any sequential simulation must be seen to have failed. Indeed:

   - A legitimate simulation will not terminate, because for one of the two processes, x will no longer be equal to 0, while
   - An illegitimate simulation will "terminate" illegally, having executed the "**then**" operations of one or both of P_0 or P_1 too late.

2. The second observation follows directly from the first. It is clear that P_0 and P_1 must be executed simultaneously for a proper outcome of the computation. The parallel (that is, two-processor) solution succeeds in accomplishing exactly this.

Finally, a word about the role of time. Real time, as mentioned earlier, is kept by a global clock and is equivalent to internal computer time. It is important to stress here that the time variable is never used explicitly by the computation C_0. Time intervenes only in the circumstance where it is needed to signal that C_0 has failed (when the "**else**" part of an "**if**" statement, either in P_0 or in P_1, is executed). In other words, time is noticed solely when time requirements are neglected.

3   Consequences

The two-process computation C_0 of Sect. 2 shows that no sequential (that is, uniprocessor) computer can ever be universal. Even if it is given an unbounded amount of memory and an unlimited amount of time (like a Turing Machine, for example), processor p_0 fails to solve the problem. Even if it is permitted interaction with the outside world (unlike a Turing Machine), p_0 fails. Finally, and most importantly for our purposes in this chapter, even if p_0 is capable of an infinite number of *sequential operations* per time unit (like an Accelerating Machine [33] or, more generally, a Supertask Machine [27, 30, 58]), it still fails to meet the requirements of the computation C_0.

Notice that the parallel (that is, multiprocessor) computer succeeded in performing C_0 satisfactorily. This demonstrates an important and often overlooked feature of parallelism: far from being simply a faster alternative to sequential computing, it is essential for the success of certain inherently parallel computations [18, 19, 46–51]. The two-process problem is *uncomputable* by a sequential computer and *computable* by a parallel one. Thus, the example not only serves to make the nonuniversality result more general and therefore stronger, it also offers a new way to distinguish computability from uncomputability via sequential and parallel computing.

Does this mean that the parallel computer is universal? Certainly not, for it is possible to construct a computation with three processes, namely, P_0, P_1, and P_2, for which a two-processor computer fails. A three-processor computer may succeed, but it will then be thwarted by a four-process computation. Such reasoning continues indefinitely. Taking this argument to its logical conclusion, only a computer capable of an infinite number of *parallel* operations per time unit can be universal.

4 Generalizations

Various options are available to generalize our result. In this section, we describe two such generalizations. Recall that in Sect. 2 we used the CREW PRAM as the parallel model of computation. In this model, several processors can read simultaneously from the same shared memory location, but no simultaneous write is allowed. Two alternative shared memory parallel models are the Exclusive Read Exclusive Write Parallel Random Access Machine (EREW PRAM) and the Concurrent-Read Concurrent-Write Parallel Random Access Machine (CRCW PRAM) [3]. In the EREW PRAM, at most one processor can gain access to a shared memory location during a time unit, either for reading or for writing. In the CRCW PRAM, a shared memory location can be accessed simultaneously by several processors during a time unit, either for reading by all of them (when executing a concurrent-read instruction) or for writing by all of them (when executing a concurrent-write instruction). In the latter case, memory conflicts are resolved in a variety of ways, including the *priority* concurrent-write instruction, where the processor with the highest writing priority succeeds in writing and all others fail, the *common* concurrent-write instruction, where the write operation succeeds if and only if all processors are attempting to write the same value, and the *combining* concurrent-write instruction, where all the values being written are combined into one (using, for example, the *arithmetic sum*, the *logical and*, the *maximum*, and so on) [3]. For our purposes in this paper, we shall use the *combining with arithmetic sum* as our write instruction.

Let C_1 and C_2 be the two generalizations of the computation C_0, to be proposed in Sects. 4.1 and 4.2, respectively. Both C_1 and C_2 use the idea hinted to in Sect. 3, whereby several processes are part of the computation to be carried out. We will show that C_1 is possible if an n-processor EREW PRAM is available, while an n-

processor CRCW PRAM is needed to execute C_2. Furthermore, both C_1 and C_2 cannot be performed successfully, neither by a RAM nor by a PRAM, of any type, equipped with fewer than n processors.

4.1   Using Several Global and Local Variables

In our first generalization of the example in Sect. 2, we assume the presence of n global variables, namely, $x_0, x_1, \ldots, x_{n-1}$, all of which are time critical, and all of which are initialized to 0. There are also n nonzero local variables, namely, $y_0, y_1, \ldots, y_{n-1}$, belonging, respectively, to the n processes $P_0, P_1, \ldots, P_{n-1}$ that make up C_1. The computation C_1 is as follows:

P_0 : **if** $x_0 = 0$ **then** $x_1 \leftarrow y_0$ **else** loop forever **end if**.

P_1 : **if** $x_1 = 0$ **then** $x_2 \leftarrow y_1$ **else** loop forever **end if**.

P_2 : **if** $x_2 = 0$ **then** $x_3 \leftarrow y_2$ **else** loop forever **end if**.

$\vdots$

P_{n-2} : **if** $x_{n-2} = 0$ **then** $x_{n-1} \leftarrow y_{n-2}$ **else** loop forever **end if**.

P_{n-1} : **if** $x_{n-1} = 0$ **then** $x_0 \leftarrow y_{n-1}$ **else** loop forever **end if**.

Suppose that the computation C_1 begins when $x_i = 0$, for $i = 0, 1, \ldots, n - 1$. For every $i, 0 \leq i \leq n - 1$, if P_i is to be completed successfully, it must be executed *while* x_i is indeed equal to 0, and not at any later time when x_i has been modified by $P_{(i-1) \bmod n}$ and is no longer equal to 0. On an EREW PRAM with n processors, namely, $p_0, p_1, \ldots, p_{n-1}$, it is possible to test all the $x_i, 0 \leq i \leq n - 1$, for equality to 0 in one time unit; this is followed by assigning to all the $x_i, 0 \leq i \leq n - 1$, their new values during the next time unit. Thus all the processes $P_i, 0 \leq i \leq n - 1$, and hence the computation C_1, terminate successfully. A RAM has but a single processor p_0 and, as a consequence, it fails to meet the time-critical requirements of C_1. At best, it can perform no more than $n - 1$ of the n processes as required (assuming it executes the processes in the order $P_{n-1}, P_{n-2}, \ldots, P_1$, then fails at P_0 since x_0 was modified by P_{n-1}), and thus does not terminate. An EREW PRAM with only $n - 1$ processors, $p_0, p_1, \ldots, p_{n-2}$, cannot do any better. At best, it too will attempt to execute at least one of the P_i when $x_i \neq 0$ and hence fail to complete at least one of the processes on time.

4.2 Using a Single Global Variable and No Local Variable

Our second generalization of the example in Sect. 2 requires the presence of only a single, time-critical, global variable x. Let $x = 0$ initially. With n processes, $P_0, P_1, \ldots, P_{n-1}$, the computation C_2 looks as follows:

P_0 : **if** $x = 0$ **then** $x \leftarrow 1$ **else** loop forever **end if**.
P_1 : **if** $x = 0$ **then** $x \leftarrow 1$ **else** loop forever **end if**.
P_2 : **if** $x = 0$ **then** $x \leftarrow 1$ **else** loop forever **end if**.

$\vdots$

P_{n-1} : **if** $x = 0$ **then** $x \leftarrow 1$ **else** loop forever **end if**.

If the computation C_2 starts when $x = 0$, it is required that the "**then** $x \leftarrow 1$" operation be performed *as soon as* it is determined that x is indeed equal to 0.

The n processors of a CRCW PRAM, namely $p_0, p_1, \ldots, p_{n-1}$, read x in parallel, find it equal to 0, and simultaneously increment x, by 1 each, resulting in $x = n$. Now all the processes, and hence the computation C_2, halt gracefully. A single-processor computer is hopeless to perform this computation, but so also is the n-processor CRCW PRAM if presented with an $n + 1$ process version of C_2; they will both run forever.

5 Conclusion

Despite considerable evidence to the contrary [15, 16, 21–23, 28, 31, 34, 55, 57, 59, 60, 62, 63], belief in the universality principle, particularly (but not exclusively) in connection with the Turing Machine, remains one of the most enduring myths in computer science (see, for example, [5, 32, 38]). In this paper we presented a new counter-example to it, the simplest such counter-example of which we are aware. Unlike previous counter-examples to computational universality, it does not rely on extraneous phenomena, such as the availability of input variables that are time varying, computational complexity that changes with time or order of execution, physical variables that interact with each other, uncertain deadlines, or mathematical conditions among the variables that must be obeyed throughout the computation. In the most basic case of the new example, all that is used is a single pre-existing global variable whose value is modified by the computation itself.

Further to its extreme simplicity, this new nonuniversality result is more powerful than earlier ones. It was previously thought, based on past counter-examples, that only a computer capable of an infinite number of basic operations per time unit could be universal [6]. We have shown in this paper that even a computer capable of an infinite number of basic *sequential* operations cannot be universal. Thus, computational universality requires an infinite number of basic *parallel* operations per time unit.

In his classic, discipline-creating paper [61], Alan Turing defined what it means for a number to be computable or uncomputable. The distinction is made by fixing a model of computation and determining whether or not that model is capable of producing a desired number. Thus, the mathematical constants π and e, for example, are computable (to a desired precision). By contrast, there are uncomputable numbers, namely, those that are the outcome of unsolvable problems such as, for example, the *Halting Problem* [25]. This conventional distinction between the computable and the uncomputable has hitherto been adopted, almost universally, with respect to the Turing Machine, as the 'ultimate' model of computation, the baseline. The present paper provides an alternative but complementary way to distinguish between computable and uncomputable numbers. While the background of Turing's distinction is a fixed model of computation, our examples exploit the fact that there exist multiple possible general-purpose computer models, not all equivalent. A number that may be uncomputable on some models, may be computable on others. For example, the number x in C_0 is computable on a parallel computer with the proper number of processors, but uncomputable otherwise (whether sequentially or in parallel). The same is true for the numbers in C_1 and C_2. Our examples, therefore, offer up parallelism as a new baseline model for computation, acknowledging that other models yet to be dreamed up will eventually replace it [46].

Our counter-example to universality also serves to illustrate the importance of parallelism in computing. Virtually the entire body of literature on parallel computation suggests that the *raison d'être* of parallel computers is to speed up sequential computations [20, 43, 53, 64]. We have shown here that parallel computing is considerably more valuable since it can make the difference between computability and uncomputability. Specifically, we have identified a problem that a parallel computer can solve while a sequential computer cannot. In other words, the set of problems solvable in parallel is a strict superset of the set of problems solved sequentially. Therefore, on the hierarchy of computational models, in which models are ranked by their power [42, 54], a parallel computer is strictly more powerful than a sequential one.

In summary, our paper offers three contributions. It strengthens the notion of nonuniversality in computation by extending the domain in which it holds to all sequential machines, even those capable of an infinite number of operations per time unit; it offers a new unconventional way to distinguish between what is computable and what is uncomputable; and it puts in sharp focus an important difference between sequential and parallel computing.

References

1. Abramsky, S., et al.: Handbook of Logic in Computer Science. Clarendon Press, Oxford (1992)
2. Akl, S.G.: The Design and Analysis of Parallel Algorithms. Prentice Hall, Englewood Cliffs, New Jersey (1989)
3. Akl, S.G.: Parallel Computation: Models and Methods. Prentice Hall, Upper Saddle River, New Jersey (1997)
4. Akl, S.G.: Superlinear performance in real-time parallel computation. J. Supercomput. **29**(1), 89–111 (2004)
5. Akl, S.G.: Universality in computation: some quotes of interest, Technical Report No. 2006-511, School of Computing, Queen's University, Kingston, Ontario, April 2006, 13 p. http://www.cs.queensu.ca/home/akl/techreports/quotes.pdf
6. Akl, S.G.: Three counterexamples to dispel the myth of the universal computer. Parallel Process. Lett. **16**(3), 381–403 (2006)
7. Akl, S.G.: Conventional or unconventional: is any computer universal? Chapter 6 In: Adamatzky, A., Teuscher, C. (eds.) From Utopian to Genuine Unconventional Computers, pp. 101–136. Luniver Press, Frome, United Kingdom (2006)
8. Akl, S.G., Gödel's incompleteness theorem and nonuniversality in computing. In: Proceedings of the Workshop on Unconventional Computational Problems, Sixth International Conference on Unconventional Computation, Kingston, Canada, August 2007, pp. 1–23 (2007)
9. Akl, S.G.: Even accelerating machines are not universal. Int. J. Unconventional Comput. **3**(2), 105–121 (2007)
10. Akl, S.G.: Unconventional computational problems with consequences to universality. Int. J. Unconventional Comput. **4**(1), 89–98 (2008)
11. Akl, S.G.: Evolving Computational Systems, Chapter 1. In: Rajasekaran, S., Reif, J.H. (eds.) Parallel Computing: Models, Algorithms, and Applications. Taylor and Francis, CRC Press, Boca Raton, Florida, pp. 1–22 (2008)
12. Akl, S.G.: Ubiquity and simultaneity: the science and philosophy of space and time in unconventional computation, Keynote address. In: Conference on the Science and Philosophy of Unconventional Computing, The University of Cambridge, Cambridge, United Kingdom (2009)
13. Akl, S.G.: Time travel: a new hypercomputational paradigm. Int. J. Unconventional Comput. **6**(5), 329–351 (2010)
14. Akl, S.G.: What is computation? Int. J. Parallel Emergent Distrib. Syst. **29**(4), 337–345 (2014)
15. Akl, S.G.: Nonuniversality in computation: fifteen misconceptions rectified. In: Adamatzky, A. (ed.) Advances in Unconventional Computing. Springer, Switzerland, pp. 1–30 (2017)
16. Akl, S.G.: Nonuniversality explained. Int. J. Parallel, Emergent Distrib. Syst **31**(3), 201–219 (2016)
17. Akl, S.G., Nagy, M.: Introduction to parallel computation, Chapter 2. In: Trobec, R., Vajteršic, M., Zinterhof, P. (Eds.) Parallel Computing: Numerics, Applications, and Trends. Springer, London, United Kingdom, pp. 43–80 (2009)
18. Akl, S.G., Nagy, M.: The future of parallel computation, Chapter 15. In: Trobec, R., Vajteršic, M., Zinterhof, P. (eds.) Parallel Computing: Numerics, Applications, and Trends. Springer, London, United Kingdom, pp. 471–510 (2009)
19. Akl, S.G., Yao, W.: Parallel computation and measurement uncertainty in nonlinear dynamical systems. J. Math. Modell. Algorithms Special Issue on Parallel Sci. Comput. Appl. **4**, 5–15 (2005)
20. Blazewicz, J., Ecker, K., Plateau, B., Trystram, D. (eds.): Handbook on Parallel and Distributed Processing. Springer, Berlin (2000)
21. Burgin, M.: Super-Recursive Algorithms. Springer, New York (2005)
22. Calude, C.S., Păun, G.: Bio-steps beyond turing. BioSystems **77**, 175–194 (2004)
23. Copeland, B.J.: Super turing-machines. Complexity **4**, 30–32 (1998)
24. Cormen, T.H., Leiserson, C.E., Rivest, R.L., Stein, C.: Introduction to Algorithms. MIT Press, Cambridge, Massachusetts (2009)

25. Davis, M.: Computability and Unsolvability. McGraw-Hill, New York (1958)
26. Davis, M.: The Universal Computer. Norton, W.W (2000)
27. Davies, E.B.: Building infinite machines. Br. J. Philos. Sci. **52**, 671–682 (2001)
28. Denning, P.J.: Reflections on a symposium on computation. Comput. J. **55**(7), 799–802 (2012)
29. Deutsch, D.: The Fabric of Reality. Penguin Books, London, United Kingdom (1997)
30. Earman, J., Norton, J.D.: Infinite pains: the trouble with supertasks. In: Morton, A., Stich, S.P. (eds.) Benacerraf and his Critics, pp. 231–261. Massachusetts, Blackwell, Cambridge (1996)
31. Etesi, G., Németi, I.: Non-turing computations via Malament-Hogarth space-times. Int. J. Theor. Phys. **41**(2), 341–370 (2002)
32. Fortnow, L.: The enduring legacy of the Turing machine. Comput. J. **55**(7), 830–831 (2012)
33. Fraser, R., Akl, S.G.: Accelerating machines: a review. Int. J. Parallel Emergent Distrib. Syst. **23**(1), 81–104 (2008)
34. Goldin, D., Wegner, P.: The Church-Turing thesis: breaking the myth. In: Proceedings of the First international conference on Computability in Europe: New Computational Paradigms. Springer, Berlin, pp. 152–168 (2005)
35. Greenlaw, R., Hoover, H.J., Ruzzo, W.L.: Limits to Parallel Computation. Oxford University Press, New York (1995)
36. Harel, D.: Algorithmics: The Spirit of Computing. Addison-Wesley, Reading, Massachusetts (1992)
37. Hillis, D.: The Pattern on the Stone. Basic Books, New York, New York (1998)
38. Hopcroft, J.E., Ullman, J.D.: Formal Languages and their Relations to Automata. Addison-Wesley, Reading, Massachusetts (1969)
39. Jájá, J.: An Introduction to Parallel Algorithms. Addison-Wesley, Reading, Massachusetts (1992)
40. Kronsjö, L.: Computational Complexity of Sequential and Parallel Algorithms. Wiley, New York (1985)
41. Leighton, F.T.: Introduction to Parallel Algorithms and Architectures. Morgan Kaufmann, San Mateo, California (1992)
42. Lewis, H.R., Papadimitriou, C.H.: Elements of the Theory of Computation. Prentice Hall, Englewood Cliffs, New Jersey (1981)
43. Lewis, T.G., El-Rewini, H.: Introduction to Parallel Computing. Prentice Hall, Englewood Cliffs, New Jersey (1992)
44. Mandrioli, D., Ghezzi, C.: Theoretical Foundations of Computer Science. Wiley, New York, New York (1987)
45. Minsky, M.L.: Computation: Finite and Infinite Machines. Prentice-Hall (1967)
46. Nagy, M., Akl, S.G.: On the importance of parallelism for quantum computation and the concept of a universal computer. In: Proceedings of the Fourth International Conference on Unconventional Computation, Sevilla, Spain, October 2005, LNCS 3699, pp. 176–190 (2005)
47. Nagy, M., Akl, S.G.: Quantum measurements and universal computation. Int. J. Unconventional Comput. **2**(1), 73–88 (2006)
48. Nagy, M., Akl, S.G.: Quantum computing: beyond the limits of conventional computation. Int. J. Parallel, Emergent Distrib. Syst. Special Issue on Emergent Comput. **22**(2), 123–135 (2007)
49. Nagy, M., Akl, S.G.: Parallelism in quantum information processing defeats the Universal Computer. In: Proceedings of the Workshop on Unconventional Computational Problems, Sixth International Conference on Unconventional Computation, Kingston, Canada, August 2007, pp. 25–52; also in: Parallel Process. Lett. Special Issue on Unconventional Computational Problems, Vol. 17, No. 3, September 2007, pp. 233–262 (2007)
50. Nagy, N., Akl, S.G.: Computations with uncertain time constraints: effects on parallelism and universality. In: Proceedings of the Tenth International Conference on Unconventional Computation, Turku, Finland, June 2011, LNCS 6714, pp. 152–163 (2011)
51. Nagy, N., Akl, S.G.: Computing with uncertainty and its implications to universality. Int. J. Parallel, Emergent Distrib. Syst. **27**(2), 169–192 (2012)
52. Penrose, R.: The Emperor's New Mind. Oxford University Press, New York (1989)

53. Rajasekaran, S., Reif, J.H. (eds.): Parallel Computing: Models, Algorithms, and Applications. Taylor and Francis, CRC Press, Boca Raton, Florida (2008)
54. Savage, J.E.: Models of Computation. Addison-Wesley (1998)
55. Siegelmann, H.T.: Neural Networks and Analog Computation: Beyond the Turing limit. Birkhäuser, Boston (1999)
56. Sipser, M.: Introduction to the Theory of Computation. PWS Publishing Company, Boston, Massachusetts (1997)
57. Stannett, M.: X-machines and the halting problem: building a super-Turing machine. Formal Aspects Comput. 2(4), 331–341 (1990)
58. Steinhart, E.: Infinitely complex machines. In: Schuster, A. (ed.) Intelligent Computing Everywhere, pp. 25–43. Springer, New York (2007)
59. Stepney, S.: Non-classical hypercomputation. Int. J. Unconventional Comput. 5(3–4), 267–276 (2009)
60. Syropoulos, A.: Hypercomputation. Springer, New York (2008)
61. Turing, A.M.: On computable numbers with an application to the Entscheidungsproblem. In: Proceedings of the London mathematical Society, Ser. 2, vol. 42, 1936, pp. 230–265; Vol. 43, 1937, pp. 544–546 (1937)
62. Van Leeuwen, J., Wiedermann, J.: The Turing machine paradigm in contemporary computing. In: Engquist, B., Schmidt, W. (eds.) Mathematics Unlimited—2001 and Beyond, pp. 1139–1156. Springer, Berlin (2000)
63. Wegner, P.: Why interaction is more powerful than algorithms. Commun. ACM 40(5), 80–91 (1997)
64. Zomaya, A.Y. (ed.): Parallel and Distributed Computing Handbook. McGraw-Hill, New York (1996)

On the Microscopic View of Time and Messages

Nicola Santoro

Abstract In distributed message-passing systems, *synchronous* computations rely on and exploit for their correctness and/or efficiency the existence of some reliable mechanism, which provides all system entities with a globally consistent view of *time*, e.g., a common global clock. Many of these computations, however, exploit time at a *macroscopic* level: they assume that transmission of an unbounded amount of information can be done in constant time. We are instead interested in the *microscopic* level of synchronous computations; that is, the study of computability and complexity when, in a constant amount of time, only a constant number of bits can be transmitted. Our general interest includes the extreme case, when a message contains only a single bit. We discuss the basics of computing at the microscopic level, describing simple but powerful computational tools, and analyzing their use.

1 Introduction

A *message-passing system* is a model of a distributed computing environment, which in turn models many artificial systems (e.g., distributed systems, communication networks, systolic architectures, etc.); it provides a language to describe its components, its behaviour, its properties; furthermore, it includes the tools for the analysis and the measurement of such an environment.

Time is an human artifact superimposed on nature in our attempt to quantify its qualitative processes. This quantification has the immediate effect of discretizing the perceived continuum and enabling its measurement.

The presence of time and the discretization it imposes are the predominant aspects in *synchronous* message-passing system. Indeed, messages and time are the two crucial components of synchronous computations, and their interplay is the determining element for both feasibility and complexity of the computations.

N. Santoro (✉)
School of Computer Science, Carleton University, Ottawa, Canada
e-mail: santoro@scs.carleton.ca

© Springer International Publishing Switzerland 2017   71
A. Adamatzky (ed.), *Emergent Computation*, Emergence, Complexity
and Computation 24, DOI 10.1007/978-3-319-46376-6_5

Many synchronous computations however view and exploit time only at a *macro-scopic* level: they assume that a unit of time is large enough for the transmission of an unbounded amount of information.

Instead, like several other researchers, we are interested in the *microscopic* analysis of synchronous computations; that is, we are interested in the study of computability and complexity when, in a unit of time, only a constant number of bits can be transmitted. Our interest covers also the most extreme case, when a message contains only a single bit. Not surprisingly, great part of the nature and beauty of synchronous computing, is revealed only under the microscope.

The aim of this chapter is to introduce the basics of computing at the microscopic level, describing simple but powerful computational tools, and analyzing their use. The terminology and notation are from [37].

1.1 Message-Passing Systems

In the language of distributed computing, a message-passing system is a collection of computational *entities* which communicate by sending and receiving bounded sequences of bits called *messages*. A binary relation, called out-neighbour, defines for each entity x the subset of the other entities, called out-neighbourhood, to which x can send a message; analogous is the definition of in-neighbourhood of an entity. If, for each entity, its in-neighbourhood coincides with its out-neighbourhood, we will use the terms neighbour and neighbourhood. The couple $G = (V, E)$ where V is the set of entities and E is the out-neighbour relation defines a graph G which describes the communication topology of the system. Hence, graph-theoretic concepts and terminology (e.g., nodes, edges, diameter, etc.) can be used to describe distributed algorithms and analyze their performance. In the following, the terms vertex, node, site, and entity will have the same meaning; analogously, the terms edge, arc, link and line will be used interchangeably. Messages received at an entity are processed there in the order they arrive; if more than one message arrives at the same entity at the same time, they will be processed in arbitrary order.

Each entity is provided with local processing and storage capabilities, and a local clock. The behaviour of the entities can be conveniently described as finite-state and event-driven; that is, each entity at any time is in a particular system state (from a finite set of states) and, when a predefined external event occurs (e.g., a message is received, the local clock is increased by one unit, etc.), it will serially perform some operations whose nature depends on the current state and on the occurred event. The operations that can be performed are local computations, transmission of messages, and changes of state. Thus, the behaviour of an entity is a set of rules of the form *State x Event → Action*, where *State* is a system state, *Event* is one of a predefined set of external events, and *Action* is an indivisible sequence of local operations. The set of rules, the same for all entities, is called a *distributed algorithm* or *protocol*. The entities might have distinguished initial values, e.g., an identity; if this is not the case, the system is said to be *anonymous*.

The basic model is based on only two simple axioms:

- *Local Orientation*: Every entity can distinguish between its (in- and out-) neighbours, and can detect from which in-neighbour a received message was sent.
- *Finite Delays*: In absence of failure, a message sent to an out-neighbour is eventually received there in its integrity.

As a consequence of the Local Orientation axiom, it can be assumed that each entity x has a distinct label associated to each out-edge (i.e., edge connecting x to an out-neighbour) and in-edge (i.e., edge connecting an in-neighbour to x).

Note that the Finite Delays axiom does not imply the existence of any bound on transmission delays; it only states that, in absence of failure, a message will arrive after a finite delay in its integrity.

Any additional restriction of the general model defines a specific submodel. For example, the following additional axiom, called Message Ordering, defines a system where the transmission of messages obeys a FIFO discipline: messages sent to the same out-neighbour, if they arrive, will do so in the same order in which they were sent. By convention, all axioms defining a submodel are common knowledge to all entities. Common restrictions usually relate to reliability, time, or communication.

1.2 Synchronous Systems

With respect to *time*, the basic model does not make any assumption on the local clocks nor (except for the fact that it is finite) on transmission time (which include both processing and queueing delays). For these reason, the systems described by the basic model are referred to as *asynchronous*, and represent one end of the spectrum of message-passing systems with respect to time. On the other end are *synchronous systems*; that is, systems defined by two assumptions about *time*:

- *Synchronized Clocks* (SC): all local clocks 'tick' simultaneously (although they might not sign the same value).
- *Bounded Transmission Delays* (BTD): there exists a known upper bound on the number of clock ticks required for message transmission (including processing and queueing delays).

Since the bound is known a priori to all entities and all local clocks tick simultaneously, the unit of time can be redefined so that the BTD axiom can be replaced (as is almost always done) by the axiom

- *Unitary Transmission Delays* (UTD): message transmission is performed in a single unit of time.

In other words, if an entity sends a message at local clock tick t to a neighbour, the message is received and processed there at time $t + 1$ (sender's time). To avoid paradoxical situations, it is assumed that at any clock tick only one message can be send to the same neighbour.

In the following, like in almost all the literature, *synchronous* means simultaneous presence of SC and UTD.

1.3 Macro Versus Micro

In a synchronous message-passing system $\mathcal{S}$, the complexity of a distributed algorithm is evaluated with respect to two basic parameters: the number of message transmissions performed during the execution, and the number of clock ticks elapsed from the time the first entity starts the execution to the time the last entity terminates its participation in the computation.

The interval of time between successive clock ticks (sometimes called a *round*[1]) is bounded by some system parameter but, by definition, is long enough so that any message sent at a clock tick t arrives and is processed at its destination at clock tick $t + 1$. As a consequence, two very different assumptions on the message size are possible, and have been made in the literature, each offering a different view of synchronous computations:

1. message size is unbounded (e.g., all the data to be transmitted always fits in a single message); this is the *macroscopic view*.
2. the message size is bounded by some system constant B; this is the *microscopic view*.

In the microscopic view, if a computation requires an entity to transmit $M > B$ bits, it is actually requiring the transmission of at least $\lceil M/B \rceil$ messages (and not one, like in the macro level). Furthermore, since at a clock tick the entity can send at most one message to the same neighbour, the transmission of $M > m$ bits will require at least $\lceil M/B \rceil$ clock ticks (and not one, like in the macro level).

These microscopic facts are clearly invisible to the commonly used macroscopic view (e.g., the **LOCAL** model of computation [34]). Indeed, as already mentioned, a macroscopic view hides great part of the nature and beauty of synchronous computing, revealed only under the microscope. From this moment on, we will consider only the microscopic viewpoint.

1.4 Under the Microscope: The Difference Time Makes

Under the microscopic view, the unique characteristics of synchronous computations appear very clear. Among the many examples and results, the best known is the one expressed by the following "folk" theorem:

---

[1]These intervals are usually assumed to have all the same length, but such a condition is not necessary.

Property 1 *Any finite sequence of bits can be communicated in S transmitting two messages, regardless of the message size.*

Proof Let α be the sequence of bits, and let u and v be the transmitting entity and its receiving neighbour, respectively. Consider the following protocol for u: 1. send a message; 2. wait $g(\alpha)$ clock ticks; and 3. send another message, where $g(\alpha)$ is the integer whose binary representation is 1α. The protocol for v is the following: 1. upon receiving the first message, set count to zero; 2. at each clock tick, if no message is received, increase count by one, otherwise stop. Obviously, when y stops, count $= g(\alpha)$. □

The above property is a striking example of the difference that computing with time makes: since the message size is irrelevant and since the string α is finite but arbitrary, and the content of the transmitted messages is irrelevant, the property states that any amount of information (e.g., several Facebook datasets) can be communicated by transmitting just two bits.

The property, as stated, is incomplete from a complexity point of view. In fact, in a synchronous system, time and transmission complexities are intrinsically related to a degree non existent in asynchronous systems. In the example above, the *constant* bit complexity is achieved at the cost of a time complexity which is *exponential* in the length of the sequence of bits to be communicated, as stated by the following reformulation of the above property:

Lemma 1 *Any finite sequence of bits α can be communicated in S transmitting two bits in time $2^{\alpha+1}$.*

1.5 Organization

In the following, we present in some details some interesting aspects of computing with time, always in the inherent interplay between time and transmissions.

In Sect. 2 we discuss the most basic distributed computation, *two-party communication*: the communication of information between two neighbouring entities; the described results are from [31]. In Sect. 3 we present a simple yet powerful technique, *waiting*, that exploits the availability of time as a computational element; the results described in Sect. 3.1 are from [35], those in Sect. 3.2 from [11]. A general technique, *guessing*, which can be used to avoid the transmission of unbounded values, is discussed in Sect. 4; the described results are from [41]. Finally, we look at another basic activity, *wakeup*, and again investigate the time versus bits tradeoffs that it offers in the case of complete network; the discussed results are from [18].

2 Two-Party Communication

In a system of communicating entities, the most basic and fundamental problem is obviously the process of efficiently and accurately communicating information between two neighbouring entities.

This problem is sometimes called *TWO-PARTY COMMUNICATION* problem, and any solution algorithm is called a TPC protocol or *communicator*. Due to the basic nature of the process, the choice of a communicator can greatly affect the overall performance of the higher-level protocols employed in the system. Associated with any communicator are clearly two related cost measures: the total number of bit transmissions and the total number of clock ticks elapsed during the communication; as we will see, the study of the two-party communication problem in synchronous networks is really the study of the *trade-off* between time and transmissions.

2.1 Basic Communicators

Consider two entities, called the *sender* and the *receiver*, connected by a direct link; at each time unit, the sender can either transmit a bit or remain silent; a bit transmitted by the sender at time t will be received and processed by the receiver at time $t + 1$ (sender's time). A *quantum of silence* (or, simply, quantum) is the number of clock ticks between two successive bit transmissions; the quantum is zero if the bits are sent at two consecutive clock ticks.

Given a countable (and possibly infinite) universe U, the *two-party communication problem* for U, denoted by TPC(U), is the problem of the sender communicating without ambiguity to the receiver arbitrary elements of U using any combination of bit transmissions and silence. Since U is countable, we will assume without loss of generality that U is a set of consecutive integers starting from 0.

As observed in Sect. 1.4, any positive integer x can be communicated transmitting only two bits. This is achieved by the well-known 2-bits Communicator C_2, to which Lemma 1 refers in the Introduction:

Communicator C_2:

- To communicate a positive integer x, the sender transmits a first bit, waits a quantum of silence $q_1 = x$, and then sends the second and final bit b_1
- To reconstruct x, the receiver simply reconstructs the quantum of silence q_1 between the two received bits.

Using this communicator, the number of bits transmitted is 2 and the time is x.

Interestingly, if we increase the number of transmissions, time becomes sublinear. Consider the following protocol C_3 that uses 3 bits:

Communicator C_3:

- To communicate a positive integer x, the sender transmits three bits in order: b_0, b_1, and b_2; the quantum of silence q_1 between the first two transmissions, and q_2 between the second and the last are: $q_1 = \lfloor \sqrt{x} \rfloor$ and $q_2 = x - \lfloor \sqrt{x} \rfloor^2$.
- To obtain x the receiver simply computes $(q_1)^2 + q_2$.

Notice that $q_1 = x - \lfloor \sqrt{x} \rfloor^2 \leq 2\sqrt{x}$; thus, protocol C_3 has time *sublinear* time complexity $\leq 3\lfloor \sqrt{x} \rfloor + 3$. The method used by protocol C_3 can be easily extended to arbitrary $k = 2^r + 1$, obtaining a communicator C_k that communicates any integer x transmitting k bits using at most $k \, x^{\frac{1}{k-1}} + k$ time units.

Notice that here, as in the rest of this section, the transmitted bits are used only as delimiters; this renders the protocols resistant to message corruptions. In corruption-free systems, the bounds can obviously be improved by using the bits to convey information [31].

2.2 Optimal Communicators

At this point the natural question is what are the optimal communicators. We first discuss lower-bounds on the time-bits trade-off for the two-party communication problem both in the worst and in the average case. The bounds apply to any solution protocol, regardless of the schemes employed for encoding, transmitting and decoding. We then describe a solution protocol whose cost matches the lowerbounds.

2.2.1 Lower Bounds

Consider $C_b(U)$; i.e., the two-party communication problem for U using exactly b bit transmissions. Observe that b time units will be required to transmit the b bits; hence, the concern is on the amount of *additional* time required by the protocol. Obviously, the time before the first transmission and after the last transmission cannot be used to convey information.

Let $c(U, b)$ denote the number of time units needed in the worst case to solve $C_b(U)$. To derive a bound on $c(U, b)$, we will consider the dual problem of determining the size $\omega(t, b)$ of the largest set $\ddot{U}$ for which $c(\ddot{U}, b) \leq t$; that is, $\ddot{U}$ is the largest set for which the two-party communication problem can always be solved using b transmissions and at most t additional time units. Notice that, with b bit transmissions, it is only possible to distinguish $k = b - 1$ quanta; hence, the dual problem can be rephrased as follows:

> Determine the largest positive integer $n = \omega(t, b)$ such that every $x \in Z_n = \{0, 1, \ldots, n\}$ can be communicated using $k = b - 1$ distinguished quanta whose total sum is at most t.

This problem has an exact solution which will enable us to establish the desired bounds. Let $\mathbf{Bin}(x, y)$ denote the binomial coefficient $\binom{x}{y}$.

Theorem 1 $\omega(t, b) = \mathbf{Bin}(t + k, k)$.

Proof Let $n = \omega(t, b)$; by definition, it must be possible to communicate any element in $Z_n = \{0, 1, \ldots, n\}$ using $k = b - 1$ distinguished quanta requiring at most time t. In other words, $\omega(t, k + 1)$ is equal to the number of distinct k-tuples $\langle t_1, t_2, \ldots, t_k \rangle$ of positive integers such that $\sum_{1 \le i \le k} t_i \le t$. Given a positive integer x, let $T_k[x]$ denote the number of compositions of x of size k; i.e.,

$$T_k[x] = |\{\langle x_1, x_2, \ldots, x_k \rangle : \sum x_j = x, x_j \in Z^+\}|$$

Since $T_k[x] = \mathbf{Bin}(x + k - 1, k - 1)$, it follows that

$$\omega(t, k + 1) = \sum_i T_k[i] = \sum_i \mathbf{Bin}(i + k - 1, k - 1) = \mathbf{Bin}(t + k, k)$$

which proves the theorem.   □

We can now establish a *worst case* lower bound. Given two positive integers x and k, let $f(x, k)$ be the smallest integer t such that $x \le \omega(t, k + 1)$.

Theorem 2 *Any solution protocol for $C_{k+1}(U)$ requires $f(|U|, k)$ time units in the worst case.*

Proof From Theorem 1, it follows that $c(U, b) = f(|U|, k)$.   □

Theorem 3 *Let $f(|U|, k) = t$. For any solution protocol P for $C_{k+1}(U)$, there exists a partition of U into $t + 1$ disjoint subsets $U_0, U_1, \ldots, U_t$ such that*

1. $|U_i| = \mathbf{Bin}(i + k - 1, k - 1), 0 \le i < t; |U_t| \le \mathbf{Bin}(t + k - 1, k - 1)$
2. *the time $P(x)$ required by P to communicate $x \in U_i$ is $P(x) \ge i$.*

Proof Since $f(|U|, k) = t$, by Theorem 1, U is the largest set for which the two-party communication problem can always be solved using $b = k + 1$ transmissions and at most t additional time units. Given a protocol P for $C_{k+1}(U)$, order the elements $x \in U$ according to the time $P(x)$ required by P to communicate them; let $\ddot{U}$ be the corresponding ordered set. Define $\ddot{U}_i$ to be the subset composed of the elements of $\ddot{U}$ whose ranking, with respect to the ordering defined above, is in the range $\sum_{0 \le j < i} \mathbf{Bin}(j + k - 1, k - 1), \sum_{0 \le j \le i} \mathbf{Bin}(j + k - 1, k - 1)$. Since $f(|U|, k) = t$, it follows that $|\ddot{U}_i| = \mathbf{Bin}(i + k - 1, k - 1)$ for $0 \le i < t$ and $|\ddot{U}_t| \le \mathbf{Bin}(t + k - 1, k - 1)$ which proves part 1 of the theorem.

We will now show that, for every $x \in \ddot{U}_i$, $P(x) \ge i$. By contradiction, let this not be the case. Let $j \le t$ be the smallest index for which there exists an $x \in \ddot{U}_i$ such that $P(x) < j$. This implies that there exists a $j' < t$ such that $|\{x \in U : P(x) = j'\}| > \mathbf{Bin}(j' + k - 1, k - 1)$. In other words, in protocol P, the number of elements which are uniquely identified using k quanta for a total of j' time is greater than the number $T_k[j'] = \mathbf{Bin}(j' + k - 1, k - 1)$ of compositions of j' of size k; a clear contradiction. Hence, for every $x \in \ddot{U}_i$, $P(x) \ge i$, proving part 2 of the theorem.   □

This gives us an *average case* lower bound:

Theorem 4 *Any solution protocol for* $C_{k+1}(U)$ *requires*

$$\frac{tm + \sum_{0 \leq i < t} i \, \mathbf{Bin}(i + k - 1, k - 1)}{|U|}$$

time on the average where $t = f(|U|, k)$ *and* $m = t(|U| - \sum_{0 \leq i < t} i \, \mathbf{Bin}(i + k - 1, k - 1))$.

Proof From Theorem 3. □

2.2.2 An Optimal Solution

We now introduce a protocol whose cost matches both the worst and the average case lower bounds; we can actually show that this communicator is optimal at any point of the time-bits tradeoff.

Given two k-tuples $q = \langle q_1, q_2, \ldots, q_k \rangle$ and $q' = \langle q_1', q_2', \ldots, q_k' \rangle$ of positive integers, we say that $q \prec q'$ if $q_j - q_j'$ for $1 \leq j < l$, and $q_l < q_l'$ for some index l, $1 \leq l \leq k + 1$. For a given k, let V_t be the ordered set of k-tuples $q = \langle q_1, q_2, \ldots, q_k \rangle$ where $q_i \in Z^+$ and $\sum_i q_i \leq t$; that is $V_t[i] < V_t[i + 1]$. Obviously, the size of V_t is $\mathbf{Bin}(t + k, k)$. Any two integers t and i, $1 \leq i \leq \mathbf{Bin}(t + k, k)$, uniquely identifies a k-tuple $V_t[i] = \langle q_1, q_2, \ldots, q_k \rangle$ where $\sum_i q_i \leq t$; conversely, any k-tuple $\langle q_1, q_2, \ldots, q_k \rangle$ uniquely identifies the integers $t = \sum_i q_i$ and i, $1 \leq i \leq \mathbf{Bin}(t + k, k)$, such that $V_t[i] = \langle q_1, q_2, \ldots, q_k \rangle$.

The solution algorithm, $P1$, is described below; it comprises of an encoding scheme, a decoding scheme, and a communication protocol.

Encoding Scheme: Given X and k,
1. Let t be the smallest integer such that $X \leq \mathbf{Bin}(t + k, k)$; i.e., $t = f(X, k)$.
2. Determine $V_t[X] = \langle q_1, q_2, \ldots, q_k \rangle$
3. Set $encoding(X) = \langle p_0, p_1, \ldots, p_{2k} \rangle$, where $p_{2i} = b \in \{0, 1\}$ and $p_{2i+1} = q_i$, $(0 \leq i < k)$.

The value X to be communicated will be encoded as a $(2k + 1)$-tuple $\langle p_0, p_1, \ldots, p_{2k} \rangle$, where the even elements $p_0, p_2, \ldots, p_{2k}$ are arbitrary bits and the odd elements $p_1, p_3, \ldots, p_{2k-1}$ form the k-tuple corresponding to the X-th element of the set $V_{f(X,k)}$; i.e., $\langle p_1, p_3, \ldots, p_{2k-1} \rangle = V_{f(X,k)}[X]$.

Once the $(2k + 1)$-tuple $\langle p_0, p_1, \ldots, p_{2k} \rangle$ corresponding to the encoding of X has been determined, the actual communication can start. The encoded information is communicated as follows: the element $p_{2i} = b \in \{0, 1\}$ is transmitted and the element $p_{2i+1} = q_i$ is communicated by waiting a quantum of silence of length q_i.

Communication Protocol
SEND(X):
    Compute $encoding(X) = \langle p_0, p_1, \ldots, p_{2k} \rangle$;
    for $0 \leq i \leq 2k$
        if $even(i)$ then transmit p_i else wait p_i time units;
    endfor
RECEIVE(Z):
    $i := 0$;
    receive(b);
    $p_0 := b$;
    Repeat until $i = k$
        wait q until receive(b);
        $p_{2i+1} := q; i := i + 1; p_{2i} := b$;
    $Z := \langle p_0, p_1, \ldots, p_{2k} \rangle$;
    Compute $decoding(Z)$;

Once the last bit p_{2k} has been received, the receiving entity has received the $(2k + 1)$-tuple $\langle p_0, p_1, \ldots, p_{2k} \rangle$ and will apply to it the decoding scheme. To decode $\langle p_0, p_1, \ldots, p_{2k} \rangle$, the receiver will extract the $(k + 1)$-tuple $\langle q_1, q_2, \ldots, q_k \rangle$ formed by the odd elements $q_i = p_{2i+1}$, $(0 \leq i < k)$ and compute $t = \sum_i q_i$; at this point X, the communicated value, is the unique integer such that $1 \leq X \leq \mathbf{Bin}(t + k, k)$ and $V_t[X] = \langle q_1, q_2, \ldots, q_k \rangle$.

Decoding Scheme: Given $Z = \langle p_0, p_1, \ldots, p_{2k} \rangle$ and k,
1. Let $Y = \langle q_1, q_2, \ldots, q_k \rangle$ where $q_i = p_{2i+1}$, $(0 \leq i < k)$; let $t = \sum_i q_i$.
2. Find X such that $V_t[X] = Y$.
3. Set $decoding(Z) = X$.

For a fixed k, let $P(X)$ denote the amount of time required by algorithm P to communicate integer X using k bit transmissions. Recall (from Sect. 3) that $f(X, k)$ is the smallest integer t such that $x \leq \omega(t, k + 1)$.

Lemma 2 *For a fixed k, $P(X) = f(X, k)$ for every integer X.*

Proof By construction.   □

Theorem 5 *P is worst-case optimal for every $Z_n = \{0, 1, \ldots, n\}$.*

Proof By Lemma 2 and Theorem 2.   □

Protocol P actually satisfies a much stronger notion of optimality. A solution protocol A is *everywhere optimal* for U if, for every solution protocol B and $\forall b \geq 2$, there exists a permutation π of the elements of U such that $\forall x \in U : A(x, b) \leq$

$B(\pi(x), b)$. In other words, for every choice of the number of transmitted bits, A requires no more time to communicate any element of U (within a relabelling) than any other solution algorithm. Obviously, everywhere optimality implies both worst-case and average-case time-bits optimality.

Theorem 6 *For a fixed k, P is everywhere optimal for every $Z_n = \{0, 1, \ldots, n\}$.*

Proof Given Z_n, let $t = f(n, k)$ be the smallest integer such that $n \leq \omega(t, k + 1)$. Assume for simplicity that $n = \mathbf{Bin}(t + k, k)$. Let $S_i = \{x \in Z_n : P1(x) = i\}$. By Lemma 2, for every $x \in Z_n, P1(x) = f(x, k) \leq t$; hence, $|S_i| = \mathbf{Bin}(i + k - 1, k - 1), 0 \leq i \leq t$. Recall that, by Theorem 2, for any solution algorithm A, there exists a partition of Z_n into $t + 1$ disjoint subsets $A_0, A_1, \ldots, A_t$ such that $|A_i| = \mathbf{Bin}(i + k - 1, k - 1)$ and $A(x) \geq i$ for every $x \in A_i$. Therefore, there exists a permutation π of Z_n such that $P1(x) \leq A(\pi(x))$ for all $x \in Z_n$, proving the theorem. □

3 Waiting

In synchronous systems, time can be used to avoid the transmission of messages of unbounded length, i.e., unbounded values. The communicators described in the previous section are an instance of a simple and direct way of exploiting time to communicate unbounded values transmitting only a constant number of bits.

In this section, we describe another technique that makes an explicit use of time and that can be efficiently used as an alternative to transmitting possibly unbounded values. The technique assumes that every entity x, in addition to its own integer value $v(x)$ (not necessarily unique), has locally available a bound w on the number n of entities and a monotonically increasing integer function f, the same for all entities With respect to this technique, an entity can be either *active, processing* or *passive*. Initially, all entities are *active*.

The technique applies to both undirected and (strongly connected) directed graphs (i.e., bidirectional and unidirectional networks). In the following, the term 'neighbours' and the phrase 'all other neighbours' are assumed to mean for digraphs 'out-neighbours' and 'all out-neighbours', respectively. The technique is as follows:

Waiting Technique
1. An *active* entity x waits $f(v(x), w)$ time units.
2. If, during this time, it receives any message, it will forward it to all its other neighbours and become *passive*; otherwise it becomes *processing* and sends a message to all its neighbours.

We will now show how to "mutate" the basic technique so to work in different environments and different problems.

3.1 Minimum Finding and Election

Consider the situation where each entity x has a positive integer value $v(x)$; values might not be distinct. *MINIMUM FINDING* is the problem of moving from an initial configuration where all entities are in the same state *available*, to a configuration where every entity whose associated value is the minimum of all the values is in a predefined state *minimum* and all others are in a different predefined state *large*.

To deal with different initiation times, a pre-processing phase is added to the basic technique so to bound the delay between distinct starting times. Following is the algorithm where $w \geq \Delta(G)$ is known to all nodes, and $f(a, b) = 2ab$.

Algorithm WaitMinElect
- **Rule 0**. If an *available* entity wants to start the algorithm or receives an ACTIVATION message, it sends an ACTIVATION message to all other neighbours and becomes *active*.
- **Rule 1**. An *active* entity x waits $f(v(x), w)$ time units, ignoring any ACTIVATION message. If, during this time it receives an END message, it forwards it to all other neighbours and becomes *large*; otherwise, it sends an END message to all its neighbours and becomes *minimum*.
- **Rule 2**. A *large* entity ignores all END messages.

Theorem 7 *The minimum value v_{min} in any synchronous graph G with n nodes and e edges can be found with at most 4e bits in at most $2wv_{min} + 2\Delta(G)$ time units, provided $w \geq \Delta(G)$ is known.*

Proof Let $t(x)$ denote the time delay, from the start of the execution of the algorithm, to the time entity x becomes *active*. Let x and y be two nodes such that $v(x) < v(y)$. Entity x will become active at time $t(x)$ and will wait $f(x, w) = 2v(x)w$ time units; a message broadcasted by x would reach y after $d(x, y)$ time units, where $d(x, y)$ denotes the length of the shortest path from x to y in G. Since (from Rule 0) $t(x) \leq t(y) + d(y, x)$, it follows that

$$t(x) + 2v(x)w + d(x, y) \leq t(y) + d(y, x) + 2v(x)w + d(x, y) < t(y) + 2v(x)w + 2\Delta(G)$$

This implies that the entities with smallest value become *active* within at most $\Delta(G)$ time units from the time the algorithm is first started; they will finish waiting before everybody else, and thus send an END message; furthermore, this message will reach every other entity while they are still waiting. Thus, any entity z with the smallest identity (i.e., $v(z) = v_{min}$) will become *minimum* while all others will become *large*. This process will require at most $2v_{min}w + 2\Delta(G)$ time units. Each edge will be traversed by at most two ACTIVATION messages and two END messages; since a single bit is sufficient to distinguish between the two types of messages, a total of at most 4e bits will be transmitted.   □

Notice that, if all initial values $v(x)$ are distinct, only one entity will become *minimum*, while all others become *large*. This means, that protocol *WaitMinElect* actually solves the *LEADER ELECTION* problem; this problem requires moving the system from an initial configuration where all entities are in the same state ("candidate"), each with a distinct value, to a final configuration where all entities are in the same predefined state ("defeated"), except one which is a distinguished state ("leader"). Hence, the unique *minimum* is the elected *leader*.

Theorem 8 *A* leader *can be elected in any synchronous graph G with n nodes and e edges with at most 4e bits in at most* $2wv_{min} + 2\Delta(G)$ *time units, where* v_{min} *is the smallest value, provided* $w \geq \Delta(G)$ *is known.*

In specific classes of graphs more specific bounds apply:

Corollary 1 *Knowing n, an election can be performed in a unidirectional* ring *exchanging* $2n$ *bits in time* $(n + 1)v_{min} + 2n - 1$, *provided that the entities are aware of being in a ring.*

Proof To prove the time, choose $f(a, b) = a(b + 1)$ and observe that, in a unidirectional ring, $d(x, y) + d(y, x) = n$ for all x and y. For the bit complexity, observe that $e = n$ and that each edge will be traversed by exactly one ACTIVATION and one entities message.

Corollary 2 *Knowing* n_1 *and* n_2, *an election can be performed in a* $n_1 \times n_2$ *mesh exchanging* $O(n)$ *bits in time* $O((n_1 + n_2)v_{min})$, *provided that the entities are aware of being in a mesh.*

Proof In a mesh of $n = n_1 \times n_2$ nodes, $\Delta(G) = n_1 + n_2$; by choosing $w = d(G)$ and $f(a, b) = a(b + 1)$ the time bounds is achieved. Since $e = O(n)$, the message result follows.

Corollary 3 *Knowing n, an election can be performed in an unlabelled* hypercube *exchanging* $O(n \log n)$ *bits in time* $O(\log n v_{min})$, *provided that the entities are aware of being in a hypercube.*

Proof In a hypercube of $n =$ nodes, $\Delta(G) = \log n$; by choosing $w = \Delta(G)$ and $f(a, b) = a(b + 1)$ the time bounds is achieved. Since $e = O(n \log n)$, the message result follows.

Corollary 4 *With simultaneous initiation, an election can be performed in a* complete graph *exchanging* $n - 1$ *bits in time* $2v_{min} + 1$, *provided the entities are aware of being in a complete graph.*

Proof Remove Rule 0 (unnecessary because of simultaneous initiation) and Rule 2; modify Rule 1 so that received END message is not forwarded, and choose $f(a, b) = 2a$. This choice of f ensures that the entity with smallest identity v_{min} will finish waiting at least two time units before everybody else; hence, all other entities will become *passive* after $2v_{min} + 1$ time units. Following these modifications, the only communication occurring in this election process will be the bit sent from the entity with smallest identity to all other entities. □

3.2 Symmetry Breaking in Rings

If the assumption on the uniqueness of the values $v(x)$ does not hold, the election problem cannot obviously be solved by an extrema-finding process. If the nodes have no identities (i.e., the system is anonymous) then no deterministic solution exists for the election problem, duly renamed *SYMMETRY BREAKING*, regardless of whether the network is synchronous or not [3]. Thus, if any solution exists, it must be a randomized algorithm.

We now shown that, using the *Waiting Technique*, symmetry can be broken in a ring with $O(n)$ bits and time units on the average without any assumption on simultaneous initiation.

The algorithm is composed of a sequence of rounds; in each round, all nodes become *awake*. In round i, upon becoming *awake*, a node x chooses a random value $v(x, i) \in \{0, 1\}$ with a biased coin: it selects 0 with probability $\frac{1}{n}$ and 1 with probability $\frac{n-1}{n}$. All nodes participate in determining whether exactly one node has chosen 0 (Situation 1), or not (Situation 2). If Situation 1 has occured, the only node that has chosen 0 becomes *leader*, all other nodes become *defeated*, and the algorithm terminates; if Situation 2 has occured, all nodes start a new round.

Initially, all nodes are in a *sleeping* state. Any *sleeping* node can spontaneously become awake at any time and start the first round. To deal with different initiation times, a pre-processing phase is added in each round so to bound the delay between distinct starting times in that round.

A detailed description of the algorithm is as follows.

Algorithm SymmBreak
- **Rule 1.** A *sleeping* node:

  1. It can become spontaneously *awake* and execute the Wake-up routine.
  2. If it receives a WAKE- UP message, it becomes *awake* and executes the Wake-up routine.

- **Rule 2.** An *awake* node:

  1. It ignores any received WAKE- UP message.
  2. If it receives a CLAIM message, it becomes *half-awake* and sends the message on.
  3. If it receives a END message, it becomes *defeated* and passes the message on. (* Situation 1 *)
  4. When clock $= n$, if no CLAIM is received (see rule 2.2) and the number it selected is 0 it becomes *candidate* and sends a CLAIM message.
  5. When clock $= 2n$, if no CLAIM is received (see rule 2.2) it executes the Wake-up routine. (* Situation 2 *)

- **Rule 3.** A *candidate* node:

  1. If it receives a CLAIM with clock $< 2n$, it becomes *awake* and executes the Wake-up routine. (* Situation 2 *)
  2. If it receives a CLAIM and its clock equals $2n$, it becomes *elected* and sends a END message. (* Situation 1 *)

- **Rule 4.** A *half-awake* node:

  1. If it receives a WAKE- UP message, it becomes *awake* and executes the Wake-up routine.
  2. If it receives a END message, it becomes *defeated* and passes the message on. (* Situation 2 *)

where the Wake-up routine is as follows

Wake-up Routine
1. choose 0 with probability $\frac{1}{n}$ and 1 with probability $\frac{n-1}{n}$;
2. set clock $:= 0$ and send a WAKE- UP message.

An important property of the algorithm is expressed by the following

Lemma 3 *(i) Every node starts its execution of a round within $n - 1$ time units from the start of that round.*
(ii) If exactly one node becomes candidate *during this round, that node becomes* elected *and all others become* defeated; *otherwise, all nodes start another round.*

Proof Call a round a *success* if Situation 1 occurs. Assume the algorithm has performed $s - 1$ unsuccessful rounds and that (i) holds at the beginning the s-th round $(s \geq 1)$. Let $t(x)$ denote the time at which node x becomes awake in this round; a node x becomes *candidate* if and only if it has choosen 0 and it has not received any CLAIM in the (global) time interval $(t(x), t(x) + n)$; furthermore, only *candidate* nodes originate CLAIM messages. Three cases are possible depending on whether exactly one, more than one, or no node becomes *candidate* in the round, respectively.

Case 1: exactly one node x becomes *candidate*. (Note: this case occurs if and only if only one node chooses 0 in this round.) In this case, x will send a CLAIM at time $t(x) + n$. This message will reach node y at time $t(x) + d(x, y) + n$; since $t(x) \leq t(y) + d(y, x)$, it follows that node y will receive the CLAIM at time $t(x) + d(x, y) + n \leq t(y) + d(y, x) + d(x, y) + n = t(y) + 2n$. Thus, by rule 2.2, y becomes *half-awake* and sends the message on. In other words, the CLAIM message originated by x will travel along the ring transforming every node (except x) into *half-awake* and will arrive at x at time $t(x) + 2n$; when this occurs, x becomes *elected* and originates an END message (rule 2.2) which will make all other nodes *defeated* (rule 4.2).

Case 2: more than one node becomes *candidate*. Let $x_1, x_2, \ldots, x_k$ become *candidate* in this round; w.l.g. assume $t(x_i) \le t(x_{i+1})$, and let $r(x_i)$ denote the *candidate* nearest to x_i clockwise. First observe that, for all *candidate* nodes x_i and x_j, $t(x_j) < t(x_i) + d(x_i, x_j)$ (otherwise $t(x_i) + d(x_i, x_j) + n \le t(x_j) + n$, and x_j would receive a CLAIM with clock $\le n$ becoming *half-awake* and not *candidate* by rule 2.2). This implies that $t(x_i) + n < t(r(x_i)) + d(r(x_i), x_i) + n$; that is,

$$t(x_i) + d(x_i, r(x_i)) + n < t(r(x_i)) + d(r(x_i), x_i) + d(x_i, r(x_i)) + n = t(r(x_i)) + 2n$$

In other words, a CLAIM from x_i will reach node $r(x_i)$ before $r(x_i)$ counts $2n$. By rule 3.1, $r(x_i)$ will then kill the CLAIM and start the next round by becoming *awake* and sending a WAKE- UP message; thus, within at most $n - 1$ additional time units from the time the first x_i becomes *awake* again, all nodes are *awake*.

Case 3: nobody becomes *candidate*. (Note: this case occurs if and only if nobody chooses 0.) In this case, no CLAIM will be sent, and each node x will start the next round by becoming *awake* at time $t(x) + 2n$ (rule 2.5).

Summarizing, if part (i) of the lemma holds for the s-th round, then part (ii) will also hold; furthermore, if the round is not a success, part (i) will hold for the $(s + 1)$-th round. Since (i) holds initially (i.e., for $s = 1$), the lemma is proved.   □

The only thing left now is to see after how many rounds a leader will be elected. Perhaps surprisingly, the process terminates after less than 3 expected rounds.

Theorem 9 *Symmetry can be broken in a unidirectional ring using $2n$ bits and $2en$ time units on the average regardless of the initiation time, where c is a constant and $e = 2.7 \ldots$ is the basis of the natural logarithm.*

Proof In any one round, each node will send exactly one WAKE- UP message. If at least one node becomes *candidate*, then each node will send or forward exactly one CLAIM message. Since there are a constant number of message types, each message will use a constant c number of bits. Thus, each round will use at most $2cn$ bits. Within $2n$ time steps when the first node on a round executed the Wake-up routine, either a unique node is *elected* or a new round is started. For any round of random selections, the probability that exactly one node selects 0 is

$$\mathbf{Bin}(n, 1)\frac{1}{n}(\frac{n-1}{n})^{n-1} = (\frac{n-1}{n})^{n-1}$$

For n large enough, this quantity is easily bounded:

$$\lim_{n \to \infty} (\frac{n-1}{n})^{n-1} = \frac{1}{e}$$

Thus, the expected number of rounds until this situation occurs is less than e; by Lemma 3, if this event occurs, the algorithm terminates.   □

In the above theorem, the factor 2 can be removed from both the time and bit complexity by allowing the nodes to immediately become *candidate* if they select 1 in the Wake-up routine, and modifying the algorithm appropriately.

4 Guessing

Another powerful technique that allows to compute functions on unbounded values without ever transmitting them is guessing. Let us consider again the *MINIMUM FINDING* problem, that is the problem of computing $v_{min} = \min\{v(x)\}$. Let us assume that all entities know n and start at the same time.

Consider the following distributed algorithm, where p is a parameter available to all entities:

Decide(p)
$clock := 0$; (* start counting *)
if $v(x) \leq p$ then
    send YES to all neighbors;
    state := *decided*;
else *state* := *undecided*;
    if (YES is received   and $clock < n$ and   *state* = *undecided*) then
        send the message to all neighbors which have not sent any message to you;
        state := *decided*;
    else ignore the message.

Note that forwarding a YES message can be done at most once by any entity since after sending it the entity becomes *decided*. Also note that, due to the synchrony in the network, this message could have been received from more than one neighbor in the current time slot, and that it is forwarded only to the other neighbors.

Lemma 4 *Let all entities know n and p, and simultaneously start the execution of* **Decide**(p) *at time 0. Then, at time n:*

1. *if all local values are greater than p, then all entities are undecided;*
2. *if there is at least one local value* $v(x) \leq p$, *then all entities are decided.*

Furthermore, the number of bits transmitted is zero in case (1), and at most 2e in case (2).

Proof At time zero entity x becomes *undecided* and sends no message iff its value $v(x)$ is greater than p. Thus, if all values are greater than p, no messages will be transmitted during the execution of **Decide**; furthermore, all entities will remain *undecided*, at time n. If entity x has local value $v(x) \leq p$, it will become *decided* at time zero and send YES messages to all its neighbours. An entity in state *decided* ignores all YES messages; an entity in state *undecided* receiving a YES message becomes *decided* and forwards the message only to the neighbours from which such a message has not yet been received; thus, at most two messages will be transmitted on each edge, for a total of at most $2e$ bits. Since the underlying communication graph is connected, it is easily shown that by time n each entity that was not decided at time zero has received at least one YES message. Since an *undecided* entity becomes *decided* as soon as it receives a message, all entities become *decided* within $n - 1$ time units.   □

Using this property, we can effectively employ **Decide** as a building block for our computations.

4.1 Minimum Finding as a Guessing Games

A technique for minimum-finding can be developed by performing a sequence of executions of **Decide** as follows. Initially, all entities choose the same initial value g_1 and simultaneously perform **Decide**(g_1). After n time units, all entities will be aware of whether the minimum value is greater than g_1 (case (1) in Lemma 4) or not (case (2) in Lemma 4); note that the latter case means that we overestimated or guessed the minimum value; this case will be called *overestimate*, even if the correct value has been guessed. Based on the outcome, a new value g_2 will be chosen by all entities, which will then simultaneously perform **Decide**(g_2). In general, based on the outcome of the execution of **Decide**(g_i), all entities will choose a value g_{i+1} and simultaneously perform **Decide**(g_{i+1}); this process is repeated until the minimum value is unambiguously determined. Depending on which strategy is employed for choosing g_{i+1} given the outcome of **Decide**(g_i), different minimum-finding algorithms will result from this technique. This technique allows to reformulate the minimum-finding problem in terms of a *number-guessing game*, as follows.

Guessing Game
1. the network is a player;
2. the minimum value in the network is a number, previously chosen and unknown to the player;
3. the player has to guess the number, by only asking questions of the type "is the number greater than g?", where each question corresponds to a simultaneous execution of **decide**(g);
4. cases (1) and (2) of Lemma 4 correspond to a "yes" and a "no" answer to the question, respectively; the latter case will be termed an *overestimate*.

First observe that, by definition, to each solution strategy for this number-guessing game corresponds a solution algorithm for the minimum-finding problem. As for the complexity of these solution algorithms recall that, by Lemma 4, each execution of

Decide (i.e., each question) requires n time units, while the number of bits transmitted is either zero or at most $2e$, depending on whether the answer is "yes" or "no", respectively. The following theorem has thus been proved.

Theorem 10 *Let S be a solution strategy for the number-guessing game which requires $b(X)$ overestimates and a total of $t(X)$ questions in the worst case, where X is the unknown number. Let v_{min} be the smallest value in the network, and assume n is known to all entities. Then*

1. *minimum-finding can be performed in an anonymous network using at most $n \cdot t(v_{min})$ time and $2 \cdot e \cdot b(v_{min})$ bits;*
2. *election in a network with distinct values can be performed using at most $n \cdot t(v_{min})$ time and $2 \cdot e \cdot b(v_{min})$ bits;*
3. *a spanning-tree in a network with distinct values can be constructed using at most $n \cdot t(v_{min})$ time and $2 \cdot e \cdot (b(v_{min}) + 2)$ bits.*

4.2   Optimal Solutions

We are interested in determining the guessing strategy that offers the best use of time for reducing the amount of bit transmissions. This means to find the strategy that solves the guessing games with the minimum number of questions (each question costs n time units) of which up to a given number b are overestimates (each overestimate costs the transmission of $2e$ bits).

Consider first the case in which the unknown number is a positive integer in the interval $[1, M]$; i.e., the values are $v(x) \leq M$. We will see later the case when the interval is unbounded, i.e., $M = +\infty$.

4.2.1   Lower Bound

We want to determine the minimum number $h(M, b)$ of questions needed to correctly guess any value in $[1, M]$ with no more than b overestimates.

To do so, we consider the "converse" problem of determining the largest integer $f(t, b)$ such that any value X known to be within the interval $[1, f(t, b)]$ can be guessed using at most t questions of which at most b are overestimates. The next theorem determines $f(t, b)$.

Theorem 11 *For every $t \geq b \geq 1$,*

$$f(t, b) = \sum_{0 \leq i \leq b} \mathbf{Bin}(t, i). \qquad (1)$$

Proof It is easy to see that $f(t, 1) = t + 1$, since the only algorithm that makes at most one overestimate is "sequential search"; i.e., using the guesses $g_1 = 1, g_2 = 2, \ldots, g_t = t$. It also trivial to see that $f(t, t) = 2^t$ since, if any question can be an overestimates, the largest possible interval is $[1, 2^t]$.

Now, for $t > b \geq 1$, suppose the first question is "is the number $> v$?". If the answer is *yes*, the unknown number is greater than v, and the player has to find it with $t - 1$ questions and b overestimates; hence, the largest interval that can be correctly searched in this case is $[v + 1, v + f(t - 1, b)]$. If the answer is *no*, then the unknown number lies in the interval $[1, v]$, to be searched using $t - 1$ questions and at most $b - 1$ overestimates. Thus, the largest value of v that allows for a correct solution in this case is $f(t - 1, b - 1)$. We therefore have

$$\forall b, t > b \geq 1, \ f(t, b) = f(t - 1, b) + f(t - 1, b - 1). \tag{2}$$

One can show now that the unique solution to (2), satisfying the boundary conditions $f(t, 1) = t + 1$ and $f(t, t) = 2^t$, is given by (1). (Another approach is to determine the generating function $F(x, y) = 1/(1 - z)(1 - y - z * y)$). $\qquad \square$

For future use we extend the definition of $f(t, b)$ to the case where $1 \leq t < b$, so as to satisfy (2) for every t and b, by

$$f(t, b) = f(t, t) \qquad for \ 1 \leq t < b. \tag{3}$$

We can now return to our original quest for determining a bound on the minimum number $h(M, b)$ of questions needed to correctly guess any value in $[1, M]$ with no more than b overestimates. We are now able to do so; in fact, by Theorem 11 we have:

Theorem 12 *Let $\hat{t}(M, b) = \min\{t : f(t, b) \geq M\}$. Then, for every $b \geq 1$,*

$$\hat{t}(M, b) \geq h(M, b) \geq \hat{t}(M, b) - 1. \tag{4}$$

Observe that, if $M = f(t, b)$, then $h(M, b) = \hat{t}(M, b) = t$.

There is no known closed-form expression for $h(M, b)$; however, it can be closely estimated as follows:

Lemma 5

$$h(M, b) = (b! N)^{\frac{1}{b}} + \epsilon b, \quad for \ some \ \epsilon = \epsilon(N, b) \ with \ -1 < \epsilon < 1. \tag{5}$$

Proof 1. By induction on (t, b) (using (2)), one can show that

$$\textbf{Bin}(t + 1, b) \leq f(t, b) \tag{6}$$

By (4) and (6) we have

$$\mathbf{Bin}(h, b) \le f(h-1, b) < N$$

and hence

$$h(h-1) \cdots (h-b+1) < b\,!\,N.$$

Thus $(h-b+1)^b < b\,!\,N$ and we have

$$h < (b\,!\,N)^{\frac{1}{b}} + b. \tag{7}$$

2. By induction on (t, b) (using (2)), one can show that

$$f(t, b) \le \mathbf{Bin}(t+b, b). \tag{8}$$

By (4) and (8) we have

$$N \le f(h, N) \le \mathbf{Bin}(h+b, b),$$

and hence

$$b\,!\,N \le (h+b)(h+b-1) \cdots (h+1).$$

Thus, using the inequality between the arithmetic and geometric means:

$$(b\,!\,N)^{\frac{1}{b}} \le [(h+b)(h+b-1) \cdots (h+1)\,]^{\frac{1}{b}}$$
$$\le \frac{1}{b}\,[(h+b) + (h+b-1) + \cdots + (h+1)\,] = h + \frac{b+1}{2}.$$

Therefore

$$h \ge (b\,!\,N)^{\frac{1}{b}} - \frac{b+1}{2}. \tag{9}$$

The lemma follows from (7) and (9). $\qquad\qquad\square$

4.2.2 Optimal Protocol

The optimal guessing strategy follows directly from the proof of Theorem 11:

Optimal Guessing Strategy To optimally search in $[1, M]$ with at most k overestimates:

1. Use as a guess $p = h(q - 1, k - 1)$, where $q \geq k$ is the smallest integer such that $M \leq h(q, k)$.
2. If p is an underestimate, then optimally search in $[p + 1, M]$ with k overestimates.
3. If it is an overestimate, then optimally search in $[1, p]$ with $k - 1$ overestimates.

This means that

Theorem 13 *The number guessing game in a bounded interval $[1, M]$ can be solved with b bits and $\hat{\imath}(M, b)$ questions.*

Worst case optimality follows from Theorem 12.

We must still consider solving the guessing game when the interval in unbounded; i.e., $M = +\infty$. A solution strategy could be to first determine a bounded interval containing X using $b' < b$ overestimates, and then use the Optimal Guessing Strategy above to find X in this interval with at most $b - b'$ overestimates.

To determine an interval containing X, we find an upperbound on X by using a monotonically increasing integer function g and proceeding through a sequence of questions "*is the number* $> g(i)$?" ($i = 1, 2, \ldots$), until we determine the value j such that $g(j - 1) < X \leq g(j)$. This approach requires exactly j questions and one overestimate.

Once this is done, we are left to determine X in an interval of size $\Delta(j) = g(j) - g(j - 1)$ with only $b - 1$ overestimates; this can be done using the Optimal Guessing Strategy above with at most $h(\Delta(j), b - 1)$ questions.

The entire process will thus require at most $j + h(\Delta(j), b - 1)$ questions. In other words:

Theorem 14 *The number guessing game in a unbounded interval can be solved with b overestimates using at most $2h(X, b) - 1$ questions, where X is the unknown number.*

Proof Choose $g(i) = f(i, b)$, for $i \geq 1$. Let $t = h(X, b)$ (i.e., the smallest integer i such that $f(i, b) \geq X$); then, following the above procedure, we stop when $j = t$. From this follows that $\Delta(j) = g(t) - g(t - 1) = f(t, b) - f(t - 1, b) = f(t - 1, b - 1)$; that is, $t - 1$ questions will suffice to solve the resulting guessing game in an interval of size $\Delta(j$ with $b - 1$ overestimates. Altogether, we determined the unknown X with a total of at most $2t - 1$ questions and b overestimates. $\square$

4.3 Improved Bounds for Distributed Problems

Using the correspondence between guessing games and minimum-finding, the results of the previous section will now be reinterpreted in the context of distributed computations. First observe that, in the distributed problem, no upper-bound is assumed on the range of the values among which the minimum must be found. Further observe that each solution strategy for the number guessing game with k overestimates corresponds to a minimum-finding algorithm requiring the transmission of $O(k \cdot e)$ bits (Theorem 10). Let C_k denote the class of such minimum-finding algorithms.

Theorem 15 *The minimum value v_{min} in a synchronous anonymous network can be determined using at most $O(k \cdot e)$ bits in time $O(k \cdot n \cdot v_{min}^{\frac{1}{k}})$ for any integer $k > 0$, provided n is known to the entities, and the entities start simultaneously. For every value of the integer k, this bound is optimal among all algorithms in C_k.*

Proof Let t be the smallest integer such that $f(t, k) \geq v_{min}$; by Theorem 13 it follows that v_{min} can be guessed using at most $2t - 1$ questions. Thus, by Theorem 10, the minimum value v_{min} can be determined using at most $(4t - 2) \cdot n$ time and $2 \cdot k \cdot e$ bits. By Lemma 5, $t < (k! \cdot i)^{\frac{1}{k}} + k$, which is approximately $\frac{i^{\frac{1}{k}}k}{e} + k$ (using Stirling's approximation), from which the bound follows. By Theorem 11 and Lemma 5, any algorithm in C_k requires at least $\Omega((\frac{k! \cdot v_{min}}{2})^{\frac{1}{k}})$ questions, from which optimality follows. $\square$

In a similar way, the following theorem can be proved.

Theorem 16 *In a synchronous network with distinct values, election and spanning-tree construction can be performed using at most $O(k \cdot e)$ bits in time $O(k \cdot n \cdot v_{min}^{\frac{1}{k}})$ for any k, where v_{min} is the smallest value in the network, provided n is known and the entities start simultaneously.*

Again, by choosing k to be any constant > 1, the theorem yields an improvement in the time complexity of using waiting (Theorem 8) without increasing the order of magnitude of the bit complexity.

5 Waking up in Complete Networks

A basic activity in distributed computing systems is that of *WAKE-UP*: initially all entities are *asleep*; one or more entities, called *initiators*, independently wake-up and send WAKEUP messages to some neighbours, starting a process to ensure that within finite time all entities become *awake*. Since the WAKEUP message contains no other information, the wakeup process is a prototypical microscopic computation.

The wake-up process is used in a variety of situations, including *initialization*, *notification*, and *reset*, e.g., to ensure that every entity in the system becomes aware of the start (or termination) of a computation. Called also *weak unison* and *distributed reset*, this process can be carried out by totally anonymous entities. It offers an interesting trade-off between time (the difference between the time the first entity wakes up and the time that all entities are awakened) and communication (number of WAKEUP messages sent in that interval) in synchronous networks.

This is indeed the case in *complete networks*. For simplicity, let denote the set of nodes by $\{0, 1, \ldots, n-1\}$, and for nodes x, y, the label of the edge $\{x, y\}$ is the integer $(y - x) \mod n$.

The obvious solution is to "flood": an initiator sends a WAKEUP message to all its neighbours. Since we are in a complete graph, this protocol requires only a single round. The number of messages will however be $k(n-1)$, where k is the number of initiators; this means that, in the worst case, $n^2 - n$ messages will be transmitted. The only way to keep down the number of messages is for initiators to send only to a subset of their neighbours. For example, if each node x sends only to node $(x + 1) \mod n$ then only n messages will be sent in total, regardless of the number of initiators. However, the time to complete wakeup will be the maximum distance between successive initiators; this means that, in the worst case (i.e., with a single initiator), the wakeup requires $n - 1$ time units.

In this type of situation, to measure the complexity of a protocol, the integrated cost measure *time* × *bits* (*TB*) is used, i.e., the number of messages times the number of steps required in the worst case for the completion of the algorithm, over all possible choice and schedule of the initiators. Notice that, for the two algorithms described above, the *TB*-complexity is the same: $O(n^2)$. The quest is for more efficient wakeup protocols.

5.1 Oblivious Protocols

We consider a special class of protocols, those where an entity sends the WAKEUP message to the same set of neighbours both when it is an initiator and when it receives a WAKEUP message while asleep. This class of protocols is called *oblivious* (because the set of neighbours does not depend on the state of the entity).

Notice that, since the network is anonymous, an oblivious protocol P can be seen as specifying a subset S of integers modulo n and requiring every entity x to send a WAKEUP message only to the subset $\{(x + j) \mod n : j \in S(P)\}$ of its neighbours. It is assumed that $1 \in S$ while $0 \notin S$.

Further notice that the set $S = \{d_0, d_1, d_2, \ldots, d_{k-1}\}$, where $d_0 = 1$ and $d_{i-1} < d_i < n$ for $1 \leq i < n$, defines a graph $R_n[d_1, d_2, \ldots, d_{k-1}]$ where the nodes are $0, 1, \ldots, n-1$ and there is an edge between nodes x, y if and only if $(x - y) \mod n \in S$. The class of graphs so defined are known as *chordal ring*.

Example 1 For $n \leq 2^k$ the chordal ring $R_n[2, 4, \ldots, 2^{k-1}]$ has diameter $\leq 2 \log n$ and degree $\log n$. This chordal ring is also called *double-cube*.

Example 2 For $n < k!$ the chordal ring $R_n[2!, 3!, \ldots, (k-1)!]$ has diameter $O(k^2)$ (use the fact that every $x < n$ can be represented in the mixed basis $1!, 2!, \ldots, (k-1)!$ as $x = x_1 + x_2 2! + \cdots + x_{k-1}(k-1)!$, with $0 \le x_i \le i$, for $i \ge 1$) and degree $n \le \log n / \log \log n$.

Summarizing, the edges on which messages are sent during the execution of an oblivious wakeup protocol P form a chordal ring R. The message and the time complexity of P will be the number of edges and the diameter of R, respectively.

We will use this observation to derive an optimal wakeup protocol.

5.2   Lower Bounds for Oblivious Protocols

We first derive a lower bound on the TB-complexity of oblivious wakeup protocols. The derivation of this lower bound is based on the following bound for chordal rings.

Lemma 6 *Let chordal ring $R_n[d_1, d_2, \ldots, d_{k-1}]$ have diameter ψ. Then*

$$k \cdot \psi = \Omega \left(\log^2 n \right),$$

As a consequence, we have the following:

Theorem 17 *Any oblivious wakeup protocol has $\Omega(n \log^2 n)$ TB-complexity.*

Proof This follows easily from Lemma 6. Since the protocol is oblivious every entity transmits a fixed number of messages in each iteration of the wakeup protocol, say k. The graph resulting from such a protocol is the chordal ring $R_n[S]$, where S is a set of size k. The time required for the WAKEUP message to reach all the entities is at least the diameter ψ of the chordal ring $R_n[S]$. Eventually all n entities are awakened. Since the protocol is oblivious every entity that receives a WAKEUP message must transmit to all its k neighbours. Hence the number of messages transmitted during the execution of the protocol is nk. It follows that the complexity is at least $nk\psi = \Omega(n \log^2 n)$. $\square$

5.3   Optimal Oblivious Wakeup

In this section we describe an optimal oblivious wakeup algorithms. To derive it, first observe that the k-dimensional mesh can be viewed as a chordal ring $R_n[S]$, for some set S of links. For example, the 2-dimensional mesh is the chordal ring $R_n[\sqrt{n}]$, while the k-dimensional mesh is the chordal ring $R_n[n^{1/k}, n^{2/k}, \ldots, n^{(k-1)/k}]$.

For each point $p = (p_1, p_2, \ldots, p_k)$ in the k-dimensional mesh, let

$$S_p^i = \{(p_1, \ldots, p_{i-1}, x_i, p_{i+1}, \ldots, p_k) : x_i < n^{1/k}\}.$$

If we define $S^i = \{(0, \ldots, 0, x_i, 0, \ldots, 0) : 0 \le x_i < n^{1/k}\}$ then we see easily that $S_p^i = p + S^i$. Let $S_p = S_p^1 \cup \cdots \cup S_p^k$, and $S = S^1 \cup \cdots \cup S^k$.

This indicates a way to design an efficient oblivious wakeup algorithm for the complete graph which terminates in k steps. In this protocol, entity p sends wakeup messages to all and only the entities in the set $p + S$.

Oblivious Chordal Wakeup (for entity p)
1. If p is an initiator then it sends a WAKEUP message to all its neighbors in the set $p + S$ and becomes awake.
2. If p receives a WAKEUP message from another entity and is not awake then it sends WAKEUP messages to all entities in the set $p + S$ and becomes awake.

Theorem 18 *For any k, if $n = q^k$ for some q, then protocol Oblivious Chordal Wakeup has TB-complexity $O(k^2 n^{(k+1)/k})$.*

Proof Let $n = q^k$; then in the execution of protocol Oblivious Chordal Wakeup, the size of each transmission is $kn^{1/k}$. A transmission from entity $p = (p_1, \ldots, p_k)$ will reach all entities of the form $p' = (p_1, \ldots, p_{i-1}, p_i', p_{i+1}, \ldots, p_k)$, where $0 \le p_i' < n, i = 1, 2, \ldots, k$. Therefore every entity will be reached after k steps. The complexity is easily seen to be as claimed.   □

More generally, by carefully choosing S, we have

Theorem 19 *For any k and any $m \le n^{1/k}$, if $n = q^k$ for some q, then protocol Oblivious Chordal Wakeup is a t-step wakeup protocol, where $t = \frac{k}{m} n^{1/k}$, and its TB-complexity is $O(m k t n) = O(k^2 n^{(k+1)/k})$.*

Proof Let $n = q^k$, $m \le n^{1/k}$, and $t = \frac{d}{m} n^{1/k}$. In the execution of Oblivious Chordal Wakeup, each entity p transmits to the set $p + S$, where $S = S_1 \cup \cdots \cup S_k$ and $S_i = \{(0, \ldots, 0, x_i, 0, \ldots, 0) : 0 \le x_i < m\}$.   □

We now have all the ingredients to make Oblivious Chordal Wakeup optimal.

Theorem 20 *Oblivious wakeup is possible with optimal TB-complexity $\Theta(n \log^2 n)$.*

Proof Consider protocol Oblivious Chordal Wakeup when the set of neighbours to which an entity p sends the WAKEUP messages defines the double-cube, i.e., the chordal ring $R_n[2, 2^2, \ldots, 2^{k-1}]$ described in Example 1. By Theorem 19, the claimed complexity follows. Optimality follows from the lower bound for oblivious protocols of Theorem 17.   □

6 Further Reading

The aim of this chapter has been to introduce the basics of synchronous computing at the microscopic level, describing some simple but powerful computational and analytical tools.

Notice that, even though everything has been expressed in terms of message-passing, the validity of what we said is independent of whether message transmission to a neighbour and its reception are implemented by using physical communication channels, or by writing to and reading from a predesigned shared register; in the latter case, the microscopic view examines synchronous computations when the size of the registers is limited by a system constant.

The reader interested in knowing more about the microscopic nature of synchrony is referred to the overview material in Chap. 6 of [37], as well as to the significant amount of investigations on the subject.

These investigations cover a wide spectrum of problems and topics, including *election* (e.g., [8, 12, 13, 25, 33, 35, 40, 41]), *extrema finding* (e.g., [1, 36, 41]), *symmetry breaking* (e.g., [11, 15, 20, 26]), *consensus* (e.g., [7]), *communicators* and their use (e.g., [5, 6, 30, 31, 38]), *shortest paths* (e.g., [28]), *unison, firing squad* and *wake-up* (e.g., [4, 10, 16, 18, 29, 32]), *matching* (e.g., [19, 42]). Also relevant are the results in the more powerful CONGEST model, on problem such as *minimum dominating sets* (e.g. [21, 23]), *coloring* and *independent sets* (e.g. [2, 15, 17, 22, 27, 39]), and *minimum-spanning-tree construction* (e.g. [9, 14, 24, 34]). For a recent investigation in a different application area see [43].

Acknowledgments This work has been supported in part by the Natural Sciences and Engineering Research Council (Canada) under the Discovery Grant program.

References

1. Alimonti, P., Flocchini, P., Santoro, N.: Finding the extrema of a distributed multiset of values. J. Parallel Distrib. Comput. **37**, 123–133 (1996)
2. Alon, N., Babai, L., Itai, A.: A fast and simple randomized parallel algorithm for the maximal independent set problem. J. Algorithms **7**(4), 567–583 (1986)
3. Angluin, D.: Local and global properties in networks of processes. In: Proceedings of the 12th ACM Symposium on Theory of Computing (STOC), pp. 82–93 (1980)
4. Arora, A., Dolev, S., Gouda, M.: Maintaining digital clocks in step. Parallel Process. Lett. **1**(1), 11–18 (1991)
5. Attiya, H., Snir, M., Warmuth, M.K.: Computing on an anonymous ring. J. ACM **35**(4), 845–875 (1988)
6. Bar-Noy, A., Naor, J., Naor, M.: One bit algorithms. Distrib. Comput. **4**(1), 3–8 (1990)
7. Bar-Noy, A., Dolev, D.: Consensus algorithms with one-bit messages. Distrib. Comput. **4**(3), 105–110 (1991)
8. Bodlaender, H.L., Tel, G.: Bit-optimal election in synchronous rings. Inf. Process. Lett. **36**(1):53–64 (1990)
9. Elkin, M.: A faster distributed protocol for constructing a minimum spanning tree. J. Comput. Syst. Sci. **72**(8), 1282–1308 (2006)

10. Even, S., Rajsbaum, S.: Unison, canon and sluggish clocks in networks controlled by a synchronizer. Math. Syst. Theor. **28**, 421–435 (1995)
11. Frederickson, G.N., Santoro, N.: Breaking symmetry in synchronous networks. In: Proceedings of 2nd International Workshop on Parallel Computing and VLSI (now SPAA), pp. 26–33 (1986)
12. Frederickson, G.N., Lynch, N.A.: Electing a leader in a synchronous ring. J. ACM **34**, 95–115 (1987)
13. Gafni, E.: Improvements in the time complexity of two message-optimal election algorithms. In: Proceedings of the 4th Conference on Principles of Distributed Computing (PODC), pp. 175–185 (1985)
14. Garay, J., Kutten, S., Peleg, D.: A sub-linear time distributed algorithm for minimum-weight spanning trees. SIAM J. Comput. **27**, 302–316 (1998)
15. Goldberg, A.V., Plotkin, S.A., Shannon, G.E.: Parallel symmetry-breaking in sparse graphs. SIAM J. Discr. Math. **1**(4), 434–446 (1988)
16. Gouda, M., Herman, T.: Stabilizing unison. Inf. Process. Lett. **35**(4), 171–175 (1990)
17. Halldórsson, M.M., Konrad, C.: Distributed large independent sets in one round on bounded-independence graphs. In: Proceedings of the 29th International Symposium on Distributed Computing (DISC), pp. 559–572 (2015)
18. Israeli, A., Kranakis, E., Krizanc, D., Santoro, N.: Time-messages tradeoffs for the weak unison problem. Nordic J. Comput. **4**, 317–329 (1997)
19. Israeli, A., Itai, A.: A fast and simple randomized parallel algorithm for maximal matching. Inf. Process. Lett. **22**, 77–80 (1986)
20. Itai, A., Rodeh, M.: Symmetry breaking in distributed networks. Inf. Comput. **88**(1), 60871 (1990)
21. Jia, L., Rajaraman, R., Suel, T.: An efficient distributed algorithm for constructing small dominating sets. Distrib. Comput. **15**(4), 193–205 (2002)
22. Kothapalli, K., Onus, M., Scheideler, C., Schindelhauer, C.: Distributed coloring in $O(\sqrt{\log n})$ bit rounds. In: Proceedings of the 20th International Parallel and Distributed Processing Symposium (IPDPS) (2006)
23. Kutten, S., Peleg, D.: Fast distributed construction of k-dominating sets and applications. J. Algorithms **28**, 40–66 (1998)
24. Lotker, Z., Patt-Shamir, B., Pavlov, E., Peleg, D.: Minimum-weight spanning tree construction in $O(\log \log n)$ communication rounds. SIAM J. Comput. **35**(1), 120–131 (2005)
25. Marchetti-Spaccamela, A.: New protocols for the election of a leader in a ring. Theor. Comput. Sci. **54**(1), 53–64 (1987)
26. Mayer, A., Ofek, Y., Ostrovsky, R., Yung, M.: Self-stabilizing symmetry breaking in constant-space. In: Proceedings of the 24th ACM Symposium on Theory of Computing (STOC), pp. 667–678 (1992)
27. Métivier, Y., Robson, J.M., Saheb-Djahromi, N., Zemmari, A.: An optimal bit complexity randomized distributed MIS algorithm. Distrib. Comput. **23**(5–6), 331–340 (2011)
28. Métivier, Y., Robson, J.M., Zemmari, A.: A distributed enumeration algorithm and applications to all pairs shortest paths, diameter. Inf. Comput. **247**, 141–151 (2016)
29. Nishitani, Y., Honda, N.: The firing squad synchronization problem for graphs. Theor. Comput. Sci. **14**, 39–61 (1981)
30. O'Reilly, U.-M., Santoro, N.: Asynchronous to synchronous transformations. In: Proceedings of the 4th International Conference on Principles of Distributed Systems (OPODIS), pp. 265–282 (2000)
31. O'Reilly, U.-M., Santoro, N.: Tight bound for synchronous communication of information using bits and silence. Discret. Appl. Math. **129**, 195–209 (2003)
32. Ostrovsky, R., Wilkerson, D.S.: Faster computation on directed networks of automata. In: Proceedings of the 14th ACM Symposium on Principles of Distributed Computing (PODC), pp. 38–46 (1995)
33. Overmars, M., Santoro, N.: Improved bounds for electing a leader in a synchronous ring. Algorithmica **18**, 246–262 (1997)
34. Peleg, D.: Distributed Computing: A Locality-Sensitive Approach. SIAM (2000)

35. Santoro, N., Rotem, D.: On the complexity of distributed elections in synchronous graphs. In: Proceedings of the 11th International Workshop on Graphtheoretic Concepts in Computer Science (WG), pp. 337–346 (1985)
36. Santoro, N.: Computing with time: temporal dimensions in distributed computing. In: Proceedings of the 28th Allerton Conference on Communication, Control and Computing, pp. 558–567 (1990)
37. Santoro, N.: Design and Analysis of Distributed Algorithms. Wiley (2007)
38. Schmeltz, B.: Optimal tradeoffs between time and bit complexity in synchronous rings. In: Proceedings of the 7th Symposium on Theoretical Aspects of Computer Science (STACS), pp. 275–284 (1990)
39. Schneider, J., Wattenhofer, R.: Trading bit, message, and time complexity of distributed algorithms. In: Proceedings of the 25th International Symposium on Distributed Computing (DISC), pp. 51–65 (2011)
40. Spirakis, P., Tampakas, B.: Efficient distributed algorithms by using the Archimedean time assumption. Informatique Théorique et Applications $23(1)$, 113–128 (1989)
41. van Leeuwen, J., Santoro, N., Urrutia, J., Zaks, S.: Guessing games and distributed computations in anonymous networks. In: Proceedings of the 14th International Colloquium on Automata, Languages and Programming (ICALP), pp. 347–356 (1987)
42. Wattenhofer, M., Wattenhofer, R.: Distributed weighted matching. In: Proceedings of the 18th International Symposium on Distributed Computing (DISC), pp. 335–348 (2004)
43. Xu, L., Jeavons, P.: Simple algorithms for distributed leader election in anonymous synchronous rings and complete networks inspired by neural development fruit flies. Int. J. Neural Syst. $25(7)$ (2015)

Descriptional Complexity of Error Detection

Timothy Ng, David Rappaport and Kai Salomaa

Abstract The neighbourhood of a language L consists of all strings that are within a given distance from a string of L. For example, additive distances or the prefix-distance are regularity preserving in the sense that the neighbourhood of a regular language is always regular. For error detection and error correction applications an important question is to determine the size of the minimal deterministic finite automaton (DFA) needed to recognize the neighbourhood of a language recognized by an n state DFA. This paper surveys recent work on the state complexity of neighbourhoods of regularity preserving distances.

1 Introduction

Distance is a fundamental concept in mathematics which gives a numerical value to express the "closeness" of two objects. How we define "closeness" depends on what the objects we want to compare are and why we want to compare them. Here the objects we are interested in are strings, or sequences of symbols. Strings are particularly important in computer science, where many different kinds of objects are often represented as sequences of symbols.

A distance between strings can be extended into a distance between sets of strings, or languages. There are various ways to extend distances from strings to languages which are motivated by a number of applications, such as specification repair [1], computational biology [24], and error detection in communication channels [16, 19]. Information on the theory and applications of error correcting codes can be found in [7, 25, 36].

T. Ng · D. Rappaport · K. Salomaa (✉)
School of Computing, Queen's University, Kingston, ON K7L 2N8, Canada
e-mail: ksalomaa@cs.queensu.ca

T. Ng
e-mail: ng@cs.queensu.ca

D. Rappaport
e-mail: daver@cs.queensu.ca

© Springer International Publishing Switzerland 2017
A. Adamatzky (ed.), *Emergent Computation*, Emergence, Complexity and Computation 24, DOI 10.1007/978-3-319-46376-6_6

101

The *Encyclopedia of Distances* by Deza and Deza [8] contains an extensive list of distances that are used across a large number of different fields, including geometry, biology, coding theory, image processing, and physics, among others. For each of these definitions, we can ask questions about the behaviour of these distances and their properties. The computational question typically considered is how hard it is to compute the distance between given languages [12, 14, 17, 29].

Suppose we are given a distance d between strings. A mathematically elegant extension of d to languages L_1 and L_2 is the *Hausdorff distance* [6, 8], which gives a good overall measure of the similarity of L_1 and L_2. On the other hand, for error detection and error correction applications when the distance function is used to measure the number of errors in strings, the natural way to define the distance between languages L_1 and L_2 is to take simply the distance between two closest strings in L_1 and L_2, respectively. If we assume that errors have unit weight, then L_1 and L_2 having distance r means that we can distinguish strings of L_1 and L_2, respectively, on a channel that introduces at most $r - 1$ errors [13, 18]. A related notion is the *inner distance* (or *self-distance*) of a language: if the distance of any two distinct strings of a language L is at least r, the language L corrects $r - 1$ errors [14, 16, 19].

The *neighbourhood* of radius r of a language L consists of all strings that are within distance at most r from a string of L. We say that a distance d is *regularity preserving* if the neighbourhood of a regular language with respect to d is always regular. This gives rise to the question how large is the deterministic finite automaton (DFA) needed to recognize the neighbourhood of a regular language. Roughly speaking, determining the optimal size of the DFA for the neighbourhood gives the *state complexity of error detection*. Note that since complementation does not change the size of a DFA, the size of the minimal DFA for the neighbourhood of L of radius r is equal to the state complexity of the set of strings that have distance at least $r + 1$ from any string in L.[1] Over the last 20 years there has been much work on the state complexity of various regularity preserving operations and the reader can find more references in the survey [11].

It is known that the neighbourhood of a regular language with respect to an *additive distance* or *additive quasi-distance* [4] or with respect to the *prefix distance* and its variants [6] is always regular. The state complexity of neighbourhoods with respect to the Hamming distance was first considered by Povarov [30]. A tight lower bound for general additive distances was given by the current authors, however, a limitation is that the alphabet size depends on the size of the original DFA.

This paper surveys algorithmic properties and descriptional complexity of commonly used distance measures between sets of strings and, in particular, recent work on the state complexity of regularity preserving distances and related questions, such as approximate pattern matching. The contents of the paper is as follows. In the next section we recall some basic definitions on finite automata and Sect. 3 discusses distances between languages and regularity preserving distances. Descriptional com-

---

[1]Strictly speaking, for incomplete DFAs the state complexity of L and the complement of L, respectively, may differ by one.

plexity of neighbourhoods of regular languages is discussed in Sect. 4 and the last section highlights some open problems and further research topics on the descriptional complexity of error channels.

2 Definitions

We assume that the reader is familiar with the basics of finite automata and regular languages and below we just fix some notations. All unexplained notions can be found e.g. in the texts by Shallit [35] or Yu [37].

In the following Σ stands always for a finite alphabet and Σ^* is the set of strings over Σ. The length of a string $w \in \Sigma^*$ is $|w|$ and ε is the empty string. For $1 \leq i \leq |w|$, w_i stands for the ith symbol of w. The reversal of a string w is $w^R = w_{|w|}w_{|w|-1} \ldots w_1$. If $w = xyz$ we say that x is a prefix, z is a suffix and y is a substring of w. Here any of the strings x, y, z may be ε.

A nondeterministic finite automaton (NFA) is a tuple $A = (\Sigma, Q, \delta, Q_0, F)$ where Σ is the input alphabet, Q is the finite set of states, $\delta \colon Q \times \Sigma \to 2^Q$ is the multivalued transition function, $Q_0 \subseteq Q$ is the set of initial states and $F \subseteq Q$ is the set of final states. In the usual way δ is extended as a function $Q \times \Sigma^* \to 2^Q$ and the language accepted by A is $L(A) = \{w \in \Sigma^* \mid \delta(Q_0, w) \cap F \neq \emptyset\}$. The automaton A is a deterministic finite automaton (DFA) if $|Q_0| = 1$ and δ is a single valued partial function. If δ is a total function, the DFA A is complete. Note that our definition allows DFAs to be incomplete, i.e., some transitions may be undefined. In cases where the transitions function is required to be always defined we use the term "complete DFA". It is well known that the deterministic and nondeterministic finite automata recognize the class of *regular languages*.

The (right) Kleene congruence of a language $L \subseteq \Sigma^*$ is the relation $\equiv_L \subseteq \Sigma^* \times \Sigma^*$ defined by setting $x \equiv_L y$ iff $\lfloor (\forall z \in \Sigma^*)\ xz \in L \Leftrightarrow yz \in L \rfloor$. The language L is regular if and only if the index of $\equiv_L$ is finite and, in this case, the index of $\equiv_L$ is equal to the size of the minimal complete DFA for L [35]. For a given regular language L, the number of states of the minimal incomplete and minimal complete DFA recognizing L differ by at most one.

By the *state complexity* of a regular language L, $\mathrm{sc}(L)$, we mean the number of states of the minimal incomplete DFA recognizing L. The *nondeterministic state complexity* of L, $\mathrm{nsc}(L)$, is the number of states of a minimal NFA recognizing L. Note that a state minimal NFA for a regular language need not be unique.

3 Distance Measures on Strings and Sets of Strings

A *distance* on strings over Σ is a function $d : \Sigma^* \times \Sigma^* \to \mathbb{Q}$ which for all strings $x, y, z \in \Sigma^*$ satisfies the following

1. $d(x, y) = 0$ if and only if $x = y$,
2. $d(x, y) = d(y, x)$ (symmetry),
3. $d(x, z) \leq d(x, y) + d(y, z)$ (triangle-inequality).

From the above conditions it follows easily that $d(x, y)$ must be non-negative for all $x, y \in \Sigma^*$.

A *quasi-distance* is a function for which the first condition is weakened from "iff" to "if", that is, a quasi-distance of two distinct strings may be zero. If d is a quasi-distance on Σ, we can define an equivalence relation $\sim_d$ on Σ by setting $x \sim_d y$ if and only if $d(x, y) = 0$. Then the mapping $d'([x]_{\sim_d}, [y]_{\sim_d}) = d(x, y)$ is a distance over $\Sigma^* / \sim_d$ [4].

A quasi-distance d is *integral* if for all strings x and y, $d(x, y) \in \mathbb{N}$. Note that a distance is a special case of a quasi-distance and all properties that hold for quasi-distances apply also to distances.

We now recall the definition of some commonly used distance measures on strings. The *Hamming distance* of two equal length strings x and y counts the number of positions in which x differs from y. Formally, the Hamming distance between strings $x, y \in \Sigma^*$ is defined as

$$d_H(x, y) = \begin{cases} |\{1 \leq i \leq |w| \mid x_i \neq y_i \}| & \text{if } |x| = |y|, \\ \text{undefined} & \text{otherwise.} \end{cases}$$

The distance is defined only when x is the same length as y.

For equal length strings Hamming distance counts the number of substitution operations needed to transform x into y. A natural extension for all pairs of strings is the *Levenshtein distance* [22], also called the *edit distance*, which counts the number of atomic substitution, insertion and deletion operations required to transform x into y. Formally the Levenshtein distance can be defined in terms of error systems considered by Kari and Konstantinidis [13] as a formalization of error in terms of formal languages, see also [15]. An error system is a formal language over the alphabet of edit operations. For an alphabet Σ, let $\mathcal{E}_\Sigma$ be the alphabet of edit operations over Σ defined by

$$\mathcal{E}_\Sigma = \{ (a/b) \mid a, b \in \Sigma \cup \{\varepsilon\}, ab \neq \epsilon \}.$$

An *error* is an edit operation (a/b) where $a \neq b$. An *edit string* is a string over $\mathcal{E}_\Sigma$. The weight $|e|_{\neq}$ of an edit string e is the number of errors in e. For an edit string $e = (a_1/b_1)(a_2/b_2) \ldots (a_n/b_n)$, we call $x = a_1 a_2 \ldots a_n$ the input part and $y = b_1 b_2 \ldots b_n$ the output part of e ($a_i, b_i \in \Sigma \cup \{\varepsilon\}$, $i = 1, \ldots, n$).

Now the edit distance between strings x and y, $d_e(x, y)$, is defined as the minimum weight edit string e having x (respectively, y) as the input (respectively, the output) part. The above definition has assigned weight one to all errors. It is possible to consider also definitions where the weights need not be equal [12, 13]. General edit distances are examples of additive quasi-distances considered in Sect. 3.2.

Example 1 An edit string to transform the string *hiphop* into *lollipop* is

$$e_1 = \frac{\varepsilon\ h\ i\ p\ h\ \varepsilon\ o\ p}{l\ o\ l\ l\ i\ p\ o\ p}.$$

The length of this edit string is 8 and its weight is 6.

The edit string e_1 is not a minimum weight edit string between the given strings. The edit distance of the string *hiphop* and *lollipop* is 5, via the edit string

$$\frac{h\ \varepsilon\ \epsilon\ \varepsilon\ i\ p\ h\ o\ p}{l\ o\ l\ l\ i\ p\ \varepsilon\ o\ p}.$$

Instead of counting the number of edit operations, the similarity of strings can be defined by way of their longest common prefix, suffix, or substring, respectively [6]. A parameterized prefix distance between regular languages has been considered by Kutrib et al. [20] for estimating the fault tolerance of information transmission applications. For example, the *prefix distance* of strings x and y is the sum of the length of the suffix of x and the suffix of y that occurs after their longest common prefix. Formally it is defined by

$$d_p(x, y) = |x| + |y| - 2 \cdot \max_{z \in \Sigma^*}\{|z| \mid x, y \in z \cdot \Sigma^*\}.$$

The definitions of the *suffix distance* and *substring distance* are analogous.[2]

Example 2 The strings *yorkdale* and *yorkville* have a prefix distance of 9 via their longest common prefix *york*. The strings *woodbine* and *guildwood* have a substring distance of 9 through their longest common substring *wood*. The strings *parkdale* and *riverdale* have a suffix distance of 9 through the longest common suffix *dale*.

3.1 Distance Between languages

If d is a distance on strings over Σ, the natural way to define the distance between a string $w \in \Sigma^*$ and a language $L \subseteq \Sigma^*$ is

$$d(w, L) = \inf\{d(w, w') \mid w' \in L\}.$$

We want to further extend the definition to measure the distance between two languages, or sets of strings. The *relative distance* [6] of the language L_1 to the language L_2 is defined as

$$d(L_1|L_2) = \sup\{d(w_1, L_2) \mid w_1 \in L_1\}.$$

---

[2]The latter is called in [6] subword distance but this term has been used also for a distance defined in terms of the longest noncontinuous subword [23].

The *bounded repair problem* [1] consists of deciding whether or not the relative edit distance of the *restriction language* L_1 to the *target language* L_2 is finite. The bounded repair problem (in the general non-streaming case) is PSPACE-complete when the restriction and target language are specified by NFAs and it is coNP-complete when the restriction language is specified by an NFA and the target language by a DFA [1]. The coNP-hardness result holds also when the restriction language is specified by a DFA. A variant of the bounded repair problem asks, roughly speaking, what is the number of edits per symbol of a string w in the restriction language that is needed to transform w to a string of the target language [2]. Chatterjee et al. [5] have studied systematically the complexity of the closely related *threshold edit distance problem* for pushdown automata and finite automata.

Note that the relative distance is not, in general, symmetric. In order to satisfy symmetry, Choffrut and Pighizzini [6] use the *Hausdorff distance* to define the distance between L_1 and L_2 by taking the maximum value

$$d_{\text{Hdorff}}(L_1, L_2) = \max\{\ d(L_1|L_2),\ d(L_2|L_1)\ \}.$$

The relative edit distance between regular languages was shown to be computable by reducing it to the limitedness problem for distance automata [6], and Leung and Podolskiy [21] have given an exponential time algorithm as well as a PSPACE-hardness lower bound for the limitedness problem. As mentioned above, a PSPACE algorithm for the relative distance between regular languages is known from the more recent work on the bounded repair problem [1].

The Hausdorff distance having a small value means, intuitively, that every string of L_1 is close to some strings of L_2 and vice versa, that is, the binary relation $L_1 \times L_2$ is "almost reflexive" with respect to the distance under consideration [6].

In the above sense the Hausdorff distance gives a good measure of similarity between languages. However, in error detection applications we want to ensure that every string of L_1 is at some minimum distance from every string of L_2 and vice versa. Thus, in the following we extend a distance from strings to languages simply by taking the smallest distance of two strings in the respective languages:

$$d(L_1, L_2) = \inf\{\ d(w_1, w_2)\ |\ w_1 \in L_1, w_2 \in L_2\ \}.$$

Unless otherwise mentioned, in the following when speaking about the edit distance (or some other distance measure) for languages, we mean the above definition. The *inner distance* of a language L [17] is

$$d(L) = \inf\{\ d(w, z)\ |\ w, z \in L,\ w \neq z\ \}.$$

The maximal error-detecting capability of a language L, in the sense defined by Konstantinidis and Silva [19], is one less than the inner distance of L.

The edit distance between two regular languages can be computed in polynomial time and the corresponding question for context-free languages is unsolvable [24].

Also the edit distance between a regular language and a context-free language can be computed in polynomial time [12].

3.2 Regularity Preserving Distances

We can consider the topological notion of neighbourhoods, or balls, of radius r with respect to a given distance d on strings. Informally, a neighbourhood of a language L is the set of strings which are at most a distance r away from some string in L according to the distance measure under consideration.

Formally, the neighbourhood of radius $r \geq 0$ of a language L under quasi-distance d is defined as

$$E(L, d, r) = \{ x \in \Sigma^* \mid (\exists y \in L) \, d(x, y) \leq r \}.$$

Suppose that the distance d measures the number of errors introduced in an information transmission channel [13, 15, 19]. Then $E(L, d, r)$ consists of all strings that a channel that introduces at most r errors can output when the input is a string of L. In other words, the complement of $E(L, d, r)$ is the unique maximal language L' such that $d(L, L') > r$.

Using the notion of neighbourhood we define the following notions on quasi-distances:

Definition 1 Let d be a quasi-distance on Σ^*. We say that d is *finite* if, for all $w \in \Sigma^*$ and $r \geq 0$ the neighbourhood $E(\{w\}, d, r)$ is finite.

We say that d is *regularity preserving* if for all regular languages L and $r \geq 0$, the neighbourhood $E(L, d, r)$ is regular.

For example, the edit, prefix, suffix and substring-distances are clearly all finite. On the other hand, it is known that finiteness of a distance d does not guarantee that d is regularity preserving [4]. Calude et al. [4] introduced a notion of additivity that is sufficient to guarantee that a quasi-distance preserves regularity. We say that a quasi-distance is *additive* if it respects composition of strings in the following sense.

Definition 2 A quasi-distance d on Σ^* is additive if for all $w_1, w_2 \in \Sigma^*$ and $r \geq 0$,

$$E(\{w_1 w_2\}, d, r) = \bigcup_{r_1 + r_2 = r} E(\{w_1\}, d, r_1) \cdot E(\{w_2\}, d, r_2), \tag{1}$$

Informally the additivity of d means that any neighbourhood of radius r of a concatenation of two strings $w_1 w_2$ consists of exactly all the language concatenations of neighbourhoods of w_1 and w_2 whose radii sum up to r. An additive distance is always finite but additive quasi-distances need not be finite [4].

It is easy to verify that the edit distance d_e is additive. For the edit distance the inclusion from left to right in (1) holds directly by definition and verifying the converse inclusion needs a short proof [4]. Also, the Hamming distance can be viewed as an additive distance if we define that the cost of deletions and insertions is infinite, that is, the distance between symbols of Σ and ε is defined to be infinite.

The additivity property is sufficient to guarantee that any neighbourhood of a regular language is regular.

Theorem 1 *([4]) An additive quasi-distance preserves regularity.*

For a given DFA A and radius $r \geq 0$, the original proof of Theorem 1 in [4] first verifies that the neighbourhoods of individual alphabet symbols are regular and then using this property and additivity of the quasi-distance d constructs an NFA for the neighbourhood $E(L(A), d, r)$. The construction is far from optimal from the state complexity point of view (since it results only in an NFA that, in general, needs to be determinized) and in the next section we will discuss constructions with better state complexity. Schulz and Mihov [34] have given a time efficient DFA construction for the neighbourhood of a single string w: if the radius is viewed as a constant, the DFA can be constructed in time linear in the length of w.

On the other hand, additivity is not necessary for a distance to preserve recognizability. It is clear that the prefix-, suffix- or substring distances are not additive. However, given a DFA A it is easy to construct an NFA B that recognizes a neighbourhood of $L(A)$ with respect to the prefix-distance by, roughly speaking, guessing the longest common prefix w_p of the input and a string in $L(A)$, and then counting the length of the remaining suffix of the input. Naturally the number of states of B has to depend on the radius, and depending on the state the simulated computation of A ends in, the NFA B can "know" the length of the shortest suffix that completes w_p to a string of $L(A)$. An analogous construction works for the suffix- and the substring distance.

Theorem 2 *([6, 28]) The prefix-, suffix- and substring distances preserve regularity.*

4   State Complexity of Neighbourhoods

A (combinatorial) channel is, in general, a binary relation on strings describing all input-output situations permitted by the channel. For more information on error channels and error detection we refer the reader e.g. to [9, 13, 16, 19]. If all substitution, insertion and deletion errors have a unit cost, the number of errors introduced by the channel is upper bounded by the edit distance of an input–output pair, and when using errors with general weights (including possibly zero weight) we can bound the number of errors by an additive quasi-distance.

Now assume that our channel C introduces at most r errors and consider a regular language L that is recognized by a DFA with n states. The the complement of the neighbourhood $E(L, d_e, 2r + 1)$ is the unique maximal language L' such that we can distinguish outputs produced by C on inputs from L and L', respectively. Complementation changes the size of a minimal incomplete DFA by at most one. This means that determining the state complexity of the neighbourhood $E(L, d_e, 2r + 1)$ as a function of n can be viewed as the state complexity of error detection on a channel with at most r errors.

Povarov [30] was the first to systematically investigate the state complexity of Hamming-neighbourhoods. Note that a Hamming-neighbourhood of radius r can be viewed as a radius r neighbourhood where the underlying edit distance assigns value $r + 1$ to all insertion and deletion operations, and in this way the Hamming distance can be interpreted as a special case of an additive distance.

We begin by recalling the NFA construction for Hamming neighbourhoods due to Povarov [30] since it is used for the first upper bound for deterministic state complexity, as well as in later constructions from [27]. An alternative proof for the upper bound of Theorem 3 based on finite transducers can be found in [31]. Recall that d_H denotes the Hamming distance.

Theorem 3 *([30, 31]) If $L \subseteq \Sigma^*$ has an NFA with n states and $r \in \mathbb{N}$, then*

$$\mathrm{nsc}(E(L, d_H, r)) \leq n \cdot (r + 1).$$

For every $r \in \mathbb{N}$ and $n > r$ there exists an n-state NFA A over a two letter alphabet such that

$$\mathrm{nsc}(E(L(A), d_H, r) = n \cdot (r + 1).$$

Proof sketch for the upper bound. Suppose L is recognized by an NFA $A = (\Sigma, Q, \delta, q_0, F_A)$. The neighbourhood $E(L(A), d_H, r)$ is recognized by an NFA

$$B = (\Sigma, Q \times \{0, 1, \ldots, r\}, \gamma, (q_0, 0), F_A \times \{0, 1, \ldots, r\}),$$

where the transitions of γ are defined by setting for $q \in Q, 0 \leq i \leq r$ and $b \in \Sigma$:

$$\gamma((q, i), b) = \begin{cases} \{(p, i) \mid p \in \delta(q, b)\} \cup \{(p, i + 1) \mid (\exists c \in \Sigma)\ p \in \delta(q, c)\} & \text{if } i < r, \\ \{(p, i) \mid p \in \delta(q, b)\} & \text{if } i = r. \end{cases}$$

The first component of the state of B simulates a computation of A on some input (possibly containing errors), and the second component keeps track of the cumulative error. Note that the definition of γ allows the possibility of increasing the value of the second component also on a transition with the correct input symbol (the case when $c = b$). These transitions, although redundant, clearly do not change the language of B. $\qquad\square$

Theorem 3 is stated in [30] using the Hamming distance and the same upper bound straightforwardly translates for any additive integral quasi-distance. In the NFA construction the set of states remains $Q \times \{0, 1, \ldots, r\}$ and, when the input symbol is b, an error transition on $c \in \Sigma \cup \{\varepsilon\}$ takes state $(q, i), (q \in Q, 0 \leq i \leq r)$ to all possible states $(p, i + d(b, c))$, where $i + d(b, c) \leq r$ and p is reached from state q on input c in the original NFA. Additionally the construction adds ε-transitions that simulate an insertion operation.

Corollary 1 *If L is recognized by an NFA with n states, d is an additive integral quasi-distance and $r \in \mathbb{N}_0$, then*

$$\text{nsc}(E(L, d, r)) \leq n \cdot (r + 1).$$

The derivation of an upper bound for the deterministic state complexity of Hamming neighbourhoods uses the NFA construction of the proof of Theorem 3 and the upper bound for the number of reachable states for the corresponding DFA makes use of the redundant transitions $(p, i + 1) \in \gamma((q, i), b)$ where $p \in \delta(q, b)$ in the definition of the NFA transition relation. Since [30] uses complete DFAs, below we translate the upper bound construction also for incomplete DFAs. The lower bound of Theorem 4 for neighbourhoods of radius one uses a construction where the minimal DFA for the Hamming neighbourhood does not have a dead state which means that the same lower bound holds when state complexity is based on incomplete DFAs.

Theorem 4 *([30])*

(i) *If A is a complete DFA with n states and $r \in \mathbb{N}_0$, then $E(L(A), d_H, r))$ has a complete DFA with at most $\frac{1}{2} \cdot n \cdot 2^{nr} + 1$ states.*

(ii) *If A is an incomplete DFA with n states and $r \in \mathbb{N}_0$, then $E(L(A), d_H, r))$ has an incomplete DFA with at most $\frac{1}{2} \cdot (n + 2) \cdot 2^{nr}$ states.*

(iii) *For all $n \geq 4$, there exists a complete DFA A with n states defined over a binary alphabet such that*

$$\text{sc}(E(L(A), d_H, 1) = \frac{3}{8}n \cdot 2^n - 2^{n-4} + n.$$

Proof The proofs of (i) and (iii) can be found in [30]. Here we just translate the former proof for incomplete DFAs to give the estimation (ii).

Suppose L is recognized by an incomplete DFA $A = (\Sigma, Q, \delta, q_0, F_A)$ where $|Q| = n$. Let $B = (\Sigma, P, \gamma, p_0, F_B)$ be the NFA constructed for the neighbourhood $E(L(A), d_H, r)$ as in the proof of Theorem 3. In particular, the set of states P is $Q \times \{0, 1, \ldots, r\}$.

Let B' be the DFA obtained from B using the standard subset construction. Since A is deterministic, from the construction of B it follows that any subset of P that is reachable as a state of B' can have at most one element of $Q \times \{0\}$ and, furthermore, the reachable subsets $X \subseteq P$ have the following property. If $X \neq \{(q_0, 0)\}$ and X contains an element $(q, 0)$, $q \in Q$, then from the definition of the transitions of the NFA B it follows that also $(q, 1) \in X$.

This means that the number of non-empty subsets of P that are reachable as states of B' is upper bounded by

$$1 + (n \cdot 2^{n-1} + 2^n) \cdot 2^{n(r-1)} - 1 = \frac{1}{2} \cdot (n + 2) \cdot 2^{nr}.$$

$\square$

The lower bound for radius one neighbourhoods is roughly within a factor of $\frac{3}{4}$ of the upper bound of Theorem 4. Significantly, the lower bound is over a binary alphabet—up to date this is the only good lower bound for the deterministic state complexity of additive neighbourhoods where the alphabet size does not depend on the size of the DFA and, furthermore, the underlying distance is just the Hamming distance.

Shamkin [33] has constructed finite languages L_n, $n \geq 4$, over a ternary alphabet such that L_n has an incomplete DFA of size n and for all $r \leq \frac{n}{2} - 1$ the state complexity of the radius r Hamming neighbourhood of L_n is at least $2^{\lfloor \frac{n}{2} - r \rfloor}$. The lower bound for Hamming neighbourhoods of radius $r \geq 1$ is proportional to 2^{-r}, that is, with a fixed number of states the lower bound decreases with increasing radius. This seems to be quite far from the upper bound.

The upper bounds of Theorem 4 could be improved by adding further redundant transitions to the original NFA construction (from Theorem 3) and then using a more detailed analysis of the number reachable states of the corresponding DFA. However, in the next subsection we get a better upper bound for the deterministic state complexity of neighbourhoods using a different approach. Instead of constructing an NFA and then determinizing it, we construct a DFA for the neighbourhood directly based on the finite automaton recognizing the original language.

4.1 Neighbourhoods of General Additive Distances

Here we consider the state complexity of neighbourhoods with respect to a general additive distance, and in the next subsection the prefix-distance and other related distance functions. If d is a regularity preserving distance, in the informal discussion we use the term *state complexity of d* to mean the state complexity of neighbourhoods with respect to d (given as a function of the state complexity of the original regular language).

In the rest of this section, without separate mention we assume that all distances and quasi-distances are integral, i.e., the range of values consists of the non-negative integers. When considering the state complexity of neighbourhoods, this is not more restrictive than using rational values. Note that an additive distance d is completely determined by the distances between elements of $\Sigma \cup \{\varepsilon\}$. Thus, if d has rational values we can find a constant k such that there is an integral distance d' that satisfies, for all strings $x, y \in \Sigma^*$, $d'(x, y) = k \cdot d(x, y)$. Consequently for any language L and radius $r \geq 0$, $E(L, d, r) = E(L, d', k \cdot r)$.

Theorem 5 *([26, 32]) Let d be an additive quasi-distance. If L has an NFA with n states and $r \geq 0$,*

$$\mathrm{sc}(E(L(A), d, r)) \leq (r + 2)^n - 1.$$

The statement of Theorem 5 in [26, 32] does not have the term "-1" because there DFAs are required to be complete. The proof (for distances and quasi-distances in [26, 32], respectively) uses a construction based on additive weighted finite automata. Below we outline a direct DFA construction for the neighbourhood of an additive distance.

Proof sketch for Theorem 5. For simplicity we assume that d is a distance. This implies that for any $w \in \Sigma^*$ and $r' \geq 0$, $E(\{w\}, d, r')$ is finite [4].

Let $A = (\Sigma, Q, \delta, q_0, F_A)$ and denote $Q = \{q_0, q_1, \ldots, q_{n-1}\}$. We construct for the neighbourhood $E(L(A), d, r)$ a DFA $B = (\Sigma, P, \gamma, p_0, F_B)$ where the set of states is

$$P = \{0, 1, \ldots, r + 1\}^n - \{(r + 1, \ldots, r + 1)\},$$

$$F_B = \{(x_0, x_1, \ldots, x_{n-1}) \in P \mid (\exists 0 \leq j \leq n - 1) \; x_j \leq r \text{ and } q_j \in F_A\},$$

and $p_0 = (0, i_1, i_2, \ldots, i_{n-1})$, where $i_j, 1 \leq j \leq n - 1$, is the minimum of the set

$$S_{\text{dist}-\varepsilon} = (\{d(\varepsilon, w) \mid q_j \in \delta(q_0, w)\} \cup \{r + 1\}).$$

Finally, the transitions of γ are defined as follows. For $b \in \Sigma$ and $(x_0, x_1, \ldots, x_{n-1}) \in P$, we define

$$\gamma((x_0, x_1, \ldots, x_{n-1}), b) = (z_0, z_1, \ldots, z_{n-1}),$$

where $z_j, 0 \leq j \leq n - 1$, is the minimum of the set

$$S_{\text{dist}-j} = \{x_i + d(b, w) \mid q_j \in \delta(q_i, w)\} \cup \{r + 1\}.$$

Since d is a distance, the neighbourhood $E(\{b\}, d, r + 1)$ is finite and the set $S_{\text{dist}-j}$ can be effectively constructed.

Intuitively, the DFA B operates as follows. Assuming that B has processed an input $u \in \Sigma^*$ and the current state is $(x_0, x_1, \ldots, x_{n-1})$, the component $x_j, 0 \leq j \leq n - 1$, is the smallest distance between u and a string $w \in \Sigma^*$ that in the original NFA A takes the state q_0 to q_j. If $x_j = r + 1$, then there is no string w with $d(u, w) \leq r$ that takes q_0 to q_j. The state $(r + 1, \ldots, r + 1)$ would correspond to the dead state of the computation and is omitted from the set of states.

The initial state p_0 is chosen to satisfy the above property. Using the additivity of d it can be verified inductively that if the state $(x_0, x_1, \ldots, x_{n-1})$ satisfies the above described property, so does $\gamma((x_0, x_1, \ldots, x_{n-1}), b), b \in \Sigma$. (The formal argument for correctness is analogous to the one used in [26, 32] for constructing a weighted finite automaton to recognize the neighbourhood.)

The choice of the set of final states guarantees that B accepts a string $u \in \Sigma^*$ iff $u \in E(L(A), d, r)$. $\square$

When r and n are at least two, the upper bound of Theorem 5 is better than the upper bound of Theorem 4. The natural question is then whether the upper bound can be reached. Below we give a positive answer to this question. However, a limitation is that the size of the alphabet depends on the size of the DFA and the used (quasi-)distance needs to be defined based on the chosen radius.

Proposition 1 *([27])*

(i) *For all $n, r \geq 1$, there exists an additive distance d_r and an NFA A_n with n states over an alphabet of size $2n - 1$ such that $\text{sc}(E(L(A_n), d, r) = (r + 2)^n - 1$.*

(ii) *For all $n, r \geq 1$, there exists an additive quasi-distance d'_r and a DFA A'_n with n states over an alphabet of size $3n - 2$ such that $\text{sc}(E(L(A'_n), d', r) = (r + 2)^n - 1$.*

Proof The constructions for (i) and (ii) are variants of each other and (ii) is presented in detail in [27]. Here we outline the construction for (i).

Choose $\Sigma_n = \{a_1, \ldots, a_{n-1}, b_1, \ldots, b_n\}$. For $r \in \mathbb{N}$, we define a distance $d_r :$ $\Sigma_n^* \times \Sigma_n^* \to \mathbb{N}_0$ by the conditions:

- $d_r(a_i, a_j) = r + 1$ for $i \neq j$,
- $d_r(b_i, b_j) = 1$ for $i \neq j$,
- $d_r(a_i, b_j) = r + 1$ for all $1 \leq i, j \leq n$,
- $d_r(\sigma, \epsilon) = r + 1$ for all $\sigma \in \Sigma$.

The above conditions specify a unique additive distance on Σ_n. Note that when we set the deletion cost to be $r + 1$, due to symmetry of d_r also the insertion cost will be $r + 1$.

We define the following family of n-state NFAs $A_n = (Q_n, \Sigma_n, \delta, 1, \{n\})$ where $Q_n = \{1, \ldots, n\}$ and the transition function δ is defined by setting

- $\delta(i, a_i) = \{i, i + 1\}$ for $1 \leq i \leq n - 1$,
- $\delta(i, a_j) = i$ for $1 \leq i \leq n - 2$ and $i + 1 \leq j \leq n - 1$,
- $\delta(i, b_j) = i$ for $1 \leq i \leq n - 1$ and $j = i - 1$ or $i + 1 \leq j \leq n$.

All transitions not listed above are undefined. The NFA A_n is depicted in Fig. 1.

As in Corollary 1 we construct an NFA $B_{n,r} = (Q'_n, \Sigma_n, \delta', q'_0, F')$, for the neighbourhood $E(L(A_n), d_r, r)$, where $Q'_n = Q_n \times \{0, 1, \ldots, r\}$, $q'_0 = (q_0, 0)$, $F' = \{n\} \times \{0, 1, \ldots, r\}$ and the transition function δ' is defined by

- $\delta'((q, j), a_q) = \{(q, j), (q + 1, j)\}$ for $1 \leq q \leq n - 1$,
- $\delta'((q, j), a_{q'}) = \{(q, j)\}$ for all $1 \leq q \leq n - 1$ and $q \leq q' \leq n - 1$,
- $\delta'((q, j), b_i) = \{(q, j + 1)\}$ for $1 \leq q \leq n$ and $i = 1, \ldots, q - 2, q$,
- $\delta'((q, j), b_i) = \{(q, j)\}$ for $1 \leq q \leq n$ and $i = q - 1, q + 1, \ldots, n$.

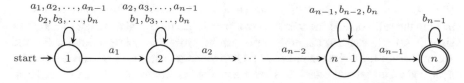

Fig. 1 The NFA A_n

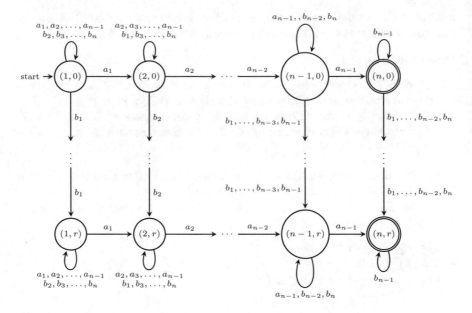

Fig. 2 The NFA $B_{n,r}$ for the neighbourhood $E(L(A_n), d_r, r)$

All transitions not listed above are undefined. The NFA $B_{n,r}$ is depicted in Fig. 2.

Note that the distance d_r associates cost one to substituting b_i with $b_j, i \neq j$, and cost $r + 1$ to all other substitutions, insertions and deletions. The "error transitions" are depicted as non-horizontal transitions in Fig. 2 and, due to the above observation, the only error transitions take a state (i, k) to $(i, k + 1)$ on symbol b_z, $1 \leq i \leq n$, $0 \leq k \leq r - 1$, where in the NFA A_n, $\delta(i, b_z)$ is undefined.

For $0 \leq k_i \leq r + 1$, $1 \leq i \leq n$, we define a string

$$w(k_1, \ldots, k_n) = a_1 b_1^{k_1} a_2 b_2^{k_2} \ldots a_{n-1} b_{n-1}^{k_{n-1}} b_n^{k_n}.$$

Claim 1. If $k_i \leq r$, then there exists a computation C_i of the NFA $B_{n,r}$ which reaches the state (i, k_i) at the end of the input $w(k_1, \ldots, k_n)$, $1 \leq i \leq n$. There is no computation of $B_{n,r}$ on $w(k_1, \ldots, k_n)$ that reaches a state (i, k_i') with $k_i' < k_i$. Furthermore, if $k_i = r + 1$, no computation of $B_{n,r}$ reaches at the end of $w(k_1, \ldots, k_n)$ a state where the first component is i.

The first part of the claim is easy to verify by direct inspection: C_i reaches state $(i, 0)$ by reading the prefix $a_1 b_1^{k_1} \ldots a_{i-1} b_{i-1}^{k_{i-1}}$. In state $(i, 0)$ the computation C_i reads a_i with the selfloop and the vertical error transitions on $b_i^{k_i}$ take the computation to state (i, k_i) where the remaining suffix can be processed using selfloops. For the second part of the claim we note that all horizontal transitions in $B_{n,r}$ are labeled by a_j's, and all horizontal transitions increment the first component of the state. Thus, the only way to reach a state (i, k_i'), would be to read each of the symbols $a_1, \ldots,$

a_{i-1} using a transition that increments the first component and then read a_i with a selfloop. Now the following k_i symbols b_i must be processed using error transitions which means that k_i' cannot be smaller than k_i.

Using Claim 1 we can now verify any two distinct strings $w(k_1, \ldots, k_n)$ and $w(k_1', \ldots, k_n')$, $0 \le k_i, k_i' \le r + 1$, are inequivalent with respect to the Kleene congruence of $E(L(A_n), d_n, r)$. Choose $1 \le j \le n$ such that $k_j < k_j'$ and define $z = b_j^{r-k_j} a_{j+1} \ldots a_{n-1}$.

By Claim 1, $w(k_1, \ldots, k_n) \cdot z \in L(B_n, r)$ because after reading $w(k_1, \ldots, k_n)$ the NFA $B_{n,r}$ can reach the state (j, k_j) and continuing the computation on z can make $r - k_j$ further error transitions. Similarly using Claim 1 we see that no computation of $B_{n,r}$ on $w(k_1', \ldots, k_n') \cdot z$ cannot reach an accepting state where the first component is n, because to do so it would need to reach on the prefix $w(k_1', \ldots, k_n')$ a state (j, ℓ) where $\ell \le k_j$. Since $k_j < k_j'$ this is impossible by Claim 1.

Since $L(B_{n,r}) = E(L(A_n), d_n, r)$ it follows that the minimal complete DFA for $E(L(A_n), d_n, r)$ has at least $(r + 2)^n$ states. The congruence class of $w(r + 1, \ldots, r + 1)$ corresponds to the dead state of the DFA and can be omitted.

The proof for the part (ii) in [27] introduces additional alphabet symbols $c_1, \ldots, c_{n-1}$ and in each nondeterministic transition of A_n on a_i, in the DFA A_n' the selfloop is labeled instead by c_i, $1 \le i \le n - 1$. The quasi-distance d_r' assigns distance zero to a_i and c_i, $1 \le i \le n - 1$. With these definitions a lower bound argument for the size of a DFA for $E(L(A_n'), d_r', r)$ similar to the one used above for (i) goes through.   □

To conclude this section we mention that the construction of Proposition 1 can be modified to yield a tight lower bound for the state complexity of approximate pattern matching. The descriptional complexity of pattern matching with mismatches was first considered by El-Mabrouk [10]. Given a pattern P of length m and a text T, the problem is to determine whether or not T contains substrings of length m having characters differing from P in at most r positions, that is, substrings having Hamming distance at most r from P. For a pattern $P = a^m$ consisting of occurrences of only one character, the state complexity was shown to be $\binom{m+1}{r+1}$ [10].

Extending the problem for a general additive quasi-distance d and a set of patterns given as a regular language $L \subseteq \Sigma^*$, we want to determine the state complexity of the set $\Sigma^* \cdot E(L, d, r) \cdot \Sigma^*$, that is, the set of strings that contain a substring within distance at most r from a string of L. The following lower bound is based on a modification of the construction used in Proposition 1.

Proposition 2 ([27]) *For $n, r \in \mathbb{N}$, there exist an additive distance d and an NFA A with n states defined over an alphabet of size $2n - 1$ such that the minimal DFA for $\Sigma^* E(L(A), d, r) \Sigma^*$ must have at least $(r + 2)^{n-2} + 1$ states.*

The authors [27] give an upper bound matching the bound of Proposition 2 which means that the state complexity of approximate pattern matching with r errors is exactly $(r + 2)^{n-2} + 1$. Brzozowski et al. [3] have shown that, for an n-state DFA language L, the worst case state complexity of the two-sided ideal $\Sigma^* L \Sigma^*$ is $2^{n-2} + 1$

which corresponds to having error radius zero in approximate pattern matching. The lower bound for the error free case is obtained with a three letter alphabet [3] whereas Proposition 2 needs a variable size alphabet.

4.2   State Complexity of Prefix Distance

Additivity is not a necessary condition for a distance to be regularity preserving. For example, by Theorem 2 the prefix, suffix, and substring distances preserve regularity while these distances clearly are not additive.

The neighbourhood $E(L, d_p, r)$ (where d_p is the prefix distance and $r \geq 0$) consists of strings w that share a "long" prefix with a string $u \in L$, more precisely, it is required that the combined length of the parts of w and u outside their longest common prefix is at most the constant r. In view of this, it seems reasonable to expect that the state complexity prefix distance neighbourhoods does not incur a similar exponential size blow-up as, for example, the edit distance.

Theorem 6 ([28]) *For $n > r \geq 0$ and a DFA A with n states, the neighbourhood $E(L(A), d_p, r)$ can be recognized by a DFA with $n \cdot (r + 1) - \frac{r(r+1)}{2}$ states.*

For $n > r \geq 0$ there exists a regular language L over an alphabet of size $n + 1$ with $\mathrm{sc}(L) = n$ such that

$$\mathrm{sc}(E(L, d_p, r)) = n \cdot (r + 1) - \frac{r(r + 1)}{2}.$$

Proof sketch. We outline the general idea only for the upper bound [28]. Suppose $A = (\Sigma, Q, \delta, q_0, F_A)$. We can construct for the neighbourhood $E(L(A), d_p, r)$ a DFA B with state set

$$P = (Q - F_A) \times \{1, \dots, r + 1\} \cup F_A \cup \{p_1, \dots, p_r\}.$$

Intuitively, B operates as follows. The computation of B simulates the computation of A and, in states $(q, j) \in (Q - F_A) \times \{1, \dots, r + 1\}$ the second component j keeps track of the minimum of the following two values: (i) the number of steps A needs q to reach a final state, and, (ii) the minimum path length in A from q to a final state that first goes one or more steps back in the current computation and then any number of steps forward (on an arbitrary input). Elements of F_A have always counter value zero and, hence, are not associated with the second component representing a counter. The details of the definition of the transitions of B that correctly update the counter value in the second component of the states can be found in [28].

If the simulated computation of A encounters an undefined transition, B performs at most r further transitions using a sequence of "error-transitions" using the states $p_1, \dots, p_r$. The number of allowable error transitions depends on the value of the counter when the undefined transition of A was encountered.

All states of B except the states $(q, r + 1), q \in Q - F$, with counter value $r + 1$, are final. Note that the states of the form $(q, r + 1), q \in Q - F$, are needed because when simulating the transitions of A in the first component the counter value may also decrease if the "forward" distance in A to a final state becomes smaller.

The state set P of B has in total $(n - |F_A|) \cdot (r + 1) + r + |F_A|$ elements. The size of P is maximized by choosing $|F_A| = 1$ and, furthermore, it can be verified that at least $\frac{r(r+1)}{2}$ elements of P must, independently of the alphabet size, be unreachable as states of B. $\qquad\square$

The lower bound construction for Theorem 6 uses an alphabet of size $n + 1$ where n is the number of states of the original DFA. It is known that the general upper bound cannot be reached using an alphabet of size $n - 2$.

Proposition 3 *([28]) Let A be a DFA with n states. If the state complexity of $E(L(A), d_p, r)$ equals $n \cdot (r + 1) - \frac{r(r+1)}{2}$, then the alphabet of A needs at least $n - 1$ letters.*

The paper [28] gives tight bounds also for the nondeterministic state complexity of neighbourhoods defined by the prefix, suffix, and substring distances. The bounds for the nondeterministic state complexity of the prefix distance and suffix distance, respectively, coincide due to the observation that $d_s(x, y) = d_p(x^R, y^R)$ for all strings x, y, and the fact that the transitions of an NFA can be reversed without changing the size of the NFA which means that, for any regular language L, $\mathrm{nsc}(L) = \mathrm{nsc}(L^R)$.

On the other hand, the deterministic state complexity of L^R is usually significantly different from the state complexity of L [11, 37]. It seems likely that constructing a DFA for the neighbourhood of an n-state DFA with respect to the suffix distance causes a much larger worst-case size blow-up than the bound for prefix distance in Theorem 6. The precise deterministic state complexity of the suffix distance remains open.

5 Conclusion and Open Problems

The precise worst-case state complexity of the radius r neighbourhood of an n state DFA language with respect to an additive quasi-distance is $(r + 2)^n - 1$. However, the lower bound construction of Proposition 1 has the following limitations:

- The construction uses an alphabet that depends linearly on the number of states of the original DFA.
- The underlying distance is defined based on the radius of the neighbourhood.
- We don't have a tight bound for state complexity defined in terms of complete DFAs. A complete DFA with $(r + 2)^n$ states can recognize a radius r neighbourhood of an n state DFA. It is not known how to construct a complete n state DFA matching this upper bound.

The main open problem consists of proving lower bounds for additive distances (or quasi-distances) using languages over a binary, or constant size, alphabet. The known good lower bound construction based on binary alphabets, due to Povarov [30], deals only with the restricted case of radius one Hamming neighbourhoods (Theorem 4). The other important improvement to the lower bound result of Proposition 1 would be to find a construction where the same distance (or quasi-distance) definition works for neighbourhoods of arbitrary radius.

Descriptional complexity questions are relevant also for the more general error channels considered by Kari and Konstantinidis [13] and Konstantinidis and Silva [19]. As briefly discussed in Sect. 3, the edit distance of two strings being at most a constant r can be defined in terms of an error system that allows at most r substitution, insertion and deletion errors. With respect to this channel, the set of possible outputs for an input belonging to a regular language L consists of the edit distance neighbourhood of L having radius r.

General error channels (or error systems) realized by rational channels [9, 13, 19] can formalize many further types of errors, such as transposition errors, or so called scattered or burst errors that are relevant for data communication applications. The set of possible outputs $C(L)$ produced by a rational error channel C corresponding to inputs belonging to a regular language L is always regular. However, the set $C(L)$ need not be a neighbourhood of L defined by a distance metric and future work can consist to determine the state complexity of $C(L)$ as a function of the state complexity of L and the size of a finite transducer realizing the error channel C. The descriptional complexity of error systems has been considered from a different point of view by Kari and Konstantinidis [13] who establish upper and lower bounds for the sizes of DFAs that recognize a given error system,

References

1. Benedikt, M., Puppis, G., Riveros, C.: Bounded repairability of word languages. J. Comput. Syst. Sci. **79**, 1302–1321 (2013)
2. Benedikt, M., Puppis, G., Riveros, C.: The per-character cost of repairing word languages. Theoret. Comput. Sci. **539**, 38–67 (2014)
3. Brzozowski, J., Jirásková, G., Li, B.: Quotient complexity of ideal languages. In: Latin American Theoretical Informatics Symposium, pp. 208–221 (2010)
4. Calude, C.S., Salomaa, K., Yu, S.: Additive distances and quasi-distances between words. J. Univers. Comput. Sci. **8**, 141–152 (2002)
5. Chatterjee, K., Henzinger, T.A., Ibsen-Jensen, R., Otop, J.: Edit distance for pushdown automata. In: 42nd ICALP, Proceedings. Part II. Lecture Notes in Computer Science, vol. 9135, pp. 121–133 (2015)
6. Choffrut, C., Pighizzini, G.: Distances between languages and reflexivity of relations. Theoret. Comput. Sci. **286**, 117–138 (2002)
7. Cohen, G., Honkala, I., Litsyn, S., Lobstein, A.: Covering Codes. Elsevier North-Holland Mathematical Library, vol. 54 (1997)
8. Deza, M.M., Deza, E.: Encyclopedia of Distances. Springer, Berlin Heidelberg (2009)
9. Dudzinski, K., Konstantinidis, S.: Formal descriptions of code properties: decidability, complexity, implementation. Int. J. Found. Comput. Sci. **23**, 67–85 (2012)
10. El-Mabrouk, N.: On the size of minimal automata for approximate string matching. Technical report, Institut Gaspard Monge, Université de Marne la Vallée, Paris (1997)

11. Gao, Y., Moreira, N., Reis, R., Yu, S.: A survey on operational state complexity. arXiv:1509.03254v1 [cs.FL], Sept 2015. (To appear in *Computer Science Review*.)
12. Han, Y.-S., Ko, S.-K., Salomaa, K.: The edit-distance between a regular language and a context-free language. Int. J. Found. Comput. Sci. **24**, 1067–1082 (2013)
13. Kari, L., Konstantinidis, S.: Descriptional complexity of error/edit systems. J. Autom. Lang. Comb. **9**(2/3), 293–309 (2004)
14. Kari, L., Konstantinidis, S., Kopecki, S., Yang, M.: An efficient algorithm for computing the edit distance of a regular language via input-altering transducers (2014)
15. Konstantinidis, S.: An algebra of discrete channels that involve combinations of three basic error types. Inform. Comput. **167**, 120–131 (2001)
16. Konstantinidis, S.: Transducers and the properties of error-detection, error-correction, and finite-delay decodability. J. Univers. Comput. Sci. **8**, 278–291 (2002)
17. Konstantinidis, S.: Computing the edit distance of a regular language. Inform. Comput. **205**, 1307–1316 (2007)
18. Konstantinidis, S., Silva, P.: Maximal error-detecting capabilities of formal languages. J. Autom. Lang. Comb. **13**(1), 55–71 (2008)
19. Konstantinidis, S., Silva, P.V.: Computing maximal error-detecting capabilities and distances of regular languages. Fund. Inform. **101**, 257–270 (2010)
20. Kutrib, M., Meckel, K., Wendlandt, M.: Parameterized prefix distance between regular languages. In: SOFSEM 2014: Theory and Practice of Computer Science, pp. 419–430 (2014)
21. Leung, H., Podolskiy, V.: The limitedness problem on distance automata: Hashiguchi's method revisited. Theoret. Comput. Sci. **310**, 147–158 (2004)
22. Levenshtein, V.I.: Binary codes capable of correcting deletions, insertions, and reversals. Sov. Phys. Dokl. **10**(8), 707–710 (1966)
23. Lothaire, M.: Applied combinatorics on words, Chapter 1 algorithms on words. In: Encyclopedia of Mathematics and It's Applications, vol. 105. Cambridge University Press, New York (2005)
24. Mohri, M.: Edit-distance of weighted automata: general definitions and algorithms. Int. J. Found. Comput. Sci. **14**(6), 957–982 (2003)
25. Morelos-Zaragoza, R.H.: The Art of Error Correcting Coding. John Wiley & Sons, Chichester, England (2006)
26. Ng, T., Rappaport, D., Salomaa, K.: Quasi-distances and weighted finite automata. In: Proceedings of DCFS 2015, Lecture Notes Computer Science, vol. 9118, pp. 209–219 (2015)
27. Ng, T., Rappaport, D., Salomaa, K.: State complexity of neighbourhoods and approximate pattern matching. In: Potapov I., (ed.) Proceedings of DLT 2015. Lecture Notes Computer Science, vol. 9168, pp. 389–400 (2015)
28. Ng, T., Rappaport, D., Salomaa, K.: State complexity of prefix distance. In: Proceedings of CIAA 2015. Lecture Notes Computer Science, vol. 9223, pp. 238–249 (2015)
29. Pighizzini, G.: How hard is computing the edit distance? Inform. Comput. **165**(1), 1–13 (2001)
30. Povarov, G.: Descriptive complexity of the Hamming neighborhood of a regular language. In: Language and Automata Theory and Applications, pp. 509–520 (2007). (An updated version available at: http://gp-sci.googlecode.com/svn/trunk/phd/.../HammingNeighborhood.pdf)
31. Povarov, G.: Finite transducers and nondeterministic state complexity of regular languages. Russ. Math. (Iz. VUZ) **54**(6), 19–25 (2010)
32. Salomaa, K., Schofield, P.: State complexity of additive weighted finite automata. Int. J. Found. Comput. Sci. **18**(6), 1407–1416 (2007)
33. Shamkin, S.: Descriptional complexity of Hamming neighbourhoods of finite languages (in Russian). M.Sc. Thesis, Ural Federal University, Ekaterinburg, Russia (2011)
34. Schulz, K.U., Mihov, S.: Fast string correction with Levenshtein automata. Int. J. Doc. Anal. Recogn. **5**, 67–85 (2002)
35. Shallit, J.: A Second Course in Formal Languages and Automata Theory, Cambridge University Press (2009)
36. van Lint, J.H.: Introduction to Coding Theory. Springer, Graduate Texts in Mathematics (1999)
37. Yu, S.: Regular languages. In: Rozenberg, G., Salomaa, A., (eds.) Handbook of Formal Languages, pp. 41–110. Springer, Berlin, Heidelberg (1997)

A Less Known Side of Quantum Cryptography

Naya Nagy, Marius Nagy and Selim G. Akl

Abstract The most tangible impact of quantum information processing that we can perceive today is undoubtedly in the area of quantum cryptography. Quantum key distribution protocols, starting with the groundbreaking BB84, are at the heart of actual physical equipment designed to ensure the security of network communications through quantum means. But, despite their practical success, some important questions related to these protocols remain insufficiently explored, even to the point where they give rise to false myths. This chapter dispels these myths showing what is the exact condition necessary to achieve genuine quantum key distribution (and not just key enhancement), how authentication can also be done quantum mechanically and how testing for possible acts of eavesdropping can also be done on qubits that were never "touched" while in transit.

1 Introduction

It is without any doubt, that from the point of view of practical applications, quantum cryptography is the most successful branch of quantum information processing. Shor's polynomial time algorithm for factorizing integers using quantum properties [21] is a remarkable milestone, proving the power of a quantum computer. However, experimental physicists are still struggling to find the best physical embodiment for a qubit that would make a scalable quantum computer a practical reality.

On the other hand, physical equipment that can ensure the security of network communications through quantum means have been available for years now. These

N. Nagy (✉)
Kingdom of Saudi Arabia, University of Dammam, Dammam, Saudi Arabia
e-mail: nmnagy@uod.edu.sa

M. Nagy
Kingdom of Saudi Arabia, Prince Mohammad Bin Fahd University, Dhahran, Saudi Arabia
e-mail: mnagy@pmu.edu.sa

S.G. Akl
Queen's University, Kingston, ON, Canada
e-mail: akl@cs.queensu.ca

© Springer International Publishing Switzerland 2017
A. Adamatzky (ed.), *Emergent Computation*, Emergence, Complexity
and Computation 24, DOI 10.1007/978-3-319-46376-6_7

technologies are based on the pioneering work of Bennett and Brassard [2]. They showed for the first time how quantum properties (specifically photon polarization) can be used to help two parties establish a shared secret key in order to communicate secretly through a public channel. The BB84 protocol was the inception of a new field of quantum key distribution, with many researchers devising new methods, analyzing security properties or extending the initial result. Little by little, an array of important results have established themselves as facts or common knowledge shaping the public perception on quantum cryptography.

This chapter intends to bring under scrutiny some of the common beliefs marking the field of quantum cryptography, either dispelling some of its myths or shedding light on some less known, unconventional results that may look surprising at first.

Two of the most important problems in cryptography are concerned with the *security* and *authenticity* of exchanged messages. There are perfectly good ways to achieve these two goals, provided the two parties (generically referred to as Alice and Bob) wishing to communicate over an insecure (public) channel share a secret key. Therefore, the *key distribution* step, allowing Alice and Bob to establish a secret key prior to exchanging any messages, is of capital importance for many areas of cryptography.

Various schemes have been proposed over time to ensure the security and authenticity of communications without resorting to a previously shared private key (Diffie-Hellman, Digital Signature Algorithm, RSA). Probably, the most successful example of such a public-key system is the RSA cryptographic system, based on the RSA algorithm [18]. The security of public-key cryptographic communication systems rests on unproved assumptions about the difficulty to compute the decryption key, even when the encryption key is known. The RSA algorithm, for example, so popular today, capitalizes on the presumed intractability of factoring large numbers in a reasonable amount of time, although nobody was able to prove that factoring is not in P.

With the advent of processing information at the quantum level, the security of cryptographic protocols was set on a firmer foundation. Quantum key distribution (QKD) schemes were proposed, whose security is guaranteed by the very laws of physics (quantum mechanics, more precisely). What really distinguishes them from the classical cryptographic protocols is that they make the difference in terms of *intrusion detection*. In a classical scheme, one can only hope that the adversary simply does not have enough computational resources to gain knowledge of the information in transit. There is no protocol that allows for the detection of an eavesdropper. The ability to copy classical information without restriction is responsible for this situation. In contrast, since arbitrary quantum bits cannot be cloned [24], it is much more difficult for an eavesdropper to spy on a quantum communication without being detected.

Several techniques exist that exploit quantum effects for key distribution [1, 2, 10]. Their aim is to maximize the intrusion detection rate, upon which the security of the protocols rests. In these protocols, Alice conveys the secret information to Bob by encoding it into some quantum properties of photons. At the other end, Bob has to subject each photon to a quantum measurement, as soon as it is received, in order to agree with Alice on a common key.

All existing protocols, except the one presented here, actually perform a key enhancement. Genuine key distribution is even deemed impossible due to the fact that the quantum protocol needs to authenticate the classical channel. The authentication of the classical channel is usually briefly mentioned, as being performed by a classical authentication scheme, or in some cases it is not mentioned explicitly. A small key authenticates the classical channel and then the quantum protocol develops a large key. But, as we show in Sect. 3, authentication need not be done by classical means. The quantum protocol itself can authenticate the communication partners. This means that all information exchanged between the partners is fully public, albeit public protected. Public protected information means public information that cannot be changed by a third party. A crucial note is justified here, namely that *all* classical schemes need public *protected* information. In essence, we show that public protected information is enough for genuine quantum key distribution.

A second myth we wish to dispel in this chapter is related to one-time pads. One-time pads are the most secure solution for encrypting messages, exactly because they are not supposed to be reused. The drawback is that the two communicating parties need to meet in person in order to exchange the keys they will be using for encryption/decryption of messages and each key must be as long as the message that will be encrypted by that key. We show in Sect. 4 that one-time pad communication can be made readily available by quantum or partially quantum messages. The one-time pads are generated as needed by the initiating party of the communication. The two parties do not need to meet in order to agree on the value of the one-time pad. The only secret information that Alice and Bob share a priori is a reading mask. This reading mask defines the quantum encryption of the one-time pad.

The idea of encrypting the encryption key and sending it along with the encrypted message is part of the cryptographic folklore: for example a DES (data encryption standard) key is used to encrypt a message, but is itself encrypted using an RSA key. Our quantum version of the idea adds to the benefits of the classic version the detectability of the intruder and the impossibility to copy the qubits. Thus, the intruder is detected when only reading the message header or body. Also, for fully quantum messages, both the encrypted key and the message body cannot be copied by an eavesdropper for later use.

Finally, it is common sense to accept that a potential eavesdropper can be detected based on how extensive her eavesdropping actions have been. The more qubits "disturbed" through eavesdropping, the higher the chances to catch her. From this common sense comes the apparently reasonable assumption that Eve (the typical eavesdropper) can only be detected if at least one of the qubits tested or verified for eavesdropping have been tampered with while in transit. However surprising as it may seem, this is not actually true. Section 5 presents the details of a QKD scheme which employs the quantum Fourier transform in order to propagate the disturbance caused by eavesdropping on a qubit to subsequent qubits in the sequence transmitted. Thus, even if the intruder tries to "hide" behind a low level of eavesdropping, our protocol has the ability to detect Eve's actions in spite of the fact that the qubits tested for eavesdropping may have not been "touched" directly by Eve.

In an effort to make the material in this chapter accessible to a broad spectrum of readers, even without a formal background in quantum mechanics, we begin our exposition with a brief introduction to the field of quantum information processing. The familiarized reader is advised to skip this preliminary section.

2 Fundamentals of Quantum Information Processing

2.1 Qubits and the Bloch Sphere

A classical bit may be in one of two states 0 and 1. The state of a qubit [17] is defined as a superposition of the classical 0 and 1. Using Dirac's notation, a qubit is $q = \alpha|0\rangle + \beta|1\rangle$, where α and β are complex numbers. Since a qubit is normalized to length one, $|\alpha|^2 + |\beta|^2 = 1$, then $|\alpha|^2$ represents the probability of the qubit to collapse to 0 when measured, and $|\beta|^2$ represents the probability of the qubit to collapse to 1. The qubit can be rewritten in polar coordinates $q = e^{i\gamma}(\cos\phi|0\rangle + e^{i\theta}\sin\phi|1\rangle)$, where $\alpha = e^{i\gamma}\cos\phi$ and $\beta = e^{i\gamma}e^{i\theta}\sin\phi$. Because the global factor $e^{i\gamma}$ is indistinguishable through measurement, it can be dropped. Thus, the qubit can be written simply as $q = \cos\phi|0\rangle + e^{i\theta}\sin\phi|1\rangle$.

Qubits can be graphically represented as arrows on a sphere, called the Bloch sphere, see Fig. 1. Consider a three dimensional sphere of radius 1, in the coordinate system Ox, Oy, and Oz. The Ox axis points to the right, Oy points away from the paper, and Oz points upwards. Figure 2 shows the representation of an arbitrary qubit. The qubit shows as an arrow of length 1. Its orientation is described by ϕ, the angle with the Oz axis, and by θ, the angle with the Ox axis measured in a counterclockwise direction. Therefore, each qubit $q = \cos\phi|0\rangle + e^{i\theta}\sin\phi|1\rangle$ has a unique representation on the Bloch sphere.

Fig. 1 The Bloch sphere has a unitary radius and is centered in the origin of the coordinate system

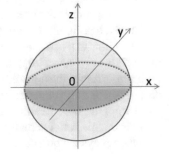

Fig. 2 The qubit $\cos\phi|0\rangle + e^{i\theta}\sin\phi|1\rangle$ is a vector on the Bloch sphere. ϕ is the angle with the Oz axis, and θ is the angle with the Ox axis measured in a counterclockwise direction

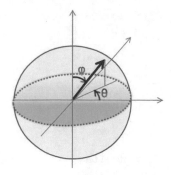

Some particular qubits are shown with their representation:

- $|0\rangle$ in Fig. 3.
- $|1\rangle$ in Fig. 4.
- $\frac{1}{\sqrt{2}}(|0\rangle + |1\rangle)$ in Fig. 5.
- $\frac{1}{\sqrt{2}}(|0\rangle + e^{i\frac{\pi}{2}}|1\rangle)$ in Fig. 6.
- $\frac{1}{\sqrt{2}}(|0\rangle - |1\rangle)$ in Fig. 7.
- $\frac{1}{\sqrt{2}}(|0\rangle - e^{i\frac{\pi}{2}}|1\rangle)$ in Fig. 8.

Note that all qubits in a balanced superposition are represented on the equator. A balanced superposition means that the qubit has an equal chance 50 % to collapse to 0 or 1.

Fig. 3 The qubit $|0\rangle$

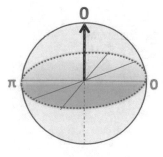

Fig. 4 The qubit $|1\rangle$

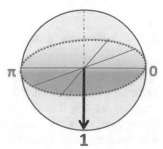

Fig. 5 The qubit
$\frac{1}{\sqrt{2}}(|0\rangle + |1\rangle)$

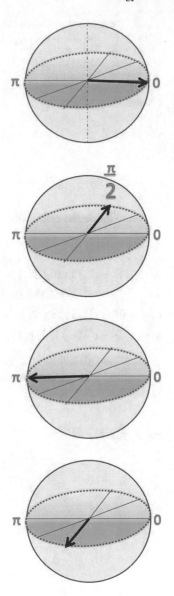

Fig. 6 The qubit
$\frac{1}{\sqrt{2}}(|0\rangle + e^{i\frac{\pi}{2}}|1\rangle)$

Fig. 7 The qubit
$\frac{1}{\sqrt{2}}(|0\rangle - |1\rangle)$

Fig. 8 The qubit
$\frac{1}{\sqrt{2}}(|0\rangle - e^{i\frac{\pi}{2}}|1\rangle)$

2.2 Quantum Gates

Quantum protocols use a small set of common gates. Three such gates are described below: the NOT gate, the phase-shift gate and the Hadamard gate. All these gates are unary gates; they transform one qubit. Nevertheless, these gates may be extended to binary gates by adding a control qubit to the primary qubit. The control qubit decides whether the gate is actually applied to its primary qubit. If the control qubit is $|1\rangle$, the primary qubit is transformed according to the gate's definition. If the control qubit is $|0\rangle$, the primary qubit passes the gate undisturbed.

2.2.1 The Controlled NOT Gate

Probably the simplest gate that can be applied to a qubit is the plain NOT gate. It reverses the $|0\rangle$ to $|1\rangle$ and vice-versa. For an arbitrary qubit, the NOT gate reverses the probabilities of obtaining the binary 0 or 1 through measurement.

$$\mathsf{NOT}(\alpha|0\rangle + \beta|1\rangle) = \beta|0\rangle + \alpha|1\rangle).$$

The NOT gate has the following transformation matrix

$$\mathsf{NOT} = \begin{bmatrix} 0 & 1 \\ 1 & 0 \end{bmatrix}.$$

A more useful variant of the gate is the Controlled NOT gate, denoted CNOT. The CNOT gate takes two input qubits and consequently also has two output qubits, see Fig. 9. The control qubit gives the option either to indeed reverse the target qubit or to leave it unchanged. If the control qubit is $|0\rangle$, the target qubit remains unchanged. If the control qubit is $|1\rangle$, the target qubit comes out negated on the second output. The first output is a copy of the control qubit. The CNOT gate is described by the following transformation matrix. Note that the matrix has size four as it applies to two qubits.

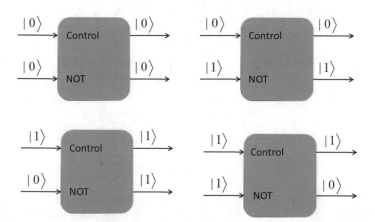

Fig. 9 The controlled NOT gate as applied to the computational basis

$$\text{CNOT} = \begin{bmatrix} 1 & 0 & 0 & 0 \\ 0 & 1 & 0 & 0 \\ 0 & 0 & 0 & 1 \\ 0 & 0 & 1 & 0 \end{bmatrix}.$$

Thus, consider an arbitrary ensemble of two qubits $q_c\, q_t = \alpha|00\rangle + \beta|01\rangle + \gamma|10\rangle + \delta|11\rangle$, where $|\alpha|^2 + |\beta|^2 + |\gamma|^2 + |\delta|^2 = 1$. q_c is the control qubit and q_t is the target (NOT) qubit. CNOT performs the following transformation

$$\text{CNOT}\, q_c\, q = \begin{bmatrix} 1 & 0 & 0 & 0 \\ 0 & 1 & 0 & 0 \\ 0 & 0 & 0 & 1 \\ 0 & 0 & 1 & 0 \end{bmatrix} \cdot \begin{bmatrix} \alpha \\ \beta \\ \gamma \\ \delta \end{bmatrix} = \begin{bmatrix} \alpha \\ \beta \\ \delta \\ \gamma \end{bmatrix}.$$

Let us consider a few particular cases. Suppose the control qubit is $q_c = |0\rangle$ and consequently the target qubit remains unchanged.

$$\text{CNOT}|00\rangle = \begin{bmatrix} 1 & 0 & 0 & 0 \\ 0 & 1 & 0 & 0 \\ 0 & 0 & 0 & 1 \\ 0 & 0 & 1 & 0 \end{bmatrix} \cdot \begin{bmatrix} 1 \\ 0 \\ 0 \\ 0 \end{bmatrix} = \begin{bmatrix} 1 \\ 0 \\ 0 \\ 0 \end{bmatrix} = |00\rangle.$$

$$\text{CNOT}|01\rangle = \begin{bmatrix} 1 & 0 & 0 & 0 \\ 0 & 1 & 0 & 0 \\ 0 & 0 & 0 & 1 \\ 0 & 0 & 1 & 0 \end{bmatrix} \cdot \begin{bmatrix} 0 \\ 1 \\ 0 \\ 0 \end{bmatrix} = \begin{bmatrix} 0 \\ 1 \\ 0 \\ 0 \end{bmatrix} = |01\rangle.$$

If the control qubit is $q_c = |1\rangle$ the target qubit is flipped.

$$\text{CNOT}|10\rangle = \begin{bmatrix} 1 & 0 & 0 & 0 \\ 0 & 1 & 0 & 0 \\ 0 & 0 & 0 & 1 \\ 0 & 0 & 1 & 0 \end{bmatrix} \cdot \begin{bmatrix} 0 \\ 0 \\ 1 \\ 0 \end{bmatrix} = \begin{bmatrix} 0 \\ 0 \\ 0 \\ 1 \end{bmatrix} = |11\rangle.$$

$$\text{CNOT}|11\rangle = \begin{bmatrix} 1 & 0 & 0 & 0 \\ 0 & 1 & 0 & 0 \\ 0 & 0 & 0 & 1 \\ 0 & 0 & 1 & 0 \end{bmatrix} \cdot \begin{bmatrix} 0 \\ 0 \\ 0 \\ 1 \end{bmatrix} = \begin{bmatrix} 0 \\ 0 \\ 1 \\ 0 \end{bmatrix} = |10\rangle.$$

The CNOT gate can be used to clone a qubit if it is encoded in the computational basis. That is to say, for a qubit that is in one of the two basis values $|0\rangle$ and $|1\rangle$, the CNOT gate can produce an exact copy. To obtain this, the qubit to be cloned is fed to the control input, and the NOT input is fed a $|0\rangle$, see Fig. 10. Such cloning may be used by the eavesdropper to obtain a copy of a qubit that is part of a secret message. If successful, the eavesdropper may remain hidden as she allows the initial

Fig. 10 The controlled NOT gate used for cloning a qubit encoding the computational basis

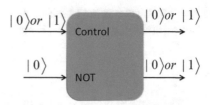

qubit to pass through to the rightful destination. This tentative eavesdropping, called translucent eavesdropping can be revealed by using two or more encoding bases, such as for example the computational basis and the Hadamard basis.

2.2.2 The Phase-Shift Gate

The phase-shift gate, R_θ, rotates the relative phase of a qubit by an angle θ. R_θ is described by the transformation matrix

$$R_\theta = \begin{bmatrix} 1 & 0 \\ 0 & e^{i\theta} \end{bmatrix}.$$

If R_θ is applied to an arbitrary qubit $q = \alpha|0\rangle + \beta|1\rangle$, where $|\alpha|^2 + |\beta|^2 = 1$, the following transformation happens

$$R_\theta \cdot \begin{bmatrix} \alpha \\ \beta \end{bmatrix} = \begin{bmatrix} \alpha \\ e^{i\theta}\beta \end{bmatrix}$$

Let us consider a few particular cases.

Rotation by a $\theta = \pi$ angle.

Using Euler's formula, $e^{i\theta} = \cos\theta + i\sin\theta$, we have $e^{i\pi} = \cos(\pi) + i\sin(\pi) = -1$ (Fig. 11). The phase-shift gate for π is also called the Pauli-Z gate, denoted

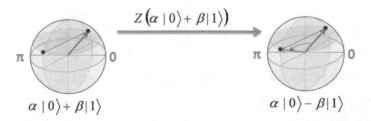

Fig. 11 The qubit on the left is an arbitrary qubit $\alpha|0\rangle + \beta|1\rangle$, with $|\alpha|^2 + |\beta|^2 = 1$. When the Z gate is applied to it, the qubit vector gets rotated counter clock wise around the z-axis by an angle of π

$Z = R_\pi$. It rotates the phase of the $|1\rangle$ by changing the sign of its coefficient. The transformation matrix of the Z gate is

$$Z = \begin{bmatrix} 1 & 0 \\ 0 & -1 \end{bmatrix}.$$

The transformation can be shown to be unitary. The adjoint transformation is obtained by transposing the matrix and each complex number is replaced by its complex conjugate. As such, $Z^\dagger$ is the same as Z. We have $Z^\dagger = \begin{bmatrix} 1 & 0 \\ 0 & -1 \end{bmatrix}$. And the multiplication of Z with its adjoint becomes $Z \cdot Z^\dagger = \begin{bmatrix} 1 & 0 \\ 0 & -1 \end{bmatrix} \cdot \begin{bmatrix} 1 & 0 \\ 0 & -1 \end{bmatrix} = \begin{bmatrix} 1 & 0 \\ 0 & 1 \end{bmatrix} = I$, which is the identity matrix.

If Z is applied on an arbitrary qubit $\alpha|0\rangle + \beta|1\rangle$, with α and β complex numbers and $|\alpha|^2 + |\beta|^2 = 1$, the following transformation happens

$$Z(\alpha|0\rangle + \beta|1\rangle) = Z \cdot \begin{bmatrix} \alpha \\ \beta \end{bmatrix} = \begin{bmatrix} 1 & 0 \\ 0 & -1 \end{bmatrix} \cdot \begin{bmatrix} \alpha \\ \beta \end{bmatrix} = \begin{bmatrix} \alpha \\ -\beta \end{bmatrix} = \alpha|0\rangle - \beta|1\rangle.$$

Z applied to the computational basis, $|0\rangle$ and $|1\rangle$ leaves the qubits unchanged.

$$Z|0\rangle = \begin{bmatrix} 1 & 0 \\ 0 & -1 \end{bmatrix} \cdot \begin{bmatrix} 1 \\ 0 \end{bmatrix} = \begin{bmatrix} 1 \\ 0 \end{bmatrix} = |0\rangle.$$

$$Z|1\rangle = \begin{bmatrix} 1 & 0 \\ 0 & -1 \end{bmatrix} \cdot \begin{bmatrix} 0 \\ 1 \end{bmatrix} = \begin{bmatrix} 0 \\ -1 \end{bmatrix} = -|1\rangle = |1\rangle.$$

We could write $-|1\rangle = |1\rangle$ because the global factor -1, having the modulus 1, is indistinguishable through measurement.

Z applied to a qubit on the equator rotates the qubit along the equator by an angle of π. For example a qubit in balanced superposition, such as $\frac{1}{\sqrt{2}}(|0\rangle + |1\rangle)$, an arrow pointing to the right on the Bloch sphere, undergoes the following Z transformation:

$$Z \cdot \frac{1}{\sqrt{2}}(|0\rangle + |1\rangle) = \begin{bmatrix} 1 & 0 \\ 0 & -1 \end{bmatrix} \cdot \frac{1}{\sqrt{2}}\begin{bmatrix} 1 \\ 1 \end{bmatrix} = \frac{1}{\sqrt{2}}\begin{bmatrix} 1 \\ -1 \end{bmatrix} = \frac{1}{\sqrt{2}}(|0\rangle - |1\rangle).$$

The resulting qubit points to the right on the Bloch sphere (Fig. 12).

Rotation by a $\theta = \frac{\pi}{2}$ angle.

Using Euler's formula, $e^{i\theta} = \cos\theta + i\sin\theta$, we have $e^{i\frac{\pi}{2}} = \cos(\frac{\pi}{2}) + i\sin(\frac{\pi}{2}) = i$.

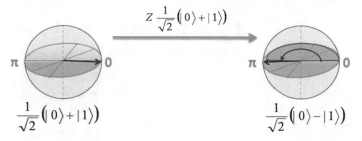

Fig. 12 The qubit on the left is the balanced superposition $\frac{1}{\sqrt{2}}(|0\rangle + |1\rangle)$. When the Z gate is applied to it, it gets rotated along the equator by an angle of π

The phase-shift gate for $\frac{\pi}{2}$ is also called the phase gate, denoted $S = R_{\frac{\pi}{2}}$. It rotates the phase of the $|1\rangle$ component by an angle of $\frac{\pi}{2}$, the equivalent of i. The transformation matrix for the S gate is

$$R_{\frac{\pi}{2}} = \begin{bmatrix} 1 & 0 \\ 0 & i \end{bmatrix}.$$

Its adjoint transformation is $R_{\frac{\pi}{2}}^{\dagger} = \begin{bmatrix} 1 & 0 \\ 0 & -i \end{bmatrix}$. The multiplication $R_{\frac{\pi}{2}} \cdot R_{\frac{\pi}{2}}^{\dagger} = \begin{bmatrix} 1 & 0 \\ 0 & i \end{bmatrix} \cdot \begin{bmatrix} 1 & 0 \\ 0 & -i \end{bmatrix} = \begin{bmatrix} 1 & 0 \\ 0 & 1 \end{bmatrix} = I$, which proves the transformation to be unitary.

$R_{\frac{\pi}{2}}$ applied to the computational basis, $|0\rangle$ and $|1\rangle$ leaves the qubits unchanged.

$$R_{\frac{\pi}{2}}|0\rangle = \begin{bmatrix} 1 & 0 \\ 0 & i \end{bmatrix} \cdot \begin{bmatrix} 1 \\ 0 \end{bmatrix} = \begin{bmatrix} 1 \\ 0 \end{bmatrix} = |0\rangle.$$

$$R_{\frac{\pi}{2}}|1\rangle = \begin{bmatrix} 1 & 0 \\ 0 & i \end{bmatrix} \cdot \begin{bmatrix} 0 \\ 1 \end{bmatrix} = \begin{bmatrix} 0 \\ i \end{bmatrix} = i|1\rangle = |1\rangle.$$

We could write $i|1\rangle = |1\rangle$ because the global factor i, having the modulus 1, is indistinguishable through measurement.

$R_{\frac{\pi}{2}}$ applied to a qubit on the equator rotates the qubit along the equator by an angle of $\frac{\pi}{2}$, counterclockwise. For example, if a qubit in balanced superposition, such as $\frac{1}{\sqrt{2}}(|0\rangle + |1\rangle)$, an arrow pointing to the right on the Bloch sphere, undergoes the following $R_{\frac{\pi}{2}}$ transformation:

$$R_{\frac{\pi}{2}} \cdot \frac{1}{\sqrt{2}}(|0\rangle + |1\rangle) = \begin{bmatrix} 1 & 0 \\ 0 & i \end{bmatrix} \cdot \frac{1}{\sqrt{2}} \begin{bmatrix} 1 \\ 1 \end{bmatrix} = \frac{1}{\sqrt{2}} \begin{bmatrix} 1 \\ i \end{bmatrix} =$$

$$\frac{1}{\sqrt{2}}(|0\rangle + i|1\rangle). \qquad (1)$$

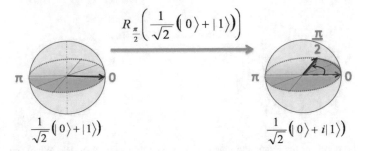

Fig. 13 The qubit on the left is the balanced superposition $\frac{1}{\sqrt{2}}(|0\rangle + |1\rangle)$. When the $R_{\frac{\pi}{2}}$ gate is applied to it, it gets rotated along the equator by an angle of $\frac{\pi}{2}$, in the counterclockwise direction. The resulting qubit points away from the page

The resulting qubit points away from the page on the Bloch sphere (Fig. 13).

If $R_{\frac{\pi}{2}}$ is applied again on the resulting qubit $\frac{1}{\sqrt{2}}(|0\rangle + i|1\rangle)$, the arrow is rotated additionally by the same angle, $\frac{\pi}{2}$, in counterclockwise direction.

$$R_{\frac{\pi}{2}} \cdot \frac{1}{\sqrt{2}}(|0\rangle + i|1\rangle) = \begin{bmatrix} 1 & 0 \\ 0 & i \end{bmatrix} \cdot \frac{1}{\sqrt{2}} \begin{bmatrix} 1 \\ i \end{bmatrix} = \frac{1}{\sqrt{2}} \begin{bmatrix} 1 \\ -1 \end{bmatrix} =$$
$$\frac{1}{\sqrt{2}}(|0\rangle - |1\rangle). \qquad (2)$$

This resulting qubit points to the left on the Bloch sphere (Fig. 14). Thus applying the $R_{\frac{\pi}{2}}$ twice is equivalent to applying the Z gate once.

Rotation by a $\theta = -\frac{\pi}{2}$ angle.

Using Euler's formula, $e^{i\theta} = \cos\theta + i\sin\theta$, we have $e^{-i\frac{\pi}{2}} = \cos(\frac{\pi}{2}) - i\sin(\frac{\pi}{2}) = -i$. This transformation performs a clockwise rotation by an angle of $\frac{\pi}{2}$, the equivalent of a counterclockwise rotation by $-\frac{\pi}{2}$.

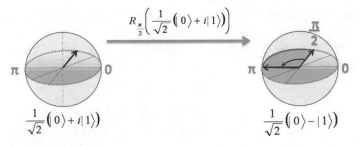

Fig. 14 The qubit on the left is the balanced superposition $\frac{1}{\sqrt{2}}(|0\rangle + i|1\rangle)$. When the $R_{\frac{\pi}{2}}$ gate is applied to it, it gets rotated along the equator by an angle of $\frac{\pi}{2}$, in the counter clock wise direction. The resulting qubit points to the right on the Bloch sphere

2.2.3 The Hadamard Gate

The Hadamard gate is described by the transformation matrix

$$H = \frac{1}{\sqrt{2}} \begin{bmatrix} 1 & 1 \\ 1 & -1 \end{bmatrix}.$$

The transformation can easily be shown to be unitary as all values in the Hadamard matrix are real and the matrix is symmetric. Therefore, the adjoint transformation is identical to the original, $H^\dagger = H$. The multiplication of H with its adjoint becomes

$H \cdot H^\dagger = \frac{1}{\sqrt{2}} \begin{bmatrix} 1 & 1 \\ 1 & -1 \end{bmatrix} \cdot \frac{1}{\sqrt{2}} \begin{bmatrix} 1 & 1 \\ 1 & -1 \end{bmatrix} = \frac{1}{2} \begin{bmatrix} 2 & 0 \\ 0 & 2 \end{bmatrix} = \begin{bmatrix} 1 & 0 \\ 0 & 1 \end{bmatrix} = I$, which is the identity

matrix.

If H is applied on an arbitrary qubit $q = \alpha|0\rangle + \beta|1\rangle$, where $|\alpha|^2 + |\beta|^2 = 1$, the following transformation happens

$$H \cdot \begin{bmatrix} \alpha \\ \beta \end{bmatrix} = \frac{1}{\sqrt{2}} \begin{bmatrix} 1 & 1 \\ 1 & -1 \end{bmatrix} \cdot \begin{bmatrix} \alpha \\ \beta \end{bmatrix} = \frac{1}{\sqrt{2}} \begin{bmatrix} \alpha + \beta \\ \alpha - \beta \end{bmatrix} =$$
$$\frac{1}{\sqrt{2}}(\alpha + \beta)|0\rangle + \frac{1}{\sqrt{2}}(\alpha - \beta)|1\rangle. \tag{3}$$

H applied to the computational basis, $|0\rangle$ and $|1\rangle$ brings the qubits into a balanced superposition.

$$H|0\rangle = \frac{1}{\sqrt{2}} \begin{bmatrix} 1 & 1 \\ 1 & -1 \end{bmatrix} \cdot \begin{bmatrix} 1 \\ 0 \end{bmatrix} = \frac{1}{\sqrt{2}} \begin{bmatrix} 1 \\ 1 \end{bmatrix} = \frac{1}{\sqrt{2}}(|0\rangle + |1\rangle). \tag{4}$$

This is a vector on the equator pointing to the right, see Fig. 15.

$$H|1\rangle = \frac{1}{\sqrt{2}} \begin{bmatrix} 1 & 1 \\ 1 & -1 \end{bmatrix} \cdot \begin{bmatrix} 0 \\ 1 \end{bmatrix} = \frac{1}{\sqrt{2}} \begin{bmatrix} 1 \\ -1 \end{bmatrix} = \frac{1}{\sqrt{2}}(|0\rangle - |1\rangle). \tag{5}$$

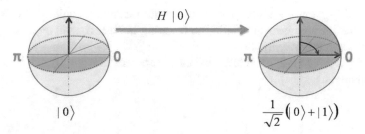

Fig. 15 The qubit on the left is classical zero. When the H gate is applied to it, it becomes the balanced superposition $\frac{1}{\sqrt{2}}(|0\rangle + |1\rangle)$

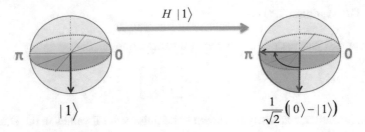

Fig. 16 The qubit on the left is classical one. When the H gate is applied to it, it becomes the balanced superposition $\frac{1}{\sqrt{2}}(|0\rangle - |1\rangle)$

This is a vector on the equator pointing to the left, see Fig. 16. The two resulting qubits, while both being balanced superpositions, differ in their phase.

As the Hadamard gate is its own inverse, $H = H^\dagger$, when the Hadamard gate is applied two times on a qubit, the qubit remains unchanged. In particular, $HH|0\rangle = H(\frac{1}{\sqrt{2}}(|0\rangle + |1\rangle)) = |0\rangle$, and $HH|1\rangle = H(\frac{1}{\sqrt{2}}(|0\rangle - |1\rangle)) = |1\rangle$.

Let us consider a few more Hadamard transformations as they will be useful in the algorithms presented in the following sections.

When Hadamard is applied to the balanced superposition $\frac{1}{\sqrt{2}}(|0\rangle + i|1\rangle)$ the following rotation happens:

$$H(\frac{1}{\sqrt{2}}(|0\rangle + i|1\rangle)) = \frac{1}{\sqrt{2}}\begin{bmatrix} 1 & 1 \\ 1 & -1 \end{bmatrix} \cdot \frac{1}{\sqrt{2}}(|0\rangle + i|1\rangle) =$$

$$\frac{1}{2}\begin{bmatrix} 1 & 1 \\ 1 & -1 \end{bmatrix} \cdot \begin{bmatrix} 1 \\ i \end{bmatrix} = \frac{1}{2}\begin{bmatrix} 1+i \\ 1-i \end{bmatrix} = \frac{1}{2}[(1+i)|0\rangle + (1-i)|1\rangle] =$$

$$\frac{1+i}{2}\left(|0\rangle + \frac{1-i}{1+i}|1\rangle\right) = \frac{1+i}{2}\left(|0\rangle + \frac{(1-i)^2}{1-i^2}|1\rangle\right) =$$

$$\frac{1+i}{2}\left(|0\rangle + \frac{1-2i+i^2}{1+1}|1\rangle\right) = \frac{1+i}{2}(|0\rangle - i|1\rangle) =$$

$$\frac{1+i}{\sqrt{2}}\frac{1}{\sqrt{2}}(|0\rangle - i|1\rangle) = e^{i\frac{\pi}{4}}\frac{1}{\sqrt{2}}(|0\rangle - i|1\rangle) = \frac{1}{\sqrt{2}}(|0\rangle - i|1\rangle). \qquad (6)$$

We could ignore the global factor $e^{i\frac{\pi}{4}}$, as its modulus is unitary and is undistinguishable through measurement. The resulting qubit is rotated by an angle of π around the equator, see Fig. 17.

When Hadamard is applied to the balanced superposition $\frac{1}{\sqrt{2}}(|0\rangle - i|1\rangle)$ a similar rotation happens:

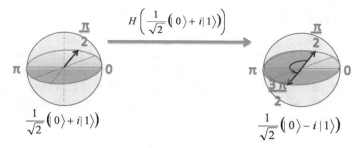

Fig. 17 The qubit on the left is the balanced superposition $\frac{1}{\sqrt{2}}(|0\rangle + i|1\rangle)$. When the H gate is applied to it, it gets rotated by an angle of π around the equator and becomes the balanced superposition $\frac{1}{\sqrt{2}}(|0\rangle - i|1\rangle)$

$$H(\frac{1}{\sqrt{2}}(|0\rangle - i|1\rangle)) = \frac{1}{\sqrt{2}}\begin{bmatrix} 1 & 1 \\ 1 & -1 \end{bmatrix} \cdot \frac{1}{\sqrt{2}}(|0\rangle - i|1\rangle) =$$

$$\frac{1}{2}\begin{bmatrix} 1 & 1 \\ 1 & -1 \end{bmatrix} \cdot \begin{bmatrix} 1 \\ -i \end{bmatrix} = \frac{1}{2}\begin{bmatrix} 1-i \\ 1+i \end{bmatrix} = \frac{1}{2}[(1-i)|0\rangle + (1+i)|1\rangle] =$$

$$\frac{1-i}{2}\left(|0\rangle + \frac{1+i}{1-i}|1\rangle\right) = \frac{1-i}{2}\left(|0\rangle + \frac{(1+i)^2}{1-i^2}|1\rangle\right) =$$

$$\frac{1-i}{2}\left(|0\rangle + \frac{1+2i+i^2}{1+1}|1\rangle\right) = \frac{1-i}{2}(|0\rangle + i|1\rangle) =$$

$$\frac{1-i}{\sqrt{2}}\frac{1}{\sqrt{2}}(|0\rangle + i|1\rangle) = e^{-i\frac{\pi}{4}}\frac{1}{\sqrt{2}}(|0\rangle + i|1\rangle) = \frac{1}{\sqrt{2}}(|0\rangle + i|1\rangle). \tag{7}$$

We could ignore the global factor $e^{-i\frac{\pi}{4}}$, as its modulus is unitary and is undistinguishable through measurement. The resulting qubit is rotated by an angle of π around the equator, see Fig. 18.

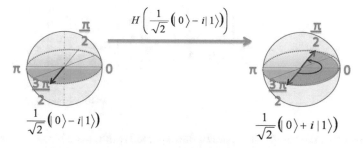

Fig. 18 The qubit on the left is the balanced superposition $\frac{1}{\sqrt{2}}(|0\rangle - i|1\rangle)$. When the H gate is applied to it, it gets rotated by an angle of π around the equator and becomes the balanced superposition $\frac{1}{\sqrt{2}}(|0\rangle + i|1\rangle)$

2.3   The Tensor Product

Two or more qubits together may be considered to form an ensemble. When a transformation is defined for one qubit, it may algebraically be described as affecting the ensemble. For example, if the first qubit is transformed by a Hadamard gate, while the second remains unchanged, the ensemble transformation can be described by the following matrix:

$$H \otimes I_2 = \frac{1}{\sqrt{2}} \begin{bmatrix} 1 & 1 \\ 1 & -1 \end{bmatrix} \otimes \begin{bmatrix} 1 & 0 \\ 0 & 1 \end{bmatrix} =$$

$$\frac{1}{\sqrt{2}} \begin{bmatrix} 1 \cdot \begin{bmatrix} 1 & 0 \\ 0 & 1 \end{bmatrix} & 1 \cdot \begin{bmatrix} 1 & 0 \\ 0 & 1 \end{bmatrix} \\ 1 \cdot \begin{bmatrix} 1 & 0 \\ 0 & 1 \end{bmatrix} & -1 \cdot \begin{bmatrix} 1 & 0 \\ 0 & 1 \end{bmatrix} \end{bmatrix} = \frac{1}{\sqrt{2}} \begin{bmatrix} 1 & 0 & 1 & 0 \\ 0 & 1 & 0 & 1 \\ 1 & 0 & -1 & 0 \\ 0 & 1 & 0 & -1 \end{bmatrix}. \tag{8}$$

Consider an arbitrary ensemble of two qubits $\alpha|00\rangle + \beta|01\rangle + \gamma|10\rangle + \delta|11\rangle$, where $|\alpha|^2 + |\beta|^2 + |\gamma|^2 + |\delta|^2 = 1$. A Hadamard gate applied on the first qubit is described below:

$$H \otimes I_2(\alpha|00\rangle + \beta|01\rangle + \gamma|10\rangle + \delta|11\rangle) =$$

$$\frac{1}{\sqrt{2}} \begin{bmatrix} 1 & 0 & 1 & 0 \\ 0 & 1 & 0 & 1 \\ 1 & 0 & -1 & 0 \\ 0 & 1 & 0 & -1 \end{bmatrix} \cdot \begin{bmatrix} \alpha \\ \beta \\ \gamma \\ \delta \end{bmatrix} = \frac{1}{\sqrt{2}} \begin{bmatrix} \alpha + \gamma \\ \beta + \delta \\ \alpha - \gamma \\ \beta - \delta \end{bmatrix} =$$

$$\frac{\alpha + \gamma}{\sqrt{2}}|00\rangle + \frac{\beta + \delta}{\sqrt{2}}|01\rangle + \frac{\alpha - \gamma}{\sqrt{2}}|10\rangle + \frac{\beta - \delta}{\sqrt{2}}|11\rangle. \tag{9}$$

For a base vector, the formula is much simplified. Consider the first base vector $|00\rangle$.

$$H \otimes I_2(|00\rangle) = \frac{1}{\sqrt{2}} \begin{bmatrix} 1 & 0 & 1 & 0 \\ 0 & 1 & 0 & 1 \\ 1 & 0 & -1 & 0 \\ 0 & 1 & 0 & -1 \end{bmatrix} \cdot \begin{bmatrix} 1 \\ 0 \\ 0 \\ 0 \end{bmatrix} = \frac{1}{\sqrt{2}} \begin{bmatrix} 1 \\ 0 \\ 1 \\ 0 \end{bmatrix} =$$

$$\frac{1}{\sqrt{2}}(|00\rangle + |10\rangle). \tag{10}$$

The result of the computation shows the expected result, namely that the first qubit is transformed to $H|0\rangle$, while the second qubit remains unchanged.

3 Quantum Authenticated Key Distribution

In cryptography, key distribution is the process whereby two parties reach an agreement on the value of a secret key. Several protocols exist in the quantum cryptography literature for the distribution of quantum keys [1, 2, 10]. These protocols achieve a higher confidence in the key's secrecy than classical methods. To date, quantum key distribution algorithms have used two communication media: a quantum channel, with quantum bits, and a classical channel, carrying classical information. The classical channel needs to be authenticated.

The algorithm presented here improves the quantum key distribution in two ways. First, there is no classical communication channel. Communication between the two parties is done solely via one insecure quantum channel. Secondly, authentication is done by the quantum algorithm itself, using two public keys. This is essentially different from previous algorithms, where authentication was done exclusively by classical means or by a trusted authority. It was previously believed that authentication is impossible by quantum means only [12]. Our protocol proves the opposite.

3.1 Public Keys: Classical Versus Quantum

There is no doubt that public key cryptosystems dominate cryptographic applications today. Their aim is to allow exchanging secret messages reliably and secretly. Public key cryptosysems offer commercially satisfactory security levels. Formally, the problem to be solved cryptographically can be formulated as two entities, Alice and Bob, that want to exchange secret messages on a classical insecure channel. A malevolent third party, Eve, may take advantage of the insecurity of the channel and listen to the message or tamper with its content. The security of the public key cryptosystem relies on the difficulty of inverting particular algebraic functions, also called "one-way" functions.

3.1.1 Protected Public Keys

Secure communication is achieved using two types of keys: a public key and a private key. If Bob wants to send a secret message to Alice, he uses the public key of Alice to encrypt the message. Alice then reads the message after using her private key for decryption. There are a few very important characteristics of the two keys implied in this communication. Alice's private key is secret, and not shared with anybody else. In particular, Bob does not need to know Alice's private key. This is a major advantage, as the private key is never seen on any communication channel and therefore, its secrecy is ensured.

By contrast, Alice's public key is available to anybody. Bob needs to know it, and also the eavesdropper, Eve, has access to it. In order for the protocol to work, the public key is guaranteed to be protected. This means, there is a consensus about the public key value. Both Bob and Alice are sure that they use the correct, same public key. Eve cannot masquerade as Alice and change the value of Alice's public key, making Bob use a false public key to encrypt his message. This feature is crucial for a public key cryptosystem to work. The public key cryptosystems need the public key to be protected, and accept it as given that such a protection of the public key is practically possible. Current public key algorithms, such as the RSA [18], need to continuously increase the length of the protected public key in order to maintain acceptable security levels.

Our quantum key distribution protocol also relies on the protectedness of public keys. The public keys used in our algorithm are regular binary numbers, but differ in meaning from the conventional public key, such as the RSA key. We will call the public keys used in our quantum algorithm *quantum generated public keys*. Alice has a protected quantum generated public key and Bob has another protected quantum generated public key. In fact these two public keys are the only protected information exchange between Alice and Bob. Exactly as in the case of classical public key cryptosystems, our algorithm requires that such public keys can be published protectedly, with the guarantee that the keys' values *are and remain* protected from masquerading. As will be seen from the algorithm itself, besides having public keys, Alice and Bob share only an insecure quantum channel.

3.2 Quantum Key Distribution Algorithms

The security of the classical public key RSA cryptosystem relies on the theoretically unproven assumption that factoring large numbers is intractable on classical computers. As described in [17], quantum computers can break some of the best public key cryptosystems.

Quantum cryptography aims to design mechanisms for secret communication with higher security than protocols based on the public key approach. Privacy of a message and its credibility is well satisfied in a private key cryptosystem setting. Alice and Bob share one and the same secret key, k_s. Bob uses the secret key for encryption and Alice consequently decrypts the message with the same key. As long as k_s is unknown to anybody else, the secrecy of the communication is satisfied. There exist various encryption/decryption functions using k_s, such that the encrypted message reveals no information whatsoever about the content of the message, provided the key k_s is unavailable.

Generally, QKD protocols involve two stages. The first one is usually a one-way communication (from Alice to Bob) over a quantum channel. In this stage, a random sequence of bits generated by Alice is transmitted over to Bob, each 0 or 1 encoded in some quantum observable (photon polarization is the natural choice). Having measured each incoming qubit in one of the pre-defined bases, Bob must now communicate with Alice over a public channel to exchange information about the encoding of each qubit and eventually agree upon a common secret key. This two-way communication between Alice and Bob over a classical channel represents the second stage of QKD protocols. The above two-stage scenario forms the backbone of all schemes developed so far in order to distribute classical keys through quantum means. They differ only in the particular quantum mechanical features or principles chosen to achieve their goal.

The first quantum protocol for key distribution was developed in 1984 by Charles Bennett and Gilles Brassard and is hence known as BB84 [2]. It is characterized by the use of two quantum alphabets (orthonormal bases) for encoding and decoding the bits transmitted. One consists of the vertical and horizontal polarization states of photons, while the other orthonormal basis corresponds to polarization directions formed respectively by 45° clockwise and counter-clockwise rotations off from the vertical. The convention used for the two quantum alphabets could be

$$\begin{cases} \text{``0''} = | \rightarrow \rangle \\ \text{``1''} = | \uparrow \rangle \end{cases}$$

in the case of the vertical/horizontal basis, and

$$\begin{cases} \text{``0''} = | \nearrow \rangle \\ \text{``1''} = | \nwarrow \rangle \end{cases}$$

for the oblique basis. In the first stage of BB84, Alice randomly chooses one of these two agreed-upon quantum alphabets for each bit transmitted. At the receiving end, Bob also selects one basis, at random, to measure each incoming photon and decode the bit carried. By comparing the alphabet used for encoding with that used for decoding, in the second stage of the protocol, Alice and Bob can reach an agreement for a common binary substring called the *raw key*, by keeping only those bits for which the encoding and decoding basis was the same and discarding all the others (roughly half of the total number of bits transmitted). Figure 19 illustrates this process.

Using a pair of conjugate (incompatible) observables, the BB84 protocol relies on Heisenberg's uncertainty principle coupled with the inevitable disturbance caused by quantum measurements to detect potential eavesdroppers. On average, 25 % of the photons that Eve (the prototypical eavesdropper) chooses to tamper with will give rise to disagreements between Alice's raw key and Bob's raw key. Things get more

Fig. 19 Quantum key distribution in the absence of eavesdropping

Alice	0	1	1	0	0	1	0	0	0	1	0	0	1	1	0
	×	+	×	+	+	+	×	×	+	+	×	+	×	×	×
	↗	↑	↖	→	→	↑	↗	↗	→	↑	↗	→	↖	↖	↗
Bob	+	+	×	+	×	×	+	×	×	+	+	+	×	+	×
	1	1	1	0	0	1	1	0	1	1	0	0	1	1	0
key		1	1	0				0		1		0	1		0

complicated when such disagreements can also be the result of imperfections or noise in the quantum channel. Consequently, Eve could adopt the strategy of gaining only partial knowledge about the key by trying to hide behind the noise level. To cope with such low levels of eavesdropping, Bennett et al. [4] have proposed the method of *privacy amplification*, a mathematical technique based on the principle of hashing functions that magnifies Eve's uncertainty over the final form of the key.

Using pairs of orthogonal polarization states as quantum alphabets for the transmitted bits is not a necessary condition. Bennett showed [1] that any two non-orthogonal quantum states can be used to achieve key distribution in a practical interferometric realization using low-intensity coherent light pulses. The protocol (known as B92) needs only one quantum alphabet, but with non-orthogonal polarization states. Therefore, Bob must be equipped with a POVM (positive operator value measure) receiver in order to interpret the incoming photons properly. As in the case of BB84, eavesdropping attempts are made apparent by an unusual error rate in Bob's raw key. Specific to B92 is the possibility of detecting eavesdroppers by an unusual erasure rate (inconclusive receptions) for Bob.

The protocols that offer the best security, at least from a theoretical viewpoint, are based on entanglement (EPR pairs). Inspired by EPR experiments designed to test Bell's inequality, Artur Ekert thought of a way of using entangled pairs for distributing cryptographic keys by quantum means [10]. In the first stage of his scheme, Alice and Bob receive entangled particles from a central source and perform independent measurements upon them. The shared secret key is established in the second stage, when Alice and Bob publicly confront the orientations they adopted for each measurement.

Similarly to BB84, the key will consist of only those bits that were measured in the same basis by both participants. Unlike the BB84 protocol, however, the remaining bits are not discarded, but the strength of their correlations is used to test for eavesdroppers. These correlations must exceed anything that is possible classically, according to Bell's theorem, if the original EPR pairs were untampered with. A related, but simpler EPR cryptographic scheme was described by Bennett et al. [3] that is proved secure without the need to invoke Bell's theorem. They also show the equivalence between their scheme and the original BB84 key distribution protocol.

Protocols resorting to EPR pairs offer a qualitatively new level of security, that becomes apparent by considering the scenario in which someone attempts to make measurements on the particles before they arrive at the legitimate receiver. For an entanglement-free protocol, such an eavesdropping strategy aims at gaining knowledge of the information encoded in the qubits transmitted. But in the case of schemes based on EPR pairs, Eve cannot elicit any information from the transiting particles simply because there is no information encoded there. The information about the secret key has yet to come into being once Alice and Bob perform their measurements.

Another advantage of entanglement-based schemes refers to the issue of privacy amplification. The limitations of the classical privacy amplification based on hashing algorithms are overcome in the quantum privacy amplification technique developed in 1996 [9]. The quantum procedure, which is applicable only to-based quantum cryptography, is in fact an entanglement purification process that can be repeatedly applied to impurely entangled particles to cleanse them of any signs of tampering by Eve.

However, these advantages of entanglement-based cryptography are rather theoretical at the moment because storing entangled particles is only possible for a fraction of a second as yet, and entanglement purification depends on quantum computational hardware that, although simple, has yet to be built. In contrast, implementations of the original BB84 protocol are well within the capabilities of current technology, reaching the point where they have become commercially viable.

Note that all quantum key distribution algorithms mentioned above require that the classical channel be authenticated. Authentication is supposed to be done by classical means. The authenticated classical channel prevents Eve from masquerading as someone else and tamper with the communication. The general view in quantum cryptographic literature (see for example [12]) is that authentication is not possible through quantum means and consequently, for any secure quantum communication a classical authentication scheme needs to be used.

As will be clear from the algorithm described in this section, authentication of a quantum communication protocol is not only possible by quantum means only, but in fact a classical channel is superfluous. The general authentication scheme has been developed in [15]. In this previous result, classical communication was still needed, though the classical channel was not authenticated. In the present improved version of the protocol, the classical channel is removed completely. The robustness of the algorithm comes also from the simplicity of the communication support available. Alice and Bob share an insecure quantum channel and two quantum generated public keys. They have an authentication step at the end of the protocol, with the help of the quantum generated public keys. Note that authentication in our algorithm is done at the end of the protocol and is derived from the quantum algorithm itself.

Shi et al. [20] also describe a quantum key distribution algorithm that does not use a classical channel. Authentication is done by a trusted authority, that provides the entangled qubits to Alice and Bob. In our protocol, such a trusted authority is not needed. The entangled qubits may come from an insecure source.

3.3 Entangled Qubits

The key distribution algorithm we present in the following subsection relies on entangled qubits. Alice and Bob, each possess one of a pair of entangled qubits. If one party, say Alice, measures her qubit, Bob's qubit will collapse to the state compatible with Alice's measurement. The qubit pair is in one of the four Bell states:

$$\frac{1}{\sqrt{2}}(|00\rangle \pm |11\rangle)$$

$$\frac{1}{\sqrt{2}}(|01\rangle \pm |10\rangle)$$

Suppose Alice and Bob share a pair of entangled qubits described by the first Bell state:

$$\frac{1}{\sqrt{2}}(|00\rangle + |11\rangle)$$

Alice has the first qubit and Bob has the second. If Alice measures her qubit and sees a 0, then Bob's qubit has collapsed to $|0\rangle$ as well. Bob will measure a 0 with certainty, that is, with probability 1. Again, if Alice measures a 1, Bob will measure a 1 as well, with probability 1. The same scenario happens if Bob is the first to measure his qubit.

Note that any measurement on one qubit of this entanglement collapses the other qubit to a *classical* state. This property is specific to all four Bell states and is then exploited by the key distribution algorithms mentioned above: If Alice measures her qubit, she *knows* what value Bob will measure. The entanglement employed by the algorithm described in this section, however, does not have this property directly.

3.3.1 Entanglement Caused by Phase Incompatibility

Let us look now at an unusual form of entanglement. Consider the following ensemble of two qubits:

$$\phi = \frac{1}{2}(-|00\rangle + |01\rangle + |10\rangle + |11\rangle)$$

The ensemble has all four components, $|00\rangle$, $|01\rangle$, $|10\rangle$, and $|11\rangle$, in its expression. And yet, this ensemble is entangled.

Consider the following proof. Suppose the ensemble ϕ is not entangled. This means ϕ can be written as a scalar product of two independent qubits:

$$\phi = \frac{1}{2}(\alpha_1|0\rangle + \beta_1|1\rangle)(\alpha_2|0\rangle + \beta_2|1\rangle)$$

Matching the coefficients from each base vector, we have the following conditions:

1. $\alpha_1\alpha_2 = -1$
2. $\alpha_1\beta_2 = 1$
3. $\alpha_2\beta_1 = 1$
4. $\beta_1\beta_2 = 1$.

The multiplication of conditions 1 and 4 yields: $\alpha_1\alpha_2\beta_1\beta_2 = -1$. On the other hand, from conditions 2 and 3, we have: $\alpha_1\alpha_2\beta_1\beta_2 = 1$. This is a contradiction.

The product $\alpha_1 \alpha_2 \beta_1 \beta_2$ cannot have two values, both $+1$ and -1. It follows that ϕ cannot be decomposed and thus the two qubits are entangled.

The entanglement of the ensemble is caused by the *signs* in front of the four base vector components. Thus, it is not that some vector is missing in the expression of the ensemble, rather it is the phases of the base vectors that keep the two qubits entangled. Let us investigate what happens to the ensemble ϕ, when the entanglement is disrupted through measurement.

If the first qubit q_1 is measured and yields $q_1 = |0\rangle = 0$ then the second qubit collapses to $q_2 = \frac{1}{\sqrt{2}}(-|0\rangle + |1\rangle)$. This is not a classical state, but a simple Hadamard gate transforms q_2 into a classical state.

Applying the Hadamard gate to an arbitrary qubit, we have $H(\alpha|0\rangle + \beta|1\rangle) = \alpha\frac{|0\rangle+|1\rangle}{\sqrt{2}} + \beta\frac{|0\rangle-|1\rangle}{\sqrt{2}}$. For our collapsed q_2, we have $H(q_2) = H(\frac{1}{\sqrt{2}}(-|0\rangle + |1\rangle)) = -|1\rangle$. This is a classical 1.

The converse happens when qubit q_1 yields 1 through measurement. In this case, q_2 collapses to $q_2 = \frac{1}{\sqrt{2}}(|0\rangle + |1\rangle)$. Applying the Hadamard gate transforms q_2 to $H(q_2) = H(\frac{1}{\sqrt{2}}(|0\rangle + |1\rangle)) = |0\rangle = 0$. Again this is a classical state 0.

It follows that by using the Hadamard gate, there is a clear correlation between the measured values of the first and second qubit. In particular, they always have opposite values.

A similar scenario can be developed, when the second qubit q_2 is measured first. In this case, the first qubit q_1, transformed by a Hadamard gate, yields the opposite value of q_2.

3.4 The Algorithm

The goal of the key distribution algorithm described below is to establish a secret key, known only to Alice and Bob. Subsequently, when Alice and Bob exchange messages, they will use this key to encrypt/decrypt their messages. One session is required to establish a binary secret key, called *secret*, such that Alice and Bob are in consensus about the value of the secret key. The secret key *secret* consists of n bits, $secret = b_1 b_2 \ldots b_n$. Technically, to perform the algorithm, Alice and Bob need an array of entangled qubit pairs, and two protected public keys. Note that Alice and Bob do not communicate on *any* classical channel.

The array of the entangled qubits has length l, it consists of l qubit pairs denoted $(q_{1A}, q_{1B}), (q_{2A}, q_{2B}), \ldots, (q_{lA}, q_{lB})$. The array is split between Alice and Bob. Alice receives the first qubit of each entangled qubit pair, namely $q_{1A}, q_{2A}, \ldots, q_{lA}$, and Bob receives the second half of the qubit pairs, $q_{1B}, q_{2B}, \ldots, q_{lB}$. The entanglement of a qubit pair is of the type described in the previous subsection, namely, phase incompatibility. The array of qubits is unprotected. There is no guarantee that the qubits of a pair are indeed entangled; indeed, Eve may have disrupted the entanglement. Also, Eve may have masqueraded as either Alice or Bob, modifying the entangled qubits, such that Alice's qubit is actually entangled with a qubit in Eve's possession rather than Bob's, and the same holds for Bob. In case Eve has disrupted the entanglement or has masqueraded, any result of the algorithm is discarded and the key distribution is attempted all over again, from the beginning.

The size n of the secret key is less than half of the length l of the initial qubit array, $n < \frac{l}{2}$. Indeed, $\frac{l}{2}$ qubits, that is half of the qubits, are discarded because the bases in which Alice and Bob measure are inconsistent 50 % of the time. From the remaining half of qubits a further arbitrary number is sacrificed for security checking. The number of qubits thus sacrificed depends on the desired degree of security.

Two public keys are needed by the algorithm. Alice has a public key key_A and Bob has a public key key_B. The two public keys key_A and key_B are independent. Alice and Bob use these public keys to exchange classical binary information and also, very importantly, for authentication. The keys, as used in this algorithm, have some characteristics that are different from the classical public keys. The keys are established *during* the computation. They are not known prior to the key distribution algorithm and are defined in value during the computation according to the measured values of some of the qubits. This means that the keys are available *after* the key distribution protocol. Consequently, the keys have to be posted after the algorithm, which is unlike the classical case, where a public key is known in advance.

Also, the two public keys key_A and key_B are valid for one session, that is, for one application of the key distribution algorithm. If Alice and Bob want to distribute a second secret key using the same algorithm, they will have to create new public keys, which are different in value from the public keys of the previous session.

The key distribution algorithm, like all quantum key distribution algorithms, develops the value of the secret key during the computation. Implicitly, the values of the public keys as well are developed *during* the computation. There exists no knowledge whatsoever about the values of the keys (secret and public) prior to running the algorithm.

Both Alice and Bob follow the same steps briefly denoted below:

1. **Measure your entangled qubits**
2. **Compute your own public key and post it**
3. **Read your partner's key and check for eavesdropping**
4. **Construct the value of the secret key**

A detailed description of the algorithm follows.

Step 1

Alice works with the array of qubits $q_{1A}, q_{2A}, \ldots, q_{lA}$. Binary information is rendered by the results of measuring. All measurements are performed in the standard computational basis. Alice has two options for processing her qubits. She either measures a qubit directly, or she transforms the qubit by a Hadamard gate and measures afterwards. For each qubit, q_{iA}, Alice decides randomly on one of the two processing options. Notably, there is no communication with Bob at this stage. To look at a concrete example, suppose Alice has 10 qubits $q_{1A}, q_{2A}, \ldots, q_{10A}$. Qubits q_{iA} transformed by the Hadamard gate are denoted Hq_{iA}; for those measured directly the notation is unchanged. Suppose that by random choice, Alice has processed her qubits as follows:

$$q_{1A}, Hq_{2A}, Hq_{3A}, q_{4A}, q_{5A}, q_{6A}, Hq_{7A}, Hq_{8A}, q_{9A}, q_{10A},$$

and suppose again, she has measured the following binary values:

$$1, 1, 1, 0, 0, 0, 0, 1, 1, 1$$

In the meantime, Bob processes his qubits $q_{1B}, q_{2B}, \ldots, q_{10B}$ following the same policy. He too, has a random choice on each qubit: to measure directly or to measure after a Hadamard transformation. Suppose again, that by random choice, Bob has obtained the following array:

$$Hq_{1B}, Hq_{2B}, q_{3B}, Hq_{4B}, q_{5B}, q_{6B}, q_{7B}, Hq_{8B}, Hq_{9B}, q_{10B},$$

with the values
$$0, 1, 0, 1, 1, 0, 1, 0, 0, 1$$

We have seen in the previous subsection that two entangled qubits $q_{iA}q_{iB} = \frac{1}{2}(-|00\rangle + |01\rangle + |10\rangle + |11\rangle)$, consistently render opposite classical bit measurements, if and only if exactly one qubit is measured directly and the other is measured after a Hadamard transformation. It is of no consequence whether the first qubit is Hadamard transformed or the second. The order of the qubits is irrelevant, the important issue is that exactly one of the qubits is passing a Hadamard gate. Thus, there are two "valid" measurement options:

1. q_{iA}, Hq_{iB} and
2. Hq_{iA}, q_{iB}

These measurement scenarios are valid in the sense that they, and only they, yield opposite classical bits after measurement. Each of Alice and Bob knows with certainty the value the other person has measured. Such qubits are considered valid by Alice and Bob and will be used to form the secret key and to check for eavesdropping.

Measurements of the form

3. q_{iA}, q_{iB} and
4. Hq_{iA}, Hq_{iB}

cannot be used by Alice and Bob. For any value measured by Alice, the value measured by Bob is still determined probabilistically. Qubits measured according to these scenarios, will unfortunately have to be discarded. As the scenarios 1, 2, 3, 4 are equally likely, 50% of the initial qubits will be discarded because of probabilistically inconsistent measurements.

As mentioned, half of the l qubits are discarded because of incompatible measurement bases. The size n of the secret key is therefore $n < \frac{l}{2}$. From the remaining qubits, depending on the desired security level, some other qubits are sacrificed for checking.

For the example of the 10 qubits given above, there are five valid qubit-pairs:

$$(q_{1A}, Hq_{1B}), (Hq_{3A}, q_{3B}), (q_{4A}, Hq_{4B}), (Hq_{7A}, q_{7B}), (q_{9A}, Hq_{9B}),$$

carrying the values
$$(1, 0), (1, 0), (0, 1), (0, 1), (1, 0)$$

Step 2

At this point Alice has no idea what measuring option Bob has employed on his qubits. She does not know that qubits 1, 3, 4, 7, and 9 are valid. Bob is in the same situation.

Therefore, Alice will publish her measuring strategy in her public key. Alice has measured $l = 10$ qubits. As such, the first l bits of Alice's public key explain which qubits have been Hadamard transformed and which were measured directly. If Alice has applied the Hadamard gate on qubit q_iA then the i-th qubit of the public key is set to 1, $key_A(i) = 1$. Otherwise, if q_{iA} has been measured directly, then the i-th qubit is 0, $key_A(i) = 0$. For the example of 10 qubits, the first ten bits of Alice's public key are

$$key_A(1..10) = 0110001100$$

The second part of Alice's public key is used for security checking. A certain fraction f, for example $f = 40\%$, of the original qubits are made public for Bob to check for eavesdropping. Alice chooses randomly 40 % of her l qubits. For each chosen qubit, Alice publishes the index of the qubit and the binary value she has measured. To continue our example, Alice chooses randomly the indices 1, 2, 9, 10. She will publish index 1 with value 1, index 2 with value 1, index 9 with value 1 and index 10 with value 1. Translated in binary this is

$$(0001)1(0010)1(1001)1(1010)1$$

Alice's final public key is the concatenation of the measuring (Hadamard/no Hadamard) information and the qubit checking information:

$$key_A = 0110001100 \quad 0001\ 1\ 0010\ 1\ 1001\ 1\ 1010\ 1$$

The length of the public key depends on the length l of the qubit array and also on the desired security level given by the fraction f. The following formula computes the length of the key:

$$length(key_A) = l + f(1 + \log l)$$

Here, l, the first term in the sum, refers to the measuring strategy; the second term, $f(1 + \log l)$, represents the part that publishes the qubits for eavesdropping checking.

Bob creates his public key following exactly the same steps. Bob's measuring strategy is encoded at the beginning of his public key. For our example, this means

$$key_B(1..10) = 1101000110$$

Suppose Bob sacrifices qubits 1, 5, 7, 8 for checking. In his public key he will publish $(0001)0(0101)1(0111)1(1000)0$. Thus, Bob's final key, the one that Alice and indeed everybody can see, is:

$$key_B = 1101000110 \quad 0001 \quad 0 \quad 0101 \quad 1 \quad 0111 \quad 1 \quad 1000 \quad 0$$

Both Alice's and Bob's keys, key_A and key_B are made public and are available to everybody, including Eve.

Step 3

At this stage, Alice and Bob, in full knowledge and consensus of each other's keys, will proceed to check for eavesdropping. Alice is looking at Bob's public key key_B and learns the values Bob has measured on the randomly sacrificed $f = 40\%$ of his qubits, here qubits 1, 5, 7, 8. Because of the various measuring options, only half of the $f = 40\%$ qubits will be useful. In our example, qubits 1 and 7 are measured with correct options, namely exactly one Hadamard gate applied to an entangled pair. Alice can find out the valid qubits by XOR-ing the measuring strategy of Bob with her own:

$$(0110001100)XOR(1101000110) = (1011001010)$$

which means qubits 1, 3, 4, 7, 9 have been measured well. Alice is left only to compare the values of qubits 1 and 7 she has measured with the values posted by Bob. With no malevolent interference, the binary values are opposite. Thus, if these values are opposite, Alice concludes that the protocol was not influenced by Eve. Otherwise, Alice discards all information and starts all over again. Bob performs the same checking. He will find the valid qubits posted by Alice 1 and 9 and will compare Alice's binary measured values with his own. Thus Bob makes his own independent decision concerning eavesdroppping. For reasonably large qubit arrays and a reasonably large number of qubits checked, Alice and Bob will reach the same conclusion concerning the validity of the measured binary data. This conclusion effectively implies the absence of eavesdropping/masquerading (assuming, of course, that the qubits were initially entangled).

Step 4

At this stage, the possibility of eavesdropping has already been eliminated. The qubits that have not been published by Alice or Bob in their public keys continue to be unknown to anybody else. These unpublished qubits form the secret key *secret*, that is, *secret* will be formed from Alice's recorded values, and Bob's complementary values. In our ten qubit example, valid unpublished qubits are qubits 4 and 9. Therefore, the secret key will be Alice's qubits 4 and 9:

$$secret = 01$$

Bob has to complement his qubits to reach the same value as Alice.

The size (length) n of the secret key depends on the initial length of the qubit array l, as well as the fraction of discarded qubits f. Alice and Bob have decided randomly which qubits to publish. In the worst case, the set of qubits published by Alice is disjoint from the set published by Bob. Thus, the fraction of unpublished qubits is $1 - 2f$. From these unpublished qubits, only half (50 %) are measured correctly. The length of the secret key is given by the formula

$$n = (1 - 2f)\frac{1}{2}l$$

For our example

$$n = (1 - 2\frac{40}{100})\frac{1}{2}10 = 1$$

The length of the secret key is 1 in the worst case. For our particular example we could use 2 bits.

3.5 Security Evaluation or Catching the Evil Eavesdropper

Let us consider the algorithm described in the previous subsection, from the point of view of the eavesdropper Eve. Eve wants to ideally gather knowledge about the value of the secret key without being noticed by either Alice or Bob. It is well known that an entangled qubit pair reveals no information whatsoever unless the qubits are measured and the entangled state collapses. Even so, the algorithm presented in this section supposes that the entanglement is not protected, only the public keys are protected. This means that the qubits are not guaranteed to be entangled. Eve may masquerade and distribute qubit arrays of her own choice. It is of no advantage to Eve to distribute entangled qubits, as she gains no knowledge about the future secret key from unmeasured entangled qubits. The best choice for Eve is to distribute classical bits, or independent qubits in a known state.

The best Eve can do is to give Alice an array of classical 0s:

$$q_{1A}q_{2A} \ldots q_{lA} = 00 \ldots 0$$

and to Bob an array of $H1$:

$$q_{1B}q_{2B}\dots q_{lB} = H1\,H1\dots H1$$

All other possible arrays Eve could send to Alice and Bob are equivalent or less advantageous than the arrays above. In particular, Eve will want to send any pair (q_{iA}, q_{iB}) that *can* be measured correctly: $(0, H1)$, $(H0, 1)$, $(1, H0)$, or $(H1, 0)$. Any such pair is equally advantageous. For simplicity we will discuss the arrays of 0s and $H1$s, respectively. For a pair $(0, H1)$, Alice and Bob apply randomly one of the four measurement options (see Sect. 3.4). The first correct measurement option (q_{iA}, Hq_{iB}) consistently yields complementary correct results, namely $(0, 1)$. The second correct measurement option (Hq_{iA}, q_{iB}) yields all four possible classical bit combinations $(0, 0)$, $(0, 1)$, $(1, 0)$, and $(1, 1)$. Moreover, these combinations are equally likely. In one-half of the cases, measurements will be $(0, 0)$ or $(1, 1)$. This cannot happen, if the qubits are entangled and untouched. This situation reveals the intervention of Eve. Thus, on any qubit checked for eavesdropping, there is a $\frac{1}{4} \times \frac{1}{2} = \frac{1}{8}$ chance of detecting Eve.

As Alice and Bob respectively check a fraction f of the original array, the expected number of times Eve is detected, that is, the *expected detection rate*, is

$$expected\_detection\_rate = \frac{1}{8} \times f \times l$$

For our example, the expected detection rate is

$$expected\_detection\_rate = \frac{1}{8} \times \frac{40}{100} \times 10 = \frac{1}{2} = 50\%$$

Eve is caught 50 % of the time. This expected detection rate is rather low given the toy example we have considered, but of course it can be increased arbitrarily by increasing f and/or l.

Suppose we have an array of 1024 qubits and work with the same fraction $f = \frac{40}{100}$. In this case, the length of the final key is

$$n = (1 - 2\frac{40}{100})\frac{1}{2}1024 \approx 100$$

This is a length that can be used in practice.

The number of qubits checked by Alice (and also by Bob) is

$$checked\_qubits = \frac{1}{2} \times \frac{40}{100} \times 1024 = 204.8$$

On each qubit, Eve can escape being caught with probability $\frac{3}{4}$. Thus Eve can escape with probability $\frac{3}{4}^{204.8} = 3.25 \times 10^{-26}$. This probability is infinitesimal for any practical purposes.

The algorithm presented above shows clearly that authentication can be done by quantum means only. Besides an insecure quantum channel, Alice and Bob have only protected quantum generated keys to communicate with. The parallel with the classical authentication scheme is simple. In classical authentication, Alice and Bob have

1. an insecure classical channel and
2. one or two standard protected public keys, posted before any communication on the channel.

In the quantum authentication scheme presented in this section, Alice and Bob equivalently have two items:

1. an insecure quantum communication channel, and
2. two quantum generated protected public keys.

An important difference concerning the two types of public keys, classical and quantum generated, is that the value of a quantum generated public key is developed during the computation and posted after any communication on the quantum channel is performed. Therefore, the quantum generated public keys depend on the specific communication session. They are not known prior to the execution of the key distribution algorithm and differ in value from one session to the next. This mirrors the behavior of the secret key to be established by the key distribution protocol. In all quantum key distribution protocols, the secret key is developed *during* the execution of the protocol.

The algorithm presented performs quantum key distribution based on entangled qubit pairs. The entanglement type is not of the generally used Bell states, but an unusual entanglement based on phase incompatibility. The advantage of this type of entanglement is that Alice and Bob perform *different* measurement steps: one is measuring the qubit directly, and the other is measuring after applying a Hadamard gate. Therefore, the measurement is not symmetric. This property, combined with random choice on the measurement steps leaves Eve with no knowledge of how to measure a tampered qubit in advance. How other protocols and algorithms may benefit from asymmetric measurement is an open problem.

The principle of checking and authenticating at the end of the protocol with quantum generated public keys, is not restricted to the algorithm described here. The same type of public keys, generated per session, posted after the execution of the main body of the algorithm, can be successfully used in authenticating other types of algorithms. This is also a direction worth investigating.

If entangled qubits are easily available, the secret key established by the algorithm can be arbitrarily long. Our algorithm can distribute a "one time pad" [23] without Alice and Bob having to meet. To use one time pads, traditionally, Alice and Bob meet in secret and exchange a long list of keys, each as long as the message it is supposed to encrypt, and each to be used exactly once.

4 One-Time Pads Without Prior Communication

If Alice and Bob *share* a set of long secret keys, the secrecy of their communication is guaranteed. They use each key for exactly one communication and the key is discarded afterwards. Needless to say, the secret keys are independent, randomly generated and thus, one key does not reveal any information about any other key. In the cryptographic world, these keys are usually referred to as one-time pads [19]. Any encryption/decryption function will be good enough to ensure secrecy, such as for example, a binary XOR of the message text with the one-time pad (assuming the text and key are binary strings). The only condition on the key is to be at least as long as the message itself, so that there are no repetitions of the key's usage.

The obvious drawback of any scheme with one-time pads is that Alice and Bob need a prior reliable agreement on the value of the secret keys. In practice, to date, the only viable solution to reaching a consensus and keeping the secrecy of the keys is for Alice and Bob to meet in advance. They need to have a secure, private meeting in which they agree on the value of *all* secret keys to be used henceforth. If, after communicating for some time, they run out of keys, Alice and Bob need another secret meeting. The basic idea is that for *any* one-time pad that Alice and Bob use, there existed a prior secret meeting of Alice and Bob and in this meeting the value of the one-time pad was defined.

By endowing messages with quantum properties, we show that encryption and decryption can be done with one-time pads and the value of the one-time pads are generated without Alice and Bob having to meet. Thus, messages are quantum messages or at least partially quantum. A message (see Fig. 20) consists of two concatenated arrays. The message header is the first part and is an array of qubits. The header renders the value of the one-time pad. The message body may be an array of classical bits or again an array of qubits. We will discuss both options. The body contains the information to be transmitted in an encrypted form.

4.1 Reading Masks

As mentioned in the introduction, the value of the one-time pad is carried by the array of qubits in the header. Some of these qubits are in a basic state, either 0 or 1,

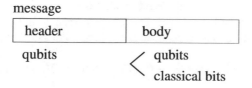

Fig. 20 A message consists of two concatenated parts. The header is an array of qubits. The body may be an array of classical bits or an array of qubits as well

while the rest of the qubits are in a balanced superposition of 0 and 1. We say that the (binary) value of the one-time pad is quantum encrypted. For each classical bit b_i of the pad, there exist four possible encryption options, resulting into the qubit q_i. These options are:

1. $q_i = b_i$. The qubit carries the exact value of the classical bit.

   - if $b_i = 0$ then $q_i = 0$.
   - if $b_i = 1$ then $q_i = 1$.

2. $q_i = NOT\ b_i$. The qubit carries the value of the complement of the classical bit.

   - if $b_i = 0$ then $q_i = 1$.
   - if $b_i = 1$ then $q_i = 0$.

3. $q_i = Hb_i$. The qubit is a superposition, obtained by applying the Hadamard gate on b_i.

   - if $b_i = 0$ then $q_i = H0 = \frac{1}{\sqrt{2}}(|0\rangle + |1\rangle)$.
   - if $b_i = 1$ then $q_i = H1 = \frac{1}{\sqrt{2}}(|0\rangle - |1\rangle)$.

4. $q_i = H\ NOT\ b_i$. The qubit is a superposition obtained by applying the Hadamard gate to the complement of b_i.

   - if $b_i = 0$ then $q_i = H\ NOT\ 0 = \frac{1}{\sqrt{2}}(|0\rangle - |1\rangle)$.
   - if $b_i = 1$ then $q_i = H\ NOT\ 1 = \frac{1}{\sqrt{2}}(|0\rangle + |1\rangle)$.

An array that explains how to encrypt each b_i, or conversely, explains how to read each q_i is called an *encryption/decryption mask* or simply a *reading mask*. Let us see how the reading mask works on an example.

Suppose the secret key is

$$secret = 00111001$$

It is eight bits long. The reading mask has to have the same length. The value of the reading mask is independent of the value of the secret key. Consider the mask to be

$$mask = *\ H\ *\ (H\ NOT)\ NOT\ H\ H\ *,$$

where $*$ means read directly and the other notations are self explanatory.

To obtain the quantum encrypted version of the secret key, each bit needs to be transformed as defined by that position in the mask. For our particular example:

$$secret\_encrypt = 0\ H0\ 1\ H0\ 0\ H0\ H0\ 1$$

4.2 One-Time Pad Communication with Classical Message

It is clear by now that Alice and Bob can exchange messages containing quantum encrypted keys provided that they share the reading mask. We consider that Alice and Bob do have a "meeting point" before they start an indefinite exchange of messages. This means that they meet and agree on a reading mask (see Fig. 21), or they may develop a secret reading mask using any quantum key distribution algorithm. Another way of viewing the sharing of the reading mask is to consider Alice and Bob as two devices, rather than two people. In this case the secret reading mask is given to both Alice and Bob before their deployment.

In addition to a secret reading mask, Alice and Bob have to share some secret information to be used for witnesses to catch the intruder. In particular, for every witness qubit two pieces of information are needed: an index describing its position in the header and the binary value it carries.

Once Alice and Bob share the reading mask, they can exchange messages. Either Alice or Bob can initiate such a message. Suppose Alice wants to send a message to Bob. She generates a random key k to be used only once. Then she encrypts the information with k and places it in the body of the message. Subsequently, Alice quantum-encrypts k using the reading mask, places it in the header, and sends the message. Figure 22 shows the steps performed by Alice.

Bob on the other end receives both the header and the body of the message. He first decrypts the header using the secret reading mask and retrieves the one-time pad k. Then Bob decrypts the message body with the one-time pad k. Figure 23 shows the steps performed by Bob.

Fig. 21 Alice and Bob share a secret reading mask

Fig. 22 Alice takes four steps to send a message to Bob

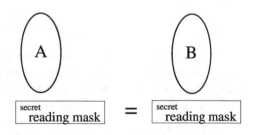

Alice

1. Generate one-time pad.

2. INFORMATION + one-time pad } classical encryption

3. ONE-TIME PAD + reading mask } quantum encrypt

| header | body |

4. Send message.

Fig. 23 Bob takes three
steps when receiving a
message from Alice

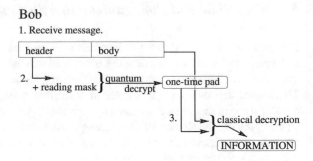

4.3 What Eve Can Do

We consider the standard setting, in which Alice and Bob are trusted. The communication channel is vulnerable to attacks from Eve. Eve may intercept a message, try to read it, change it, or send spurious messages. Suppose Eve intercepts a message. She is interested to decrypt the message body that contains the information being transmitted. Looking at the encrypted message body does not reveal anything about the decryption key. In order to get the secret key, Eve has to look at the header of the message. Because the header of the message is quantum encrypted, Eve has no way of knowing how to read the header.

The best Eve can do to retrieve the one-time pad through measurement of the head, is to guess the reading strategy, meaning that she guesses the value of the reading mask. Suppose Eve has chosen to read all qubits simply in the computational basis. This means the reading mask would be

$$mask = *\ \ *\ \ *\ldots$$

Actually all other guesses are equivalent in terms of the performance of the guess. Note that on average 25 % of the header qubits are indeed encoded with $*$, 25 % of the qubits are encoded with NOT, 25 % with H, and 25 % with $H\ NOT$. Therefore Eve's guessed key will have the following performance:

1. On the 25 % $*$ qubits, all guesses are correct.
2. On the 25 % NOT qubits, all guesses are wrong.
3. On the 25 % H qubits, half of the guesses are correct.
4. On the 25 % $H\ NOT$ qubits, half of the guesses are correct.

Overall, 50 % of the secret key bits are guessed correctly and the rest are wrong. Yet this is exactly as if Eve guesses the secret key directly, by tossing a coin, without bothering to read the header at all. It means that without any information about the reading mask, Eve cannot get any information about the secret key.

Eve's intervention, on the other hand, is detectable by Bob. If Eve has touched a witness qubit of the header of some message through reading, the state of the qubit will be detectably changed. Remember that Eve does not know whether a qubit is in a simple state or in a superposition.

Suppose the qubit is in a simple state: $|0\rangle$ or $|1\rangle$. If Eve chooses to read the qubit in the computational basis, her intervention remains hidden. Yet, if Eve chooses to rotate the qubit with a Hadamard gate first and then measures, she will have rotated the qubit into a superposition: $H0$ or $H1$. Measurement collapses the superposition to any binary value with 50 % probability. Thus, there is a 50 % probability that the qubit has collapsed to the wrong value. As Eve has a 50 % probability to choose applying an H gate, the probability of altering the value of the qubit is 25 %. A similar scenario happens if the qubit was initially in a superposition $H0$ or $H1$. If Eve decides to measure in the computational basis, the result of the measurement has a 50 % probability of being wrong. If Eve applies a H gate, measures and then rotates the qubit back with a H gate, her intervention remains hidden. Overall, Eve has altered the value of the qubit with a chance of 25 %.

For Bob to notice the change in the value of the qubit, he has to expect the qubit to have a certain value. For this, Alice has to set the witness qubits to the known values. Bob checks the witnesses and if they match the expected values, he concludes that Eve has not touched the header.

As such, any reading intervention of Eve will be of no advantage to her and will be probabilistically detectable by Bob. The probability to detect Eve's intervention grows exponentially with the number of qubits touched by Eve.

The only "information" Eve can see is the encoded message body. While the encoded version does not reveal anything about the information content, it is still a sequence of encoded information that Eve can read and copy without leaving any trace of her intervention. We will see in the next subsection that fully quantum messages will restrict Eve even in this action.

4.4 Fully Quantum Messages

Consider now that qubits are easily available and there are no technical impediments to using fully quantum messages. That is, both the header and the body of the message are arrays of qubits.

In this case, we may consider the body of the message to be quantum encrypted. As such

- the header contains the one-time pad, quantum encrypted with the secret encryption mask, and
- the body contains the intended information, quantum encrypted with the one-time pad.

When Alice wants to send a message to Bob, she goes through the same steps as before. Alice randomly generates a one-time pad that gets quantum encrypted with the secret reading mask and represents the header of the message. Then Alice quantum encrypts the information with the one-time pad. This is done by interpreting the one-time pad as a reading mask. The one-time pad is divided into groups of two bits. Every group of two bits encodes the reading strategy ($*$, NOT, H, H NOT) of one qubit. Then Alice places the encrypted message in the message body. Bob, on receiving the message performs the same steps reversed. The only difference to the partially quantum communication is that the body is quantum encrypted.

4.4.1 Discussion

Fully quantum message exchange offers at least all advantages of the partially quantum message exchange. In addition, whenever Eve attempts to read the message body, she changes the states of the superpositions of the qubits she reads. If the message body contains witness qubits as described in Sect. 4.3, Bob can detect her intervention. This means that Eve is detectable now whenever she reads a part of the message.

When Eve reads the message body, she has absolutely no benefit from the classical binary array she obtains. This array is not even some classical encryption of the information transmitted. The array Eve gets is actually not even uniquely determined, as it depends on how the superpositions collapse.

Therefore, what Eve reads is nothing but garbage and has no clear connection to the binary information to be transmitted. It is not even fully predictable and its predictability depends on Eve's reading strategy for each qubit, whether Eve measures directly or applies a Hadamard gate.

4.5 More Messages Means Safer Messages

Stepping now a few steps back for an overview of the implications of this work, we can see that the use of quantum cryptography brings about a fundamental change in the way we think about secret communication. This refers to a change in the philosophy of cryptography, of some core beliefs taken for granted. For example, one such belief of classical cryptography and quantum cryptography to date was stated as follows:

> The more Alice and Bob communicate secretly in a certain cryptographic setting, the less secure that setting becomes and it needs to be updated or reinitialized from outside of the system.

From this principle comes the advice to change passwords regularly. If a password has been used for a long time, it has been exposed to attacks all along and *some* of its content might have leaked out. Also, security systems tend to be checked after a while, in order to determine whether they still work reliably. This check is done from *outside* of the security system.

A fully quantum security system may work in a fundamentally different way. Authentication can be done with quantum cryptography [16]. An effectively secret key can be distributed using protected *public* information only [13]. Using this secret key as a reading mask, Alice and Bob may now communicate secretly with one-time pad encrypted messages. Moreover, after communicating for some time, Alice and Bob will know whether Eve was present in the slightest way during their communication, as Eve's intervention, such as simple reading is detectable. This means that the more Alice and Bob use the quantum communication system, the more their confidence in the secrecy of their communication increases. Implicitly, the amount of effectively secret information shared by Alice and Bob increases over time. This may include the change or lengthening of the reading mask, or the change of the encryption/decryption strategy. That is, Alice may communicate secretly to Bob the intention of changing the reading mask, and after Bob agrees by an equally secret message, Alice may encrypt and send a new, longer encryption mask. Note that these revisions of the security scheme are now made from the *inside* of the system. A check from the *outside* of the system is no longer needed. The following short story summarizes the quantum philosophy of cryptography:

> Alice and Bob, two cryptographic entities unknown to each other, meet in a public place. The public place may be a crowded cafeteria. Here, a formal introduction takes place. Alice and Bob present themselves with some public names, called public keys. Once the introduction is over, Alice and Bob leave the cafeteria to go each on a different life-path. From then on, Alice and Bob can communicate secretly "forever". As time passes, their "acquaintance" develops into a "friendship" and as such the trust in the privacy of their messages increases.

As this heart-warming story suggests, the quantum cryptographic system is self-sustainable and moreover builds itself up over time.

Because the one time pads are randomly generated, even if Eve would possess a series of quantum-encrypted one-time pads, this would not reveal anything about the reading mask. In addition, as Eve's intervention is detectable, Alice and Bob will know how many message headers or fully quantum messages have been captured by Eve. Therefore, Alice and Bob will know whether Eve has in her possession a series of header readings and may act accordingly.

The unprecedented advantage of using fully quantum messages is that Eve is clearly detectable in all her actions. If Alice and Bob have communicated undisturbed for some time, the amount of shared secret information increases. They may agree on a longer or a new reading mask, using their previous message exchanges only. As such, the confidence and security of the communication increases over time.

5 Quantum Key Distribution Using Disturbance Amplification

At an abstract level, a QKD protocol could be described in terms of qubits transmitted over a quantum channel. For practical implementations, the physical realization usually chosen to embody a qubit is the photon. Since it travels at the speed of light and its polarization can be easily manipulated, the photon is naturally suited for transmitting information. Still, in some cases, other realizations are equally possible, like manipulating the spin of an electron, for instance. Regardless of their possible implementations, all QKD protocols share one basic constraint: qubits are measured individually as soon as they are received. Storing the incoming qubits for later processing and/or measurement is not taken into consideration. This is quite intuitive, especially if we think about photons, which, by their nature, are made to travel and not to store information locally, in a static fashion.

This section investigates the opportunities created by the relaxation of the aforementioned constraint. More explicitly, we are interested in what benefits can be gained and at what cost, if we allow the qubits transmitted over the quantum channel to be stored for a determined amount of time by the receiving party. In the following, we motivate the feasibility of this assumption, even if the qubits are realized as photons.

Two of the main proposals for building a practical quantum computer are based on "ion traps" and cavity QED (quantum electrodynamics), respectively. In the ion trap scheme imagined by Cirac and Zoller [6], a quantum memory register would be physically realized by using "fences" of electromagnetic fields to trap a number of ions within the central region of an evacuated chamber. Each imprisoned ion embodies a qubit, with the ground state representing $|0\rangle$ and a metastable state representing $|1\rangle$. The operation of a quantum gate is effected by shining a pulse of light from a laser beam of the appropriate frequency onto the target ion. Although very simple quantum algorithms have been implemented on an ion trap quantum computer [11], the technology's main drawback remains scalability.

In the other proposal, which goes by the name of "flying qubit"-based quantum computer, quantum information is encoded in the polarization states of photons. The interaction necessary to emulate the functionality of a controlled-NOT quantum gate can be mediated by a drifting cesium atom, when the photons are placed inside a small cavity with highly reflecting walls. Quantum-phase gates based on cavity QED have been successfully realized experimentally [8, 22], yet again, it is a very challenging endeavor to extend this technology to complicated quantum circuits.

One of the ideas that emerged in order to overcome the scalability problem is a hybrid approach that combines the advantages of both ion trap and cavity QED technologies. In this approach, ion traps of limited size each would be interconnected through fiber optics, forming a quantum network. Thus, photons could be used to transfer quantum information between distant trapped atoms, while each of the multibit ion traps is responsible for storing information and local processing. The cavity QED interactions can provide the necessary methods for exchanging quantum information between the two different carriers [7]. Alternatively, the same goal can be achieved by using entanglement between a trapped atom and a photon [5].

The techniques proposed to implement a quantum network can also be applied in a cryptographic context. The qubits "flying" through the quantum channel will still be realized as photons, but whenever the receiving party wishes to store them (until it has better knowledge about their encoding, for example), the information they carry is transferred to a local ion trap quantum register, which is much more

Fig. 24 Schematics of random phase shift protocol for QKD

suited for storing information over an extended length of time. Hence, our working assumption is motivated practically by the advancements made on the way toward building a quantum computer.

The immediate benefit of storing qubits during a quantum protocol for a more "intelligent" processing/measurement is an important reduction in the communication volume required, both quantum and classical. In the case of BB84, for instance, if Bob can safely store the qubits received from Alice until the second stage of the protocol, when he is informed of the exact encoding for each of them, then an appropriate measurement can be performed for each qubit. In this way, no qubit has to be discarded due to a mismatch between the encoding and decoding alphabet. For a shared secret key of a specified length, this leads to a 50 % reduction in the total number of qubits that have to be transmitted. With fewer qubits transmitted, the volume of the classical communication in stage 2 of the protocol is reduced too. The fact that an eavesdropper may gain knowledge about the correct measurement basis for each qubit is of no advantage to her, since the qubits are no longer in her possession.

But reducing the amount of communication between Alice and Bob is not the only advantage offered by temporarily storing qubits. This possibility opens the door for designing new QKD schemes that have higher rates of intrusion detection and are therefore more secure. In the next two subsections we show explicitly how storing qubits for a limited time can be exploited to enhance security.

5.1 Random $\frac{\pi}{2}$ Phase Shift Protocol

We first describe a BB84 equivalent protocol that we will use as a building block in designing a QKD scheme based on the quantum Fourier transform. The main idea of the protocol described in this subsection is to encode each transmitted bit (0 or 1) into the relative phase between the $|0\rangle$ and $|1\rangle$ components of a balanced superposition and then encrypt the resulting qubit by applying a random phase shift gate, as depicted in Fig. 24. The Hadamard gate provides the encoding alphabet

$$
\begin{cases}
\text{"0"} \mapsto \frac{1}{\sqrt{2}}(|0\rangle + |1\rangle) \\[2mm]
\text{"1"} \mapsto \frac{1}{\sqrt{2}}(|0\rangle - |1\rangle)
\end{cases}
$$

and the R_θ gate rotates the relative phase with an angle θ

$$R_\theta = \begin{bmatrix} 1 & 0 \\ 0 & e^{i\theta} \end{bmatrix}, \ \theta \in \{0, \frac{\pi}{2}\}.$$

Note that R_0 does not affect the state of the qubit onto which the gate is applied, while $R_{\pi/2}$ rotates the qubit halfway between the two symbols of the encoding alphabet. The protocol conforms to the generic two-stage structure, as described in the following.

Random $\frac{\pi}{2}$ phase shift protocol for QKD

Stage 1:   Communication over a quantum channel

>   Step 1.   Alice flips a fair coin to generate a random binary sequence that she intends to share with Bob.
>   Step 2.   For each bit j in the sequence, Alice chooses, again at random, an angle $\theta = 0$ or $\theta = \pi/2$. She then prepares, accordingly, a qubit in the state $|\psi\rangle = R_\theta H |j\rangle$ that she sends over to Bob.
>   Step 3.   Bob applies the necessary procedures for safely storing the qubits received from Alice until the second stage of the protocol, when he gains knowledge of which qubits have been phase shifted.

Stage 2:   Communication over a public channel

>   Phase 1.   Raw key extraction
>   Step 1.   Alice informs Bob about her choice of θ for each transmitted bit.
>   Step 2.   Knowing the relative phase shift θ for each stored qubit $|\psi\rangle$, Bob recovers the original bit transmitted, by computing $|j\rangle = H R_\theta^\dagger |\psi\rangle$ and then measuring $|j\rangle$ in the normal computational basis $\{|0\rangle, |1\rangle\}$. Following this procedure, Bob obtains a binary sequence that should be identical to the one randomly generated by Alice, provided no eavesdropping or errors interfered with the quantum transmission.
>   Phase 2.   Error estimation
>   Step 1.   Over the public channel, Alice and Bob compare portions of their raw keys to estimate the error rate Err. The bits tested are deleted from their raw keys. If $Err = 0$ the remaining bits form their final secret key.
>   Step 2.   If $Err > 0$, but still sufficiently small, Alice and Bob may decide to apply privacy amplification techniques to minimize Eve's knowledge about their final secret key. Otherwise, if Err exceeds a certain threshold, they discard the whole sequence and start all over again.

The analogy with BB84 becomes apparent if we assimilate the encoding alphabet with the horizontal/vertical basis and the $\pi/2$ relative phase shift with the oblique basis. What are Eve's chances to break the above protocol and find a loophole that may allow her to elicit information about the secret key? In what follows, we analyze two main eavesdropping strategies that Eve may adopt.

5.1.1 Opaque Eavesdropping

The most straightforward way in which Eve could spy on the quantum communication between Alice and Bob would be to intercept Alice's information carriers and measure them in some appropriate basis. If she could undo the rotation (with angle θ) applied by Alice, then she could measure the intercepted qubit using the basis $\{\frac{1}{\sqrt{2}}(|0\rangle + |1\rangle), \frac{1}{\sqrt{2}}(|0\rangle - |1\rangle)\}$. Such a measurement is carried out by first passing the qubit through a Hadamard gate and then measuring it in the normal computational basis $\{|0\rangle, |1\rangle\}$.

Since Eve has no information about θ, trying to rotate the qubit back with $\pi/2$ (see Fig. 25) is in no way a better strategy than applying the Hadamard gate directly. Without loss of generality, consider what happens if the qubit intercepted by Eve encodes the bit 0 (the other case proceeds in an analogous way yielding a symmetric result). Before the qubit is acted upon, its state is given by

$$|\psi^0\rangle = R_\theta H|0\rangle = \frac{1}{\sqrt{2}}|0\rangle + \frac{1}{\sqrt{2}}e^{i\theta}|1\rangle. \tag{11}$$

Eve is assumed to have knowledge of the encoding alphabet, so she reverses the effect of the Hadamard gate by also applying a Hadamard gate (which is its own inverse):

$$H|\psi^0\rangle = \frac{1}{2}(|0\rangle + |1\rangle) + \frac{e^{i\theta}}{2}(|0\rangle - |1\rangle) = \frac{1+e^{i\theta}}{2}|0\rangle + \frac{1-e^{i\theta}}{2}|1\rangle. \tag{12}$$

Fig. 25 Bit encoding in the random $\frac{\pi}{2}$ phase shift protocol

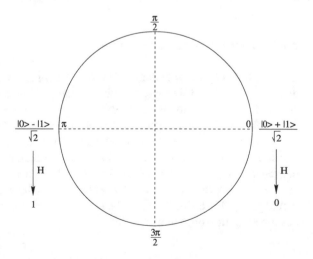

Upon observing the above state, Eve will see a 0 with probability

$$p^0_{Eve} = |\frac{1+e^{i\theta}}{2}|^2 = \frac{1+\cos\theta}{2}.$$ (13)

and a 1 with probability

$$p^1_{Eve} = |\frac{1-e^{i\theta}}{2}|^2 = \frac{1-\cos\theta}{2}.$$ (14)

where θ is either 0 or $\pi/2$ (see Fig. 25). Next, Eve uses the Hadamard transform again to prepare a qubit in an encoded state compatible with the measurement's outcome and sends it to Bob. Bob keeps the qubit untouched until Alice informs him of the correct rotation angle θ. Then, he applies the $R^{\dagger}_{\theta}$ gate, thus inducing a relative phase of $-\theta$, since he received the qubit from Eve and not from Alice. Finally, Bob measures the qubit in the Hadamard basis, obtaining a 0 with the following probability:

$$p^0_{Bob} = p^0_{Eve} \cdot |\frac{1+e^{-i\theta}}{2}|^2 + p^1_{Eve} \cdot |\frac{1-e^{-i\theta}}{2}|^2 = \frac{1+\cos^2\theta}{2}.$$ (15)

For $\theta = 0$, $p^0_{Bob} = 1$ and Eve gets undetected, but if $\theta = \pi/2$, p^0_{Bob} is only $1/2$, so, on average, there is a 25 % probability of detecting Eve for each qubit she chooses to eavesdrop on (same as BB84). Of course, this probability can be made arbitrarily close to 1 by testing a sufficiently large number of qubits. In turn, this requires a large number of qubits to be transmitted through the quantum channel. If Bob can store these qubits until the second stage of the protocol, the cost of the total communication (both quantum and classical) is effectively halved. Parity checking techniques, to avoid discarding bits when testing for eavesdropping, are also applicable.

5.1.2 Translucent Eavesdropping

In order to avoid the inevitable disturbance caused by a measurement, Eve could decide for a more subtle eavesdropping technique. She could choose, for instance, to entangle Alice's information carrier with her own probe, sending half of the entangled pair to Bob while keeping the other half for herself. Then, upon finding about the correct θ angle, by listening in on the conversation between Alice and Bob on the classical channel, Eve can apply the $R^{\dagger}_{\theta}$ and Hadamard gates to the qubit in her possession, hoping to unlock the information hidden within it. We focus again on the case when Alice encodes a 0, with the observation that the analysis for the other case would proceed in a similar way. The eavesdropping operation, performed by Eve, is described by the following equation:

$$CNOT((\frac{1}{\sqrt{2}}|0\rangle + \frac{e^{i\theta}}{\sqrt{2}}|1\rangle) \otimes |0\rangle) = \frac{1}{\sqrt{2}}|00\rangle + \frac{e^{i\theta}}{\sqrt{2}}|11\rangle.$$ (16)

where *CNOT* denotes the application of a controlled-*NOT* operation, with the qubit intercepted from Alice acting as the control qubit. When Alice discloses to Bob whether she applied the $\pi/2$ relative phase shift or not, Eve can proceed to effect the $R_\theta^\dagger$ and Hadamard transformations on the qubit that remained in her possession. This will change the state of the ensemble Eve-Bob as follows:

$$H \otimes I(R_\theta^\dagger \otimes I(\tfrac{1}{\sqrt{2}}|00\rangle + \tfrac{e^{i\theta}}{\sqrt{2}}|11\rangle))) = H \otimes I(\tfrac{1}{\sqrt{2}}|00\rangle + \tfrac{1}{\sqrt{2}}|11\rangle)$$

$$= \frac{1}{2}(|00\rangle + |01\rangle + |10\rangle - |11\rangle). \tag{17}$$

Similarly, if a bit with the value 1 would have been transmitted by Alice, the state of the entangled ensemble would have been

$$\frac{1}{2}(|00\rangle - |01\rangle + |10\rangle + |11\rangle). \tag{18}$$

Although distinguishing among states (17) and (18) is possible by applying a two-qubit gate on the whole ensemble, no information can be elicited by acting only on one qubit. In particular, a quantum measurement in the normal computational basis will yield a 0 or a 1 with equal probability.

The description and analysis of the protocol assumed an error-free quantum channel. The issue of noise can be addressed by introducing an additional phase to the second stage of the protocol. During this phase, Alice and Bob remove all errors from their tentative final key, producing a common error-free key, called *reconciled key* (see [12], Chap. III, for details).

We conclude the analysis of the random $\pi/2$ phase shift protocol with a few observations that, although formulated for the protocol presented in this subsection, can be generalized, in a suitable form, to probably any existing QKD scheme. For each qubit Eve decides to tamper with, there is a certain chance (25 % in our case, as well as for BB84) that she will be caught. It is important to emphasize that this probability is independent of the actions performed on the other qubits transmitted through the quantum channel. The only way Eve can be detected is to test one of the qubits she decided to spy on. In half of the cases, when she is lucky, the quantum state retransmitted to Bob is identical to the one intercepted from Alice, so she gains knowledge of the bit transmitted without any possibility of being detected. On the other hand, if she gets unlucky, then her uncertainty about the bit transmitted is total and, in addition, she disturbs the state of the qubit, introducing an error rate in Bob's raw key.

Consequently, Eve could settle for a low level of eavesdropping, trying to gain only partial knowledge of the secret key, while minimizing the chances of being detected. She could even take advantage of the imperfections in the quantum channel, trying to hide behind the "noise". In the next subsection, we propose a conceptually new kind of QKD scheme that aims to maximize Eve's uncertainty about the bits she eavesdropped on, even after the public discussion between Alice and Bob, while giving Bob higher chances of detecting Eve, even for a smaller number of bits tested. The main idea of the protocol is to propagate the disruption caused by Eve when measuring a qubit to other qubits in the sequence as well. To this end we take advantage of the data dependencies introduced by the application of the quantum Fourier transform.

5.2 QKD Scheme Based on the Fourier Transform

The quantum Fourier transform (QFT) is a very powerful tool, allowing the design of quantum algorithms that are exponentially faster than their best classical counterparts, as in the case of Shor's quantum algorithms for factoring integers and computing discrete logarithms. We show herein that the QFT and its inverse can also be successfully used to build quantum key distribution protocols that offer improved eavesdropping detection rates while maximizing the eavesdropper's uncertainty about the binary sequence transmitted.

The QFT is a linear operator whose action on any of the computational basis vectors $|0\rangle, |1\rangle, \ldots, |2^n - 1\rangle$ associated with an n-qubit register is described by the following transformation:

$$|j\rangle \longrightarrow \frac{1}{\sqrt{2^n}} \sum_{k=0}^{2^n-1} e^{2\pi i jk/2^n} |k\rangle, \ 0 \le j \le 2^n - 1. \tag{19}$$

Equation (19) can be rewritten as a tensor product of the n qubits involved, as follows:

$$|j_1 j_2 \ldots j_n\rangle \longrightarrow \frac{(|0\rangle + e^{2\pi i 0.j_n}|1\rangle) \otimes (|0\rangle + e^{2\pi i 0.j_{n-1}j_n}|1\rangle) \otimes \cdots \otimes (|0\rangle + e^{2\pi i 0.j_1 j_2 \ldots j_n}|1\rangle)}{2^{n/2}}. \tag{20}$$

Equation (20) provides the blueprint for devising a circuit implementing the QFT that requires only $\Theta(n^2)$ elementary quantum gates (see Fig. 26).

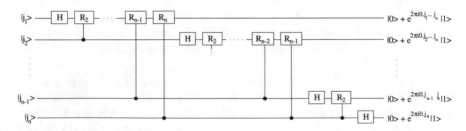

Fig. 26 Quantum circuit performing the discrete Fourier transform

Note that each Fourier transformed qubit is in a balanced superposition of $|0\rangle$ and $|1\rangle$. They differ from one another only in the relative phase between the $|0\rangle$ and the $|1\rangle$ components. For the first qubit in the tensor product, j_n will introduce a phase shift of 0 or π, depending on whether its value is 0 or 1, respectively. The phase of the second qubit is determined (controlled) by both j_n and j_{n-1}. It can amount to $\pi + \pi/2$, provided j_{n-1} and j_n are both 1. This dependency on the values of all the previous qubits continues up to (and including) the last term in the tensor product. When $|j_1\rangle$ gets Fourier transformed, the coefficient of $|1\rangle$ in the superposition involves all the digits in the binary expansion of j.

In the case of each qubit, the 0 or π phase induced by its own binary value is implemented through a Hadamard gate. The dependency on the previous qubits is reflected in the use of controlled phase shifts, as depicted in Fig. 26. Reversing each gate in Fig. 26 gives us an efficient quantum circuit (depicted in Fig. 27) for performing the inverse Fourier transform.

Because of the interdependencies introduced by the controlled rotations, the procedure computing the inverse quantum Fourier transform must start by computing $|j_n\rangle$ and then work its way up to $|j_1\rangle$. The value of $|j_n\rangle$ is needed in the computation of $|j_{n-1}\rangle$. Both $|j_n\rangle$ and $|j_{n-1}\rangle$ are required in order to obtain $|j_{n-2}\rangle$. This continues in the same manner, until finally, the values of all the higher rank bits are used to determine $|j_1\rangle$ precisely.

This fixed order of execution can be exploited to design secure QKD schemes. The protocol that we describe in the following can be seen as a generalization of the random $\pi/2$ phase shift protocol, both relying on encapsulating information in the relative phase between the two components in a superposition. However, the Fourier transform brings into play the *rank* of a qubit in the sequence, thus giving a *context* to each qubit transmitted.

Employing the Fourier transform instead of the random $\pi/2$ phase shift as the encryption method does not alter the main structure of the protocol, so we will just point out the differences relative to the description we provided in the previous subsection. In step 2 of the quantum communication stage, Alice applies the QFT to the binary sequence generated in the previous step, by passing it through the quantum circuit depicted in Fig. 26. Then, she scrambles the resulting qubit sequence by choosing an arbitrary permutation of the qubits and sends them to Bob.

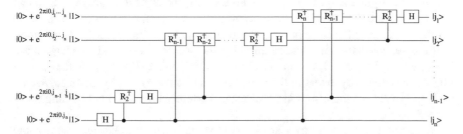

Fig. 27 Quantum circuit performing the inverse Fourier transform

In stage 2 of the protocol, Alice informs Bob of the correct order in which he must place the received qubits (in other words, the *rank* of each qubit is disclosed). Consequently, the raw key extraction step can proceed with Bob applying the inverse Fourier transform to the properly re-arranged qubit sequence. In the absence of any eavesdropping or transmission errors, Bob must end up with the same bit sequence that Alice randomly produced at the outset of the protocol.

When Eve decides to spy on an arbitrary qubit in the sequence, she doesn't know its rank and is therefore ignorant of the influence exerted on it by the previous qubits in the ordered sequence. Without access to this additional information (the qubit's context), Eve can have no confidence in the outcome of an eventual measurement in the Hadamard basis pointing to a 0 or a 1.

Example.

Suppose that the bit string that Alice wants to convey to Bob is 10011010, so that $j_1 = 1$ and $j_8 = 0$. Consider what happens if Eve intercepts the qubit of rank 6 and measures it in the Hadamard basis. Since its state is

$$|0\rangle + e^{2\pi i 0.010}|1\rangle = |0\rangle + e^{\frac{\pi}{2}i}|1\rangle, \tag{21}$$

exactly halfway between $|0\rangle$ and $|1\rangle$ (relative phase $\pi/2$), there is an equal probability for either outcome to be realized. Consequently, even after learning it's context, Eve's uncertainty over this bit is total. Following her measurement, Eve can either send $H|0\rangle$ or $H|1\rangle$ to Bob. In any case, Bob will undo the $\pi/2$ rotation supposedly caused by $j_7 = 1$, therefore having a 50 % chance of detecting Eve, provided he and Alice choose to test bit j_6. But if Bob measures bit j_6 as 1, then the error introduced by Eve's action is still detectable, even if the qubit whose state she disturbed is not checked by Alice and Bob. Thus, when applying the inverse Fourier transform on the qubit of rank 5, its quantum state becomes

$$|0\rangle + e^{(\pi + \frac{\pi}{4} - \frac{\pi}{4} - \frac{\pi}{2})i}|1\rangle \tag{22}$$

and in 50 % of the cases Alice and Bob will discover a mismatch in their values for this bit. An erroneous bit j_6 will continue to influence the outcome of the following bits, up to j_1. The strength of this influence decreases with the rank and probably becomes negligible in a few steps. Nevertheless, if the error in j_6 propagates to one of its neighbors, then this bit acts as a new source of error, creating the mechanism for the initial disturbance to propagate indefinitely. So, unlike other QKD schemes, in this case, eavesdropping on one qubit has the potential to introduce a large number of errors. In general, for an arbitrary qubit of rank k ($0 < k \leq n$), the relative phase shift caused by errors in the previous bits (from n to $k + 1$) varies between 0 and $\sum_{i=1}^{n-k} \pi/2^i$, as the errors induced may interfere with each other, adding up or canceling out.

Since Eve's uncertainty over an observed value is based on her ignorance about the context involved, it appears that the weak spot of the protocol lies in the high rank qubits. The highest rank qubit, for instance, is context-free (having no predecessors), so Eve can be certain of its value, provided she has performed a measurement on it. But because she doesn't know the ranks of the qubits transmitted during the quantum communication stage, she must eavesdrop on many qubits to increase her chances of learning the value of j_n. This, in turn, will cause more disturbance and therefore increase the risk of being detected.

In our example, by learning that the value of j_8 equals 0, Eve also becomes aware that j_8 has no influence on j_7, so her measurement on j_7 (if performed) must have yielded its true value. However, since $j_7 = 1$, there is an equal probability that a hypothetical measurement on j_6 has revealed the correct or incorrect value. For an arbitrary bit string $j_1 \ldots j_n$, Eve can end up knowing the values of the last k bits, where $j_{n-k+1} = 1$ and $j_{n-k+2}, \ldots, j_{n-1}, j_n$ are all zeroes, assuming that she performed all the necessary measurements on the qubits in transit. In practice, since the binary sequence transmitted is chosen at random, the probability of it ending in more than two or three consecutive zeroes is very low.

One immediate solution is for Alice and Bob to discard those bits from their raw keys. Alternatively, the protocol described above, and based on the Fourier transform, could be combined with the random $\pi/2$ phase shift protocol presented in the previous subsection. In this way, each qubit may get an additional $\pi/2$ relative phase shift, increasing Eve's uncertainty about the trailing bits in the sequence while maintaining the uncertainty level for the others.

One important idea that we brought forward in this section is to harness the dependencies between qubits created by the quantum Fourier transform in order to obtain a protocol with superior performance. When compared with existing QKD schemes, the protocol using the QFT offers better eavesdropping detection rates by propagating the disruption caused to one qubit to the following qubits in the sequence. This makes the protocol more efficient in terms of the number of bits that have to be tested in order to achieve a certain level of security. Also, the lack of knowledge over a qubit's context, at the time of eavesdropping, maximizes Eve's uncertainty about the information encoded within its quantum state, thus making the protocol more secure.

These benefits come at the cost of a more complex processing required at both ends of the link. However, the computational power assumed to be available for Alice and Bob is not that of a quantum computer. Computing the QFT and its inverse in the special case of a sequence made up of classical bits requires only the application of single-qubit gates. Although all the phase shift gates in Figs. 26 and 27 are controlled-rotations, the control qubit is in fact always classical. Consequently, the net effect of such a controlled-gate is the application of the phase shift rotation onto the target qubit, if the control is 1, or no transformation at all, if the control is 0. Therefore, Alice and Bob need only to be able to perform Hadamard and phase shift rotations of single-qubit quantum states. Parallel processing can also be applied in order to avoid decoherence [14].

The protocol for QKD developed in this section demonstrates that the QFT is a versatile tool, with important applications not only in quantum algorithms, but also in quantum cryptography. It allows for the design of new QKD schemes with clear advantages over the existing ones, especially for low levels of eavesdropping. Furthermore, the results obtained herein suggest that the role of QFT in the general area of data security is much more important than previously believed.

References

1. Bennett, C.H.: Quantum cryptography using any two nonorthogonal states. Phys. Rev. Lett. **68**(21), 3121–3124 (1992)
2. Bennett, C.H., Brassard, G.: Quantum cryptography: public key distribution and coin tossing. In: Proceedings of IEEE International Conference on Computers, Systems and Signal Processing, pp. 175–179. IEEE, New York (1984). Bangalore, India, December 1984
3. Bennett, C.H., Brassard, G., Mermin, N.D.: Quantum cryptography without Bell's theorem. Phys. Rev. Lett. **68**(5), 557–559 (1992)
4. Bennett, C.H., Brassard, G., Robert, J.M.: Privacy amplification by public discussion. SIAM J. Comput. **17**(2), 210–229 (1988)
5. Blinov, B.B., Moehring, D.L., Duan, L.M., Monroe, C.: Observation of entanglement between a single trapped atom and a single photon. Nature **428**, 153–157 (2004)
6. Cirac, I., Zoller, P.: Quantum computations with cold trapped ions. Phys. Rev. Lett. **74**, 4091–4094 (1995)
7. Cirac, I., Zoller, P., Kimble, H.J., Mabuchi, H.: Quantum state transfer and entanglement distribution among distant nodes in a quantum network. Phys. Rev. Lett. **78**(16), 3221–3224 (1997). http://arxiv.org/abs/quant-ph/9611017
8. Davidovich, L., et al.: Quantum switches and nonlocal microwave fields. Phys. Rev. Lett. **71**(15), 2360–2363 (1993)
9. Deutsch, D., Ekert, A., Jozsa, R., Macchiavello, C., Popescu, S., Sanpera, A.: Quantum privacy amplification and the security of quantum cryptography over noisy channels. Phys. Rev. Lett. **77**, 2818–2821 (1996). http://arxiv.org/abs/quant-ph/9604039
10. Ekert, A.: Quantum cryptography based on Bell's theorem. Phys. Rev. Lett. **67**, 661–663 (1991)
11. Gulde, S., et al.: Implementation of the Deutsch-Jozsa algorithm on an ion-trap quantum computer. Nature **421**, 48–50 (2003)
12. Lomonaco Jr., S.J. (ed.): Quantum Computation: a grand mathematical challenge for the twenty-first century and the millennium, In: Proceedings of Symposia in Applied Mathematics, vol. 58. American Mathematical Society, Short Course, Washington, DC (2000)
13. Nagy, N., Akl, S.G.: Authenticated quantum key distribution without classical communication. Parallel Process. Lett., Spec. Issue Unconv. Comput. Prob. **17**(3), 323–335 (2007)
14. Nagy, M., Akl, S.G.: Coping with decoherence: parallelizing the quantum Fourier transform. Parallel Process. Lett., Spec. Issue Adv. Quantum Comput. **20**(3), 213–226 (2010)
15. Nagy, N., Akl, S.G.: Quantum authenticated key distribution. In: Proceedings of International Conference on Unconventional Computation. Lecture Notes in Computer Science, vol. 4618, pp. 127–136. Springer, Heidelberg (2007)
16. Nagy, N., Nagy, M., Akl, S.G.: Key distribution versus key enhancement in quantum cryptography. Parallel Process. Lett., Spec. Issue Adv. Quantum Comput. **20**(03), 239–250 (2010)
17. Nielsen, M.A., Chuang, I.L.: Quantum Computation and Quantum Information. Cambridge University Press, Cambridge, UK (2000)
18. Rivest, R.L., Shamir, A., Adleman, L.M.: A method of obtaining digital signatures and public-key cryptosystems. Commun. ACM **21**(2), 120–126 (1978)
19. Shannon, C.: Communication theory of secrecy systems. Bell Syst. Tech. J. **28**(4), 656–715 (1949)
20. Shi, B.S., Li, J., Liu, J.M., Fan, X.F., Guo, G.C.: Quantum key distribution and quantum authentication based on entangled states. Phys. Lett. A **281**(2–3), 83–87 (2001)

21. Shor, P.W.: Polynomial-time algorithms for prime factorization and discrete logarithms on a quantum computer. Spec. Issue Quantum Comput. SIAM J. Comput. **26**(5), 1484–1509 (1997)
22. Turchette, Q., et al.: Measurement of conditional phase shifts for quantum logic. Phys. Rev. Lett. **75**(25), 4710–4713 (1995)
23. Vaudenay, S.: A Classical Introduction to Cryptography: Applications for Communications Security. Springer (2006)
24. Wootters, W.K., Zurek, W.H.: A single quantum cannot be cloned. Nature **299**, 802–803 (1982)

Emergence in Context-Free Parallel Communicating Grammar Systems: What Does and Does not Make a Grammar System More Expressive Than Its Parts

Stefan D. Bruda and Mary Sarah Ruth Wilkin

Abstract We investigate the emergent factors that affect the expressiveness of parallel communicating grammar systems (PCGS) with context-free components. It is already known that synchronization is a significant such a factor. In addition we show that serving multiple queries from multiple components simultaneously (broadcast communication) is not an emergent factor, but serving multiple queries from a single component is. We further identify a notion of interference that has significant emergent consequences. In the process we introduce several potentially useful techniques for the analysis of PCGS with context free components. In particular we introduce the notion of PCGS parse trees, and also some techniques such as "copycat" components and "reset" components that are potentially useful in developing an algorithm for the elimination of broadcast communication.

1 Introduction

Parallel Communicating Grammar Systems (PCGS) have been introduced as a language-theoretic treatment of concurrent (or more general, multi-agent) systems [10, 24]. A PCGS consists of several component grammars that work in parallel on separate strings. The components also communicate with each other: One component grammar may request strings generated by others, and several components may make queries at the same time. Several variants of PCGS can be defined based on various synchronization and communication assumptions. Emergent behaviour was expected and indeed did not fail to manifest itself: because of the synchronization and communication facilities, PCGS whose components are of a certain type are generally (though not always) more powerful than a single grammar of the same type [5, 7, 10, 24, 25].

S.D. Bruda (✉) · M.S.R. Wilkin
Department of Computer Science, Bishop's University, Sherbrooke,
QC J1M 1Z7, Canada
e-mail: stefan@bruda.ca

M.S.R. Wilkin
e-mail: swilkin@cs.ubishops.ca

© Springer International Publishing Switzerland 2017   171
A. Adamatzky (ed.), *Emergent Computation*, Emergence, Complexity
and Computation 24, DOI 10.1007/978-3-319-46376-6_8

In this paper we present a thorough analysis of the expressiveness of PCGS whose components are context-free grammars (context-free PCGS henceforth). It was shown earlier [6, 7, 11] that context-free PCGS are Turing complete, yet these results are established under a certain communication model (called broadcast communication) which is assumed in many papers on the matter yet is not implied by the original definition of PCGS (which uses however implicitly the one-step communication model). We therefore wonder whether such a result holds for the original definition. It turns out that this is indeed the case, though the construction that establishes Turing completeness in the one-step communication case (Sect. 4) is considerably more complex. While the construction itself is large and unyielding, it introduces several general techniques including "copycat" grammars (that behave identically to other grammars in the system so that different components can query different grammars without interfering with each other) and "reset" grammars (that reset other components at precise moments in the derivation). We believe that these techniques are interesting, general, and useful. In particular we believe that they can be eventually used algorithmically, so that one does not need to repeat the manual construction from Sect. 4, but perform such a conversion algorithmically and for any context-free PCGS using broadcast communication instead. One way or another, we establish that broadcast communication is not an emergent factor in PCGS with context-free components.

Having established the Turing completeness of the most powerful version of context-free PCGS, we turn our attention to the expressiveness of weaker variants of these systems. We use for this investigation the concept of parse tree, one of the most important constructs in the realm of context-free languages. We first introduce the notion of parse trees for context-free PCGS (Sect. 5) as a natural extension of context-free parse trees, showing in the process that any context-free PCGS derivation has an associated parse tree. Based on parse trees we then identify a particular notion of "interference" (Sect. 6) and we find that one cannot fully characterize PCGS derivations using parse trees (in the sense that each derivation has a parse tree and also the other way around) whenever our notion of interference is present. Complete characterization of derivations by parse trees thus only holds for one very restricted context-free PCGS variant (Sect. 7) which will turn out to be called unsynchronized, returning, unique-query context-free PCGS. One consequence of the existence of such a characterization is in the realm of generative power. More precisely, this variant of context-free PCGS is the weakest of them all (Sect. 8).

The end result of our investigation is the assessment of the various emergent factors in the behaviour of PCGS with context-free components. We already knew that synchronization has a big impact, but we find that a weaker notion of interference has quite a significant impact as well. The simultaneity of queries does and does not have an emergent impact: On one hand it does not matter whether we allow multiple queries from different components to be serviced simultaneously (broadcast communication); we can produce the same result even if we serve one component at a time (one-step communication). On the other hand, whether we allow multiple simultaneous queries in a component or not makes quite a bit of a difference in the expressiveness of the system.

On a practical note, we believe that context-free PCGS have a promising future in formal methods, where more expressive does not necessarily mean better. Therefore the various emergent factors presented above will determine between other things the most promising flavour of PCGS for such an application.

We will offer further details on emergence and its practical consequences only after all the concepts have been defined and all the results have been established. A detailed discussion on the matter will therefore be postponed until Sect. 9.

2 Preliminaries

For some alphabet (i.e., finite set) V, some word $x \in V^*$, and a set $U \subseteq V$, $|x|$ denotes the length of x and $|x|_U$ denotes the number of occurrences of elements of U in x. By abuse of notation we write $|x|_a$ instead of $|x|_{\{a\}}$ for singleton sets $U = \{a\}$. The empty string (and only the empty string) is denoted by ε.

A grammar [22] is a quadruple $G = (N, \Sigma, S, R)$. Σ is the finite, nonempty set of terminals. N is the finite, nonempty set of nonterminals, and is disjoint from Σ. $S \in N$ is a designated nonterminal referred to as the start symbol or axiom. R is a finite set of rewriting rules, of the form $A \to u$ where $A \in (\Sigma \cup N)^* N (\Sigma \cup N)^*$ and $u \in (\Sigma \cup N)^*$ (A and u are strings of terminals and nonterminals but A contains at least one nonterminal). Given a grammar G, the $\Rightarrow_G$ binary operator on strings from the alphabet $W = (\Sigma \cup N)^*$ is defined as follows: $w_1 A w_2 \Rightarrow_G w_1 u w_2$ if and only if $A \to u \in R$ and $w_1, w_2 \in (\Sigma \cup N)^*$. The language generated by a grammar $G = (\Sigma, N, S, R)$ is $\mathcal{L}(G) = \{w \in \Sigma^* : S \Rightarrow_G^* w\}$, where $\Rightarrow_G^*$ denotes as usual the reflexive and transitive closure of $\Rightarrow_G$. We often omit the subscript G when no ambiguity is thus introduced.

Languages generated by (unrestricted) grammars are referred to as recursively enumerable (RE). A grammar G is called context sensitive if each rewriting rule $A \to u$ in R satisfies $|A| \le |u|$; the languages generated by these grammars are referred to as context sensitive (CS). G is context free if every rewriting rule $A \to u$ in R satisfies $|A| = 1$ (meaning that A is a single nonterminal); these grammars generate context-free languages (CF). A special type of context-free grammars are linear grammars, generating linear languages (LIN), where no rewriting rule is allowed to have more that one nonterminal symbol on its right hand side. Finally, grammars are regular and generate regular languages (REG) if their rewriting rules have one of the following forms: $A \to cB$, $A \to c$, $A \to \varepsilon$, or $A \to B$ where A, B are nonterminals and c is a terminal [17, 20].

Definition 1 [22] A *parse tree* for a context-free grammar $G = (N, \Sigma, R, S)$ is a tree whose nodes are labelled with symbols from the set $N \cup \Sigma$. It is defined inductively as follows (based on Fig. 1a–d on p. XX): For every $a \in N \cup \Sigma$ the tree depicted in Fig. 1a is a parse tree with *yield* a; for every $A \to \varepsilon \in R$ the tree from Fig. 1b is a parse tree with yield ε. Suppose that the n trees from Fig. 1c are parse trees with yields $y_1, y_2, \ldots, y_n$ and that $A \rightarrow A_1 A_2 \ldots A_n \in R$; then the tree shown in Fig. 1d is a parse tree with yield $y_1 y_2 \ldots y_n$.

Note that the yield of a parse tree is the sequence of leaf labels as obtained by an inorder traversal of the tree. For every parse tree with root A and yield y there exists a derivation $A \Rightarrow^* y$ (and the other way around) in a natural way [22].

2.1 Parallel Communicating Grammar Systems

A Parallel Communicating Grammar System (PCGS) provides a theoretical proto-type that combines the concepts of grammars with parallelism and communication. A PCGS consists of a number of grammars that communicate with each other and thus cooperate in the generation of strings.

Definition 2 [10]: A PCGS of degree n for some $n \geq 1$ is an $(n + 3)$ tuple $\Gamma = (N, K, \Sigma, G_1, \ldots, Gn)$ where N is a nonterminal alphabet, Σ is a terminal alphabet, and K is the set of query symbols, $K = \{Q_1, Q_2, \ldots, Q_n\}$. The sets N, Σ, and K are mutually disjoint. $G_i = (N \cup K, \Sigma, R_i, S_i)$, $1 \leq i \leq n$ are grammars; they represent the components of the system. The indices $1, \ldots, n$ of the symbols in K point to $G_1, \ldots, G_n$, respectively.

A derivation in a PCGS consists of a series of communication and rewriting steps. A rewriting step is not possible if communication is requested (which happens whenever a query symbol appears in one of the components of a configuration).

Definition 3 [10]: Let $\Gamma = (N, K, \Sigma, G_1, \ldots, Gn)$ be a PCGS as above, and $(x_i, x_2, \ldots, x_n)$ and $(y_i, y_2, \ldots, y_n)$ be two n-tuples with $x_i, y_i \in (N \cup K \cup \Sigma)^*$, $1 \leq i \leq n$. We write $(x_i, \ldots, x_n) \Rightarrow_\Gamma (y_i, \ldots, y_n)$ iff one of the following two cases holds:

1. $|x_i|_K = 0$, $1 \leq i \leq n$, and for each i, $1 \leq i \leq n$, we have $x_i \Rightarrow_{G_i} y_i$ (in the gram-mar G_i), or $[x_i \in \Sigma^*$ and] $x_i = y_i$.
2. $|x_i|_K > 0$ for some $1 \leq i \leq n$; let $x_i = z_1 Q_{i_1} z_2 Q_{i_2} \ldots z_t Q_{i_t} z_{t+1}$, with $t \geq 1$ and $z_j \in (N \cup \Sigma)^*$, $1 \leq j \leq t + 1$. Then $y_i = z_1 x_{i_1} z_2 x_{i_2} \ldots z_t x_{i_t} z_{t+1}$ [and $y_{i_j} = S_{i_j}$, $1 \leq j \leq t$] whenever $|x_{i_j}|_K = 0$, $1 \leq j \leq t$. If on the other hand $|x_{i_j}|_K \neq 0$ for some $1 \leq j \leq t$, then $y_i = x_i$. For all $1 \leq k \leq n$, $y_k = x_k$ whenever y_k was not specified above.

The presence of "[and $y_{i_j} = S_{i_j}$, $1 \leq j \leq t$]" in the definition makes the PCGS *returning*. The PCGS is *non-returning* if the phrase is eliminated.

We use $\Rightarrow_\Gamma$ for both component-wise and communication steps. A sequence of interleaved rewriting and communication steps will be denoted by $\Rightarrow_\Gamma^*$, the reflex-ive and transitive closure of $\Rightarrow_\Gamma$. As usual we omit the subscript Γ whenever no ambiguity is introduced by such an omission.

The first case in Definition 3 above is called a component-wise derivation step and the second a communication step. Informally, an n-tuple $(x_1, \ldots, x_n)$ yields $(y_1, \ldots, y_n)$ as follows:

1. If there is no query symbol in $x_1, \ldots, x_n$, then we have a component-wise derivation $(x_i \Rightarrow_{G_i} y_i, 1 \le i \le n$, which means that one rule is used per component G_i), unless x_i is all terminals $(x_i \in T^*)$ in which case it remains unchanged $(y_i = x_i)$.
2. If we have query symbols then a communication step is required. When this occurs each query symbol Q_j in x_i is replaced by x_j, if and only if x_j does not contain query symbols. In other words, a communication step involves the query symbol Q_j being replaced by the string x_j; the result of this replacement is referred to as Q_j being *satisfied* (by x_j). Once the communication step is complete the grammar G_j continues processing from its axiom, unless the system is non-returning. Communication steps always have priority over rewriting steps; if not all the query symbols are satisfied during a communication step, then they can still be satisfied during subsequent communication steps.

A tuple $(x_1, x_2, \ldots, x_n) \in ((N \cup K \cup \Sigma)^*)^n$ as above is called a *configuration* of the system. We call x_i a component of the configuration or only a component if the reference to the configuration is understood from the context. Rules $Q_j \to \alpha$ are never used so we can assume that no such rules are present.

The derivation in a PCGS is blocked if no component-wise derivation can be applied to a nonterminal symbol in some component, or circular queries appear. The latter happens when G_{i_1} introduces Q_{i_2}, G_{i_2} introduces $Q_{i_3}, \ldots, G_{i_{k-1}}$ introduces Q_{i_k} and G_{i_k} introduces Q_{i_1}; in such a case no rewriting step is possible (as communication has priority), but no communication steps are possible either.

A string generated by a PCGS Γ is the result of a derivation that starts from the tuple of axioms $(S_1, S_2, \ldots, S_n)$. A number of rewriting and/or communication steps are performed until G_1 produces a terminal string (we do not restrict the form of, or indeed care about the rest of the components of the final configuration). The language generated by Γ consists of exactly all the strings generated by Γ:

Definition 4 [10]: The language generated by a PCGS Γ is $\mathcal{L}(\Gamma) = \{w \in \Sigma^* : (S_1, S_2, \ldots, S_n) \Rightarrow_\Gamma^* (w, \sigma_2, \ldots, \sigma_n), \sigma_i \in (N \cup K \cup \Sigma)^*, 2 \le i \le n\}$.

Several variants of PCGS can be defined. If a component-wise derivation requires that any component containing nonterminals be rewritten (i.e., the bracketed phrase "$[x_i \in \Sigma^*$ and]" is present in the first case of Definition 3) then the system is *synchronized* (note however that any component that is already a string of terminals will not be rewritten under any circumstance). Otherwise we have an *unsynchronized* system, that allows components to either perform derivations or stay put.

A PCGS is called *returning* (to the axiom) if, after communication, a component which has communicated a string resumes the work from its axiom as described by the phrase "[and $y_{i_j} = S_{i_j}, 1 \le j \le t$]" in the second case of Definition 3. A PCGS is called *non-returning* if components continue working using the current string after a query (i.e., the bracketed phrase above is erased from the definition).

Finally, a PCGS is called *centralized* if only the first component grammar G_1 can control the communication, meaning that only G_1 can introduce query symbols; in the absence of such a restriction the system is non-centralized.

The family of languages generated by a non-centralized, returning PCGS with n components of type X (where X is an element of the Chomsky hierarchy) will be denoted by $PC_n(X)$. The language families generated by centralized PCGS will be represented by $CPC_n(X)$. The fact that the PCGS is non-returning will be indicated by the addition of an N, thus obtaining the classes $NPC_n(X)$ and $NCPC_n(X)$. Let M be a class of PCGS, $M \in (PC, CPC, NPC, NCPC)$; then we define:

$$M(X) = M_*(X) = \bigcup_{n \geq 1} M_n(X)$$

In the case of a returning PCGS a consequence of Definition 3 is that the communicated components are reset to their respective axioms as soon as all the query symbols are satisfied in the component that requested the communication. However, several papers e.g., [6, 7, 11] imply a different reset model, where the communicated components reset to their axioms only after all the queries (from all the other components) have been satisfied. We refer to the communication model from Definition 3 as *one-step communication*, while the other model will be called *broadcast communication*.

3   Previous Work

In this section we summarize the existing results regarding the expressiveness of the most commonly studied PCGS. One will notice that not all structural variations have been studied in this respect. Most of the existing results are about centralized systems, and even then not all of the centralized variants have been studied thoroughly. We pay particular attention to PCGS with context-free components, the object of our discussion.

CS and RE are the two most powerful PCGS and grammar types. Somehow surprisingly their behavior is quite similar. It is immediate that a RE grammar is just as powerful as a PCGS with RE components: $RE = Y_n(RE) = Y_*(RE), n \geq 1$, for all $Y \in \{PC, CPC, NPC, NCPC\}$ [10]; PCGS of this type are thus not very interesting since they are just as powerful as a PCGS with one component. The same holds to some degree for PCGS with context-sensitive components versus context-sensitive languages: $CS = Y_n(CS) = Y_*(CS), n \geq 1$, for $Y \in \{CPC, NCPC\}$ [10]. Note however that this result describes the centralized case; we would expect the non-centralized case to be more powerful, so presumably this result does not hold in the non-centralized case.

One should note that PCGS with CS components are computationally expensive, which limits their usefulness. As is the case with normal grammars, the most useful classes are the simple ones. The results in the area of PCGS with regular or context-free components are therefore much more interesting.

We note first that the class of languages generated by centralized returning PCGS with regular components is a proper subset of the class of languages generated by non-centralized, returning PCGS with regular components: $CPC_n(REG) \subsetneq PC_n(REG)$, $n > 1$ [26]. This indicates that the generative power of a PCGS is greater than the generative power of a single grammar component, and that the more communication facilities we have the more powerful the resulting system is. A similar result was found for PCGS with context free components; however in this case increased communication does not necessarily make the system more powerful: $CPC_*(CF) \subseteq PC_*(CF)$ [12].

Generally, a centralized PCGS is a particular case of a non-centralized PCGS. As a consequence the class of languages generated by a centralized PCGS of any type can be generated by a non-centralized PCGS of the same type: $CPC_n(X) \subseteq PC_n(X)$ for any $n \geq 1$. The generative power of a PCGS is greater that of a single grammar component because of communication facilities, and once this parameter is restricted the generative power is also restricted.

The following two results further demonstrate that there are limitations to the generative power of PCGS. When we have only two regular components the languages generated by centralized PCGS are all context free. Even the non-centralized variant is limited to generating context-free languages: $CPC_2(REG) \subsetneq CF$ and $PC_2(REG) \subseteq CF$ [10].

Another way to increase the generative power of a system is to increase the number of components in the system. We have shown that this does not change the generative capacity in the RE and (to some degree) CS case. However if we examine classes that are lower in the hierarchy we notice that an increase in the number of components generally increases the generative capacity of the system [10].

1. There exists a language generated by PCGS with 2 or more REG components that cannot be generated by a linear grammar: $Y_n(REG) \setminus LIN \neq \emptyset$ for $n \geq 2$, $Y \in \{PC, CPC, NPC, NCPC\}$.
2. There exists a language generated by a PCGS with 3 or more REG components that cannot be generated by a context free grammar: $Y_n(REG) \setminus CF \neq \emptyset$ for $n \geq 3$ (and $n \geq 2$ for non-returning PCGS), $Y \in \{PC, CPC, NPC, NCPC\}$.
3. There exists a language generated by a PCGS with 2 or more linear components that cannot be generated by a context free grammar: $Y_n(LIN) \setminus CF \neq \emptyset$, $n \geq 2$, $Y \in \{PC, CPC, NPC, NCPC\}$.
4. There exists a language generated by a non-returning PCGS with 2 or more regular components that cannot be generated by a context free grammar: $Y_n(REG) \setminus CF \neq \emptyset$, $n \geq 2$, $Y \in \{NPC, NCPC\}$.

Obviously an increase in the power of the components will generally increase the power of a PCGS. This holds strictly in the centralized case for REG versus LIN versus CF components: $CPC_n(REG) \subsetneq CPC_n(LIN) \subsetneq CPC_n(CF)$, $n \geq 1$, [10]. Presumably the same relationship would hold for the non-centralized case, but this has not been investigated.

We already mentioned the number of components as an important factor in the generative power of PCGS. It therefore makes sense to consider the hierarchies generated by this factor. Some of these hierarchies are in fact infinite, namely $CPC_n(REG)$ and $CPC_n(LIN)$, $n \geq 1$ [10].

Some hierarchies however collapse. We have already mentioned that $CPC_n(CS)$ and $NCPC_n(CS)$, $n \geq 1$, do not give infinite hierarchies, for all of these classes coincide with CS. Lower classes also produce collapsing hierarchies; for instance non-centralized context-free PCGS with 11 components can generate the whole class RE [7]:

$$RE = PC_{11}(CF) = PC_*(CF).\tag{1}$$

A later paper found that a context-free PCGS with only 5 components can generate the entire class of RE languages by creating a PCGS that has two components that track the number of nonterminals and use the fact that for each RE language L there exists and Extended Post Correspondence problem P [18] such that $\mathcal{L}(P) = L$ [11]:

$$RE = PC_5(CF) = PC_*(CF).\tag{2}$$

Other papers have examined the size complexity of returning and non-returning CF systems even further. It has been shown that every recursively enumerable language can be generated by a returning context-free PCGS, where the number of nonterminals in the system is less than or equal to a natural number k [6].

This all being said, the earlier results showing the Turing completeness of returning context-free PCGS [6, 7, 11] are valid only under the broadcast communication model [27]. Whether a variant of these results exists for the one-step communication model will be discussed in the next section.

Turing completeness was also shown for non-returning systems [8, 23]. In particular, if $k \geq 2$ and $L \subseteq \{a_1, \ldots, a_k\}^+$ is a recursively enumerable language, then there exists a non-returning context-free PCGS without ε-rules (meaning without rules of the form $A \to \varepsilon$) that generates L [8]. If we consider that non-returning systems can be simulated by returning systems with the help of assistance grammars holding intermediate strings [14], these results [8, 23] also apply to returning systems (though the number of components necessary for this to happen does not remain the same).

4   Broadcast Communication Is not Emergent

For the reminder of this paper we will focus on context-free PCGS. In particular in this section we will consider the synchronized, returning variant. We already mentioned that this variant is Turing complete under the broadcast communication model [6, 7, 11], but that the original proofs no longer hold when one-step communication is used instead. As it turns out, context-free PCGS are Turing complete under any communication model, though the one-step communication construction is considerably more complex. We have:

Theorem 1 [27] $RE = \mathcal{L}(PC_{95}(CF)) = \mathcal{L}(PC_*(CF))$.

Proof (sketch) Similarly to one of the proofs for the broadcast communication model [7] we use a context-free PCGS to simulate an arbitrary 2-counter Turing machine. We use all of the components used originally in their construction, but with modified labels. However, we now have to ensure that the components can work together under one-step communication without stumbling over each other.

Let $M = (\Sigma \cup \{Z, B\}, E, R)$ be a 2-counter Turing machine [16]. M has a tape alphabet $\Sigma \cup \{Z, B\}$, a set of internal states E with $q_0, q_F \in E$ and a set of transition rules R. The 2-counter machine has a read only input tape and two counters that are semi-infinite storage tapes. The alphabet of the storage tapes contains two symbols Z and B, while the input tape has the alphabet $\Sigma \cup \{B\}$. The transition relation is defined as follows: if $(x, q, c_1, c_2, q', e_1, e_2, g) \in R$ then $x \in \Sigma \cup \{B\}$, $q, q' \in E$, $c_1, c_2 \in \{Z, B\}$, $e_1, e_2 \in \{-1, 0, +1\}$, and $g \in \{0, +1\}$. The starting and final states of M are denoted by q_0 and q_F, respectively. A transition of the 2-counter machine $(x, q, c_1, c_2, q', e_1, e_2, g) \in R$ is then enabled by the current state q, the symbol currently scanned on the input tape x, and the current value of the two counters c_1 and c_2 (which can be either Z for zero or B for everything else). The effect of such a transition is that the state of the machine is changed to q'; the counter $k \in \{1, 2\}$ is decremented, unchanged, or incremented whenever the value of e_k is -1, 0, or $+1$, respectively; and the input head is advanced if $g = +1$ and stays put if $g = 0$. When the input head scans the last non-blank symbol on the input tape and the machine M is in the accepting state q_F then the input string is accepted by the machine. $\mathcal{L}(M)$ be the language of exactly all the input strings accepted by M.

The following PCGS with 95 components will simulate a given 2-counter Turing machine:

$$\Gamma = (N, K, \Sigma \cup \{a\}, G_{m_{original}}, \ldots G_{m_{29}}, G_{P_1}^{C_1}, \ldots, G_{P_{1_5}}^{C_1}, G_{P_2}^{c_1}, G_{P_3}^{c_1}, G_{P_1}^{C_1}, \ldots$$
$$G_{P_{1_5}}^{C_1}, G_{P_2}^{c_2}, G_{P_3}^{c_2}, G_{Pa_1} \ldots, G_{Pa_{1_5}} G_{a_2}, G_{resetGMPa1}^{14}, G_{resetP_1}^{4} \ldots G_{resetP_4}^{4})$$

where

$$N = \{[x, q, c_1, c_2, e_1, e_2], [e_1]', [e_2]', [I], [I]', \langle I \rangle, \langle x, q, c_1, c_2, e_1, e_2 \rangle |$$
$$x \in \Sigma, q \in E, C_1, c_2 \in \{Z, B\}, e_1, e_2 \in \{-1, 0, +1\}\} \cup$$
$$\{S, S_1, S_2, S_3, S_4, S_4^{(1)}, S_4^{(2)}, S^{(1)}, S^{(2)}, S^{(3)}, S^{(4)}\} \cup \{A, C\}$$

The component definitions from the original system have the word *original* in their label to differentiate them from the helper grammars that were added in order to accommodate the requirements of a on step-communication (returning) system.

For clarity the labels l of the query symbols Q_l will no longer be purely numerical. We group newly introduced rules in sets labeled $\mathfrak{N}$. Those components that do not have an equivalent in the original construction have all their rules in the set $\mathfrak{N}$.

The new master contains the same rewriting rules and communications steps as it had in the original construction [7]. The primary role of the master is to maintain its relationship with the P_{a_1} component grammar. In addition, we copy the functionality of the master in five more helper grammars designed to handle queries from $P_1^{c_1}$, $P_2^{c_1}$, $P_3^{c_1}$, $P_4^{c_1}$, $P_1^{c_2}$, $P_2^{c_2}$, $P_3^{c_2}$, and $P_4^{c_2}$ (these components will all be described in detail later).

$$
\begin{aligned}
P_{GM_{Original}} = \{ & S \to [I], [I] \to C, C \to Q_{a_1} \} \cup \\
& \{ \langle I \rangle \to [x, q, Z, Z, e_1, e_2] | (x, q_0, Z, Z, e_1, e_2, 0) \in R, x \in \Sigma \} \cup \\
& \{ \langle I \rangle \to x[y, q, Z, Z, e_1, e_2] | (x, q_0, Z, Z, q, e_1, e_2, +1) \in R, x, y \in \Sigma \} \cup \\
& \{ \langle x, q, c_1', c_2', e_1', e_2' \rangle \to [x, q', c_1, c_2, e_1, e_2] | x \in \Sigma, c_1', c_2' \in \{Z, B\}, \\
& \quad (x, q, c_1, c_2, q', e_1, e_2, 0) \in R, e_1', e_2' \in \{-1, 0, +1\} \} \cup \\
& \{ \langle x, q, c_1', c_2', e_1', e_2' \rangle \to x[y, q', c_1, c_2, e_1, e_2], \\
& \quad \langle x, q_F, c_1', c_2', e_1', e_2' \rangle \to x | (x, q, c_1, c_2, q', e_1, e_2, +1) \in R, \\
& \quad c_1', c_2' \in \{Z, B\}, e_1', e_2' \in \{-1, 0, +1\}, x, y \in \Sigma \}
\end{aligned}
$$

The following 5 helper grammars simulate rules from the new master but each component is designed to work with different components in $P_1^{c_1}$, including $P_{1_{originalS_1}}^{c_1}$ its four newly defined helpers. The components below work with the $P_1^{c_1}$ grammars as the single grammar version would have in the original construction but the labels of the query symbols have been modified to reflect the labels of their matching component grammar.

$$
\begin{aligned}
P_{GM_{S1}}^{c_1} = \{ & S \to [I], [I] \to C \} \cup \mathfrak{N} = \{ C \to Q_{a_1 P_{a_1} S_1}^{c_1} \} \cup \\
& \{ \langle I \rangle \to [x, q, Z, Z, e_1, e_2] | (x, q_0, Z, Z, e_1, e_2, 0) \in R, x \in \Sigma \} \cup \\
& \{ \langle I \rangle \to x[y, q, Z, Z, e_1, e_2] | (x, q_0, Z, Z, q, e_1, e_2, +1) \in R, x, y \in \Sigma \} \cup \\
& \{ \langle x, q, c_1', c_2', e_1', e_2' \rangle \to [x, q', c_1, c_2, e_1, e_2] | (x, q, c_1, c_2, q', e_1, e_2, 0) \in R, \\
& \quad x \in \Sigma, c_1', c_2' \in \{Z, B\}, e_1', e_2' \in \{-1, 0, +1\} \} \cup \\
& \{ \langle x, q, c_1', c_2', e_1', e_2' \rangle \to x[y, q', c_1, c_2, e_1, e_2], \langle x, q_F, c_1', c_2', e_1', e_2' \rangle \to x | \\
& \quad (x, q, c_1, c_2, q', e_1, e_2, +1) \in R, c_1', c_2' \in \{Z, B\}, e_1', e_2' \in \{-1, 0, +1\}, \\
& \quad x, y \in \Sigma \}
\end{aligned}
$$

$$
\begin{aligned}
P_{GM_{S1H2(S4)}}^{c_1} = \{ & S \to [I], [I] \to C \} \cup \\
& \mathfrak{N} = \{ C \to Q_{a_1 P_{a_1} S_1 H_2(S4)}^{c_1}, S \to Q_{a_1 P_{a_1} S_1 H_2(S4)}^{c_1} \} \cup \\
& \{ \langle I \rangle \to [x, q, Z, Z, e_1, e_2] | (x, q_0, Z, Z, e_1, e_2, 0) \in R, x \in \Sigma \} \cup \\
& \{ \langle I \rangle \to x[y, q, Z, Z, e_1, e_2] | (x, q_0, Z, Z, q, e_1, e_2, +1) \in R, x, y \in \Sigma \} \cup \\
& \{ \langle x, q, c_1', c_2', e_1', e_2' \rangle \to [x, q', c_1, c_2, e_1, e_2] | x \in \Sigma, c_1', c_2' \in \{Z, B\}, \\
& \quad (x, q, c_1, c_2, q', e_1, e_2, 0) \in R, e_1', e_2' \in \{-1, 0, +1\} \} \cup \\
& \{ \langle x, q, c_1', c_2', e_1', e_2' \rangle \to x[y, q', c_1, c_2, e_1, e_2], \langle x, q_F, c_1', c_2', e_1', e_2' \rangle \\
& \quad \to x | (x, q, c_1, c_2, q', e_1, e_2, +1) \in R, c_1', c_2' \in \{Z, B\}, \\
& \quad e_1', e_2' \in \{-1, 0, +1\}, x, y \in \Sigma \}
\end{aligned}
$$

$$P^{c_1}_{GM_{S1H3(S4)}} = \{S \to [I], [I] \to C\} \cup \mathfrak{N} = \{C \to Q^{c_1}_{a_1 P_{a_1} S_1 H_3(S_4)}, S \to Q^{c_1}_{a_1 P_{a_1} S_1 H_3(S_4)}\} \cup$$
$$\{\langle I \rangle \to [x, q, Z, Z, e_1, e_2] | (x, q_0, Z, Z, e_1, e_2, 0) \in R, x \in \Sigma\} \cup$$
$$\{\langle I \rangle \to x[y, q, Z, Z, e_1, e_2] | (x, q_0, Z, Z, q, e_1, e_2, +1) \in R, x, y \in \Sigma\} \cup$$
$$\{\langle x, q, c'_1, c'_2, e'_1, e'_2 \rangle \to [x, q', c_1, c_2, e_1, e_2] | (x, q, c_1, c_2, q', e_1, e_2, 0) \in R,$$
$$x \in \Sigma, c'_1, c'_2 \in \{Z, B\}, e'_1, e'_2 \in \{-1, 0, +1\}\} \cup$$
$$\{\langle x, q, c'_1, c'_2, e'_1, e'_2 \rangle \to x[y, q', c_1, c_2, e_1, e_2], \langle x, q_F, c'_1, c'_2, e'_1, e'_2 \rangle$$
$$\to x | (x, q, c_1, c_2, q', e_1, e_2, +1) \in R, c'_1, c'_2 \in \{Z, B\},$$
$$e'_1, e'_2 \in \{-1, 0, +1\}, x, y \in \Sigma\}$$

$$P^{c_1}_{GM_{S1(S2)}} = \{S \to [I], [I] \to C\} \cup \mathfrak{N} = \{C \to Q^{c_1}_{a_1 P_{a_1} S_1(S_2)}\} \cup$$
$$\{\langle I \rangle \to [x, q, Z, Z, e_1, e_2] | (x, q_0, Z, Z, e_1, e_2, 0) \in R, \dot{x} \in \Sigma\} \cup$$
$$\{\langle I \rangle \to x[y, q, Z, Z, e_1, e_2] | (x, q_0, Z, Z, q, e_1, e_2, +1) \in R, x, y \in \Sigma\} \cup$$
$$\{\langle x, q, c'_1, c'_2, e'_1, e'_2 \rangle \to [x, q', c_1, c_2, e_1, e_2] | (x, q, c_1, c_2, q', e_1, e_2, 0) \in R,$$
$$x \in \Sigma, c'_1, c'_2 \in \{Z, B\}, e'_1, e'_2 \in \{-1, 0, +1\}\} \cup$$
$$\{\langle x, q, c'_1, c'_2, e'_1, e'_2 \rangle \to x[y, q', c_1, c_2, e_1, e_2], \langle x, q_F, c'_1, c'_2, e'_1, e'_2 \rangle \to x |$$
$$(x, q, c_1, c_2, q', e_1, e_2, +1) \in R, c'_1, c'_2 \in \{Z, B\}, e'_1, e'_2 \in \{-1, 0, +1\},$$
$$x, y \in \Sigma\}$$

$$P^{c_1}_{GM_{S1(S3)}} = \{S \to [I], [I] \to C\} \cup \mathfrak{N} = \{C \to Q^{c_1}_{a_1 P_{a_1} S_1(S_3)}\} \cup$$
$$\{\langle I \rangle \to [x, q, Z, Z, e_1, e_2] | (x, q_0, Z, Z, e_1, e_2, 0) \in R, x \in \Sigma\} \cup$$
$$\{\langle I \rangle \to x[y, q, Z, Z, e_1, e_2] | (x, q_0, Z, Z, q, e_1, e_2, +1) \in R, x, y \in \Sigma\} \cup$$
$$\{\langle x, q, c'_1, c'_2, e'_1, e'_2 \rangle \to [x, q', c_1, c_2, e_1, e_2] | (x, q, c_1, c_2, q', e_1, e_2, 0) \in R,$$
$$x \in \Sigma, c'_1, c'_2 \in \{Z, B\}, e'_1, e'_2 \in \{-1, 0, +1\}\} \cup$$
$$\{\langle x, q, c'_1, c'_2, e'_1, e'_2 \rangle \to x[y, q', c_1, c_2, e_1, e_2], \langle x, q_F, c'_1, c'_2, e'_1, e'_2 \rangle \to x |$$
$$(x, q, c_1, c_2, q', e_1, e_2, +1) \in R, c'_1, c'_2 \in \{Z, B\}, e'_1, e'_2 \in \{-1, 0, +1\},$$
$$x, y \in \Sigma\}$$

We only need one $P^{c_1}_2$ component. The grammar below will simulate rules from the master grammar and will work indirectly with $P^{c_1}_{2_{OriginalS2}}$ holding intermediate strings.

$$P^{c_1}_{GM_{S2}} = \{S \to [I], [I] \to C\} \cup \mathfrak{N} = \{C \to Q^{c_1}_{a_1 P_{a_1} S_2}\} \cup$$
$$\{\langle I \rangle \to [x, q, Z, Z, e_1, e_2] | (x, q_0, Z, Z, e_1, e_2, 0) \in R, x \in \Sigma\} \cup$$
$$\{\langle I \rangle \to x[y, q, Z, Z, e_1, e_2] | (x, q_0, Z, Z, q, e_1, e_2, +1) \in R, x, y \in \Sigma\} \cup$$
$$\{\langle x, q, c'_1, c'_2, e'_1, e'_2 \rangle \to [x, q', c_1, c_2, e_1, e_2] | (x, q, c_1, c_2, q', e_1, e_2, 0) \in R,$$
$$x \in \Sigma, c'_1, c'_2 \in \{Z, B\}, e'_1, e'_2 \in \{-1, 0, +1\}\} \cup$$
$$\{\langle x, q, c'_1, c'_2, e'_1, e'_2 \rangle \to x[y, q', c_1, c_2, e_1, e_2], \langle x, q_F, c'_1, c'_2, e'_1, e'_2 \rangle \to x |$$
$$(x, q, c_1, c_2, q', e_1, e_2, +1) \in R, c'_1, c'_2 \in \{Z, B\}, e'_1, e'_2 \in \{-1, 0, +1\},$$
$$x, y \in \Sigma\}$$

Similar to the $P_2^{c_1}$ we only need one $P_3^{c_1}$, which will will work indirectly with $P_{3_{OriginalS3}}^{c_1}$.

$$P_{GM_{S3}}^{c_1} = \{S \to [I], [I] \to C\} \cup \mathfrak{N} = \{C \to Q_{a_1 P_{a_1} S_3}^{c_1}\} \cup$$
$$\{\langle I \rangle \to [x, q, Z, Z, e_1, e_2] | (x, q_0, Z, Z, e_1, e_2, 0) \in R, x \in \Sigma\} \cup$$
$$\{\langle I \rangle \to x[y, q, Z, Z, e_1, e_2] | (x, q_0, Z, Z, q, e_1, e_2, +1) \in R, x, y \in \Sigma\} \cup$$
$$\{\langle x, q, c_1', c_2', e_1', e_2' \rangle \to [x, q', c_1, c_2, e_1, e_2] | (x, q, c_1, c_2, q', e_1, e_2, 0) \in R,$$
$$x \in \Sigma, c_1', c_2' \in \{Z, B\}, e_1', e_2' \in \{-1, 0, +1\}\} \cup$$
$$\{\langle x, q, c_1', c_2', e_1', e_2' \rangle \to x[y, q', c_1, c_2, e_1, e_2], \langle x, q_F, c_1', c_2', e_1', e_2' \rangle \to x|$$
$$(x, q, c_1, c_2, q', e_1, e_2, +1) \in R, c_1', c_2' \in \{Z, B\}, e_1', e_2' \in \{-1, 0, +1\},$$
$$x, y \in \Sigma\}$$

The following 7 helper grammars imitate P_{a_1}. The first 5 work with $P_{1_{original}}^{c_1}$ and four of its helpers, while the remaining 2 work with $P_{2_{original}}^{c_1}$ and $P_{3_{original}}^{c_1}$ holding intermediate strings during derivations. A new rule allows these grammars to reset themselves by querying their new helper component defined later in the "reset" section.

$$P_{GM_{PA1S1}}^{c_1} = \{S \to [I], [I] \to C\} \cup$$
$$\mathfrak{N} = \{C \to Q_{Reset_{GM_{Pa1_{S1}^{c_1}}}}\} \cup$$
$$\{\langle I \rangle \to [x, q, Z, Z, e_1, e_2] | (x, q_0, Z, Z, e_1, e_2, 0) \in R, x \in \Sigma\} \cup$$
$$\{\langle I \rangle \to x[y, q, Z, Z, e_1, e_2] | (x, q_0, Z, Z, q, e_1, e_2, +1) \in R, x, y \in \Sigma\} \cup$$
$$\{\langle x, q, c_1', c_2', e_1', e_2' \rangle \to [x, q', c_1, c_2, e_1, e_2] | (x, q, c_1, c_2, q', e_1, e_2, 0) \in R,$$
$$x \in \Sigma, c_1', c_2' \in \{Z, B\}, e_1', e_2' \in \{-1, 0, +1\}\} \cup$$
$$\{\langle x, q, c_1', c_2', e_1', e_2' \rangle \to x[y, q', c_1, c_2, e_1, e_2], \langle x, q_F, c_1', c_2', e_1', e_2' \rangle \to x|$$
$$(x, q, c_1, c_2, q', e_1, e_2, +1) \in R, c_1', c_2' \in \{Z, B\}, e_1', e_2' \in \{-1, 0, +1\},$$
$$x, y \in \Sigma\}$$

$$P_{GM_{PA1S1H2}}^{c_1} = \{S \to [I], [I] \to C\} \cup$$
$$\mathfrak{N} = \{C \to Q_{Reset_{GM_{Pa1S1H2(S4)}^{c_1}}}\} \cup$$
$$\{\langle I \rangle \to [x, q, Z, Z, e_1, e_2] | (x, q_0, Z, Z, e_1, e_2, 0) \in R, x \in \Sigma\} \cup$$
$$\{\langle I \rangle \to x[y, q, Z, Z, e_1, e_2] | (x, q_0, Z, Z, q, e_1, e_2, +1) \in R, x, y \in \Sigma\} \cup$$
$$\{\langle x, q, c_1', c_2', e_1', e_2' \rangle \to [x, q', c_1, c_2, e_1, e_2] | (x, q, c_1, c_2, q', e_1, e_2, 0) \in R,$$
$$x \in \Sigma, c_1', c_2' \in \{Z, B\}, e_1', e_2' \in \{-1, 0, +1\}\} \cup$$
$$\{\langle x, q, c_1', c_2', e_1', e_2' \rangle \to x[y, q', c_1, c_2, e_1, e_2], \langle x, q_F, c_1', c_2', e_1', e_2' \rangle \to x|$$
$$(x, q, c_1, c_2, q', e_1, e_2, +1) \in R, c_1', c_2' \in \{Z, B\}, e_1', e_2' \in \{-1, 0, +1\},$$
$$x, y \in \Sigma\}$$

$$P^{c_1}_{GM_{PA1S1H3}} = \{S \to [I], [I] \to C\} \cup$$

$$\mathfrak{N} = \{C \to Q_{Reset_{GM^{c_1}_{Pa1S1H3(S4)}}}\} \cup$$

$$\{\langle I \rangle \to [x, q, Z, Z, e_1, e_2] | (x, q_0, Z, Z, e_1, e_2, 0) \in R, x \in \Sigma\} \cup$$

$$\{\langle I \rangle \to x[y, q, Z, Z, e_1, e_2] | (x, q_0, Z, Z, q, e_1, e_2, +1) \in R, x, y \in \Sigma\} \cup$$

$$\{\langle x, q, c'_1, c'_2, e'_1, e'_2 \rangle \to [x, q', c_1, c_2, e_1, e_2] | (x, q, c_1, c_2, q', e_1, e_2, 0) \in R,$$
$$x \in \Sigma, c'_1, c'_2 \in \{Z, B\}, e'_1, e'_2 \in \{-1, 0, +1\}\} \cup$$

$$\{\langle x, q, c'_1, c'_2, e'_1, e'_2 \rangle \to x[y, q', c_1, c_2, e_1, e_2], \langle x, q_F, c'_1, c'_2, e'_1, e'_2 \rangle \to x|$$
$$(x, q, c_1, c_2, q', e_1, e_2, +1) \in R, c'_1, c'_2 \in \{Z, B\}, e'_1, e'_2 \in \{-1, 0, +1\},$$
$$x, y \in \Sigma\}$$

$$P^{c_1}_{GM_{PA1S1(S2)}} = \{S \to [I], [I] \to C\} \cup$$

$$\mathfrak{N} = \{C \to Q_{Reset_{GM^{c_1}_{Pa1S1(S2)}}}\} \cup$$

$$\{\langle I \rangle \to [x, q, Z, Z, e_1, e_2] | (x, q_0, Z, Z, e_1, e_2, 0) \in R, x \in \Sigma\} \cup$$

$$\{\langle I \rangle \to x[y, q, Z, Z, e_1, e_2] | (x, q_0, Z, Z, q, e_1, e_2, +1) \in R, x, y \in \Sigma\} \cup$$

$$\{\langle x, q, c'_1, c'_2, e'_1, e'_2 \rangle \to [x, q', c_1, c_2, e_1, e_2] | (x, q, c_1, c_2, q', e_1, e_2, 0) \in R,$$
$$x \in \Sigma, c'_1, c'_2 \in \{Z, B\}, e'_1, e'_2 \in \{-1, 0, +1\}\} \cup$$

$$\{\langle x, q, c'_1, c'_2, e'_1, e'_2 \rangle \to x[y, q', c_1, c_2, e_1, e_2], \langle x, q_F, c'_1, c'_2, e'_1, e'_2 \rangle \to x|$$
$$(x, q, c_1, c_2, q', e_1, e_2, +1) \in R, c'_1, c'_2 \in \{Z, B\}, e'_1, e'_2 \in \{-1, 0, +1\},$$
$$x, y \in \Sigma\}$$

$$P^{c_1}_{GM_{PA1S1(S3)}} = \{S \to [I], [I] \to C\} \cup \mathfrak{N} = \{C \to Q_{Reset_{GM^{c_1}_{Pa1S1(S3)}}}\} \cup$$

$$\{\langle I \rangle \to [x, q, Z, Z, e_1, e_2] | (x, q_0, Z, Z, e_1, e_2, 0) \in R, x \in \Sigma\} \cup$$

$$\{\langle I \rangle \to x[y, q, Z, Z, e_1, e_2] | (x, q_0, Z, Z, q, e_1, e_2, +1) \in R, x, y \in \Sigma\} \cup$$

$$\{\langle x, q, c'_1, c'_2, e'_1, e'_2 \rangle \to [x, q', c_1, c_2, e_1, e_2] | (x, q, c_1, c_2, q', e_1, e_2, 0) \in R,$$
$$x \in \Sigma, c'_1, c'_2 \in \{Z, B\}, e'_1, e'_2 \in \{-1, 0, +1\}\} \cup$$

$$\{\langle x, q, c'_1, c'_2, e'_1, e'_2 \rangle \to x[y, q', c_1, c_2, e_1, e_2], \langle x, q_F, c'_1, c'_2, e'_1, e'_2 \rangle \to x|$$
$$(x, q, c_1, c_2, q', e_1, e_2, +1) \in R, c'_1, c'_2 \in \{Z, B\}, e'_1, e'_2 \in \{-1, 0, +1\},$$
$$x, y \in \Sigma\}$$

$$P^{c_1}_{GM_{PA1S2}} = \{S \to [I], [I] \to C\} \cup \mathfrak{N} = \{C \to Q_{Reset_{GM^{c_1}_{Pa1S2}}}\} \cup$$

$$\{\langle I \rangle \to [x, q, Z, Z, e_1, e_2] | (x, q_0, Z, Z, e_1, e_2, 0) \in R, x \in \Sigma\} \cup$$

$$\{\langle I \rangle \to x[y, q, Z, Z, e_1, e_2] | (x, q_0, Z, Z, q, e_1, e_2, +1) \in R, x, y \in \Sigma\} \cup$$

$$\{\langle x, q, c'_1, c'_2, e'_1, e'_2 \rangle \to [x, q', c_1, c_2, e_1, e_2] | (x, q, c_1, c_2, q', e_1, e_2, 0) \in R,$$
$$x \in \Sigma, c'_1, c'_2 \in \{Z, B\}, e'_1, e'_2 \in \{-1, 0, +1\}\} \cup$$

$$\{\langle x, q, c'_1, c'_2, e'_1, e'_2 \rangle \to x[y, q', c_1, c_2, e_1, e_2], \langle x, q_F, c'_1, c'_2, e'_1, e'_2 \rangle \to x|$$
$$(x, q, c_1, c_2, q', e_1, e_2, +1) \in R, c'_1, c'_2 \in \{Z, B\}, e'_1, e'_2 \in \{-1, 0, +1\},$$
$$x, y \in \Sigma\}$$

$$P^{c_1}_{GM_{PA1S3}} = \{S \to [I], [I] \to C\} \cup \mathfrak{N} = \{C \to Q_{Reset_{GM^{c_1}_{Pa_1S3}}}\} \cup$$

$$\{\langle I \rangle \to [x, q, Z, Z, e_1, e_2] | (x, q_0, Z, Z, e_1, e_2, 0) \in R, x \in \Sigma\} \cup$$

$$\{\langle I \rangle \to x[y, q, Z, Z, e_1, e_2] | (x, q_0, Z, Z, q, e_1, e_2, +1) \in R, x, y \in \Sigma\} \cup$$

$$\{\langle x, q, c'_1, c'_2, e'_1, e'_2 \rangle \to [x, q', c_1, c_2, e_1, e_2] | (x, q, c_1, c_2, q', e_1, e_2, 0) \in R,$$
$$x \in \Sigma, c'_1, c'_2 \in \{Z, B\}, e'_1, e'_2 \in \{-1, 0, +1\}\} \cup$$

$$\{\langle x, q, c'_1, c'_2, e'_1, e'_2 \rangle \to x[y, q', c_1, c_2, e_1, e_2], \langle x, q_F, c'_1, c'_2, e'_1, e'_2 \rangle \to x|$$
$$(x, q, c_1, c_2, q', e_1, e_2, +1) \in R, c'_1, c'_2 \in \{Z, B\}, e'_1, e'_2 \in \{-1, 0, +1\},$$
$$x, y \in \Sigma\}$$

The following 5 helpers simulate rules from the new master and are designed to work with a different component in the $P^{c_2}_1$ family.

$$P^{c_2}_{GM_{S1}} = \{S \to [I], [I] \to C\} \cup \mathfrak{N} = \{C \to Q^{c_2}_{a_1 P_{a_1} S_1}\} \cup$$

$$\{\langle I \rangle \to [x, q, Z, Z, e_1, e_2] | (x, q_0, Z, Z, e_1, e_2, 0) \in R, x \in \Sigma\} \cup$$

$$\{\langle I \rangle \to x[y, q, Z, Z, e_1, e_2] | (x, q_0, Z, Z, q, e_1, e_2, +1) \in R, x, y \in \Sigma\} \cup$$

$$\{\langle x, q, c'_1, c'_2, e'_1, e'_2 \rangle \to [x, q', c_1, c_2, e_1, e_2] | (x, q, c_1, c_2, q', e_1, e_2, 0) \in R,$$
$$x \in \Sigma, c'_1, c'_2 \in \{Z, B\}, e'_1, e'_2 \in \{-1, 0, +1\}\} \cup$$

$$\{\langle x, q, c'_1, c'_2, e'_1, e'_2 \rangle \to x[y, q', c_1, c_2, e_1, e_2], \langle x, q_F, c'_1, c'_2, e'_1, e'_2 \rangle \to x|$$
$$(x, q, c_1, c_2, q', e_1, e_2, +1) \in R, c'_1, c'_2 \in \{Z, B\}, e'_1, e'_2 \in \{-1, 0, +1\},$$
$$x, y \in \Sigma\}$$

$$P^{c_2}_{GM_{S1H2(S4)}} = \{S \to [I], [I] \to C\} \cup \mathfrak{N} = \{C \to Q^{c_2}_{a_1 P_{a_1} S_1 H_2(S_4)}, S \to Q^{c_2}_{a_1 P_{a_1} S_1 H_2(S_4)}\} \cup$$

$$\{\langle I \rangle \to [x, q, Z, Z, e_1, e_2] | (x, q_0, Z, Z, e_1, e_2, 0) \in R, x \in \Sigma\} \cup$$

$$\{\langle I \rangle \to x[y, q, Z, Z, e_1, e_2] | (x, q_0, Z, Z, q, e_1, e_2, +1) \in R, x, y \in \Sigma\} \cup$$

$$\{\langle x, q, c'_1, c'_2, e'_1, e'_2 \rangle \to [x, q', c_1, c_2, e_1, e_2] | (x, q, c_1, c_2, q', e_1, e_2, 0) \in R,$$
$$x \in \Sigma, c'_1, c'_2 \in \{Z, B\}, e'_1, e'_2 \in \{-1, 0, +1\}\} \cup$$

$$\{\langle x, q, c'_1, c'_2, e'_1, e'_2 \rangle \to x[y, q', c_1, c_2, e_1, e_2], \langle x, q_F, c'_1, c'_2, e'_1, e'_2 \rangle \to x|$$
$$(x, q, c_1, c_2, q', e_1, e_2, +1) \in R, c'_1, c'_2 \in \{Z, B\}, e'_1, e'_2 \in \{-1, 0, +1\},$$
$$x, y \in \Sigma\}$$

$$P^{c_2}_{GM_{S1H3(S4)}} = \{S \to [I], [I] \to C\} \cup \mathfrak{N} = \{C \to Q^{c_2}_{a_1 P_{a_1} S_1 H_3(S_4)}, S \to Q^{c_2}_{a_1 P_{a_1} S_1 H_3(S_4)}\} \cup$$

$$\{\langle I \rangle \to [x, q, Z, Z, e_1, e_2] | (x, q_0, Z, Z, e_1, e_2, 0) \in R, x \in \Sigma\} \cup$$

$$\{\langle I \rangle \to x[y, q, Z, Z, e_1, e_2] | (x, q_0, Z, Z, q, e_1, e_2, +1) \in R, x, y \in \Sigma\} \cup$$

$$\{\langle x, q, c'_1, c'_2, e'_1, e'_2 \rangle \to [x, q', c_1, c_2, e_1, e_2] | (x, q, c_1, c_2, q', e_1, e_2, 0) \in R,$$
$$x \in \Sigma, c'_1, c'_2 \in \{Z, B\}, e'_1, e'_2 \in \{-1, 0, +1\}\} \cup$$

$$\{\langle x, q, c'_1, c'_2, e'_1, e'_2 \rangle \to x[y, q', c_1, c_2, e_1, e_2], \langle x, q_F, c'_1, c'_2, e'_1, e'_2 \rangle \to x|$$
$$(x, q, c_1, c_2, q', e_1, e_2, +1) \in R, c'_1, c'_2 \in \{Z, B\}, e'_1, e'_2 \in \{-1, 0, +1\},$$
$$x, y \in \Sigma\}$$

$P^{c_2}_{GM_{S1(S2)}} = \{S \rightarrow [I], [I] \rightarrow C\} \cup \mathfrak{N} = \{C \rightarrow Q^{c_2}_{a_1 P_{a_1} S_1(S_2)}\} \cup$

$\{\langle I \rangle \rightarrow [x, q, Z, Z, e_1, e_2] | (x, q_0, Z, Z, e_1, e_2, 0) \in R, x \in \Sigma\} \cup$

$\{\langle I \rangle \rightarrow x[y, q, Z, Z, e_1, e_2] | (x, q_0, Z, Z, q, e_1, e_2, +1) \in R, x, y \in \Sigma\} \cup$

$\{\langle x, q, c'_1, c'_2, e'_1, e'_2 \rangle \rightarrow [x, q', c_1, c_2, e_1, e_2] | (x, q, c_1, c_2, q', e_1, e_2, 0) \in R,$

$\quad x \in \Sigma, c'_1, c'_2 \in \{Z, B\}, e'_1, e'_2 \in \{-1, 0, +1\}\} \cup$

$\{\langle x, q, c'_1, c'_2, e'_1, e'_2 \rangle \rightarrow x[y, q', c_1, c_2, e_1, e_2], \langle x, q_F, c'_1, c'_2, e'_1, e'_2 \rangle \rightarrow x|$

$\quad (x, q, c_1, c_2, q', e_1, e_2, +1) \in R, c'_1, c'_2 \in \{Z, B\}, e'_1, e'_2 \in \{-1, 0, +1\},$

$\quad x, y \in \Sigma\}$

$P^{c_2}_{GM_{S1(S3)}} = \{S \rightarrow [I], [I] \rightarrow C\} \cup \mathfrak{N} = \{C \rightarrow Q^{c_2}_{a_1 P_{a_1} S_1(S_2)}\} \cup$

$\{\langle I \rangle \rightarrow [x, q, Z, Z, e_1, e_2] | (x, q_0, Z, Z, e_1, e_2, 0) \in R, x \in \Sigma\} \cup$

$\{\langle I \rangle \rightarrow x[y, q, Z, Z, e_1, e_2] | (x, q_0, Z, Z, q, e_1, e_2, +1) \in R, x, y \in \Sigma\} \cup$

$\{\langle x, q, c'_1, c'_2, e'_1, e'_2 \rangle \rightarrow [x, q', c_1, c_2, e_1, e_2] | (x, q, c_1, c_2, q', e_1, e_2, 0) \in R,$

$\quad x \in \Sigma, c'_1, c'_2 \in \{Z, B\}, e'_1, e'_2 \in \{-1, 0, +1\}\} \cup$

$\{\langle x, q, c'_1, c'_2, e'_1, e'_2 \rangle \rightarrow x[y, q', c_1, c_2, e_1, e_2], \langle x, q_F, c'_1, c'_2, e'_1, e'_2 \rangle \rightarrow x|$

$\quad (x, q, c_1, c_2, q', e_1, e_2, +1) \in R, c'_1, c'_2 \in \{Z, B\}, e'_1, e'_2 \in \{-1, 0, +1\},$

$\quad x, y \in \Sigma\}$

There is only one $P^{c_2}_2$ and one $P^{c_2}_3$, as in the original system. The master helpers below work indirectly with them.

$P^{c_2}_{GM_{S2}} = \{S \rightarrow [I], [I] \rightarrow C\} \cup \mathfrak{N} = \{C \rightarrow Q^{c_2}_{a_1 P_{a_1} S_2}\} \cup$

$\{\langle I \rangle \rightarrow [x, q, Z, Z, e_1, e_2] | (x, q_0, Z, Z, e_1, e_2, 0) \in R, x \in \Sigma\} \cup$

$\{\langle I \rangle \rightarrow x[y, q, Z, Z, e_1, e_2] | (x, q_0, Z, Z, q, e_1, e_2, +1) \in R, x, y \in \Sigma\} \cup$

$\{\langle x, q, c'_1, c'_2, e'_1, e'_2 \rangle \rightarrow [x, q', c_1, c_2, e_1, e_2] | (x, q, c_1, c_2, q', e_1, e_2, 0) \in R,$

$\quad x \in \Sigma, c'_1, c'_2 \in \{Z, B\}, e'_1, e'_2 \in \{-1, 0, +1\}\} \cup$

$\{\langle x, q, c'_1, c'_2, e'_1, e'_2 \rangle \rightarrow x[y, q', c_1, c_2, e_1, e_2], \langle x, q_F, c'_1, c'_2, e'_1, e'_2 \rangle \rightarrow x|$

$\quad (x, q, c_1, c_2, q', e_1, e_2, +1) \in R, c'_1, c'_2 \in \{Z, B\}, e'_1, e'_2 \in \{-1, 0, +1\},$

$\quad x, y \in \Sigma\}$

$P^{c_2}_{GM_{S3}} = \{S \rightarrow [I], [I] \rightarrow C\} \cup \mathfrak{N} = \{C \rightarrow Q^{c_2}_{a_1 P_{a_1} S_3}\} \cup$

$\{\langle I \rangle \rightarrow [x, q, Z, Z, e_1, e_2] | (x, q_0, Z, Z, e_1, e_2, 0) \in R, x \in \Sigma\} \cup$

$\{\langle I \rangle \rightarrow x[y, q, Z, Z, e_1, e_2] | (x, q_0, Z, Z, q, e_1, e_2, +1) \in R, x, y \in \Sigma\} \cup$

$\{\langle x, q, c'_1, c'_2, e'_1, e'_2 \rangle \rightarrow [x, q', c_1, c_2, e_1, e_2] | (x, q, c_1, c_2, q', e_1, e_2, 0) \in R,$

$\quad x \in \Sigma, c'_1, c'_2 \in \{Z, B\}, e'_1, e'_2 \in \{-1, 0, +1\}\} \cup$

$\{\langle x, q, c'_1, c'_2, e'_1, e'_2 \rangle \rightarrow x[y, q', c_1, c_2, e_1, e_2], \langle x, q_F, c'_1, c'_2, e'_1, e'_2 \rangle \rightarrow x|$

$\quad (x, q, c_1, c_2, q', e_1, e_2, +1) \in R, c'_1, c'_2 \in \{Z, B\}, e'_1, e'_2 \in \{-1, 0, +1\},$

$\quad x, y \in \Sigma\}$

The following 7 grammars work with the $P^{c_2}_{a_1}$ components; the first 5 work with the $P^{c_2}_1$ helper grammars, and the other 2 work with $P^{c_2}_{2_{Original S_2}}$ and $P^{c_2}_{3_{Original S_3}}$. A new rule has been added to these components which allows them to reset themselves by querying their matching reset component (defined later).

$$P^{c2}_{GM_{PA1S1}} = \{S \to [I], [I] \to C\} \cup \mathfrak{N} = \{C \to Q_{Reset_{GM^{c2}_{Pa1S1}}}\} \cup$$

$$\{\langle I \rangle \to [x, q, Z, Z, e_1, e_2] | (x, q_0, Z, Z, e_1, e_2, 0) \in R, x \in \Sigma\} \cup$$
$$\{\langle I \rangle \to x[y, q, Z, Z, e_1, e_2] | (x, q_0, Z, Z, q, e_1, e_2, +1) \in R, x, y \in \Sigma\} \cup$$
$$\{\langle x, q, c'_1, c'_2, e'_1, e'_2 \rangle \to [x, q', c_1, c_2, e_1, e_2] | (x, q, c_1, c_2, q', e_1, e_2, 0) \in R,$$
$$\quad x \in \Sigma, c'_1, c'_2 \in \{Z, B\}, e'_1, e'_2 \in \{-1, 0, +1\}\} \cup$$
$$\{\langle x, q, c'_1, c'_2, e'_1, e'_2 \rangle \to x[y, q', c_1, c_2, e_1, e_2], \langle x, q_F, c'_1, c'_2, e'_1, e'_2 \rangle \to x|$$
$$\quad (x, q, c_1, c_2, q', e_1, e_2, +1) \in R, c'_1, c'_2 \in \{Z, B\}, e'_1, e'_2 \in \{-1, 0, +1\},$$
$$\quad x, y \in \Sigma\}$$

$$P^{c2}_{GM_{PA1S1H2}} = \{S \to [I], [I] \to C\} \cup \mathfrak{N} = \{C \to Q_{Reset_{GM^{c2}_{Pa1S1H2(S4)}}}\} \cup$$

$$\{\langle I \rangle \to [x, q, Z, Z, e_1, e_2] | (x, q_0, Z, Z, e_1, e_2, 0) \in R, x \in \Sigma\} \cup$$
$$\{\langle I \rangle \to x[y, q, Z, Z, e_1, e_2] | (x, q_0, Z, Z, q, e_1, e_2, +1) \in R, x, y \in \Sigma\} \cup$$
$$\{\langle x, q, c'_1, c'_2, e'_1, e'_2 \rangle \to [x, q', c_1, c_2, e_1, e_2] | (x, q, c_1, c_2, q', e_1, e_2, 0) \in R,$$
$$\quad x \in \Sigma, c'_1, c'_2 \in \{Z, B\}, e'_1, e'_2 \in \{-1, 0, +1\}\} \cup$$
$$\{\langle x, q, c'_1, c'_2, e'_1, e'_2 \rangle \to x[y, q', c_1, c_2, e_1, e_2], \langle x, q_F, c'_1, c'_2, e'_1, e'_2 \rangle \to x|$$
$$\quad (x, q, c_1, c_2, q', e_1, e_2, +1) \in R, c'_1, c'_2 \in \{Z, B\}, e'_1, e'_2 \in \{-1, 0, +1\},$$
$$\quad x, y \in \Sigma\}$$

$$P^{c2}_{GM_{PA1S1H3}} = \{S \to [I], [I] \to C\} \cup \mathfrak{N} = \{C \to Q_{Reset_{GM^{c2}_{Pa1S1H3(S4)}}}\} \cup$$

$$\{\langle I \rangle \to [x, q, Z, Z, e_1, e_2] | (x, q_0, Z, Z, e_1, e_2, 0) \in R, x \in \Sigma\} \cup$$
$$\{\langle I \rangle \to x[y, q, Z, Z, e_1, e_2] | (x, q_0, Z, Z, q, e_1, e_2, +1) \in R, x, y \in \Sigma\} \cup$$
$$\{\langle x, q, c'_1, c'_2, e'_1, e'_2 \rangle \to [x, q', c_1, c_2, e_1, e_2] | (x, q, c_1, c_2, q', e_1, e_2, 0) \in R,$$
$$\quad x \in \Sigma, c'_1, c'_2 \in \{Z, B\}, e'_1, e'_2 \in \{-1, 0, +1\}\} \cup$$
$$\{\langle x, q, c'_1, c'_2, e'_1, e'_2 \rangle \to x[y, q', c_1, c_2, e_1, e_2], \langle x, q_F, c'_1, c'_2, e'_1, e'_2 \rangle \to x|$$
$$\quad (x, q, c_1, c_2, q', e_1, e_2, +1) \in R, c'_1, c'_2 \in \{Z, B\}, e'_1, e'_2 \in \{-1, 0, +1\},$$
$$\quad x, y \in \Sigma\}$$

$$P^{c2}_{GM_{PA1S1S2}} = \{S \to [I], [I] \to C\} \cup \mathfrak{N} = \{C \to Q_{Reset_{GM^{c2}_{Pa1S1(S2)}}}\} \cup$$

$$\{\langle I \rangle \to [x, q, Z, Z, e_1, e_2] | (x, q_0, Z, Z, e_1, e_2, 0) \in R, x \in \Sigma\} \cup$$
$$\{\langle I \rangle \to x[y, q, Z, Z, e_1, e_2] | (x, q_0, Z, Z, q, e_1, e_2, +1) \in R, x, y \in \Sigma\} \cup$$
$$\{\langle x, q, c'_1, c'_2, e'_1, e'_2 \rangle \to [x, q', c_1, c_2, e_1, e_2] | (x, q, c_1, c_2, q', e_1, e_2, 0) \in R,$$
$$\quad x \in \Sigma, c'_1, c'_2 \in \{Z, B\}, e'_1, e'_2 \in \{-1, 0, +1\}\} \cup$$
$$\{\langle x, q, c'_1, c'_2, e'_1, e'_2 \rangle \to x[y, q', c_1, c_2, e_1, e_2], \langle x, q_F, c'_1, c'_2, e'_1, e'_2 \rangle \to x|$$
$$\quad (x, q, c_1, c_2, q', e_1, e_2, +1) \in R, c'_1, c'_2 \in \{Z, B\}, e'_1, e'_2 \in \{-1, 0, +1\},$$
$$\quad x, y \in \Sigma\}$$

$$P^{c_2}_{GM_{PA1S1S3}} = \{S \to [I], [I] \to C\} \cup \mathfrak{N} = \{C \to Q_{Reset_{GM^{c_2}_{Pa1S1(S3)}}}\} \cup$$

$$\{\langle I \rangle \to [x, q, Z, Z, e_1, e_2] | (x, q_0, Z, Z, e_1, e_2, 0) \in R, x \in \Sigma\} \cup$$

$$\{\langle I \rangle \to x[y, q, Z, Z, e_1, e_2] | (x, q_0, Z, Z, q, e_1, e_2, +1) \in R, x, y \in \Sigma\} \cup$$

$$\{\langle x, q, c'_1, c'_2, e'_1, e'_2 \rangle \to [x, q', c_1, c_2, e_1, e_2] | (x, q, c_1, c_2, q', e_1, e_2, 0) \in R,$$

$$x \in \Sigma, c'_1, c'_2 \in \{Z, B\}, e'_1, e'_2 \in \{-1, 0, +1\}\} \cup$$

$$\{\langle x, q, c'_1, c'_2, e'_1, e'_2 \rangle \to x[y, q', c_1, c_2, e_1, e_2], \langle x, q_F, c'_1, c'_2, e'_1, e'_2 \rangle \to x|$$

$$(x, q, c_1, c_2, q', e_1, e_2, +1) \in R, c'_1, c'_2 \in \{Z, B\}, e'_1, e'_2 \in \{-1, 0, +1\},$$

$$x, y \in \Sigma\}$$

$$P^{c_2}_{GM_{PA1S2}} = \{S \to [I], [I] \to C\} \cup \mathfrak{N} = \{C \to Q_{Reset_{GM^{c_2}_{Pa1S2}}}\} \cup$$

$$\{\langle I \rangle \to [x, q, Z, Z, e_1, e_2] | (x, q_0, Z, Z, e_1, e_2, 0) \in R, x \in \Sigma\} \cup$$

$$\{\langle I \rangle \to x[y, q, Z, Z, e_1, e_2] | (x, q_0, Z, Z, q, e_1, e_2, +1) \in R, x, y \in \Sigma\} \cup$$

$$\{\langle x, q, c'_1, c'_2, e'_1, e'_2 \rangle \to [x, q', c_1, c_2, e_1, e_2] | (x, q, c_1, c_2, q', e_1, e_2, 0) \in R,$$

$$x \in \Sigma, c'_1, c'_2 \in \{Z, B\}, e'_1, e'_2 \in \{-1, 0, +1\}\} \cup$$

$$\{\langle x, q, c'_1, c'_2, e'_1, e'_2 \rangle \to x[y, q', c_1, c_2, e_1, e_2], \langle x, q_F, c'_1, c'_2, e'_1, e'_2 \rangle \to x|$$

$$(x, q, c_1, c_2, q', e_1, e_2, +1) \in R, c'_1, c'_2 \in \{Z, B\}, e'_1, e'_2 \in \{-1, 0, +1\},$$

$$x, y \in \Sigma\}$$

$$P^{c_2}_{GM_{PA1S3}} = \{S \to [I], [I] \to C\} \cup \mathfrak{N} = \{C \to Q_{Reset_{GM^{c_2}_{Pa1S3}}}\} \cup$$

$$\{\langle I \rangle \to [x, q, Z, Z, e_1, e_2] | (x, q_0, Z, Z, e_1, e_2, 0) \in R, x \in \Sigma\} \cup$$

$$\{\langle I \rangle \to x[y, q, Z, Z, e_1, e_2] | (x, q_0, Z, Z, q, e_1, e_2, +1) \in R, x, y \in \Sigma\} \cup$$

$$\{\langle x, q, c'_1, c'_2, e'_1, e'_2 \rangle \to [x, q', c_1, c_2, e_1, e_2] | (x, q, c_1, c_2, q', e_1, e_2, 0) \in R,$$

$$x \in \Sigma, c'_1, c'_2 \in \{Z, B\}, e'_1, e'_2 \in \{-1, 0, +1\}\} \cup$$

$$\{\langle x, q, c'_1, c'_2, e'_1, e'_2 \rangle \to x[y, q', c_1, c_2, e_1, e_2], \langle x, q_F, c'_1, c'_2, e'_1, e'_2 \rangle \to x|$$

$$(x, q, c_1, c_2, q', e_1, e_2, +1) \in R, c'_1, c'_2 \in \{Z, B\}, e'_1, e'_2 \in \{-1, 0, +1\},$$

$$x, y \in \Sigma\}$$

$P^{c_1}_{1_{originalS_1}}$ contains the same rewriting rules and communication steps as the component $P^{c_1}_1$ in the original system [7], with some re-labelling of queries so that the components query their corresponding helper grammars in the other sections of the system. We now have 4 new helper grammars to ensure that $P^{c_1}_2$, $P^{c_1}_3$, and $P^{c_1}_4$ have their own unique component grammars to communicate with.

$$P^{c_1}_{1_{originalS_1}} = \mathfrak{N} = \{S_1 \to Q^{c_1}_{GM_{S_1}}, S_1 \to Q^{c_1}_{4S_{1original}}, C \to Q^{c_1}_{GM_{S_1}}\} \cup$$

$$[x, q, c_1, c_2, e_1, e_2] \to [e_1]', [+1]' \to AAC, [0]' \to AC, [-1]' \to C|$$

$$x \in \Sigma, q \in E, c_1, c_2 \in \{Z, B\}, e_1, e_2 \in \{-1, 0, +1\}\} \cup$$

$$\{[I] \to [I]', \quad [I]' \to AC\}$$

$$P^{c_1}_{1_{S_1 H_2(S_4)}} = \mathfrak{N} = \{S_1 \to Q^{c_1}_{GM_{S_1 H_2(S_4)}}, S_1 \to Q^{c_1}_{4S_1 H_2(S_4)}, C \to Q_{GM_{S_1 H_2(S_4)}},$$

$$C \to W\} \cup$$

$$[x, q, c_1, c_2, e_1, e_2] \to [e_1]', [+1]' \to AAC, [0]' \to AC, [-1]' \to C|$$

$$x \in \Sigma, q \in E, c_1, c_2 \in \{Z, B\}, e_1, e_2 \in \{-1, 0, +1\} \cup$$
$$\{[I] \rightarrow [I]', [I]' \rightarrow AC\}$$
$$P_{1_{S_1 H_3(S_4)}}^{c_1} = \mathfrak{N} = \{S_1 \rightarrow Q_{GM_{S_1 H_3(S_4)}}^{c_1}, S_1 \rightarrow Q_{4S_1 H_3(S_4)}^{c_1}, C \rightarrow Q_{GM_{S_1 H_3(S_4)}},$$
$$C \rightarrow W\} \cup$$
$$\{[x, q, c_1, c_2, e_1, e_2] \rightarrow [e_1]', [+1]' \rightarrow AAC, [0]' \rightarrow AC, [-1]' \rightarrow C|$$
$$x \in \Sigma, q \in E, c_1, c_2 \in \{Z, B\}, e_1, e_2 \in \{-1, 0, +1\} \cup$$
$$\{[I] \rightarrow [I]', [I]' \rightarrow AC\}$$

The following two $P_1^{c_1}$ helpers will ensure the proper derivation of $P_{2_{OriginalS2}}^{c_1}$ and $P_{3_{OriginalS3}}^{c_1}$. They work by communicating with their corresponding helper grammars and their designated special helper in the $P_4^{c_1}$ section.

$$P_{1_{S_1}(S_2)}^{c_1} = \mathfrak{N} = \{S_1 \rightarrow Q_{GM_{S_1(S_2)}}^{c_1}, S_1 \rightarrow Q_{4SpecialHelper1_{S1S2}}^{c_1}, C \rightarrow Q_{GM_{S_1(S_2)}},$$
$$S_4 \rightarrow S_4^{(1)}, S_4^{(1)} \rightarrow Q_{P_{1_{S_1 H_2(S_4)}}^{c_1}}\} \cup$$
$$\{[x, q, c_1, c_2, e_1, e_2] \rightarrow [e_1]', [+1]' \rightarrow AAC, [0]' \rightarrow AC, [-1]' \rightarrow C|$$
$$x \in \Sigma, q \in E, c_1, c_2 \in \{Z, B\}, e_1, e_2 \in \{-1, 0, +1\}\} \cup$$
$$\{[I] \rightarrow [I]', [I]' \rightarrow AC\}$$
$$P_{1_{S_1}(S_3)}^{c_1} = \mathfrak{N} = \{S_1 \rightarrow Q_{GM_{S_1(S_3)}}^{c_1}, S_1 \rightarrow Q_{4SpecialHelper1_{S1S3}}^{c_1}, C \rightarrow Q_{GM_{S_1(S_3)}}^{c_1},$$
$$S_4 \rightarrow S_4^{(1)}, S_4^{(1)} \rightarrow Q_{P_{1_{S_1 H_3(S_4)}}^{c_1}}\} \cup$$
$$\{[x, q, c_1, c_2, e_1, e_2] \rightarrow [e_1]', [+1]' \rightarrow AAC, [0]' \rightarrow AC, [-1]' \rightarrow C|$$
$$x \in \Sigma, q \in E, c_1, c_2 \in \{Z, B\}, e_1, e_2 \in \{-1, 0, +1\}\} \cup$$
$$\{[I] \rightarrow [I]', [I]' \rightarrow AC\}$$

The grammars $P_2^{c_1}$ and $P_3^{c_1}$ have been renamed and labels have been modified to ensure that they work with their matching helper components.

$$P_{2_{OriginalS2}}^{c_1} = \mathfrak{N} = \{S_2 \rightarrow Q_{GMS_2}^{c_1}, S_2 \rightarrow Q_{4S_2}^{c_1}, C \rightarrow Q_{GMS_2}^{c_1}, A \rightarrow A\} \cup$$
$$\{[x, q, Z, c_2, e_1, e_2] \rightarrow [x, q, Z, c_2, e_1, e_2], [I] \rightarrow [I]|x \in \Sigma, q \in E,$$
$$c_2 \in \{Z, B\}, e_1, e_2 \in \{-1, 0, +1\}\}$$
$$P_{3_{OriginalS3}}^{c_1} = \mathfrak{N} = \{S_3 \rightarrow Q_{GMS_3}^{c_1}, S_3 \rightarrow Q_{4S_3}^{c_1}, C \rightarrow Q_{GMS_3}^{c_1}\} \cup$$
$$\{[x, q, Z, c_2, e_1, e_2] \rightarrow a, [x, q, B, c_2, e_1, e_2] \rightarrow [x, q, B, c_2, e_1, e_2]$$
$$[I] \rightarrow [I]|x \in \Sigma, q \in E, c_2 \in \{Z, B\}, e_1, e_2 \in \{-1, 0, +1\}\}$$

The component $P_{4_{OriginalS4}}^{c_1}$, needs extra helper grammars to ensure that components defined in other sections have their own unique $P_4^{c_1}$ component to query.

$$P^{c_1}_{4_{Original S_4}} = \{S_4 \to S_4^{(1)}, S_4^{(1)} \to S_4^{(2)}\} \cup \mathfrak{N} = \{S_4^{(2)} \to Q^{c_1}_{P_1 S_1}\} \cup \{A \to a\}$$

$$P^{c_1}_{4_{S_1 H_2(S_4)}} = \{S_4 \to S_4^{(1)}, S_4^{(1)} \to S_4^{(2)}\} \cup \mathfrak{N} = \{S_4^{(2)} \to Q^{c_1}_{P_1 S_1 H_2(S_4)}, S_4^{(2)} \to S_4^{(2)}\} \cup$$
$$\{A \to a\}$$

$$P^{c_1}_{4_{S_1 H_3(S_4)}} = \{S_4 \to S_4^{(1)}, S_4^{(1)} \to S_4^{(2)}\} \cup \mathfrak{N} = \{S_4^{(2)} \to Q^{c_1}_{P_1 S_1 H_3(S_4)}, S_4^{(2)} \to S_4^{(2)}\} \cup$$
$$\{A \to a\}$$

$$P^{c_1}_{4_{S_2}} = \{S_4 \to S_4^{(1)}, S_4^{(1)} \to S_4^{(2)}\} \cup \mathfrak{N} = \{S_4^{(2)} \to Q^{c_1}_{P_1 S_2}\} \cup \{A \to a\}$$

$$P^{c_1}_{4_{S_3}} = \{S_4 \to S_4^{(1)}, S_4^{(1)} \to S_4^{(2)}\} \cup \mathfrak{N} = \{S_4^{(2)} \to Q^{c_1}_{P_1 S_3}\} \cup \{A \to a\}$$

$$P^{c_1}_{4_{Special Helper 1_{S1S2}}} = P^{c_1}_{4_{Special Helper 2_{S1S3}}} = \mathfrak{N} = \{S_4 \to S_4\}$$

$P^{c_2}_{1_{Original S_1}}$ is similar to the original $P^{c_2}_1$. It also needs 4 new helper grammars.

$$P^{c_2}_{1_{Original S_1}} = \mathfrak{N} = \{S_1 \to Q^{c_2}_{GM S_1}, S_1 \to Q^{c_2}_{P_4 S_1}, C \to Q^{c_2}_{GM S_1}\} \cup$$
$$\{[x, q, c_1, c_2, e_1, e_2] \to [e_2]', [+1]' \to AAC, [0] \to AC, [-1] \to C|$$
$$x \in \Sigma, q \in E, c_1, c_2 \in \{Z, B\}, e_1, e_2 \in \{-1, 0, +1\}\} \cup$$
$$\{[I] \to [I]', [I]' \to AC\}$$

$$P^{c_2}_{1_{S_1 H_2(S_4)}} = \mathfrak{N} = \{S_1 \to Q^{c_2}_{GM S_1 H_2(S_4)}, S_1 \to Q^{c_2}_{P_4 S_1 H_2(S_4)}, C \to Q^{c_2}_{GM S_1 H_2(S_4)}, C \to W\} \cup$$
$$\{[x, q, c_1, c_2, e_1, e_2] \to [e_2]', [+1]' \to AAC, [0] \to AC, [-1] \to C|$$
$$x \in \Sigma, q \in E, c_1, c_2 \in \{Z, B\}, e_1, e_2 \in \{-1, 0, +1\}\} \cup \{[I] \to [I]',$$
$$[I]' \to AC\}$$

$$P^{c_2}_{1_{S_1 H_3(S_4)}} = \mathfrak{N} = \{S_1 \to Q^{c_2}_{GM S_1 H_3(S_4)}, S_1 \to Q^{c_2}_{P_4 S_1 H_3(S_4)}, C \to Q^{c_2}_{GM S_1 H_3(S_4)},$$
$$C \to W\} \cup$$
$$\{[x, q, c_1, c_2, e_1, e_2] \to [e_2]', [+1]' \to AAC, [0] \to AC, [-1] \to C|$$
$$x \in \Sigma, q \in E, c_1, c_2 \in \{Z, B\}, e_1, e_2 \in \{-1, 0, +1\}\} \cup$$
$$\{[I] \to [I]', [I]' \to AC\}$$

$$P^{c_2}_{1_{S_1(S_2)}} = \mathfrak{N} = \{S_1 \to Q^{c_2}_{GM S_1(S_2)}, S_1 \to Q^{c_2}_{4 Special Helper 1_{S1S2}}, C \to Q^{c_2}_{GM S_1(S_2)},$$
$$S_4 \to S_4^{(1)}, S_4^{(1)} \to Q_{P^{c_2}_{1_{S_1 H_2(S_4)}}}\} \cup$$
$$\{[x, q, c_1, c_2, e_1, e_2] \to [e_2]', [+1]' \to AAC, [0] \to AC, [-1] \to C|$$
$$x \in \Sigma, q \in E, c_1, c_2 \in \{Z, B\}, e_1, e_2 \in \{-1, 0, +1\}\} \cup$$
$$\{[I] \to [I]', [I]' \to AC\}$$

$$P^{c_2}_{1_{S_1(S_3)}} = \mathfrak{N} = \{S_1 \rightarrow Q^{c_2}_{GM_{S_1(S_3)}}, S_1 \rightarrow Q^{c_2}_{4SpecialHelper1_{S1S3}}, C \rightarrow Q^{c_2}_{GM_{S_1(S_3)}},$$

$$S_4 \rightarrow S_4^{(1)}, S_4^{(1)} \rightarrow Q_{P^{c_2}_{1_{S_1 H_3(S_4)}}}\} \cup$$

$$\{[x, q, c_1, c_2, e_1, e_2] \rightarrow [e_2]', [+1]' \rightarrow AAC, [0] \rightarrow AC, [-1] \rightarrow C|$$

$$x \in \Sigma, q \in E, c_1, c_2 \in \{Z, B\}, e_1, e_2 \in \{-1, 0, +1\}\} \cup$$

$$\{[I] \rightarrow [I]', [I]' \rightarrow AC\}$$

The components $P_2^{c_2}$ and $P_3^{c_2}$ are the same as in the original system, except for modified labels. They do not need any helper.

$$P^{c_2}_{2_{OriginalS_2}} = \mathfrak{N} = \{S_2 \rightarrow Q^{c_2}_{GM_{S_2}}, S_2 \rightarrow Q^{c_2}_{P_4 S_2}, C \rightarrow Q^{c_2}_{GM_{S_2}}\} \cup \{A \rightarrow A\} \cup$$

$$\{[x, q, c_1, Z, e_1, e_2] \rightarrow a, [x, q, c_1, B, e_1, e_2] \rightarrow [x, q, c_1, B, e_1, e_2],$$

$$[I] \rightarrow [I]| \quad x \in \Sigma, q \in E, c1 \in \{Z, B\}, e_1, e_2 \in \{-1, 0, +1\}\}$$

$$P^{c_2}_{3_{OriginalS3}} = \mathfrak{N} = \{S_3 \rightarrow Q^{c_2}_{GM_{S_3}}, S_3 \rightarrow Q^{c_2}_{P_4 S_2}, C \rightarrow Q^{c_2}_{GM_{S_3}}\} \cup$$

$$\{[x, q, c_1, Z, e_1, e_2] \rightarrow a, [x, q, c_1, B, e_1, e_2] \rightarrow [x, q, c_1, B, e_1, e_2]$$

$$[I] \rightarrow [I]|x \in \Sigma, q \in E, c1 \in \{Z, B\}, e_1, e_2 \in \{-1, 0, +1\}\}$$

The component $P^{c_2}_{4_{OriginalS4}}$ on the other hand requires 6 additional helper components.

$$P^{c_2}_{4_{OriginalS4}} = \{S_4 \rightarrow S_4^{(1)}, S_4^{(1)} \rightarrow S_4^{(2)}\} \cup \mathfrak{N} = \{S_4^{(2)} \rightarrow Q^{c_2}_{P_1 S_1}\} \cup \{A \rightarrow a\}$$

$$P^{c_2}_{4_{S_1 H_2(S_4)}} = \{S_4 \rightarrow S_4^{(1)}, S_4^{(1)} \rightarrow S_4^{(2)}\} \cup$$

$$\mathfrak{N} = \{S_4^{(2)} \rightarrow Q^{c_2}_{P_1 S_1 H_2(S_4)}, S_4^{(2)} \rightarrow S_4^{(2)}\} \cup \{A \rightarrow a\}$$

$$P^{c_2}_{4_{S_1 H_3(S_4)}} = \{S_4 \rightarrow S_4^{(1)}, S_4^{(1)} \rightarrow S_4^{(2)}\} \cup$$

$$\mathfrak{N} = \{S_4^{(2)} \rightarrow Q^{c_2}_{P_1 S_1 H_3(S_4)}, S_4^{(2)} \rightarrow S_4^{(2)}\} \cup \{A \rightarrow a\}$$

$$P^{c_2}_{4_{S_2}} = \{S_4 \rightarrow S_4^{(1)}, S_4^{(1)} \rightarrow S_4^{(2)}\} \cup \mathfrak{N} = \{S_4^{(2)} \rightarrow Q^{c_2}_{P_1 S_2}\} \cup \{A \rightarrow a\}$$

$$P^{c_2}_{4_{S_3}} = \{S_4 \rightarrow S_4^{(1)}, S_4^{(1)} \rightarrow S_4^{(2)}\} \cup \mathfrak{N} = \{S_4^{(2)} \rightarrow Q^{c_2}_{P_1 S_3}\} \cup \{A \rightarrow a\}$$

$$P^{c_2}_{4SpecialHelper1_{S1S2}} = P^{c_2}_{4SpecialHelper2_{S1S3}} = \mathfrak{N} = \{S_4 \rightarrow S_4\}$$

The original P_{a_1} grammar remains as it was in the original system. In order for component grammars in sections $P_1^{c_1}, P_2^{c_1}, P_3^{c_1}, P_4^{c_1}, P_1^{c_2}, P_2^{c_2}, P_3^{c_2}$, and $P_4^{c_2}$ to derive correctly 14 additional P_{a_1} helpers have been added to the system. Their names and labels reflect the components they will work with during a derivation.

$$P_{a_1 Original} = \mathfrak{N} = \{S \rightarrow Q_{GM_{original}}\} \cup$$
$$\{[I] \rightarrow \langle I \rangle, [x, q, c_1, c_2, e_1, e_2] \rightarrow \langle x, q, c_1, c_2, e_1, e_2 \rangle,$$
$$\langle x, q, c_1, c_2, e_1, e_2 \rangle \rightarrow \langle x, q, c_1, c_2, e_1, e_2 \rangle, \langle I \rangle \rightarrow \langle I \rangle |$$
$$x \in \Sigma, q \in E, c_1, c_2 \in \{Z, B\}, e_1, e_2 \in \{-1, 0, +1\}\}$$

$$P_{a_1 GMS_1}^{c_1} = \mathfrak{N} = \{S \rightarrow Q_{GMPA1S_1}^{c_1}, C \rightarrow C\} \cup$$
$$\{[I] \rightarrow \langle I \rangle, [x, q, c_1, c_2, e_1, e_2] \rightarrow \langle x, q, c_1, c_2, e_1, e_2 \rangle,$$
$$\langle x, q, c_1, c_2, e_1, e_2 \rangle \rightarrow \langle x, q, c_1, c_2, e_1, e_2 \rangle, \langle I \rangle \rightarrow \langle I \rangle |$$
$$x \in \Sigma, q \in E, c_1, c_2 \in \{Z, B\}, e_1, e_2 \in \{-1, 0, +1\}\}$$

$$P_{a_1 GMS_1 H_2(S_4)}^{c_1} = \mathfrak{N} = \{S \rightarrow Q_{GMPA1S_1 H_2(S_4)}^{c_1}, C \rightarrow C\} \cup$$
$$\{[I] \rightarrow \langle I \rangle, [x, q, c_1, c_2, e_1, e_2] \rightarrow \langle x, q, c_1, c_2, e_1, e_2 \rangle,$$
$$\langle x, q, c_1, c_2, e_1, e_2 \rangle \rightarrow \langle x, q, c_1, c_2, e_1, e_2 \rangle, \langle I \rangle \rightarrow \langle I \rangle |$$
$$x \in \Sigma, q \in E, c_1, c_2 \in \{Z, B\}, e_1, e_2 \in \{-1, 0, +1\}\}$$

$$P_{a_1 GMS_1 H_3(S_4)}^{c_1} = \mathfrak{N} = \{S \rightarrow Q_{GMPA1S_1 H_3(S_4)}^{c_1}, C \rightarrow C\} \cup$$
$$\{[I] \rightarrow \langle I \rangle, [x, q, c_1, c_2, e_1, e_2] \rightarrow \langle x, q, c_1, c_2, e_1, e_2 \rangle,$$
$$\langle x, q, c_1, c_2, e_1, e_2 \rangle \rightarrow \langle x, q, c_1, c_2, e_1, e_2 \rangle, \langle I \rangle \rightarrow \langle I \rangle |$$
$$x \in \Sigma, q \in E, c_1, c_2 \in \{Z, B\}, e_1, e_2 \in \{-1, 0, +1\}\}$$

$$P_{a_1 GMS_1(S_2)}^{c_1} = \mathfrak{N} = \{S \rightarrow Q_{GMS_1(S_2)}^{c_1}, C \rightarrow C\} \cup$$
$$\{[I] \rightarrow \langle I \rangle, [x, q, c_1, c_2, e_1, e_2] \rightarrow \langle x, q, c_1, c_2, e_1, e_2 \rangle,$$
$$\langle x, q, c_1, c_2, e_1, e_2 \rangle \rightarrow \langle x, q, c_1, c_2, e_1, e_2 \rangle, \langle I \rangle \rightarrow \langle I \rangle |$$
$$x \in \Sigma, q \in E, c_1, c_2 \in \{Z, B\}, e_1, e_2 \in \{-1, 0, +1\}\}$$

$$P_{a_1 GMS_1(S_3)}^{c_1} = \mathfrak{N} = \{S \rightarrow Q_{GMS_1(S_3)}^{c_1}, C \rightarrow C\} \cup$$
$$\{[I] \rightarrow \langle I \rangle, [x, q, c_1, c_2, e_1, e_2] \rightarrow \langle x, q, c_1, c_2, e_1, e_2 \rangle,$$
$$\langle x, q, c_1, c_2, e_1, e_2 \rangle \rightarrow \langle x, q, c_1, c_2, e_1, e_2 \rangle, \langle I \rangle \rightarrow \langle I \rangle |$$
$$x \in \Sigma, q \in E, c_1, c_2 \in \{Z, B\}, e_1, e_2 \in \{-1, 0, +1\}\}$$

$$P_{a_1 GMS_2}^{c_1} = \mathfrak{N} = \{S \rightarrow Q_{GMPA1S_2}^{c_1}, C \rightarrow C\} \cup$$
$$\{[I] \rightarrow \langle I \rangle, [x, q, c_1, c_2, e_1, e_2] \rightarrow \langle x, q, c_1, c_2, e_1, e_2 \rangle,$$
$$\langle x, q, c_1, c_2, e_1, e_2 \rangle \rightarrow \langle x, q, c_1, c_2, e_1, e_2 \rangle, \langle I \rangle \rightarrow \langle I \rangle |$$
$$x \in \Sigma, q \in E, c_1, c_2 \in \{Z, B\}, e_1, e_2 \in \{-1, 0, +1\}\}$$

$$P_{a_1 GMS_3}^{c_1} = \mathfrak{N} = \{S \rightarrow Q_{GMPA1S_3}^{c_1}, C \rightarrow C\} \cup$$
$$\{[I] \rightarrow \langle I \rangle, [x, q, c_1, c_2, e_1, e_2] \rightarrow \langle x, q, c_1, c_2, e_1, e_2 \rangle,$$
$$\langle x, q, c_1, c_2, e_1, e_2 \rangle \rightarrow \langle x, q, c_1, c_2, e_1, e_2 \rangle, \langle I \rangle \rightarrow \langle I \rangle |$$
$$x \in \Sigma, q \in E, c_1, c_2 \in \{Z, B\}, e_1, e_2 \in \{-1, 0, +1\}\}$$

$$P_{a_1 GMS_1}^{c_2} = \mathfrak{N} = \{S \rightarrow Q_{GMPA1S_1}^{c_2}, C \rightarrow C\} \cup$$
$$\{[I] \rightarrow \langle I \rangle, [x, q, c_1, c_2, e_1, e_2] \rightarrow \langle x, q, c_1, c_2, e_1, e_2 \rangle,$$
$$\langle x, q, c_1, c_2, e_1, e_2 \rangle \rightarrow \langle x, q, c_1, c_2, e_1, e_2 \rangle, \langle I \rangle \rightarrow \langle I \rangle |$$

$$x \in \Sigma, q \in E, c_1, c_2 \in \{Z, B\}, e_1, e_2 \in \{-1, 0, +1\}\}$$

$$P^{c_2}_{a_1 GMS_1 H_2(S_4)} = \mathfrak{N} = \{S \to Q^{c_2}_{GMPA1S_1 H_2(S_4)}, C \to C\} \cup$$

$$\{[I] \to \langle I \rangle, [x, q, c_1, c_2, e_1, e_2] \to \langle x, q, c_1, c_2, e_1, e_2 \rangle,$$

$$\langle x, q, c_1, c_2, e_1, e_2 \rangle \to \langle x, q, c_1, c_2, e_1, e_2 \rangle, \langle I \rangle \to \langle I \rangle |$$

$$x \in \Sigma, q \in E, c_1, c_2 \in \{Z, B\}, e_1, e_2 \in \{-1, 0, +1\}\}$$

$$P^{c_2}_{a_1 GMS_1 H_3(S_4)} = \mathfrak{N} = \{S \to Q^{c_2}_{GMPA1S_1 H_3(S_4)}, C \to C\} \cup$$

$$\{[I] \to \langle I \rangle, [x, q, c_1, c_2, e_1, e_2] \to \langle x, q, c_1, c_2, e_1, e_2 \rangle,$$

$$\langle x, q, c_1, c_2, e_1, e_2 \rangle \to \langle x, q, c_1, c_2, e_1, e_2 \rangle, \langle I \rangle \to \langle I \rangle |$$

$$x \in \Sigma, q \in E, c_1, c_2 \in \{Z, B\}, e_1, e_2 \in \{-1, 0, +1\}\}$$

$$P^{c_2}_{a_1 GMS_1(S_2)} = \mathfrak{N} = \{S \to Q^{c_2}_{GMS_1(S_2)}, C \to C\} \cup$$

$$\{[I] \to \langle I \rangle, [x, q, c_1, c_2, e_1, e_2] \to \langle x, q, c_1, c_2, e_1, e_2 \rangle,$$

$$\langle x, q, c_1, c_2, e_1, e_2 \rangle \to \langle x, q, c_1, c_2, e_1, e_2 \rangle, \langle I \rangle \to \langle I \rangle |$$

$$x \in \Sigma, q \in E, c_1, c_2 \in \{Z, B\}, e_1, e_2 \in \{-1, 0, +1\}\}$$

$$P^{c_2}_{a_1 GMS_1(S_3)} = \mathfrak{N} = \{S \to Q^{c_2}_{GMS_1(S_3)}, C \to C\} \cup$$

$$\{[I] \to \langle I \rangle, [x, q, c_1, c_2, e_1, e_2] \to \langle x, q, c_1, c_2, e_1, e_2 \rangle,$$

$$\langle x, q, c_1, c_2, e_1, e_2 \rangle \to \langle x, q, c_1, c_2, e_1, e_2 \rangle, \langle I \rangle \to \langle I \rangle |$$

$$x \in \Sigma, q \in E, c_1, c_2 \in \{Z, B\}, e_1, e_2 \in \{-1, 0, +1\}\}$$

$$P^{c_2}_{a_1 GMS_2} = \mathfrak{N} = \{S \to Q^{c_2}_{GMPA1S_2}, C \to C\} \cup$$

$$\{[I] \to \langle I \rangle, [x, q, c_1, c_2, e_1, e_2] \to \langle x, q, c_1, c_2, e_1, e_2 \rangle,$$

$$\langle x, q, c_1, c_2, e_1, e_2 \rangle \to \langle x, q, c_1, c_2, e_1, e_2 \rangle, \langle I \rangle \to \langle I \rangle |$$

$$x \in \Sigma, q \in E, c_1, c_2 \in \{Z, B\}, e_1, e_2 \in \{-1, 0, +1\}\}$$

$$P^{c_2}_{a_1 GMS_3} = \mathfrak{N} = \{S \to Q^{c_2}_{GMPA1S_3}, C \to C\} \cup$$

$$\{[I] \to \langle I \rangle, [x, q, c_1, c_2, e_1, e_2] \to \langle x, q, c_1, c_2, e_1, e_2 \rangle,$$

$$\langle x, q, c_1, c_2, e_1, e_2 \rangle \to \langle x, q, c_1, c_2, e_1, e_2 \rangle, \langle I \rangle \to \langle I \rangle |$$

$$x \in \Sigma, q \in E, c_1, c_2 \in \{Z, B\}, e_1, e_2 \in \{-1, 0, +1\}\}$$

The original component grammar P_{a_2} remains unchanged and works as it did in the original system.

$$P_{a_2 Original} = \{S \to S^3, S^{(1)} \to S^{(2)}, S^{(2)} \to S^{(3)}, S^{(3)} \to S^{(4)}\} \cup$$

$$\mathfrak{N} = \{S^{(4)} \to Q^{c_1}_{P_{2 original S_2}} Q^{c_1}_{P_{3 original S_3}} Q^{c_2}_{P_{2 Original S_2}} Q^{c_2}_{P_{3 original S_3}} S^{(1)}\}.$$

Now we define the grammars that are used to reset the P_{a1} helpers. They will send the nonterminal $\langle I \rangle$ to their matching component grammar, which will allow their derivation to restart. None of the components below are part of the original system.

$$Reset_{GM^{c1}_{Pal_{S1}}} = Reset_{GM^{c1}_{Pal_{S1H2(S4)}}} = Reset_{GM^{c1}_{Pal_{S1H3(S4)}}} =$$

$$Reset_{GM^{c1}_{Pal_{S1(S2)}}} = Reset_{GM^{c1}_{Pal_{S1(S3)}}} = Reset_{GM^{c1}_{Pal_{S2}}} =$$

$$Reset_{GM^{c1}_{Pal_{S3}}} = Reset_{GM^{c2}_{Pal_{S1}}} = Reset_{GM^{c2}_{Pal_{S1H2(S4)}}} =$$

$$Reset_{GM^{c2}_{Pal_{S1H3(S4)}}} = Reset_{GM^{c2}_{Pal_{S1(S2)}}} = Reset_{GM^{c2}_{Pal_{S1(S3)}}} =$$

$$Reset_{GM^{c2}_{Pal_{S2}}} = Reset_{GM^{c2}_{Pal_{S3}}} = \mathfrak{N} = \{S \to \langle I \rangle, \langle I \rangle \to \langle I \rangle\}$$

The components below will reset $P^{c1}_{1_{S_1 H_2(S_4)}}$, $P^{c1}_{1_{S_1 H_3(S_4)}}$, $P^{c2}_{1_{S_1 H_2(S_4)}}$, and $P^{c2}_{1_{S_1 H_3(S_4)}}$.

$$U_s = \{ U \to U_1, U_1 \to U_2, U_2 \to U_3, U_3 \to U_4, U_4 \to U_5, U_6 \to U_7 \}$$

$$Reset_{P^{c1}_{1_{S_1 H_2(S_4)}}} = \mathfrak{N} = (U_s \cup \{U_7 \to Q_{P^{c1}_{1_{S_1 H_2(S_4)}}} U_4\})$$

$$Reset_{P^{c1}_{1_{S_1 H_3(S_4)}}} = \mathfrak{N} = (U_s \cup \{U_7 \to Q_{P^{c1}_{1_{S_1 H_3(S_4)}}} U_4\})$$

$$Reset_{P^{c2}_{1_{S_1 H_2(S_4)}}} = \mathfrak{N} = (U_s \cup \{U_7 \to Q_{P^{c1}_{1_{S_1 H_2(S_4)}}} U_4\})$$

$$Reset_{P^{c2}_{1_{S_1 H_3(S_4)}}} = \mathfrak{N} = (U_s \cup \{U_7 \to Q_{P^{c1}_{1_{S_1 H_3(S_4)}}} U_4\})$$

The following grammars will reset $P^{c1}_{4_{S_1 H_2(S_4)}}$, $P^{c1}_{4_{S_1 H_3(S_4)}}$, $P^{c2}_{4_{S_1 H_2(S_4)}}$, and $P^{c2}_{4_{S_1 H_3(S_4)}}$.

$$T_s = \{ T \to T_1, T_1 \to T_2, T_2 \to T_3, T_3 \to T_4, T_4 \to T_5, T_6 \to T_7 \}$$

$$Reset_{P^{c1}_{4_{S_1 H_2(S_4)}}} = \mathfrak{N} = (T_s \cup \{T_7 \to Q_{P^{c1}_{4_{S_1 H_2(S_4)}}} T_4\})$$

$$Reset_{P^{c1}_{4_{S_1 H_3(S_4)}}} = \mathfrak{N} = (T_s \cup \{T_7 \to Q_{P^{c1}_{4_{S_1 H_3(S_4)}}} T_4\})$$

$$Reset_{P^{c2}_{4_{S_1 H_2(S_4)}}} = \mathfrak{N} = (T_s \cup \{T_7 \to Q_{P^{c2}_{4_{S_1 H_2(S_4)}}} T_4\})$$

$$Reset_{P^{c2}_{4_{S_1 H_3(S_4)}}} = \mathfrak{N} = (T_s \cup \{T_7 \to Q_{P^{c2}_{4_{S_1 H_3(S_4)}}} T_4\})$$

In order for our construction to hold it is enough for the grammars that represent the original components to terminate the derivation with the same strings as in the original 11-component derivation.

The master grammar will control the derivation. The string $[x, q, c_1, c_2, e_1, e_2]$ present in the master component, where $x \in \Sigma$, $q \in E$, $c_1, c_2 \in \{Z, B\}$, $e_1, e_2 \in \{-1, 0, +1\}$ means that the 2-counter machine M is in state q, the input head proceeds to scan x onto the input tape and c_1, c_2 on the two storage (counter) tapes, respectively, and then the heads of the storage tapes are moved according to values in e_1, and e_2. The number of A symbols in the strings of the c_1, c_2 component grammars keep track of the value of the counters of M, meaning that these numbers should always match the value stored in the counters of M or else the system will block.

The components defined as "original" will work with the Turing machine M simulating the steps of M in their derivation. The system will change its configuration in sync with the state of M and according to the value of the string derived so far in the master component (which will correspond at the end of the derivation with an input accepted by M). The interested reader is referred to the original proof [27] for the details of the simulation.   □

5   Context-Free PCGS Parse Trees and Meta-Trees

The notion of parse trees can be naturally extended to context-free PCGS (often just PCGS henceforth, with the understanding that all grammars from now on are context-free unless otherwise stated). In fact we use almost the same construction as in Definition 1, though we also need to account for communication between components. We also need to keep track of which portion of which tree has been generated by which component grammar.

Definition 5 (*PCGS Parse Trees*) Let $\Gamma = (N, K, \Sigma, G_1, \ldots, G_n)$ be a PCGS with context-free components.

A parse tree for some component $G_i = (N \cup K, \Sigma, R_i, S_i)$, $1 \leq i \leq n$, of Γ is defined inductively as follows: For every $a \in N \cup K \cup \Sigma$ the tree depicted in Fig. 1a is a parse tree with *yield a*, and for every $A \rightarrow \varepsilon \in R_i$ the tree depicted in Fig. 1b is a parse tree with yield ε. Suppose that the n trees from Fig. 1c are parse trees with yields $y_1, y_2, \ldots, y_n$ and that $A \rightarrow A_1 A_2 \ldots A_n \in R_i$; then the tree shown in Fig. 1d is a parse tree with yield $y_1 y_2 \ldots y_n$. If the tree depicted in Fig. 1e is a parse tree of G_j, then the tree from Fig. 1f is a parse tree for G_i, $1 \leq i, j \leq n, i \neq j$.

The yield of a parse tree continues to be the sequence of leaf labels as obtained by an inorder traversal of that parse tree.

Note that parse trees are identified as belonging to a specific component G_i in the PCGS. They are therefore constructed using only rewriting rules from that component, until a query node Q_j is generated. When this happens, the sub-tree rooted at the subsequent S_j is only allowed to use rewriting rules from G_j, and so on.

Similar with the context-free case, we can create parse trees that correspond to derivations in a PCGS. However now we have n components so logically we should

Fig. 1 Parse trees for context-free grammars as well as PCGS (**a, b, c, d**); supplementary parse trees for context-free PCGS (**e, f**)

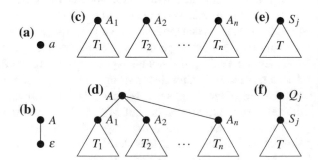

construct simultaneously n parse trees; the n parse trees are collectively referred to as a *parse forest*.

Definition 6 (*PCGS Parse Forest*)   A parse forest for Γ is an n-tuple $\mathcal{T} = (T_1, T_2, \ldots, T_n)$, with T_i a parse tree for G_i as in Definition 5, $1 \leq i \leq n$. The first component T_1 of a parse forest $\mathcal{T}$ is called the master parse tree (of $\mathcal{T}$).

As it turns out Definition 6 will only have a temporary role, as we will eventually show that the master parse tree alone is enough to characterize a PCGS derivation. For the time being however the following establishes the usefulness of parse forests.

Lemma 1 *Every derivation resulting in a configuration $(x_1, x_2, \ldots, x_n)$ in a PCGS with context-free components has an equivalent parse forest; the yields of the parse trees in that forest are $x_1, x_2, \ldots, x_n$, respectively*

Proof Let $\Gamma = (N, K, \Sigma, G_1, \ldots, G_n)$ be a context-free PCGS with $G_i = (N \cup K, \Sigma, R_i, S_i)$, $1 \leq i \leq n$. We construct the trees in the parse forest simultaneously in a natural way as follows:

1. All the n trees are initialized as single nodes labelled S_i, $1 \leq i \leq n$, respectively.
2. If no leaf node labelled with a query symbol exists in any of the n parse trees, then each component i chooses a leaf labelled with a nonterminal A such that $A \to w_1 w_2 \ldots w_k \in R_i$, $w_j \in N \cup K \cup \Sigma$. The respective (former) leaf gains k children labelled w_1, w_2, ..., w_k (or just one child labelled ε whenever $k = 0$). If Γ is synchronized then every component must perform such an expansion as long as a nonterminal leaf is present in its parse tree; otherwise, a component can either perform the expansion or leave its parse tree unchanged. In all cases, if there is no leaf labelled by a nonterminal in the parse tree for some component, then that component does not alter its parse tree.
3. If at least one component has a leaf node labelled with a query symbol, then the following process takes place:
   For a non-returning system Γ, a leaf labelled with a query symbol (say, Q_j) in some component (say, i) is chosen, such that the parse tree for component j has no leafs labelled with query symbols. The leaf labelled with Q_j gains one child labelled S_j which is the root of a copy of the current parse tree of component j. This process is continued for as long as there are leaves labelled with query symbols in the forest.
   Suppose now that Γ is a returning PCGS. Let now T_i be some component tree with leafs labelled with query symbols, and let Q_{i_1}, ..., Q_{i_k} be exactly all the nodes labelled with query symbols in T_i (note that by abuse of notation we henceforth refer to a node by its label; note further that the labels Q_{i_1}, ..., Q_{i_k} above are not necessarily different from each other). Then for all $1 \leq j \leq k$ the leaf Q_{i_j} gains one child labelled S_{i_j} which is the root of a copy of the current parse tree of component i_j. If the one-step communication model is used, the parse trees of components i_j are then reset to single nodes labelled S_{i_j}, $1 \leq j \leq k$.
   The process outlined in the previous paragraph is performed for as long as there are leafs labelled with query symbols in the forest. If the broadcast communication

model is used, then immediately after this process completes, all the component trees that have been queried are reset to single nodes labelled with their respective axioms.

Showing that given a derivation such a construction exists proceeds as follows: We consider the component strings x_i, $1 \leq i \leq n$, as they are rewritten, together with the corresponding parse trees T_i, $1 \leq i \leq n$, in the forest as they are constructed during the same derivation. We proceed by induction over the number m of derivation steps performed.

For $m = 0$ the component strings are all initialized with the respective axioms, while the parse trees have all one (root) node labelled with the axiom; their yields are obviously identical with the component strings.

The component strings at step m are then rewritten to obtain the component strings at step $m + 1$ as follows:

Suppose that no component string contains query symbols, so we have a component-wise derivation step. By inductive assumption it follows that there is no leaf labelled with a query symbol in any of the corresponding parse trees. The rewriting that takes place in x_i picks a nonterminal A (such that $x_i = x_i' A x_i''$) to be rewritten using a rule $A \rightarrow \alpha_1 \alpha_2 \ldots \alpha_k \in R_i$, with $\alpha_i \in N \cup \Sigma \cup K$, $1 \leq i \leq k$. Then A is replaced in the string by $\alpha_1 \alpha_2 \ldots \alpha_k$ (such that x_i becomes $x_i' \alpha_1 \alpha_2 \ldots \alpha_k x_i''$). By definition the corresponding node A in T_i gains k children labelled α_1, α_2, ..., α_k. The yield of T_i was $x_i' A x_i''$ (by inductive assumption) and is changed as follows: it contains x_i' (since nothing changes to the left of the node labelled A), followed by $\alpha_1 \alpha_2 \ldots \alpha_k$ (A disappears from the yield since the node is now internal; its place is taken by the labels of its children according to the inorder traversal), followed by x_i'' (since nothing changes to the right of the node labelled A), as desired.

Whenever x_i contains only terminals (case in which x_i does not change), the yield of T_i contains only terminals by inductive assumption, which means that all the leafs of T_i are labelled with terminals. In such a case no expansion can take place, so the tree remains unchanged, again as desired.

We note that whether some components can remain unchanged in the current step even if they can be rewritten depends on whether the system is synchronized or not; the same dependency is specified in the definition for the respective forest.

Suppose now that query symbols are present in some components, so a communication step takes place. Let component x_i contain Q_{i_1}, ..., Q_{i_k} (so that $x_i = w_1 Q_{i_1} w_2 Q_{i_2} \cdots w_k Q_{i_k} x'$); then after the communication step x_i becomes $w_1 x_{i_1} w_2 x_{i_2} \ldots w_k x_{i_k} x'$. The yield of T_i was by inductive assumption $w_1 Q_{i_1} w_2 Q_{i_2} \cdots w_k Q_{i_k} x'$. Now each leaf Q_{i_j}, $1 \leq j \leq k$ becomes internal (so it disappears from the yield). It's place is taken by the yield of the subtree now rooted at Q_{i_j} (by the definition of inorder traversal); this yield is however x_{i_j} (indeed, the tree rooted at Q_{i_j} is a copy of T_{i_j}, which has yield x_{i_j} by inductive assumption) and so the yield of T_i becomes $w_1 x_{i_1} w_2 x_{i_2} \cdots w_k x_{i_k} x'$ (again by the definition of inorder traversal), as desired. T_{i_j} are reset to their initial form iff x_{i_j} is reset to the axiom in the PCGS (namely, whenever the system is returning and the communication model being used requires the reset of x_{i_j}). That circular queries block equally the derivation and the construction of the parse trees is also immediate from definitions. $\qquad\square$

In order for parse forests to be useful we must also be able to get a parse forest and reconstruct some derivation that gave birth to it. It turns out that this is possible. We can actually do even better and reconstruct the derivation starting solely from the master parse tree of the forest. In order to do this we will find the following notion of meta-tree useful.

Definition 7 (*Meta-tree*)   Let $\Gamma = (N, K, \Sigma, G_1, \ldots, G_n)$ be a context-free PCGS and let T_k, $1 \leq k \leq n$ be a tree in the parse forest of some derivation $(S_1, S_2, \ldots, S_n) \Rightarrow_\Gamma^* (x_1, x_2, \ldots, x_n)$.

A *meta-node* (of T_k) is then a maximal region of T_k which (a) has all the edges produced by applications of rules from a single component G_i of Γ, and (b) is rooted at S_i, the axiom of G_i.

The *meta-tree* $\mu(T_k)$ of T_k is then a the tree of meta-nodes constructed using a function μ defined recursively (and naturally) such that for some parse tree T $\mu(T)$ produces the following meta-tree:

1. The root r of $\mu(T)$ is the meta-node rooted at the root of T. It has one child for every leaf labelled with a query symbol Q_j in r (zero children if no such a label exists in r).
2. For each edge (Q_j, S_j) in T that originates from a leaf of r labelled Q_j the respective child of r is $\mu(T')$, where T' is the tree rooted at S_j.

The yield of $\mu(T)$ is defined as being the same as the yield of the underlying parse tree T.

The meta-tree $\mu(T_k)$ of T_k is a tree of meta-nodes such that there exists an edge in $\mu(T_k)$ for exactly all the edges (Q_j, S_j) in T_k; this edge connects the meta-node that contains Q_j (the parent) with the meta-node rooted at S_j (the child). To illustrate the concept of meta-node and meta-tree intuitively refer to the sample parse tree depicted in Fig. 2a. This parse tree has four meta-nodes $\mathbf{M_1}$, $\mathbf{M_2}$, $\mathbf{M_3}$, and $\mathbf{M_4}$ which are shown in Fig. 2b. The meta-tree $\mu(\mathbf{T})$ (Fig. 2c) is rooted at $\mathbf{M_1}$, which in turn has $\mathbf{M_2}$ and $\mathbf{M_3}$ as its children; $\mathbf{M_2}$ is a leaf (it has zero children since there are no nodes labelled with query symbols in this meta-node), and the sole child of $\mathbf{M_3}$ is $\mathbf{M_4}$ (which is also a leaf).

Lemma 2   *There exists a meta-tree $\mu(T)$ for every parse tree T from any PCGS parse forest. Furthermore $\mu(T)$ covers all the nodes from T.*

Proof That $\mu(T)$ is a tree is immediate by the definition of a parse tree. Indeed, component-wise derivations build a meta-node; the introduction of a query symbol is followed immediately by the connection of that symbol with the respective axiom, which then starts a new meta-node that becomes the child of the initial meta-node, and so on.

A simple inductive argument over the depth of the nodes in T further shows that $\mu(T)$ covers all the nodes of T. Indeed, the root S_k of T (at depth 1) is evidently the

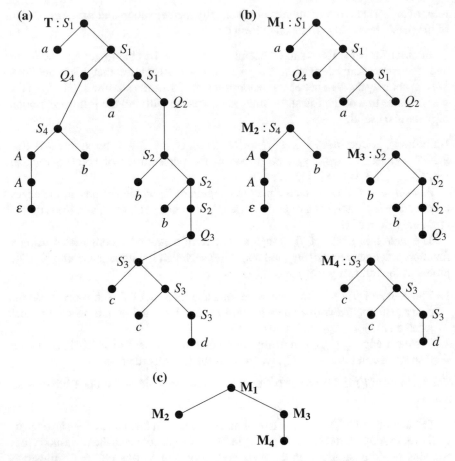

Fig. 2 A parse tree **T** (**a**), its meta-nodes M_1, M_2, M_3, and M_4 (**b**), and the associated meta-tree (**c**)

root of a meta-node (and so inside some meta-node). Then if a node A (at depth d) with children A_1, A_2, …, A_k (at depth $d + 1$) is already in a meta-node rooted at S_i, then the rewriting rule that generates its children must come from G_i by the definition of a PCGS parse tree and so all A_1, A_2, …, A_k are in the same meta-node. If a query Q_j (depth d) is in some meta-node, then its only child must be S_j (again by the definition of a PCGS parse tree), which means that S_j (depth $d + 1$) is the beginning of a new meta-node (and so belong to some, although different meta-node).   □

Overall we established a natural bijection μ between every PCGS parse tree and its meta-tree. In passing, the definition of μ (in Definition 7) or alternatively the inductive argument used in the proof of Lemma 2 effectively establish an algorithm for computing μ.

Now we can show as promised that a valid derivation can be constructed from every master parse tree.

Lemma 3 *Given a master parse tree T with yield w for some PCGS Γ, one can reconstruct a derivation in Γ that produces w in its first component, provided that T was constructed based on a valid derivation in Γ.*

Proof Let $\Gamma = (N, K, \Sigma, G_1, \ldots, G_n)$ and let $\mu(T)$ be the meta-tree of T. The length of a derivation $A \Rightarrow w_1 \Rightarrow w_2 \Rightarrow \cdots \Rightarrow w_m$ is m, the number of steps in that derivation. We proceed with our proof using a structural induction over $\mu(T)$.

There are no query symbols in any of the leaves of $\mu(T)$, so any derivation consistent with the respective leaf will do. Such a valid derivation can be obtained out of the meta-node using the standard technique used for context-free grammars and their parse trees [22]. At least one such a derivation must exist since the whole tree T comes from a valid derivation in Γ, so the base case is established. It is worth noting additionally that the length of any derivation corresponding to a leaf is equal to the number of internal nodes in that leaf (since each of these internal nodes correspond to the application of one rewriting rule and so with one step in the derivation).

Consider now the inductive step that is, a meta-node N with κ internal nodes containing exactly all the query symbols $Q_{N_1}, Q_{N_2}, \ldots, Q_{N_p}$ (which are all necessarily leafs of the meta-node). Let the sub-trees rooted at the roots of the children of N be called $N_1, N_2, \ldots, N_p$, respectively, and let their number of internal nodes be $\kappa_1, \kappa_2, \ldots, \kappa_p$, again respectively.

The proof of this step proceeds more conveniently if split it into one case for Γ being synchronized and another for Γ being unsynchronized.

Synchronized: Recall that none of the meta-nodes $N, N_1, N_2, \ldots, N_p$ feature queries as internal nodes by definition, so they were all created by a sequence of component-wise derivation steps. In addition, this sequence of steps is synchronized between all the trees (Γ being synchronized), meaning that each time an internal node was added to N, one internal node is also added to each of its children $N_1, N_2, \ldots, N_p$. Therefore the child N_i has κ_i internal nodes iff the respective query symbol Q_{N_i} (than connects it to N) was introduced in N after precisely κ_i component-wise derivation steps. Out of all the possible derivations that generate N we thus choose one derivation that introduce every Q_{N_i} after κ_i steps. Such a derivation must exist since the whole tree T was generated by a valid derivation to begin with.

The derivation we thus choose happens in the component given by the root of N. The other components' derivations are given by the inductive assumption. Putting all of these together and then satisfying the p query symbols in the usual way (communication and if applicable reduction to axiom for the communicated component) we obtain a whole derivation for N as a parse tree, as desired.

Unsynchronized: We proceed as in the synchronized case, except that now we no longer need to introduce the query symbols after any precise number of steps. We therefore just choose a derivation that introduces all the query symbols present in N. We then proceed in the same manner to construct the complete derivation for the parse tree N. □

Note that the algorithm implied by the construction of Lemma 3 has a high time complexity (since multiple derivations must be tried for most meta-nodes) and can likely be improved. At this stage however just having an algorithm suffices.

Now we can finally state the main result that establishes the usefulness of parse trees for PCGS with context-free components. Indeed, we showed that every derivation in a PCGS has a corresponding parse forest (Lemma 1) and so a corresponding parse tree (the master parse tree of that forest). Once a master parse tree is given, one derivation that generates it can be determined (Lemma 3); this essentially makes the rest of the forest unnecessary. Putting these two points together we have:

Theorem 2 [4] *Every derivation in a PCGS Γ with context-free components that produces a string w (in the sense of the language generated by a PCGS that is, by the first component grammar) has an equivalent parse tree with yield w. Conversely, given a master parse tree T with yield w that has been constructed according to a valid derivation in Γ, one can reconstruct a derivation in Γ that produces w.* □

We essentially showed that a given derivation in a PCGS is characterized by a single parse tree. Whether this goes the other way around (that is, whether any parse tree corresponds to a derivation) has a more complex answer that will be studied in the next couple of sections.

6   Interference, PCGS Derivations, and Parse Trees

The usefulness of parse trees for context-free grammars is that they characterize exactly all the derivations in a grammar, meaning that every derivation has an equivalent parse tree but also every parse tree corresponds to at least one derivation in the grammar. The first property is already established for PCGS and their parse trees in Theorem 2 above. PCGS parse trees are a relatively straightforward extension of the concept of (context-free) parse trees, so they have the potential of having the second property as well.

However tempting (and useful) this might be, we will eventually show that this is most of the time not the case. More precisely, the existence of the second property will turn out to be dependent on the following concept of interference:

Definition 8 (*Checkpoint and Interference*) Let $\Gamma = (N, K, \Sigma, G_1, \ldots, G_n)$ be a context-free PCGS. We use $\gamma_0 \Rightarrow_\Gamma \gamma_1 \Rightarrow_\Gamma \cdots \Rightarrow_\Gamma \gamma_p$ to refer to any complete derivation in Γ, meaning that there is no γ_{p+1} such that $\gamma_p \Rightarrow_\Gamma \gamma_{p+1}$. We further put

$\gamma_k = (\gamma_{k1}, \gamma_{k2}, \ldots, \gamma_{kn})$, meaning that we use γ_{ki} to refer to the i-th component of the configuration γ_k, for $0 \le k \le p$ and $1 \le i \le n$.

A *checkpoint* of (component) G_i by G_j, $1 \le i, j \le n$ during some derivation in Γ is either p (the end of the derivation) or some $1 \le k < p$ such that $|\gamma_{kj}|_{Q_i} \ne 0$ (the event of G_j querying G_i). Note that during a particular derivation there may be multiple checkpoints of G_i by G_j.

A component G_j *interferes* with another component G_i in some derivation whenever there exists a checkpoint C of G_i by G_j and a string w_i (the *interference string*) such that

1. $S_i \Rightarrow^*_{G_i} w_i$ (w_i can be produced by grammar G_i if it acts alone outside Γ), and
2. $\gamma_{Ci} \ne w_i$ (w_i cannot be produced by G_i at step C in the respective derivation of Γ).

In order to illustrate the concepts of checkpoint and interference we use the following non-returning PCGS (which incidentally speaking we shall meet again in Corollary 1): $\Gamma = (\{S_1, S_2, S_2'\}, \{Q_1, Q_2\}, \{a, b\}, G_1, G_2)$, where $G_1 = (\{S_1, S_2'\}, \{a, b\}, R_1, S_1)$, $G_2 = (\{S_2, S_2'\}, \{a, b\}, R_2, S_2)$, and

$$R_1 = \{S_1 \rightarrow aS_1, S_1 \rightarrow Q_2, S_2' \rightarrow \varepsilon\}$$
$$R_2 = \{S_2 \rightarrow S_2', S_2' \rightarrow bS_2'\}$$

Consider now the following derivation:

$$\begin{aligned}
\gamma_0 &= (S_1, S_2) &\Rightarrow \\
\gamma_1 &= (aS_1, S_2') &\Rightarrow \\
\gamma_2 &= (aaS_1, bS_2') &\Rightarrow \\
\gamma_3 &= (aaQ_2, bbS_2') &\Rightarrow \\
\gamma_4 &= (aabbS_2', bbbS_2') &\Rightarrow \\
\gamma_5 &= (aabb, bbbbS_2')
\end{aligned}$$

In this derivation 3 and 5 are both checkpoints of G_2 by G_1. Indeed, Q_2 is introduced in γ_{31} (and so γ_3 defines a checkpoint) and γ_5 is the end of the derivation (and so it also defines a checkpoint). Furthermore G_1 interferes with G_2 at checkpoint 3, as follows: $S_2 \Rightarrow_{G_2} S_2' \Rightarrow_{G_2} bS_2'$ and so $S_2 \Rightarrow^*_{G_2} bS_2'$ (if the grammar G_2 acts alone), yet $\gamma_{32} = bbS_2' \ne bS_2'$. The interference string in this case is thus bS_2'.

Clearly the components of a PCGS must somehow "interfere" with each other in a general sense (else the resulting system will not serve any purpose not already served by its master grammar alone). Our notion of interference is more restricted and is meant to only identify those cases in which one component controls another in between queries. As it turns out, this notion of interference makes the difference for whether parse trees or forests characterize completely PCGS derivations (second property mentioned at the beginning of this section).

Theorem 3 [4] *Every (master) parse tree of a PCGS Γ with root S_1 and yield w corresponds to a (not necessarily unique) derivation $(S_1, S_2, \ldots, S_n) \Rightarrow_\Gamma^* (w, x_2, \ldots, x_n)$ for some $x_i \in N \cup \Sigma, 2 \le i \le n$ iff there is no interference in Γ.*

Proof Let $\Gamma = (N, K, \Sigma, G_1, \ldots, G_n)$ and let T be some master parse tree with $\mu(T)$ its corresponding meta-tree. (Recall that the notion and properties of a meta-tree were introduced in Definition 7 and Lemma 2.)

We consider first the case in which no interference is present and we proceed by structural induction over the (structure of) $\mu(T)$.

A leaf meta-node rooted at $S_i, 0 \le i \le n$ has no queries anywhere inside it; moreover the rewriting rules that create the leaf come all from a single component. It is thus equivalent to some component-wise derivation starting from the axiom S_i in the respective component grammar G_i. In the absence of interference multiple components can create any possible combination of such leafs concurrently, including the ones corresponding to the actual leafs of the given meta-tree. Indeed we note first that all the components start from their axioms. The fact that there is no interference in Γ means that the components can reach any combination of individual outcomes, including the ones corresponding to the leafs being considered. We can then simply choose a component-wise derivation in Γ that produces a configuration which includes the yields of all the leafs, as desired; the base case is established.

Consider now some meta-node rooted at $S_i, 0 \le i \le n$ together with all its children rooted at $S_{i_1}, S_{i_2}, S_{i_k}$ (and so introduced by the query symbols $Q_{i_1}, Q_{i_2}, Q_{i_k}$, respectively). By the same argument as above this meta-node corresponds to a component-wise derivation in the respective component. In the absence of interference the components corresponding to its children have the time to reach any combination of configurations starting from their respective axioms (by induction hypothesis), including the configurations that correspond to the actual children of the node in discussion. The parent node is immaterial in the process as it does not perform any query and so does not interfere with its children. Our meta-node corresponds to a derivation in the respective component resulting in some string that includes the query symbols $Q_{i_1}, Q_{i_2}, \ldots, Q_{i_k}$ (by the definition of a parse tree). These query symbols must then become roots to trees corresponding to derivations in their respective components (again by the definition of a parse tree). The actual children do corresponds to such derivations (by induction hypothesis). Therefore the whole tree corresponds to a derivation, as desired. The induction is complete.

Consider now the checkpoint C of G_i by G_j such that G_j interferes with G_i at C in an otherwise successful derivation of Γ with parse tree T. Let w_i be the interference string. The checkpoint C corresponds in T to a node labelled Q_i having S_i as sole child (which in turn is the root of some subtree). Replace then the aforementioned tree rooted at S_i with the tree corresponding to the derivation $S_i \Rightarrow_{G_i}^* w_i$. We still have a parse tree, yet such a tree cannot correspond to any derivation in Γ since this would imply that w_i is communicated to G_j at checkpoint C (an impossibility since the interference string w_i is not available at that point). $\qquad\Box$

7 When All PCGS Parse Trees Correspond to Derivations and When They Do Not

Theorem 3 together with Definition 8 allows us to easily see which (if any) PCGS variant feature complete characterization of their derivation based on parse trees.

Consider first synchronized PCGS. Such a system can perform "inter-component counting," in the sense that the number of steps being the same in two (or more) components allows for the generation of comparable numbers of symbols in both (all three, etc.). In such a case the number of nodes in the component parse trees is kept synchronized, so taking an arbitrary parse tree from one component and plugging in into the master parse tree will not do (since this synchronization is no longer taken into account). More to the point such an "inter-component counting" is a clear interference by Definition 8 so we cannot have a complete characterization by parse trees:

Corollary 1 *There exists a master parse tree of some synchronized PCGS that does not correspond to any derivation in that PCGS. There exists a parse forest of some synchronized PCGS that does not correspond to any derivation in that PCGS. This all holds for both returning and non-returning PCGS.*

Proof Once the existence of a master parse tree not corresponding to any derivation is established, the existence of a parse forest with the same property is immediate (just take the master parse tree thus found and add some arbitrary but valid $n - 1$ parse trees for the remaining components).

Interference between different components of a synchronized PCGS is clearly present (since the components are synchronized and therefore restrict each other in the number of derivation steps available). Theorem 3 establishes the desired result.

A constructive proof of this result by finding an actual interference string in some PCGS is also easy to establish. We also present this proof in order to better illustrate the concept of interference, especially in conjunction with the "inter-component counting" effect mentioned above.

Let $\Gamma = (\{S_1, S_2, S_2'\}, \{Q_1, Q_2\}, \{a, b\}, G_1, G_2)$ be a PCGS, where $G_1 = (\{S_1, S_2'\}, \{a, b\}, R_1, S_1)$ and $G_2 = (\{S_2, S_2'\}, \{a, b\}, R_2, S_2)$ such that

$$R_1 = \{S_1 \rightarrow aS_1, S_1 \rightarrow Q_2, S_2' \rightarrow \varepsilon\}$$
$$R_2 = \{S_2 \rightarrow S_2', S_2' \rightarrow bS_2'\}$$

A derivation in Γ can only proceed as follows: Once the first component queries the second the derivation is almost finished. Indeed, the only possible rewriting will erase S_2' (communicated from the second component), thus reaching a string of terminals. The only successful derivations thus involve some $p > 0$ component-wise derivation steps followed by a communication step, followed by one final component-wise derivation step.

Whenever Q_2 is introduced by the first component-wise derivation ($p = 1$) we have:

$$(S_1, S_2) \Rightarrow (Q_2, S_2') \Rightarrow (S_2', \sigma) \Rightarrow (\varepsilon, \sigma')$$

with σ being either S_2 or S_2' and so σ' being S_2' or bS_2' depending on whether the system is returning or not. In either case the result of the derivation is ε.

If Q_2 is introduced later ($p > 1$) we start with $(S_1, S_2) \Rightarrow (aS_1, S_2')$. Each subsequent $p - 1$ derivations introduce one a in the first component and one b in the second, so after p steps we obtain:

$$(S_1, S_2) \Rightarrow^* (aa^{p-1}S_1, b^{p-1}S_2') = (a^p S_1, b^{p-1}S_2')$$

A final component-wise derivation introduces Q_2 in the first component while the second gains another b, so we have:

$$(S_1, S_2) \Rightarrow^* (a^p Q_2, bb^{p-1}S_2') = (a^p Q_2, b^p S_2')$$

The second component is then communicated to the first, followed by the erasure of S_2' in the first component. Therefore:

$$(S_1, S_2) \Rightarrow^* (a^p b^p S_2', \sigma) \Rightarrow (a^p b^p, \sigma')$$

where σ is either S_2 or $b^p S_2'$ (depending on whether Γ is returning or not) and so σ' is either S_2' or $b^{p+1} S_2'$, respectively. Clearly whether Γ is returning or not is immaterial as far as the string produced by the derivation is concerned.

As argued earlier there is no other possible derivation. We thus conclude that $\mathcal{L}(\Gamma) = \{a^p b^p : p \geq 0\}$.

Consider now the tree depicted in Fig. 3. It is a master parse tree for Γ according to Definition 5, but its yield is $aab \notin \mathcal{L}(\Gamma)$ and so cannot have any equivalent derivation. Incidentally, this tree was constructed specifically as a counterexample using the pertinent portion of the proof of Theorem 3 (bS_2' being an interference string of G_2 by G_1). □

The unsynchronized, non-returning PCGS do not offer such strong synchronization, so it is reasonable to believe that parse trees might after all describe exactly all their derivations. Unfortunately, we still have interference and so this is yet again not the case:

Corollary 2 *There exists a master parse tree of some non-returning, unsynchronized PCGS that does not correspond to any derivation in that PCGS. There exists a parse forest of some synchronized PCGS that does not correspond to any derivation in that PCGS.*

Proof Again once the existence of a master parse tree not corresponding to any derivation is established, the existence of a parse forest with the same property is immediate (just take the master parse tree thus found and add some arbitrary but valid $n - 1$ parse trees for the remaining components).

Fig. 3 A master parse tree
that cannot correspond to
any derivation in the
respective PCGS

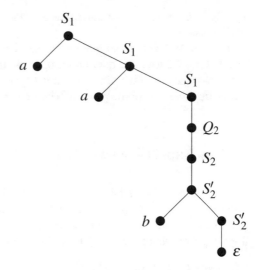

We will establish interference using a system with two components having the
following sets of rewriting rules, respectively:

$$R_1 = \{S_1 \rightarrow aQ_2, S_2 \rightarrow bQ_2, S_2 \rightarrow c\}$$
$$R_2 = \{S_2 \rightarrow xS_2\}.$$

We have $(S_1, S_2) \Rightarrow^* (aQ_2, x^k S_2) \Rightarrow (ax^k S_2, x^k S_2)$. This first checkpoint does not
feature any interference. If the derivation continues however using $S_2 \rightarrow bQ_2$ rather
than $S_2 \rightarrow c$, we get $(ax^k S_2, x^k S_2) \Rightarrow^* (ax^k Q_2, x^{k+p} S_2)$ for some $p \geq 0$. Clearly
the second component is forbidden to produce any string $x^q S_2$ with $q \leq n$ even if it is
perfectly capable of doing so if left to its own devices. The interference is established
and so the result is an immediate consequence of Theorem 3.   □

The unsynchronized, returning case is slightly more interesting, though the result
is still mostly negative:

Definition 9 (*Unique-query PCGS*) A unique-query PCGS is a PCGS in which no
rewriting rule contains two or more occurrences of the same query symbol.

Corollary 3 *Every parse tree with root S_1 and yield w of an unsynchronized, returning PCGS Γ corresponds to a (not necessarily unique) derivation $(S_1, S_2, \ldots, S_n)$
$\Rightarrow^* (w, x_2, \ldots, x_n)$ in Γ for some $x_i \in N \cup \Sigma, 2 \leq i \leq n$ iff Γ is a unique-query
PCGS.*

Proof If (by contrapositive): If Γ is not unique query then interference happens as
follows: Let $A \rightarrow \sigma_1 Q_i \sigma_2 Q_i \sigma_3$ be one of the rules of component G_j than makes Γ
non-unique-query. G_j then interferes with G_i since G_j imposes a fixed number of
steps (zero!) on G_i between the satisfaction of the first and the second occurrence of

Q_i. By Theorem 3 at least one parse tree that does not correspond to any derivation in Γ exists.

Only if: When Γ is unique query then no interference is possible. Indeed, any checkpoint reduces the queried component to its axiom, and then the respective component has as much time as needed to derive by itself any possible string. The whole positive part of the proof of Theorem 3 applies literally to this case.   □

8   The Expressiveness of Unique-Query Context-Free PCGS

Between other things, the complete characterization of derivations in unique-query context-free PCGS by parse trees implies that this variant is by far the least powerful PCGS with context-free components possible. Indeed, in such a case we do not gain anything from using a PCGS; a simple context-free grammar will do just as well.

Theorem 4 [4] *Exactly all the languages generated by unique-query, unsynchronized, returning context-free PCGS are context free.*

Proof That every context-free language can be generated by a unique-query, unsynchronized, returning context-free PCGS is immediate since a context-free grammar is a special case of PCGS (with one component and no use for query symbols).

Let now $\Gamma = (N, K, \Sigma, G_1, \ldots, G_n)$ be a unique-query, unsynchronized, returning context-free PCGS with $G_i = (N \cup K, \Sigma, R_i, S_i), 1 \leq i \leq n$. Let σ_i be a renaming such that $\sigma_i(x)$ is the string x in which all the occurrences $\eta \in N$ are replaced by (η, i). We extend naturally σ_i to rewriting rules as $\sigma_i(A \to \alpha) = \sigma_i(A) \to \sigma_i(\alpha)$, and to sets of rewriting rules as $\sigma_i(\rho) = \{\sigma_i(r) : r \in \rho\}$.

Consider then the following context-free grammar: $G = ((N \times \{1, 2, \ldots, n\}) \cup K, \Sigma, R, (S_1, 1))$, where $R = \bigcup_{0 \leq i \leq n} R'_i$ with $R'_0 = \{Q_j \to (S_j, j) : 1 \leq j \leq n\}$ and $R'_i = \sigma_i(R_i), 1 \leq i \leq n$. We find that there is a natural bijection β between the set of parse trees of Γ and the set of parse trees of G such that the yield of T is identical with the yield of $\beta(T)$ and thus we complete the proof. Indeed, this establishes that $w \in \mathcal{L}(\Gamma) \implies w \in \mathcal{L}(G)$ by Theorem 2, that $w \in \mathcal{L}(G) \implies w \in \mathcal{L}(\Gamma)$ by Corollary 3, and so that $\mathcal{L}(\Gamma) = \mathcal{L}(G)$, as desired.

Intuitively, a derivation in G simulates the derivation in Γ component by component while the respective parse tree is constructed (and so generates an isomorphic parse tree). The nonterminals are the old nonterminals in Γ with an added associated index to keep track which rewriting (i.e., node expansion) happens in which component. The occurrence of a query symbol will change this index in G, meaning that in Γ whatever happens afterward is the result of a derivation in a different component. This all follows faithfully the definition of PCGS parse trees (Definition 5).

Formally, given some parse tree T of Γ we construct $\beta(T)$ in a natural way by considering every meta-node N in the meta-tree $\mu(T)$ and relabeling all the nodes in N with root S_i from ν to $\sigma_i(\nu)$. (Recall that the notion and properties of a meta-tree were introduced in Definition 7 and Lemma 2.)

That β is one-to-one is immediate given that σ_i is one-to-one and that only the node labels are changed by β while the structure of the tree remains the same.

Let now T' be a parse tree of G. T' is rooted at (S_i, i) with $i = 1$, the axiom of G and so $\beta^{-1}(T')$ is rooted at S_i for $i = 1$ (the axiom of G_1). Then every node labeled (A, i) is expanded using some rule from $\sigma_i(R_i)$ (only these are usable given the second component of the label), meaning that the corresponding node in $\beta^{-1}(T')$ is expanded using some rule from R_i. When a node labeled Q_j is encountered, it can only be expanded using the rule $Q_j \rightarrow (S_j, j)$, so the label of its sole child is (S_j, j), and so from then on only rules from $\sigma_j(R_j)$ will be applicable (same reason as above). Therefore in $\beta^{-1}(T')$ a node labeled Q_j can only have one child labeled S_j, and from then on only rules from R_j will be applicable. As far as $\beta^{-1}(T')$ is concerned the above description matches exactly Definition 5 and so $\beta^{-1}(T')$ is a parse tree for Γ. The function β is thus onto.

The fact that β is a bijective relabeling of nodes in a tree which does not change terminal symbols established immediately that the yields of T and $\beta(T)$ are the same for any T. $\qquad\square$

It is also worth noting that for synchronized context-free PCGS non-returning systems were first found to be (not necessarily strictly) weaker than their returning counterparts [14]. Subsequently returning and non-returning context-free PCGS turned out to be equivalent (and also Turing complete) [8, 9, 23]. The initial find in the synchronized case [14] turns out to be reversed in the unique-query, unsynchronized case (for the returning, unique-query PCGS are the weakest of them all). On the other hand, the subsequent find for the synchronized case (that returning and non-returning PCGS are equivalent [8, 9, 23]) does not hold for the unique-query, unsynchronized case. Overall we have:

Corollary 4 *Any unsynchronized, returning, unique-query context-free PCGS can be simulated by an unsynchronized, non-returning, unique-query context-free PCGS, but not the other way around. Indeed, the non-returning variant of such systems is strictly stronger than the returning variant.*

Proof An unsynchronized, non-returning, unique-query context-free PCGS T can simulate an unsynchronized, returning, unique-query context-free PCGS S in a trivial manner: we construct the context-free grammar corresponding to S as in Theorem 4 above and we just call that T (that is, a non-returning unique-query context-free PCGS). T is the lowest common denominator of all the context-free PCGS (having one component and no use for queries) so it can be any kind of context-free PCGS.

The following example will then show that non-returning, unsynchronized unique-query context-free PCGS are strictly stronger than their returning version:

Let $N = \{S_1, S_2, S_3, S_4, A, B\}$ and $\Sigma = \{a, b\}$. Let then $\Gamma = (N, K, \Sigma, (N \cup K, \Sigma, R_1, S_1), (N \cup K, \Sigma, R_2, S_2), (N \cup K, \Sigma, R_3, S_3), (N \cup K, \Sigma, R_4, S_4))$ be an unsynchronized non-returning unique-query context-free PCGS with

$$R_1 = \{S_1 \rightarrow Q_3Q_4, B \rightarrow \varepsilon\}$$
$$R_2 = \{S_2 \rightarrow aS_2, S_2 \rightarrow bS_2, S_2 \rightarrow A, A \rightarrow A\}$$
$$R_3 = \{S_3 \rightarrow Q_2, A \rightarrow B\}$$
$$R_4 = \{S_4 \rightarrow Q_2, A \rightarrow B\}$$

Clearly, the first component must query to have any chance of producing a terminal string. Both the second and third components must query before the first component queries, for if this is not the case then either S_3 or S_4 (or both) will find their way into the first component, which cannot erase them and so the derivation cannot be successful.

S_2 can generate any string in $\{a, b\}^*$ in the second component before being rewritten to A; once such a rewriting happens, the second component will not change anymore (since only the rule $A \rightarrow A$ will be applicable henceforth). The rewriting of S_2 to A must furthermore happen before any query takes place; if this is not so, then S_2 will find its way into the first component via the third and fourth components but cannot be erased and so the first component cannot produce a terminal string.

In all a successful derivation in Γ can only proceed as follows before any query takes place: For an arbitrary $w \in \{a, b\}^*$,

$$(S_1, S_2, S_3, S_4) \Rightarrow^* (S_1, wA, S_3, S_4)$$

As we argued above, both the third and the fourth component must query before the first component does so. The third and fourth component may query at different times, but the timing does not matter since from now on the second component will not change anymore. Any potentially successful continuation of the derivation is therefore equivalent to the following one:

$$(S_1, wA, S_3, S_4) \Rightarrow (S_1, wA, Q_3, Q_4) \Rightarrow (S_1, wA, wA, wA)$$

Now before the first component queries the A nonterminals in the third and fourth component must both be rewritten to B; if this is not the case then A fill find its way into the first component where it cannot be erased. The derivation will therefore continue along the following line (or equivalent):

$$(S_1, wA, wA, wA) \Rightarrow (Q_3Q_4, wA, wB, wB)$$
$$\Rightarrow (wBwB, wA, wB, wB) \Rightarrow^* (ww, wA, wB, wB)$$

As explained throughout the above argument no other derivation is possible, and so $\mathcal{L}(\Gamma) = \{ww : w \in \{a, b\}^*\}$. $\mathcal{L}(\Gamma)$ is not context free [22] and so Γ cannot be simulated by any unsynchronized, returning, unique-query context-free PCGS since the latter only generate context-free languages (as shown in Theorem 4). $\square$

We are not aware of any results in the intermediate case (namely, unsynchronized PCGS that are not necessarily unique query).

9 Conclusions

Emergence in teams of context-free constructs is by no means unexpected. A trivial argument (which is also a classical class exercise [22]) makes a two-stack pushdown automaton Turing complete. The way in which emergence appears and the factors that enable it are however interesting and worth investigating. Formal languages after all do not live in a vacuum and need to model real life phenomena in order to be useful.

There are various features of context-free PCGS that realize emergence. One well-studied such a feature is synchronization. The other emergent feature is the capability of introducing multiple queries in the same component string simultaneously (the lack of the unique-query feature). The main result of this paper is that these are the *only* emergent features. Indeed, when all these features are eliminated then we obtain unique-query, unsynchronized, returning context-free PCGS, a formalism which is no more expressive than any one of its components (that is, a context-free grammar).

Another potential emergent feature is broadcast communication. We found however that broadcast communication is not really an emergent feature, since it can be readily simulated using the one-step communication model. The construction used to show this is manual and tedious, but is worth remembering because it introduces a number of general patterns, as follows:

1. Copycat components contain rules similar to the original components, and derive the same strings during the same steps as the original components. Such a construction allows for each of the original grammars to request the same string at the same time without the need to query the same component (that is, without needing broadcast communication).
2. Reset components reset some of the copycat grammars at precise steps in the derivation in order to fix synchronization issues.
3. Waiting rules ensure that communication steps would only be triggered at certain points in the derivation.
4. Selective rewriting rules were used in conjunction with blocking, thus allowing certain rewriting rules to be successful only at specific steps and ensuring that no undesired strings are created.

We believe that the above techniques are applicable not only to our construction but in a more general environment. That is, they appear to be useful for eliminating broadcast communication in general. We further believe that our transformation can be accomplished algorithmically. Whether this is indeed the case and if so in what circumstances is an interesting open question.

9.1 *Incidental Results*

We introduced the concept of parse forests and parse trees as a natural extension of parse trees for context-free grammars. More precisely, some of our results were obtained using the intermediate step of a parse forest, which in the end got reduced to a single parse tree. While parse forests do not convey any more information than their master parse trees, they may be needed as an intermediate step in any process of actually constructing parse trees. Indeed, the only immediate way arising from our results of actually constructing a parse tree for a context-free PCGS is to construct the parse forest first and then discard all but the first tree of that forest. It also turned out to be the case that parse trees and parse forests are of limited theoretical utility for context-free PCGS, but they nonetheless may prove to be useful in practical applications.

The concept of splitting a PCGS parse tree into regions thus obtaining a meta-tree is natural, very simple, and worth remembering. It has proven extremely useful in our effort and might have further utility elsewhere.

A final nod goes to an interesting difference in the power of returning versus non-returning context-free PCGS. In the synchronized case it has been known that the non-returning and returning variants are equivalent [8, 9, 23]. When it comes to unique-query, unsynchronized PCGS this equivalence no longer holds (see Theorem 4). It would be interesting to know how is the wind blowing in the intermediate case (unsynchronized PCGS that are not necessarily unique query); given the known results the simulation can essentially go either way, or it may even be that the two variants are not comparable (though we do believe that this is unlikely).

9.2 PCGS in Formal Methods?

The motivation of PCGS is claimed to be the study of concurrent systems. Relatively recently some, however strenuous links with practice have beet attempted [19], but overall virtually no work has been performed on actually linking PCGS with any practical field. PCGS with context-free components in particular (which we believe to have the most practical utility) were ignored almost completely. Instead, PCGS have been studied relatively extensively (though not completely) with respect to theoretical properties such as generative power and then their study has largely stalled.

One of our research interest is the formal specification and verification of recursive, concurrent systems. We find context-free PCGS particularly enticing for this purpose. Indeed, on one hand context-free grammars (or equivalently pushdown automata) model naturally the control flow of sequential computation in typical programming languages with nested, and potentially recursive invocations of program modules such as procedures and methods. On the other hand, the communication facilities offered by PCGS are a good model for inter-process communication. In other words, the context-free components are an excellent model of recursive subsystems, while the communication between these components (which is done in a "remote procedure call"-like fashion) seems particularly suitable for putting these subsystems together.

Many non-regular properties are required for the verification of complex, recursive and concurrent systems. One needs to specify and verify properties such as "if p holds when a module is invoked, the module must return, and q must hold upon return" [1]. Non-regular properties however generate an infinite state space, which cannot be handled by finite-state process algebrae or by standard verification techniques such as model checking. Context-free process algebrae such as basic process algebra or BPA [2] can specify such context-free properties. Still, most of the software use many parallel components (such as multiple threads). In addition, many conformance-testing techniques (such as may/must testing [13]) use test cases that run in parallel with the process under test. Concurrency is therefore required for software verification, but cannot be provided by simple context-free process algebrae since context-free languages are not closed under intersection [22].

On the automata side the class of multi-stack visibly pushdown languages (MVPL) has been introduced to address such a need [3, 21]; no equivalent mechanism is known on the grammatical (and thus process-algebraic) side. In addition, permitting concurrency in a compositional manner is done in MVPL using constructs that do not seem to be naturally portable to the grammatical (or process-algebraic) side; it appears that any specification formalism based on MVPL requires an excessive exposition of implementation details (which should normally be hidden in the specification phase). We wonder whether context-free PCGS can be used as an underlying model that would remedy such a lack of abstraction.

The attention received by PCGS so far was from the formal languages community and so people have focused on the "more powerful is better" facet of these systems. It did not take long to identify Turing complete variants, the apex of such a pursuit. Power (and Turing-completeness in particular) are essential properties for identifying those real-life phenomena (computational or otherwise) that can be modelled using the respective formalism.

When it comes to formal methods however the goal is less ambitious. Instead of trying to model general computational phenomena, we are aiming at modelling only the interactions of (albeit complex) computing systems with their environment, so that such interactions can be specified formally and then verified. In this context the facet of interest of any underlying formalism becomes "less powerful is better," as long as the properties of interest can still be modelled. In this respect our results suggest strongly that synchronized context-free PCGS are unnecessarily powerful, and we also note that unsynchronized PCGS appear to be more amenable to practical applications, they being less powerful but still expressive enough to model complex, potentially recursive systems. Unfortunately however unsynchronized PCGS have received little attention so far: They have been found to be weaker in terms of generative power compared to their synchronized counterparts, and then they have been effectively ignored.

Other variants have also been proposed, including centralized PCGS [10] or PCGS with terminal transmission [15] (in which only terminal strings can be communicated). They further simplify the formalism which is in general a good thing, but when it comes to formal verification we believe that these further simplifications reduce too much the ability of the formalism to model real-life phenomena. In all, we thus believe that unsynchronized context-free PCGS are in the sweet spot when it comes to formal methods.

We further note that the unique query variant is the only one that can make full use of parse trees, though whether the use of parse trees is indeed required for our purposes remains to be seen. More interestingly, the computational power of the unique-query variant does not vary with the number of components, which is appealing from the point of view of modelling parallel systems with arbitrary (and possibly dynamic) number of concurrent threads of execution. Whether the unique query restriction is a reasonable restriction remains however to be seen.

In general, before starting to use them in formal methods (or indeed any practical domain), unsynchronized PCGS need to be analyzed thoroughly, especially with respect to generative capacity and closure properties. Such an analysis is as we already mentioned missing almost completely. Doing this is included in our short-term interests. At the same time the possible ways of modelling the behaviour of complex application software using PCGS need to be investigated, and this is yet another of our immediate interests. Indeed, we moved tentatively our interest from MVPL to PCGS simply because of the necessarily awkward form of an MVPL-based specification; should a PCGS-based specification be equally awkward, our pursuit becomes substantially less interesting.

Acknowledgments This research was supported by Bishop's University. Part of this research was also supported by the Natural Sciences and Engineering Research Council of Canada. We are indebted to an anonymous reviewer of one of our papers, who offered an extension of one of our results free of charge! Indeed, the second part of Corollary 4 as well as the respective portion of the subsequent proof are essentially the work of that reviewer.

References

1. Alur, R., Etessami, K., Madhusudan, P.: A temporal logic of nested calls and returns. In: Proceedings of the 10th International Conference on Tools and Algorithms for the Construction and Analysis of Systems (TACAS 04), Lecture Notes in Computer Science, vol. 2988, pp. 467–481. Springer (2004)
2. Bergstra, J.A., Klop, J.W.: Process theory based on bisimulation semantics. In: Linear Time, Branching Time and Partial Order in Logics and Models for Concurrency, Lecture Notes in Computer Science, vol. 354, pp. 50–122. Springer (1988)
3. Bruda, S.D., Bin Waez, M.T.: Unrestricted and disjoint operations over multi-stack visibly pushdown languages. In: Proceedings of the 6th International Conference on Software and Data Technologies (ICSOFT 2011), vol. 2, pp. 156–161. Seville, Spain (2011)
4. Bruda, S.D., Wilkin, M.S.R.: Parse trees and unique queries in context-free parallel communicating grammar systems. Technical Report 2013-001, Department of Computer Science, Bishop's University (2013)
5. Cai, L.: The computational complexity of linear PCGS. Comput. AI **15**(2–3), 199–210 (1989)
6. Csuhaj-Varjú, E.: On size complexity of context-free returning parallel communicating grammar systems. In: Martin-Vide, C., Mitrana, V. (eds.) Where Mathematics, Computer Scients, Linguistics and Biology Meet, pp. 37–49. Springer (2001)
7. Csuhaj-Varjú, E., Vaszil, G.: On the computational completeness of context-free parallel communicating grammar systems. Theor. Comput. Sci. **215**(1–2), 349–358 (1999)
8. Csuhaj-Varjú, E., Vaszil, G.: On the size complexity of non-returning context-free PC grammar systems. In: 11th International Workshop on Descriptional Complexity of Formal Systems (DCFS 2009), pp. 91–100 (2009)

9. Csuhaj-Varjú, E., Vaszil, G.: On the number of components and clusters of nonreturning parallel communicating grammar systems. In: Holzer, M., Kutrib, M., Pighizzing, G. (eds.) 13th International Workshop on Descriptional Complexity of Formal Systems (DCFS 2011), pp. 121–134. Gieen/Limburg, Germany (2011)
10. Csuhaj-Varjú, E., Dassow, J., Kelemen, J., Paun, G.: Grammar Systems: A Grammatical Approach to Distribution and Cooperation. Gordon and Breach, London (1994)
11. Csuhaj-Varjú, E., Gheorghe, P., Vaszil, G.: PC grammar systems with five context-free components generate all recursively enumerable languages. Theor. Comput. Sci. **299**, 785–794 (2003)
12. Dassow, J., Paun, G., Rozenberg, G.: Grammar systems. In: Handbook of Formal Languages—Volume 2: Linear Modeling: Background and Applications, pp. 155–213. Springer (1997)
13. De Nicola, R., Hennessy, M.C.B.: Testing equivalences for processes. Theor. Comput. Sci. **34**, 83–133 (1984)
14. Dumitrescu, S.: Non-returning PC grammar systems can be simulated by returning systems. Theor. Comput. Sci. **165**, 463–474 (1996)
15. Fernau, H.: Parallel communicating grammar systems with terminal transmission. Acta Inform. **37**, 511–540 (2001)
16. Fischer, P.C.: Turing machines with restricted memory access. Inform. Comput. **9**, 364–379 (1966)
17. Garey, M.R., Johnson, D.S.: Computers and Intractability a Guide to the Theory of NP-Completeness. Macmillan Higher Education (1979)
18. Geffert, V.: Context-free-like forms for the phrase-structure grammars. In: Mathematical Foundations of Computer Science, Lecture Notes in Computer Science, vol. 324, pp. 309–317. Springer (1988)
19. Grando, M.A., Mitrana, V.: A possible connection between two theories: grammar systems and concurrent programming. Fundam. Inform. **76**, 325–336 (2007)
20. Katsirelos, G., Maneth, S., Narodytska, N., Walsh, T.: Restricted global grammar contraints. In: Principles and Practice of Constraint Programming (CP 2009), Lecture Notes in Computer Science, vol. 5732, pp. 501–508 (2009)
21. La Torre, S., Madhusudan, P., Parlato, G.: A robust class of context-sensitive languages. In: Proceedings of the 22nd Annual IEEE Symposium on Logic in Computer Science (LICS 07), pp. 161–170. IEEE Computer Society, Washington, DC (2007)
22. Lewis, H.R., Papadimitriou, C.H.: Elements of the Theory of Computation, 2nd edn. Prentice-Hall (1998)
23. Mandache, N.: On the computational power of context-free PCGS. Theor. Comput. Sci. **237**, 135–148 (2000)
24. Paun, G., Santean, L.: Parallel communicating grammar systems: the regular case. Analele Universitatii din Bucuresti, Seria Matematica-Informatica **38**(2), 55–63 (1989)
25. Paun, G., Santean, L.: Further remarks on parallel communicating grammar systems. Int. J. Comput. Math. **34**, 187–203 (1990)
26. Santean, L.: Parallel communicating grammar systems. Bull. EATCS (Formal Lang. Theory Column) **1**(2) (1990)
27. Wilkin, M.S.R., Bruda, S.D.: Parallel communicating grammar systems with context-free components are Turing complete for any communication model. Technical Report 2014-003, Department of Computer Science, Bishop's University (2014)

Structural Properties of Generalized Exchanged Hypercubes

Eddie Cheng, Ke Qiu and Zhizhang Shen

Abstract It has been shown that, when a linear number of vertices are removed from a Generalized Exchanged Hypercube (GEH), a generalized version of the interesting exchanged hypercube, its surviving graph consists of a large connected component and smaller component(s) containing altogether a rather limited number of vertices. In this chapter, we further apply the above connectivity result to derive several fault-tolerance related structural parameters for GEH, including its restricted connectivity, cyclic vertex-connectivity, component connectivity, and its conditional diagnosability in terms of the comparison diagnosis model.

1 Introduction

It is certainly unavoidable that some of the processing nodes within a multi-processor system become faulty, leading to a faulty system. To have an effective system to work with, we are naturally interested in the fault tolerance properties of these systems, seeking answers to such questions as how many faulty nodes will disrupt such a system, or disconnect its associated graph in graph theoretical terms; and how disrupted the *surviving system (graph)* will become when a certain number of nodes and/or links become faulty, thus effectively removed. For example, will the surviving graph completely break apart, or are most of its nodes still connected in a component? We might also be interested in knowing more about the details, e.g.,

E. Cheng (✉)
Department of Mathematics and Statistics, Oakland University,
Rochester, MI 48309, USA
e-mail: echeng@oakland.edu

K. Qiu
Department of Computer Science, Brock University, St. Catharines,
ON L2S 3A1, Canada
e-mail: kqiu@brocku.ca

Z. Shen
Department of Computer Science and Technology, Plymouth State University,
Plymouth, NH 03264, USA
e-mail: zshen@plymouth.edu

© Springer International Publishing Switzerland 2017
A. Adamatzky (ed.), *Emergent Computation*, Emergence, Complexity
and Computation 24, DOI 10.1007/978-3-319-46376-6_9

215

the relationship between the maximum number of the faulty nodes and the minimum number of components in such a surviving graph.

A related issue is that, once processing nodes become faulty, could we know exactly which ones are faulty so that the fault-free status of the system can be restored? The number of such detectable faulty nodes in a system certainly depends on its topology, the restriction placed on such a faulty set, as well as the modeling assumptions, and the maximum number of detectable faulty nodes in such a system is called its *diagnosability*. One major modeling approach to this regard is called the *comparison diagnosis model* [14, 26, 27, 33], where each processing node performs a diagnosis by sending the same input to each and every pair of its distinct neighbors, and then comparing their responses. Based on such comparison results made by all the nodes, the faulty status of the whole system can be determined. Various efficient algorithms to detect such faulty sets have also been proposed, e.g., [33, 35, 47].

To address the unlikelihood that all the neighbors of a certain node in such a system will fail at the same time, the notion of the *conditional diagnosability* of a graph G was introduced in [18], defined as the maximum number of detectable faulty nodes in G, assuming that no faulty set contains all the neighbors of any node in G. Such a faulty set is henceforth referred to as a *conditional faulty set*. This more realistic notion leads to an improved measurement of the fault tolerance capability of network structures and is thus of great interest [1, 4, 5, 7, 18].

Answers to the aforementioned fault tolerance related questions are often expressed in terms of connectivity related properties of a graph underlying such a surviving structure [1, 13, 15, 24, 41–43]. In particular, a general connectivity result has been demonstrated in [9] for the *generalized exchanged hypercube* structure that, when a linear number of vertices are removed from such a structure, the surviving graph is either connected or consists of a large connected component and small components containing a small number of vertices. The results as reported in this chapter can be seen as a companion work of [9]: We apply the above general result to further derive for this topological structure several fault tolerance related measurements, including its (i) restricted connectivity, i.e., the size of a minimum vertex cut such that the degree of every vertex in the surviving graph will have a guaranteed lower bound; (ii) cyclic vertex-connectivity, i.e., the size of a minimum vertex cut such that at least two components in the surviving graph contain a cycle; (iii) component connectivity, i.e., the size of a minimum vertex cut whose removal leads to multiple components in its surviving graph; as well as (iv) conditional diagnosability in terms of the comparison diagnostic model.

The rest of this chapter proceeds as follows: We briefly review the exchanged hypercube [3, 22] and the class of generalized exchanged hypercubes [9] in the next section; our exposition is based on [9]. We state the general result obtained in [9] in Sect. 3. We then apply the aforementioned general connectivity property as associated with the generalized exchanged hypercube to derive various parameters that generalize the concept of connectivity, namely, restricted connectivity and cyclic vertex-connectivity in Sect. 4, component connectivity and conditional diagnosability in Sects. 5 and 6, respectively. We conclude this chapter with some final remarks in Sect. 7.

2 The Exchanged Hypercube and its Generalization

The n-dimensional hypercube [13], often referred to as the n-cube and denoted by Q_n, is perhaps one of the most studied and utilized interconnection structures, as it possesses many desirable properties such as vertex and edge symmetry, high connectivity, and small diameter thus lower communication cost, as well as the existence of a simple routing algorithm. More specifically, an n-cube has 2^n nodes $0, 1, 2, \ldots, 2^n - 1$ where (u, v) is an edge (arc) if u's and v's binary representations differ in exactly one position, i.e., $u = u_{n-1}u_{n-2} \cdots u_{i+1}u_i u_{i-1} \cdots u_1 u_0$ and $v = u_{n-1}u_{n-2} \cdots u_{i+1}\overline{u_i}u_{i-1} \cdots u_1 u_0, 0 \le i \le n - 1$. Figure 1 shows a 3-cube.

Several hypercube variants have since been suggested, including augmented cubes, crossed cubes, enhanced cubes, folded hypercube, Möbius cubes, and twisted cubes.

The exchanged hypercube was proposed in [3, 22] as another edge removal variant of the hypercube, where about half of the edges are systematically removed [3, Theorem 2]. With such a significantly reduced complexity, besides addressing a scaling issue as associated with the hypercube structure, the exchanged hypercube still manages to inherit several attractive properties of the hypercube such as incremental expandability [3], bipancyclicity [23], connectivity and super connectivity [25], and existence of a fault tolerant routing algorithm [22]. With essentially the same diameter and eccentricity, but reduced maximum degree and Wiener index [17], the bounds of its domination number, as well its surface area and average distance, have also been established in [16, 17], respectively. We will further study some of its fault tolerance related connectivity properties in this chapter.

The exchanged hypercube, denoted by $EH(s, t)$, where $s, t \ge 1$, is defined as an undirected graph (V, E), where V is the collection of all the binary strings of length $s + t + 1$. Hence, $|V(EH(s, t))| = 2^{s+t+1}$. A vertex u of an exchanged hypercube $EH(s, t)$ is denoted by $A(u)B(u)C(u)$, where $A(u) = a_{s-1} \cdots a_0$, $B(u) = b_{t-1} \cdots b_0$, and $C(u) = c$. $C(u)$ is sometimes referred to as the C *bit* of

Fig. 1 A 3-Cube

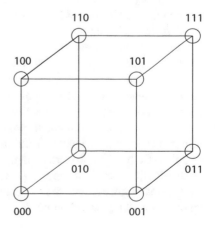

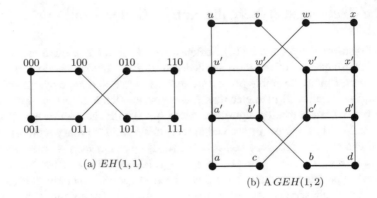

(a) $EH(1,1)$

(b) A $GEH(1,2)$

Fig. 2 Simple exchanged hypercubes

u henceforth. Let $u, v \in V(EH(s,t))$, $(u,v) \in E$ if and only if it falls into one of the following three mutually exclusive cases: E_1: $C(u) \neq C(v)$, but $A(u) = A(v)$ and $B(u) = B(v)$; E_2: $C(u) = C(v) = 0$, $A(u)$ and $A(v)$ differ in exactly one bit in position $p \in [0,s)$, while $B(u) = B(v)$; and E_3: $C(u) = C(v) = 1$, $A(u) = A(v)$, but $B(u)$ and $B(v)$ differ in exactly one bit in position $p \in [0,t)$.

Figure 2a shows $EH(1,1)$, where $(000,001)$, $(000,100)$, and $(001,011)$ are examples of E_1, E_2, and E_3 edges, respectively.

Each collection of 2^s vertices, sharing the same B segment and 0 as their common C bit, forms a Q_s, referred to as a *Class-0 cluster*, via the associated E_2 edges. Clearly, there are a total of 2^t such hypercubes in $EH(s,t)$. Similarly, each collection of 2^t vertices, sharing the same A segment and 1 as their common C bit, forms a Q_t, a *Class-1 cluster*, via the associated E_3 edges. There are a total of 2^s such hypercubes in $EH(s,t)$. We thus refer to both E_2 and E_3 edges collectively as *cube edges*.

Class-0 clusters and Class-1 clusters are referred to as *clusters of opposite class* of each other. Clearly, each vertex u in a cluster is adjacent to a unique vertex in a cluster of opposite class via an E_1 edge, denoted by u' in the rest of this chapter. By definition, $A(u)B(u) = A(u')B(u')$ but $C(u') = \overline{C(u)}$, namely, the complement of the C *bit* of u. Since these E_1 edges connect vertices belonging to different clusters, we refer to them as *cross edges*.

A key structural property of the exchanged hypercube is that, let u, v be two vertices of the same cluster C in $EH(s,t)$, $s, t \geq 1$, then u' and v' belong to two different clusters of a class opposite to that of C, via cross edges. Here the set of cross edges are chosen specifically. One may wonder the role of the specific set of cross edges chosen among all possible sets. In terms of shortest path, it plays an important role in ensuring the resulting graph has a small diameter. However, in terms of connectivity type properties, there may be no differences among different set of cross edges. Indeed, in the recursive definition of the hypercube, one can replace the specific matching between the two smaller hypercubes by any perfect matching. This leads to a wider class of networks, which leads to the even more general matching composition networks. We can apply the same type of generalization to the exchanged hypercube, that is, although the existence of a perfect matching between vertices via the cross edges is structurally essential, the specifics of such a matching, i.e., details such as which vertices are matched with each other, is not. We will show that such generalized networks also have strong connectivity type results. Of course, the proof will be more involved due to the generality.

We thus generalized this class of exchanged hypercubes in [9] as follows: A generalized exchanged hypercube, denoted by $GEH(s, t, f), s, t \geq 1$, consists of two classes of hypercubes: One class contains 2^t s-cubes, each labeled with the shared B segment, and referred to collectively as the *Class-0 clusters*; and the other contains 2^s t-cubes, each labeled with the shared A segment, and referred to collectively as the *Class-1 clusters*. Class-0 and Class-1 clusters will be referred to as clusters of *opposite class* of each other, *same class* otherwise, and collectively as *clusters*, $C_i, i \in [0, 2^s + 2^t)$, when their categories are irrelevant to the issue. When $s = t$, we simply refer to one of the classes of hypercubes as Class-0 clusters, and the other as Class-1 clusters. Set E_h, the *cube edges*, collects all the usual $(s + t)2^{s+t-1}$ edges in the hypercubes of both classes.

The function f is a bijection between vertices of Class-0 clusters and those of Class-1 clusters such that, for u, v, two vertices of the same cluster, $f(u)$ and $f(v)$ belong to two different clusters, as observed in the aforementioned structural property of the exchanged hypercubes. We naturally refer to such an edge $(u, f(u))$ as a *cross edge*. Set E_c collects all the 2^{s+t} cross edges in between the clusters of opposite classes. Such a bijection f ensures the existence, but ignores the specifics, of a perfect matching between vertices of Class-0 clusters and those in the Class-1 clusters.

By its definition, in a generalized exchanged hypercube, all of the 2^s distinct vertices in a specific Class-0 cluster, a Q_s, out of 2^t of them, are adjacent, via cross edges, to 2^s vertices, each of which is located in a unique Class-1 cluster, a Q_t; and all of the 2^t distinct vertices in a specific Class-1 cluster, out of 2^s of them, are adjacent to 2^t vertices, each of which is located in a unique Class-0 cluster. As an example, Fig. 2b shows one example of $GEH(1, 2)$, where there are four Class-0 clusters, (u, v), (w, x), (a, c), and (b, d), each being an edge, technically a Q_1; and two Class-1 clusters, (a', b', u', w') and (c', d', v', x'), both being Q_2. Each of the two vertices in an edge is adjacent to a unique vertex in a Q_2, and each of the four vertices in a Q_2 is adjacent to a unique vertex in an edge.

The above observation motivates us to further define a labeled *structure graph*, $G(s, t, \omega)$, *associated with* $GEH(s, t, f)$, where $V(G(s, t, \omega))$ collects all the clusters in $GEH(s, t, f)$. Each vertex in this structure graph, sometimes also referred to as a *cluster*, corresponding to a Class-0 cluster, is adjacent to 2^s vertices, each corresponding to a Class-1 cluster; and conversely, each vertex corresponding to a Class-1 cluster, is adjacent to 2^t vertices, each corresponding to a Class-0 cluster. Each edge, e, in $G(S, t, \omega)$, corresponding to a cross edge $(u, f(u))$ in $GEH(s, t, f)$, is labeled with $\omega(e)$ $(= (u, f(u)))$. It is clear that such a structure graph, $G(s, t, \omega)$, is isomorphic to a complete bipartite graph $K_{2^s, 2^t}$. When f and/or ω are irrelevant to the issue in the discussion, we may choose to exclude them in the notation. In particular, by $GEH(s, t)$, we mean $GEH(s, t, f)$ for some appropriate perfect matching f;

and, by $G(s, t)$, we mean a structure graph $G(s, t, \omega)$ associated with a generalized exchanged hypercube $GEH(s, t, f)$, where ω is induced by f.

An exchanged hypercube is certainly a generalized exchanged hypercube, where the cross edges are specified with E_1; while the class of generalized exchanged hypercubes is strictly more general than that of the exchanged hypercubes since we have a lot more freedom in choosing the cross edges between the clusters of opposite classes: Any perfect matching between the vertices of clusters of opposite classes will do.

Obviously, some topological properties (such as the distance between a specific pair of vertices) may vary wildly depending on the specifics of such a matching, but others do not. For example, as shown in [22, Theorem 1], $EH(s, t)$ is isomorphic to $EH(t, s)$. This property also holds for a generalized exchanged hypercube since, in the above definition of $GEH(s, t)$, the roles as played by Class-0 clusters and Class-1 clusters are symmetric to each other. As a result, we assume $1 \leq s \leq t$, when addressing $GEH(s, t)$, in the rest of this chapter. Furthermore, as we will expose in the rest of this chapter, several other structure properties, and fault tolerance related measurements, are also independent of this perfect matching between the vertices of opposite clusters. Such an observation reveals the naturalness and robustness of the generalized exchanged hypercube.

3 A Connectivity Result Associated with Linearly Many Faults

Let G be a graph, and let $S \subset V(G)$, we use $N_G(S)$ to refer to the *open neighbors of all the vertices of S in G*, excluding those in S. (We often omit the subscript G from this notation, and others, when the context is clear.) Such a graph G is *r-regular* if the degree of every vertex in $V(G)$ is r.

The *vertex connectivity* of a non-complete graph G, denoted by $\kappa(G)$, refers to the minimum size of a vertex cut F, $F \subset V(G)$, such that the surviving graph $G - F$ is disconnected, which is obtained from G by deleting all the vertices in F from G, together with edges incident to at least one vertex in F. By convention, the vertex connectivity of a complete graph K_n is $n - 1$. On the other hand, the *edge connectivity* of a graph G, denoted by $\kappa'(G)$, refers to the minimum size of an edge cut D, $D \subset E(G)$, such that the surviving graph $G - D$ is disconnected, which is obtained from G by removing all the edges as contained in D. Let $\delta(G)$ be the minimum degree among those of all the vertices in a graph G, clearly, $\delta(GEH(s, t)) = s + 1$, when $1 \leq s \leq t$. Indeed the following well-known result relates the vertex connectivity, the edge connectivity, and the minimum degree of a *simple graph* G, where there is at most one edge between any two vertices.

Lemma 1 *[37, Theorem 4.1.9] Let G be a simple graph, then $\kappa(G) \leq \kappa'(G) \leq \delta(G)$.*

Naturally, it is desirable for a graph G to have the property that $\kappa(G) = \delta$. A non-complete graph G with at least $r + 1$ vertices is *r-connected* if deleting any set of at most $r - 1$ vertices results in a connected graph. A complete graph with $r + 1$ vertices, denoted by K_{r+1}, is k-connected for all $k \leq r$. An r-regular graph is *maximally connected* if it is r-connected. A maximally connected r-regular graph is also *tightly super-connected* if, for every $F \subset V(G)$ with $|F| = r$, the graph $G - F$ is either connected or it consists of two components, one being a singleton. Clearly, in a tightly super-connected graph, all the neighbors of the aforementioned singleton fall in such a set F. When used as an interconnection network, an r-regular tightly super-connected structure is more preferable than an r-regular maximally connected graph, as when up to r vertices become faulty, the surviving graph of such a tightly super-connected graph, except one vertex, is still connected, thus functioning. We observe that a maximally connected graph does not need to be tightly super-connected. For example, in a given $K_{3,3} (= (V_1, V_2, E))$, $K_{3,3} - V_1 = V_2$, i.e., three singletons, thus $K_{3,3}$ is not tightly super-connected, although it is maximally connected. On the other hand, it is well known that Q_n is tightly super-connected [41, Theorem 3.3], thus maximally connected.

Noticing that the generalized exchanged hypercube $GEH(s, t)$, $1 \leq s \leq t$, is not regular, except when $s = t$, we thus slightly generalize the above notions as follows: We say that G is δ-*maximally connected* if, for all $F \subset V(G)$, $|F| < \delta(G)$, $G - F$ is connected; and G is δ-*tightly super-connected* if it is δ-maximally connected, and, for all $F \subset V(G)$, $|F| \leq \delta(G)$, $G - F$ is either connected or it consists of one large (connected) component plus one singleton. Clearly, $K_{m,n}$, $1 \leq m \leq n$, is δ-maximally connected [37, Example 4.1.2], although it is not δ-tightly super-connected, while Q_n is. For $GEH(s, t)$ to be useful as an interconnection network, it should be δ-tightly super-connected. In fact, an even stronger statement is true.

Theorem 1 *Let $s \in [1, t]$, and let $k \in [1, s]$, then*

1. *there is $F \subset V(GEH(s, t))$, $|F| = ks - \frac{k(k-1)}{2} + 1$, such that $GEH(s, t) - F$ contains a component of size k; and*
2. *for all $F \subset V(GEH(s, t))$, $|F| \leq ks - \frac{k(k-1)}{2}$, $GEH(s, t) - F$ is either connected or it consists of a large component and small components containing at most $k - 1$ vertices.*

The proof of Theorem 1 is given in [9]. (We note that the proof for the case of $s = 3$ for Part 2 was omitted due to space constraint. In the appendix, we give a proof for this case.) For example, if we set $k = 1$ in Part 2 of Theorem 1, we have that, for all F, $|F| \leq s$, $GEH(s, t) - F$ is connected, that is, it is maximally connected. On the other hand, if we set $k = 1$ in Part 1 of Theorem 1, we have that for some F, $|F| = s + 1$, $GEH(s, t) - F$ contains a singleton. Furthermore, if we then set $k = 2$ in Part 2 of Theorem 1, we have that, when $|F| \leq 2s - 1$, $GEH(s, t) - F$ is either connected or it contains a large component plus a singleton.

The following result is immediate by Theorem 1, and will be made use of in the next section, when we address the component connectivity of $GEH(s, t)$.

Corollary 1 *Let* $F \subset V(GEH(s,t)), s \in [1,t]$. *If* $GEH(s,t) - F$ *consists of a large component and other components that contain at least* k ($\in [1,s]$) *vertices, then*

$$|F| \geq ks - \frac{(k-1)k}{2} + 1.$$

On the other hand, if we set k to 3, where $k \in [1,s]$, in Part 2 of Theorem 1, the following result plays a critical role when we derive the conditional diagnosability in Sect. 6.

Corollary 2 *Let* $F \subset V(GEH(s,t)), s \in [3,t]$. *If* $|F| \leq 3s - 3$, *then* GEH $(s,t) - F$ *is either connected or it consists of a large component and small components that contain at most two vertices altogether.*

4   The Restricted and Cyclic Vertex-Connectivity

Given a non-complete graph $G(V,E), F \subset V$ is a g-disconnecting set of G if $G - F$ is disconnected and every vertex in $G - F$ has degree at least g (≥ 0). The restricted connectivity of order g of G, denoted as $\kappa_g(G)$, is defined as the size of a minimum g-disconnecting set of G [10, 11].

While $\kappa_0(G)$ coincides with the traditional vertex connectivity $\kappa(G)$, $\kappa_g(G)$, $g \geq 1$, is often used to characterize other fault tolerance properties, such as the g-good-neighbor conditional diagnosability, of various network structures, including the hypercube [30, 31, 39], the m-ary n-dimensional hypercube [38, 45]. In particular, the following general result is derived in [20, Theorem 3.3].

Theorem 2 *For* $1 \leq s \leq t$, *and* $g \in [0,s]$, $\kappa_g(EH(s,t)) = (s - g + 1)2^g$.

We now initiate the study of this measurement for $GEH(s,t)$.

Theorem 3 *Let* $3 \leq s \leq t$, $\kappa_1(GEH(s,t)) = 2s$.

Proof Let $k = 2$ in Theorem 1, we have that if $|F| \leq 2s - 1$, $GEH(s,t) - F$ is either connected or it has two components, one of which is a singleton. Thus $\kappa_1(GEH(s,t)) \geq 2s$.

Let u and v be two adjacent vertices in a Class-0 cluster in $GEH(s,t)$. Clearly, $|N(\{u,v\})| = 2s$ as $GEH(s,t)$ is triangle-free by definition. Now let $k = 3$ in Theorem 1, we have that if $|F| \leq 3s - 3$, $GEH(s,t) - F$ has a large component and small components with at most two vertices in total. (This includes the case when $GEH(s,t) - F$ is connected.) Since $2s \leq 3s - 3$, when $s \geq 3$, $GEH(s,t) - N(\{u,v\})$ has two components, one of which is a K_2, i.e., (u,v), while none of the vertices in the large component is isolated, thus each having a degree at least 1. Hence, $\kappa_1(GEH(s,t)) = 2s$. $\square$

We comment that $\kappa_1(G)$ is referred to as the super connectivity of G in [25, 39], i.e., the survival graph contains no isolated vertex when such a minimum vertex cut is removed. Theorem 3 immediately leads to the super connectivity of $EH(s, t), 3 \leq s \leq t$, one of the main results in [25].

The proof for the following observation is straightforward.

Lemma 2 *Let $n \geq 4$ and let C_4 be a 4-cycle, then the degree of every vertex in $Q_n - N(C_4)$ is at least 2.*

We are now ready to prove the following result.

Theorem 4 $\kappa_2(GEH(s, t)) = 4s - 4$, $s \in [6, t]$.

Proof Let $k = 4$ in Theorem 1, we have that, if $|F| \leq 4s - 6$, $GEH(s, t) - F$ has a large component and small components with at most three vertices in total. Since $GEH(s, t)$ is triangle-free, the three vertices in small components cannot form a triangle. Thus $\kappa_2(GEH(s, t)) \geq 4s - 5$. We now claim that this number is at least $4s - 4$. Suppose the size of a minimum 2-disconnecting set of $GEH(s, t)$ is $4s - 5$ and let S be such a set. Let $k = 5$ in Theorem 1, we have that, if $|F| \leq 5s - 10$, $GEH(s, t) - F$ has a large component and small components with at most four vertices in total. Since $4s - 5 \leq 5s - 10$, for $s \geq 5$, the statement holds for S. Furthermore, as S is a 2-disconnecting set, and the graph is triangle-free, the small component of $GEH(s, t) - S$ must contain exactly four vertices, which form a 4-cycle. To isolate this 4-cycle, we need to delete at least $4(s - 1) (= 4s - 4)$ vertices, a contradiction.

To show that $4s - 4$ suffices, let A be the vertex-set of a 4-cycle in a Class-0 cluster C of $GEH(s, t)$. It is clear that $|N(A)| = 4s - 4$. Now apply Theorem 1 with $k = 5$ again, we can conclude that $GEH(s, t) - N(A)$ contains a large component and small components with at most four vertices, as $4s - 4 \leq 5s - 10$, when $s \geq 6$. Since the large component contains at least $2^{s+t+1} - 4s$ vertices, and $2^{s+t+1} - 4s \geq 2^{2s+1} - 4s > 4$, $s \geq 6$, we conclude that the surviving graph contains one large component and the prescribed 4-cycle. We claim that every vertex u in this large component of $GEH(s, t) - N(A)$ is of degree at least 2. If u is a vertex of C, then it has degree at least 2 by Lemma 2; otherwise, the degree of u in $GEH(s, t) - N(A)$ is at least the degree of u in $GEH(s, t)$-1, thus at least 2. Therefore $N(A)$ is a 2-disconnecting set. □

It is clear that both Theorems 3 and 4 agree with Theorem 2 when setting g to 1, and 2, respectively.

Let G be a graph, we refer to F $(\subset V(G))$ a *cyclic vertex-cut* of G if $G - F$ is disconnected and at least two components in $G - F$ contain a cycle. The *cyclic vertex-connectivity* of a graph G is then defined as the size of a minimum cyclic vertex-cut in G. This notion was originally introduced to study the Four Color problem [36], and has since been applied to study other graph theory problems, including that of the Integer Flow Conjectures [46]. Recently, the cyclic vertex-connectivity results of several interconnection networks have also been reported in literature, e.g., [6, 44].

By following the arguments as we made in proving Theorem 4, we can similarly show that the cyclic vertex-connectivity of $GEH(s, t)$ is $4s - 4$.

It is pointed out in [16, pp. 159] that DC_n, the dual-cube-like network [1], which generalizes the dual-cube structure [21], is isomorphic to $EH(n-1, n-1)$, a special case of $GEH(n-1, n-1)$. Hence, we immediately have the following result:

Corollary 3 *For* $n \geq 3$, $\kappa_1(DC_n) = 2n - 2$. *For* $n \geq 7$, $\kappa_2(DC_n) = 4n - 8$, *which is also the value of its cyclic vertex-connectivity.*

5 The Component Connectivity

Component connectivity of a graph characterizes the size of a minimum vertex cut whose removal leaves its surviving graph in a certain number of components. This notion, as introduced in [2, 32] and further addressed in, e.g., [19, 28, 29], is to overcome the deficiency of the ordinary notion of vertex connectivity when used to measure the fault tolerance of interconnection networks. Indeed, with two graphs of same vertex connectivity, when a corresponding vertex cut is removed, their respective surviving graphs could have quite different number of components. For example, as pointed out in [19], the vertex connectivity of both $K_{1,n}$ and the path graph $P_{n+1}, n \geq 2$, is 1, but, when a cut vertex is removed, the surviving graph of $K_{1,n}$ consists of n singletons, while that of the path graph consists of just two components.

It is worth pointing out that there exists yet another alternative generalization of this vertex connectivity concept as proposed in [12]. The *k-tree connectivity* of a graph G is defined as the minimum k such that internally disjoint Steiner trees exist on all the k-subsets of $V(G)$. For a connection between the component connectivity and this latter tree based generalization, readers are referred to [19] and the references cited within.

Let G be a non-complete graph, an *r-component cut* of $G, r \geq 2$, refers to a set of vertices whose removal results in a surviving graph with at least r components. The *r-component connectivity,* or simply r-connectivity [19], denoted by $\bar{\kappa}_r(G)$, of G refers to the size of a minimum r-component cut of G (If there is no r-component cut of G, we simply define $\bar{\kappa}_r(G)$ to be ∞.). Clearly, $\bar{\kappa}_2(G)$ is just the usual vertex connectivity of G. It is also easy to see, by definition, that $\bar{\kappa}_m(G) \leq \bar{\kappa}_{m+1}(G), m \geq 2$.

As mentioned earlier, $\bar{\kappa}_n(K_{1,n}) = 1$. For P_{2n+1}, if we remove every other vertex, n vertices in total, the surviving graph consists of $n+1$ singletons. Thus, $\bar{\kappa}_{n+1}(P_{2n+1}) \leq n$. Clearly, $\bar{\kappa}_2(P_3) \geq 1$, and an inductive argument shows that $\bar{\kappa}_{n+1}(P_{2n+1}) \geq n$. Hence, $\bar{\kappa}_{n+1}(P_{2n+1}) = n$. The same idea also applies to P_{2n}, except that removing the last vertex will not increase the number of singletons. Thus, we only need to remove the first $n-1$ vertices and the surviving graph ends up with $n-1$ singletons and one component of size 2, n components altogether. We thus have $\bar{\kappa}_n(P_{2n}) = n - 1$.

The above analysis shows that, although both $K_{1,n}$ and P_{n+1} share the same vertex connectivity, it just takes out one cut vertex to break a $K_{1,n}$ into n pieces, but it has to take out about half of the vertices to achieve the same effect in a path graph that contains twice as many vertices. As a result, we may conclude that a path graph is more resilient as compared with a star graph from this perspective. Hence, this measure of component connectivity characterizes more faithfully the degree of an interconnection network to stay intact, when a number of processing nodes become faulty.

The following result on the $(r + 1)$-component connectivity of the hypercube Q_n, $n \geq 2$, has been derived in [15, Theorem 2.1].

Theorem 5 *For all $n \geq 3$, $k \in [1, n]$, $\bar{\kappa}_{r+1}(Q_n) = rn - \frac{r(r+1)}{2} + 1$.*

We now derive $\bar{\kappa}_{r+1}(GEG(s, t))$, the component connectivity of a generalized exchanged hypercube $GEH(s, t)$, $1 \leq s \leq t$, for $r \in [2, s]$.

Theorem 6 *Let $1 \leq s \leq t$. For $r \in [1, s]$, $\bar{\kappa}_{r+1}(GEH(s, t)) = rs - \frac{r(r-1)}{2} + 1$.*

Proof Let u be a vertex in C_0, a Class-0 cluster of $GEH(s, t)$, $1 \leq s \leq t$, S be a collection of r ($\in [1, s + 1]$) neighbors of u in $GEH(s, t)$, and let u' be the unique neighbor of u in a Class-1 cluster, C'_1, via a cross edge. Depending on whether $u' \in S$, we can construct the open neighbor set $N_r(S)$ of S, where $|S| = r$, in two ways, referred to as $N_r^1(S)$ ($N_r^2(S)$), respectively.

- Assume that $u' \notin S$. Then, for those r ($\in [1, s]$) neighbors of u in C_1, each has $s + 1$ neighbors in $GEH(s, t)$, a total of $r(s + 1)$ vertices. But, for each of them, u is counted once as its neighbor, although it should be counted just once in $N(S)$. Moreover, every common neighbor shared by any two of these neighbors of u is counted twice, while each of them should also be counted only once in $N(S)$. As a result, we have

$$|N_r^1(S)| = r(s + 1) - (r - 1) - \binom{r}{2} = rs - \binom{r}{2} + 1$$

as a hypercube has no $K_{2,3}$ as a subgraph. We notice that $N_r^1(S)$ is only defined when $r \in [1, s]$. As an example, in $GEH(1, 2)$, as shown in Fig. 2b, we have that $s = 1$, $t = 2$, thus $r = 1$. If we pick $S = \{v\}$, then, $N_1^1(S) = \{u, v'\}$. On the other hand, the above result gives us $|N_1^1(S_1)| = 2$.
- Alternatively, when $u' \in S$, then each of the $r - 1$ ($\in [0, s]$) neighbors of u in C_0 has s neighbors, plus another one via a cross edge; and u' has $t + 1$ neighbors in C'_0, a total of $(r - 1)(s + 1) + (t + 1)$ vertices. Similar to the previous case, u is counted once for each of these r neighbors of u, a total of r times, while it should be included just once in $N_r^2(S)$; and, each vertex adjacent to any two of these $r - 1$ neighbors of u in C_0 is counted twice, but it should be counted just once. (Notice that only u is adjacent to both u' and those $r - 1$ neighbors in C_0. Just assume there is another vertex v adjacent to both u' and u_1, a neighbor of u in C_0. By definition, v cannot be in a cluster of Class 0, as then it won't be adjacent to u, a vertex in a Class-0 cluster. Thus, v is either located in C_0 or in a Class-1 cluster C'. Assume that v occurs in C_0, then, because cross edges form a perfect matching, there is only one cross edge in between C_0 and C'_0, since (u, u')

is already part of the matching, v ($\neq u$) cannot be adjacent to u' via another cross edge. By the same token, because of the existence of the edge (u, u'), v cannot be located in a Class-1 cluster, either.) Removing all these redundancies, the size of this *alternative* construction can be calculated as follows:

$$|N_r(S_2)| = (r - 1)(s + 1) + (t + 1) - (r - 1) - \binom{r - 1}{2}$$

$$= rs - \binom{r - 1}{2} + (t - s + 1)$$

$$= |N_r(S_1)| + (t - s + r - 1) \geq |N_r(S_1)|, \tag{1}$$

when $r \in [1, s + 1]$, as $1 \leq s \leq t$, by assumption. Clearly, $N_r(S_1) = N_r(S_2)$ when $s = t$ and $r = 1$.

To continue with our previous example, for $r = 1$, if we now pick $S = \{u'\}$, we have $N_1^2(S) = \{u, a', w'\}$, while Eq. 1 gives $|N_1^2(S)| = 3$. In this case, $|N_1(S_1)| < |N_1(S_2)|$, since $s \neq t$, although $r = 1$. Moreover, for $r = 2$, although $N_2^1(S)$ is not defined, when we set $S = \{v, u'\}$, the alternative construction gives us that, $N_2^2(S) = \{u, a', w', v'\}$, while $|N_2^2(S)| = 4$ by Eq. 1.

It is easy to see that, once $N_r^1(S)$ (respectively, $N_r^2(S)$) is removed, all the neighbors of vertices in S are removed and none of these vertices in S are adjacent since $GEH(s, t)$ is bipartite. Hence, $GEH(s, t) - N_r^1(S)$ (respectively, $GEH(s, t) - N_r^2(S)$) contains at least $r + 1$ components, including at least r singletons. Thus, $N_r^1(S)$ (respectively, $N_r^2(S)$) is a $(r + 1)$-component cut. As a result, when $r \in [1, s]$,

$$\bar{\kappa}_{r+1}(GEH(s, t)) \leq \min\{N_r(S_1), N_r(S_2)\} = N_r(S_1) = rs - \binom{r}{2} + 1.$$

Let F be a minimum $(r + 1)$-component cut. Then $GEH(s, t) - F$ has at least $r + 1$ components. Thus, it has one large components and r "smaller" components. Clearly, these "smaller" components collectively has at least r vertices. By Corollary 1, for $1 \leq s \leq t$, $r \in [1, s]$, $\bar{\kappa}_{r+1}(GEH(s, t)) \geq |F| \geq rs - \frac{r(r-1)}{2} + 1$.

Thus, for $1 \leq s \leq t$, $r \in [1, s]$, $\bar{\kappa}_{r+1}(GEH(s, t)) = rs - \frac{r(r-1)}{2} + 1$. $\square$

The following result is based on the relationship between the dual-cube-like network and that of the generalized exchanged hypercube, as pointed out in the last section.

Corollary 4 *For all* $n \geq 3$, *if* $r \in [1, n - 1]$, $\bar{\kappa}_{r+1}(DC_n) = r(n - 1) - \frac{r(r-1)}{2} + 1$.

6 The Conditional Diagnosability

The conditional diagnosability of interconnection networks has been studied by using a number of ad-hoc methods [18, 47]. Recently, gathering various ad-hoc methods developed in the last decade, an unified approach was developed [4, 14], which has been applied to find the conditional diagnosability of many interconnection networks, e.g., [4, 5, 7]. We give a brief overview here and refer readers to the aforementioned literature for further details.

According to the comparison diagnosis model [26, 27, 33], a comparator, $w \in G$, sends the same input to each and every pair of its neighbors, v and x in G, and generates a result, which tells if v and x are faulty, assuming w is not. A collection of all such results is called a syndrome of the diagnosis. Since a faulty comparator can lead to unreliable results, a set of faulty vertices may also produce different syndromes. Two distinct faulty sets F_1 and F_2 are indistinguishable if and only if they are compatible with at least one syndrome, distinguishable otherwise. Hence, $t_c(G)$, the conditional diagnosability of G, equals the maximum number d such that for all distinct pairs of conditional faulty sets, (F_1, F_2), $|F_1| \leq d$, $|F_2| \leq d$, F_1 and F_2 are distinguishable.

We notice that the central structure of the above comparison diagnosis model is a length two path, $p_2(v, w, x)$, centered at a vertex w. Clearly, any vertex in a viable interconnection network should have at least one neighbor outside the neighborhood of such a length two path centered at w, an arbitrary but fixed vertex. Otherwise, this length two path will immediately turn into a bottleneck, and make the network fault-intolerant. This observation motivates the following notion of a good length two path [5]: Let G be a graph, we call $p_2(v, w, x)$, a path of length 2 in G, a *good path* if, for every vertex $z \notin N(\{w\}) \cup \{w\}$, $N(\{z\}) \nsubseteq N(\{v, w, x\}) \cup \{v, x\}$.

By definition, to show that, for a given graph G, $t_c(G) \leq d$, we only need to construct a pair of distinct conditional faulty sets (F_1, F_2), $|F_1| \leq d + 1$, $F_2 \leq d + 1$, such that (F_1, F_2) is indistinguishable. The following result [34] provides such an upper bound of $t_c(G)$.

Proposition 1 *Let G be a graph where $p_2(v, w, x)$ forms a good path of length two in G. Then $t_c(G) \leq |N_G(\{v, w, x\})|$.*

It seems that, to get an upper bound for $t_c(G)$ by applying Proposition 1, we have to minimize $|N_G(\{v, w, x\})|$ over all good paths of length two in G, which may not be easy. On the other hand, as we will show, there is often a good candidate for a minimizer. We should also point out that the above result does not imply that a conditional faulty set obtained via a length two path is always a minimizer of such an upper bound. In fact, such an upper bound is sometimes obtained through a four cycle [40].

Given $GEH(s, t)$, $2 \leq s \leq t$, we select a four cycle $C_4 = (v, w, x, u, v)$ in C_0, a Class-0 cluster, and consider $p_2(v, w, x)$. Let any vertex $z \notin N(\{w\}) \cup \{w\}$. Such a z

must exist since C_0 ($\equiv Q_s$) contains at least four vertices when $s \geq 2$. By definition, z is adjacent to vertex z' in a cluster uniquely associated with C_0, which is also different from those corresponding to either v, w or x by definition. Thus, z' cannot be either v or x, and z' cannot be adjacent to either v, w or x. In other words, $p_2(v, w, x)$ is a good path.

Both v and x have $s - 1$ neighbors that are not on $p_2(v, w, x)$, while w has only $s - 2$ of them. Moreover, u is a neighbor of both v and x thus gets over counted once. Finally, all three of v, w, and x are adjacent to a unique vertex in their respectively associated cluster. Hence, $|N_{GEH(s,t)}(\{v, w, x\})| = 3s - 2$.

By Proposition 1, we have achieved the following upper bound result.

Lemma 3 *For all* $2 \leq s \leq t$, $t_c(GEH(s, t)) \leq 3s - 2$.

The issue now becomes how to verify this upper bound is also a lower bound, thus an exact bound, of $t_c(GEH(s, t))$. In general, this is quite challenging since we have to show that, for all conditional faulty set pairs (F_1, F_2), $|F_1| \leq d$, $|F_2| \leq d$, they are distinguishable. Fortunately, as previously mentioned, several general results to this regard have recently emerged, one of which is the following [4].

Theorem 7 *Let G be a graph, $\delta(G) \geq 3$, such that (1) for any $T \subset V(G)$, $|T| \leq d$, $G - T$ contains a large component and smaller components which contain at most two vertices in total; and (2) $|V(G)| > (\Delta(G) + 2)d + 4$, where $\Delta(G)$ refers to the maximum degree of vertices in G. Then $t_c(G) \geq d + 1$.*

When $2 \leq s \leq t$, $\delta(GEH(s, t)) = s + 1 \geq 3$, $\Delta(GEH(s, t)) = t + 1$, and $|V(GEH(s, t))| = 2^{s+t+1}$. Condition 1 of Theorem 7, for the generalized exchanged hypercube, immediately follows from Corollary 2, when $3 \leq s \leq t$. What is left for us to do is to check Condition 2 of Theorem 7, when $d = 3s - 3$, namely,

$$2^{s+t+1} = |V(G)| > (\Delta(G) + 2)d + 4 = 3(t + 3)(s - 1) + 4. \tag{2}$$

We only need to show that $2^{s+t-1} > (t + 3)(s - 1) + 1$, which holds when $2^{s+t-1} > (t + 4)(s - 1)$, since $s \geq 2$. This last inequality holds if $2^{s-1} > s - 1$ and $2^t \geq t + 4$. The first part certainly holds when $s \geq 2$, while the second part holds for $t \geq 3$. Finally, setting $t = 2$ in Eq. 2, we have $2^{s+3} > 15(s - 1) + 4 = 15s - 11$, which certainly holds for all $s \geq 2$.

Hence, Condition 2 holds for all $s, t \geq 2$. By Corollary 2, Lemma 3, and Theorem 7, we have obtained the following result.

Theorem 8 *For $3 \leq s \leq t$, $t_c(GEH(s, t)) = 3s - 2$.*

Last but not least, the diagnosability of the dual-cube-like network DC_n has been derived in [5, Theorem 7.2] to be $3n - 5$, $n \geq 4$, that certainly coincides with Theorem 8, when setting $s = t = n - 1 \geq 3$.

7 Concluding Remarks

In this chapter, we applied a general connectivity result to further derive several fault tolerance measurements for the generalized exchanged hypercube, including its restricted connectivity, cyclic vertex-connectivity, component connectivity, and its conditional diagnosability, in terms of the comparison diagnosis model.

These results show that the generalized exchanged hypercube is a natural and robust interconnection topology and the general connectivity result is truly general and useful, which might be applied to derive other interesting connectivity related results.

We comment that similar connectivity results have also been reported in the literature [8] for the complete cubic networks with its underlying structure graph being a complete graph.

Appendix

In this section, we give a proof of $s = 3$ for Part 2 of Theorem 1. We first state a number of preliminary results from [9]. (The proof of Lemma 5 was omitted but it is similar to the one for Lemma 4.)

Lemma 4 *[9, Lemma 3.3]* $GEH(s, t), 1 \leq s \leq t$, *is δ-maximally connected.*

Lemma 5 *[9, Lemma 3.4]* $GEH(s, t), 2 \leq s \leq t$, *is δ-tightly super-connected.*

Lemma 6 *[9, Lemma 4.1] Let $F \subset V(GEH(s, t)), s \in [2, t], |F| \leq ks - \frac{k(k-1)}{2}$, there exists Y, a connected component of $GEH(s, t) - F$, such that, for all $i \in [0, 2^s + 2^t)$, if $C_i - F_i$ is connected, it is a subgraph of Y.*

We note that Lemma 5 does not hold for $s = 1$ as $GEH(1, t)$ contains 2^t Class-0 clusters, each of which is an edge, and two Class-1 clusters, each isomorphic to a Q_t. (Cf. Fig. 2b). Let (u, v) be one of these edges. When $\{u', v'\} \subseteq F$, $GEH(1, t) - F$ contains (u, v) and other components containing a total of $2^{t+2} - 2 \geq 6$ vertices.

We are now ready to prove $s = 3$ for Part 2 of Theorem 1. When $s = 3$, $k \in [1, 3]$. We notice that, when $k = 1$, $|F| \leq s$, $GEH(s, t) - F$ is then connected, by Lemma 4. We thus only need to consider the cases of $k = 2$ and $k = 3$.

For the case of $k = 2$, thus $|F| \leq 5$ by Part 2, we need to show that $GEH(3, t) - F, t \geq 3$, is either connected or contains a large component together with a singleton. By Lemma 5, when $|F| \leq 4$, $GEH(3, t) - F$ is either connected or it consists of a large component and one singleton. Thus, we only need to consider the case of $|F| = 5$.

Let $F_i = F \cap V(C_i), i \in [0, 2^s + 2^t)$. If, for some $l, |F_l| = 5$, then all the other clusters contain no faulty vertices, thus they are all connected. Clearly $GEH(s, t) - F_l$ will be connected, as well, since every vertex in $C_l - F_l$ is adjacent, via a cross edge, to a vertex located in a connected cluster. If for some $l, |F_l| = 4$, and the remaining faulty vertex f falls into another cluster, then all the clusters, other than C_l, are connected. $GEH(s, t) - F$ is then either connected or contains a large component and a singleton u ($\in V(C_l) \setminus F_l$), when C_l is isomorphic to Q_3, u is adjacent to f via a cross edge, and all the three neighbors of u in C_l fall into F_l. We now assume that $|F_l| = 3$, when the other clusters collectively hold two faulty vertices, thus all connected by the maximum connectivity of hypercubes, as $s = 3$. If $C_l - F_l$ is connected, so is $GEH(s, t) - F$ by Lemma 6. Otherwise, if $C_l - F_l$ is disconnected, then C_l is isomorphic to Q_3, and $C_l - F_l$ contains a $K_{1,3}$ and a singleton f. Since the other clusters jointly hold two faulty vertices, this $K_{1,3}$ must be part of the large component of $GEH(3, t) - F$, as at least one of its four vertices is adjacent to a non-faulty vertex in this large component. Then, $GEH(3, t), 3 \le t$, is either connected or contains a large component and one singleton u when u is adjacent to one of the two faulty vertices, while all its three neighbors in C_l form F_l. The other cases are symmetric to the above.

We now turn to the case of $k = 3$, i.e., $|F| \le 6$, when we have to show that $GEH(3, t) - F, t \ge 3$, is either connected or contains a large component and small components altogether with at most two vertices. In light of the previous case, we only need to consider the case of $|F| = 6$.

If for some l, $|F_l| \ge 4$, then other clusters, sharing at most two faulty vertices, must be individually connected in the resulting graph by the assumption of $s = 3$ and Lemma 4, and belong to the same component, say Y, in the resulting graph by Lemma 6. By definition, those non-faulty vertices of C_l are part of Y. Hence, $GEH(s, t) - F$ is either connected, or contains a large component and smaller ones with at most two vertices, when the remaining up to two vertices in $V(C_l) \setminus C_l$ are adjacent to the faulty vertices in $F \setminus F_l$ via cross edges, while sharing their faulty neighbors in F_l.

We now consider the case when, for all l, F_l contains at most three of these vertices. Since for all l, C_l is isomorphic to a cube $Q_m, m \ge s$ ($= 3$), when $m \ge 4$, all such $C_l - F_l$'s are connected by Lemma 4, and so is $GEH(s, t) - F$, by Lemma 6. We thus only need to consider the case when C_l is isomorphic to Q_3, where $|F_l| = 3$.

If for some l, $|F_l| = 3$, and for $j \ne l, |F_j| < 3$, then $C_j - F_j, j \ne l$, will all be connected by Lemma 6. If $C_l - F_l$ is connected, so is $GEH(s, t) - F$ by Lemma 6. Now assume that $C_l - F_l$ is not connected. Notice that C_l is isomorphic to a Q_3, and its surviving graph contains a singleton u and a $K_{1,3}$. Since there are only three faulty vertices located outside C_l, and $K_{1,3}$ contains four vertices, it must be part of a large connected component. Thus, in this case, $GEH(3, t), t \ge 3$, is either connected or contains a large component and a singleton u, when u is adjacent to one of these remaining faulty vertices in $F \setminus F_l$, and all its three neighbors are contained in F_l.

We finally consider the subcase that $|F_l| = |F_{l'}| = 3$, where both C_l and $C_{l'}$ are isomorphic to Q_3, when, for $j \notin \{l, l'\}$, F_j is empty. If both $C_l - F_l$ and $C_{l'} - F_{l'}$ are connected, then $GEH(s, t)$ is also connected by Lemma 6. We now assume, without loss of generality, $C_l - F_l$ is connected, but $C_{l'} - F_{l'}$ is not, when it contains a singleton u' and a $K_{1,3}$. $GEH(s, t) - F$, in this case, is either connected or contains a large component and a singleton u' when it is adjacent to a vertex in F_l and all its three neighbors in C_l' constitute F_l'. For the remaining case, when neither of them is connected, namely, $C_l - F_l$ (respectively, $C_{l'} - F_{l'}$) contains a singleton u (respectively, u') and a $K_{1,3}$. By the same token, $GEH(s, t) - F$ is either connected or it contains a large component and smaller component(s) with at most two faulty vertices u and u', when u' (respectively, u) is adjacent to a vertex in F_l (respectively, $F_{l'}$) and all its three neighbors in C_l' (respectively, C_l) fall into F_l' (respectively, F_l).

References

1. Angjeli, A., Cheng, E., Lipták, L.: Linearly many faults in dual-cube-like networks. Theor. Comput. Sci. **472**, 1–8 (2013)
2. Chartrand, G., Kapoor, S.F., Lesniak, L., Lick, D.R.: Generalized connectivity in graphs. Bull. Bombay Math. Colloq. **2**, 1–6 (1984)
3. Chen, Y.-W.: A comment on "The exchanged hypercube". IEEE Trans. Parallel Dist. Syst. **18**, 576 (2007)
4. Cheng, E., Lipták, L.L., Qiu, K., Shen, Z.: On deriving conditional diagnosability of interconnection networks. Inf. Process. Lett. **112**, 674–677 (2012)
5. Cheng, E., Lipták, L.L., Qiu, K., Shen, Z.: A unified approach to the conditional diagnosability of interconnection networks. J. Interconnect. Netw. **13**, 1250007 (19 pages) (2012)
6. Cheng, E., Lipták, L.L., Qiu, K., Shen, Z.: Cyclic vertex connectivity of Cayley graphs generated by transposition trees. Graphs Comb. **29**(4), 835–841 (2013)
7. Cheng, E., Qiu, K., Shen, Z.: On the conditional diagnosability of matching composition networks. Theor. Comput. Sci. **557**, 101–114 (2014)
8. Cheng, E., Qiu, K., Shen, Z.: Connectivity results of complete cubic network as associated with linearly many faults. J. Interconnect. Netw. **15**(1 & 2), 1550007 (23 pages) (2015)
9. Cheng, E., Qiu, K., Shen, Z.: A strong connectivity property of the generalized exchanged hypercube. Discrete Appl. Math. http://dx.doi.org/10.1016/j.dam.2015.11.014
10. Esfahanian, A.H.: Generalized measures of fault tolerance with application to n-cube networks. IEEE Trans. Comput. **38**(11), 1586–1591 (1989)
11. Esfahanian, A.H., Hakimi, S.L.: On computing a conditional edge-connectivity of a graph. Inf. Process. Lett. **27**, 195–199 (1988)
12. Hager, M.: Pendant tree-connectivity. J. Comb. Theory **38**, 179–189 (1985)
13. Harary, F., Hayes, J.P., Wu, H.-J.: A survey of the theory of hypercube. Comput. Math. Appl. **15**(4), 277–289 (1988)
14. Hong, W.-S., Hsieh, S.-Y.: Strong diagnosability and conditional diagnosability of augmented cubes under the comparison diagnosis model. IEEE Trans. Reliab. **61**, 140–148 (2012)
15. Hsu, L.-H., Cheng, E., Lipták, L., Tan, J.J.M., Lin, C.-K., Ho, T.-Y.: Component connectivity of the hypercubes. Int. J. Comput. Math. **89**(2), 137–145 (2012)
16. Klavžar, S., Ma, M.: The domination number of exchanged hypercubes. Inf. Process. Lett. **114**, 159–162 (2014)
17. Klavžar, S., Ma, M.: Average distance, surface area, and other structural properties of exchanged hypercubes. J. Supercomput. **69**, 306–317 (2014)
18. Lai, P.-L., Tan, J.J.M., Chang, C.-P., Hsu, L.-H.: Conditional diagnosability measures for large multiprocessor systems. IEEE Trans. Comput. **54**, 165–175 (2005)
19. Li, X., Mao, Y.: The generalized 3-connectivity of lexicographic product graphs. Discrete Math. Theor. Comput. Sci. **16**(1), 339–354 (2014)
20. Li, X.-J., Xu, J.-M.: Generalized measures of fault tolerance in exchanged hypercubes. Inf. Process. Lett. **113**, 533–537 (2013)
21. Li, Y., Peng, S., Chu, W.: Efficient collective communications in dual-cube. J. Supercomput. **28**, 71–90 (2004)

22. Loh, P.K.K., Hsu, W.J., Pan, Y.: The exchanged hypercube. IEEE Trans. Parallel Dist. Syst. **16**, 866–874 (2005)
23. Ma, M., Liu, B.: Cycles embedding in exchanged hypercubes. Inf. Process. Lett. **110**, 71–76 (2009)
24. Ma, M.: The connectivity of exchanged hypercubes. Discrete Math. Algorithms Appl. **2**, 213–220 (2010)
25. Ma, M., Zhu, L.: The super connectivity of exchanged hypercubes. Inf. Process. Lett. **111**, 360–364 (2011)
26. Maeng, J., Malek, M.: A comparison connection assignment for self-diagnosis of multiprocessor systems. In: Proceedings of 11th International Symposium on Fault-Tolerant Computing, pp. 173–175 (1981)
27. Malek, M.: A comparison connection assignment for diagnosis of multiprocessor systems. In: Proceedings of 7th International Symposium on Computer Architecture, pp. 31–35 (1980)
28. Oellermann, O.R.: On the l-connectivity of a graph. Graphs Comb. **3**, 285–299 (1987)
29. Oellermann, O.R.: A note on the l-connectivity function of a graph. Congessus Numerantium **60**, 181–188 (1987)
30. Oh, A.D., Choi, H.-A.: Generalized measures of fault tolerance in n-cube networks. IEEE Trans. Parallel Dist. Syst. **4**, 702–703 (1993)
31. Peng, S.L., Lin, C.K., Tan, J.J.M., Hsu, L.H.: The g-good-neighbor conditional diagnosability of hypercube under the PMC model. Appl. Math. Comput. **218**(21), 10406–10412 (2012)
32. Sampathkumar, E.: Connectivity of a graph— A generalization. J. Comb. Inform. Syst. Sci. **9**, 71–78 (1984)
33. Sengupta, A., Dahbura, A.T.: On self-diagnosable multiprocessor systems: diagnosis by the comparison approach. IEEE Trans. Comput. **41**, 1386–1396 (1992)
34. Stewart, I.: A general technique to establish the asymptotic conditional diagnosability of interconnection networks. Theor. Comput. Sci. **452**, 132–147 (2012)
35. Sullivan, G.F.: A polynomial time algorithm for fault diagnosability. In: Proceedings of 25th Annual Symposium Foundations Computer Science, IEEE Computer Society, pp. 148–156 (1984)
36. Tai, P.G.: Remarks on the coloring of maps. Proc. R. Soc. Edinb. **10**, 501–503 (1880)
37. West, D.B.: Introduction to Graph Theory, 2nd edn. Prentice Hall, Upper Saddle River, NJ (2001)
38. Wu, J., Guo, G.: Fault tolerance measures for m-ary n-dimensional hypercubes based on forbidden faulty sets. IEEE Trans. Comput. **47**, 888–893 (1998)
39. Xu, J.-M., Wang, J.-W., Wang, W.-W.: On super and restricted connectivity of some interconnection networks. Ars Combinatoria **94**, 25–32 (2010)
40. Yang, M.-C.: Conditional diagnosability of balanced hypercubes under the MM^* model. J. Supercomput. **65**, 1264–1278 (2013)
41. Yang, X., Evans, D.J., Megson, G.M.: On the maximal connected component of a hypercube with faulty vertices. Int. J. Comput. Math. **81**(5), 515–525 (2004)
42. Yang, X., Evans, D.J., Megson, G.M.: On the maximal connected component of a hypercube with faulty vertices II. Int. J. Comput. Math. **81**(10), 1175–1185 (2004)
43. Yang, X., Evans, D.J., Megson, G.M.: On the maximal connected component of a hypercube with faulty vertices III. Int. J. Comput. Math. **83**(1), 27–37 (2006)
44. Yu, Z., Liu, Q., Zhang, Z.: Cyclic vertex connectivity of star graphs. In: Proceedings of Fourth Annual International Conference on Combinatorial Optimization and Applications (COCOA'2010) Dec. 18–20, 2010, Kailua-Kona, HI, USA, Springer LNCS 6508(Part I), pp. 212–221 (2010)
45. Yuan, J., Liu, A., Ma, X., Liu, X., Qin, X., Zhang, J.: The g-good-neighbor conditional diagnosability of k-ary n-cubes under the PMC model and MM* model. Int. J. Parallel, Emergent Distrib. Syst. **26**(4), 1165–1177 (2015)
46. Zhang, C.Q.: Integer Flows and Cycle Covers of Graphs. Marcel Dekker Inc., New York (1997)
47. Zhu, Q.: On conditional diagnosability and reliability of the BC networks. J. Supercomput. **45**, 173–1184 (2008)

Enumerated BSP Automata

Gaetan Hains

Abstract Parallel software needs formal descriptions and mathematically verified tools for its core concepts that are data-distribution and inter-process exchanges or synchronizations. Existing formalisms are either non-specific, like process algebras, or unrelated to standard Computer Science, like algorithmic skeletons or parallel design patterns. This has negative effects on undergraduate training, general understanding and mathematically-verified software tools for scalable programming. To fill a part of this gap, we adapt the classical theory of finite automata to bulk-synchronous parallel computing (BSP) by defining BSP words, BSP automata and BSP regular expressions. BSP automata are built from vectors of finite automata, one per computational unit location. In this first model the vector of automata is enumerated, hence the adjective *enumerated* in the title. We also show symbolic (intensional) notations to avoid this enumeration. The resulting definitions and properties have applications in the areas of data-parallel programming and verification, scalable data-structures and scalable algorithm design.

1 Introduction and Background

This paper introduces a new theory of bulk-synchronous parallel computing (BSP), by adapting classical automata theory to BSP. It attempts to provide the simplest possible mathematical description of BSP computations. With maximal reuse of existing Computer Science it is hoped that this theory will find its way into more complex formalisms for parallel programming tools, language designs and software-engineering.

BSP is a theory of parallel computing introduced by Valiant in the late 1980s [20] and developed by McColl [14]. Unlike theories of concurrency that generalize sequential computation, BSP retains the deterministic and predictable behaviour of sequential machines of the Von Neumann type, while taking advantage of concurrent

G. Hains (✉)
Huawei France R&D Center, 20 quai du Point du Jour, 92100 Boulogne-Billancourt, France
e-mail: gaetan.hains@huawei.com

© Springer International Publishing Switzerland 2017
A. Adamatzky (ed.), *Emergent Computation*, Emergence, Complexity and Computation 24, DOI 10.1007/978-3-319-46376-6_10

execution for accelerating computations. **Concurrency theory is related to BSP in the way microscopes are related to telescopes: they are built from similar components but look in opposite directions** (Fig. 1).

Just as parallel algorithms are a *special case* of sequential algorithms (they realize a sub- complexity class of problems i.e. NC $\subseteq$ P), BSP machines are close relatives of sequential machines whose instruction cycles are built from vectors of asynchronous sequential computations and are called *supersteps*. When the vector's elements terminate, they are globally synchronized, which guarantees determinism and allows predictable performance. This global sequence is called a superstep: launch a vector of asynchronous and independent sequential computations, wait until they all terminate and synchronize them. A *BSP computation* is a sequence of supersteps realized by a BSP computer, that is a vector of Von Neuman sequential computers linked by a global synchronization device. As we will see by adapting finite automata theory to BSP

- BSP automata are special cases of finite-state machines because they are finitely-defined systems, but
- the correspondence is not trivial and both the finite-alphabet hypothesis and the classical theory of product automata have to be adapted to account for the two-level nature of BSP computation.

In the usual definition and application of BSP, the sequential elements also *exchange* data during synchronization and there is a simple linear model of time complexity that estimates the delay for synchronization and data exchange in units of sequential computation. The model defined in the present paper can be extended to represent communication, as in the BSPCCS process algebra [16].

2 Bulk-Synchronous Words and Languages

Automata theory is both an elementary and standard part of Computer Science and an area of advanced research through topics such as tree automata, pattern matching and concurrency theory. It is also universally used by computing system through lexical analysis, text processing and similar operations. We are interested here in the core elementary theory of automata as described for example in [15, 21] or in the initial chapters of graduate textbooks such as [9, 18].

Let Σ be a finite alphabet and $p > 0$ an integer constant. Elements of Σ represent inputs to the automaton, or more generally events it takes part into. Constant p represents the assumption of a vector of parallel computation units executing the events or receiving them as signals. The local sequential computers or computations are indexed by $[p] = \{0, 1, \ldots, p-1\}$ and variable i will be assumed to range over $[p]$. In programming systems like BSPlib [7] this variable i is called `pid` for "Processor ID". A value $i \in [p]$ is sometimes called a *processor*, or an *explicit process* [11] or simply a *location*. Throughout the paper, all vectors will be assumed to be indexed by $[p]$.

Concurrent systems & theories	Bulk-synchronous parallelism
Arbitrary asynchronism	Structured asynchronism
High-complexity	Low-complexity
Distributed computing	Parallel computing
Unpredictable performance	Predictable performance
Endless processes like servers	Finite processes like algorithms
Not scalable	Massively scalable
Pairwise synchronizations	Collective synchronizations
Implicit shared memory	Explicit distributed memory
Implicit processes	Explicit processes (pid variable)

Fig. 1 Concurrency versus bulk-synchronous parallelism

Our whole theory of BSP computation is parametrized over this constant p, which is thus "static" and fixed for a given application of the theory. There is now a large body of research about BSP and many generalizations have been studied but for the sake of generality we model here only the standard core of BSP.

Our first definition represents the asynchronous part of supersteps: vectors of sequential computations, which automata theory sees as p-vectors of traces or *word-vectors*.

Definition 1 Elements of $(\Sigma^*)^p$ will be called *word-vectors*. A *BSP word* over Σ is a sequence of word-vectors i.e. a sequence of $((\Sigma^*)^p)^*$. A *BSP language* over Σ is a set of BSP words over Σ.

Remark 1 The word-vector $< \epsilon, \ldots, \epsilon >$ is not equivalent to an empty BSP word ϵ as the former will trigger a global synchronization, while the latter will not. In other words, $< \epsilon, \ldots, \epsilon >$ has length one and ϵ has length zero.

In our examples we will assume that $\Sigma = \{a, b\}$, ϵ is the empty "scalar" word and $p = 4$ without loss of generality.

For example $\mathbf{v}_1 = < ab, a, \epsilon, ba >$ and $\mathbf{v}_2 = < bbb, aa, b, a >$ are word-vector and $\mathbf{w} = \mathbf{v}_1\mathbf{v}_2$ is a BSP word. It is understood that $\mathbf{w}$ represents two successive supersteps and that

$$\mathbf{w} = \mathbf{v}_1\mathbf{v}_2 \neq < abbbb, aaa, b, baa >$$

that is: concatenation of BSP words is not the same as pointwise concatenation of word-vectors. Concatenation of BSP words represents phases of collective communications and barrier synchronizations (see Fig. 2, where vectors are drawn vertically). Concatenation of BSP words accordingly means concatenation of (sequences of) word vectors:

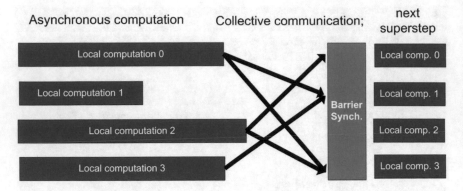

Fig. 2 A BSP superstep

$$\mathbf{w} = \mathbf{v}_1\mathbf{v}_2 = < ab, a, \epsilon, ba > < bbb, aa, b, a > .$$

Let $\mathbf{v}_a = < \epsilon, a, aa, aaa >$ respectively $\mathbf{v}_b = < \epsilon, b, bb, bbb >$ be word-vectors whose local words are a^i and b^i respectively. Then $L_a = \{\mathbf{v}_a\}$, $L_b = \{\mathbf{v}_b\}$ and $L_2 = \{\epsilon, \mathbf{v}_a, \mathbf{v}_b, \mathbf{v}_a\mathbf{v}_b\}$ are finite BSP languages and $L_3 = \{\epsilon, \mathbf{v}_a, \mathbf{v}_a\mathbf{v}_b, \mathbf{v}_a\mathbf{v}_b\mathbf{v}_a, \ldots\}$ is an infinite BSP language.

3 Finite Versus Infinite Alphabet

A BSP word is built from an infinite alphabet: even when Σ is finite, the set of word-vectors will be infinite. This part of the model illustrates the fact that a BSP computer is two-level: it is built from sequential computers, whose computations are finite but of unlimited length. But the infinite-alphabet property is not caused by the (finitely-many) computing elements, it would still hold if $p = 1$. It is rather a consequence of the fact that synchronization barriers are cooperative and not pre-emptive. Individual local computations have to *terminate* before a superstep ends with synchronization.

In his famous paper [19] Turing gives sketches several arguments for the choice of a finite alphabet. One is physical-topological: infinite alphabets realized by a finite physical device would require infinite precision of the device reading a symbol from working memory. He also gives another argument against infinite alphabets:

> compound symbols [such as arabic numerals], if they are too lengthy, cannot be observed at a glance

and even mentions, less convincingly, the case of Chinese ideograms as an attempt

> to have an enumerable infinity of symbols.

So our notion of BSP computation would appear to be incoherent with classic Church-Turing models: it is built from an infinite alphabet of symbols. However that would only be the case if we chose to use BSP languages as a model for decidability, which they are not intended to be. The BSP model was invented to model parallel *algorithms*, not arbitrary parallel computations. All local computations are therefore assumed to terminate and so is the global sequence of supersteps.

The best point of view on this question of infinite-vs-finite alphabet for BSP is that **BSP languages are sets of traces having a series-parallel structure representing the behaviour of parallel computers that all synchronize periodically**.

4 Bulk-Synchronous Automata

We now define BSP automata as acceptance machines for BSP words.

Definition 2 A BSP automaton **A** is a structure

$$(\{Q^i\}_{i\in[p]}, \Sigma, \{\delta^i\}_{i\in[p]}, \{q_0^i\}_{i\in[p]}, \{F^i\}_{i\in[p]}, \Delta)$$

such that for every i, $(Q^i, \Sigma, \delta^i, q_0^i, F^i)$ is a deterministic finite automaton (DFA),[1] and $\Delta : \mathbf{Q} \to \mathbf{Q}$ is called the *synchronization function* where $\mathbf{Q} = (Q^0 \times \ldots \times Q^{(p-1)})$ is called the set of *global states*.

In other words a BSP automaton is a vector of sequential automata A^i over the same alphabet Σ, together with a synchronization function that maps state-vectors to state-vectors.

Observe that the synchronization function is finite, like the transition functions, and that its value depends on a whole vector of local states. Because of it, a BSP automaton is more than the *product* [6] of its local automata (see Appendix 1 for an explanation).

Let Q^i be a set of local states at location i, $\delta^i : Q^i \times \Sigma \to Q^i$ a local transition function on those states and $\delta^i* : Q^i \times \Sigma^* \to Q^i$, the extended transition function on Σ-words. Right-application notation is sometimes convenient: $\delta^i * (q, w)$ can be written qw e.g. $qab = \delta(\delta(q, a), b)$.

Define a *transition function* δ on word-vectors as follows. For $\mathbf{q} \in \mathbf{Q}$ and $\mathbf{w} = < w^0, \ldots, w^{p-1} >$ a word-vector

$$\mathbf{q}w = \Delta(< q^0 w^0, \ldots, q^{p-1} w^{p-1} >) \tag{1}$$

i.e. "synchronization" of the result of application of local transition functions to local words. Function Δ is the model of a synchronization barrier because its local results depend on the whole vector of asynchronous results.

A BSP word is a sequence of word vectors. It is read by a BSP automaton as follows (Fig. 3):

---

[1] Q^i is the finite set of states, δ the transition function, $q^i \in Q^i$ the initial state and $F^i \subseteq Q^i$ the non-empty set of accepting states.

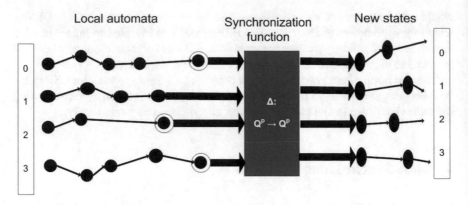

Fig. 3 A BSP automaton

1. If the sequence of word vectors is empty, the vector state remains the vector of local initial states; otherwise continue.
2. If $< w^0, \ldots, w^{p-1} >$ is the first word vector. Local automaton i applies w^i to its initial state and transition function to reach some state q^i, **not necessarily an accepting state**.
3. The synchronization function maps $\Delta :< q^0, \ldots, q^{p-1} > \rightarrow < q'^0, \ldots, q'^{p-1} >$.
4. If there are no more word vectors, and $\forall i. \ q'^i \in F^i$, the BSP word is accepted.
5. If there are no more word vectors, and $\exists i. \ q'^i \notin F^i$, the BSP word is rejected.
6. If there are more word vectors, control returns to step 2. but with local automaton i in state q'^i, for every location i.

Finite automata have no explicit notion of variables and values but states can be used to encode them e.g. $q_x 1 = \{(x, 1)\}$, $q_x 2 = \{(x, 2)\}, \ldots$. As a result, the synchronization function Δ can encode the communication of values between locations i, j, although this is not explicit in the general theory.

Proposition 1 *A BSP automaton is equivalent to a deterministic automaton over (the infinite alphabet of) word-vectors.*

Proof If all p finite automata are deterministic, then the transition function on word-vectors is a total and well-defined function of type $\mathbf{Q} \times (\Sigma^*)^p \rightarrow \mathbf{Q}$. The following structure built from $\mathbf{A}$ is a deterministic automaton by construction:

$$(\mathbf{Q}, (\Sigma^*)^p, \delta, < q_0^0, \ldots, q_0^{p-1} >, (F^0 \times \ldots \times F^{p-1})).$$

The automaton is deterministic because δ is well-defined and total because of 1 and the fact that local automata are deterministic. □

As Proposition 1 states, the BSP automaton is a deterministic automaton but its alphabet is infinite. The synchronization function Δ finite (can be enumerated) but enumerating the transition function δ is impossible: it is a table over $Q^p \times (\Sigma^*)^p$

whose second component is infinite. So δ is infinite, but it has an obvious finite representation: the vector of finite transition functions δ^i. As a result, a BSP automaton is practically equivalent to a DFA modulo the above syntactic changes.

Definition 3 As shown in the proof of Proposition 1, a BSP automaton $\mathbf{A}$ is a DFA on word-vectors. A BSP-word $\mathbf{w}$ is *accepted* by $\mathbf{A}$ if the reflexive-transitive closure of δ takes initial state $\mathbf{q}_0 = < q_0^0, \ldots, q_0^{p-1} >$ to an accepting state of $(F^0 \times \ldots \times F^{p-1})$ when applied to $\mathbf{w}$. The *language* of $\mathbf{A}$ is its set of accepted BSP-words.

We now give BSP automata to recognize BSP languages from Sect. 2.

Let A_a^i and A_b^i be the unique minimal DFA to recognize $\mathbf{v}_a$ and $\mathbf{v}_b$. Define A_a as the BSP automaton $(< A_a^0, A_a^1, A_a^2, A_a^3 >, \Delta)$ where Δ is the identity function. Then, for word-vector $\mathbf{a}$, the local transition functions of A_a will lead to a vector of accepting states, which the synchronization function Δ will leave unchanged. For any other word-vector $\mathbf{w}$, the local transition functions will lead to a vector of non-accepting states, unchanged by synchronization. As a result A_a accepts exactly language $L_a = \{\mathbf{v}_a\}$. A similar construction with letter b gives a BSP automaton A_b to accept L_b.

We now define a BSP automaton to accept

$$L_2 = \{\epsilon, \mathbf{v}_a, \mathbf{v}_b, \mathbf{v}_a \mathbf{v}_b\}.$$

Let A_{a+b}^i be a DFA that accepts language $\{\epsilon, a^i, b^i\}$ with exactly three accepting states: q_0^i initial state for accepting ϵ, q_{Fa}^i for accepting a^i and q_{Fb}^i for accepting b^i. Let $A_{\epsilon+b}^i$ be a DFA that accepts language $\{\epsilon, b^i\}$ with initial (accepting) state q_b^i. Define the BSP automaton as

$$A_{a+b} = (< A_{a+b}^i \cup A_{\epsilon+b}^i : i = 0, 1, 2, 3 >, \Delta)$$

where the local automaton has the union of accepting states and initial state q_0^i. Define also

$$\Delta < q_{Fa}^0, q_{Fa}^1, q_{Fa}^2, q_{Fa}^3 > = < q_b^0, q_b^1, q_b^2, q_b^3 >$$

and Δ is the identity function on all other vector-states. Then A_{a+b} on ϵ leads to $\Delta(\epsilon) = \epsilon$ which is by definition accepting. Automaton A_{a+b} applied to word-vector $\mathbf{a}$ leads to

$$\Delta < q_{Fa}^0, q_{Fa}^1, q_{Fa}^2, q_{Fa}^3 > = < q_b^0, q_b^1, q_b^2, q_b^3 >$$

an accepting state. Automaton A_{a+b} applied to word-vector $\mathbf{b}$ leads asynchronously to

$$< q_{Fb}^0, q_{Fb}^1, q_{Fb}^2, q_{Fb}^3 >$$

unchanged by Δ and that is an accepting state. Automaton A_{a+b} applied to word-vector $\mathbf{ab}$ leads through $\mathbf{a}$ and synchronization to $< q_b^0, q_b^1, q_b^2, q_b^3 >$ and from there asynchronously to accepting states of $A^i_{\epsilon+b}$ that the second synchronization preserves. So **ab** is also accepted and it can be checked that any other sequence of word-vectors is not accepted.

5 Non-determinism and Empty Transitions

A non-deterministic finite automaton (NFA) is a finite automaton whose transition function has type $Q \times \Sigma \rightarrow \mathcal{P}(Q)$ i.e. zero, one or more transitions $\delta(q, a)$ can exist for a given symbol a. The closure of its transition function is the union of all possible paths defined by δ for an input word.

A non-deterministic finite automaton with empty transitions (ϵ-NFA) is an NFA over alphabet $\Sigma \cup \{\epsilon\}$ where ϵ does not denote the empty word but a special "internal" symbol that represents "spontaneous" state changes happening without input. The closure of its transition function is the union of all possible NFA transitions on the input word interleaved with an arbitrary number of ϵ symbols.[2]

The languages recognized by NFA and by ϵ-NFA are same regular languages generated by regular expressions and recognized by DFA [15]. This holds because of:

1. a polynomial-time algorithm to remove ϵ-transitions without changing the language, and
2. an exponential-time algorithm to convert an NFA into an equivalent DFA.

The former transformation is called the subset algorithm because it generates a DFA whose states are subsets of the NFA states.

Definition 4 A non-deterministic BSP automaton (NBSPA) is a BSP automaton whose local automata are of type $Q \times \Sigma \rightarrow \mathcal{P}(Q)$ and whose synchronization function has type $\Delta : \mathbf{Q} \rightarrow \mathcal{P}(\mathbf{Q})$.

Definition 5 A non-deterministic BSP automaton with empty transitions (ϵ-NBSPA) is a NBSPA whose local automata are ϵ-NFA (Fig. 4).

Remark that the definition of empty transitions for BSP automata leaves the synchronization function Δ unchanged.

A (standard, deterministic) BSP automaton is by definition a special case of NBSPA and of ϵ-NBSPA but we need to verify whether the latter encode the same class of languages. The answer is positive and given by the next propositions.

---

[2]This notion of empty transitions is convenient but theoretically delicate. In the case of finite automata it preserves all elementary properties but that is not the case for communicating automata. For example Milner's CCS process algebra [17] uses a spontaneous-transition symbol τ with a similar property, but this changes the so-called bisimulation semantics of *communicating* automata. A more conservative process algebra can be built by replacing τ with an explicit clock-tick symbol Θ [1]. The resulting algebra of processes combines a simple bisimulation semantics with the algebraic simplicity (e.g. distributive law) similar to regular languages.

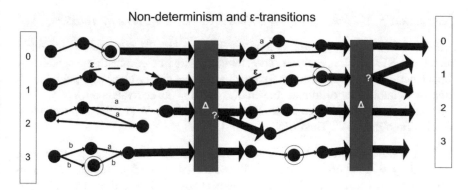

Fig. 4 An ϵ-NBSPA

Proposition 2 *The language of a NBSPA can be accepted by a deterministic BSP automaton.*

Proof Let N be a NBSPA defined by $(< N^0, \ldots, N^{p-1} >, \Delta)$ where the N^i are NFA and $\Delta : \mathbf{Q} \to \mathcal{P}(\mathbf{Q})$. Let Q^i be the set of states of N^i.

By the subset algorithm there exists p DFA D^i accepting the same (scalar) languages as the N^i and whose states are parts of $\mathcal{P}(Q^i)$. Define $\Delta' : \mathcal{P}(\mathbf{Q}) \to \mathcal{P}(\mathbf{Q})$ by

$$\Delta'\{\mathbf{q^1}, \ldots, \mathbf{q^n}\} = \bigcup_{i=1}^{n} \Delta(\mathbf{q^i})$$

so Δ' sends a set of possible vector states to a set of vector states (a non-deterministic choice of synchronization transition). Define D as the deterministic automaton $D = (< D^0, \ldots, D^{p-1} >, \Delta')$. Then we can verify that $L(N) \subseteq L(D)$ by induction on the number of supersteps S in an accepted BSP word.

- ($S = 0$) If ϵ is accepted by N that is because the initial vector-state in N, $\mathbf{q}^0$ is accepting. By definition of the subset algorithm, the accepting vector state of D is built from local accepting states, and the initial state of D is just $\mathbf{q}^0$. As a result, $\mathbf{q}^0$ is also accepting in D and so D accepts ϵ.
- ($S = 1$) If a word-vector $\mathbf{w} =< w^0, \ldots, w^{p-1} >$ is accepted by N then one of the paths in N^i applied to w^i leads to a state q'^i such that $\Delta < q'^0, \ldots, q'^{p-1} >$ contains an accepting state-vector. By the subset algorithm, $q'^i \in Q'^i$ where Q'^i is a state of D^i and by the definition of Δ' then $\Delta' < Q'^0, \ldots, Q'^{p-1} >$ contains an accepting state-vector.
- ($S \geq 2$) If a BSP word $\mathbf{w}^1; \ldots; \mathbf{w}^{n-1}; \mathbf{w}^n$ is accepted by N then N applied to $\mathbf{w}^1; \ldots; \mathbf{w}^{n-1}$ leads to a set of vector-states among which one $\mathbf{Q}$ can be chosen as initial vector state from which N would accept w^n. By construction D contains a vector-state containing $\mathbf{Q}$. Apply then the above one-superstep proof from $\mathbf{Q}$ to show that if N leads to acceptance, so does D. □

Proposition 3 *The language of an ϵ-NBSPA can be recognized by a NBSPA.*

Proof An ϵ-NBSPA N' is simply a non-deterministic BSP automaton built from a vector of ϵ-NFA N'^i. Its synchronization function is non-deterministic but contains no "spontaneous" empty supersteps. Standard automata theory gives us a polynomial-time ϵ-reachability algorithm to convert every N'^i into an equivalent NFA N^i without ϵ-transitions. Define N to be the NBSPA built from the N^i and the same synchronization function as N'. Then $L(N) = L(N')$.   □

As a result, non-determinism and ϵ-transitions do not change the languages accepted by BSP automata. Just as with sequential "scalar" automata, those syntactic extensions can be used at an exponential cost in time and number of states. Depending on the complexity of the synchronization functions, the blow-up factor may also depend on p.

6   Sequentialization

Every parallel computation can be simulated sequentially and the theory of BSP automata expresses this fact by a transformation from BSP automata to classical finite automata. A word $u = a_1 \ldots a_n$ of Σ^* is *localized* to i as follows: $u@i = (a_1, i) \ldots (a_n, i)$. A word-vector $\mathbf{w} \in (\Sigma^*)^p$ is *sequentialized* to a word Seq($\mathbf{w}$) on alphabet $\Sigma \times [p]$ by the transformation:

$$\text{Seq}(\mathbf{w}) = \mathbf{w}^0@0 \ldots \mathbf{w}^{p-1}@(p-1).$$

In other words, Seq($\mathbf{w}$) concatenates the words of word-vector $\mathbf{w}$ after having labelled them by their locations (any interleaving of the localized words would satisfy our purpose, but ordered concatenation is simpler). For example if $\mathbf{w} = < b, \epsilon, bb, aa >$ then Seq($\mathbf{w}$) $= \mathbf{w}^0@0 \ldots \mathbf{w}^3@3 = (b, 0)(b, 2)(b, 2)(a, 3)(a, 3)$.

Definition 6 A BSP word on $(\Sigma^*)^p$ is *sequentialized* to a word on $(\Sigma \times [p]) \cup \{; \}$ as follows (Fig. 5):

$$\text{Seq}(\epsilon) = \epsilon$$
$$\text{Seq}(\mathbf{v}_1 \ldots \mathbf{v}_n) = \text{Seq}(\mathbf{v}_1); \ldots ; \text{Seq}(\mathbf{v}_n);$$

A BSP language L is sequentialized to Seq(L) by sequentializing every one of its BSP words.

The following remarks should be kept in mind because they are much more than a syntactic detail. 1. The sequentialization of a BSP word is either empty or contains at least one semicolon, and 2. function Seq has one of two possible types.

- To sequentialize word vectors Seq : $(\Sigma^*)^p \to (\Sigma \times [p])^*$.
- To sequentialize *BSP words* Seq : $((\Sigma^*)^p)^* \to ((\Sigma \times [p]) \cup \{; \})^*$.

BSP element: type $\longrightarrow$ local / sequential element
$\epsilon : \Sigma^* \xrightarrow{@i} \epsilon$
$a : \Sigma^* \xrightarrow{@i} (a, i)$
$abaa : \Sigma^* \xrightarrow{@i} (a, i)(b, i)(a, i)(a, i)$
$\epsilon =< \epsilon, \epsilon, \epsilon, \epsilon >: (\Sigma^*)^p \xrightarrow{\text{Seq}} \epsilon$
$\mathbf{v}_1 =< aba, b, bbb, a >: (\Sigma^*)^p \xrightarrow{\text{Seq}} (a, 0)(b, 0)(a, 0)(b, 1)(b, 2)(b, 2)(a, 3)$
$\mathbf{v}_2 =< a, \epsilon, bbb, \epsilon >: (\Sigma^*)^p \xrightarrow{\text{Seq}} (a, 0)(b, 2)(b, 2)(b, 2)$
$\epsilon =< \epsilon, \epsilon, \epsilon, \epsilon >: (\Sigma^*)^p \xrightarrow{\text{Seq}} \epsilon$
$\epsilon : ((\Sigma^*)^p)^* \xrightarrow{\text{Seq}} \epsilon$
$\epsilon =< \epsilon, \epsilon, \epsilon, \epsilon >: ((\Sigma^*)^p)^* \xrightarrow{\text{Seq}} (\epsilon;) = ;$
$\mathbf{v}_2 \epsilon : ((\Sigma^*)^p)^* \xrightarrow{\text{Seq}} (a, 0)(b, 2)(b, 2)(b, 2); ;$
$\epsilon \, \mathbf{v}_2 : ((\Sigma^*)^p)^* \xrightarrow{\text{Seq}} ; (a, 0)(b, 2)(b, 2)(b, 2);$
$< \epsilon, a, \epsilon, a >< b, b, b, b >: ((\Sigma^*)^p)^* \xrightarrow{\text{Seq}} (a, 1)(a, 3); (b, 0)(b, 1)(b, 2)(b, 3);$

Fig. 5 Localization and sequentialization

For simplicity we denote both by the same symbol Seq but the first one is only an auxiliary part of the definition of the second one.

So if $\epsilon =< \epsilon, \ldots, \epsilon >$ is considered to be a word-vector, then it is sequentialized to the empty word. But as a BSP word it is sequentialized to the one-symbol word ";" (Definition 6). This is the theoretical representation that even an "empty" BSP algorithm (whose every local process has an empty execution trace) must end by a synchronization barrier that propagates the coherent information "end execution" to every location. In terms of BSP automata this means that even if Δ is the identity function, and it follows a vector of empty computations, it still must be applied once.

A finite automaton $A = (Q, \Sigma, \delta, q_0, F)$ of alphabet Σ can be *localized* to $i \in [p]$ and becomes automaton $A@i$ by the transformation

$$A@i = (Q \times \{i\}, \Sigma \times \{i\}, \delta@i, (q_0, i), F \times \{i\})$$

where $(\delta@i)((q, i), (a, i)) = (\delta(q, a), i)$.

Proposition 4 *For any BSP automaton A on Σ, there exists a finite automaton $Seq(A)$ on $(\Sigma \times [p]) \cup \{; \}$ such that $Seq(L(A)) = L(Seq(A))$.*

Proof Let $A = (< A^0, \ldots, A^{p-1} >, \Delta)$ with $A^i = (Q^i, \Sigma, \delta^i, q_0^i, F^i)$.

Define vector states $Q = \prod_{i=0}^{p-1} Q^i$ for the sequential automaton i.e. all vectors of local states.

Define localized transition function $\delta_a(\mathbf{q}, a) = \mathbf{q}[i := \delta^i(\mathbf{q}_i, a)]$ i.e. the local asynchronous transition at i for any letter a localized at i.

Define a vector of initial state $q_0 =< (q_0^0, 0), (q_0^1, 1) \ldots, (q_0^{p-1}, p-1) >$ with the local initial states.

Define also the set of unanimously-accepting vector states $F = \prod_{i=0}^{p-1} F^i$.

Then $A_a = (Q, \Sigma \times [p], \delta_a, q_0, F)$ is a DFA that can simulate the application of A to any word-vector $\mathbf{w} = < w^0, \ldots, w^{p-1} >$ as follows.

Let $w = \text{Seq}(\mathbf{w})$ then $\delta_a(q_0, w) = < \delta^0(q_0^0, \mathbf{w}^0), \ldots, \delta^{p-1}(q_0^{p-1}, \mathbf{w}^{p-1}) >$.

As a result, the asynchronous automaton A_a simulates A in the absence of synchronizations. That covers the trivial case of accepting the empty BSP word whose sequentialization is $\text{Seq}(\epsilon) = \epsilon$. Indeed if $\epsilon \in L(A)$ that is because $\forall i.q_0^i \in F^i$ and then by definition $q_0 \in F$ so $\epsilon \in L(A_a)$. But even a single word-vector (single superstep) involves the synchronization function when it is considered as a BSP word.

To simulate its effect with the sequential automaton, transform A_a to a DFA $A_;$ on $(\Sigma \times [p]) \cup \{; \}$ as follows.

Let $\delta_;$ be the extension of δ_a with transitions on symbol semicolon ; that simulate the effect of the synchronization function Δ. For any state vector $\mathbf{q}$ define $\delta_;(\mathbf{q}, ;) = \Delta(\mathbf{q})$. Since the synchronization function is total, this ensures that $\delta_;$ is a total function and that $A_;$ is a DFA. Consider a non-empty BSP word of length one i.e. a word vector $\mathbf{v}$ (which *could* be a vector of empty words). The effect of $A_;$ on $\text{Seq}(\mathbf{v}) = (\text{Seq}(\mathbf{v}));$ is the same as the effect of A on $\mathbf{v}$. Therefore $\mathbf{v} \in L(A)$ iff $\text{Seq}(\mathbf{v}) \in L(A_;)$.

A trivial induction argument shows that this is also the case for a BSP word of any length. We therefore define $\text{Seq}(A) = A_;$ and conclude that

$$\mathbf{v}_1 \ldots \mathbf{v}_n \in L(A) \quad \Leftrightarrow \quad \text{Seq}(\mathbf{v}_1); \ldots ; \text{Seq}(\mathbf{v}_n); \in L(A_;)$$

i.e. $\text{Seq}(L(A)) = L(A_;) = L(\text{Seq}(A))$. □

7 Parallelization

We have seen in Sect. 6 the sequentialization of word-vectors by localization of their words, one symbol at a time $\text{Seq} : (\Sigma^*)^p \to (\Sigma \times [p])^*$. It is easy to invert this transformation and define $\text{Par} : (\Sigma \times [p])^* \to (\Sigma^*)^p$ so that $\text{Par}(\text{Seq}(\mathbf{w})) = \mathbf{w}$.

Let $\epsilon[i:=u]$ be the word vector that is empty everywhere except for word u at position i. Let $\mathbf{u} \cdot \mathbf{v}$ be the pointwise concatenation of word-vectors i.e.

$$< u^0, \ldots, u^{p-1} > \cdot < v^0, \ldots, v^{p-1} > = < u^0 v^0, \ldots, u^{p-1} v^{p-1} > .$$

Define $\text{Par} : \Sigma \times [p] \to (\Sigma^*)^p$ by

$$\text{Par}(a, i) = \epsilon[i:=a]$$

so that for example $\text{Par}(u@i) = \epsilon[i:=u]$. Define then Par on sequentialized words of $(\Sigma \times [p])^*$ by

$$\text{Par}((a, i)(b, j) \ldots) = \text{Par}(a, i) \cdot \text{Par}(b, j) \ldots$$

and in particular $Par(\epsilon) = \epsilon$ the vector of empty words (or "empty-word vector" not to be confused with the empty BSP word $\epsilon \in (\Sigma^*)^p)^*$). For example

$$Par((a, 0)(b, 1)(b, 0)(b, 3)) =< ab, b, \epsilon, b > .$$

The following follows directly from the definition of Seq on word-vectors.

Lemma 1 *Parallelization is the left-inverse of sequentialization on word-vectors* $(\Sigma^*)^p$:

$$Par(Seq(\mathbf{v})) = \mathbf{v}.$$

In fact, any permutation π of $Seq(\mathbf{w})$ that does not reorder co-located letters would also preserve the parallelization $Par(\pi(Seq(\mathbf{w})))$ but we will not expand on this for it is not essential to our developments.

Function Par has one of three possible types.

- To parallelize localized letters $Par : (\Sigma \times [p]) \to (\Sigma^*)^p$.
- To parallelize semicolon-free words $Par : (\Sigma \times [p])^* \to (\Sigma^*)^p$.
- To parallelize localized words with semicolons $Par : (\Sigma \times [p]) \cup \{; \})^* \to ((\Sigma^*)^p)^*$.

Again, this can lead to ambiguity if the input type is unknown: the semicolon-free word is mapped to the empty-word vector, but the empty general word of type $((\Sigma \times [p]) \cup \{; \})^*)$ is mapped to the empty BSP word (Fig. 6). This ambiguity is of course only a convenience for notation but, as we have seen earlier, the difference between empty-word vector and empty BSP word in fundamental.

The following straightforward consequence of our definitions shows that Par is a non-injective function.

Proposition 5 *If w is a word of $(\Sigma \times [p])^*$ and π is a permutation of w that does not exchange co-located letters, then $Par(\pi(w)) = Par(w)$.*

For this reason, Seq is not the left-inverse of Par:

$$\exists w \in (\Sigma \times [p])^*. \ Seq(Par(w)) \neq w.$$

local / sequential element: type $\longrightarrow$ vector/BSP element: type
$(a, 1) : \Sigma \times [p] \xrightarrow{\text{Par}} < \epsilon, a, \epsilon, \epsilon >: (\Sigma^*)^p$
$\epsilon : (\Sigma \times [p])^* \xrightarrow{\text{Par}} < \epsilon, \epsilon, \epsilon, \epsilon >: (\Sigma^*)^p$
$(a, 1)(b, 3)(a, 1) : (\Sigma \times [p])^* \xrightarrow{\text{Par}} < \epsilon, aa, \epsilon, b >: (\Sigma^*)^p$
$(a, 0)(b, 0)(a, 0)(b, 1)(b, 2)(b, 2)(a, 3) \xrightarrow{\text{Par}} < aba, b, bbb, a >: (\Sigma^*)^p$
$(a, 0)(b, 0)(b, 2)(a, 3); (a, 0)(b, 1)(b, 2)(b, 2); \xrightarrow{\text{Par}} < ab, \epsilon, b, a >< a, b, bb, \epsilon >: ((\Sigma^*)^p)^*$

Fig. 6 Parallelization

For example if $w = (a, 0)(b, 3)(a, 1)$ then $\text{Par}(w) = < aa, \epsilon, \epsilon, b >$ and
$\text{Seq}(Par(w)) = (a, 0)(a, 1)(b, 3) \neq w$. But Seq $\circ$ Par is clearly a normal form for
words of $(\Sigma \times [p])^*$: it sorts their letters in increasing order of locations.

Proposition 6 *Reduction to normal form* $\cong=$ *Seq $\circ$ Par is a congruence for con-*
catenation on $(\Sigma \times [p])^*$ *and* $(\Sigma \times [p])^*/\cong$ *is isomorphic to* $(\Sigma^*)^p$.

Proof Taking the normal form by $\cong$ preserves the value of Par, and Par is surjective. Taking the i-subword of any $w \in (\Sigma \times [p])^*$ is a homomorphism for concatenation. Therefore Par is a homomorphism from word concatenation to word-
vector concatenation. As a result Par is injective on $(\Sigma \times [p])^*/\cong$, surjective and
homomorphic. $\square$

Concurrency theories like process algebras [17] ignore the notion of localization
and simply consider interleavings π that forget the locations i. That is why they are
models of *shared-memory* computers and that was one of the reasons for inventing
theories like BSP that do not abstract from distributed-memory.

As the semicolon symbol ; encodes synchronization barriers i.e. the end of supersteps, it is natural to extend parallelization to all words on $(\Sigma \times [p]) \cup \{; \}$.

Definition 7 Let $\alpha = \alpha_0; \ldots; \alpha_n$; where $\alpha_i \in (\Sigma \times [p])^*$.
Then $\text{Par}(\alpha) = \text{Par}(\alpha_0) \ldots \text{Par}(\alpha_n)$.

For example $\text{Par}((a, 0)(b, 1)(b, 0)(b, 3); (a, 2)(a, 2)(b, 3); ; (a, 0))$ is the BSP word:

$$< ab, b, \epsilon, b >< \epsilon, \epsilon, aa, b >< \epsilon, \epsilon, \epsilon, \epsilon >< a, \epsilon, \epsilon, \epsilon > .$$

The inversion property on word-vectors then follows from our definitions.

Lemma 2 *Parallelization is the left-inverse of sequentialization on BSP words*
$((\Sigma^*)^p)^*$:
$$Par(Seq(\mathbf{w})) = \mathbf{w}.$$

The reasoning in the other direction, about Seq $\circ$ Par applies to BSP words identically
and yields the same result as for word vectors (individual BSP supersteps): $\cong=$
Seq $\circ$ Par sorts inter-semicolon sequences in increasing order of location, it is a
congruence for concatenation on $((\Sigma \times [p]) \cup \{; \})^*$ and leads to a parallel-sequential
isomorphism.

Reduction to normal form $\cong=$ Seq $\circ$ Par is a congruence for concatenation on
sequential words but with the important exclusion of non-empty semicolon-free
sequential words that are meaningless for BSP.

Definition 8 $\Sigma_{p;} = (((\Sigma \times [p])^*);)$

In other words, $\Sigma_{p;}^*$ is the set of sequential localized words, without non-empty
semicolon-free words. We find that $\Sigma_{p;}^*/\cong$ is isomorphic to the BSP words $((\Sigma^*)^p)^*$.

Proposition 7 *Reduction to normal form* $\cong=$ *Seq* $\circ$ *Par is a congruence for concatenation on* $\Sigma_{p;}^*$ *and* $\Sigma_{p;}^* / \cong$ *is isomorphic to* $((\Sigma^*)^p)^*$.

Proof The proof is almost identical to that of Proposition 6. The only (key) difference is that Par would not be a bijection if applied to the whole of $((\Sigma \times [p]) \cup \{; \})^*$. $\square$

Definition 9 The parallelization Par(L) of a language on $(\Sigma \times [p]) \cup \{; \}$ is {Par(α) : $\alpha \in L$} and the sequentialization Seq(L') of a BSP language is {Seq($\mathbf{w}$) : $\mathbf{w} \in L'$}.

We now give results about inverting the sequentialization of BSP automata.

The first result is about inverting the sequentialization of "incomplete superstep" BSP words. Such words correspond to sequentialized words on $\Sigma_{p;}$ i.e. words of the form $((\Sigma \times [p])^*);)$. It would appear that such words contain all the necessary information to be recognized by a BSP automaton. One word at a time this is true, but it does not hold of regular languages of this type. Take for example the regular language of expression $((a, 0) + (b, 1))^*$ then its parallelized language can be recognized by a BSP automaton whose language is $< a^*, b^*, \epsilon, \epsilon >$, essentially because the number of a events is independent from the number of b events. But a language like that of $((a, 0)(b, 1))^*$ is parallelized to language $\{< a^n, b^n, \epsilon, \epsilon > \mid n \geq 0\}$ which cannot be recognized by a BSP automaton because the local automata at locations 0 and 1 would need to keep synchronised without the help of the synchronization function. However if the sequentialized language is given extra synchronization semicolons, then it can be recognized by a BSP automaton. In the above example, the language of expression $((a, 0)(b, 1);)^*$ is parallelized to language $\{< a, b, \epsilon, \epsilon >^n \mid n \geq 0\} =< a, b, \epsilon, \epsilon >^*$ for which a BSP automaton exists. The process of adding semicolons to a sequential word or language will be called *over-synchronization*.

Definition 10 For $w \in ((\Sigma \times [p])^*) \cup \{; \}$, we say that w' over-sychronizes w and write $w \leq_; w'$ if w' is obtained by interleaving w with a word of the form $;^*$. A language L' over-synchronizes language L, written $L \leq_; L'$, if there is a bijection from L to L' which is an over-synchronization. An automaton A' over-synchronizes automaton A, written $A \leq_; A'$ if $L(A) \leq_; L(A')$.

Lemma 3 *For any automaton A on $(\Sigma \times [p])$ there is a sequential automaton $A' \geq_; A$, and a BSP automaton Par(A) on Σ such that $L(Par(A)) = Par(L(A'))$.*

Proof Let r_A be a regular expression such that $L(r_A) = L(A)$ (Appendix 2). We show by induction on the syntax of r_a that there exists a BSP automaton to recognize the parallelization of an over-synchronization of $L(A)$.

If $r_A = \emptyset$ then Par($L(r_A)$) = $\emptyset$ and the BSP automaton can be any one that has empty sets of accepting states. If $r_A = \epsilon$ then $L(r_A) = \{\epsilon\}$ and Par($L(r_A)$) = $\{\epsilon\}$ so the BSP automaton should recognize nothing but the empty BSP word. To obtain this, define its local automata as accepting $\{\epsilon\}$ and the synchronization function is the identity.

If $r_A = r_0^*$ then by induction there is a BSP automaton A_0 to recognize $\text{Par}(L(r_0))$. Add new unique accepting states q_F^i to its local automata and ϵ-transitions from their (previously) accepting states to the q_F^i. Add to A_0's synchronization function the mapping from $< q_F^0, \ldots, q_F^{p-1} >$ to the initial states of all finite automata. Call this new ϵ-NBSPA A_1. Then $L(A_1) = \text{Par}(L((r_0;)^*))$ i.e. the over-sychronization $(r_0;)^*$ has A_1 as accepting BSP automaton.

If $r_A = r_1 + r_2$ then by induction there are BSP automata A_j and $r_j' \geq_; r_j$ such that $L(A_j) = \text{Par}(L(r_j'))$ for $j = 1, 2$. Then build an NBSPA A_0 whose local automata have: the union of local states from A_1, A_2 with an added new initial state with an ϵ-transition leaving it to each of the (previously) initial states from A_1^i and A_2^i, transition function that are the union of local transition functions from A_1, A_2 and a new final state q_F^i. The synchronization function of A_0 is the union of synchronization functions of A_1, A_2 with the added mappings from all state vectors that are uniformly ($\forall i$) accepting for r_1' or uniformly accepting for r_2' to $< q_F^0, \ldots, q_F^{p-1} >$. Then $L(A_0) = L(A_1) \cup L(A_2) = \text{Par}((r_1' + r_2');)$ which is an over-synchronization of r_A.

If $r_A = r_1 r_2$ then a similar construction leads to a BSP automaton accepting the parallelization of $r_1; r_2$. $\qquad\square$

The second result follows about inverting the sequentialization of all BSP words.

Theorem 1 *For any automaton A on $(\Sigma \times [p]) \cup \{; \}$ there is a sequential automaton $A' \geq_; A$, and a BSP automaton $\text{Par}(A)$ on Σ such that $L(\text{Par}(A)) = \text{Par}(L(A'))$.*

Proof Let r_A be a regular expression such that $L(r_A) = L(A)$ (Appendix 2). We will show by induction on the syntax of r_A that there exists a BSP automaton $\text{Par}(A)$ to recognize $\text{Par}(L(r_A)) = \text{Par}(L(A))$.

If $r_A = \emptyset$ then $\text{Par}(L(r_A)) = \emptyset$ and the BSP automaton can be any one that has empty sets of accepting states.

If $r_A = \epsilon$ then $L(r_A) = \{\epsilon\}$ and $\text{Par}(L(r_A)) = \{\epsilon\}$ so the BSP automaton should recognize nothing but the empty BSP word. To obtain this, define its local automata as accepting $\{\epsilon\}$ and the synchronization function can be arbitrary because it does not get applied on the empty BSP word.

It is not possible to have $r_A = (a, i)$ then it is easy to build a BSP automaton to recognize $\text{Par}(\{(a, i); \})$.

If $r_A =;$ then $L(r_A) = \{; \} = \{\epsilon; \}$ so $\text{Par}(L(r_A)) = \{\text{Par}(\epsilon)\} = \{\epsilon\}$. Define then A^i as a finite automaton with initial state q_0^i, a single accepting state equal to q_0^i and transition function to accept only the empty word. Let $\mathbf{q}_0 = < q_0^0, \ldots, q_0^{p-1} >$ and define $\Delta(\mathbf{q}_0) = \mathbf{q}_0$ and a different value of $\Delta(\mathbf{q})$ for all other $\mathbf{q}$. Define the BSP automaton $\text{Par}(A) = (< A^0, \ldots, A^{p-1} >, \Delta)$. Then applying $\text{Par}(A)$ to ϵ leads to $\mathbf{q}_0$ vacuously on the local automata and then by one application of Δ, so ϵ is accepted. Applying $\text{Par}(A)$ to any other BSP word leads to non-acceptance so $L(\text{Par}(A)) = \{\epsilon\}$.

If $r_A = r_0^*, r_A = r_1 + r_2$ or $r_A = r_1 r_2$ then the corresponding induction steps used in the proof of Lemma 3 apply directly. $\qquad\square$

Moreover, as seen in Sect. 5 there exists a deterministic BSP automaton A equivalent to the ϵ-NBSPA constructed in the proof of Theorem 1.

8   Bulk-Synchronous Regular Expressions

In this section it is shown how to adapt regular expressions (Appendix 2) to BSP
languages.

A BSP regular expression is an expression R from the following grammar:

$$R ::= \emptyset \mid \epsilon \mid < r^0, \ldots, r^{p-1} > \mid R; R \mid R* \mid R + R$$

where r^i is any (scalar) regular expression. The set of BSP regular expressions is
BSPRE and the language any BSP regular expression is defined by $L :$ BSPRE $\rightarrow$
$\mathcal{P}(((\Sigma^*)^p)^*)$ as:

R	$L(R)$
$\emptyset$	$\{\ \}$
ϵ	$\{\epsilon\}$
$< r^0, \ldots, r^{p-1} >$	$L(r^0) \times \ldots \times L(r^{p-1})$
$R_1; R_2$	$L(R_1)L(R_2)$
R^*	$L(R)^*$
$R_1 + R_2$	$L(R_1) \cup L(R_2)$

We now show that Kleene's equivalence theorems (Appendix 2 and [10]) can be
adapted to the two-level BSP regular expressions and automata.

Theorem 2 *For $R \in$ BSPRE there exists a BSP automaton A_R such that $L(A_R) =$
$L(R)$.*

Proof We proceed by induction on the syntax of R. If $R = \emptyset$ the BSP automaton
simply needs to have empty (local) sets of accepting states. If $R = \epsilon$ the BSP automata
should have as unique accepted BSP word the empty one. That is obtained by having
accepting (local) start states and all transitions leading to different (non-accepting
states), with an indentity synchronization function.

If $R =< r^0, \ldots, r^{p-1} >$ then there exist classical automata A^i on Σ such that
$L(A^i) = L(r^i)$ (Appendix 2). The BSP automaton is then simply the collection of
those automata with identity synchronization function.

If $R = R_1; R_2$ then by induction there exists BSP automata A_1, A_2 such that
$L(A_j) = L(R_j)$ for $j = 1, 2$. Define the BSP automaton A whose states is the disjoint union of those of the A_j, whose accepting states are those of A_2, whose initial
vector state is that of A_1, whose (partial) Δ_A is the union of the synchronization
functions of the A_j with an added ϵ-transition from all accepting state vectors in A_1
to the (previously) initial state vector of A_2. The resulting A is an ϵ-NBSPA accepting
language $L(A_1)L(A_2) = L(R)$.

If is of the form $R = R_0^*$ or $R = R_1 + R_2$, similar constructions lead to ϵ-NBSPA
whose language is R. $\square$

Theorem 3 *For A a BSP automaton there exists $R_A \in$ BSPRE such that $L(R_A) = L(A)$.*

Proof Assume

$$A = (\{Q^i\}_{i \in [p]}, \Sigma, \{\delta^i\}_{i \in [p]}, \{q_0^i\}_{i \in [p]}, \{F^i\}_{i \in [p]}, \Delta) \quad \text{and} \quad A^i = (Q^i, \Sigma, \delta^i, q_0^i, F^i).$$

Let $Q = \bigcup_i Q^i$ be the union of all states in the local automata, then the states $\mathbf{Q}$ of A are all in Q^p. Similarly, let $\mathbf{F} \subseteq \mathbf{Q}$ be the accepting states of Q.

Let $q_1, q_2 \in Q$ be any local states. Then by Kleene's theorem there exists $r(q_1, q_2)^i \in$ RE such that $L(r(q_1, q_2)^i)$ is the set of Σ words that lead from q_1 to q_2 with δ^i. Let RE_A be the *finite set* of all such regular expressions over all pairs of states and all location $i \in [p]$. Let $\Sigma_A = (\mathrm{RE}_A)^p \cup \{;\}$, a large but finite "alphabet" for the following construction of a "vector-automaton" $\mathbf{A}$ equivalent to A. Define $\Delta_A : Q \times \Sigma_A \to Q$ by $\Delta_A(\mathbf{q}_1, \mathbf{r}) = \mathbf{q}_2$ where $\mathbf{q}_1, \mathbf{q}_2$ are vectors of states linked at every location by the local projection of $\mathbf{r}$. Define also $\Delta_A(\mathbf{q}_1, ;) = \mathbf{q}_2$ iff $\Delta(\mathbf{q}_1) = \mathbf{q}_2$. Define at last $\mathbf{A} = (\mathbf{Q}, \Sigma_A, \Delta_A, \mathbf{F})$.

This NFA can be applied to BSP words by applying the vectors of regular expressions to the word vectors pointwise, and traversing any semicolon-edge when there is a change of word. By defining transition in this manner, $\mathbf{A}$ is an acceptance mechanism for BSP word whose accepted language is precisely $L(A)$.

By Kleene's theorem applied to $\mathbf{A}$, there is a (normal) regular expression built from alphabet $\Sigma_A = (\mathrm{RE}_A)^p \cup \{;\}$ whose language is $L(\mathbf{A}) = L(A)$. By construction, such a regular expression is precisely a BSPRE whose language is that of A. $\qquad \Box$

It is convenient to write $r = r'$ in RE (respectively $R = R'$ in BSPRE) when the two regular expressions (resp. BSP reg. expr.) have the same language.

Proposition 8 *(Sect. 9.3.1 of [15]) For $r, r_1, r_2, r_3 \in RE$:*
$\epsilon r = r\epsilon = r \quad r_1(r_2 r_1)^* = (r_1 r_2)^* r_1$
$\emptyset r = r\emptyset = \emptyset \quad (r_1 \cup r_2)^* = (r_1^* r_2^*)^*$
$\epsilon^* = \emptyset^* = \epsilon \quad r_1(r_2 \cup r_3) = r_1 r_2 \cup r_1 r_3$

Classical equivalences such as the above hold also for BSP regular expressions R, R_1, R_2, R_3 because they involve no interactions between the two levels of BSP syntax.

9   Minimization

For a given DFA A there exists a so-called *minimal* DFA M_A [21]: $L(M_A) = L(A)$, the number of states of M_A is minimal amongst all automata of equal language. Moreover the minimal automaton M_A is unique: it is isomorphic to any other M_A' of equal language and of the same size. The computation $A \mapsto M_A$ is called *minimization* and can be realized by sequential algorithms of worst-case quadratic time in the number of states of A.

Let us recall the state congruence relation used for the invariant of those algorithms, and its very compact formulation by Benzaken (Chap. 2, Sect. 6.3 of [2]).

Definition 11 Let $A = (Q, \Sigma, q_0, \delta, F)$ be a DFA and $k \geq 0$ an integer. For $q \in Q$, define A_q to be the language accepted by A starting from q i.e. $A_q = L(A[q_0 := q])$. For $p, q \in Q$ define $p \simeq_k q$ or "p, q are k-equivalent" to mean $L_k(A_p) = L_k(A_q)$ where $L_k(\ )$ denotes the sub-language of words no longer than k.

Then k-equivalence $\simeq_k$ is clearly an equivalence relation on Q and:

- $\simeq_0$ has only two equivalence classes of states, namely F and $Q - F$: $L_0(p) = L_0(q)$ precisely when the empty word is accepted from either state. That is true when both are in F and false otherwise.
- $p \simeq_{k+1} q$ iff $(p \simeq_0 q)$ and $\forall a \in \Sigma.(\delta(p, a) \simeq_k \delta(q, a))$: two states define the same language of length $\leq k + 1$ iff 1 they are both accepting/non-accepting and 2 any pair of transitions $\delta(p, a)$ and $\delta(q, a)$ leaving them on the same symbol a, leads to k-equivalent states.

By definition, $(k + 1)$-equivalence is a (non-strict) refinement of k-equivalence, so its equivalence classes $Q/ \simeq_{k+1}$ are obtained by splitting some equivalence classes of $Q/ \simeq_k$. Moreover, if for some k we have $Q/ \simeq_{k+1} = Q/ \simeq_k$ then all $Q/ \simeq_i$ are equal for $i = k, k + 1, k + 2 \ldots$. Observe then that in the series of partitions $Q/ \simeq_0, Q/ \simeq_1, \ldots Q/ \simeq_i \ldots$, the number of equivalence classes is non-decreasing, yet by definition it cannot be greater than the number of states $| Q |$. It follows that $A_p = A_q$ iff $p \simeq_{|Q|} q$ and it can be proved that $A/ \simeq_{|Q|}$ is the unique minimal DFA equivalent to A. Sequential algorithms for computing it can be derived from the above construction, among them Hopcroft's algorithm [8] of time complexity $O(n \log n)$ where $n = | Q |$.

The above ideas have been generalized by D'Antoni and Veanes to so-called symbolic finite automata (SFA) whose alphabets are logical formulae rather than elementary letters [5]. They generalize the above DFA minimization method to SFA and find that the key requirement is to check rapidly for satisfiability of $\phi \wedge \psi$ when considering transitions of the form $\delta(p, \phi)$ and $\delta(q, \psi)$. From the construction used in the proof of Theorem 3 it appears that our BSP automata are a special case of SFA and that those results [5] apply to them. But we will not make use of this general result for the sake of simplicity and to keep this paper self-contained. Efficient algorithms for BSP automata will also benefit from our elementary presentation that only builds from DFAs, REs and vectors, as there is no guarantee that excessively general methods lead to efficient algorithms for the class of BSP automata.

Let Min be the minimization function on DFA that results from applying Hopcroft's algorithm. Observe that it is not sufficient to minimize a BSP automaton by minimizing its local automata: we must account for the synchronization function. The states of a BSP automaton are the state-vectors of $\prod_i Q^i$. But if we apply the classical method to state-vectors and alphabet Σ_p, then all minimization properties and methods apply.

Fig. 7 Automaton A_a

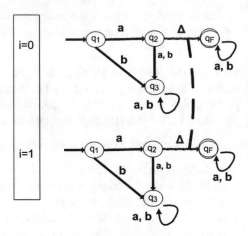

Proposition 9 *If A is a deterministic BSP automaton on Σ then there exists a sequential automaton $Min(Seq(A))$ that accepts the same $Seq(L(A))$ and is of minimal size.*

Proof Consider A as a special notation for $Seq(A)$: an automaton on $(\Sigma \times [p]) \cup \{; \}$ i.e. with vector-states but single-symbol local transitions or global transitions on ; defined by the synchronization function Δ. Then clearly A is deterministic so it is a DFA that can be minimized. Apply sequential minimization to obtain the result. $\square$

Minimizing BSP automata is considerably more complex than minimizing DFA. The reason is that pointwise minimization of the local automata, without reference to the synchronization function, may change the accepted BSP language. Let us illustrate this property by an example.

Example 1 Let A_a be the DFA with four states q_1, q_2, q_3, q_F, initial state q_1, unique accepting state q_F, and transitions as shown in Fig. 7 (ignoring Δ). Then clearly $L(A_a) = \emptyset$ because q_F is unreachable from q_1.

Define also the BSP automaton $\mathbf{A}_a = (<A_a, \ldots, A_a>, \Delta)$ with the following synchronization function:

$\mathbf{q}$	$\Delta(\mathbf{q})$
$\mathbf{q}_2 = < q_2, \ldots, q_2 > \xrightarrow{\Delta}$	$\mathbf{q}_F = < q_F, \ldots, q_F >$
$\mathbf{q}_F = < q_F, \ldots, q_F > \xrightarrow{\rightarrow}$	$\mathbf{q}_{1F} = < q_1, q_F, \ldots, q_F >$
any other $\mathbf{q}$ $\xrightarrow{\rightarrow}$	$\mathbf{q}_{1F} = < q_1, q_F, \ldots, q_F >$

Since $\mathbf{q}_F$ is the only accepting vector state for $\mathbf{A}_a$, and since the initial state is $< q_1, \ldots, q_1 >$ it follows that the empty BSP word is not accepted by $\mathbf{A}_a$. Any BSP word of $L(\mathbf{A}_a)$ is therefore of length one or more, so must trigger one or more applications of Δ. By definition, the only such application leading to acceptance is $\Delta(\mathbf{q}_2)$. By definition of A_a, the only word-vectors leading to $\mathbf{q}_2$ is $\mathbf{a} =< a, a, \ldots, a >$. So the BSP word $\mathbf{a} \in L(\mathbf{A}_a)$.

Fig. 8 Locally minimal automaton $\text{Min}(A_a)$

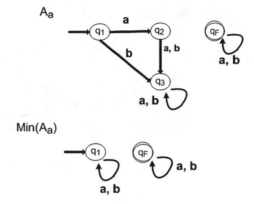

Any longer BSP words are not accepted, because 1 by definition of A_a, local transitions will only lead from $\mathbf{q}_F$ to itself and 2 synchronization Δ will then lead to $\mathbf{q}_{1F}$ which is not accepted, and similarly for a BSP word of length more than two.

As a result $L(\mathbf{A}_a) = \mathbf{a} = < a, a, \ldots, a >$.

Consider now *local* minimization of the BSP automaton $\mathbf{A}_a$ of Example 1. That yields the BSP automaton $(< \text{Min}(A_a), \ldots, \text{Min}(A_a) >, \Delta)$ where $\text{Min}(A_a)$ is the minimal DFA for accepting the empty language i.e. the two-state DFA of Fig. 8. Local state q_1 in $\text{Min}(A_a)$ is actually the equivalence class $\{q_1, q_2, q_3\}$ in A_a so the synchronization function would send $< q_1, \ldots, q_1 >$ to $\Delta(q_2, \ldots, q_2 >) = \mathbf{q}_F$ so that any BSP word of length one would be accepted. The result would then be a BSP automaton whose language is $< (a + b)^*, \ldots, (a + b)^* > \neq L(\mathbf{A}_a)$. The above remarks show that local minimization alone does not preserve the BSP language.

The application of Min $\circ$ Seq as in Proposition 9 has a disadvantage: it produces an automaton whose parallelization is not obvious. Sequentialization can then be reversed but only at the cost of over-synchronization (by Theorem 1).

In other words, if we apply Min $\circ$ Seq and then Par $\circ \leq$; in the hope of minimization, the resulting BSP automaton may have a reduced number of states but an increased number of synchronizations. In practical terms that means that the BSP automaton's implementation will consume less space, and process BSP words in the same number of local transitions, but require an increased number of global barriers. Proposition 9 is thus a first but insufficient step towards BSP automata minimization.

Figure 9 illustrates the minimization of $\text{Seq}(\mathbf{A}_a)$ (with $p = 2$, sufficient for illustrating the computation). The circled groups of state vectors are provisional congruence classes to be refined until the Min algorithm reaches its fixed point. They strongly depend on the structure of Δ and as we have seen, the resulting (sequential) automaton on $\Sigma_{p;}$ can only be re-parallelized at the expense of extra synchronizations.

More important for our purpose, the objective of bulk-sychronous parallelism is to provide realistic and predictable *parallel speed-up*. BSP theory includes a cost-model that relates sequential time, the number of processes p and global synchronization

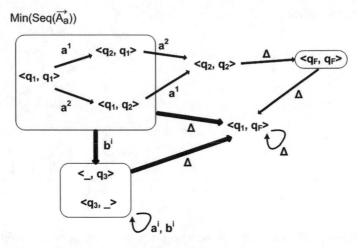

Fig. 9 Sequential minimization of BSP automaton $\mathbf{A}_a$

(and communication) delays. Automaton minimization is directly related to space complexity, memory consumption, but as seen in this section it can lead to higher synchronization costs hence more time complexity. As a step in this direction, we now adapt the cost-model of BSP theory to BSP automata and show how it can be used as objective function in the search for fast parallel versions of sequential automata.

10   Cost-Model

Words from regular languages can be recognized by (classical) finite automata in time proportional to their length. Being models of parallel algorithms, BSP automata are meant to accelerate this process. Ideally a word from a regular language could be recognized p times faster by a BSP automaton. This is certainly possible but, in general, parallel recognition requires more than one superstep so that the BSP automaton's operations require a BSP word of length more than one. Moreover, BSP theory and systems have show that the synchronization function Δ's implementation incurs costs that may be larger than the speed-up of parallelism.

In this section we show how to accelerate the recognition of regular languages, and define a detailed version of the BSP cost model to quantify the time-space cost of doing this.

The first auxiliary notion is concatenation-factorization on sequential words.

Definition 12 A *factorization function* on Σ words is a function $\Phi : \Sigma^* \to (\Sigma^+)^*$ such that

$$\Phi(\epsilon) = \epsilon$$
$$|w| > 0 \implies |\Phi(w)| > 0$$
$$\Phi(w) = w_1, w_2 \ldots, w_n \implies w_1 w_2 \ldots w_n = w$$

By definition, a factorization function sends the empty word to itself and sends a non-empty w to a non-empty sequence of non-empty words whose concatenation is w itself.

Next we define the distribution of sequential words to (BSP) locations. Recall that $(\Sigma_{p;})^*$ is the set of sequentialized BSP words $((\Sigma \times [p])^*;)^*$.

Definition 13 Given a factorization function Φ on Σ words, a *distribution function* based on Φ is a $D_\Phi : \Sigma^* \to (\Sigma_{p;})^*$ such that

$$D_\Phi(\epsilon) = \epsilon$$
$$\Phi(w) = w_1, w_2 \ldots, w_n \implies D_\Phi(w) = w_1'; w_2'; \ldots w_n';$$
$$w_t = a_1 \ldots a_k \implies w_t' = (a_1, i_1) \ldots (a_k, i_k)$$
$$i_1, \ldots, i_k \in [p]$$

The distribution of a language on Σ is the set of distributions of its words i.e. $D_\Phi(L) = \{D_\Phi(w) \mid w \in L\}$.

This definition is such that a distribution $D(w)$ is the sequential image of a BSP word and $\mathrm{Seq}(\mathrm{Par}(D_\Phi(w))) \cong D_\Phi(w)$ (Sect. 7).

For example if

$$w = aaabba$$

one possible factorization is

$$\Phi(w) = aaab, ba$$

and one possible associated distribution is

$$D_\Phi(w) = (a, 3)(a, 2)(a, 2)(b, 0); (b, 1)(a, 1);$$

with

$$\mathrm{Par}(D_\Phi(w)) = < b, \epsilon, aa, a > < \epsilon, ba, \epsilon, \epsilon >$$

and

$$\mathrm{Seq}(\mathrm{Par}(D_\Phi(w))) = (b, 0)(a, 2)(a, 2)(a, 3); (b, 1)(a, 1); \cong D_\Phi.$$

The definition of distribution function is flexible enough to allow any word and any language to be distributed to a BSP word or BSP language. The existence of distributions is a trivial fact of no interest in itself. What matters is optimization: the discovery of distributions with minimal parallel execution time. To define this we need to define the cost of a BSP automaton's computations. The synchronization

cost is an experimental constant that depends on the physical machine executing one
of our BSP automata.

Definition 14 Let $\mathbf{v} \in (\Sigma^*)^p$ be a word vector. Its *BSP cost* $\text{cost}(\mathbf{v}) = \max_i |v^i|$ is
the length of its longest element. Define also $l \in \mathbf{N}^+$, the *barrier synchronization
cost* constant. For a BSP word $w = \mathbf{v}_1 \ldots \mathbf{v}_S \in ((\Sigma^*)^p)^*$, its BSP cost is

$$\text{cost}(w) = \Sigma_{t=1}^S (\text{cost}(\mathbf{v}_t) + l) = Sl + \Sigma_{t=1}^S \text{cost}(\mathbf{v}_t).$$

The reader familiar with BSP theory will have noticed that our cost function covers
local sequential computation and global synchronization but not communication.
This is indeed a simplification and assumes, not that communication is "free" but
that an implementation always uses all-to-all communications and that its usual BSP
cost of $p \times g$ is here hidden in the l constant.

More detailed presentations of BSP automata will refine this, for example by
taking into account the actual dependencies in the synchronization function: a purely
local Δ actually costs less than one whose values (output states) depend on all the
input states. The above-defined cost model is a pessimistic upper-bound for this.
We now explain how BSP automata encode the elements of BSP algorithm design
namely load balancing and minimal synchronization.

Definition 15 For a given distribution function D_Φ of factorization Φ, the BSP cost
of a sequential word $w \in \Sigma^*$ with respect to D_Φ is defined as the BSP cost of the
parallelization of its distribution:

$$\text{cost}_{D_\Phi}(w) = \text{cost}(\text{Par}(D_\Phi(w)))$$

For example in the above example with $w = aaabba$ we had

$$\text{Par}(D_\Phi(w)) = <b, \epsilon, aa, a> <\epsilon, ba, \epsilon, \epsilon>$$

and so

$$\text{cost}_{D_\Phi}(w) = \text{cost}(\text{Par}(D_\Phi(w))) = 2 + l + 2 + l = 4 + 2l.$$

A direct consequence of the cost model is that the cost of a word with respect to
D_Φ is least if the factorization $\Phi(w)$ produces a minimal number of factors (hence
minimal number of BSP supersteps) while the distribution of each factor $D_\Phi(w_t)$
has the least maximal local length (hence the most balanced distribution). This bi-
objective cost function is the basis of BSP algorithm design: for a given amount of
parallelism, balance the lengths of local computations while minimizing the number
of supersteps.

Problem 1 BSP-PARALLELIZE-WORDWISE
Input: *A regular language L given by a regular expression r or DFA A.*
Goal: *Find a distribution D_Φ and BSP automaton A_D such that $L(A) = \text{Par}(D_\Phi(L))$
and $|A_D| \in O(|A|)$.*

Subject to: $\forall w \in \Sigma^*.\ cost_{D_\Phi}(w)$ *is minimal over* $\{(\Phi, D_\Phi, A_D) \mid L(A) = Par(D_\Phi(L))\}$.

Minimization for every individual w is not a standard formulation. A better one is:

Problem 2 BSP-PARALLELIZE
Input: *A regular language L given by a regular expression or DFA.*
Goal: *Find a distribution D_Φ and BSP automaton A_D such that $L(A) = Par(D_\Phi(L))$ and $|A_D| \in O(|A|)$.*
Subject to: $T_{D_\Phi}(n) = \max\{cost_{D_\Phi}(w) \mid |w| = n\}$ *is minimal over* $\{(\Phi, D_\Phi, A_D) \mid L(A) = Par(D_\Phi(L))\}$, *for all $n \geq 0$.*

Theoretical work can concentrate on $\lim_{n \to \infty} T_{D_\Phi}(n)$ while certain applications could consider only fixed-size input words i.e. a single value of n. The former is clearly a general algorithm-design problem and the latter is more likely to have an algorithmic solution. The present formulation of BSP automata leave open both theoretical and practical explorations: depending on the space of factorization and distribution functions that is considered, the BSP-PARALLELIZE problem could have widely different complexities.

In the next section we explore an important subproblem: finding BSP automata parallelizations for the block-wise distribution function $D_{\div p}$. The cost is then equal to l times the number of supersteps and BSP-PARALLELIZE amounts to minimizing the number of supersteps. But as specified in the problem definition ($|A_D| \in O(|A|)$) this should not be at the cost of an explosion in the number of states. We present elements of both lower- and upper-bound for this parameter.

11 Parallel Acceleration

Problem BSP-PARALLELIZE sets the goal of finding the fastest possible tuple (factorization, distribution, BSP automaton) of dimension p to recognize a given regular language L. *Fastest* refers to the cost of the BSP words once they are factorized and distributed for the BSP automaton. As a first step towards such optimal solutions, we will adapt the experimental notion of parallel speedup and show some parallelizations measured thus.

Definition 16 Let L be a regular language and (Φ, D_Φ, A_D) a factorization, distribution and BSP automaton for L i.e. $Par(D_\Phi(L))$. The parallel *speedup* obtained by (Φ, D_Φ, A_D) on a given word size n is the ratio

$$speedup(\Phi, D_\Phi, A_D, n) = \min\{n/cost_{D_\Phi}(w) \mid |w| = n\}$$

The n term in the denominator is $|w|$, the cost of sequential recognition by a DFA. On first inspection, the definition of speedup does not appear to depend on the language L being recognized. But it actually does. A speedup value is only possible by virtue of a BSP automaton recognizing L with the given factorization (supersteps) and distribution (data placement).

We take three examples of simple regular languages to parallelize.

- $L_1 = L(a^*)$
- $L_2 = L(a^*b^*)$
- $L_3 = L((a + b)^*bbb(a + b)^*)$

Example 2 **Parallel recognition of L_1.** Sequential recognition of a^* amounts to reading a word $w \in \Sigma^*$ sequentially with a DFA for this language. The simplest DFA A_{a^*} has two states q_1, q_2, starts from q_1, accepting states $F = \{q_1\}$, and transitions $\delta(q_2, x) = q_2, \delta(q_1, a) = q_1, \delta(q_1, b) = q_2$.

A simple and efficient parallelization for L_1 is $(\Phi_1, D_\circ, A_1)$ defined as follows. The factorization function keeps the input word into a single superstep word: $\Phi_1(w) = (w)$.

The "remainder p" distribution function sends letters to locations in cyclic fashion: $D_\circ(u_0 \ldots u_{n-1}) = (u_0, 0 \mod p) \ldots (u_{n-1}, (n-1) \mod p)$;

The BSP automaton $A_1 = (< A_{a^*}, \ldots, A_{a^*} >, \text{Id})$ has a copy of the DFA for accepting a^* at every location so any input word containing letter b will put one location into non-accepting local state. The synchronization function is the identity on state vectors. As a result, the BSP words accepted are those that only contain letter a, i.e. $L(A_1) = \text{Par}(D_\circ(L))$.

By construction $\text{cost}_{D_\circ}(w) = \text{cost}(w) + l$ i.e. the cost of the distributed word vector + the cost of one barrier. Cyclic distribution is known to have cost n/p or $(n/p) + 1$ because no location receives more than that many of the letters. As a result the speedup is $\frac{n}{(n/p)+l}$ which tends to p (ideal speedup) for large input sizes.

The above construction is an asymptotic solution to BSP-PARALLELIZE for this language because any BSP automaton costs one l term on non-empty input, and processing the whole input word is both necessary and requires at least parallel cost n/p.

Example 3 **Parallel recognition of L_2.** Sequential recognition of a^*b^* is done by a DFA having states q_0, q_1, q_2 of which q_0 is initial, accepting states q_0, q_1 and transition function $q_0 \xrightarrow{a} q_0, q_0 \xrightarrow{b} q_1$,

$q_1 \xrightarrow{b} q_1, q_1 \xrightarrow{a} q_2$,

$q_2 \xrightarrow{x} q_2$.

Let us call this automaton A_2.

Consider again parallelization with factorization function Φ_1 i.e. BSP words of length one superstep. Take as distribution function the "div-p" function that sends to each location a block of length $k \geq p$ except one possibly shorter block at the end: $D_{\div p}(u_0 u_1 \ldots u_{n-1}) = (u_0, 0/p)(u_1, 1/p) \ldots (u_{n-1}, (n-1)/p)$

For example

$$D_{\div 4}(u_0 u_1 \ldots u_8) = (u_0, 0)(u_1, 0)(u_2, 0)(u_3, 0)(u_4, 1)(u_5, 1)(u_6, 1)(u_7, 1)(u_8, 2).$$

We now show that a BSP automaton can be built for Φ_1 and $D_{\div p}$ to accept L_2. Its parallel speedup will then be the same as in Example 2. Consider the BSPRE $R = R_0 + R_1 + R_2 + R_3$ where

$R_3 = < a^*, a^*, a^*, a^*b^* >$,
$R_2 = < a^*, a^*, a^*b^*, b^* >$,
$R_1 = < a^*, a^*b^*, b^*, b^* >$,
$R_0 = < a^*b^*, b^*, b^*, b^* >$.

By construction $L(R) = \mathrm{Par}(D_{\div p}(L_2))$ because the words of L_2 split into four equal-length blocks and parallelized are precisely of one of the four forms specified by R.

It is not sufficient for our purpose to apply Theorem 2 to R because the constructive proof given there introduces unnecessary synchronization. To obtain a one-superstep BSP automaton for L_2 based on R proceed as follows. Build a DFA A'_2 with states q_0, q_1, q_2, q_3 such that q_0, q_1, q_3 are accepting, state q_0 accepts $L(a^+)$, state q_1 accepts $L(a^+b^*)$, state q_3 accepts $L(b^*)$ and words leading to q_2 are not from the union of those languages (which equals L_2). Let the BSP automaton have A'_2 as local automaton at every location and define the synchronization function as follows:

$\Delta(< (q_0 + q_1), (q_0 + q_1), (q_0 + q_1), (q_0 + q_1) >) =$ an accepting vector state; to accept R_3,

$\Delta(< (q_0 + q_1), (q_0 + q_1), (q_0 + q_1), q_3 >) =$ an accepting vector state; to accept R_2,

$\Delta(< (q_0 + q_1), (q_0 + q_1), q_3, q_3 >) =$ an accepting vector state; to accept R_1,

$\Delta(< (q_0 + q_1), q_3, q_3, q_3 >) =$ an accepting vector state; to accept R_0,

Δ sends any other state vector to a non-accepting vector state.

It then follows that the BSP automaton accepts L_2 in one superstep for the given distribution. This completes the example.

Example 4 **Parallel recognition of L_3.** Sequential recognition of $(a + b)^*bbb(a + b)^*$ amounts to searching for the first sequence bbb in a given word. A simple manner of obtaining a DFA for this is to start for a NFA with a sequence of 4 states from initial to accepting, each one related to the next by a unique $\delta(q_j, b) = q_{j+1}$ transition, and then apply the NFA-to-DFA transformation. Another method is to retain the four states and add all missing transitions to obtain a DFA. Let us call it A_3.

A_3, and thus L_3 can be parallelized to a one-superstep BSP automaton by a construction similar to that of Example 3 above. The parallelization uses factorization Φ_1 and distribution $D_{\div p}$: it sends the first $n/p = |w|/p$ elements of w to location 0, the next n/p to location 1, etc. with a single superstep symbol $;$ at the end.

To do this we consider the three factors $w = w_1w_2$ of any $w \in L_3$ where $w_1 \in L((a + b)^*) - \{bbb\}$, $w_2 \in L(bbb(a + b)^*)$ i.e. w_2 begins with the first occurrence of bbb in w. Then we consider all the p possible positions for the first letter of w_2. Each one corresponds through $D_{\div p}$ to a BSPRE. For example $|w_1bbb| \leq n/p$ iff $\mathrm{Par}(D_{\div p}(w)) \in L(< ((a + b)^*) - \{bbb\})bbb, (a + b)^*, (a + b)^*, \ldots >)$. A BSP automaton A_0 can be derived from this BSP regular expression: by definition it operates in one superstep. Similar BSP automata A_i can be derived from the hypothesis that the first b symbol of the first bbb sequence in w starts at a certain point in w. It follows that $A_0 + A_1 + A_2 + \ldots$ is a BSP automaton for L_3. Moreover it is possible to combine those BSP automata by a purely local process: add (create the disjuction) of all local DFA, and then build the combined synchronization function Δ by operating independently on the accepting states every local part A_j^i. The resulting BSP automaton accepts L_3 in a single superstep. Its speedup is the same as for Example 3.

Warning In our parallelization examples above it is assumed that an input word is split into regular blocks before being input to a custom-built BSPA. If processing time is understood as the time required to accept/refuse a given input word in each language, then our constructions indeed provide a $p\times$ speedup over the initial "sequential" DFA. But the reader should be aware that the BSPA are in general non-deterministic (NBSPA) and that to obtain this speedup in practice requires to transform them into equivalent deterministic BSPA. This pre-processing is amortized over the whole language but may have an exponential cost in space and time.

The construction of Examples 3 and 4 can clearly be applied to the general word recognition problem: for any given $x \in L((a + b)^*)$, one can construct a one-superstep BSP automaton A (i.e. based on Φ_1 and $D_{\div p}$) that paralellizes the language $(a + b)^*x(a + b)^*$. This A is the sum (language union) of $O(\max(p, |x|))$ BSP automata whose local DFA are minor modifications of A_x, the minimal DFA for accepting x.

All examples shown above provide candidate solutions to BSP-PARALLELIZE: they parallelize the given regular language L_j in one superstep with a BSP automaton whose size is linear in the size of a minimal DFA for L_j. All three examples are regular language of star-height one, and in general it is not clear whether such a parallelization is always possible.

Problem 3 *OPEN PROBLEM: does every instance of BSP-PARALLELIZE have a one-superstep solution?*

The answer would be positive if the number of states in the BSP automaton solution were allowed to grow exponentially. However the construction for showing this is very different from that of our above examples.

Proposition 10 *Every regular language L of regular expression r has a one-superstep parallelization $(\Phi_1, D_{\div p}, A)$ that can be constructed in time exponential in $|r|$ and such that $|A|$ is also exponential in $|r|$.*

Proof We show how $L = L(r)$ can be parallelized to a 1-superstep BSP automaton. Define $L_n = L \cap L((a + b)^n)$ and apply the following steps to build $(\Phi_1, D_{\div p}, A)$ such that $L(A) = \text{Par}(D_{\div p}(L))$. Assume without loss of generality that $p = 2$ (if $p > 2$ the construction can be extended by induction).

1. Compute $L^0 = L \cap (a + b)^{n/2}$. Those words are the ones location 0 should accept in A: the first half of L_n's words i.e. L's words for a given length input length n. Let A^0 be a DFA and r^0 a regular expression for L^0.
2. Compute $L_n = L \cap (a + b)^n$ as a regular expression r_n.

3. For every one of the 2^n words $x \in L_{n/2}$, compute the Brzozowski differential $D_x(r_n)$ whose language is known to equal $x \backslash L_n = \{y \mid xy \in L_n\}$. This computation is a simple but exponential time-size converging normalization on the regular expression [4, 21].
4. Let $L' = \sum_{x \in (a+b)^{n/2}} (x \cap L_{n/2}) \backslash L_n$. Let A' be a DFA for accepting L'.
5. Define $A = (< A^0, A' >, \Delta)$ with Δ mapping to an accepting state vector, only those pairs of accepting states that correspond to the same x prefix.

By construction A will accept at location 0 precisely the first halves of words in L_n, and at location 1 their corresponding suffixes. The local automata are a sum (union) of all such possibilities and the synchronization function Δ recombines them in the correct way. $\qquad\qquad\square$

In this section we have begun exploring parallelizations of regular languages. We have only shown one-superstep examples because there are trivial n-supersteps parallelizations that are of no interest either theoretically or practically. On simple examples of star-height one, space-efficient one-supersteps parallelizations have been constructed. It has also been proved that any regular language can have a one-superstep parallelization if exponential space (number of states) is allowed. It remains to explore intermediate solutions and how their complexity relates to star-height of the input regular language.

12 Intensional Notation and Application to Programming

In the theory presented up to this point, parallel vectors are enumerated but this is not a scalable point of view on parallel programming. It is more usual and convenient to represent vectors as functions from position i to the local element. This was the basis for the λ-calculus in [12] whose primitives are now implemented in BSML (BSP-OCaml). We show how to improve our theory of BSP automata in this manner so that vectors are not enumerated but defined by a simple symbolic notation.

Assume that the locations $i \in [p]$ are written in binary notation $0, 1, 10, 11 \ldots$. Define a binary regular expression (BRE) by the following grammar:

$$b ::= \emptyset \mid 0 \mid 1 \mid bb \mid b + b \mid b^*$$

Notice that BRE cannot encode the empty word. This notation is used to encode sets of locations. For example $b_1 = (0 + 1)^* 1$ is the set of odd-rank locations, $b_2 = 0(0 + 1)(0 + 1)$ represents the first four locations when $p = 8$, and $b_3 = 010$ $(0 + 1)(0 + 1)(0 + 1)$ the third 8-position block of positions when $p = 32$ i.e. positions 16–23 over 6 binary digits. It would be possible to make this notation symbolic over p but that would require additional syntax and here we only explain how to make it symbolic over the position integers for a fixed p.

To avoid enumerating BSP vectors, replace the enumeration $< r_0, r_1, \ldots, r_{p-1} >$ by a grammar clause for *intensional vectors* of regular expressions:

$$R ::= \ < r@b >$$

where $r \in$ RE and $b \in$ BRE. The meaning of $< r@b >$ is the vector of regular expressions whose local value is r at locations pid $\in L(b)$ and $\emptyset$ at other locations. For example if $p = 8$, the BSPRE $< (a+b)^+ @b_2 >$ represents, in enumerated form,

$$< (a+b)^+, \ (a+b)^+, \ (a+b)^+, \ (a+b)^+, \ \emptyset, \ \emptyset, \ \emptyset, \ \emptyset >$$

i.e. the BSP language of one-superstep BSP words with non-empty local traces at positions 0–3 but empty traces at positions 4–7.

It is also possible to create BSP vectors by superposition $\|$ of multiple $r@b$ expressions. For example if $p = 4$, the BSPRE $< a@(0 + 1)^*0 \ \| \ b@(0 + 1)^*1 >$ corresponds to the enumeration

$$< a, b, a, b > .$$

With this new notation, redefine the BSP regular expressions:

$$R ::= \emptyset \ | \ \epsilon \ | < V > | \ R; R \ | \ R* \ | \ R + R.$$

using a new sub-grammar for BSP vectors:

$$V ::= r@b \ | \ V \ \| \ V$$

where $r \in$ RE and $b \in$ BRE. The language of those intensional BSP regular expressions is defined with new rules for intensional vectors:

R	$L(R)$
$\emptyset$	$\{ \}$
ϵ	$\{\epsilon\}$
$< r@b >$	$\prod_{i=0}^{i=p-1} \begin{cases} L(r) \text{ if } i \in L(b) \\ \{\emptyset\} \quad \text{else} \end{cases}$
$< r_1@b_1 \ \| \ \ldots \ \| \ r_k@b_k >$	$\prod_{i=0}^{i=p-1} \bigcup \{L(r_j) \mid i \in L(b_j), 1 \leq j \leq k\}$
$R_1; R_2$	$L(R_1)L(R_2)$
R^*	$L(R)^*$
$R_1 + R_2$	$L(R_1) \cup L(R_2)$

This new notation is "scalable" in the sense that its parallel implementations can slice it into local parts of the $< r@b >$ sub-expressions and simply combine their local values as (regular) functions of the location number pid. This is similar to what data-parallel programming languages provide. But its restriction to regular expressions has a major advantage: one location can compute the set of locations that hold a certain value. For a parallel implementation this amounts to inverting the communication relation, without specific source-code information to that effect.

We illustrate this kind of application on a simple but meaningful example: converting "get" requests for remote data into "put" operations for sending data. Assume we are programming a one-million core machine $p = 2^{20}$ in a high-level BSP language and a global parallel instruction (purely functional, for simplicity) of the form

$$\text{get datavector from indexvector}$$

whose input types are a

$$\text{datavector : float}^p; \quad \text{indexvector : (int set)}^p$$

and whose output type is
$$\text{(float set)}^p.$$

Let datavector be $< d_0, \ldots, d_{p-1} >$ and indexvector be $< I_0, \ldots, I_{p-1} >$ and assume that the get-from instruction realizes a global BSP operation whose resulting value is the vector $< A_0, \ldots, A_{p-1} >$ whose local values are

$$A_i = \{d_j \mid j \in I_i\}$$

. In other words get-from moves the elements of datavector as if every processor i sends a request for local data to processors whose positions j are listed in the local table I_i of indexvector. Consider now three successively improved data-parallel implementations for this operation.

12.1 2-Phases Implementation

A straightforward implementation is to use two BSP supersteps. The first one sends a set of requests from every processor i to processors $j \in I_i$. The second superstep sends back the requested data i.e. processor j communicates back with all requesting processors $\{i' \mid j \in I_{i'}\}$. The disadvantage of this scheme is that its BSP costs includes two global barriers (i.e. twice the global latency) and implementors wish to avoid it by "converting get into put" using one of the two following methods.

12.2 1-Phase $O(p)$ Inversion

The SPMD paradigm for data-parallel programming ensures that the source program is common to every local processor and thus the code for our instruction is known at every position i, only data d_i and I_i are local. We can consider the I_i to be (finite) languages of 20-bit words ($\log p$-bit words) and improve the get-from instruction's syntax as follows: indexvector is given as a BSPRE e.g. $< r@b >$ where r

encodes the I_i. As a result of this language construct, every processor j can directly compute its set of target processors for sending data $\{i' \mid j \in I_{i'}\}$ by simply running every 20-bit word i' through a finite automaton for r: if accepted and if $j \in L(b)$ then processor j should send its d_j to processor i'. This can be done in time proportional to $2^{20} = p$. Moreover it does not require two BSP supersteps but only one: the "get" is implemented directly by a "send", thanks to the simplicity of the sub-language on integer sets for I_i.

12.3 1-Phase $O(\log P)$ Inversion

An even more efficient implementation of the 1-phase implementation is possible due to the simplicity of BSPRE. Every processor j can simply enumerate $L(r)$ because it is a regular language. This can be done in time proportional to the size of this set times the length of the words in it: that is $O(\log p)$ time the number of messages [13]. In our example, if processor j has a small number of requests to satisfy e.g. 3, that would prevent it from executing p or one million instructions.

12.4 Other Intensional Notations

All p-indexed vectors in the theory of BSP automata can be manipulated with similar regular-indexing notations. For example the factorization Φ and distribution D functions on sequential regular languages can likewise be restricted to intensional notations. The result would be to automate the inversion of D, and from there compute a BSPRE directly from a sequential regular expression.

Moreover, partition, distribution and synchronization are enumerated functions whose implementation may not be obvious. Defining a regular notation for those functions improves their ease of programming, makes expressions "scalable" (parametric on p) and leads to useful inversion algorithms e.g. inverse distribution. For example BSPLib [7] and many other "SPMD" data-parallel programming systems present the local code (which corresponds to our local sequential automata) as a function of the location number called `pid`.

All the advantages of an intensional notation can be obtained by an extended notation for BSPRE that we define below. Moreover as we will now explain, the low complexity of regular languages allows us to automate the inversion of the (location $\rightarrow$ value) map, a useful operation for parallel algorithms that is rarely provided by parallel languages.

13 Conclusions and Future Work

We have defined and begun exploring a BSP variant of elementary automata theory. Some key observations are that BSP automata are more than product automata, their natural alphabet is the set of regular expressions, and their state-space is exponential in the number of parallel locations. BSP automata and BSP languages preserve all the classical closure properties: non-determinism, ϵ-transitions and determinization, but break the classical properties of minimization. The interaction between state-minimization and BSP cost optimization remains to be understood. Compact symbolic notations can be designed for the parallel-vector components of BSP automata and BSP regular-expressions. BSP automata can help automate bulk-synchronous parallel programming e.g. as a declarative language for connection supersteps, defining communication structures and cost optimization.

Future work will explore (a) BSP regular grammars and their generalization to BSP context-free languages, (b) the application of BSP automata to parallel text processing and parsing, (c) applications to pattern matching and to parallel data structure (tries etc.) (d) generalizations of BSP automata to heterogeneous and hierarchical architectures.

BSP automata constitute a clear and easily-understood basis for teaching, specifying and writing parallel programs. They can be used to combine the control- and communication structure of BSP programs, analyze or optimize that structure. BSP regular expressions are useful for declarative programming of parallel operations with explicit data placement and synchronizations.

Acknowledgments The author thanks Frédéric Loulergue, Thibaut Tachon, Arnaud Lallouet and Chong Li for their insightful remarks and help with proofreading. This work has been supported by the author's position at Huawei/CSI-FRC.

Appendix 1

1.1 BSP Automaton Versus Product of Automata

The theory of products of automata is developed by Gécseg in [6]. It describes decompositions of finite automata as products of simpler ones, and is closely related to the theory of semigroup decompositions.

A BSP computation is more than a vector of sequential computations, and this is reflected by the fact that a BSP automaton is more than a vector of DFA. This is relatively obvious but we make it here completely explicit by comparing that definition (Definition 2) with that of a product automaton.

Let $A^i = (Q^i, \Sigma, \delta^i, q_0^i, F^i), i \in [p]$ be a vector of DFA and $A = (A^0, \ldots, A^{p-1}, \Delta)$ a BSP automaton built from them. Gécseg's definition (Definition 4.2 in [6]) of *machine product* applies to Mealy machines i.e. DFA with an output function added. For the purpose of language recognition the output functions are not necessary, so we consider the machine product $\prod_i A^i$ as the Gécseg without outputs. According to Definition 4.2 in [6], $\prod_i A^i$ is a state machine with vector states $\prod_i Q^i$, just like the BSP automaton, a new externally-defined alphabet X, and a special transition function δ_ψ based on the externally-given function

$$\phi : (\prod_i Q^i) \times X \to \Sigma^p$$

such that

$$\delta_\psi(< q^0, \ldots, q^{p-1} >, x) = < \delta^0(q^0, x^0), \ldots, \delta^{p-1}(q^{p-1}, x^{p-1}) >$$

where $< x^0, \ldots, x^{p-1} >= \phi((< q^0, \ldots, q^{p-1} >, x)$.

It is trivial to show that the automaton product can simulate the asynchronous parts of a BSP computation. But the structure of δ_ψ is not the same as a synchronization function Δ. The product automaton could simulate the BSP automaton but at the expense of an unnatural encoding e.g. $X = (\Sigma^*)^p \times \prod_i Q^i$ to let ϕ distinguish asynchronous versus synchronous applications of Δ.

But an alphabet which contains states and trace histories is hardly a natural (and low-complexity) encoding. Following this theoretical direction would defeat the purpose of BSP automata that is not the study of algebraic decompositions, or decidability, but rather to investigate programming notations having BSP implementations.

Appendix 2

2.1 Regular Expressions

Regular expressions are a well-known notation for the languages of finite automata. The definitions and properties we state below can be found in every textbook on finite automata for example Chap. 3 of [21]. The languages denoted by regular expressions are called regular, and that class of languages is the same as those recognized by a DFA or its equivalent, non-deterministic variants.

A regular expression is an expression r from the following grammar:

$$r ::= \emptyset \mid \epsilon \mid a \mid rr \mid r* \mid r + r$$

where $a \in \Sigma$ is any symbol from the alphabet. We write RE for the set of regular expressions . The language of a regular expression is defined by $L : \text{RE} \to \mathcal{P}(\Sigma^*)$ where function L translates $\emptyset$ to the empty language, ϵ (resp. a) to a singleton empty word (resp. singleton one-symbol word), $r_1 r_2$ to the concatenation of languages, r^* to $L(r)^* = \bigcup_{n \geq 0} L(r)^n$ and $r_1 + r_2$ to the union of the two languages. The union, concatenation, *-closure of two regular languages is regular. The complement of a regular language is also regular [15].

For $r \in RE$, there exists a NFA A such that $L(A) = L(r)$. A time-optimal quadratic time algorithm for this transformation is described in [3]. It has been improved to a linear-space and parallelisable algorithm in [22]. Both use the *Glushkov* automation of r whose states are the positions in r's syntax tree.

Inversely, for A a finite automaton, there exists $r \in RE$ such that $L(r) = L(A)$. The two equivalence properties are called Kleene's theorems [10].

References

1. Anantharaman, S., Hains, G.: A synchronous bisimulation-based approach for information flow analysis. In: Third Workshop on Automated Verification of Critical Systems: (AVOCS'03), Southampton, (UK) (2003)
2. Benzaken, C.: Systèmes formels: introduction à la logique et à la théorie des langages. Masson (1991)
3. Brüggemann-Klein, A.: Regular expressions into finite automata. Theoret. Comput. Sci. **120**(2), 197–213 (1993)
4. Brzozowski, J.A.: Derivatives of regular expressions. J. ACM **11**(4), 481–494 (1964)
5. D'Antoni, L., Veanes, M.: Minimization of symbolic automata. In: Proceedings of the 41st ACM SIGPLAN-SIGACT Symposium on Principles of Programming Languages, POPL '14, pp. 541–553. ACM (2014)
6. Gécseg, F.: Products of Automata. Springer (1986)
7. Hill, J.M.D., McColl, B., Stefanescu, D.C., Goudreau, M.W., Lang, K., Rao, S.B., Suel, T., Tsantilas, T., Bisseling, R.H.: BSPlib: The BSP programming library. Parallel Comput. **24**(14) (1998)
8. Hopcroft, J.: An n log n algorithm for minimizing states in a finite automaton. Technical Report No. STAN-CS-70-190, Stanford University, Department of Computer Science (1971)
9. John, M.: Howie. Automata and languages. Clarendon Press, Oxford (1991)
10. Kleene, S.C.: Representation of events in nerve nets and finite automata. Automata Stud., 3–41 (1956)
11. Loulergue, F., Hains, G.: Functional parallel programming with explicit processes: Beyond SPMD. Lecture Notes in Computer Science, vol. 1300. Springer (1997)
12. Loulergue, F., Hains, G., Foisy, Ch.: A calculus of recursive-parallel BSP programs. Sci. Comput. Programm. (2000)
13. Mäkinen, E.: On lexicographic enumeration of regular and context-free languages. Acta Cybern. **13**(1), 55–61 (1997)
14. McColl, W.F.: Scalable computing. In: van Leeuwen, J. (ed.) Computer Science Today. LNCS, vol. 1000. Springer (1995)
15. McNaughton, Robert: Elementary Computability, Formal Languages and Automata. Prentice-Hall, Englewood Cliffs, NJ (1982)
16. Merlin, A., Hains, G.: A bulk-synchronous parallel process algebra. Comput. Lang. Syst. Struct. **33**(3), 111–133 (2007)
17. Milner, R.: Communication and Concurrency. Prentice Hall, New York (1989)
18. Pin, J.-É.: Variétés de Langages Formels. Masson, Paris (1984)
19. Turing, A.: On computable numbers with an application to the entscheidungs problem. Proc. Lond. Math. Soc. **2**(42), 230–265 (1936)
20. Valiant, L.G.: A bridging model for parallel computation. CACM **33**(8), 103 (1990)
21. Wood, D.: Theory of Computation. Wiley (1987)
22. Ziadi, D., Ponty, J.-L., Champarnaud, J.-M.: Passage d'une expression rationnelle à un automate fini non-déterministe. Bull. Belgian Math. Soc. Simon Stevin **4**(1), 177 (1997)

Coping with Silent Errors in HPC Applications

Guillaume Aupy, Anne Benoit, Aurlien Cavelan, Massimiliano Fasi, Yves Robert, Hongyang Sun and Bora Uçar

Abstract This chapter describes a unified framework for the detection and correction of silent errors, which constitute a major threat for scientific applications at extreme-scale. We first motivate the problem and explain why checkpointing must be combined with some verification mechanism. Then we introduce a general-purpose technique based upon computational patterns that periodically repeat over time. These patterns interleave verifications and checkpoints, and we show how to determine the pattern minimizing expected execution time. Then we move to application-specific techniques and review dynamic programming algorithms for linear chains of tasks, as well as ABFT-oriented algorithms for iterative methods in sparse linear algebra. **Thanks to Selim Akl, by Yves Robert**—I have a vivid souvenir of Selim's visit to Lyon in the early 90s. Selim had obtained a Louis Néel fellowship devoted to promote exchanges between Canada and the Rhône-Alpes area in France, and he spent 6 months in Lyon with his family. Michel Cosnard was the

G. Aupy (✉)
Penn State University, State College, USA
e-mail: guillaume.aupy@ens-lyon.org

A. Benoit · A. Cavelan · Y. Robert
ENS Lyon, Lyon, France
e-mail: anne.benoit@ens-lyon.fr

A. Cavelan
e-mail: aurelien.cavelan@ens-lyon.fr

M. Fasi
The University of Manchester, Manchester, UK
e-mail: massimiliano.fasi@manchester.ac.uk

Y. Robert
University of Tennessee, Knoxville, USA
e-mail: yves.robert@ens-lyon.fr

H. Sun
ENS Lyon & INRIA, Lyon, France
e-mail: hongyang.sun@ens-lyon.fr

B. Uçar
CNRS & ENS Lyon, Lyon, France
e-mail: bora.ucar@ens-lyon.fr

© Springer International Publishing Switzerland 2017
A. Adamatzky (ed.), *Emergent Computation*, Emergence, Complexity
and Computation 24, DOI 10.1007/978-3-319-46376-6_11

head of the LIP laboratory at that time. Selim gave a course on parallel algorithms, mainly sorting and PRAM, that sparkled a lot of interest among both our students and the researchers in the lab. During his stay, Selim initiated several collaborations with Jean Duprat, Afonso Ferreira and Pierre Fraigniaud. Although I never collaborated with him, I would like to thank him for his vision. I was then a young professor in LIP, and I felt like meeting a star, but a very kind one. His two books, *Parallel Sorting Algorithms* and *The Design and Analysis of Parallel Algorithms*, had a huge influence on many researchers at LIP (including myself), as they helped shape our view of parallel complexity. Later on we all took different research directions (PRAM, hypercubes, systolic arrays, scheduling, routing, …) but Selim laid the foundations of the field for us, and we are grateful to him.

1   Introduction

For High-Performance Computing (HPC) applications, scale is a major opportunity. Massive parallelism with 100,000+ nodes is the most viable path to achieving sustained petascale performance. Future platforms will exploit even more computing resources to enter the exascale era.

Unfortunately, scale is also a major threat, because resilience becomes a key challenge. Even if each node provides an individual MTBF (Mean Time Between Failures) of, say, one century, a machine with 100,000 such nodes encounters on average a failure every 9 h, an interval much shorter than the execution time of many HPC applications. Note that (i) a one-century MTBF per node is an optimistic figure, given that each node features several hundreds of cores; and (ii) in some scenarios for the path to exascale computing [15], one envisions platforms including up to one million such nodes, whose MTBF will decrease to 52 min.

Several kinds of errors need to be considered when computing at scale. In the recent years, the HPC community has traditionally focused on fail-stop errors, such as hardware failures. The de facto general-purpose technique to recover from fail-stop errors is checkpoint/restart [11, 17]. This technique employs checkpoints to periodically save the state of a parallel application, so that when an error strikes some process, the application can be restored into one of its former states. There are several families of checkpointing protocols, but they share a common feature: each checkpoint forms a consistent recovery line, i.e., when an error is detected, one can rollback to the last checkpoint and resume execution, after a downtime and a recovery time. Many models are available to understand the behavior of the checkpointing and restarting techniques [8, 14, 31, 37].

While the picture is quite clear for fail-stop errors, the community has yet to devise an efficient approach to cope with silent errors, primary source of silent data corruptions. Such errors must also be accounted for when executing HPC applications [28, 30, 39–41]. They may be caused, for instance, by soft errors in L1 cache,

Fig. 1 Error and detection
latency

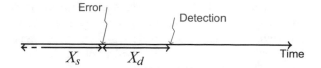

arithmetic errors in the ALU (Arithmetic and Logic Unit), or bit flips due to cosmic radiation. The main issue is that the impact of silent errors is not immediate, since they do not manifest themselves until the corrupted data impact the result of the computation (see Fig. 1), leading to a failure. If an error striking before the last checkpoint is detected after that checkpoint, then the checkpoint is corrupted, and cannot be used to restore the application. If only fail-stop failures are considered, a checkpoint cannot contain a corrupted state, because a process subject to failure cannot create a checkpoint or participate to the application: failures are naturally contained to failed processes. When dealing with silent errors, however, faults can propagate to other processes and checkpoints, because processes continue to participate and follow the protocol during the interval that separates the occurrence of the error from its detection.

In Fig. 1, X_s and X_d are random variables that represent the time until the next silent error and its detection latency, respectively. We usually assume that silent errors strike according to a Poisson process of parameter λ, so that X_s has the distribution of an exponential law of parameter λ and mean $1/\lambda$. On the contrary, it is very hard to make assumptions on the distribution of X_d. To alleviate the problem of detection latency, one may envision to keep several checkpoints in memory, and to restore the application from the last *valid* checkpoint, thereby rolling back to the last *correct* state of the application [25]. This multiple-checkpoint approach has three major drawbacks. First, it is demanding in terms of storage: each checkpoint typically represents a copy of the entire memory footprint of the application, which may well correspond to several terabytes. The second drawback is the possibility of fatal failures. Indeed, if we keep k checkpoints in memory, the approach requires that the last checkpoint still kept in memory to precede the instant when the error currently detected struck. Otherwise, all live checkpoints would be corrupted, and one would have to re-execute the entire application from scratch. The probability of a fatal failure for various error distribution laws and values of k can be evaluated [1]. The third and most serious drawback of this approach applies even without memory constraints, i.e., if we could store an infinite number of checkpoints in memory. The critical point is to determine which checkpoint is the last valid one, information which is necessary to recover from a valid application state. However, because of the detection latency (which is unknown), we do not know when the silent error has indeed occurred, hence we cannot identify the last valid checkpoint, unless some verification mechanism is enforced.

We introduce such verification mechanisms in this chapter. In Sect. 2, we discuss several approaches to validation (recomputation, checksums, coherence tests, orthogonalization checks, etc.). Then in Sect. 3 we adopt a general-purpose approach, which is agnostic of the nature of the verification mechanism. We consider a divisible-load application (which means that we can take checkpoints at any instant), and we partition the execution into computational patterns that repeat over time. The simplest pattern is represented by a work chunk followed by a verified checkpoint, which corresponds to performing a verification just before taking each checkpoint. If the verification succeeds, then one can safely store the checkpoint. If the verification fails, then a silent error has struck since the last checkpoint, and one can safely recover from it to resume the execution of the application. We compute the optimal length of the work chunk in the simplest pattern in Sect. 3.1, which amounts to revisiting Young and Daly's formula [14, 37] for silent errors. While taking a checkpoint without verification seems a bad idea (because of the memory cost, and of the risk of saving corrupted data), a validation step not immediately followed by a checkpoint may be interesting. Indeed, if silent errors are frequent enough, verifying the data in between two (verified) checkpoints, will reduce in expectation the detection latency and thus the amount of work to be re-executed due to possible silent errors. The major goal of Sect. 3 is to determine the best pattern composed of m work chunks, where each chunk is followed by a verification and the last chunk is followed by a verified checkpoint. We show how to determine m and the length of each chunk so as to minimize the *makespan*, that is the total execution time.

Then we move to application workflows. In Sect. 4, we consider application workflows that consist of a number of parallel tasks that execute on a platform, and that exchange data at the end of their execution. In other words, the task graph is a linear chain, and each task (except maybe the first and the last one) reads data from its predecessor and produces data for its successor. This scenario corresponds to a high-performance computing application whose workflow is partitioned into a succession of (typically large) tightly-coupled computational kernels, each of them being identified as a task by the model. At the end of each task, we can either perform a verification on the task output, or perform a verification followed by a checkpoint. We provide dynamic programming algorithms to determine the optimal locations of checkpoints and verifications.

The last technique that we illustrate is application-specific. In Sect. 5, we deal with sparse linear algebra kernels, and we show how to combine ABFT (Algorithm Based Fault Tolerance) with checkpointing. In a nutshell, ABFT consists in adding checksums to application data, and to view them as extended data items. The application performs the same computational updates on the original data and on the checksums, thereby avoiding the need to recompute the checksums after each update. The salient feature of this approach is *forward recovery*: ABFT is used both as an error verification and error correction mechanism: whenever a single error strikes, it can be corrected via ABFT and there is no need to rollback for recovery. Finally, we outline main conclusions and directions for future work in Sect. 6.

2 Verification Mechanisms

Considerable efforts have been directed at error-checking to reveal silent errors. Error detection is usually very costly. Hardware mechanisms, such as ECC (Error Correcting Code) memory, can detect and even correct a fraction of errors, but in practice they are complemented with software techniques. General-purpose techniques are based on replication [18, 21, 34, 38]. Indeed, performing the operation twice and comparing the results of the replicas makes it possible to detect a single silent error. With Triple Modular Redundancy [26] (TMR), errors can also be corrected by means of a voting scheme. Another approach, proposed by Moody et al. [29], is based on checkpointing and replication and enables detection and fast recovery of applications from both silent errors and hard errors.

Coming back to verification mechanisms, application-specific information can be helpful in designing ad hoc solutions, which can dramatically decrease the cost of detection. Many techniques have been advocated. They include memory scrubbing [24], but also ABFT techniques [7, 23, 35], such as coding for the SpMxV (Sparse Matrix-Vector multiplication) kernel [35], and coupling a higher-order with a lower-order scheme for Ordinary Differential Equation [6]. These methods can only detect an error but not correct it. Self-stabilizing corrections after error detection in the conjugate gradient method are investigated by Sao and Vuduc [33]. Also, Heroux and Hoemmen [22] design a fault-tolerant GMRES algorithm capable of converging despite silent errors, and Bronevetsky and de Supinski [9] provide a comparative study of detection cost for iterative methods. Elliot et al. [16] combine partial redundancy and checkpointing, and confirm the benefit of dual and triple redundancy. The drawback is that twice the number of processing resources is required (for dual redundancy).

A nice instantiation of the checkpoint and verification mechanism that we study in this chapter is provided by Chen [12], who deals with sparse iterative solvers. Consider a simple method such as the Preconditioned Conjugate Gradient (PCG) method: Chen's approach performs a periodic verification every d iterations, and a periodic checkpoint every $d \times c$ iterations, which is a particular case, with equi-spaced validations, of the approach presented later in Sect. 3.2. For PCG, the verification amounts to checking the orthogonality of two vectors and to recomputing and checking the residual. The cost of the verification is small if compared to the cost of an iteration, especially when the preconditioner requires many more flops than a SpMxV. As already mentioned, the approach presented in Sect. 3 is agnostic of the underlying error-detection technique and takes the cost of verification as an input parameter to the model.

3 Patterns for Divisible Load Applications

In this section we explain how to derive the optimal pattern of interleaving check-points and verifications. An extended presentation of the results is available in [2, 4, 10].

3.1 Revisiting Young and Daly's Formula

Consider a divisible-load application, i.e., a (parallel) job that can be interrupted at any time for checkpointing, for a nominal cost C. To deal with fail-stop failures, the execution is partitioned into same-size chunks followed by a checkpoint, and there exist well-known formulae by Young [37] and Daly [14] to determine the optimal checkpointing period.

To deal with silent errors, the simplest protocol (see Fig. 2) would be to perform a verification (at a cost V) just before taking each checkpoint. If the verification succeeds, then one can safely store the checkpoint and mark it as *valid*. If the verification fails, then an error has struck since the last checkpoint, which is correct having been verified, and one can safely recover (which takes a time R) from that checkpoint to resume the execution of the application. This protocol with verifications zeroes out the risk of fatal errors that would force to restart the execution from scratch.

To compute the optimal length of the work chunk W^*, we first have to define the objective function. The aim is to find a pattern P (with a work chunk of length W followed by a verification of length V and a checkpoint of length C) that minimizes the expected execution time of the application. Let W_{base} denote the base execution time of an application without any overhead due to resilience techniques (without loss of generality, we assume unit-speed execution). The execution is divided into periodic patterns, as shown in Fig. 2. Let $\mathbb{E}(P)$ be the expected execution time of the pattern. For large jobs, the expected makespan W_{final} of the application when taking failures into account can then be approximated by

$$W_{\text{final}} \approx \frac{\mathbb{E}(P)}{W} \times W_{\text{base}} = W_{\text{base}} + H(P) \cdot W_{\text{base}}$$

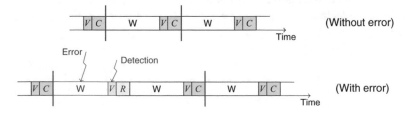

Fig. 2 The simplest pattern: a work chunk W followed by a verification V and a checkpoint C

where

$$H(P) = \frac{\mathbb{E}(P)}{W} - 1$$

is the expected *overhead* of the pattern. Thus, minimizing the expected makespan is equivalent to minimizing the pattern overhead $H(P)$. Hence, we focus on minimizing the pattern overhead. We assume that silent errors are independent and follow a *Poisson process* with arrival rate λ. The probability of having at least a silent error during a computation of length w is given by $p = 1 - e^{-\lambda w}$. We assume that errors cannot strike during recovery and verification. The following proposition shows the expected execution time of a pattern with a fixed work length W.

Proposition 1 *The expected execution time of a pattern* P *with work length W is*

$$\mathbb{E}(P) = W + V + C + \lambda W^2 + \lambda W(V + R) + O(\lambda^2 W^3). \tag{1}$$

Proof Let $p = 1 - e^{-\lambda W}$ denote the probability of having at least one silent error in the pattern. The expected execution time obeys the recursive formula

$$\mathbb{E}(P) = W + V + p(R + \mathbb{E}(P)) + (1 - p)C. \tag{2}$$

Equation (2) can be interpreted as follows: we always execute the work chunk and run the verification to detect silent errors, whose occurrence requires not only a recovery but also a re-execution of the whole pattern. Otherwise, if no silent error strikes, we can proceed with the checkpoint. Solving the recursion in Eq. (2), we obtain

$$\mathbb{E}(P) = e^{\lambda W}(W + V) + (e^{\lambda W} - 1)R + C.$$

By approximating $e^{\lambda x} = 1 + \lambda x + \frac{\lambda^2 x^2}{2}$ up to the second-order term, we can further simplify the expected execution time and obtain Eq. (1). □

The following theorem gives a first-order approximation to the optimal work length of a pattern.

Theorem 1 *A first-order approximation to the optimal work length* W^* *is given by*

$$W^* = \sqrt{\frac{V + C}{\lambda}}. \tag{3}$$

The optimal expected overhead is

$$H^*(P) = 2\sqrt{\lambda(V + C)} + O(\lambda). \tag{4}$$

Proof From the result of Proposition 1, the expected overhead of the pattern can be computed as

$$H(\mathrm{P}) = \frac{V+C}{W} + \lambda\, W + \lambda\, (V+R) + O(\lambda^2 W^2)\,. \tag{5}$$

Assume that the MTBF of the platform $\mu = 1/\lambda$ is large if compared to the resilience parameters. Then consider the first two terms of $H(\mathrm{P})$ in Eq. (5): the overhead is minimal when the pattern has length $W = \Theta(\lambda^{-1/2})$, and in that case both terms are in the order of $\lambda^{1/2}$, so that we have

$$H(\mathrm{P}) = \Theta(\lambda^{1/2}) + O(\lambda)\,.$$

Indeed, the last term $O(\lambda)$ becomes also negligible when compared to $\Theta(\lambda^{1/2})$. Hence, the optimal pattern length W^* can be obtained by balancing the first two terms in Eq. (5), which gives Eq. (3). Then, by substituting W^* back into $H(\mathrm{P})$, we get the optimal expected overhead in Eq. (4).   □

We observe from Theorem 1 that the optimal work length W^* of a pattern is in $\Theta(\lambda^{-1/2})$, and the optimal overhead $H^*(\mathrm{P})$ is in $\Theta(\lambda^{1/2})$. This allows us to express the expected execution overhead of a pattern as $H(\mathrm{P}) = \frac{o_{\mathrm{ef}}}{W} + o_{\mathrm{rw}}\, W + O(\lambda)$, where o_{ef} and o_{rw} are two key parameters that characterize two different types of overheads in the execution, and they are defined below.

Definition 1 For a given pattern, o_{ef} denotes the *error-free* overhead due to the resilience operations (e.g., verification, checkpointing), and o_{rw} denotes the *re-executed work* overhead, in terms of the fraction of re-executed work due to errors.

In the simple pattern we analyze above, these two overheads are given by $o_{\mathrm{ef}} = V + C$ and $o_{\mathrm{rw}} = \lambda$, respectively. The optimal pattern length and the optimal expected overhead can thus be expressed as

$$W^* = \sqrt{\frac{o_{\mathrm{ef}}}{o_{\mathrm{rw}}}}\,,$$

$$H^*(\mathrm{P}) = 2\sqrt{o_{\mathrm{ef}} \cdot o_{\mathrm{rw}}} + O(\lambda)\,.$$

We see that minimizing the expected execution overhead $H(\mathrm{P})$ of a pattern becomes equivalent to minimizing the product $o_{\mathrm{ef}} \times o_{\mathrm{rw}}$ up to the dominating term. Intuitively, including more resilient operations reduces the re-executed work overhead but adversely increases the error-free overhead, and vice versa. This requires a resilience protocol that finds the optimal tradeoff between o_{ef} and o_{rw}. We make use of this observation in the next section to derive the optimal pattern in a more complicated protocol where patterns are allowed to include several chunks.

3.2 Optimal Pattern

If the verification cost is small when compared to the checkpoint cost, there is room for optimization. Consider the pattern illustrated in Fig. 3 with three verifications

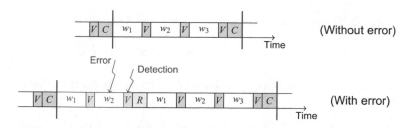

Fig. 3 Pattern with three verifications per checkpoint

per checkpoint. There are three chunks of size w_1, w_2, and w_3, each followed by a verification. Every third verification is followed by a checkpoint.

To understand the advantages of such a pattern, assume $w_1 = w_2 = w_3 = W/3$ for now, so that the total amount of work is the same as in the simplest pattern. As before, a single checkpoint needs to be kept in memory, and each error leads to re-executing the work since the last checkpoint. But detection occurs much more rapidly in the new pattern, because of the intermediate verifications. If the error strikes during the first of the three chunks, it is detected by the first verification, and only the first chunk is re-executed. Similarly, if the error strikes the execution of the second chunk (as illustrated in the figure), it is detected by the second verification, and the first two chunks are re-executed. The entire frame of work needs to be re-executed only if the error strikes during the third chunk. Under the first-order approximation as in the analysis of Theorem 1, the average amount of work to re-execute is $(1 + 2 + 3)w/3 = 2w = 2W/3$, that is, the re-executed work overhead becomes $o_{rw} = 2\lambda/3$. On the contrary, in the first pattern of Fig. 2, the amount of work to re-execute is always W, because the error is never detected before the end of the pattern. Hence, the second pattern leads to a 33 % gain in the re-execution time. However, this comes at the price of three times as many verifications, that is, the error-free overhead becomes $o_{ef} = 3V + C$. This overhead is paid in every error-free execution, and may be an overkill if the verification mechanism is too costly.

This example shows that finding the best trade-off between error-free overhead (what is paid due to the resilience method, when there is no failure during execution) and execution time (when errors strike) is not a trivial task. The optimization problem can be stated as follows: given the cost of checkpointing C, recovery R, and verification V, what is the optimal pattern to minimize the (expectation of the) execution time? A pattern is composed of several work chunks, each followed by a verification, and the last chunk is always followed by both a verification and a checkpoint. Let m denote the number of chunks in the pattern, and let w_j denote the length of the j-th chunk for $1 \leq j \leq m$. Let $W = \sum_{j=1}^{m} w_j$. We define $\beta_j = w_j/W$ be the relative length of the j-th chunk so that $\beta_j \geq 0$ and $\sum_{j=1}^{m} \beta_j = 1$. We let $\boldsymbol{\beta} = [\beta_1, \beta_2, \ldots, \beta_m]$. The goal is to determine the pattern work length W, the number of chunks m as well as the relative length vector $\boldsymbol{\beta}$.

Proposition 2 *The expected execution time of the above pattern is*

$$\mathbb{E}(P) = W + mV + C + \left(\lambda \boldsymbol{\beta}^T \mathbf{A} \boldsymbol{\beta}\right) W^2 + O(\sqrt{\lambda}), \tag{6}$$

where $\mathbf{A}$ *is an* $m \times m$ *matrix whose diagonal coefficients are equal to 1 and whose other coefficients are all equal to* $\frac{1}{2}$.

Proof Let $p_j = 1 - e^{-\lambda w_j}$ denote the probability of having at least one silent error in chunk j. To derive the expected execution time of the pattern, we need to know the probability q_j that the chunk j actually gets executed in the current attempt.

The first chunk is always executed, so we have $q_1 = 1$. Consider the second chunk, which is executed if no silent error strikes the first chunk, hence $q_2 = 1 - p_1$. In general, the probability that the j-th chunk gets executed is

$$q_j = \prod_{k=1}^{j-1}(1 - p_k).$$

Now, we are ready to compute the expected execution time of the pattern. The following gives the recursive expression:

$$\mathbb{E}(P) = \left(\prod_{k=1}^{m}(1 - p_k)\right) C$$
$$+ \left(1 - \prod_{k=1}^{m}(1 - p_k)\right)(R + \mathbb{E}(P))$$
$$+ \sum_{j=1}^{m} q_j(w_j + V). \tag{7}$$

Specifically, line 1 of Eq. (7) shows that the checkpoint at the end of the pattern is performed only when there has been no silent error in any of the chunks. Otherwise, we need to re-execute the pattern, after a recovery, as shown in line 2. Finally, line 3 shows the condition for each chunk j to be executed. By simplifying Eq. (7) and approximating the expression up to the second-order term, as in the proof of Proposition 1, we obtain

$$\mathbb{E}(P) = W + mV + C + \lambda f W^2 + O(\sqrt{\lambda}),$$

where $f = \sum_{j=1}^{m} \beta_j \left(\sum_{k=j}^{m} \beta_k\right)$, and it can be concisely written as $f = \boldsymbol{\beta}^T \mathbf{M} \boldsymbol{\beta}$, where $\mathbf{M}$ is the $m \times m$ matrix given by

$$m_{i,j} = \begin{cases} 1 & \text{for } i \leq j \\ 0 & \text{for } i > j \end{cases}.$$

By replacing $\mathbf{M}$ by its symmetric part $\mathbf{A} = \frac{\mathbf{M}+\mathbf{M}^T}{2}$, which does not affect the value of f, we obtain the matrix $\mathbf{A}$ whose diagonal coefficients are equal to 1 and whose other coefficients are all equal to $\frac{1}{2}$, and the expected execution time in Eq. (6).   □

Theorem 2 *The optimal pattern has m^* equal-length chunks, total length W^* and is such that:*

$$W^* = \sqrt{\frac{m^* V + C}{\frac{1}{2}\left(1 + \frac{1}{m^*}\right)\lambda}} , \tag{8}$$

$$\beta_j^* = \frac{1}{m^*} \text{ for } 1 \le j \le m^* , \tag{9}$$

where m^* is either $\max(1, \lfloor \bar{m}^* \rfloor)$ or $\lceil \bar{m}^* \rceil$ with

$$\bar{m}^* = \sqrt{\frac{C}{V}} . \tag{10}$$

The optimal expected overhead is

$$H^*(\mathrm{P}) = \sqrt{2\lambda\, C} + \sqrt{2\lambda\, V} + O(\lambda) . \tag{11}$$

Proof Given the number of chunks m with $\sum_{j=1}^{m} \beta_j = 1$, the function $f = \boldsymbol{\beta}^T \mathbf{A} \boldsymbol{\beta}$ is shown to be minimized [10, Theorem 1 with $r = 1$] when $\boldsymbol{\beta}$ follows Eq. (9), and its minimum value is given by $f^* = \frac{1}{2}\left(1 + \frac{1}{m}\right)$. We derive the two types of overheads as follows:

$$o_{\mathrm{ef}} = mV + C ,$$

$$o_{\mathrm{rw}} = \frac{1}{2}\left(1 + \frac{1}{m}\right)\lambda .$$

The optimal work length $W^* = \sqrt{\frac{o_{\mathrm{ef}}}{o_{\mathrm{rw}}}}$ for any fixed m is thus given by Eq. (8). The optimal number of chunks $\bar{m}^*$ shown in Eq. (10) is obtained by minimizing $F(m) = o_{\mathrm{ef}} \times o_{\mathrm{rw}}$. The number of chunks in a pattern can only be a positive integer, so m^* is either $\max(1, \lfloor \bar{m}^* \rfloor)$ or $\lceil \bar{m}^* \rceil$, since $F(m)$ is a convex function of m. Finally, substituting Eq. (10) back into $H^*(\mathrm{P}) = 2\sqrt{o_{\mathrm{ef}} \times o_{\mathrm{rw}}} + O(\lambda)$ gives rise to the optimal expected overhead as shown in Eq. (11).   □

4   Linear Workflows

For an application composed of a chain of tasks, the problem of finding the optimal checkpoint strategy, i.e., of determining which tasks to checkpoint, in order to minimize the expected execution time when subject to fail-stop failures, has been solved by Toueg and Babaoglu [36], using a dynamic programming algorithm. We revisit the problem for silent errors by exploiting verification in addition to checkpoints. An extended presentation of the results is available in [3, 5].

4.1 Setup

To deal with silent errors, resilience is provided through the use of checkpointing coupled with an error detection (or verification) mechanism. When a silent error is detected, we roll back to the nearest checkpoint and recover from there. As in Sect. 3.1, let C denote the cost of checkpointing, R the cost of recovery, and V the cost of a verification.

We consider a chain of tasks $T_1, T_2, \ldots, T_n$, where each task T_i has a weight w_i corresponding to the computational load. For notational convenience, we also define $W_{i,j} = \sum_{k=i+1}^{j} w_k$ to be the time to execute tasks T_{i+1} to T_j for any $i \leq j$. Once again we assume that silent errors occur following a *Poisson process* with arrival rate λ and that the probability of having at least one error during the execution of $W_{i,j}$ is given by $p_{i,j} = 1 - e^{-\lambda W_{i,j}}$.

We enforce that a verification is always taken immediately before each checkpoint, so that all checkpoints are valid, and hence only one checkpoint needs to be maintained at any time during the execution of the application. Furthermore, we assume that errors only strike the computations, while verifications, checkpoints, and recoveries are failure-free.

The goal is to find which task to verify and which task to checkpoint in order to minimize the expected execution time of the task chain. To solve this problem, we derive a two-level dynamic programming algorithm. For convenience, we add a virtual task T_0, which is always checkpointed, and whose recovery cost is zero. This accounts for the fact that it is always possible to restart the application from scratch at no extra cost. In the following, we describe the general scheme when considering both verifications and checkpoints.

4.2 Dynamic Programming

Figures 4 and 5 illustrate the idea of the algorithm, which contains two dynamic programming levels, responsible for placing checkpoints and verifications, respectively, as well as an additional step to compute the expected execution time between two verifications. The following describes each step of the algorithm in detail.

Placing checkpoints. The first level focuses on the placement of verified checkpoints, i.e., checkpoints preceded immediately by a verification. Let $E_{ckpt}(c_2)$ denote the expected time to successfully execute all the tasks from T_1 to T_{c_2}, where T_{c_2} is verified and checkpointed. Now, to find the last verified checkpoint before T_{c_2}, we try all possible locations from T_0 to T_{c_2-1}. For each location, say c_1, we call the function

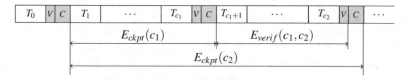

Fig. 4 First level of dynamic programming (E_{ckpt})

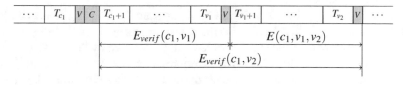

Fig. 5 Second level of dynamic programming (E_{verif}) and computation of expected execution time between two verifications (E)

recursively with $E_{ckpt}(c_1)$ (for placing checkpoints before T_{c_1}), and compute the expected time to execute the tasks from T_{c_1+1} to T_{c_2}. The latter is done through $E_{verif}(c_1, c_2)$, which also decides where to place additional verifications between T_{c_1+1} and T_{c_2}. Finally, we add the checkpointing cost C (after T_{c_2}) to $E_{ckpt}(c_2)$. Overall, we can express $E_{ckpt}(c_2)$ as follows:

$$E_{ckpt}(c_2) = \min_{0 \leq c_1 < c_2} \{E_{ckpt}(c_1) + E_{verif}(c_1, c_2) + C\}.$$

Note that a location $c_1 = 0$ means that no further checkpoints are added. In this case, we simply set $E_{ckpt}(0) = 0$, which initializes the dynamic program. The total expected time to execute all the tasks from T_1 to T_n is thus given by $E_{ckpt}(n)$.

Placing additional verifications. The second level decides where to insert additional verifications between two tasks with verified checkpoints. The function is initially called from the first level between two checkpointed tasks T_{c_1} and T_{c_2}, each of which also comes with a verification. Therefore, we define $E_{verif}(c_1, v_2)$ as the expected time to successfully execute all the tasks from T_{c_1+1} to T_{v_2}, knowing that the last checkpoint is right after task T_{c_1}, and there is no additional checkpoint between T_{c_1+1} and T_{v_2}. Note that $E_{verif}(c_1, v_2)$ accounts only for the time required to execute and verify these tasks. As before, we try all possible locations for the last verification between T_{c_1} and T_{v_2} and, for each location v_1, we call the function recursively with $E_{verif}(c_1, v_1)$. Furthermore, we add the expected time needed to successfully execute the tasks T_{v_1+1} to T_{v_2}, denoted by $E(c_1, v_1, v_2)$, given the position c_1 of the last checkpoint. Overall, we can express $E_{verif}(c_1, v_2)$ as follows:

$$E_{verif}(c_1, v_2) = \min_{c_1 \leq v_1 < v_2} \{E_{verif}(c_1, v_1) + E(c_1, v_1, v_2)\}. \tag{12}$$

Again, the case $v_1 = c_1$ means that no further verification is added, so we initialize the dynamic program with $E_{verif}(c_1, c_1) = 0$. Note that the verification cost V at the end of task T_{v_2} will be accounted for in the function $E(c_1, v_1, v_2)$.

Computing expected execution time between two verifications. Finally, to compute the expected time to successfully execute several tasks between two verifications, we need the position of the last checkpoint c_1, as well as the positions of the two verifications v_1 and v_2.

First, we pay W_{v_1, v_2} by executing all the tasks from T_{v_1+1} to T_{v_2}, followed by the cost of verification V after T_{v_2}. During the execution, there is a probability $p_{v_1, v_2} = 1 - e^{-\lambda W_{v_1, v_2}}$ of having a silent error, which will be detected by the verification after T_{v_2}. In this case, we need to perform a recovery from the last checkpoint after T_{c_1} with a cost R (set to 0 if $c_1 = 0$), and re-execute the tasks from there by calling the function $E_{verif}(c_1, v_1)$ followed by $E(c_1, v_1, v_2)$. Therefore, we can express $E(c_1, v_1, v_2)$ as follows:

$$E(c_1, v_1, v_2) = W_{v_1, v_2} + V + p_{v_1, v_2} \left(R + E_{verif}(c_1, v_1) + E(c_1, v_1, v_2) \right). \quad (13)$$

Simplifying Eq. (13), we get

$$E(c_1, v_1, v_2) = e^{\lambda W_{v_1, v_2}} \left(W_{v_1, v_2} + V \right) + \left(e^{\lambda W_{v_1, v_2}} - 1 \right) \left(R + E_{verif}(c_1, v_1) \right).$$

Complexity. The complexity is dominated by the computation of the expected completion time table $E_{verif}(c_1, v_2)$, which contains $O(n^2)$ entries, and each entry depends on at most n other entries that are already computed. All tables are computed in a bottom-up fashion, from the left to the right of the task chain. Hence, the overall complexity of the algorithm is $O(n^3)$.

5   ABFT and Checkpointing for Linear Algebra Kernels

In this section we introduce ABFT (Algorithm Based Fault Tolerance) as an application-specific technique which allows for both error detection and correction. We streamline our discussion on the CG method, however, the techniques that we describe are applicable to any iterative solver that uses sparse matrix vector multiplies and vector operations. This list includes many of the non-stationary iterative solvers such as CGNE (Conjugate Gradient on Normal Equations), BiCG (Bi-Conjugate Gradient), BiCGstab (Bi-Conjugate Gradient Stabilized), where sparse matrix transpose vector multiply operations also take place. Preconditioned variants of these solvers with an approximate inverse preconditioner (applied as an SpMxV, or two SpMxVs) can also be made fault-tolerant with the proposed scheme. The extension to PCG is described in [19].

In Sect. 5.1, we first provide a background on the CG method and give an overview of both Chen's stability tests [12] and ABFT protection schemes. Then we detail ABFT techniques for the SpMxV kernel.

Algorithm 1 The Conjugate Gradient algorithm for a positive definite matrix $\mathbf{A}$.

Input: $\mathbf{A} \in \mathbb{R}^{n \times n}, \mathbf{b}, \mathbf{x}_0 \in \mathbb{R}^n, \varepsilon \in \mathbb{R}$
Output: $\mathbf{x} \in \mathbb{R}^n : \|\mathbf{A}\mathbf{x} - \mathbf{b}\| \leq \varepsilon$
1: $\mathbf{r}_0 \leftarrow \mathbf{b} - \mathbf{A}\mathbf{x}_0$;
2: $\mathbf{p}_0 \leftarrow \mathbf{r}_0$;
3: $i \leftarrow 0$;
4: **while** $\|\mathbf{r}_i\| > \varepsilon (\|\mathbf{A}\| \cdot \|\mathbf{r}_0\| + \|\mathbf{b}\|)$ **do**
5: $\quad \mathbf{q} \leftarrow \mathbf{A}\mathbf{p}_i$;
6: $\quad \alpha_i \leftarrow \|\mathbf{r}_i\|^2 / \mathbf{p}_i^\mathsf{T} \mathbf{q}$;
7: $\quad \mathbf{x}_{i+1} \leftarrow \mathbf{x}_i + \alpha \mathbf{p}_i$;
8: $\quad \mathbf{r}_{i+1} \leftarrow \mathbf{r}_i - \alpha \mathbf{q}$;
9: $\quad \beta \leftarrow \|\mathbf{r}_{i+1}\|^2 / \|\mathbf{r}_i\|^2$;
10: $\quad \mathbf{p}_{i+1} \leftarrow \mathbf{r}_{i+1} + \beta \mathbf{p}_i$;
11: $\quad i \leftarrow i + 1$;
12: **end while**
13: **return** $\mathbf{x}_i$;

5.1 CG and Fault Tolerance Mechanisms

The code for the CG method is shown in Algorithm 1. The main loop features a sparse matrix-vector multiply, two inner products (for $\mathbf{p}_i^\mathsf{T}\mathbf{q}$ and $\|\mathbf{r}_{i+1}\|^2$), and three vector operations of the form $axpy$.

Chen's stability tests [12] amount to checking the orthogonality of vectors $\mathbf{p}_{i+1}$ and $\mathbf{q}$, at the price of computing $(\mathbf{p}_{i+1}^\mathsf{T}\mathbf{q})/(\|\mathbf{p}_{i+1}\| \|\mathbf{q}_i\|)$, and to checking the residual at the price of an additional SpMxV operation $\mathbf{A}\mathbf{x}_i - \mathbf{b}$. The dominant cost of these verifications is the additional SpMxV operation.

We investigate three fault tolerance mechanisms. The first one is ONLINE- DETECTION; this is Chen's original approach modified to save the matrix $\mathbf{A}$ in addition to the current iteration vectors. This is needed when a silent error is detected: if this error comes for a corruption in data memory, we need to recover with a valid copy of the data matrix $\mathbf{A}$. The second one is ABFT- DETECTION, which detects errors and restarts from the most recent checkpoint. The thirds one is ABFT-CORRECTION, which detects errors and corrects if there was only one, otherwise restarts from the last checkpoint. The three methods under the study keep a valid copy of $\mathbf{A}$ and have exactly the same checkpoint cost.

We now introduce the ingredients of our own protection and verification mechanisms ABFT-DETECTION and ABFT-CORRECTION. We use ABFT techniques to protect the SpMxV, its result (hence the vector $\mathbf{q}$), the matrix $\mathbf{A}$ and the input vector

$\mathbf{p}_i$. As ABFT methods for vector operations is as costly as a repeated computation, we use TMR for them for simplicity. That is we do not protect $\mathbf{p}_i$, $\mathbf{q}$, $\mathbf{r}_i$, and $\mathbf{x}_i$ of the ith loop beyond the SpMxV at line 5 with ABFT, but we compute the dots, norms and $axpy$ operations in resilient mode.

Although theoretically possible, constructing ABFT mechanism to detect up to k errors is practically not feasible for $k > 2$. The same mechanism can be used to correct up to $\lfloor k/2 \rfloor$ errors. Therefore, we focus on detecting up to two errors and correcting single errors. That is, we detect up to two errors in the computation $\mathbf{q} \leftarrow \mathbf{A}\mathbf{p}_i$ (two entries in $\mathbf{q}$ are faulty), or in $\mathbf{p}_i$, or in the sparse representation of the matrix $\mathbf{A}$. With TMR, we assume that the errors in the computation are not overly frequent so that two results out of three are correct (we assume errors do not strike the vector data here). Our fault-tolerant CG versions thus have the following ingredients: ABFT to detect up to two errors in the SpMxV and correct up to one; TMR for vector operations; and checkpoint and roll-back in case errors are not corrected. In the rest of this section, we discuss the proposed ABFT method for the SpMxV (combining ABFT with checkpointing is later in Sect. 5.3).

5.2 ABFT-SpMxV

The overhead of the standard single error correcting ABFT technique is too high for the sparse matrix-vector product case. Shantaram et al. [35] propose a cheaper ABFT SpMxV algorithm that guarantees detection of single errors striking either the computation or the memory representation of the two input operands (matrix and vector). As their results depend on the sparse storage format adopted, throughout this section we assume that sparse matrices are stored in the compressed storage format by rows (CSR) format [32, Sect. 3.4], that is by means of three distinct arrays, namely $Colid \in \mathbb{N}^{nnz(\mathbf{A})}$, $Val \in \mathbb{R}^{nnz(\mathbf{A})}$ and $Rowidx \in \mathbb{N}^{n+1}$.

Shantaram et al. can protect $\mathbf{y} \leftarrow \mathbf{A}\mathbf{x}$, where $\mathbf{A} \in \mathbb{R}^{n \times n}$ and $\mathbf{x}, \mathbf{y} \in \mathbb{R}^n$. To perform error detection, they rely on a column checksum vector $\mathbf{c}$ defined by

$$c_j = \sum_{i=1}^{n} a_{i,j} \tag{14}$$

and an auxiliary copy $\mathbf{x}'$ of the $\mathbf{x}$ vector. After having performed the actual SpMxV, to validate the result it suffices to compute $\sum_{i=1}^{n} y_i$, $\mathbf{c}^\mathsf{T}\mathbf{x}$ and $\mathbf{c}^\mathsf{T}\mathbf{x}'$, and to compare their values. It can be shown [35] that in the case of no errors, these three quantities carry the same value, whereas if a single error strikes either the memory or the computation, one of them must differ from the other two. Nevertheless, this method requires $\mathbf{A}$ to be strictly diagonally dominant, that seems to restrict too much the practical applicability of their ABFT scheme. Shantaram et al. need this condition to ensure the detection of errors striking an entry of $\mathbf{x}$ corresponding to a zero checksum column of $\mathbf{A}$. We further analyze that case and show how to overcome the issue without imposing any restriction on $\mathbf{A}$.

A nice way to characterize the problem is expressing it in geometrical terms. Let us consider the computation of a single entry of the checksum as

$$(\mathbf{w}^\mathsf{T}\mathbf{A})_j = \sum_{i=1}^{n} w_i a_{i,j} = \mathbf{w}^\mathsf{T}\mathbf{A}^j,$$

where $\mathbf{w} \in \mathbb{R}^n$ denotes the weight vector and $\mathbf{A}^j$ the j-th column of $\mathbf{A}$. Let us now interpret such an operation as the result of the scalar product $\langle \cdot, \cdot \rangle : \mathbb{R}^n \times \mathbb{R}^n \to \mathbb{R}$ defined by $\langle \mathbf{u}, \mathbf{v} \rangle \mapsto \mathbf{u}^\mathsf{T}\mathbf{v}$. It is clear that a checksum entry is zero if and only if the corresponding column of the matrix is orthogonal to the weight vector. In (14), we have chosen $\mathbf{w}$ to be such that $w_i = 1$ for $1 \le i \le n$, in order to make the computation easier. Let us see now what happens without this restriction.

The problem reduces to finding a vector $\mathbf{w} \in \mathbb{R}^n$ that is not orthogonal to any vector out of a basis $\mathscr{B} = \{\mathbf{b}_1, \ldots, \mathbf{b}_n\}$ of $\mathbb{R}^n$—the rows of the input matrix. Each one of these n vectors is perpendicular to a hyperplane h_i of $\mathbb{R}^n$, and $\mathbf{w}$ does not verify the condition

$$\langle \mathbf{w}, \mathbf{b}_i \rangle \ne 0, \tag{15}$$

for any i, if and only if it lies on h_i. As the Lebesgue measure in $\mathbb{R}^n$ of an hyperplane of $\mathbb{R}^n$ itself is zero, the union of these hyperplanes is measurable with $m_n \left(\bigcup_{i=1}^{n} h_i \right) = 0$, where m_n denotes the Lebesgue measure of $\mathbb{R}^n$. Therefore, the probability that a vector $\mathbf{w}$ randomly picked in $\mathbb{R}^n$ does not satisfy condition (15) for any i is zero.

Nevertheless, there are many reasons to consider zero checksum columns. First of all, when working with finite precision, the number of elements in $\mathbb{R}^n$ one can have is finite, and the probability of randomly picking a vector that is orthogonal to a given one could be bigger than zero. Moreover, a coefficient matrix usually comes from the discretization of a physical problem, and the distribution of its columns cannot be considered as random. Finally, using a randomly chosen vector instead of $(1, \ldots, 1)^\mathsf{T}$ increases the number of required floating point operations, causing a growth of both execution time and rounding errors. Therefore, we would like to keep $\mathbf{w} = (1, \ldots, 1)^\mathsf{T}$ as the vector of choice, in which case we need to protect SpMxV with matrices having zero column sums. There are many matrices with this property, for example the Laplacian matrices of graphs [13, Chap. 1].

Algorithm 2 ABFT-protected SpMxV, detection of 2 errors, correction of 1 error

---

Input: $\mathbf{A} \in \mathbb{R}^{n \times n}$ (as $Val \in \mathbb{R}^{nnz(\mathbf{A})}, Colid \in \mathbb{N}^{nnz(\mathbf{A})}, Rowidx \in \mathbb{R}^n$), $\mathbf{x} \in \mathbb{R}^n$
Output: $\mathbf{y} = \mathbf{A}\mathbf{x}$, correction of single error or detection of double error
1: global $\mathbf{W}^\mathsf{T} \leftarrow [\begin{smallmatrix} 1 & 1 & \cdots & 1 \\ 1 & 2 & \cdots & n \end{smallmatrix}] \in \mathbb{R}^{2 \times n}$;
2: global $\underline{\mathbf{W}}^\mathsf{T} \leftarrow [\mathbf{W}^\mathsf{T} \begin{smallmatrix} 1 \\ n+1 \end{smallmatrix}] \in \mathbb{R}^{2 \times n+1}$;
3: $\mathbf{x}' \leftarrow \mathbf{x}$;
4: $[\mathbf{C}, \mathbf{M}, \mathbf{c}_r, \mathbf{c}_x] = \text{COMPUTECHECKSUMS}(Val, Colid, Rowidx)$;
5: **return** SPMxV($Val, Colid, Rowidx, \mathbf{x}, \mathbf{x}', \mathbf{M}, \mathbf{c}_r, \mathbf{c}_x$);

6: **function** COMPUTECHECKSUMS($Val, Colid, Rowidx$)
7:     $\mathbf{C}^\mathsf{T} \leftarrow \mathbf{W}^\mathsf{T}\mathbf{A}$;
8:     $\mathbf{M} \leftarrow \mathbf{W} - \mathbf{C}$;
9:     $\mathbf{c}_r \leftarrow \underline{\mathbf{W}}^\mathsf{T} Rowidx$;
10:    $\mathbf{c}_x \leftarrow \mathbf{W}^\mathsf{T}\mathbf{x}$;
11:    **return** $\mathbf{C}, \mathbf{M}, \mathbf{c}_r, \mathbf{c}_x$;

12: **function** SPMxV($Val, Colid, Rowidx, \mathbf{x}, \mathbf{x}', \mathbf{C}, \mathbf{M}, \mathbf{c}_r, \mathbf{c}_x$)
13:    $\mathbf{s}_r \leftarrow 0 \in \mathbb{R}^{2 \times 1}$;
14:    **for** $i \leftarrow 1$ to n **do**
15:       $y_i \leftarrow 0$;
16:       $\mathbf{s}_r \leftarrow \mathbf{s}_r + [\begin{smallmatrix} w_{1,i} \\ w_{2,i} \end{smallmatrix}] Rowidx_i$;
17:       **for** $j \leftarrow Rowidx_i$ to $Rowidx_{i+1} - 1$ **do**
18:          $ind \leftarrow Colid_j$;
19:          $y_i \leftarrow y_i + Val_j \cdot x_{ind}$;
20:    $\mathbf{d}_r = \mathbf{c}_r - \mathbf{s}_r$;
21:    $\mathbf{d}_x = \mathbf{W}^\mathsf{T}\mathbf{y} - \mathbf{C}^\mathsf{T}\mathbf{x}$;
22:    $\mathbf{d}_{x'} = \mathbf{W}^\mathsf{T}(\mathbf{x}' - \mathbf{y}) - \mathbf{M}^\mathsf{T}\mathbf{x}$;
23:    **if** $\mathbf{d}_r = 0 \wedge \mathbf{d}_x = 0 \wedge \mathbf{d}_{x'} = 0$ **then**
24:       **return** $\mathbf{y}$;
25:    **else**
26:       CORRECTERRORS($Val, Colid, Rowidx, \mathbf{x}, \mathbf{x}', \mathbf{C}, \mathbf{M}, \mathbf{d}_r, \mathbf{d}_x, \mathbf{d}_{x'}, \mathbf{c}_r, \mathbf{c}_x$);

In Algorithm 2, we propose an ABFT SpMxV method that uses weighted checksums and does not require the matrix to be strictly diagonally dominant. The idea is to compute the checksum vector and then shift it by adding to all of its entries a constant value chosen so that all of the elements of the new vector are different from zero. We give the result in Theorem 3 for the simpler case of single error detection without correction, in which case Algorithm 2 has $\mathbf{W} = (1, \ldots, 1)^\mathsf{T}$ at line 1 and raises an error at line 26 (instead of correcting the error) if the tests at line 23 are not passed. The cases of multiple error detection and single error correction are proved in a technical report [20, Sect. 3.2].

Theorem 3 (Correctness of Algorithm 2 for error detection) *Let $\mathbf{A} \in \mathbb{R}^{n \times n}$ be a square matrix, let $\mathbf{x}, \mathbf{y} \in \mathbb{R}^n$ be the input and output vector respectively, and let $\mathbf{x}' = \mathbf{x}$. Let us assume that the algorithm performs the computation*

$$\widetilde{\mathbf{y}} \leftarrow \widetilde{\mathbf{A}}\widetilde{\mathbf{x}}, \tag{16}$$

where $\widetilde{\mathbf{A}} \in \mathbb{R}^{n \times n}$ *and* $\widetilde{\mathbf{x}} \in \mathbb{R}^{n}$ *are the possibly faulty representations of* $\mathbf{A}$ *and* $\mathbf{x}$ *respectively, while* $\widetilde{\mathbf{y}} \in \mathbb{R}^{n}$ *is the possibly erroneous result of the sparse matrix-vector product. Let us also assume that the encoding scheme relies on*

1. *an auxiliary checksum vector* $\mathbf{c} = \left[\sum_{i=1}^{n} a_{i,1} + k, \ldots, \sum_{i=1}^{n} a_{i,n} + k \right]$, *where* k *is such that* $\sum_{i=1}^{n} a_{i,j} + k \neq 0$ *for* $1 \leq j \leq n$,
2. *an auxiliary checksum* $y_{n+1} = k \sum_{i=i}^{n} \widetilde{x}_i$,
3. *an auxiliary counter* $\mathbf{s}_r$ *initialized to 0 and updated at runtime by adding the value of the hit element each time the Rowidx array is accessed*,
4. *an auxiliary checksum* $\mathbf{c}_r = \sum_{i=1}^{n} Rowidx_i \in \mathbb{N}$.

Then, a single error in the computation of the SpMxV causes one of the following conditions to fail:

    i. $\mathbf{c}^{\mathsf{T}} \widetilde{\mathbf{x}} = \sum_{i=1}^{n+1} \widetilde{y}_i$, *difference is in* $\mathbf{d}_x$ *at line 21*,
    ii. $\mathbf{c}^{\mathsf{T}} \mathbf{x}' = \sum_{i=1}^{n+1} \widetilde{y}_i$, *difference is in* $\mathbf{d}_{x'}$ *at line 22;*
    iii. $\mathbf{s}_r = \mathbf{c}_r$, *difference is in* $\mathbf{d}_r$ *at line 20.*

The proof of this theorem is technical and is available elsewhere [20, Theorem 1].

The function COMPUTECHECKSUM in Algorithm 2 requires just the knowledge of the matrix. Hence in the common scenario of many SpMxVs with the same matrix, it is enough to invoke it once to protect several matrix-vector multiplications. This observation will be crucial when discussing the performance of the checksumming techniques.

Extensions to $k \geq 2$ errors are discussed elsewhere [20, Section 3.2], where the following are detailed. The method just described can be extended to detect up to a total of k errors anywhere in the computation, in the representation of $\mathbf{A}$, or in the vector $\mathbf{x}$. Building up the necessary structures requires $\mathcal{O}(knnz(\mathbf{A}))$ time, and the overhead per SpMxV is $O(kn)$. For the particular case of $k = 2$ a result similar to that in Theorem 3 is also shown.

We now discuss error correction. If at least one of the tests at line 23 of Algorithm 2 fails, the algorithm invokes CORRECTERRORS in order to determine whether just one error struck either the computation or the memory and, in that case, to correct it. Indeed, whenever a single error is detected, disregarding its location (i.e., computation or memory), it can be corrected by means of a succession of various steps, as explained below; if need be, partial recomputations of the result are performed.

Specifically, we proceed as follows. To detect errors striking *Rowidx*, we compute the ratio d of the second component of $\mathbf{d}_r$ to the first one, and check whether its distance from an integer is smaller than a certain threshold parameter ε. If it is so, the algorithm concludes that the d-th element of *Rowidx* is faulty, performs the correction by subtracting the first component of $\mathbf{d}_r$ to *Rowidx*$_d$, and recomputes y_d and y_{d-1}, if the error in *Rowidx*$_d$ is a decrement; or y_{d+1} if it was an increment. Otherwise, it just emits an error.

The correction of errors striking *Val, Colid* and the computation of y are corrected together. Let now d be the ratio of the second component of $\mathbf{d}_x$ to the first one. If d is near enough to an integer, the algorithm computes the checksum matrix

$\mathbf{C}' = \mathbf{W}^\mathsf{T}\mathbf{A}$ and considers the number $z_{\widetilde{\mathbf{C}}}$ of non-zero columns of the difference matrix $\widetilde{\mathbf{C}} = | \mathbf{C} - \mathbf{C}' |$. At this stage, three cases are possible:

- If $z_{\widetilde{\mathbf{C}}} = 0$, then the error is in the computation of y_d, and can be corrected by simply recomputing this value.
- If $z_{\widetilde{\mathbf{C}}} = 1$, then the error concerns an element of *Val*. Let us call f the index of the non-zero column of $\widetilde{\mathbf{C}}$. The algorithm finds the element of *Val* corresponding to the entry at row d and column f of A and corrects it by using the column checksums much like as described for *Rowidx*. Afterwards, y_d is recomputed to fix the result.
- If $z_{\widetilde{\mathbf{C}}} = 2$, then the error concerns an element of *Colid*. Let us call f_1 and f_2 the index of the two non-zero columns and m_1, m_2 the first and last elements of *Colid* corresponding to non-zeros in row d. It is clear that there exists exactly one index m^* between m_1 and m_2 such that either $Colid_{m^*} = f_1$ or $Colid_{m^*} = f_2$. To correct the error it suffices to switch the current value of $Colid_{m^*}$, i.e., putting $Colid_{m^*} = f_2$ in the former case and $Colid_{m^*} = f_1$ in the latter. Again, y_d has to be recomputed.
- if $z_{\widetilde{\mathbf{C}}} > 2$, then errors can be detected but not corrected, and an error is emitted.

To correct errors striking $\mathbf{x}$, the algorithm computes d, that is the ratio of the second component of $\mathbf{d}_{x'}$ to the first one, and checks that the distance between d and the nearest integer is smaller than ε. Provided that this condition is verified, the algorithm computes the value of the error $\tau = \sum_{i=1}^{n} x_i - cx_1$ and corrects $x_d = x_d - \tau$. The result is updated by subtracting from $\mathbf{y}$ the vector $\mathbf{y}^\tau = \mathbf{A}\mathbf{x}^\tau$, where $\mathbf{x}^\tau \in \mathbb{R}^{n \times n}$ is such that $x_d^\tau = \tau$ and $x_i^\tau = 0$ otherwise.

Finally, note that double errors could be shadowed when using Algorithm 2, but the probability of such an event is negligible. Still, there exists an improved version which avoids this issue by adding a third checksum [20, Sect. 3.2].

5.3 Performance Model

The performance model is a simplified instance of the one discussed in Sect. 4, and we instantiate it for the three methods that we are considering, namely ONLINE-DETECTION, ABFT-DETECTION and ABFT-CORRECTION. We have a linear chain of identical tasks, where each task corresponds to one or several CG iterations. We execute T units of work followed by a verification, which we call a *chunk*, and we repeat this scheme s times, i.e., we compute s chunks, before taking a checkpoint. We say that the s chunks constitute a *frame*. The whole execution is then partitioned into frames. We assume that the checkpoint, recovery and verification operations are error-free. For each method below, we let C, R and V be the respective cost of these operations. Finally, and as before, assume a Poisson process for errors and let q be the probability of successful execution for each chunk: $q = e^{-\lambda T}$, where λ is the fault rate.

5.3.1 ONLINE-DETECTION

For Chen's method [12], we have the following parameters:

- We have chunks of d iterations, hence $T = dT_{iter}$, where T_{iter} is the raw cost of a CG iteration without any resilience method.

- The verification time V is the cost of the operations described in Sect. 5.1.

- As for silent errors, the application is protected from arithmetic errors in the ALU, as in Chen's original method, but also for corruption in data memory (because we also checkpoint the matrix **A**). Let λ_a be the rate of arithmetic errors, and λ_m be the rate of memory errors. For the latter, we have $\lambda_m = M\lambda_{word}$ if the data memory consists of M words, each susceptible to be corrupted with rate λ_{word}. Altogether, since the two error sources are independent, they have a cumulated rate of $\lambda = \lambda_a + \lambda_m$, and the success probability for a chunk is $q = e^{-\lambda T}$. The optimal values of d and s can be computed by the same method as in Sect. 4.

5.3.2 ABFT-DETECTION

When using ABFT techniques, we detect possible errors every iteration, so a chunk is a single iteration, and $T = T_{iter}$. For ABFT-DETECTION, V is the overhead due to the checksums and redundant operations to detect a single error in the method.

ABFT-DETECTION can protect the application from the same silent errors as ONLINE- DETECTION, and just as before the success probability for a chunk (a single iteration here) is $q = e^{-\lambda T}$.

5.3.3 ABFT-CORRECTION

In addition to detection, we now correct single errors at every iteration. Just as for ABFT-DETECTION, a chunk is a single iteration, and $T = T_{iter}$, but V corresponds to a larger overhead, mainly due to the extra checksums needed to detect two errors and correct a single one.

The main difference lies in the error rate. An iteration with ABFT-CORRECTION is successful if zero or one error has struck during that iteration, so that the success probability is much higher than for ONLINE- DETECTION and ABFT-DETECTION. We compute that value of the success probability as follows. We have a Poisson process of rate λ, where $\lambda = \lambda_a + \lambda_m$ as for ONLINE- DETECTION and ABFT-DETEC-TION. The probability of exactly k errors in time T is $\frac{(\lambda T)^k}{k!}e^{-\lambda T}$ [27], hence the probability of no error is $e^{-\lambda T}$ and the probability of exactly one error is $\lambda Te^{-\lambda T}$, so that $q = e^{-\lambda T} + \lambda Te^{-\lambda T}$.

5.4 Experiments

Comprehensive tests were performed and reported in the technical report [20]. The main observation is that ABFT-CORRECTION outperforms both ONLINE- DETECTION and ABFT-DETECTION for a wide range of fault rates, thereby demonstrating that combining checkpointing with ABFT correcting techniques is more efficient than pure checkpointing for most practical situations.

6 Conclusion

Both fail-stop errors and silent data corruptions are major threats to executing HPC applications at scale. While many techniques have been advocated to deal with fail-stop errors, the lack of an efficient solution to handle silent errors is a real issue.

We have presented both a general-purpose solution and application-specific techniques to deal with silent data corruptions, with a focus on minimizing the overhead. For a divisible load application, we have extended the classical bound of Young/Daly to handle silent errors by combining checkpointing and verification mechanisms. For linear workflows, we have devised a polynomial-time dynamic programming algorithm that decides the optimal checkpointing and verification positions. Then, we have introduced ABFT as an application-specific technique to both detect and correct silent errors in iterative solvers that use sparse matrix vector multiplies and vector operations.

Our approach only addresses silent data corruptions. While several techniques have been developed to cope with either type of errors, few approaches are devoted to addressing both of them simultaneously. Hence, the next step is to extend our study to encompass both fail-stop and silent data corruptions in order to propose a comprehensive solution for executing applications on large scale platforms.

References

1. Aupy, G., Benoit, A., Hérault, T., Robert, Y., Vivien, F., Zaidouni, D.: On the combination of silent error detection and checkpointing. In: Proceedings of the 2013 International Symposium on Dependable Computing, pp. 11–20 (2013)
2. Bautista-Gomez, L., Benoit, A., Cavelan, A., Raina, S.K., Robert, Y., Sun, H.: Which verification for soft error detection? In: Proceedings of the 2015 International Conference on High Performance Computing (HiPC'2015). IEEE Computer Society Press (2015)
3. Benoit, A., Cavelan, A., Robert, Y., Sun, H.: Assessing general-purpose algorithms to cope with fail-stop and silent errors. In: Proceedings of the 5th International Workshop on Performance Modeling, Benchmarking and Simulation of High Performance Computer Systems (PMBS) (2014)
4. Benoit, A., Cavelan, A., Robert, Y., Sun, H.: Optimal resilience patterns to cope with fail-stop and silent errors. Research report RR-8786, INRIA (2015). http://graal.ens-lyon.fr/~yrobert/rr8786.pdf

5. Benoit, A., Cavelan, A., Robert, Y., Sun, H.: Two-level checkpointing and partial verifications for linear task graphs. In: Proceedings of the 6th International Workshop on Performance Modeling, Benchmarking and Simulation of High Performance Computer Systems (PMBS) (2015)

6. Benson, A.R., Schmit, S., Schreiber, R.: Silent error detection in numerical time-stepping schemes. Int. J. High Perform. Comput. Appl. (2014). doi:10.1177/1094342014532297

7. Bosilca, G., Delmas, R., Dongarra, J., Langou, J.: Algorithm-based fault tolerance applied to high performance computing. J. Parallel Distrib. Comput. 69(4), 410–416 (2009)

8. Bougeret, M., Casanova, H., Rabie, M., Robert, Y., Vivien, F.: Checkpointing strategies for parallel jobs. In: Proceedings of the 2011 International Conference for High Performance Computing, Networking, Storage and Analysis (SC), pp. 1–11 (2011)

9. Bronevetsky, G., de Supinski, B.: Soft error vulnerability of iterative linear algebra methods. In: Proceedings of the 2008 International Conference on Supercomputing (ICS), pp. 155–164 (2008)

10. Cavelan, A., Raina, S.K., Robert, Y., Sun, H.: Assessing the impact of partial verifications against silent data corruptions. In: Proceedings of the 44th International Conference on Parallel Processing (ICPP) (2015)

11. Chandy, K.M., Lamport, L.: Distributed snapshots: determining global states of distributed systems. ACM Trans. Comput. Syst. 3(1), 63–75 (1985)

12. Chen, Z.: Online-ABFT: An online algorithm based fault tolerance scheme for soft error detection in iterative methods. In: Proceedings of the 18th Symposium on Principles and Practice of Parallel Programming, pp. 167–176 (2013)

13. Chung, F.R.K.: Spectral Graph Theory. American Mathematical Society (1997)

14. Daly, J.T.: A higher order estimate of the optimum checkpoint interval for restart dumps. Future Gener. Comput. Syst. 22(3), 303–312 (2006)

15. Dongarra, J., et al.: The international exascale software project: a call to cooperative action by the global high-performance community. Int. J. High Perform. Comput. Appl. 23(4), 309–322 (2009)

16. Elliott, J., Kharbas, K., Fiala, D., Mueller, F., Ferreira, K., Engelmann, C.: Combining partial redundancy and checkpointing for HPC. In: Proceedings of the 2012 IEEE International Conference on Distributed Computing Systems (ICDCS), pp. 615–626 (2012)

17. Elnozahy, E.N.M., Alvisi, L., Wang, Y.-M., Johnson, D.B.: A survey of rollback-recovery protocols in message-passing systems. ACM Comput. Surv. 34, 375–408 (2002)

18. Engelmann, C., Ong, H.H., Scorr, S.L.: The case for modular redundancy in large-scale highh performance computing systems. In: Proceeding of the 8th IASTED Infernational Conference on Parallel and Distributed Computing and Networks (PDCN), pp. 189–194 (2009)

19. Fasi, M., Langou, J., Robert, Y., Uçar, B.: A backward/forward recovery approach for the preconditioned conjugate gradient method. Research report RR-8826, INRIA, 2015. http://graal.ens-lyon.fr/~yrobert/rr8826.pdf

20. Fasi, M., Robert, Y., Uçar, B.: Combining Algorithm-based Fault Tolerance and Checkpointing for Iterative Solvers. Research Report RR-8675, INRIA, 2015. Short version appears in the proceedings of PDSEC'2015

21. Ferreira, K., Stearley, J., Laros, J.H.I., Oldfield, R., Pedretti, K., Brightwell, R., Riesen, R., Bridges, P.G., Arnold, D.: Evaluating the viability of process replication reliability for exascale systems. In: Proceedings of the 2011 International Conference for High Performance Computing, Networking, Storage and Analysis, pp. 44:1–44:12 (2011)

22. Heroux, M., Hoemmen, M.: Fault-tolerant iterative methods via selective reliability. Research report SAND2011-3915 C, Sandia National Laboratories (2011)

23. Huang, K.-H., Abraham, J.A.: Algorithm-based fault tolerance for matrix operations. IEEE Trans. Comput. 33(6), 518–528 (1984)

24. Hwang, A.A., Stefanovici, I.A., Schroeder, B.: Cosmic rays don't strike twice: understanding the nature of DRAM errors and the implications for system design. ACM SIGARCH Comput. Archit. News 40(1), 111–122 (2012)

25. Lu, G., Zheng, Z., Chien, A.A.: When is multi-version checkpointing needed? In: Proceedings of the 3rd Workshop on Fault-tolerance for HPC at extreme scale (FTXS), pp. 49–56 (2013)
26. Lyons, R.E., Vanderkulk, W.: The use of triple-modular redundancy to improve computer reliability. IBM J. Res. Dev. **6**(2), 200–209 (1962)
27. Mitzenmacher, M., Upfal, E.: Probability and Computing: Randomized Algorithms and Probabilistic Analysis. Cambridge University Press (2005)
28. Moody, A., Bronevetsky, G., Mohror, K., B.R.d. Supinski. Design, modeling, and evaluation of a scalable multi-level checkpointing system. In: Proceedings of the 2010 ACM/IEEE International Conference for High Performance Computing, Networking, Storage and Analysis (SC'10) (2010)
29. Ni, X., Meneses, E., Jain, N., Kalé, L.V.: ACR: Automatic checkpoint/restart for soft and hard error protection. In: Proceedings of the 2013 ACM/IEEE International Conference for High Performance Computing, Networking, Storage and Analysis (SC'13). ACM (2013)
30. O'Gorman, T.: The effect of cosmic rays on the soft error rate of a DRAM at ground level. IEEE Trans. Electron Devices **41**(4), 553–557 (1994)
31. Ozaki, T., Dohi, T., Okamura, H., Kaio, N.: Distribution-free checkpoint placement algorithms based on min-max principle. IEEE Trans. Dependable Secure Comput. **3**(2), 130–140 (2006)
32. Saad, Y.: Iterative Methods for Sparse Linear Systems, 2nd edn. SIAM Press (2003)
33. Sao, P., Vuduc, R.: Self-stabilizing iterative solvers. In: Proceedings of the Workshop on Latest Advances in Scalable Algorithms for Large-Scale Systems (ScalA) (2013)
34. Schroeder, B., Gibson, G.: Understanding failures in petascale computers. J. Phys. Conf. Ser. **78**(1) (2007)
35. Shantharam, M., Srinivasmurthy, S., Raghavan, P.: Fault tolerant preconditioned conjugate gradient for sparse linear system solution. In: Proceedings of the 2012 International Conference on Supercomputing, pp. 69–78 (2012)
36. Toueg, S., Babaoglu, Ö.: On the optimum checkpoint selection problem. SIAM J. Comput. **13**(3), 630–649 (1984)
37. Young, J.W.: A first order approximation to the optimum checkpoint interval. Commun. ACM **17**(9), 530–531 (1974)
38. Zheng, Z., Lan, Z.: Reliability-aware scalability models for high performance computing. In: Proceedings of the 2009 IEEE Conference on Cluster Computing (2009)
39. Ziegler, J., Muhlfeld, H., Montrose, C., Curtis, H., O'Gorman, T., Ross, J.: Accelerated testing for cosmic soft-error rate. IBM J. Res. Dev. **40**(1), 51–72 (1996)
40. Ziegler, J., Nelson, M., Shell, J., Peterson, R., Gelderloos, C., Muhlfeld, H., Montrose, C.: Cosmic ray soft error rates of 16-Mb DRAM memory chips. IEEE J. Solid-State Circuits **33**(2), 246–252 (1998)
41. Ziegler, J.F., Curtis, H.W., Muhlfeld, H.P., Montrose, C.J., Chin, B.: IBM experiments in soft fails in computer electronics. IBM J. Res. Dev. **40**(1), 3–18 (1996)

Parallel Sorting for GPUs

Frank Dehne and Hamidreza Zaboli

Abstract Selim Akl has been a ground breaking pioneer in the field of parallel sorting algorithms. His 'Parallel Sorting Algorithms' book [12], published in 1985, has been a standard text for researchers and students. Here we discuss recent advances in parallel sorting methods for many-core GPUs. We demonstrate that parallel *deterministic* sample sort for GPUs (GPU BUCKET SORT) is not only considerably faster than the best comparison-based sorting algorithm for GPUs (THRUST MERGE) but also as fast as *randomized* sample sort for GPUs (GPU SAMPLE SORT). However, *deterministic* sample sort has the advantage that bucket sizes are guaranteed and therefore its running time does not have the input data dependent fluctuations that can occur for *randomized* sample sort.

1 Introduction

Selim Akl has been a ground breaking pioneer in the field of parallel sorting algorithms. His 'Parallel Sorting Algorithms' book [12], published in 1985, has been a standard text for researchers and students. Here we discuss recent advances in parallel sorting methods for many-core GPUs.

Modern graphics processors (*GPUs*) have evolved into highly parallel and fully programmable architectures. Current many-core GPUs can contain hundreds of processor cores on one chip and can have an astounding performance. However, GPUs are known to be hard to program and current general purpose (i.e. non-graphics) GPU applications concentrate typically on problems that can be solved using fixed and/or regular data access patterns such as image processing, linear algebra, physics simulation, signal processing and scientific computing (see e.g. [7]). The design of efficient GPU methods for discrete and combinatorial problems with

F. Dehne (✉)
Carleton University, Ottawa, Canada
e-mail: frank@dehne.net

H. Zaboli
IBM, Ottawa, Canada
e-mail: hamedzaboli@gmail.com

© Springer International Publishing Switzerland 2017
A. Adamatzky (ed.), *Emergent Computation*, Emergence, Complexity
and Computation 24, DOI 10.1007/978-3-319-46376-6_12

data dependent memory access patterns is still in its infancy. The comparison-based THRUST MERGE method [11] by Nadathur Satish, Mark Harris and Michael Garland of nVIDIA Corporation was considered the best sorting method for GPUs. Nikolaj Leischner, Vitaly Osipov and Peter Sanders [9] recently published a *randomized* sample sort method for GPUs (referred to as GPU SAMPLE SORT) that significantly outperforms THRUST MERGE. However, a disadvantage of the *randomized* sample sort method is that its performance can vary for different input data distributions because the data is partitioned into buckets that are created via *randomly* selected data items. Here we demonstrate that *deterministic* sample sort for GPUs, referred to as GPU BUCKET SORT, has the same performance as the *randomized* sample sort method (GPU SAMPLE SORT) in [9].

The remainder of this paper is organized as follows. Section 2 reviews some features of GPUs that are important in this context. Section 3 reviews recent GPU based sorting methods. Section 4 outlines GPU BUCKET SORT and discusses some details of our CUDA [1] implementation. In Sect. 5, we present an experimental performance comparison between our GPU BUCKET SORT implementation, the randomized GPU SAMPLE SORT implementation in [9], and the THRUST MERGE implementation in [11].

2   Review: GPU Architectures

As in [9, 11], we will focus on nVIDIA's unified graphics and computing platform for GPUs [10] and associated *CUDA* programming model [1]. However, the discussion applies more generaly to GPUs that support the OpenCL standard [2]. A GPU consists of an array of streaming processors called *Streaming Multiprocessors (SMs)*. Each SM contains several processor cores and a small size low latency local *shared memory* that is shared by its processor cores. All SMs are connected to a *global* DRAM *memory* through an interconnection network. The global memory is arranged in independent memory partitions and the interconnection network routes the read/write memory requests from the processor cores to the respective global memory partitions, and the results back to the cores. Each global memory partition has its own queue for memory requests and arbitrates among the incoming read/write requests, seeking to maximize DRAM transfer efficiency by grouping read/write accesses to neighboring memory locations (referred to as *coalesced* global memory access). Memory latency to global DRAM memory is optimized when parallel read/write operations can be grouped into a minimum number of sub-arrays of contiguous memory locations.

It is important to note that data accesses from processor cores to their SM's local shared memory are at least an order of magnitude faster than accesses to global memory. This is our main motivation for using a sample sort based approach. An important property of sample sort is that the number of times the data has to be accessed in global memory is a small fixed constant. At the same time, *deterministic* sample sort provides a partitioning into independent parallel workloads and also gives guarantees for the sizes of those workloads. For GPUs, this implies that we are able to utilize the local shared memories efficiently and that the number of data transfers between gloabl memory and the local shared memories is a small fixed constant.

Another critical issue for the performance of CUDA implementations is conditional branching. CUDA programs typically execute very large numbers of threads. In fact, a large number of threads is required for hiding latencies of global memory accesses. The GPU has a hardware thread scheduler that is built to manage tens of thousands and even millions of concurrent threads. All threads are divided into blocks, and each block is executed by an SM. An SM executes a thread block by breaking it into groups called *warps* and executing them in parallel. The cores within an SM share various hardware components, including the instruction decoder. Therefore, the threads of a warp are executed in SIMT (single instruction, multiple threads) mode, which is a slightly more flexible version of the standard SIMD (single instruction, multiple data) mode. The main problem arises when the threads encounter a conditional branch such as an IF-THEN-ELSE statement. Depending on their data, some threads may want to execute the code associated with the "true" condition and some threads may want to execute the code associated with the "false" condition. Since the shared instruction decoder can only handle one branch at a time, different threads can not execute different branches concurrently. They have to be executed in sequence, leading to performance degradation. Recent GPUs provide a small improvement through an instruction cache at each SM that is shared by its cores. This allows for a "small" deviation between the instructions carried out by the different cores. For example, if an IF-THEN-ELSE statement is short enough so that both conditional branches fit into the instruction cache then both branches can be executed fully in parallel. However, a poorly designed algorithm with too many and/or large conditional branches can result in serial execution and very low performance.

3   GPU Sorting Methods

Early sorting algorithms for GPUs include *GPUTeraSort* [6] based on bitonic merge, and *Adaptive Bitonic Sort* [8] based on a method by Bilardi et al. [3]. *Hybrid Sort* [14] used a combination of bucket sort and merge sort, and Cederman et al. [4] proposed a quick sort based method for GPUs. Both methods [4, 14] suffer from load balancing problems. Until recently, the comparison-based THRUST MERGE method [11] by Nadathur Satish, Mark Harris and Michael Garland of nVIDIA Corporation was considered the best sorting method for GPUs. THRUST MERGE uses a combination of odd-even merge and two-way merge, and overcomes the load balancing problems mentioned above. Satish et al. [11] also presented an even faster GPU radix sort method for the special case of integer sorting. Yet, a recent paper by Nikolaj Leischner, Vitaly Osipov and Peter Sanders [9] presented a randomized sample sort method for GPUs (GPU SAMPLE SORT) that significantly outperforms THRUST MERGE [11]. However, as also discussed in Sect. 1, the fact that GPU

SAMPLE SORT is a *randomized* method implies that its performance can vary with the distribution of the input data because buckets are created through randomly selected data items. For example, the performance analysis presented in [9] measures the runtime of GPU SAMPLE SORT for several input data distributions to document the performance variations observed for different input distributions.

4 GPU BUCKET SORT: *Deterministic* Sample Sort For GPUs

In this section we outline GPU BUCKET SORT, a *deterministic* sample sort algorithm for GPUs. An overview of GPU BUCKET SORT is shown in Algorithm 1 below. It consists of a local sort (Step 1), a selection of samples that define balanced buckets (Steps 3–5), moving all data into those buckets (Steps 6–8), and a final sort of each bucket (Step 9). In our implementation of GPU BUCKET SORT we introduced several adaptations to the structure of GPUs, in particular the two level memory hierarchy, the large difference in memory access times between those two levels, and the small size of the local shared memories. We experimented with several bucket sizes and number of samples in order to best fit them to the GPU memory structure. For sorting the selected sample and the bottom level sorts of the individual buckets, we experimented with several existing GPU sorting methods such as bitonic sort, adaptive bitonic sort [8] based on [3], and parallel quick sort.

The following discussion of our implementation of GPU BUCKET SORT will focus on GPU performance issues related to shared memory usage, coalesced global memory accesses, and avoidance of conditional branching. Consider an input array A with n data items in global memory and a typical local shared memory size of $\frac{n}{m}$ data items.

In **Steps 1 and 2** of Algorithm 1, we split the array A into m sublists of $\frac{n}{m}$ data items each and then locally sort each of those m sublists. More precisely, we create m thread blocks of 512 threads each, where each thread block sorts one sublist using one SM. Each thread block first loads a sublist into the SM's local shared memory using a coalesced parallel read from global memory. Note that, each of the 512 threads is responsible for $\frac{n}{m}/512$ data items. The thread block then sorts a sublist of $\frac{n}{m}$ data items in the SM's local shared memory. We tested different implementations for the local shared memory sort within an SM, including quicksort, bitonic sort, and adaptive bitonic sort [3]. In our experiments, bitonic sort was consistently the fastest method, despite the fact that it requires $O(n \log^2 n)$ work. The reason is that, for Step 2 of Algorithm 1, we always sort a small fixed number of data items, independent of n. For such a small number of items, the simplicity of bitonic sort, it's small constants in the running time, and it's perfect match for SIMD style parallelism outweigh the disadvantage of additional work.

In **Step 3** of Algorithm 1, we select s equidistant samples from each sorted sublist. (The implementation of Step 3 is built directly into the final phase of Step 2 when the sorted sublists are written back into global memory.) Note that, the sample size s is a free parameter that needs to be tuned. With increasing s, the sizes of buckets created in Step 8 decrease and the time for sorting those buckets (Step 9) decreases as well. However, the time for managing the buckets (Steps 3–7) grows with increasing s. This trade-off will be studied in Sect. 5 where we show that $s = 64$ provides the best performance. In **Step 4**, we sort all sm selected samples in global memory, using all available SMs in parallel. Here, we compared GPU bitonic sort [6], adaptive bitonic sort [8] based on [3], and GPU SAMPLE SORT [9]. Our experiments indicate that for up to 16 M data items, simple bitonic sort is still faster than even GPU SAMPLE SORT [9] due to its simplicity, small constants, and complete avoidance of conditional branching. Hence, Step 4 was implemented via bitonic sort. In **Step 5**, we again select s equidistant *global samples* from the sorted list of sm samples. Here, each thread block/SM loads the s global samples into its local shared memory where they will remain for the next step.

Input: An array A with n data items stored in global memory.
Output: Array A sorted.

1. Split the array A into m sublists A_1, ..., A_m containing $\frac{n}{m}$ items each where $\frac{n}{m}$ is the shared memory size at each SM.
2. *Local Sort*: Sort each sublist A_i ($i=1,..., m$) locally on one SM, using the SM's shared memory as a cache.
3. *Local Sampling*: Select s equidistant samples from each sorted sublist A_i ($i=1,..., m$) for a total of sm samples.
4. *Sorting All Samples*: Sort all sm samples in global memory, using all available SMs in parallel.
5. *Global Sampling*: Select s equidistant samples from the sorted list of sm samples. We will refer to these s samples as *global samples*.
6. *Sample Indexing*: For each sorted sublist A_i ($i=1,..., m$) determine the location of each of the s global samples in A_i. This operation is done for each A_i locally on one SM, using the SM's shared memory, and will create for each A_i a partitioning into s buckets A_{i1},..., A_{is} of size a_{i1},..., a_{is}.
7. *Prefix Sum*: Through a parallel prefix sum operation on a_{11},..., a_{m1}, a_{12},..., a_{m2}, ..., a_{1s},..., a_{ms} calculate for each bucket $A_{ij}(1 \leq i \leq m, 1 \leq j \leq s,)$ its starting location l_{ij} in the final sorted sequence.
8. *Data Relocation*: Move all sm buckets $A_{ij}(1 \leq i \leq m, 1 \leq j \leq s)$ to location l_{ij}. The newly created array consists of s sublists B_1, ..., B_s where $B_j = A_{1j} \cup A_{2j} \cup ... \cup A_{mj}$ for $1 \leq j \leq s$.
9. *Sublist Sort*: Sort all sublists B_j, $1 \leq j \leq s$, using all SMs.

Algorithm 1: GPU BUCKET SORT (Deterministic Sample Sort For GPUs)

In **Step 6**, we determine for each sorted sublist A_i ($i=1, ..., m$) of $\frac{n}{m}$ data items the location of each of the s global samples in A_i. For each A_i, this operation is done locally by one thread block on one SM, using the SM's shared memory, and will create for each A_i a partitioning into s buckets A_{i1},..., A_{is} of size a_{i1},..., a_{is}. Here, we apply a parallel binary search algorithm to locate the global samples in A_i. More precisely, we first take the $\frac{s}{2}$-th global sample element and use one thread to perform a binary search in A_i, resulting in a location $l_{s/2}$ in A_i. Then we use two threads to perform two binary searches in parallel, one for the $\frac{s}{4}$-th global sample element in the part of A_i to the left of location $l_{s/2}$, and one for the $\frac{3s}{4}$-th global sample element in the part of A_i to the right of location $l_{s/2}$. This process is iterated $\log s$ times until all s global samples are located in A_i. With this, each A_i is split into s buckets $A_{i1},...,$ A_{is} of size $a_{i1},..., a_{is}$. Note that, we do not simply perform all s binary searches fully in parallel in order to avoid memory contention within the local shared memory [1].

Step 7 uses a prefix sum calculation to obtain for all buckets their starting location in the final sorted sequence. The operation is illustrated in Fig. 1 and can be implemented with coalesced memory accesses in global memory. Each row in Fig. 1 shows the $a_{i1},..., a_{is}$ calculated for each sublist. The prefix sum is implemented via a parallel column sum (using all SMs), followed by a prefix sum on the columns sums (on one SM in local shared memory), and a final update of the partial sums in each column (using all SMs).

In **Step 8**, the sm buckets are moved to their correct location in the final sorted sequence. This operation is perfectly suited for a GPU and requires one parallel coalesced data read followed by one parallel coalesced data write operation. The newly created array consists of s sublists $B_1, ..., B_s$ where each $B_j = A_{1j} \cup A_{2j} \cup ... \cup A_{mj}$ has at most $\frac{2n}{s}$ data items [13]. In **Step 9**, we sort each B_j using the same bitonic sort implementation as in Step 4. We observed that for our choice of s, each B_j contains at most $4M$ data items. For such small data sets, simple bitonic sort is again the fastest sorting algorithm for each B_j due to bitonic sort's simplicity, small constants, and complete avoidance of conditional branching.

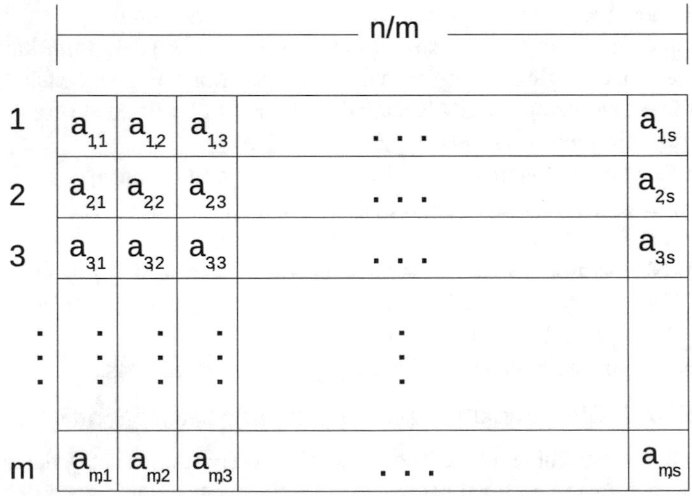

Fig. 1 Illustration of step 7 in Algorithm 1

5 Experimental Results and Discussion

Figure 2 shows in detail the time required for the individual steps of Algorithm 1 when executed on an NVIDIA Fermi GTX 480 GPU. We observe that *sublist sort* (Step 9) and *local sort* (Step 2) represent the largest portion of the total runtime of GPU BUCKET SORT. This is very encouraging in that the "overhead" involved to manage the deterministic sampling and generate buckets of guaranteed size (Steps 3–7) is small. We also observe that the *data relocation* operation (Step 8) is very efficient and a good example of the GPU's great performance for data parallel access when memory accesses can be coalesced.

Figures 3 and 4 show a comparison between GPU BUCKET SORT and the current best GPU sorting methods, *randomized* GPU SAMPLE SORT [9] and THRUST MERGE [11] on a NVIDIA Tesla C1060 GPU. For GPU BUCKET SORT, all runtimes are the averages of 100 experiments, with less than 1 ms observed variance. For *randomized* GPU SAMPLE SORT and THRUST MERGE, the runtimes shown are the ones reported in [9, 11]. For THRUST MERGE, performance data is only available for up to $n = 16$ M data items. For larger values of n, the current THRUST MERGE code shows memory errors [5]. As reported in [9], the current *randomized* GPU SAMPLE SORT code can sort up to 128 M data items on a Tesla C1060. Our GPU BUCKET SORT implementation appears to be more memory efficient. GPU BUCKET SORT can sort up to $n = 512$ M data items on a Tesla C1060. Figures 3 and 4 show the performance comparison with higher resolution for up to $n = 128$ M and for the entire range up to $n = 512$ M, respectively. We observe that, as reported in [9], *randomized* GPU SAMPLE SORT [9] significantly outperforms THRUST MERGE [11]. Most importantly, we observe that *randomized* sample sort (GPU SAMPLE SORT) [9] and *deterministic* sample sort (GPU BUCKET SORT) show nearly identical performance.

The data sets used for the performance comparison in Figs. 3 and 4 were uniformly distributed, random data items. The data distribution does not impact the performance

Fig. 2 Performance of deterministic sample sort for GPUs (GPU BUCKET SORT). Total runtime and runtime for individual steps of Algorithm 1 for varying number of data items

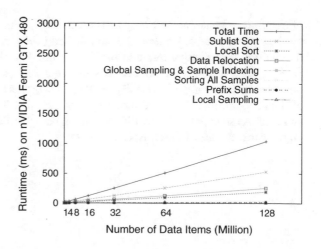

Fig. 3 Comparison between deterministic sample sort (GPU Bucket Sort), Randomized Sample Sort (GPU Sample Sort) [9] and Thrust Merge [11]. Total runtime for varying number of data items up to 128,000,000. (*Note* [11] and [9] provided data only for up to 16 and 128 M data items, respectively.)

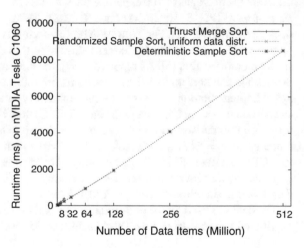

Fig. 4 Comparison between deterministic sample sort (GPU Bucket Sort), Randomized sample sort (GPU Sample Sort) [9] and Thrust Merge [11]. Total runtime for varying number of data items up to 512,000,000. (*Note* [11] and [9] provided data only for up to 16 and 128 M data items, respectively.)

of *deterministic* sample sort (GPU Bucket Sort) but has a considerable impact on the performance of *randomized* sample sort (GPU Sample Sort) [9]. In fact, the uniform data distribution used for Figs. 3 and 4 is a *best case* scenario for *randomized* sample sort where all bucket sizes are nearly identical. Figure 5 shows that our deterministic sample sort (GPU Bucket Sort) is stable under different types of data distribution. We tested three types of data distribution: Uniform, Gaussian, and Zipf. As seen in the figure, different input data distributions have little influence on the performance of GPU Bucket Sort.

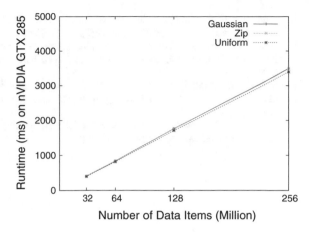

Fig. 5 Performance of deterministic sample sort (GPU BUCKET SORT) for different input data distributions

6 Conclusions

In this paper, we presented a *deterministic* sample sort algorithm for GPUs, called GPU BUCKET SORT. Our experimental evaluation indicates that GPU BUCKET SORT is considerably faster than THRUST MERGE [11], the best comparison-based sorting algorithm for GPUs, and it is exactly as fast as *randomized* sample sort for GPUs (GPU SAMPLE SORT) [9] when the input data sets used are uniformly distributed, which is a *best case* scenario for randomized sample sort. However, as observed in [9], the performance of *randomized* GPU SAMPLE SORT fluctuates with the input data distribution whereas GPU BUCKET SORT does not show such fluctuations.

References

1. NVIDIA CUDA Programming Guide. nVIDIA Corporation. www.nvidia.com
2. The OpenCL Specification 1.0. Khronos OpenCL Working Group (2009)
3. Bilardi, G., Nicolau, A.: Adaptive bitonic sorting. An optimal parallel algorithm for shared-memory machines. SIAM J. Comput. **18**(2), 216–228 (1989)
4. Cederman, D., Tsigas, P.: A practical quicksort algorithm for graphics processors. In: Proceedings of European Symposium on Algorithms (ESA), vol. 5193 of LNCS, pp. 246–258 (2008)
5. Garland, M.: Private Communication. nVIDIA Corporation (2010)
6. Govindaraju, N., Gray, J., Kumar, R., Manocha, D.: GPUTeraSort: high performance graphics co-processor sorting for large database management. In: Proceedings of International Conference on Management of Data (SIGMOD), pp. 325–336 (2006)
7. GPGPU.ORG. General-purpose computation on graphics hardware
8. Greb, A., Zachmann, G.: GPU-ABiSort: optimal parallel sorting on stream architectures. In: Proceedings of International Parallel and Distributed Processing Symposium (IPDPS) (2006)
9. Leischner, N., Osipov, V., Sanders, P.: GPU sample sort. In: Proceedings of International Parallel and Distributed Processing Symposium (IPDPS), pp. 1–10 (2010)

10. Lindholm, E., Nickolls, J., Oberman, S., Montrym, J.: NVIDIA Tesla: a unified graphics and computing architecture. IEEE Micro. **28**(2), 39–55 (2008)
11. Satish, N., Harris, M., Garland, M.: Designing efficient sorting algorithms for manycore GPUs. In: Proceedings of International Parallel and Distributed Processing Symposium (IPDPS) (2009)
12. Selim, G.A.: Parallel Sorting Algorithms. Academic Press (1985)
13. Shi, H., Schaeffer, J.: Parallel sorting by regular sampling. J. Par. Dist. Comp. **14**, 362–372 (1992)
14. Sintorn, E., Assarsson, U.: Fast parallel GPU-sorting using a hybrid algorithm. J. Parallel Distrib. Comput. **68**(10), 1381–1388 (2008)

Mining for Functional Dependencies Using Shared Radix Trees in Many-Core Multi-Threaded Systems

Joel Fuentes, Claudio Parra, David Carrillo and Isaac D. Scherson

Abstract We consider the problem of mining for functional dependencies in relational databases. Intermediate data structures, although simple, explode in size and a solution is proposed using radix trees to reduce memory utilization. Parallelism is further applied in a Multi-Core computer to further speedup the process. Because bit-permutations are the basis of the construction of a binary intermediate matrix, radix trees reduce the memory usage 10 times. Multi-Threading the construction and processing of the intermediate data leads to a concurrent computing average-over-time of 63 % on an equivalent speedup of 6.3 on a system with 12 cores, 256 GB of memory and 1 TB SSD.

1 Introduction

With the advent of computing systems that use silicon devices with many CPUs per chip, also known as Many-Core or Multi-Core systems, new challenges are posed to programming applications that attempt to use parallelism to achieve a significant computational improvement. Typical Many-Core computers use chips that contain two, four or more cores each. These Multi-Core chips also include an on-chip hierarchical shared cache system that provides local and shared caches. An interface to a large common main DRAM storage completes the solid state memory hierarchy. These systems are normally programmed using threads that are managed by

J. Fuentes (✉)
Universidad del Bío-Bío, Chillán, Chile
e-mail: jfuentes@ubiobio.cl

C. Parra
Universidad Católica del Maule, Talca, Chile
e-mail: parra.claudio.alejandro@gmail.com

D. Carrillo · I.D. Scherson
University of California, Irvine, USA
e-mail: dcarril@ics.uci.edu

I.D. Scherson
e-mail: isaac@ics.uci.edu

© Springer International Publishing Switzerland 2017
A. Adamatzky (ed.), *Emergent Computation*, Emergence, Complexity and Computation 24, DOI 10.1007/978-3-319-46376-6_13

the operating system to execute in the available cores attempting to use as much parallelism as possible. The main problem that arises is the management of shared data structures in the shared hierarchical memory to guarantee synchronized access to the data structures, avoiding deadlock and providing a correct access sequence as required by the program.

Herlihy and Shavit [7] wrote a book that discusses the methodologies used to properly program Multi-Core systems. Exploiting parallelism depends very much on the synchronization mechanisms available to avoid shared data conflicts. Their book has become a classic and has been adopted to teach MultiProgramming courses.

Experience with practical programs shows that in addition to deadlock avoidance, programmers need to be very careful about write through latencies that are bound to create incorrect value reads if threads access variables before the write through mechanism updates values throughout the memory hierarchy. It seems that even when sequentializing shared data accesses, the write through mechanism may get in the way of correct execution, and might lead to a slowdown in program execution.

In this paper we consider an actual practical problem encountered when trying to discover functional dependencies (FDs) in relational data bases using a recently introduced technique based on the generation of refutations [6]. Given a relational database with n records of k attributes each, the method starts by generating a refutation matrix that represents exhaustively all attribute groups where no dependencies can be found. It is shown that the size of this matrix can explode beyond the storage capabilities of the computer and a need is identified to represent it using radix trees. The generation of this refutation matrix is almost embarrassingly parallel and a discussion is presented on how to gain performance by exploiting concurrency both with a straight forward matrix data structure as well as with a radix tree representation.

It will be shown that a big bottleneck is an intermediate data structure whose size may overcome the available storage in the computer. Radix trees are suggested to reduce the demand on memory and are shown to yield a reduction of 10 times on average for the worst case. Parallelization of the intensive phase of the procedure is done on a Multi-Core engine using Multi-threaded programming. Recognizing the inherent sequentialization present in shared memory/data programs, a figure of merit is used to determine what percentage of the execution time threads are allowed to run in parallel. A companion speedup is also calculated. For a 12-core computer, with 256 GB of DRAM and 1 TB disk, experimentation shows that on average a speedup of 6.3 is achieved with parallelism observable 63 % of the time.

2 Algorithm for Mining Functional Dependencies

Consider a relational database where R is a relation with a set of attributes $A = \{a_i | i = 1 \ldots k\}$ and r is an instance of R where each attribute in each tuple can assume a value in some domain. We denote by $t(a_i)$ the value of attribute a_i

in tuple t. A functional dependency (FD) is an expression of the form $X \to a_i$, where $X \subseteq \{A - a_i\}$. X is called the *determinant set* and a_i is called the *dependent attribute*. The FD $X \to a_i$ is *valid* in the instance r if and only if for every pair of rows (tuples) $t, t' \in r$, whenever $t[a_j] = t'[a_j]$ for all $a_j \in X$, it is also the case that $t[a_i] = u[a_i]$.

Many direct algorithms have been proposed to find FDs in relational databases [2, 4, 5, 8, 10]. This work is based on a novel approach that first prunes the search space by determining which attributes cannot depend on others. The idea is to generate first "refutations" by exhaustively searching all tuple pairs in the relational database to identify which subsets of attributes cannot determine another attribute.

To facilitate the description of the operations in Refutation-Based FD mining (RB-FD) algorithm [6], let us give the following definitions:

1. Let $A = \{a_i \mid i = 1 \dots k\}$ be a set of attributes where each attribute can assume a value in some domain.
2. A relation R over A describes all possible tuples of values of attributes in A.
3. An instance r of R is a subset of tuples of R. We denote by $t(a_i)$ the value of attribute a_i in tuple t.
4. A refutation $[A - \{a_i\}] \nrightarrow a_i$ holds if and only if for $t, t' \in r$, $t(a_i) \neq t'(a_i) \wedge \exists a_j, j \neq i, t(a_j) = t'(a_j)$.

A refutation is found when two different values for $t(a_i)$ and $t'(a_i)$ correspond to tuples where some subset of the remainder $[A - \{a_i\}]$ attributes are equal.

The result of finding refutations can be kept on a binary matrix H where a row corresponds to the comparison of a pair of tuples t, t' in r. If $t(a_i) \neq t'(a_i)$, the row is generated with 1s at attribute positions where $t(a_j) = t'(a_j)$, $i \neq j$, and 0s elsewhere. If $t(a_i) = t'(a_i)$, no row is generated.

Note that if a_i does not depend on some set of other attributes it does not depend on any subset of those attributes. We conclude that when looking for refutations, we only need to retain those with the maximum number of attributes other than a_i.

If two refutations are generated such that the one with more 1s, say $\underline{h}$, has 1s in all the positions where other $\underline{h'}$ has 1s, then only the one with more 1s ($\underline{h}$) needs to be retained. We say that the former ($\underline{h}$) contains the latter ($\underline{h'}$). The test for containment can be summarized as follows:

$$\text{If } (\underline{h} \wedge \underline{h'}) \oplus \underline{h} = \underline{0} \Rightarrow \text{ retain } \underline{h'}$$
$$\text{If } (\underline{h} \wedge \underline{h'}) \oplus \underline{h'} = \underline{0} \Rightarrow \text{ retain } \underline{h}$$

The procedure for generating the matrix H for an attribute a_i in r is shown in pseudocode in Algorithm 2 below.

Input : relation r, set of attributes A
Output: set of refutations H

```
for t ∈ r do
    for t' ∈ r do
        if t(aᵢ)! = t'(aᵢ) then
            h = ∅;
            for aⱼ ∈ {A − aᵢ} do
                if t(aⱼ) == t'(aⱼ) then
                    │  h = h ∪ aⱼ;
                end
            end
            for h' ∈ H do
                if (h ∧ h') ⊕ h = 0 then
                    │  continue next t';
                end
                if (h ∧ h') ⊕ h = 0 then
                    │  H = H \ {h'};
                end
            end
            H = H ∪ h;
        end
    end
end
```

Algorithm 2: Generation of refutations H for attribute a_i.

Example. Given a 6-attribute relation with $A = \{a, b, c, d, e, f\}$ such that the domain for $a = \{a_1, a_2, a_3, a_4, a_5\}$, for $b = \{b_1, b_2, b_3, b_4\}$ and so on for c, d, e and f (see Fig. 1), and the instance r shown in Fig. 1 we produce the matrix H for attribute 6 as its refuted attribute. We observe that the tuples 1 and 2 produce the refutation $\{a, b\} \nrightarrow f$ (represented by the vector $\underline{h} = 11000$) since the values of attribute f in tuples 1 and 2 are distinct and for attributes a and b are equal; tuples 1 and 3 produce the refutations $\{b, d\} \nrightarrow f$ ($\underline{h} = 01010$); tuples 1 and 7 produce the refutations $\{a\} \nrightarrow f$ ($\underline{h} = 10000$) and so on. Keeping only the maximal refutations we discard the refutation $\{a\} \nrightarrow f$ and finally obtain the refutations $\{a, b\}$ and $\{b, d\}$ represented by the matrix shown below:

Fig. 1 Relation instance with $A = \{a, b, c, d, e, f\}$

Tuple ID	a	b	c	d	e	f
1	a_1	b_3	c_2	d_1	e_4	f_1
2	a_1	b_3	c_3	d_3	e_1	f_2
3	a_2	b_3	c_5	d_1	e_5	f_4
4	a_3	b_3	c_2	d_3	e_3	f_1
5	a_4	b_2	c_2	d_8	e_2	f_1
6	a_5	b_4	c_4	d_1	e_3	f_1
7	a_1	b_1	c_3	d_7	e_6	f_2

$$
\begin{array}{ccccc}
a & b & c & d & e
\end{array}
$$
$$
H = \begin{pmatrix} 1 & 1 & 0 & 0 & 0 \\ 0 & 1 & 0 & 1 & 0 \end{pmatrix}
$$

Phase 1 of the RB-FD algorithm is the generation of the refutation matrix H just described.

As seen in [6], Phase 2 consists in finding the minimal transversals of a hypergraph represented by a binary matrix $\overline{H}$, the bit-wise complement of H. A minimal transversal τ_m of the hypergraph H' is a $(k-1)$-bit binary vector that contains the minimal number of 1s such that $\tau_m \wedge h_i$ contains at least one 1 for all rows i of H'.

The procedure is supposed to generate all possible distinct minimal transversals for each matrix H (for each attribute). The detail of the minimal transversals generation is not within the main scope of this paper and are left to the interested to read in [6].

Example. From the refutation matrix found from r in Fig. 1 in the previous example, we continue now by obtaining the hypergraph of the complements of the hyperedges in H, which is $H' = \{\{c, d, e\}, \{a, c, e\}\}$, represented by the binary vectors $H' = \{00111, 10101\}$. Finally the minimal transversals correspond to the functional dependencies in minimal form:

$$
\begin{array}{ccccc}
a & b & c & d & e
\end{array}
$$
$$
H = \begin{pmatrix} 1 & 0 & 0 & 1 & 0 \\ 0 & 0 & 1 & 0 & 0 \\ 0 & 0 & 0 & 0 & 1 \end{pmatrix}
$$

That is, we have discovered the functional dependencies $\{a, d\} \to f$, $\{c\} \to f$ and $\{e\} \to f$.

Even though the generation of the refutation matrices H for each attribute has polynomial complexity, it can become dominant due to the large size of the database r (large n and k) and also because the size of H may explode as it will be seen in the next section.

3 The Size of Refutation Matrices

Consider as above a n-tuple instance r of a relation R. It was shown that to generate a refutation matrix H, all pairs of tuples in r need to be compared to obtain refutation vectors that are inserted in a running matrix H following the containment rule.

To find the number of rows in H, that is the number of "non-contained" refutations, out of the $O(n^2)$ generated by comparing all possible pairs of tuples in r, assume that some pair of tuples generated a refutation with a maximum number of 1s. It is obvious that the only non-contained other refutations that could be added to H are those binary vectors that are some permutation of this maximum 1s refutation. All other refutations will be contained in the maximum one, or in one that is permutation of the maximum one.

If q is the number of 1s in the refutation vector with the maximum number of ones, the maximum number of non-contained refutations (rows) in H is:

$$max \ rows \ in \ H = \binom{k}{q} = \frac{(k-1)!}{q!(k-1-q)!} \qquad k, q \in \mathbb{N}$$

which has a maximum when $q = \frac{k-1}{2}$.

The proof of this maximum can be obtained by induction and we omit it as it is simple and does not add substantial knowledge to this work.

Observation: If $n^2 \geq \frac{k-1!}{q!(k-1-q)!}$ and all permutations of a binary vector with $q = \frac{k-1}{2}$ 1s are generated, the maximum size of H is:

$$|H| = \frac{k-1!}{\frac{k-1}{2}!(k-1-\frac{k-1}{2})!} \tag{1}$$

Example. Table 1 illustrates the growth of memory utilization needed to store refutation matrix H when increasing the number of attributes in the worst case.

It is clear that there is a need for a compact representation of refutations in main memory. In [9] Knuth shows different forms to represents all the permutations and introduces the concept of path for these sequences. Table 2 illustrates all the combinations that corresponds to a worst case when finding $\binom{6}{3} = 20$ maximal refutations from a relation with 6 attributes. From the second to the fourth column corresponds to different forms of representing these binary strings that can also be seen as compact representations. The second column corresponds to the dual combination $b_p \ldots b_1$ that lists the position of zeros. The third column represents the primal combination $c_p \ldots c_1$ that lists the positions of the ones. The fourth column corresponds to the multicombination $d_p \ldots d_1$ that lists the number of 0s to the right of each 1.

Furthermore, Table 2 presents the path of each binary string. Each binary string is equivalent to a path of length $k - 1$ from the corner to corner of an $q \times (k - 1 - q)$ grid, because such a path contains q vertical steps and $k - 1 - q$ horizontal steps.

Table 1 Memory utilization for refutation matrix H

# Attributes	Memory
16	25.7 KB
24	8.1 MB
32	2.4 GB
40	689 GB
48	195 TB

Table 2 Representations of maximal refutations for $k = 6$ and $q = 3$

$a_5a_4a_3a_2a_1a_0$	$b_3b_2b_1$	$c_3c_2c_1$	$d_3d_2d_1$	Path
000111	543	210	000	
001011	543	310	100	
001101	541	320	110	
001110	540	321	111	
010011	532	410	200	
010101	531	420	210	
010110	530	421	211	
011001	521	430	220	
011010	520	431	221	
011100	510	432	222	
100011	432	510	300	
100101	431	520	310	
100110	430	521	311	
101001	421	530	320	
101010	420	531	321	
101100	410	532	322	
110001	321	540	330	
110010	320	541	331	
110100	310	542	332	
111000	210	543	333	

The concept of path in this sense is useful to define a new data structure to store refutations. It can be seen that the worst case presents common prefix sequences in a half of the total of refutations.

4   Reducing the Size of Intermediate Data Structures

The refutation matrix H represents a simple structure to store refutations but it produces high costs to keep the maximal refutations in it. For example, when a refutation is found, a review step over the entire matrix H must be done by removing all its subsets (contained refutations). The other important cost is the memory usage, where this structure can explode if the worst case is presented.

To deal with these issues the radix tree data structure is introduced. A radix tree is an ordered tree data structure that is used to store a dynamic set or associative array where the keys are usually strings. In a regular radix tree each edge is assigned with some symbol. Thus, any route from a tree root to one of its leaves defines precisely only one string. As the refutations are represented as a binary strings, for the radix tree only the symbols 0 and 1 are considered.

When inserting these refutation in the radix tree and based on the property that all the refutations have the same length, the refutations are represented as paths from the root to every leaf. Figure 2 on the right illustrates the refutations from H inserted in the radix tree. To make the radix tree smaller, it is possible compress it. Such form means that a number of bits B per node is defined allowing that nodes with the same bits in the same level are packed.

Unlike balanced trees, radix trees permit lookup, insertion, and deletion in $O(k)$ time rather than logarithmic. However for our radix tree implementation, a fixed number of bits B is defined as the strict number of bits per node. This value will also represent the maximum common number of bits for different refutations that can be compacted in a node.

Formally, let B be the number of bits in each node. The worst case for the radix tree when looking up for a refutation is similar to the matrix H, $O(|H|)$. It occurs when having defined the value B there are no common prefixes of size B for the set of refutations. For instance, if there exist $|H|$ different refutations and they are different from each other in the first B bits, $\{X_1, \ldots, X_B\}$, and the rest of bits $\{X_{B+1}, \ldots, X_k\}$ for every refutation are similar, then there is no possible compression. However, from the previous section it was shown that if the worst case scenario for the number of refutations occurs, then common prefixes exist (common initial paths exist from one corner to the opposite on the diagonal).

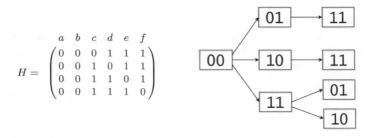

$$H = \begin{array}{c} \begin{array}{cccccc} a & b & c & d & e & f \end{array} \\ \begin{pmatrix} 0 & 0 & 0 & 1 & 1 & 1 \\ 0 & 0 & 1 & 0 & 1 & 1 \\ 0 & 0 & 1 & 1 & 0 & 1 \\ 0 & 0 & 1 & 1 & 1 & 0 \end{pmatrix} \end{array}$$

Fig. 2 Maximal refutations stored in the matrix H (*left*) and the radix tree (*right*) with $B = 2$

The average case can be calculated as follows. Let n be the number of refutations with k bits. Let $M = 2^B$ the number of possible combinations given B bits, and m_i the i-th combination of bits. Let r_a and r_b be refutations. Finally, let $bits(B, r_a)$ be the first B bits of the refutation r_a.

The probability of having two refutations with the same first B bits is:

$$Pr(bits(B, r_a) == bits(B, r_b)) = \sum_{i=1}^{M} Pr(bits(B, r_a) = m_i \ AND \ bits(B, r_b) = m_i)$$

$$= M * Pr(bits(B, r_a) = m_i) * Pr(bits(B, r_b) = m_i)$$

$$= M * (\frac{1}{M})^2$$

$$= \frac{1}{M}$$

From above, $Pr(bits(B, r_a) \neq bits(B, r_b)) = 1 - \frac{1}{M}$. Then the probability that $n - 1$ refutations are different to r_a is $(1 - \frac{1}{M})^{(n-1)}$. This means r_a is unique in its first B bits. If all the refutations are unique in their first B bits, then the expected number of different refutations is:

$$E[D] = \sum_{i=1}^{n} Pr(r_i \ is \ unique) = n * (1 - \frac{1}{M})^{(n-1)}$$

The number of refutations that share the first B bits is $n(1 - (1 - \frac{1}{M})^{(n-1)})$

For the first level of the tree there are in average $n * (1 - \frac{1}{M})^{(n-1)} * B$ bits, and if it is compared with the matrix H the memory usage is reduced by $(1 - \frac{1}{2^B})^{(n-1)}$. On the following levels the considerations are the same.

Experimental results show the important reduction in the memory usage by using the radix tree. For instance, Fig. 4 illustrates a set of experiments comparing the memory usage by the matrix H and the radix tree when increasing the number of refutations. This experiment corresponds to a sequential execution of the algorithm using both data structures.

5 Parallel Generation of Refutation Sets

The goal of the parallelization of the RB-FD on multicore systems is to minimize the processing time with the computing power being efficiently utilized. Thus, the parallelization plus the new data structure proposed would allow mining big datasets in shorter time.

From results presented in [6] it can be seen that first step of RB-FD represents the most time-consuming task. This part of the method involves two tasks:

1. Obtain the set of maximal refutations H with the form $[A - a_i] \nrightarrow a_i$ by comparing every pairs of tuples for every attribute. That is, $\forall a_i \in A$, the set of pairs of tuples $(t, t') \in r \times r$ such that $(\forall a_j \in [A - a_i])$ $t[a_j] = t'[a_j] \wedge t[a_i] \neq t'[a_i]$. This task is exactly $\Theta(k^2 \frac{n(n-1)}{2})$ with n the number of tuples and k the number of attributes.

2. Keep only the maximal refutations in H. For every new refutation, check its maximality with the existing refutations in H.

The complexity of the first task does not seem to involve a problematic issue even when having a big number of tuples. However the verification of maximal refutations when inserting a new one in H would produce an upper bound of $O(|H|)$ whose complexity becomes the hardest with the presence of the worst case scenario. Thus, we can have $O(|H|) = \frac{k!}{p!(k-q)!}$ as a verification for every new refutation found.

The *Dynamic Multithreading* model described in [3] allows programmers to specify parallelism in applications without worrying about communication protocols, load balancing, and other vagaries of static-thread programming. The model represents a multithreaded computation as a directed acyclic graph $G = (V, E)$ whose vertices are instructions and $(u, v) \in E$ if u must be executed before v. The time T_p needed to execute the computation on p cores depends on two parameters of the computation: its work T_1 and its span T_∞. The work is the running time on a single core, that is, the number of nodes (i.e., instructions) in G, assuming each instruction takes constant time. Since p cores can execute only p instructions at a time, we have $T_p = \Omega(T_1/p)$. The span is the length of the longest path in G. Since the instructions on this path need to be executed in order, we also have $T_p = \Omega(T_\infty)$. Together, these two lower bounds give $T_p = \Omega(T_\infty + T_1/p)$. The degree to which an algorithm can take advantage of the presence of $p > 1$ cores is captured by its speed-up T_1/T_p and its parallelism T_1/T_∞. In the absence of cache effects, the best possible speed-up is p, known as linear speed-up. Parallelism provides an upper bound on the achievable speed-up.

The proposed parallel solutions adopt the *Dynamic Multithreading* model. Following this model, two important features are defined to reflect the parallel behavior: nested parallelism and parallel loops. Nested parallelism allows a subroutine to be spawned, allowing the caller to proceed while the spawned subroutine is computing its result. A parallel loop (*parallel for* in Algorithms 3 and 4) is like an ordinary *for* loop, except that the iterations of the loop can execute concurrently.

Two parallel alternatives are presented in this section using radix trees.

5.1 Parallelism Through Attributes

It is easy to see that the first step of the studied algorithm is embarrassing parallel. The objective of this step is to obtain the set of maximal refutations for every attribute in A, it means that for every attribute there exists a set of refutations that is independent from the sets generated for other attributes.

Input : relation r, set of attributes A
Output: sets of refutations $H_1, \ldots, H_k$

parallel for $a \in A$ **do**
$\quad | \quad H_a = $ findRefutations(a);
end

function *findRefutations(a)*
$\quad |$ **for** $t \in r$ **do**
$\quad \quad |$ **for** $t' \in r$ **do**
$\quad \quad \quad |$ **if** $t(a)! = t'(a)$ **then**
$\quad \quad \quad \quad |$ $h = \emptyset$;
$\quad \quad \quad \quad |$ **for** $a_j \in \{A - \{a\}\}$ **do**
$\quad \quad \quad \quad \quad |$ **if** $t(a_j) == t'(a_j)$ **then**
$\quad \quad \quad \quad \quad \quad |$ $h = h \cup a_j$;
$\quad \quad \quad \quad \quad |$ **end**
$\quad \quad \quad \quad |$ **end**
$\quad \quad \quad \quad |$ addToRadixTree(H_a, h);
$\quad \quad \quad |$ **end**
$\quad \quad |$ **end**
$\quad |$ **end**
$\quad |$ return H_a;
end

Algorithm 3: Parallel RB-FD through the attributes usign radix trees (RB-FD-rtree-att).

Given the set of attributes $A = \{a_1, a_2, \ldots, a_k\}$ and an instance of relation r, this alternative associates independent threads to the search of refutations for each attribute A_i. It results in a set of concurrent threads accessing (read operations) the relation r to find refutations in it. The refutations are kept in radix trees, meaning that each thread has its own radix tree which is independent from another thread's radix tree.

Each process computes exactly $\Theta(k\frac{n(n-1)}{2})$ operations in finding refutations. The Algorithm 3 presents the parallel solution through the set of attributes. The set of refutations for each attribute (the right part in the refutation) are represented by H_{a_i} and corresponds to a radix tree.

5.2 Parallelism Through Tuples

This alternative is based on the fact that sometimes a search of FDs for a specific attribute can be needed. For instance, find all the refutations with attribute a_i as the dependent attribute. The approach consists in generating threads that go over a well-defined range on the tuples of the relation r. Thus, every refutation found by a thread will need to be stored in a common radix tree.

Formally, given a set of attributes $A = \{a_1, a_2, \ldots, a_k\}$, an instance of relation r and a number of threads p; a range b is defined as $b = |r|/p$, where $0 \leq b < |r|$.

A global radix tree is also defined with restricted access for writing and shared access for reading. Each process is given a range of contiguous tuples that will be its scope of search. The search of refutations starts when the main process takes a tuple as a pivot and sends this tuple to each thread that has the labor of finding refutations within its range. The radix tree is modified by getting exclusive access if and only if the refutation to add is maximal.

Input : relation r, set of attributes A, number of processes P
Output: sets of refutations $H_1, \ldots, H_k$

$i = 0$;
$j = 0$;
$range = |r|/P$;
for $a \in A$ **do**
   $H_a = \emptyset$;
   parallel for $t \in r$ **do**
      $i = j + 1$;
      $j = j + range$;
      findRefutations(a, t, i, j);
   end
end

function *findRefutations(a, t, i, j)*
   for $t' \in (i, j)$ **do**
      if $t(a)! = t'(a)$ **then**
         $h = \emptyset$;
         for $a_j \in \{A - \{a\}\}$ **do**
            if $t(a_j) == t'(a_j)$ **then**
               $h = h \cup a_j$;
            end
         end
         if h *is maximal* **then**
            lock();
            addToRadixTree(H_a, h);
            unlock();
         end
      end
   end
end

Algorithm 4: Parallel RB-FD through tuple ranges using shared radix trees (RB-FD-rtree-tup).

For an attribute D and a range b each thread computes exactly $\Theta(k * n * b)$ operations in finding refutations. Finding the refutations for all the attributes the complexity is $\Theta(k^2 * n * b)$. The Algorithm 4 presents the parallel solution through the set of attributes.

6 Experimental Results

A measurement of the average-over-time of the number of threads simultaneously running is introduced as the figure of merit that characterizes the efficiency of a Multi-threaded execution in a Many-Core computer. The objective is to distinguish, for a certain problem and execution, what is the proportion of the total running time for which threads execute in parallel and achieve some progress in their work.

Consider that blocking mechanisms are used to guarantee exclusive access to certain memory location. These mechanisms usually have three main states:

- Acquiring permission: The thread asks for permission to access the protected memory location. This state can consider blocking.
- Performing actions with permission: The thread performs actions on the protected memory location.
- Releasing the permission: The thread releases the permission after finishing its actions on the protected memory location.

It can be seen that the only blocking state is the first one, meaning that there can be no progress on the actions that the thread has to perform. For instance, the Lock blocking mechanism considers that only one writer thread can have the lock at a time. Thus the blocking state consists on repeatedly and unsuccessfully atomic operations by the thread to change the lock state from *unlock* to *lock*.

Formally, let Γ be the total running time of a thread and let β be the time the thread spends in blocking state, with a total of p threads running concurrently, the average of parallel running time $(APRT)$ is defined as follows:

$$APRT = \frac{\sum_{i=1}^{p} \Gamma_i - \beta_i}{p}$$

$APRT$ and the level of speedup are used to analyze the performance of the parallel algorithms described previously. We carried out a set of experiments using both of the proposals with the introduction of the radix tree.

Algorithms were implemented in the C++11 programming language. The experiments were carried out on a Dual 12 Core Xeon Haswell, with a total of 24 physical cores running at 2.60 GHz. Hyperthreading was disabled. The computer runs Linux 3.19.0-26-generic, in 64-bit mode. This machine has per-core L1 and L2 caches of sizes 32 and 256 KB, respectively and a per-processor shared L3 cache of 30 MB, with a 256 GB DDR RAM memory and 1 TB SSD. Algorithms were compared in terms of running times using the usual high-resolution (nanosecond) C functions in *time.h*.

The datasets (input) used are from UCI Machine Learning Repository [1], in particular we use the PAMAP2 Physical Activity Monitoring datasets.

6.1   Radix Tree Performance

One of the first improvement for RB-FD that was introduced in the previous section
was the use of a radix tree instead of a matrix to store maximal refutations. This new
data structure allows refutation compression by using the similar refutations' prefix
as a single node.

A number of 12 relation instances where used to carry out some experiments to
measure the performance with the goal of seeing what is the improvement achieved
by using this new data structure. Figure 3 shows the running time (in seconds) of
RB-FD using the original matrix H and the new version using a radix tree (RB-
FD-rtree). It is clear to see that RB-FD-rtree presents a significant less running time
when increasing the number of tuples in the datasets. A similar behavior occurs
when increasing the number of attributes. Furthermore, as analyzed previously, the
memory usage is reduced notoriously when using the radix tree. For instance, when
storing 30,000 maximal refutations the memory utilization is 10 times smaller than
using an array representation for the matrix H (Fig. 4).

Fig. 3 Running time of
RB-FD and RB-FD-rtree

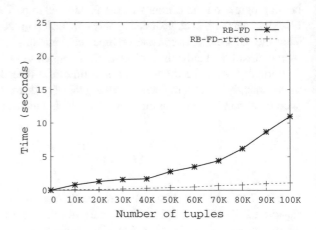

Fig. 4 Memory usage of
Matrix H and radix tree
storing refutations

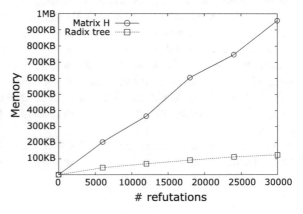

6.2   Parallel Alternatives

Table 3 shows the times achieved by RB-FD on its three versions: RB-FD-rtree (sequential algorithm) and the parallel versions RB-FD-rtree-att (parallelization through attributes) and RB-FD-rtree-tup (parallelization through tuples). This experiment corresponds to the finding of functional dependencies on a datasets with 502,182 tuples and 32 attributes. Figure 5 shows the speedup from the sequential version RB-FD-rtree when increasing the number of cores for RB-FD-rtree-att and RB-FD-rtree-tup. Up to 12 cores, it can be seen that the speedup of the parallelization through tuples grows slowly, then it stops growing and decreases its speedup a bit and continues fluctuating between a speedup of 2 and 3. The main reason for this phenomenon is that increasing the number of threads in RB-FD-rtree-tup produces data contention and more lock operations on the shared radix tree. The data contention does not occur with RB-FD-rtree-att, which presents a constant growth, since there is no shared radix tree, due to each thread has its own radix tree.

The parallelization over tuples (RB-FD-rtree-tup) works with a simple Lock as a mechanism for allowing exclusive access to the radix tree when a new maximal

Table 3 Running times (in seconds) varying the number of cores

# cores	RB-FD rtree	RB-FD rtree-att	RB-FD rtree-tup
1	4.41	4.41	4.41
4	–	2.08	1.7
8	–	1.94	1.04
12	–	1.46	0.76
16	–	1.86	0.66
20	–	1.62	0.56
24	–	1.90	0.44

Fig. 5 Speedup of parallel alternatives varying the number of cores

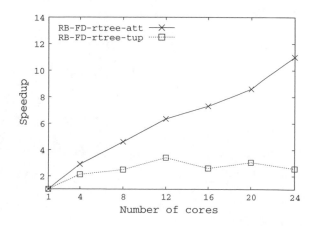

Table 4 Average of blocking and total time for threads running RB-FD-rtree-tup

# cores	Blocking time	Total time	APRT
4	0.018	0.22	0.91
8	0.030	0.156	0.79
12	0.05	0.132	0.63
16	0.048	0.106	0.57
20	0.064	0.114	0.43
24	0.082	0.126	0.35

refutation is found. Therefore it sounds interesting to see what is the APRT achieved when different number of threads are finding FDs in a dataset.

Table 3 shows the blocking and total times achieved by the algorithm when threads work on the same attribute (right-part of the refutation). As explained in Algorithm 4 the number of tuples is divided by the number of threads and each thread has to find refutations on its own range. These executions correspond to the same from the previous experiments, but focusing only on RB-FD-rtree-tup and the blocking and total times (Table 4).

According to the results, when increasing the number of cores and threads shorter running time is achieved. However after 12 cores, the running time stops decreasing and keeps stable. This behavior is explained by the addition of more blocking time as the number of cores and threads are increased. Therefore smaller APRT values are obtained and the greater blocking times are added. In other words, with more threads they have to perform less work but they suffer of blocking and waiting times (Fig. 6).

Fig. 6 Running time characterized by APRT and blocking time

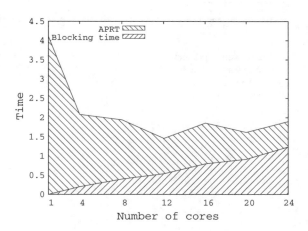

Acknowledgments We are grateful to Prof. Gilberto Gutiérrez and Dr. Pablo Sáez for their valuable guidance into the problem of finding functional dependencies. Thanks also to EMC Inc. in Irvine for lending us computing resources.

References

1. Bache, K., Lichman, M.: UCI Machine Learning Repository (2013)
2. Baixeries, J.: A formal concept analysis framework to mine functional dependencies. Workshop on Mathematical Methods for Learning. Como, Italy (2004)
3. Cormen, T.H., Leiserson, C.E., Rivest, R.L., Stein, C.: Introduction to Algorithms, 3rd edn. The MIT Press (2009)
4. Distel, F., Sertkaya, B.: On the complexity of enumerating pseudo-intents. Discrete Appl. Math. **159**(6), 450–466 (2011)
5. Flach, P., Savnik, I.: Database dependency discovery: a machine learning approach. AI Commun. **12**(3), 139–160 (1999)
6. Fuentes, J., Sáez, P., Gutiérrez, G., Scherson, I.D.: A method to find functional dependencies through refutations and duality of hypergraphs. Comput. J. bxu047 (2014)
7. Herlihy, M., Shavit, N.: The Art of Multiprocessor Programming. Morgan Kaufmann Publishers Inc., San Francisco, CA, USA (2008)
8. Huhtala, Y., Karkkainen, J., Porkka, P., Toivonen, H.: Tane: an efficient algorithm for discovering functional and approximate dependencies. Comput. J. **42**(2), 100 (1999)
9. Knuth, D.E.: The art of computer programming, vol. 3, 2nd edn. In: Sorting and Searching. Addison Wesley Longman Publishing Co., Inc, Redwood City, CA, USA (1998)
10. Novelli, N., Cicchetti, R., Fun: An efficient algorithm for mining functional and embedded dependencies. In: Database Theory—ICDT, vol. 1973. Springer, London, United Kingdom, 189–203 (2001)

Cellular Automata and Wireless Sensor Networks

Salimur Choudhury

Abstract Wireless sensor networks and cellular automata both are unconventional computing models. We can use cellular automaton based localized algorithms to solve various optimization problems of wireless sensor networks. In this chapter, we consider two very well known optimization problems and discuss various cellular automaton based algorithms for these two problems.

1 Introduction

Due to the advancement of sensor communication technology, wireless sensor networks have been used in many applications. The sensors are the main components of a *wireless sensor network*. A *sensor* is a very low cost small device that has limited battery power, short communication range, limited processing power and limited memory. A wireless sensor network (WSN) forms a distributed information processing system that gathers and processes different attributes of the network, for example, humidity, temperature, etc. A traditional wireless sensor network also includes single or multiple base stations that gather data from the sensors [53]. Each sensor of a sensor network has a sensing radius and a communication radius. A sensor can sense or monitor the region that falls within its sensing radius and communicates with other sensors that are within its communication radius. The sensors with whom a sensor can communicate are called the neighbors of the sensor. A typical wireless sensor network consists of hundreds, or even thousands, of sensors. These sensors are deployed in the monitored area and typically centralized controls are absent on these sensors [33]. These nodes are also unattended due to typical applications of sensor networks, it is not possible to have a human operator to directly attend to individual sensors. The sensors use each other (multi-hop communication) to route the information that they sense to the base stations for further processing. Habitat monitoring is an important application of wireless sensor networks. Mainwaring et al. [38] designed a wireless sensor network on Great Duck Island, Maine, USA to

S. Choudhury (✉)
Algoma University, Sault Ste. Marie, ON, Canada
e-mail: salimur.choudhury@algomau.ca

© Springer International Publishing Switzerland 2017
A. Adamatzky (ed.), *Emergent Computation*, Emergence, Complexity
and Computation 24, DOI 10.1007/978-3-319-46376-6_14

321

monitor the behavior of storm petrel. Some other habitat monitoring applications are considered in [8, 10, 27, 52]. The other important applications of wireless sensor networks are health monitoring systems [39, 45], home applications [46], etc.

Unlike traditional networking technology, energy is one of the major constraints that limits the usability of wireless sensor networks [53]. The energy of a sensor decays with time. A typical sensor can be in one of the two modes, asleep and awake. When a sensor is in the awake mode it can sense its region and also can communicate with its neighboring sensors while in the asleep mode, it can not sense or communicate. However the energy of a sensor decays in both modes. A sensor spends much more energy in the awake state than in the asleep state and the amount of energy needed for the purpose of communication increases with the increase of radius. When a sensor loses all of its energy we consider that sensor as dead. Designing energy aware techniques for different applications of wireless sensor networks is a well studied research area [26].

As the sensors spend more energy in communication and they use multi-hop communication, the sensors that are closer to the base station can be overloaded and can die early. This problem can be solved using mobile sinks [2]. When a component of a wireless sensor network (either a sensor or a sink) has moving capability then we call the network a *Mobile Wireless Sensor Network* (MWSN). Moreover, in different applications, for example, military field, mobile objects monitoring, etc., the sensors cannot be deployed in their desired positions by a centralized (or human) operator. In these cases, we can deploy the sensors densely and they can autonomously move within the network to improve the performance of the network if they have a moving capability [37].

In most of the applications of a WSN or a MWSN, two metrics are considered to measure the performance of the network [18]. One is the coverage and the other is the network lifetime. Coverage is defined as the area monitored by the network. The definition of network lifetime varies with the application. In some cases, we define a network lifetime as the time elapsed until a sensor of the network dies, while in some cases network lifetime refers to the time until a given fraction of the sensors die. The other definition of network lifetime is the time elapsed "until all the sensors are dead" [18].

Underwater sensor networks (UWSN) are relatively new applications of wireless sensor networks [1, 42]. In a typical UWSN, sensors are used to monitor different aspects of seas, lakes, etc. The sensors of a UWSN can be either static or mobile. Differing from a land based MWSN, the sensors of a mobile UWSN move in three dimensions. Deployment is one of the main challenges of the effective use of wireless sensor networks. In a typical deployment, a number of metrics are considered. One of the main metrics is the energy of the sensors. As the energy of the sensors is limited and they decay with time we need to deploy the sensors in a way that enables us to extend their lifetime as long as possible. There are other metrics often considered at the deployment, for example, connectivity, coverage, etc. [48].

Most of the algorithms found in the literature related to the sensor network deployment are either global or distributed [44]. In a real environment it is not feasible to implement a global algorithm. On the other hand, a distributed algorithm needs more communication and message passing which makes the algorithm complex. A local algorithm can be a good candidate solution for this type of optimization problems. Even the local algorithms that have been designed for WSN are quite complex in terms of processing and memory needed [47].

In this chapter we discuss cellular automaton based algorithms [14, 15] for two of the most important optimization problems of wireless sensor networks.

2 Cellular Automata

The *Cellular Automaton* is a biologically inspired model that has been widely used to model different physical systems [11, 21, 25]. We can model a two dimensional cellular automaton as a two dimensional grid. Each cell of the grid is in one of a finite number of states. The cells change state synchronously at discrete time steps and the new state depends always only on states of a local neighborhood.

One of the major benefits of using a cellular automaton based model is that it needs very limited local information to compute the solution. We can also perform extensive simulations if we use cellular automaton models as cellular automata are easy to implement. The main challenge is to design good local rules that give a satisfactory global outcome for different optimization problems. In the following subsections we describe and discuss some optimization problems and related challenges.

Typically, a large number of sensors are deployed in the environment to sense the data. Since the sensors are densely deployed, multiple sensors can sense or cover the same region. The sensor spends most of its energy to sense the environment and dies as soon as it loses all its energy. Different techniques have been applied to preserve the energy for as long as possible. One of these techniques is to let a part of the sensors sleep for a certain amount of time and let them awake only when necessary.

Much research has been done on such scheduling problems. One of the common techniques is to make a domatic partition of the sensors [40, 49]. In a domatic partition (set), some representative sensors are selected in a way that either a sensor is in that set or at least one of its neighbors is in the set. The idea is to let this set of sensors stay awake for a certain period of time so that the area of the network is covered and the life time of the network is also increased. As the domatic partition problem is NP-hard [20], many approximation algorithms have been proposed [3, 12, 19, 31]. In such algorithms, it is assumed that the network is connected. Once the network is disconnected the algorithm is of no use. Different strategies to solve energy efficient coverage problems can be found in [9].

There are some algorithms to solve the sleep-wake scheduling problem of a WSN using cellular automata [7, 18, 32]. In every case, they consider a radius 1 neighborhood of the sensors. In this chapter, we consider a radius 2 neighborhood to solve the sleep-wake scheduling problem of a wireless sensor network. Moreover, we have used two different energy levels for sensing and communicating among neighbors. To make the comparison with the earlier radius 1 algorithms realistic, in our algorithm the sensing radius remains one and the larger radius two neighborhood is used only for communicating with neighbors. Naturally, the energy used for communication depends on the square of the communication radius, and this is taken into account when calculating the energy use of sensors. Furthermore, our algorithm solves the sudden falls of the coverage of [18]. We developed a CA simulator and compare the different rules to implement the CA model. Recently a variant of our algorithm is considered in [30].

Object detecting in a WSN is one of the well studied research problems [4, 5, 28, 36]. In an animal behavior monitoring environment or in a military field, sensors are usually deployed to detect animals or enemies, respectively. To this end, algorithms are proposed that increase the probability of success. In such a situation network life time is also an important factor. As soon as the network dies, we cannot detect the object any more. We also find that our CA model increases the network lifetime and can detect more objects than the earlier algorithms based on CA models.

3   The Coverage and the Object Detection Problems

In the following, we extend the CA algorithms considered in Cunha et al. [18]. We represent the sensors as the cells of a two dimensional grid. Each cell can have three states: awake, asleep and dead. A real value is associated to each cell representing the remainder of the available battery for that sensor. At each time point, the value is decreased correspondingly, depending on whether the sensor is awake or asleep. When the value reaches 0 (or a negative value), the sensor enters permanently into the dead state. Initially, after the deployment, we set the states of a subset of cells as awake. In the awake state, the sensor can sense its region (radius 1 neighborhood) and communicates with its neighbors. For communication with neighbors, a sensor has, depending on the algorithm used, either radius 1 [18] or radius 2 neighborhood (our algorithm). The energy consumption when using the radius 2 neighborhood is larger (square of the energy consumption for the radius 1 neighborhood). While using a large communication radius increases the energy consumption, having more information about the states of nearby sensors may enable the network to make better decisions on which set of cells should be kept awake.

In practice, the sensing range and the communication range of a sensor can be different in a WSN [9]. In the asleep state, a sensor cannot sense and communicate with any of its neighbors. However, as explained below, a sensor in the asleep state periodically wakes up and communicates with its neighborhood to determine whether or not the sensor should go to the awake state. A sensor that runs out of energy goes permanently to the dead state.

We use only rules that count the number of awake sensors. The CA transition rule is applied only periodically. For $i, j \geq 0$, we define i/j rules be setting:

- A cell in the awake state that has at least i neighbors awake, goes to the asleep state; otherwise it remains awake.

- A cell in the asleep state that has less than j neighbors awake, wakes up; otherwise it stays asleep.

Thus, at periodic intervals, when the transition rules in a particular cell are applied, the cell wakes up and checks its neighborhood. For this time period the cell consumes energy (the energy consumption rate depending on the neighborhood radius as described later in the simulation and results section).

Note that, when counting the number of awake cells in the neighborhood, it does not make a difference whether the remaining cells are asleep or dead. Since a sensor in the asleep state is not using energy for communication, it appears to its neighborhood as a dead sensor. Thus, in order to keep the algorithm realistic, the CA rules cannot distinguish between sensors that are asleep and those that are dead. Though the boundary cells can always have less than i or j sensors for some rules (for example, $3/3$, $3/4$, $4/4$, etc.), for simplicity we use similar rules for all the cells.

The algorithm terminates when all sensors are in the dead state.

In this section, we describe two problems for which we apply our algorithm: coverage and object detection.

In a WSN, the area covered by the awake sensors is called the coverage of the network. A sensor in the awake state is considered to cover the sensors that are within its sensing radius (radius 1 neighborhood in our paper). Sensors in the asleep or dead states do not provide any coverage. A network, where a large number of cells is awake, provides good coverage, but due to the increased energy consumption many cells start to die out which then makes it impossible to maintain the coverage level. The goal of a coverage algorithm is to provide coverage at an acceptable threshold (measured as a percentage of the area covered) with as few sensors awake as possible. The decisions on which cells are asleep/awake need to be made locally, which makes CA a good model for this problem. What is an acceptable threshold naturally depends on the particular applications we have in mind. In general, the solutions to the coverage problem can be evaluated by considering the resulting coverage percentage as a function over time.

In the object detection problem, the network is used to detect various objects (such as wild animals, or enemy soldiers) that enter the network area. While the type of the movement may depend very much on the particular applications, here we assume that the objects move randomly. In one set of experiments, the objects change direction at each time step, and we consider another variant where the objects move in one direction a fixed amount of time before making a random choice of a new direction. As far as we know, CA algorithms have not been previously considered for the object detection problem, and naturally for more detailed results, further work can consider objects whose movement is not random.

Note that, if we require that the object be observed at all times, the problem setting would become more similar to the coverage problem. Furthermore, when discussing the object detection problem, we assume that the sensors have sensing radius zero, that is, an object is detected only when it comes into the same grid point with an awake sensor. Choosing the sensing radius is just a matter of scaling. If we use a larger sensing radius, almost all objects will be detected even with very conservative algorithms that keep only a small portion of cells awake.

We identify the best transition rules for the coverage problem and the object detection problem both with respect to the radius 1 and radius 2 neighborhood is (1/1) and compare these with each other we find that radius 2 algorithm outperforms radius 1 algorithm in all cases.

4 Dispersion of Mobile Sensor Networks

Mobile wireless sensor networks have a wide range of applications [33, 37] and there has been much research on this topic. The mobile sensors have the same general characteristics as in a sensor of a sensor network. Additionally, each mobile sensor has locomotion capability. One of the main reasons for using the mobile sensors is to improve the coverage of the networks. However, this leads to a very important research question. How one sensor can move so that it can maximize the coverage of the network?

Typically, a sensor is deployed in a network to cover some region of the network. The coverage of a network is defined as the area that is covered by its sensors. In different applications of MWSN, it is not possible to deploy the sensors deterministically so that they can maximize the coverage. More commonly, they are deployed randomly and the sensors are required to disperse autonomously using algorithms to maximize the coverage of the network [37]. In different applications, along with the coverage, the preservation of the connectivity is also a major concern. Once the sensors try to move by themselves, they can break the connectivity. Connectivity is an important aspect of the network as it is used to route the data among the nodes. We call this problem *the Mobile Dispersion Problem*.

We first discuss a cellular automaton based algorithm that maximizes the coverage of a MWSN very quickly, involves fewer sensor movements when compared with an earlier cellular automaton algorithm [48], and is simpler than other existing techniques [17, 22, 51]. Though we have not considered the energy constraints on the movement of the sensors explicitly, the number of sensor movements gives a good approximate measure of energy use and gives a good measure of the usefulness of the algorithms in practice.

In the first set-up the sensors are restricted to an area. This means that, as long as the sensors are sufficiently dispersed (the algorithm works probabilistically), most of the network stays connected. We also consider a scenario where a number of sensors are initially densely deployed and the area is (at least potentially) unbounded. Thus, the algorithm needs to explicitly ensure that the connectivity of the network is preserved.

Connectivity preservation is an important aspect of different wireless sensor networks as it is necessary to route information within the network. In the second part of this section we also consider different CA based algorithms for connectivity preserving deployment of mobile sensors [13, 16]. In our experimental set-up we first consider a "regular" $n \times n$ square formation as an initial configuration of the network and find an optimal solution using a deterministic algorithm. Finally, we study systematically more realistic random initial deployments of sensors and for the realistic initial deployments it is not equally easy to obtain an optimal solution. We need to add features to the algorithm that try to prevent the creation of "holes" in the network and also, in this context, it turns out to be useful to consider randomized variants of the algorithm.

There is another difference between the two variants of the problem. In the first variant, each cell can contain only one sensor at any time period. On the other hand, in the second variant a cell can contain more than one sensor.

Most of the solutions that have been proposed for this mobile dispersion problem are either global or distributed algorithms [17, 22, 29, 43, 51].

Vector based approach is a well known approach for deploying mobile sensors to maximize the coverage of the mobile sensor networks [23, 24, 41, 50, 54]. They all are inspired by the different physical models. In these cases, a sensor node determines its movement direction based on the force received from its neighbors. Howard et al. [24] first propose a virtual force algorithm which is localized but it is not applicable for a discrete model and computing the directions is quite complex compared to our cellular automaton algorithms. Zuo et al. [54] propose a distributed virtual force algorithm for the dispersion of the mobile sensors. They use a combination of attractive and repulsive forces to determine the movement of the sensors.

Cortes et al. [17] propose an algorithm for an analogues problem in robotics. In their algorithm, each robot (sensor in our algorithms) draws a Voronoi diagram and it moves to minimize its local uncovered areas by aligning its sensing range with the Voronoi region as much as possible. Some other Voronoi diagram based algorithms are proposed in [22, 51].

Barriere et al. [6] propose a local algorithm for the uniform dispersion of autonomous mobile robots in a grid. However, their algorithm is quite complex and the algorithm does not work for the network having communication radius less than 4.

There are other local algorithms proposed for a variant of coverage problem [34, 35] called "focused coverage" where some regions of the network have more priority to be covered. The definition of this problem is quite different than the problem that we consider in this part of the thesis.

A simple approach has been proposed in [48] based on cellular automata to maximize the coverage of the mobile sensor networks. It does not explicitly consider the connectivity constraint and there are no techniques considered in the algorithm to preserve the connectivity. We discuss this algorithm in some more details when we describe our algorithm in Sect. 4.2.

4.1 System Model of Mobile Sensor Networks

A set of n mobile sensors are deployed in a network. They are dispersed densely within the network. The sensors are homogeneous, i.e. they have same sensing (R_s) and

Fig. 1 An example of
mobile wireless sensor
network

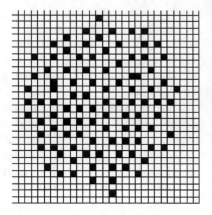

communication (R_c) radii, where $R_c > R_s$. The sensing radius refers to the monitoring function of the network and a sensor is said to cover the cells within its sensing radius. On the other hand, the sensors can communicate with other sensors that are within their communication radius. The network is connected if any two sensors can be connected by a path of sensors $m_1, \ldots, m_k$ where m_{i+1} is within the communication radius of m_i, $i = 1, \ldots, k - 1$. The coverage of the network is the area covered by the largest connected component of the network. All the sensors move with the same speed. In our cellular automaton based algorithms, the sensors can move a maximum of one cell at a time. A random deployment of a mobile sensor network is shown in Fig. 1. The small circles are the sensors.

4.2 Algorithms for Maximizing Coverage

We briefly discuss our algorithm for the first variant of the problem, i.e. maintaining connectivity of the network is not the main concern of this algorithm. The main goal of the algorithm is to increase the coverage. In this case, the sensors are deployed in a closed area and there are sufficiently many sensors so that when they are evenly distributed the majority of the sensors remain connected.

We consider a 2-D cellular automaton where the states of the cells indicate the presence or the absence of mobile sensors. The movement of the sensors is modeled by state changes. We consider different (R_c, R_s) pairs and the goal of the algorithms is to position the sensors in a way that maximizes the coverage.

The sensors can communicate and get information about the locations of nearby sensors up to R_c. During a given time period, a sensor spends much less energy to communicate with its neighbors than in monitoring its environment. So it is reasonable to assume that the communication radius is larger than the sensing radius even though the energy consumption increases with the radius.

In each time step, a sensor can move one cell in any direction. The algorithms use the information about the positions of the nearby sensors to determine the movement of the current sensor in the next time step. The goal of the algorithms is two-fold:

- to position the sensors in a way that maximizes the coverage,
- to minimize the movements of the sensors in order to conserve energy.

We propose the following cellular automaton model for this particular problem.

We consider an R_c neighborhood for each cell. We divide the neighborhood of a cell into the North East (NE), North West (NW), South East (SE) and South West (SW) quadrants. As indicated in Fig. 2 for $R_c = 4$, each quadrant consists of 20 cells. Now we describe the algorithm that is used to solve the problem.

1. Initially the sensors are placed in different ways (described in the simulation and results section) within the network. The state of the cell that contains a sensor is set to 1, otherwise 0.
2. Each time period is divided into two phases: odd and even (similar to [48]).

   - Odd Phase: The algorithms use two parameter values k' and k''. In this phase, each sensor determines whether it should move (and where) or not.
     - For each of the four quadrants $x \in \{NE, NW, SW, SE\}$ of its current location, the sensor calculates a weight, t_x that is based on the number of sensors seen in that quadrant. The algorithm gives higher weight for sensors that are closer to the current sensor. For example, in case of $R_c = 3$, the neighbors that are distance 1 away are assigned weight 3. On the other hand the neighbors that are distance 3 away are assigned weight 1.
     - Each sensor determines the maximum and the minimum value from t_x, where $x \in \{NE, NW, SW, SE\}$.

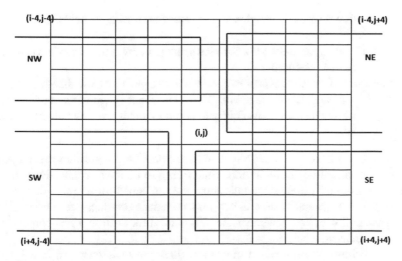

Fig. 2 The four quadrants of a cell (i, j) When R_c is 4

- If the minimum value is less than k' and the maximum value is more than k'', then the sensor chooses a position randomly from the quadrant (one of the two positions of the quadrant that are adjacent to the current location of the sensor) that has the minimum value. In case of multiple minimum values, the sensor chooses a random quadrant. In the even phase, the sensor tries to move into that position.

- Even Phase: In this phase, a sensor moves to the cell that it has chosen in the odd phase if the following conditions are met:
  - The cell where it is moving should be empty.
  - No other sensor tries to move to that cell. Randomly one sensor is chosen if there are multiple candidate sensors for that cell. A radius $R_c = 2$ communication is enough to determine who are the candidates as the sensors move one cell at a given time.

- Once the sensor decides on its next location, the state of that location is set to 1 and the state of the sensor's previous location is set to 0.

By the (k'/k'') algorithm we mean the algorithm where the sensor movement condition requires that the minimum (respectively, maximum) weight computed for the quadrants is less than k' (respectively, greater than k''). We try different (k'/k'') rules, where k' and k'' are integers, $1 \le k', k'' \le 5$. We consider three different scenarios for the grid of size 125 and compare the best rules of our model with an earlier algorithm [48] for which we use the name "COUNT". The COUNT algorithm determines the movement of the sensors based only on a numerical quantity obtained by counting the numbers of nearby sensors at different distances and does not incorporate directional information into the numerical value.

Before explaining our simulation results, we briefly mention the differences between our algorithm and the algorithm from [48] (that we call the "COUNT" algorithm):

- Our algorithm divides the neighborhoods into four quadrants but "COUNT" does not.

- "COUNT" calculates only one k for each sensor, whereas our algorithm calculates four, one for each quadrant.

- Our algorithm forces the sensor to choose a future position from that quadrant that has minimum weighted value, while "COUNT" chooses a random one from all of its neighbors. So there is a possibility to choose a position where already enough sensors exist.

In this section, we give a high-level description of our algorithm that disperses sensors from an initial configuration while trying to maintain connectivity. In particular, for simplicity, below we talk about the movement of an individual sensor. However, in the general case, one cell may contain more than one sensor and the state of the cell needs to remember the information for each sensor it contains. We consider a scenario where $R_c > R_s$.

The algorithm determines the movement direction of a sensor m_i based on the weighted number of neighbors of m_i in the positive and in the negative x-direction (respectively, y-direction). The weights assigned to neighbors in case of $R_c = 3$ are as follows: The weights for neighbors at distance 1, 2, 3 are 4, 2, 1 respectively. In case of $R_c = 2$, the weights for neighbors at distance 1, 2 are 2, 1 respectively.

In an ideal case, one sensor can have neighbors at distance 3 away in case of $R_c = 3$ and $R_s = 1$ to maximize the coverage while maintaining connectivity. For this reason we assign the weights to be inversely proportional to the distance.

The state representing an individual sensor is a pair (x, y), $x, y \in \{-1, 0, 1\}$. The state remembers the last move of the sensor. The direction of the movement is stored in the state because, in order to avoid infinite loops, the algorithm preserves the current movement direction. For example, a pair $(0, 1)$ means that in the last time step the sensor did not move in the x-direction and moved upwards along the y-direction. The next movement step of a sensor m_i is determined by the weighted neighborhood of m_i and the previous movement direction of m_i that is stored in the pair of integers representing m_i. When a cell has more than one sensor, each represented by a pair (x, y), $x, y \in \{-1, 0, 1\}$, the algorithm computes the potential movement direction for each of these sensors. (The movement direction depends on (x, y) and, thus, may be different for different sensors in the same cell.)

First we describe how the algorithm determines the direction of the next movement step. Then we describe blocking rules that are used to prevent loss of connectivity and, finally, we describe additional move-back rules that are used when the blocking rules fail.

The algorithms use a parameter (Multiplier), $M \geq 2$ that serves to encourage the sensor to keep moving in the direction of its previous movement step. We define the movement rule as follows.

Suppose a sensor m_i is represented by pair (s_x, s_y). Suppose that w_1 is the sum of weights of neighbors of m_i in the negative x-direction (to the left of m_i). Suppose that w_2 is the sum of basic weights of neighbors of m_i in the positive x-direction (to the right of m_i). The potential movement of m_i in x-direction depends on the value of s_x, i.e. by remembering the last move, the sensor tries to keep moving in the same direction, unless there is a really good reason to change direction. So the movement of a sensor along the x-direction depends on its neighbors in the positive and negative x-direction and the current value of s_x. If we use a multiplier M, then the decision of movement to the x-direction is determined by a value x-move(m_i) defined as

- if $s_x = 0$, x-move$(m_i) = w_2 - w_1$
- if $s_x = -1$, x-move$(m_i) = (M * w_2) - w_1$
- if $s_x = 1$, x-move$(m_i) = w_2 - (M * w_1)$

Now, if x-move$(m_i) = 0$ then the sensor does not move in the x-direction. If x-move$(m_i) \geq 1$ and x-move$(m_i) \leq -1$ then the sensor moves to the negative and positive direction, respectively. Movement in the y-direction is determined analogously.

If we do not have multipliers, that is, set $M = 1$, the movement rules defined by the weighted neighborhood of a sensor together with the move back rules can lead to infinite cycles.

We introduce rules that attempt to prevent the network from losing connectivity. Below we consider a movement step in the positive x-direction. The same rules apply to the 3 other directions. If a sensor m_i does not see any neighbors within distance $R_c - 1$ in the negative x-direction, a movement step in positive x-direction is blocked.

Consider a cell with n' sensors $m_1, m_2, ..., m_{n'}$. Each sensor determines independently whether it should move or not. If all the sensors move to a positive x-direction then sensor m_1 checks within distance $R_c - 1$ in the negative x-direction for the connectivity and if there is no sensor within distance $R_c - 1$ then only sensor m_1 will not move to the positive x-direction. In case all sensors in the cell try to move in one of the other three directions, a similar control step is done in the opposite direction.

There are some additional blocking rules too.

The above blocking conditions still do not guarantee the preservation of connectivity because the sensors that are within distance $R_c - 1$ of each other can move to the opposite directions. For this reason we introduce the following "move back" rules. At any given time period t, before moving to the positive x-direction, a sensor s remembers whether or not it has a neighbor in two different quadrants in the negative x-direction. At the next time period $t + 1$, if the sensor finds that there is no sensor in one of these quadrants but there was at least one sensor in the same quadrant at period t, then the sensor moves back in the negative x-direction. The same rule applies to 3 other directions.

If we are concerned only about the loss of connectivity, the move back rule can be simplified by remembering only whether or not there were neighbors in the direction opposite to the current movement, that is, in that case quadrants Q_1 and Q_2 can be combined together. We call this simplified rule the 180°-*move back rule*. The purpose of the quadrant-rules is to prevent the creation of holes in the network. Large holes increase the hop-distance between individual sensors and, thus, worsen the strong connectivity of the network.

If we have a cell with n' sensors and x-move(m_i) > 0 for all $i = 1 ... l$, where $l \leq n'$, then in case of 180°-move back rule, only sensor m_1 checks cells for a distance R_c in the negative direction and if it finds that there is no sensor in the negative direction then it moves back in the negative direction. All other sensors move according to the other rules. On the other hand, in the case of quadrant move back, before moving to a new direction each sensor remembers whether there is any sensor in the quadrants of the opposite direction.

At any period t, m_1 checks the quadrants of the opposite direction and if it finds that there is no sensor in one of the quadrants at that time period but there was at least one in that quadrant at time $t - 1$, then only sensor m_1 moves back in negative x-direction. Other sensors continue with other rules. These rules are applied similarly in the other three directions.

In a deterministic version of the algorithm, one sensor checks the rules at each time period to decide whether or not it should move from the current location. We also consider the probabilistic versions of the algorithms where at each time period, each sensor verifies the rules with some probability. Though, in a probabilistic version, it takes more time than the deterministic one to reach a final position but we find that the randomized algorithm often gives a better result and the probability of being in a cycle is much less than in the deterministic cases.

5 Conclusion

In this chapter we describe some well known cellular automaton based algorithms for two well known optimization problems in wireless sensor networks. Wireless sensor networks can be well modeled using cellular automata. A cell can represent a sensor. Sensors can decide locally their acts to accomplish some tasks to obtain a global solution for an optimization problems. Different results suggest that cellular automata can be used to model such problems of the sensor networks. We can do some extensive simulations and these algorithms can also be worked as benchmarks for other algorithms. Cellular automaton based algorithms exist in literature only consider square grids. However, considering hexagonal grids can be very interesting which is much more realistic for wireless communications.

References

1. Akyildiz, I., Pompili, D., Melodia, T.: Underwater acoustic sensor networks: research challenges. Ad Hoc Netw. (Elsevier) **3**, 257–279 (2005)
2. Alsalih, W., Hassanein, H., Akl, S.: Placement of multiple mobile data collectors in wireless sensor networks. Ad Hoc Netw. **8**(4), 378–390 (2010)
3. Alzoubi, K.M., Wan, P.J., Frieder, O.: New distributed algorithm for connected dominating set in wireless ad hoc networks. In: Proceedings of the 35th Annual Hawaii International Conference on System Sciences (HICSS), pp. 3849–3855. IEEE (2002)
4. Arora, A., Dutta, P., Bapat, S., Kulathumani, V., Zhang, H., Naik, V., Mittal, V., Cao, H., Demirbas, M., Gouda, M.: A line in the sand: a wireless sensor network for target detection, classification, and tracking. Comput. Netw. **46**(5), 605–634 (2004)
5. Aslam, J., Butler, Z., Constantin, F., Crespi, V., Cybenko, G., Rus, D.: Tracking a moving object with a binary sensor network. In: Proceedings of the 1st International Conference on Embedded Networked Sensor Systems, pp. 150–161. ACM (2003)
6. Barriere, L., Flocchini, P., Mesa-Barrameda, E., Santoro, N.: Uniform scattering of autonomous mobile robots in a grid. In: Proceedings of the IEEE International Symposium on Parallel Distributed Processing, pp. 1–8, May (2009)
7. Baryshnikov, Y.M., Coffman, E.G., Kwak, K.J.: High performance sleep-wake sensor systems based on cyclic cellular automata. In: Proceedings of the International Conference on Information Processing in Sensor Networks, IPSN'08, pp. 517–526. IEEE (2008)
8. Biagioni, E.S., Bridges, K.W.: The application of remote sensor technology to assist the recovery of rare and endangered species. Int. J. High Perform. Comput. Appl. **16** (2002)
9. Cardei, M., Wu, J.: Energy-efficient coverage problems in wireless ad-hoc sensor networks. Comput. Commun. **29**(4), 413–420 (2006)
10. Cerpa, A., Elson, J., Hamilton, M., Zhao, J., Estrin, D., Girod, L.: Habitat monitoring: application driver for wireless communications technology. In: Workshop on Data Communication in Latin America and the Caribbean, SIGCOMM LA'01, pp. 20–41, New York, NY, USA. ACM (2001)
11. Chaudhuri, P.P.: Additive cellular automata: theory and applications. In: Additive Cellular Automata: Theory and Applications. IEEE Computer Society Press (1997)
12. Cheng, X., Huang, X., Li, D., Wu, W., Du, D.Z.: A polynomial-time approximation scheme for the minimum-connected dominating set in ad hoc wireless networks. Networks **42**(4), 202–208 (2003)

13. Choudhury, S., Akl, S.G., Salomaa, K.: Cellular automaton based motion planning algorithms for mobile sensor networks. In: To Appear in Proceedings of the First International Conference on the Theory and Practice of Natural Computing, Tarragona, Spain, October, 2012

14. Choudhury, S., Salomaa, K., Akl, S.G.: A cellular automaton model for wireless sensor networks. J. Cellular Automata **7**(3), 223–241 (2012)

15. Choudhury, S., Salomaa, K., Akl, S.G.: Cellular automaton-based algorithms for the dispersion of mobile wireless sensor networks. IJPEDS **29**(2), 147–177 (2014)

16. Choudhury, S., Salomaa, K., Akl, S.G.: A cellular automaton model for connectivity preserving deployment of mobile wireless sensors. In: 2nd IEEE International Workshop on Smart Communication Protocols and Algorithms (ICC'12 WS—SCPA), Ottawa, Canada, June 10–15 (2012)

17. Cortes, J., Martinez, S., Karatas, T., Bullo, F.: Coverage control for mobile sensing networks. IEEE Trans. Robot. Autom. **20**(2), 243–255 (2004). April

18. Cunha, R.O., Silva, A.P., Loreiro, A.A.F., Ruiz, L.B.: Simulating large wireless sensor networks using cellular automata. In: Proceedings of 38th Annual Simulation Symposium, 2005, pp. 323–330. IEEE (2005)

19. Dai, F., Wu, J.: An extended localized algorithm for connected dominating set formation in ad hoc wireless networks. IEEE Trans. Parallel Distrib. Syst. **15**(10), 908–920 (2004)

20. Garey, M.R., Johnson, D.S.: Computers and intractability. A guide to the theory of NP-completeness. A Series of Books in the Mathematical Sciences. WH Freeman and Company, San Francisco, Calif (1979)

21. Garzon, M.: Models of massive parallelism, Analysis of cellular automata and neural networks. In: Texts in Theoretical Computer Science. Springer (1995)

22. Heo, N., Varshney, P.K.: A distributed self spreading algorithm for mobile wireless sensor networks. In: Wireless Communications and Networking, 2003. WCNC 2003. 2003 IEEE, vol. 3, pp. 1597–1602 (2003)

23. Heo, N., Varshney, P.K.: Energy-efficient deployment of intelligent mobile sensor networks. IEEE Trans. Syst. Man Cybernet. Part A: Syst. Humans **35**(1), 78–92 (2005)

24. Howard, A., Mataric, M.J., Sukhatme, G.S.: Mobile sensor network deployment using potential fields: a distributed, scalable solution to the area coverage problem. In: Proceedings of the 6th International Symposium on Ditributed Autonomous Robotics Systems, pp. 299–308 (2002)

25. Ilachinski, A.: Cellular Automata: A Discrete Universe. World Scientific

26. Islam, K.: Energy aware techniques for certain problems in wireless sensor networks. Ph.D. thesis, School of Computing, Queen's University, ON, Canada (2010)

27. Juang, P., Oki, H., Wang, Y., Martonosi, M., Peh, L.S., Rubenstein, D.: Energy-efficient computing for wildlife tracking: design tradeoffs and early experiences with zebranet. SIGPLAN Not. **37**(10), 96–107 (2002). October

28. Juang, P., Oki, H., Wang, Y., Martonosi, M., Peh, L.S., Rubenstein, D.: Energy-efficient computing for wildlife tracking: design tradeoffs and early experiences with zebranet. ACM SIGOPS Oper. Syst. Rev. **36**(5), 96–107 (2002)

29. Kar, K., Banerjee, S.: Node placement for connected coverage in sensor networks. In: Proceedings of WiOpt, vol. 3. Citeseer (2003)

30. Ko, S., Lee, H., Han, Y.: An Improved Cellular Automaton Model for Wireless Sensor Networks. Submitted (2012)

31. Kuhn, F., Wattenhofer, R.: Constant-time distributed dominating set approximation. Distrib. Comput. **17**(4), 303–310 (2005)

32. Li, W., Zomaya, A.Y., Al-Jumaily, A.: Cellular automata based models of wireless sensor networks. In: Proceedings of the 7th ACM International Symposium on Mobility Management and Wireless Access, pp. 1–6. ACM (2009)

33. Li, X.: Improving area coverage by mobile sensor networks. Ph.D. thesis, SCS, Carleton University, ON, Canada (2008)

34. Li, X., Frey, H., Santoro, N., Stojmenovic, I.: Focused-coverage by mobile sensor networks. In: IEEE 6th International Conference on Mobile Adhoc and Sensor Systems, 2009. MASS'09, pp. 466–475, Oct (2009)

35. Li, X., Frey, H., Santoro, N., Stojmenovic, I.: Strictly localized sensor self-deployment for optimal focused coverage. IEEE Trans. Mobile Comput. **10**(11), 1520–1533 Nov (2011)
36. Lin, C.Y., Peng, W.C., Tseng, Y.C.: Efficient in-network moving object tracking in wireless sensor networks. IEEE Trans. Mobile Comput. pp. 1044–1056 (2006)
37. Liu, B., Brass, P., Dousse, O., Nain, P., Towsley, D.: Mobility improves coverage of sensor networks. In: Mobihoc (2005)
38. Mainwaring, A., Culler, D., Polastre, J., Szewczyk, R., Anderson, J.: Wireless sensor networks for habitat monitoring. In: Proceedings of the 1st ACM International Workshop on Wireless Sensor Networks and Applications, WSNA'02, pp. 88–97, New York, NY, USA. ACM (2002)
39. Malan, D., Fulford-jones, T., Welsh, M., Moulton, S.: Codeblue: an ad hoc sensor network infrastructure for emergency medical care. In: In International Workshop on Wearable and Implantable Body Sensor Networks (2004)
40. Min, M., Du, H., Jia, X., Huang, C.X., Huang, S.C.H., Wu, W.: Improving construction for connected dominating set with Steiner tree in wireless sensor networks. J. Global Optimization **35**(1), 111–119 (2006)
41. Pac, M.R., Erkmen, A.M., Erkmen, I.: Scalable self-deployment of mobile sensor networks: a fluid dynamics approach. In: 2006 IEEE/RSJ International Conference on Intelligent Robots and Systems, pp. 1446–1451, Oct (2006)
42. Partan, J., Kurose, J., Levine, B.N.: A survey of practical issues in underwater networks. In: Proceedings of the 1st ACM International Workshop on Underwater Networks, WUWNet'06, pp. 17–24, New York, NY, USA. ACM (2006)
43. Poduri, S., Sukhatme, G.S.: Constrained coverage for mobile sensor networks. In: 2004 IEEE International Conference on Robotics and Automation, 2004. Proceedings. ICRA '04, vol. 1, pp. 165–171, April-1–May (2004)
44. Sahni, S., Xu, X.: Algorithms for wireless sensor network. Int. J. Distrib. Sensor Netw. **1**, 35–56 (2005)
45. Schwiebert, L., Gupta, S.K.S., Weinmann, J.: Research challenges in wireless networks of biomedical sensors. In: Proceedings of the 7th Annual International Conference on Mobile Computing and Networking, MobiCom'01, pp. 151–165, New York, NY, USA. ACM (2001)
46. Srivastava, M., Muntz, R., Potkonjak, M.: Smart kindergarten: sensor-based wireless networks for smart developmental problem-solving environments. In: Proceedings of the 7th Annual International Conference on Mobile Computing and Networking, MobiCom'01, pp. 132–138, New York, NY, USA. ACM (2001)
47. Suomela, J.: Optimisation problems in wireless sensor networks: local algorithms and local graphss. Ph.D. thesis, Department of Computer Science, University of Helsinki, Finland (2009)
48. Torbey, S.: Towards a framework for intuitive programming of cellular automata. M.Sc thesis, School of Computing, Queen's University, ON, Canada (2007)
49. Wan, P.J., Alzoubi, K.M., Frieder, O.: Distributed construction of connected dominating set in wireless ad hoc networks. In: Proceedings of the Twenty-First Annual Joint Conference of the IEEE Computer and Communications Societies (INFOCOM). IEEE, vol. 3, pp. 1597–1604. IEEE (2002)
50. Wang, G., Cao, G., La Porta, T.: Movement-assisted sensor deployment. IEEE Trans. Mobile Comput. **5**(6), 640–652 (2006). June
51. Wang, G., Wang, T., Jia, W., Guo, M., Li, J.: Adaptive location updates for mobile sinks in wireless sensor networks. J. Supercomput. **47**(2), 127–145 (2009)
52. Wang, H., Elson, J., Girod, L., Estrin, D., Yao, K., Vanderberge, L.: Target classification and localization in habitat monitoring. In: IEEE Proceedings of the International Conference on Speech and Signal Processing, pp. II–597–II–600 April (2003)
53. Yick, J., Mukherjee, B., Ghosal, D.: Wireless sensor network survey. Comput. Netw. **52**, 2292–2330 (2008)
54. Zou, Y., Chakrabarty, K.: Sensor deployment and target localization based on virtual forces. In: INFOCOM 2003. Twenty-Second Annual Joint Conference of the IEEE Computer and Communications. IEEE Societies, vol. 2, pp. 1293–1303, April (2003)

Connectivity Preserving Network Transformers

Othon Michail and Paul G. Spirakis

Abstract The Population Protocol model is a distributed model that concerns systems of very weak computational entities that cannot control the way they interact. The model of Network Constructors is a variant of Population Protocols capable of (algorithmically) constructing abstract networks. Both models are characterized by a *fundamental inability to terminate*. In this work, we investigate the minimal strengthenings of the latter model that could overcome this inability. Our main conclusion is that *initial connectivity of the communication topology* combined with the ability of the protocol to *transform the communication topology* and the ability of a node to *detect when its degree is equal to a small constant*, plus either a *unique leader* or the ability of *detecting common neighbors*, are sufficient to guarantee not only *termination* but also the *maximum computational power that one can hope for in this family of models*. In particular, the model, under these minimal assumptions, *computes with termination any symmetric predicate computable by a Turing Machine of space* $\Theta(n^2)$.

Supported in part by (i) the project "Foundations of Dynamic Distributed Computing Systems" (FOCUS) which is implemented under the "ARISTEIA" Action of the Operational Programme "Education and Lifelong Learning" and is co-funded by the European Union (European Social Fund) and Greek National Resources and (ii) the FET EU IP project MULTIPLEX under contract no 317532. The full paper on which this chapter is based, can be found at http://arxiv.org/abs/1512.02832.

O. Michail (✉)
Department of Computer Science, University of Liverpool, Liverpool, UK
e-mail: Othon.Michail@liverpool.ac.uk

O. Michail · P.G. Spirakis
Computer Technology Institute and Press "Diophantus" (CTI), Patras, Greece
e-mail: P.Spirakis@liverpool.ac.uk

© Springer International Publishing Switzerland 2017
A. Adamatzky (ed.), *Emergent Computation*, Emergence, Complexity and Computation 24, DOI 10.1007/978-3-319-46376-6_15

1 Introduction

A dynamic distributed computing system is a system composed of distributed computational processes in which the structure of the communication network between the processes changes over time. In one extreme, the processes cannot control and cannot accurately predict the modifications of the communication topology. Typical such examples are mobile distributed systems in which the mobility is *external* to the processes and is usually provided by the environment in which the system operates. For example, it could be a system of cell phones following the movement of the individuals carrying them or a system of nanosensors flowing in the human circulatory system. This type of mobility is known as *passive* (see e.g. [4]). On the other extreme, dynamicity may be a sole outcome of the algorithm executed by the processes. Typical examples are systems in which the processes are equipped with some *internal* mobility mechanism, like mobile robotic systems and, in general, any system with the ability to algorithmically modify the communication topology. This type of mobility is known as *active* mobility (see e.g. [20] for active self-assembly, [11, 12, 19] for mobile robots, and [1] for reconfigurable (nano)robotics under physical constraints). Recently, there is an interest in *intermediate* (or *hybrid*) systems. One such type, consists of systems in which the processes are passively mobile but still they are equipped with an internal active mechanism that allows them to have a partial (algorithmic) control of the system's dynamicity.

The intermediate model that guides our study here, is the *network constructors* model introduced in [17]. In this model, there are n extremely weak processes, computationally equivalent to anonymous finite automata, that usually have very limited knowledge of the system (e.g. they do not know its size). The processes move passively and interact in pairs whenever two of them come sufficiently close to each other. This part of the system's dynamicity is not controlled and cannot be (completely) predicted by the processes and is modeled by assuming an adversary scheduler that in every step selects a pair of processes to interact. The adversary is typically restricted to be *fair* so that it cannot forever block the system's progress (e.g. by keeping two parts of the system forever disconnected). Fairness is sufficient for analyzing the correctness of protocols for specific tasks. If additionally an estimate of the running time is desired, a typical assumption is that the scheduler is a uniform random one (which is fair with probability 1 [8] and also corresponds to the dynamicity patterns of well-mixed solutions). But in this model, there is also an internal source of dynamicity. In particular, the processes can algorithmically connect and disconnect to each other during their pairwise interactions. This can be viewed either as a physical bonding mechanism, as e.g. in reconfigurable robotics and molecular (e.g. DNA) self-assembly, or as a virtual record of local connectivity, as e.g. in a social network where a participant keeps track of and can regularly update the set of his/her associates. This allows the processes to control the construction and maintenance of a network or a shape in an uncontrolled and unpredictable dynamic environment.

The network from which the scheduler picks interactions between processes and develops the uncontrolled interaction pattern is called the *interaction network*. At the same time, the processes, by connecting and disconnecting to each other, develop another network, the *(algorithmically) constructed network*, which is a subnetwork of the interaction network. In the most abstract setting, the interaction network is the clique K_n throughout the execution, no matter what the protocol does (e.g. no matter how the protocol modifies the constructed network). In this case, the scheduler can in every step (throughout the course of the protocol) pick any possible pair of processes to interact, independently of the constructed network.[1] This is precisely the setting of [17] and also the one that we will consider in the present work.[2] But even if the interaction network is always a clique independently of the constructed network, the ability of the processes to construct a network may still allow them to counterbalance the adversary's power. For example, if the processes manage somehow to self-organize into a spanning network G, then it might be possible for them to ignore all interactions that occur over the non-links of G and thus force the actual communication pattern to be consistent with the constructed network.

The existing literature on distributed network construction [14, 17] has almost absolutely focused on the setting in which all processes are initially disconnected and the goal is for them to algorithmically self-organize into a desired (usually spanning or of size at least some required function of n) stable network or shape. In [17], the authors presented simple and efficient direct constructors and lower bounds for several basic network construction problems such as spanning line, spanning ring, and spanning star and also generic constructors capable of constructing a large class of networks by simulating a Turing Machine (abbreviated "TM" throughout). One of the main results was that for every graph language L that is decidable by a $O(\sqrt{l})$-space ($l + O(\sqrt{l})$, resp.) TM, where $l = \Theta(n^2)$ is the binary length of the input of the simulated TM, there is a protocol that constructs L equiprobably with useful space $\lfloor n/2 \rfloor$ ($\lfloor n/3 \rfloor$, resp.), where the *useful space* is defined as a lower bound on the order of the output network (the rest of the nodes being used as auxiliary and thrown away eventually as *waste*). In [14], a geometrically constrained variant was studied, where the formed network and the allowable interactions must respect the structure of the 2-dimensional (or 3-dimensional) grid network. The main result was a *terminating* protocol counting the size n of the system with high probability (abbreviated "w.h.p." throughout). This protocol was then used as a subroutine of universal constructors, establishing that the nodes can self-assemble w.h.p. into arbitrarily complex shapes while still being capable to terminate after completing the construction.

---

[1] A convenient way to think of this setting is to imagine a clique graph with its edges labeled from $\{0, 1\}$. Then, in this case, the clique is the interaction network while its subgraph induced by the edges labeled 1 is the constructed network.

[2] On the other hand, it is possible, and plausible w.r.t. several application scenarios, that the set of available interactions at a given step actually depends on the constructed network. Such a case was considered in [14], where the constructed network is always a subnetwork of the grid network and two processes can only interact if a connection between them would preserve this requirement. So, in that case, the set of available interactions is, in every step, constrained by the network that has been constructed by the protocol so far.

1.1 Our Approach and Contribution

The main goal of this work is to investigate minimal strengthenings of the population protocol and network constructors models that can maximize their computational power, also rendering them capable to terminate. To this end, we consider (for the first time in network constructors) the case in which the initial configuration of the edges is not the one in which all edges are *inactive* (i.e. those that are in state 0). In particular, we assume that the initial configuration of the edges can be any configuration in which the *active* (i.e. those that are in state 1) edges form *a connected graph spanning the set of processes.*[3] The initial configuration of the nodes is either, as in [17], the one in which all nodes are initially in the same state, e.g. in an initial state q_0, or (whenever needed) the one in which all nodes begin from q_0 apart from a pre-elected unique leader that begins from a distinct initial leader-state l_0. This choice is motivated by the fact that without some sort of bounded initial disconnectivity we can only hope for global computations and constructions that are *eventually stabilizing* (and not *terminating*), because a component can guess neither the number of components not encountered yet nor an upper bound on the time needed to interact with another one of them ([18] overcomes this by assuming that the nodes know some upper bound on this time, while [14] overcomes this by assuming a uniform random scheduler and a unique leader and by restricting correctness to be w.h.p.).

Next, observe that if the protocol is not allowed to modify the state of the edges, then the assumption of initial connectivity alone does not add any computational power to the model (in the worst case). For if we ignore for a while the ability of the model to modify the state of the edges, what we have is a model equivalent to classical population protocols [4] on a restricted interaction graph [3] (observe that the model can ignore the interactions that occur over inactive edges). Though there are some restricted interaction graphs, like the spanning line, that dramatically increase the computational power of the model (in this case making it equivalent to a TM of linear space), still there others, like the spanning star, on which the power of the model is as low as the power of classical population protocols on a clique interaction graph [10], which, in turn, is equal to the rather small class of *semilinear predicates* [5]. As we have allowed any possible connected initial set of active edges, the spanning star inclusive, the initial configuration of the edges alone (without any edge modifications) is not sufficient for strengthening the model.

Our discussion so far, suggests to consider at the same time initial connectivity (or, more generally, bounded initial disconnectivity) and the ability of the protocol to modify the state of the edges, with the hope of increasing the computational power. Unfortunately, even with this additional assumption, non-trivial terminating computation is still impossible (this is proved in Proposition 3, in Sect. 3.3). An immediate way to appreciate this, is to notice that a clique does not provide more information than an empty network about the size of the system. Even worse, if a node's initial active degree is unbounded (as e.g. is the case for the center of a spanning

---

[3] Active and inactive *edges* are not to be confused with active and passive *mobility*. An edge is said to be *active* if its state is 1 and it is said to be *inactive* if its state is 0.

star), then it is not clear even whether the stabilizing constructors that assume initial disconnectivity (as in [17]) can be adapted to work. Actually, it could be the case, that *without additional assumptions* initial connectivity may even decrease the power of the model (we leave this as an interesting open problem). For example, it could be simpler to construct a spanning line if the initial active network is empty (i.e. all edges are inactive) than if it is a clique (i.e. all edges are active). Even if it would turn out that the model does not become any weaker, we still cannot avoid the aforementioned impossibility of termination and the maximum that we can hope for is an *eventually stabilizing* universal constructor, as the one of [17].

We now add to the picture a very minimal and natural, but extremely powerful, additional assumption that, combined with our assumptions so far, will lead us to a stronger model. In particular, we equip the nodes with the ability to detect some small local degrees. For a concrete example, assume that a node can detect when its active degree is equal to 0 (otherwise it only knows that its degree is at least 1). A first immediate gain, is that we can now directly simulate any constructor that assumes an empty initial network (e.g. the constructors of [17]): every node initially deactivates the active edges incident to it until its local active degree becomes for the first time 0, and only when this occurs the node starts participating in the simulation. So, even though a node does not know its initial degree (which is due to the fact that a node in this model is a finite automaton with a state whose size is independent of the size of the system), it can still detect when it becomes equal to 0. At that point, the node does not have any active edges incident to it, therefore it can start executing the constructor that assumes an empty initial network.

Our main finding in this work, is that the initial connectivity guarantee together with the ability to modify the network and to detect small local degrees (combined with either a pre-elected leader or a natural mechanism that allows two nodes to tell whether they have a neighbor in common), are sufficient to obtain the *maximum computational power* that one can hope for in this family of models. In particular, the resulting model can compute *with termination* any symmetric predicate[4] computable by a *TM of space* $\Theta(n^2)$, and no more than this, i.e. it is an exact characterization. The symmetricity restriction can only be dropped by UIDs or by any other means of knowing and maintaining an ordering of the nodes' inputs. This power is maximal because the distributed space of the system is $\Theta(n^2)$, so we cannot hope for computations exploiting more space. The substantial improvement compared to [15, 17] is that the universal computations are now *terminating* and not just *eventually stabilizing*. It is interesting to point out that the additional assumptions and mechanisms are minimal, in the sense that the removal of each one of them leads to either an impossibility of termination or to a substantial decrease in the computational power.

In Sect. 2, we discuss further related literature. Section 3 brings together all definitions and basic facts that are used throughout the chapter. In particular, in Sect. 3.1 we formally define the model of network constructors under consideration, Sect. 3.2 formally defines the transformation problems that are considered in this work, and

---

[4]Essentially, a predicate in this type of models is called *symmetric* (or *commutative*) if permuting the input symbols does not affect the predicate's outcome.

Sect. 3.3 provides some basic impossibility results and a lower bound on the time needed to transform any network to a spanning line. In Sect. 4, we study the case in which there is a pre-elected unique leader and give two protocols for the problem, the Online-Cycle-Elimination protocol and the time-optimal Line-Around-a-Star protocol. Then, in Sect. 5, we try to drop the unique leader assumption. First, in Sect. 5.1 we show that, without additional assumptions, dropping the unique leader leads to a strong impossibility result. In face of this negative result, in Sect. 5.2 we minimally strengthen the model with a common neighbor detection mechanism and give a correct terminating protocol. Finally, in Sect. 6 we conclude and give further research directions that are opened by our work.

2 Further Related Work

The model considered in this chapter belongs to the family of population protocol models. The population protocol model [4] was originally developed as a model of highly dynamic networks of simple sensor nodes that cannot control their mobility. The first papers focused on the computational capabilities of the model which have now been almost completely characterized. In particular, if the interaction network is complete, i.e. one in which every pair of processes may interact, then the computational power of the model is equal to the class of the *semilinear predicates* (and the same holds for several variations) [5]. Semilinearity persists up to $o(\log \log n)$ local space but not more than this [9]. If additionally the connections between processes can hold a state from a finite domain (note that this is a stronger requirement than the active/inactive that the present work assumes) then the computational power dramatically increases to the commutative subclass of **NSPACE**(n^2) [15]. The latter constitutes the mediated population protocol (MPP) model, which was the first variant of population protocols to allow for states on the edges. For introductory texts to these models, the interested reader is encouraged to consult [6] and [16].

Based on the MPP model, [17] restricted attention to binary edge states and regarded them as a physical (or virtual, depending on the application) bonding mechanism. This gave rise to a "hybrid" self-assembly model, the network constructors model, in which the actual dynamicity is passive and due to the environment but still the protocol can construct a desired network by activating and deactivating appropriately the connections between the nodes. The present chapter essentially investigates the computational power of the network constructors model under the assumption that a connected spanning active topology is provided initially and also initiates the study of the *distributed network reconfiguration problem*. Recently, [14] studied a geometrically constrained variant of network constructors in which the interaction network is not complete but rather it is constrained by the existing shapes (every shape that can be formed being a sub-network of the 2D or 3D grid network). Interestingly, apart from being a model of computation, population protocols are also closely related to chemical systems. In particular, Doty [13] has recently demonstrated their formal equivalence to *chemical reaction networks* (CRNs), which model chemistry in a *well-mixed solution*.

3 Preliminaries

3.1 The Model

The model under consideration is the network constructors model of [17] with the only essential difference being that in [17] the initial configuration was always (apart from the network replication problem) the one in which all edges are inactive, while in this work the initial configuration can be any configuration in which the active edges form a spanning connected network. Still, we give a detailed presentation of the model for self-containment.

Definition 1 A *Network Constructor* (NET) is a distributed protocol defined by a 4-tuple $(Q, q_0, Q_{out}, \delta)$, where Q is a finite set of *node-states*, $q_0 \in Q$ is the *initial node-state*, $Q_{out} \subseteq Q$ is the set of *output node-states*, and $\delta : Q \times Q \times \{0, 1\} \to Q \times Q \times \{0, 1\}$ is the *transition function*. When required, also a special *initial leader-state* $l_0 \in Q$ may be defined.

If $\delta(a, b, c) = (a', b', c')$, we call $(a, b, c) \to (a', b', c')$ a *transition* (or *rule*) and we define $\delta_1(a, b, c) = a'$, $\delta_2(a, b, c) = b'$, and $\delta_3(a, b, c) = c'$. A transition $(a, b, c) \to (a', b', c')$ is called *effective* if $x \neq x'$ for at least one $x \in \{a, b, c\}$ and *ineffective* otherwise. When we present the transition function of a protocol we only present the effective transitions. Additionally, we agree that the *size* of a protocol is the number of its states, i.e. $|Q|$.

The system consists of a population V_I of n distributed *processes* (also called *nodes* when clear from context). In the generic case, there is an underlying *interaction graph* $G_I = (V_I, E_I)$ specifying the permissible interactions between the nodes. Interactions in this model are always pairwise. In this work, unless otherwise stated, G_I is a *complete undirected interaction graph*, i.e. $E_I = \{uv : u, v \in V_I \text{ and } u \neq v\}$, where $uv = \{u, v\}$. When we say that all nodes in V_I are initially *identical*, we mean that all nodes begin from the initial node-state q_0. In case we assume the existence of a unique leader, then there is a $u \in V_I$ beginning from the initial leader-state l_0 and all other $v \in V_I \setminus \{u\}$ begin from the initial node-state q_0 (which in this case may also be called the *initial nonleader-state*).

A central assumption of the model is that edges have binary states. An edge in state 0 is said to be *inactive* while an edge in state 1 is said to be *active*. In almost all problems studied in [17] (apart from the replication problem), all edges were initially inactive. Though we shall also consider this case in the present chapter, our main focus is on a different setting in which the protocol begins its execution on a precomputed set of active edges provided by some adversary. Formally, there is an input set of edges $E \subseteq E_I$, such that all $e \in E$ are initially active and all $e' \in E_I \setminus E$ are initially inactive. The set E defines the *input graph* $G = (V_I, E)$, also called the *initial active topology/graph*. Throughout this work, unless otherwise stated, we assume that the initial active topology is *connected*, which means that the active edges form a connected graph spanning V_I. This is a restriction imposed on the adversary selecting the input. In particular, the adversary is allowed to choose any initial set of active edges E (in a worst-case manner), subject to the constraint that E defines a connected graph on the whole population.

Execution of the protocol proceeds in discrete steps. In every step, a pair of nodes uv from E_I is selected by an *adversary scheduler* and these nodes interact and update their states and the state of the edge joining them according to the transition function δ. In particular, we assume that, for all distinct node-states $a, b \in Q$ and for all edge-states $c \in \{0, 1\}$, δ specifies either (a, b, c) or (b, a, c). So, if a, b, and c are the states of nodes u, v, and edge uv, respectively, then the unique rule corresponding to these states, let it be $(a, b, c) \rightarrow (a', b', c')$, is applied, the edge that was in state c updates its state to c' and if $a \neq b$, then u updates its state to a' and v updates its state to b', if $a = b$ and $a' = b'$, then both nodes update their states to a', and if $a = b$ and $a' \neq b'$, then the node that gets a' is drawn equiprobably from the two interacting nodes and the other node gets b'.

A *configuration* is a mapping $C : V_I \cup E_I \rightarrow Q \cup \{0, 1\}$ specifying the state of each node and each edge of the interaction graph. Let C and C' be configurations, and let u, υ be distinct nodes. We say that C *goes to* C' *via encounter* $e = u\upsilon$, denoted $C \xrightarrow{e} C'$, if $(C'(u), C'(v), C'(e)) = \delta(C(u), C(v), C(e))$ or $(C'(v), C'(u), C'(e)) = \delta(C(v), C(u), C(e))$ and $C'(z) = C(z)$, for all $z \in (V_I \backslash \{u, v\}) \cup (E_I \backslash \{e\})$. We say that C' *is reachable in one step from* C, denoted $C \rightarrow C'$, if $C \xrightarrow{e} C'$ for some encounter $e \in E_I$. We say that C' *is reachable from* C and write $C \rightsquigarrow C'$, if there is a sequence of configurations $C = C_0, C_1, \ldots, C_t = C'$, such that $C_i \rightarrow C_{i+1}$ for all $i, 0 \leq i < t$.

An *execution* is a finite or infinite sequence of configurations $C_0, C_1, C_2, \ldots$, where C_0 is an initial configuration and $C_i \rightarrow C_{i+1}$, for all $i \geq 0$. A *fairness condition* is imposed on the adversary to ensure the protocol makes progress. An infinite execution is *fair* if for every pair of configurations C and C' such that $C \rightarrow C'$, if C occurs infinitely often in the execution then so does C'. In what follows, every execution of a NET will by definition considered to be fair.

We define the *output of a configuration* C as the graph $G(C) = (V, E)$ where $V = \{u \in V_I : C(u) \in Q_{out}\}$ and $E = \{uv : u, v \in V, u \neq v, \text{ and } C(uv) = 1\}$. In words, the output-graph of a configuration consists of those nodes that are in output states and those edges between them that are active, i.e. the active subgraph induced by the nodes that are in output states. The output of an execution $C_0, C_1, \ldots$ is said to *stabilize* (or *converge*) to a graph G if there exists some step $t \geq 0$ s.t. $G(C_i) = G$ for all $i \geq t$, i.e. from step t and onwards the output-graph remains unchanged. Every such configuration C_i, for $i \geq t$, is called *output-stable*. The *running time* (or *time to convergence*) of an execution is defined as the minimum such t (or ∞ if no such t exists). Throughout the chapter, whenever we study the running time of a NET, we assume that interactions are chosen by a *uniform random scheduler* which, in every step, selects independently and uniformly at random one of the $|E_I| = n(n-1)/2$ possible interactions. In this case, the running time on a particular n and an initial set of active edges E becomes a random variable (abbreviated "r.v.") $X_{n,E}$ and our goal is to obtain bounds on $\max_{n,E}\{e[X_{n,E}]\}$, where $e[X]$ is the *expectation* of the r.v. X. That is, the running time of a protocol is defined here as the maximum (also called *worst-case*) expected running time over all possible initial configurations. Note that the uniform random scheduler is fair with probability 1.

Definition 2 We say that an execution of a NET on n processes *constructs a graph* (or *network*) G, if its output stabilizes to a graph isomorphic to G.

Definition 3 We say that a NET $\mathcal{A}$ *constructs a graph language L with useful space* $g(n) \leq n$, if $g(n)$ is the greatest function for which: (i) for all n, every execution of $\mathcal{A}$ on n processes constructs a $G \in L$ of order at least $g(n)$ (provided that such a G exists) and, additionally, (ii) for all $G \in L$ there is an execution of $\mathcal{A}$ on n processes, for some n satisfying $|V(G)| \geq g(n)$, that constructs G. Equivalently, we say that $\mathcal{A}$ *constructs L with waste* $n - g(n)$.

In this work, we shall also be interested in NETs that construct a graph language and additionally always *terminate*.

Definition 4 We call a NET $\mathcal{A}$ *terminating* (or say that $\mathcal{A}$ *always terminates*) if every execution of $\mathcal{A}$ reaches a *halting* configuration, that is one in which every node is in a state q_h from a set of halting states Q_{halt}, where $(q_h, q, s) \to (q_h, q, s)$ (i.e. is ineffective) for every $q_h \in Q_{halt}, q \in Q$, and $s \in \{0, 1\}$.

Finally, in order to consider TM simulations, we denote by $\mathbf{SSPACE}(f(n))$ the symmetric subclass of the complexity class $\mathbf{SPACE}(f(n))$.

3.2 Problem Definitions

Acyclicity. Let $G = (V, A)$ be the subgraph of G_I consisting of V and the active edges between nodes in V, that is $A = \{e \in E_I : C(e) = 1\}$. The initial G is connected. The goal is for the processes to stably transform G to an acyclic graph spanning V without ever breaking the connectivity of G.

Line Transformation. Let $G = (V, A)$ be the subgraph of G_I consisting of V and the active edges between nodes in V, that is $A = \{e \in E_I : C(e) = 1\}$. The initial G is connected. The goal is for the processes to stably transform G to a spanning line.

Terminating Line Transformation. The same as Line Transformation with the additional requirement that all processes must terminate.

3.3 Fundamental Inabilities

We now give a few basic impossibility results that justify the necessity of minimally strengthening the network constructors model in order to be able to solve the above main problems.

The following proposition (which is a well-known fact in the relevant literature but we include here a proof for self-containment) states that if the system does not involve edge states (i.e. the original population protocol model with transition function δ : $Q \times Q \to Q \times Q$), then a protocol cannot decide with termination whether there is a single a in the population (mainly because a node does not know how much time it has to wait to meet every other node). Though the result is not directly applicable to our model, still we believe that it might help the reader's intuition w.r.t. to the computational difficulties in this family of models.

Proposition 1 (PPs Impossibility of Termination) *There is no population protocol that can compute with termination the predicate* $(N_a \geq 1)$ *(i.e. whether there exists an a in the input assignment).*

Proof Consider a population of size n and let the nodes be $u_1, u_2, \ldots, u_n$. It suffices to prove the impossibility for the variation in which there is a unique leader, initially in state l, and all other nodes are non-leaders, initially in state q_a if their input is a and q_b if their input is not a, and all interactions are between the leader and the non-leaders. This is w.l.o.g. because this model is not weaker than the original population protocol model, which means that an impossibility for this model also transfers to the original population protocol model. Indeed, this model can easily simulate the original model as follows. Interactions between two non-leaders can be simulated via the leader: the leader first collects the state q_1 of a node u, which it marks, and then waits to interact with another node v. When this occurs, if the state of v is q_2, rule $(q_1, q_2) \to (q_1', q_2')$ is applied to the state stored by the leader and to the state of v. Then the leader waits to meet u again (which can be detected since it has been marked) in order to update its state to q_1'. When this occurs, the leader drops the stored information and starts a new simulation round.

So, let u_1 be the initial leader. Let A be a protocol that computes $(N_a \geq 1)$ and terminates on every n and every input assignment. Consider now the input assignment in which all inputs are b, that is there is no a and thus all non-leaders begin with initial state q_b. Clearly, it must hold that in every fair execution the leader terminates in a finite number of steps and says "no". These steps are interactions between the leader and the non-leaders so any such execution can be represented by a sequence of u_js from $\{u_2, \ldots, u_n\}$. Let now $s = v_1, v_2, \ldots, v_k$ be any such finite execution in which the leader says "no". $v_i \in \{u_2, \ldots, u_n\}$ is simply the node with which the leader interacted at step i.

Consider now a population of size $n + 1$. The only difference to the previous setting is that now we have added a node u_{n+1} with input a. Since now the predicate evaluates to 1, in every fair execution, A should terminate in a finite number of steps and say "yes". Take any fair execution $s' = s, v_{k+1}, \ldots, v_h$, that is s' has s as an "unfair" prefix. As s contains the same nodes as before with the same input assignment, the leader in s' terminates in precisely k steps saying "no" without knowing that an additional node with input a exists in this case. This contradicts the existence of protocol A. We should mention that the leader has no means of guessing the existence of node u_{n+1} because its termination only depends on the protocol stored in its memory which is by definition finite and independent of n (suffices to consider the longest chain of rules that leads to termination with output "no" and which corresponds to at least one feasible execution). □

Moreover, even if the system is initially connected, there are some very symmetric topologies that do not allow for strong computations. For example, if the topology is a star with the leader at the center, then the system is equivalent to population protocols on a complete interaction graph and can compute only semilinear predicates on input assignments, again only in an eventually-stabilizing way (i.e. no termination). This is captured by the following proposition.

Proposition 2 (Structure versus Computational Power) *There are initial topologies in which the computational power of population protocols is as limited as in the case of no structure at all.*

The above expose the necessity of additional assumptions, such as topology modifications, in order to hope for terminating computations and surpass the computational power of classical population protocols. So, we turn our attention again to our model, i.e. where the edges have binary states and the protocol can modify them, and consider the case in which the initial topology is always connected.

Proposition 3 (NETs Impossibility of Termination) *There is no protocol that can compute with termination the predicate $(N_a \geq 1)$, even if the initial topology is connected and even if there is a pre-elected unique leader.*

So, connectivity of the initial topology alone, even if the protocol is allowed to transform the topology, is not sufficient for non-trivial terminating computations. In the rest of the chapter we shall naturally try to overcome this by adding to the model minimal and realistic extra assumptions. Interestingly, it will turn out that there are some very plausible such assumptions that allow for: (i) termination and (ii) computation of all predicates on input assignments that can be computed by a TM in quadratic space ($O(n^2)$, where n is the number of nodes).

One of the assumptions that we will keep throughout is that the nodes are capable of detecting some small local degrees. For example, in Sect. 4 we will assume that a node can detect that it has local degree 1 or 2, otherwise it knows that it has degree in $\{0, 3, 4, ..., n-1\}$ without being able to tell its precise value. We will complement this local degree detection mechanism with either a unique leader or a common neighbor detection mechanism in order to arrive at the above strong characterization.

Keep in mind that we want to give protocols for Acyclicity and Terminating Line Transformation. In Acyclicity, the protocol begins from any connected active topology and has to transform it to an acyclic network without ever breaking the connectivity, while in Terminating Line Transformation the protocol does not necessarily have to preserve connectivity but it has to satisfy the additional requirements its constructed network to be a spanning line and to always terminate. Still, even for the Terminating Line Transformation problem we shall mostly focus on protocols that perform the transformation without ever breaking connectivity. A justification of this choice, is that arbitrary connectivity breaking could render the protocol unable to terminate even if the protocol is equipped with all the additional mechanisms mentioned above. This is made formal in Lemma 2 of Sect. 5.1. One way to appreciate this is to consider a protocol in which a leader breaks in some execution the network into an unbounded number of components. Then the leader can no longer distinguish an execution in which one of these components is being concealed from an execution that it is not. For example, if the leader is trying to construct a spanning line, then it has no means of distinguishing a spanning line on all nodes but the concealed ones from one on all nodes. Of course, this does not exclude protocols that perform some controlled connectivity breakings, e.g. a leader breaking a spanning line at one point and then waiting to reconnect the two parts. So, in principle, our problems could have been defined independently of whether connectivity is preserved or not, as they can also be solved in some cases by protocols that do not always preserve connectivity. However, in this work, for simplicity and clarity of presentation, we have chosen to focus only on those protocols that always preserve connectivity.

Before starting to present our protocols for above problems and the upper bounds on time provided by them, we give a lower bound on the time that any protocol needs in order to solve the Line Transformation problem.

Lemma 1 (Line Transformation Lower Bound) *The running time of any protocol that solves the Line Transformation problem is* $\Omega(n^2 \log n)$.

4   Transformers with a Unique Leader

We begin from the simplest case in which there is initially a pre-elected unique leader that handles the transformation. Recall that the initial active topology is connected. The goal is for the protocol to transform the active topology to a spanning line and when this occurs to detect it and terminate (i.e. solve the Terminating Line Transformation problem). Ideally, the transformation should preserve connectivity of the active topology during its course (or break connectivity in a controlled way, because, as we already discussed in the previous section, uncontrolled/arbitrary connectivity breaking may render termination impossible). Moreover, as a minimal additional assumption to make the problem solvable (in order to circumvent the impossibility of Proposition 3), we assume that a node can detect whether it has local degree 1 or 2 (otherwise it knows that it has degree in $\{0, 3, 4, ..., n-1\}$ without being able to tell its precise value). We first give a straightforward solution, with a complete presentation of its transitions and an illustration showing them in action. Though that protocol is correct, it is rather slow and it mainly serves as a demonstration of the model and the problem under consideration. Then we follow a different approach and arrive at a time-optimal protocol for the problem.

The idea of the first protocol is simple. The leader begins from its initial node and starts forming an arbitrary line by expanding one endpoint of the line towards unvisited nodes. Every such expansion either occurs over an edge that was already active from the very beginning or over an inactive edge which the protocol activates. Apart from expanding its active line with the goal of making it spanning after $n - 1$ expansions, the leader must also guarantee that eventually no cycles will have remained. One idea would be to first form a spanning line and then start eliminating all unnecessary cycles, however there is, in general, no way for the protocol to detect that the line is indeed spanning, due to the possible presence of non-line active edges joining nodes of the line. This is resolved by eliminating line-internal cycles "online" after every expansion of the line. This guarantees that when the last expansion occurs and the protocol deactivates the last cycles, the active topology will be a spanning line. Now the protocol can easily detect this by traversing the line from left to right and comparing the observed active degree sequence to the target degree sequence $1, 2, 2, \ldots, 2, 1$ (i.e. the degree sequence of a spanning line). We next give the detailed description of the protocol.

Protocol Online-Cycle-Elimination. The leader marks its initial node as "left endpoint" e_l and picks an arbitrary next node for the line (for the first step it could be from its active neighbors, because there is at least one such node due to initial connectivity) and marks that node as "right endpoint" e_r. Then the leader moves to e_r, finds an arbitrary next node which is not part of the current line, if the edge is inactive it activates it and marks that node as e_r and the previous e_r is converted to i (for "internal node" of the line). Observe that the active line is always in special states, which makes its nodes detectable.

After every such expansion, the leader starts a cycle elimination phase. In particular, the leader deactivates all edges that introduce a cycle inside its active line. To do this, it suffices after every expansion to deactivate the cycles introduced by the new right endpoint e_r. First, the leader moves to e_l (e.g. by direct communication or by traversing the active line to the left). Every time, the leader waits to meet e_r, in order to check the status of the edge; if it is active, it deactivates it and then moves on step to the right on the line. When the leader arrives at the left neighbor of e_r, all line-internal cycles have been eliminated and the leader just moves to e_r. If the degree sequence observed during the traversal to the right (the degree of a node is checked *after* checking and possibly modifying the status of its edge to e_r) was of the form $1, 2, 2, \ldots, 2, 1$ then the line is spanning and the leader terminates. Otherwise, the line is not spanning yet and the leader proceeds to the next expansion.

The code of the protocol is presented in Protocol 5. For readability, we only present the code for the expansion and cycle elimination phases and we have excluded the termination detection subroutine (it is straightforward to extend the code to also take this into account). An illustration showing what are the roles of the various states and transitions during the expansion and cycle elimination phases, is given in Fig. 1.

Theorem 1 *By assuming a pre-elected unique leader and the ability to detect local degrees 1 and 2, Protocol Online-Cycle-Elimination solves the Terminating Line Transformation problem in $\Theta(n^4)$ time.*

Proof We prove the following invariant: "For all $1 \leq i \leq n - 1$, after the ith expansion and cycle elimination phases, the leader lies on the e_r endpoint of an active

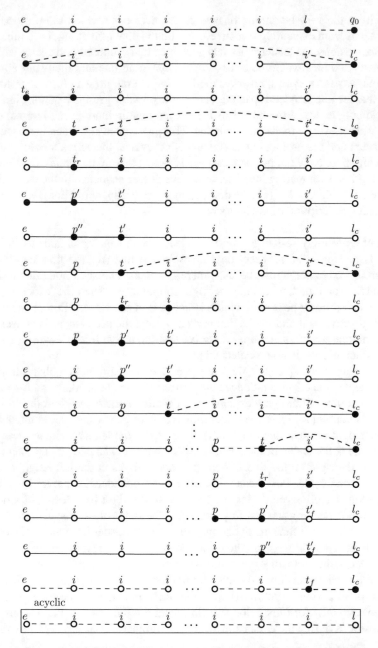

Fig. 1 An illustration of all transitions involved in Protocol Online-Cycle-Elimination during expansion of the line and elimination of the newly introduced internal cycles. The *line* at the *top* shows an expansion of the current acyclic line, the intermediate steps show the the process of eliminating cycles, and the *line* at the *bottom* is the new *acyclic line*. In every step, the two interacting nodes are colored *black* and joined by a bold edge. *Dashed edges* could be either active or inactive. The *dashed edge* of an expansion (*top line*) is activated no matter what its previous state was, while all other *dashed edges* in the figure, that correspond to (potential) cycle eliminations, are deactivated no matter what their state was

---

Algorithm 5 Online-Cycle-Elimination

$Q = \{l_0, l_1, l_c, l'_c, l, q_0, e, i, i', t, t_e, t_f, t'_f, t_r, t', p, p', p''\}$, initially the unique leader is in state l_0 and all other nodes are in state q_0

δ:

$(l_0, q_0, 1) \rightarrow (e, l_1, 1)$	$(t_e, i, 1) \rightarrow (e, t, 1)$	$(p'', t', 1) \rightarrow (p, t, 1)$
$(l_1, q_0, \cdot) \rightarrow (i', l'_c, 1)$	$(t, l_c, \cdot) \rightarrow (t_r, l_c, 0)$	$(t_r, i', 1) \rightarrow (p', t'_f, 1)$
$(e, l'_c, \cdot) \rightarrow (t_e, l_c, 0)$	$(t_r, i, 1) \rightarrow (p', t', 1)$	$(p'', t'_f, 1) \rightarrow (i, t_f, 1)$
$(t_e, i', 1) \rightarrow (e, t_f, 1)$	$(e, p', 1) \rightarrow (e, p'', 1)$	$(l, q_0, \cdot) \rightarrow (i', l'_c, 1)$
$(t_f, l_c, 1) \rightarrow (i, l, 1)$	$(p, p', 1) \rightarrow (i, p'', 1)$	

// All transitions that do not appear have no effect

// The logical structure is better followed if the transitions are read from top to bottom

---

line of (edge-)length i without line-internal cycles (still any node of the line may have active edges to the rest of the graph) and the active topology is connected". This implies that for $i < n - 1$ there is at least one node of the line that has an edge to a node not belonging to the line and that for $i = n - 1$ the active topology is a spanning line (without any other active edges).

First observe that connectivity never breaks, because whenever the protocol deactivates an edge $e = uv$, both u and v are nodes belonging to the active line formed so far (in particular, at least one of them is the e_r endpoint of the line). As e is an edge forming a cycle on the active line after its deactivation connectivity between u and v still exists by traversing the line.

We prove by induction the rest of the invariant. It holds trivially for $i = 1$. Given that it holds for any $1 \le i \le n - 2$ we prove that it holds for $i + 1$. By hypothesis, when expansion $i + 1$ occurs, the only possible line-internal cycles are between the new e_r and the rest of the line. During the cycle elimination phase the protocol eliminates all these cycles, and as a result by the end of phase $i + 1$ the active line has now length $i + 1$, it has no internal cycles and is still connected to the rest of the graph.

It remains to show that the leader terminates just after phase $n - 1$ and never at a phase $i < n - 1$. For the first part, after phase $n - 1$ the active topology is a spanning line, thus the observed degree sequence when the leader traverses it from left to right is of the form $1, 2, 2, \ldots, 2, 1$ which triggers termination. For the second part, after any phase $i < n - 1$ there is at least one node of the line having an active edge leading outside the line. In case that node is an endpoint, its active degree is at least 2 and in case it is an internal node its active degree is at least 3 (after eliminating a possible cycle of that node with e_r). So, in this case the observed degree sequence is not of the form $1, 2, 2, \ldots, 2, 1$ and, as required, the leader does not terminate.

For the running time, the worst case is when the initial active topology is a clique. In this case, the protocol must deactivate $\Theta(n^2)$ edges to transform the clique to a line. Every edge deactivation is performed by placing a mark on each endpoint of the edge and waiting for the scheduler to pick that edge for interaction. This takes time $\Theta(n^2)$, so the total time for deactivating $\Theta(n^2)$ edges is $\Theta(n^4)$. $\square$

We should mention that due to the unique-leader guarantee, it suffices to only have detection of whether the degree is equal to 1 (i.e. the detection of degree equal to 2 can be dropped). The reason is that the leader can every time break the line at some point while marking the two endpoints of the edge and then check whether one of these nodes has degree 1. If yes, then its previous degree was 2 and the leader waits for the two marked nodes to interact again in order to reconnect them, now knowing their degree.

A drawback of the above protocol is that it is rather slow. In the paper on which this chapter is based, we have developed another protocol, based on a different transformation technique, which is time-optimal. That protocol is called Line-Around-a-Star.

Theorem 2 *By assuming a pre-elected unique leader and the ability to detect local degree 1, Protocol Line-Around-a-Star solves the Terminating Line Transformation problem. Its running time is $\Theta(n^2 \log n)$, which is optimal.*

5   Transformers with Initially Identical Nodes

An immediate question, given the optimal Line-Around-a-Star protocol, is whether the unique leader assumption can be dropped and still have a correct and possibly also optimal protocol for Terminating Line Transformation. At a first sight it might seem plausible to expect that the problem is solvable. The reason is that the nodes can execute a leader election protocol (e.g. the standard pairwise elimination protocol; see e.g. [6]) guaranteeing that eventually a single leader will remain in the system which can from that point on handle the execution of one of the leader-based protocols of the previous section. The only additional guarantee is to ensure that nothing can go wrong as long as there are more than one leaders in the population. Typically, this is achieved in the population protocol literature by the reinitialization technique in which the configuration of the system is reinitialized/restored every time another leader is eliminated so that when the last leader remains a final reinitialization gives a correct system configuration for the leader to work on. In fact, this technique and others have been used in the population protocol literature to show that most population protocol models do not benefit in terms of computational power from the existence of a unique leader (still they are known to benefit in terms of efficiency).

In contrast to this intuition, we shall see in this section (see Corollary 2) that if all nodes are initially identical, Terminating Line Transformation becomes impossible to solve (with the modeling assumptions we have made so far). In particular, we will show that any protocol that makes the active topology acyclic, may disconnect it in some executions in $\Theta(n)$ components (see Corollary 1). As already discussed in Sect. 3.3, such a worst-case disconnection is severe for any terminating protocol, because, in this case, a component has no means of determining when it has interacted with (or heard from) all other components in the network.

Observation 1 *For a protocol to transform any topology to a line (or in general to an acyclic graph) without breaking connectivity, it must hold that the protocol deactivates an edge only if the edge is part of a cycle. Because deleting an edge e of an undirected graph does not disconnect the graph iff e is part of a cycle.*

There are several ways to achieve this when there is a unique leader. However, it will turn out that this is not the case when all nodes are initially identical.

5.1 Impossibility Results

An immediate question is whether there is a protocol with initially identical nodes that decides the existence of small cycles and additionally always terminates. We shall now show that this is not the case.

Theorem 3 (Strong Impossibility) *For every connected graph G with at least one cycle, there is an infinite family of graphs $\mathcal{G}$ such that for every $G' \in \mathcal{G}$ every protocol (beginning from identical states on all nodes) that makes G acyclic may disconnect G' in some executions.*

The above strong result states that *every* connected graph G has a corresponding infinite family of graphs (in most cases disjoint to the families of other graphs) such that Acyclicity cannot be solved at the same time on G and on a G' from the family. This means that it does not just happen for Acyclicity to be unsolvable in a few specific inconvenient graphs. *All graphs* are in some sense inconvenient for Acyclicity when studied together with the families that we have defined.

Corollary 1 (Acyclicity Impossibility) *If all nodes are initially identical, then any protocol that always makes the active topology acyclic may disconnect it in some executions in $\Theta(n)$ active components (i.e. in a worst-case manner).*

Lemma 2 *If a protocol breaks in some executions the active topology into $\Theta(n)$ components, then such a protocol cannot solve the Terminating Line Transformation problem.*

Corollary 2 (Terminating Line Transformation Impossibility) *If all nodes are initially identical, there is no protocol for Terminating Line Transformation.*

5.2 The Common Neighbor Detection Assumption

In light of the impossibility results of the previous section, we naturally ask whether some minimal strengthening of the model could make the problems solvable. To this end, we give to the nodes the ability to detect whether they have a neighbor in common. In particular, we assume that whenever two nodes interact, they can tell whether they have at that time a common neighbor (over active edges). Clearly, this mechanism can be used to safely deactivate an edge in case it happens that the two nodes are indeed part of a 3-cycle. If the two nodes are part only of longer cycles they still cannot deactivate the edge with certainty. Observe that the common neighbor detection mechanism is very local and easily implementable by almost any plausible system. For example, it only requires local names and at least 2-round local communication before neighborhood changes. Moreover, it is also an inherent capability of the variation of population protocols in which the nodes interact in triples instead of pairs (see e.g. [4, 7]). Interestingly, we shall see in this section that this minimal extra assumption overcomes the impossibility results both of Corollaries 1 and 2. In particular, both Acyclicity and Terminating Line Transformation become now solvable.

Proposition 4 *By assuming that nodes are equipped with the common neighbor detection mechanism, there is a protocol, called Star-Transformer, that solves Acyclicity in the setting in which all nodes are initially identical. In particular, the final acyclic active topology is always a spanning star.*

We now exploit the common neighbor detection assumption and the Star-Transformer protocol to give a correct and efficient protocol for the Terminating Line Transformation problem. The protocol, called Line-Transformer, assumes (as did the protocols of Sect. 4) the ability to detect whether the local active degree of a node is equal to 1 or 2.

Protocol Line-Transformer. We give here a high-level description. All nodes are initially leaders in state l. When two leaders interact, one of them becomes a peripheral in state p and the edge is activated. Every leader is connected to all ps that it encounters. Two ps deactivate an active edge joining them only if at the time of interaction they have a neighbor in common. When a peripheral has active local degree equal to 1, its local state is p_1, otherwise it is p or one of some other states that we will describe in the sequel.

When a leader first sees one of its own p_1s (i.e. via an active edge), it initiates the formation of a line over its p_1 peripherals (observe that the set of p_1 peripherals of a leader does not remain static, as e.g. a p_1 becomes p when some other leader connects to it). In particular, as in Protocol Line-Around-a-Star, the line will have as its "left" endpoint the center of the local star, which will be in a new state e_l, and it will start expanding over the available local p_1 peripherals over its right endpoint in state l'.

The new center e_l keeps connecting to new peripherals but it cannot become eliminated any more by other leaders. Pairwise eliminations only occur via any combination of l and l'. A local line expands over the local p_1s as follows. When the right endpoint l' encounters a p_1, which can occur only via an inactive edge, it expands on it only if the two nodes have a common neighbor (which can only be the center of the local star). If this is satisfied, the l' takes the place of the p_1, leaving behind an i (for "internal node" of the line) and the edge becomes activated.

Moreover, the center e_l deactivates every active edge it has with an i but not with the first peripheral of the line (so the first peripheral that the line uses must always be in a distinguished state i_1) and not with the l' right endpoint (because that edge is always needed for common neighbor detection during the next expansion).

Now, if the l' endpoint ever meets either another l' endpoint or an l center then one of them becomes deactivated. When it meets an l center we can always prefer to deactivate the l center because no line backtracking is required in this case. When an l center is deactivated, l simply becomes p and the edge becomes activated (it can never be a p_1 immediately but this is minor).

The most interesting case is when an l' loses from another l'. In this case, the eliminated l' becomes f. The role of f is to backtrack the whole local line construction by simply converting one after the other every i on its left to p and finally converting e_l again to p (again it cannot be a p_1 at the time of conversion). This backtracking process cannot fail because f has always a single i (or i_1) active neighbor, always the one on its left, while its right neighbor is no one initially and a p in all subsequent steps, so it knows which direction to follow. When the backtracking process ends, all the nodes of the local star are either p or p_1 so they can be attracted by the stars that are still alive.

The protocol terminates, when for the first time it holds that an e_l has local degree equal to 2 after its line has for the first time length at least 3 (nodes). When this occurs, a spanning ring has been formed and e_l can deactivate the edge (e_l, l') between the two endpoints to make it a spanning line. This completes the description of the protocol (Fig. 2).

Theorem 4 *By assuming that nodes are equipped with a common neighbor detection mechanism and have the ability to detect local degrees 1 and 2, Protocol Line-Transformer solves the Terminating Line Transformation problem in the setting in which all nodes are initially identical. Its running time is $O(n^3)$.*

Table 1 summarizes all protocols that we developed for the Terminating Line Transformation problem, both for the case of a pre-elected unique leader (Protocols Online-Cycle-Elimination and Line-Around-a-Star in Sect. 4) and for the case of identical nodes (Protocol Line-Transformer in the present section).

Finally, in the paper on which this chapter is based, we have shown how the spanning line formed with termination by *Line-Transformer* can be used to establish that the class of computable predicates is the maximum that one can hope for in this family of models.

Theorem 5 (Full Computational Power) *Let the initial active topology be connected, all nodes be initially identical, and let the nodes be equipped with degree in $\{1, 2\}$ detection and common neighbor detection. Then for every predicate $p \in \mathbf{SSPACE}(n^2)$ there is a terminating NET that computes p.*

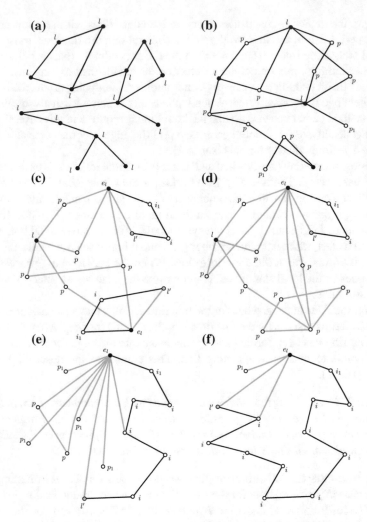

Fig. 2 An example execution of Protocol Line-Transformer. In all subfigures, *black* and *gray* edges are active and missing edges are inactive. *Black* and *gray* are used together whenever we want to highlight some subnetwork of the active network. **a** Initially, all nodes are leaders and the topology is connected. **b** Most leaders have been converted to peripherals, some leaders have attracted new peripherals, and some peripherals have disconnected from each other. **c** Two of the survived leaders have started to form lines over their p_1 peripherals. The centers of these stars are now in state e_l (*black nodes*), the other endpoint of their lines is in state l' (*gray nodes*), and the lines are drawn by *black edges*. **d** The l' endpoints of the two previous lines interacted and one of them was backtracked. **e** A single line has remained. **f** The line is almost spanning

Table 1 All protocols developed in this work for the Terminating Line Transformation problem. For each of these protocols (OCE: Online-Cycle-Elimination, LAS: Line-Around-a-Star, and LT: Line-Transformer), the table shows whether it makes use of a pre-elected unique leader, what local degree detection it uses (DD), whether it uses common neighbor detection (CND), and also its expected running time under the uniform random scheduler

Protocol	Leader	DD	CND	Expected time	Lower bound
OCE	Yes	1	No	$\Theta(n^4)$	$\Omega(n^2 \log n)$
LAS	Yes	1	No	$\Theta(n^2 \log n)$ (opt)	$\Omega(n^2 \log n)$
LT	No	1, 2	Yes	$O(n^3)$	$\Omega(n^2 \log n)$

The last column shows the best known lower bound for the problem

6 Conclusions and Further Research

There are many open problems related to the findings of the present work. We have shown that initial connectivity of the active topology combined with the ability of the protocol to transform the topology yield, under some additional minimal and local assumptions, an extremely powerful model. We managed to show this by developing protocols that transform the initial topology to a convenient one (in our case the spanning line) while always *preserving the connectivity of the topology*. Though arbitrary connectivity breaking makes termination impossible, still we have not excluded the possibility that some protocol performs some "controlled" connectivity breaking during its course, being always able to correctly reassemble the disconnected parts and terminate.

Another issue has to do with the underlying interaction model. Throughout this work we have assumed that the underlying interaction graph is the clique K_n and all of our protocols largely exploit this. Though this model is a convenient starting point to understand the basic principles of algorithmic transformations of networks, it is obvious that it is highly non-local. Realistic implementations would probably require more local or geometrically constrained models (like the one of [14]), for example, one in which, at any given time, a node can only communicate with nodes at active distance at most 2. It is also valuable to consider the Terminating Line Transformation and Acyclicity problems in models of computationally weak (and probably also anonymous) robots moving in the plane.

There are also some more technical intriguing open questions. The most prominent one is whether protocol Line-Transformer is time-optimal. Recall that its running time was shown to be $O(n^3)$. First of all, it is not clear whether the analysis is tight. The subroutine that dominates the running time is the one that tries to form a spanning line over the peripheral nodes, which is restricted by the fact that the partial lines of "sleeping" stars have to either be backtracked (which is what our solution does) or merged somehow with the lines of "awake" stars. We should mention that the spanning line subroutine that backtracks many "sleeping" lines in parallel is an immediate improvement of the best spanning line protocol of [17], called Fast-Global-Line. The improvement is due to the fact that instead of having the awake leader backtrack node-by-node sleeping lines, we now have any sleeping line backtrack itself, so that many backtrackings occur in parallel. We also have experimental evidence showing a small improvement [2] but still we do not have a proof of whether this is also an asymptotic improvement. For example, is it the case that the running time of this improvement is $O(n^3/\log n)$ (or even smaller)? This question is open. There is also room for lower bounds. Apart from the obvious lower bound for the Terminating Line Transformation problem with identical nodes, one could also focus on the spanning line construction problem with initially disconnected nodes (i.e. the Spanning Line problem of [17]). The reason is that an improvement to this problem would probably imply an improvement for Terminating Line Transformation by using the protocol as an improved subroutine of Line-Transformer for forming the lines over the peripherals of the star. The best lower bound known for Spanning Line is $\Omega(n^2)$. Some first attempts suggest that it might be non-trivial to improve this to $\Omega(n^2 \log n)$.

References

1. Aloupis, G., Collette, S., Damian, M., Demaine, E.D., Flatland, R., Langerman, S., O'Rourke, J., Pinciu, V., Ramaswami, S., Sacristán, V., Wuhrer, S.: Efficient constant-velocity reconfiguration of crystalline robots. Robotica **29**(01), 59–71 (2011)
2. Amaxilatis, D., Logaras, M., Michail, O., Spirakis, P.G.: NETCS: a new simulator of population protocols and network constructors. arXiv:1508.06731 (2015)
3. Angluin, D., Aspnes, J., Chan, M., Fischer, M.J., Jiang, H., Peralta, R.: Stably computable properties of network graphs. In: 1st IEEE International Conference on Distributed Computing in Sensor Systems (DCOSS), vol. 3560, LNCS, pp. 63–74. Springer (2005)
4. Angluin, D., Aspnes, J., Diamadi, Z., Fischer, M.J., Peralta, R.: Computation in networks of passively mobile finite-state sensors. Distrib. Comput. **18**(4), 235–253 (2006)
5. Angluin, D., Aspnes, J., Eisenstat, D., Ruppert, E.: The computational power of population protocols. Distrib. Comput. **20**(4), 279–304 (2007)
6. Aspnes, J., Ruppert, E.: An introduction to population protocols. In: Garbinato, B., Miranda, H., Rodrigues, L. (eds.) Middleware for Network Eccentric and Mobile Applications, pp. 97–120. Springer (2009)
7. Beauquier, J., Burman, J., Rosaz, L., Rozoy, B.: Non-deterministic population protocols. In: 16th International Conference on Principles of Distributed Systems (OPODIS), LNCS, pp. 61–75. Springer (2012)
8. Chatzigiannakis, I., Dolev, S., Fekete, S.P., Michail, O., Spirakis, P.G.: Not all fair probabilistic schedulers are equivalent. In: 13th International Conference on Principles of Distributed Systems (OPODIS), vol. 5923, Lecture Notes in Computer Science, pp. 33–47. Springer (2009)
9. Chatzigiannakis, I., Michail, O., Nikolaou, S., Pavlogiannis, A., Spirakis, Paul G.: Passively mobile communicating machines that use restricted space. Theor. Comput. Sci. **412**(46), 6469–6483 (2011)
10. Chatzigiannakis, Ioannis, Michail, Othon, Nikolaou, Stavros, Spirakis, Paul G.: The computational power of simple protocols for self-awareness on graphs. Theor. Comput. Sci. **512**, 98–118 (2013)
11. Cornejo, A., Kuhn, F., Ley-Wild, R., Lynch, N.: Keeping mobile robot swarms connected. In: Proceedings of the 23rd International Symposium on Distributed Computing (DISC), LNCS, pp. 496–511. Springer (2009)
12. Das, S., Flocchini, P., Santoro, N., Yamashita, M.: Forming sequences of geometric patterns with oblivious mobile robots. Distrib. Comput. **28**(2), 131–145 (2015)

13. Doty, D.: Timing in chemical reaction networks. In: Proceedings of the 25th Annual ACM-SIAM Symposium on Discrete Algorithms (SODA), pp. 772–784 (2014)

14. Michail, O.: Terminating distributed construction of shapes and patterns in a fair solution of automata. In: Proceedings of the 34th ACM Symposium on Principles of Distributed Computing (PODC), pp. 37–46. ACM (2015)

15. Michail, O., Chatzigiannakis, I., Spirakis, P.G.: Mediated population protocols. Theor. Comput. Sci. **412**(22), 2434–2450 (2011)

16. Michail, O., Chatzigiannakis, I., Spirakis, P.G.: New models for population protocols. In: Lynch, N.A. (ed.) Synthesis Lectures on Distributed Computing Theory. Morgan & Claypool (2011)

17. Michail, O., Spirakis, P.G.: Simple and efficient local codes for distributed stable network construction. In: Proceedings of the 33rd ACM Symposium on Principles of Distributed Computing (PODC), pp. 76–85. ACM (2014). (Also in Distributed Computing. doi:10.1007/s00446-015-0257-4, 2015)

18. Michail, O., Spirakis, P.G.: Terminating population protocols via some minimal global knowledge assumptions. J. Parallel Distrib. Comput. (JPDC) **81**, 1–10 (2015)

19. Suzuki, I., Yamashita, M.: distributed anonymous mobile robots: formation of geometric patterns. SIAM J. Comput. **28**(4), 1347–1363 (1999)

20. Woods, D., Chen, H.-L., Goodfriend, S., Dabby, N., Winfree, E., Yin, P.: Active self-assembly of algorithmic shapes and patterns in polylogarithmic time. In: Proceedings of the 4th Conference on Innovations in Theoretical Computer Science, pp. 353–354. ACM (2013)

Operating Secure Mobile Healthcare Services over Constrained Resource Networks

Anne Kayem, Patrick Martin, Khallid Elgazzar and Christoph Meinel

Abstract Constrained resource (lossy) networks are characterized by low-power and low-processing capability devices. Using lossy networks for service provision is a good approach to overcoming the technological constraints that characterize remote and rural areas of developing regions. In this chapter, we propose a framework for provisioning Healthcare-as-a-mobile-service over lossy networks, as a cost-effective approach to healthcare management in remote and rural areas of developing regions. The framework allows Healthcare providers collect and share patient data obtained via body sensors that periodically report health status information. Queries can be run locally on a mobile device when the stored copy, say on a cloud, is not accessible. Since mobile storage capacity limitations can be problematic for maintaining data consistency, we propose a storage management framework based on data fragmentation and caching. The fragmentation scheme classifies data according to confidentiality and affinity. While caching optimizes mobile storage by prioritizing data fragments in terms of frequency of access. Our experimental results demonstrate that our data fragmentation and caching algorithms are scalable to varied bandwidth and download rates. The battery power consumption is reasonable with respect to bandwidth variations.

A. Kayem (✉)
Internet Technologies and Systems Group, Hasso-Plattner-Institut für
Softwaresystemtechnik GmbH, Campus Griebnitzsee, Postfach 900460,
14440 Potsdam, Germany
e-mail: anne.kayem@hpi.uni-potsdam.de

P. Martin
School of Computing, Queen's University, Kingston, ON K7L 3N6, Canada
e-mail: martin@cs.queensu.ca

K. Elgazzar
School of Computer Science, Carnegie Mellon, 5000 Forbes Ave.,
GHC 9108, Pittsburgh, PA 15213, USA
e-mail: elgazzar@cs.cmu.edu

C. Meinel
Hasso-Plattner-Institut für Softwaresystemtechnik GmbH,
Campus Griebnitzsee, Postfach 900460, 14440 Potsdam, Germany

© Springer International Publishing Switzerland 2017   361
A. Adamatzky (ed.), *Emergent Computation*, Emergence, Complexity
and Computation 24, DOI 10.1007/978-3-319-46376-6_16

1   Introduction

The concept of "healthcare-as-a-mobile-service" offers a cost effective strategy to addressing health care needs, particularly in regions where access to healthcare providers is impeded by factors that include, intermittent power connectivity, lack of on-site healthcare experts, and intermittent Internet access. In regions where standard approaches to accessing the Internet are negatively impacted by bandwidth limitations, studies indicate that cheap mobile devices offer a lightweight approach to service provisioning [1]. As a consequence, health care organizations have explored mobile health (m-health) solutions as a cost-effective strategy to healthcare management and distribution in these regions [2–4].

Proposed m-health solutions suggest using a combination of smart devices (e.g. sensors to measure heart rate, blood sugar levels, etc.) and the mobile Internet to manage and distribute healthcare information. Mobile platforms have the advantage of providing increased data accessibility while granting users (patients) more control over the management of their data. The basic idea behind m-health data management is that in the case of an emergency, a healthcare practitioner might be able to avoid a misdiagnosis by retrieving the patient's medical history from the patient's mobile device. Both patients and health care providers can benefit from this infrastructure. For instance, patients can access portions of their records on demand while health care providers can access whole copies of the patient's data in addition to contacting experts for help with handling difficult cases.

1.1   Problem Statement

The idea of integrating mobile devices on healthcare distribution platforms raises the question of reliable and dependable mobile storage management in addition to data security and privacy. While storing a copy of the data on the mobile device offers the advantage of portability, mobile device storage limitations can make maintaining a complete copy of a patient's healthcare data on the mobile device challenging. Existing studies [5–8] inspired by the outsourced data paradigm propose maintaining frequent mobile to cloud updates, but this is expensive both battery and security performance-wise for the mobile device.

1.2   Contribution

In this paper, we extend our previous work [9], to provide a complete storage management framework for multi-tier scenarios in which a healthcare provider employs some form of external storage (e.g. a cloud storage provider) to store patient health data. While a patient keeps a "summarized" version of his/her health data on a mobile device. For instance, the healthcare provider may want to keep records that are quite a few years old but that are not necessarily useful for the patient's regular hospital visits. In this case, it makes sense to transfer the storage of these records to a third-party storage provider and keep only the most relevant data on the mobile device's database. This serves to ensure that the patient's medical history is easily accessible in emergency situations.

Our proposed storage management framework works by requiring the mobile device(s) to collect information pertaining to a patient's vital signs. A relevant copy of the data is maintained on the mobile device while the rest of the data is periodically offloaded to the storage provider. The storage provider integrates readings obtained from multiple sources such as hospitals, laboratories, and imaging centers, to maintain data consistency. Determining which portions of the data to store on the mobile devices happens periodically based on a data size threshold, event, or update. Finally we note that the summary data on the mobile device might include data from other sources, not necessarily only vital signs that the device collects.

We handle the problem of deciding how to offload data from the mobile device to the cloud storage provider with two algorithms that are based on the concepts of fragmentation and caching. We note however, that the problem of composing fragments to optimize query execution and that of optimizing mobile storage management are both NP-Hard [10]. Therefore, the solutions that we propose are by necessity based on heuristics. As an added extension, we provide proofs to support our claims of the NP-hardness of both the problem of composing fragments to optimize query execution and that of optimizing mobile storage management. We do this by demonstrating that these problems are in fact reducible to the problem of computing a vertex cover of minimum size and the knapsack problem, respectively.

The fragmentation algorithm creates data fragments from a patient's healthcare data, by relying on two criteria (heuristics) namely, confidentiality, and affinity. Confidentiality expresses the privacy requirements of the healthcare provider with respect to the data while affinity is useful in forming fragments with similar information to facilitate query processing. The caching algorithm supports the fragmentation algorithm by employing a prioritization mechanism to organize the created data fragments to form a tree data structure. The prioritization mechanism orders data fragments according to relevance and frequency of access, and automatically outsources the least relevant and/or frequently accessed ones. The most frequently accessed data fragments are used to form a "summarized" version of the data that is stored locally on the hospital's database and on the patient's mobile device.

The advantage of our proposed storage management framework is twofold, first we protect the mobile data from malicious tampering by only storing the most vital information on the mobile device, and second, we ensure that the e-healthcare provider has control over the data even when this is outside the provider's security domain. Our experimental results demonstrate that our data fragmentation and caching algorithms are scalable to varied bandwidth and download rates. Furthermore, the battery power consumptions incurred when supporting mobile to cloud transfers (and vice versa) are reasonable with respect to bandwidth variations and user expectations.

1.3   Outline

The rest of this chapter is structured as follows: in Sect. 2 we discuss related work on preserving security and privacy of outsourced data and how this relates to the problems evoked in Sect. 1.1. Securing outsourced data is in fact quite similar to protecting, from unauthorized access, the fragments of data that cannot be stored locally on a mobile device. So in fact, the owner of the mobile device and by default the data assumes the role of the data owner while the service provider could be the patient's home hospital or another hospital. In Sect. 3, we build on the discussion in Sect. 2 and describe our proposed storage management framework. Section 4, presents experimental results. We offer concluding remarks and directions for future work in Sect. 5.

2   Related Work

The idea of storing data electronically to facilitate access, guarantee consistency, and reduce data management costs has existed for a while in the healthcare domain [2, 4, 11, 12]. In general, electronic healthcare management systems offer the potential for better access to records when they are needed and have been evolved in modern times to incorporate access via stable Internet connections to support efficient sharing of healthcare information among patients and healthcare providers. Electronic health-care management systems typically implement a role-based access control (RBAC) scheme to handle authentication and authorization. The RBAC scheme usually places the trust at the server-end in order to protect patient's records. Typical examples of authentication-based electronic healthcare systems that use RBAC include Indivo and PCASSO [3, 13, 14].

Indivo is an open-source, and professionally developed personally controlled health care management system that allows users ownership and management of personal healthcare data [3, 14]. The Patient-Centered Access Control Secure System Online (PCASSO) platform also relies on role-based access control to handle access to the data but is more user-centric as opposed to provider-centric [11]. Until recently, the PCASSO model received little attention because in practice, securing healthcare data requires a combination of authentication, encryption, and digital signing to ensure confidentiality, and integrity of the data on the user's end [15].

Another drawback that electronic health information management systems like Indivo and PCASSO face is the lack of mechanisms to facilitate data portability. Data portability becomes an issue when users (patients) decide to switch healthcare providers and/or move to a remote area with intermittent Internet connections. In situations like this it can be impractical to rely on exporting the data from the previous healthcare provider to the new provider [4]. Therefore, existing systems need to be extended to enable data consistency and storage on servers accessible to external parties via the web. Furthermore, to cope with situations of intermittent Internet connections, users could maintain copies of their healthcare records locally on a device that is used to access the healthcare management system.

Frequently, organizations and individuals, find it secure and cost effective to transfer the management of data to external service providers. Similarly, in the health care domain, increasingly patient records are stored in repositories that can be made accessible on demand to other healthcare providers particularly in situations of emergency. In these scenarios, it is important to guarantee security and privacy because the storage service providers cannot be trusted completely.

The literature on the concept of data outsourcing began with Damiani et al.'s work [16, 17] on metadata management in outsourced encrypted databases. The authors show that to respond to queries metadata is integral to information retrieval from outsourced data. Damiani et al. also built on earlier work on securing outsourced data in untrustworthy environments to propose an access control mechanism to reduce the cost of server side authentication [18, 19]. The proposed access control mechanism selectively encrypts metadata and only grants access to users in possession of a cryptographic key capable of decrypting the data. This is in contrast to previous work where all authentication requests to the server needed to be transmitted to the data owners for verification before access could be granted or denied.

In subsequent work, Ciriani et al., proposed fragmenting the data at the service provider in order to improve the efficiency of query executions on encrypted data [20–22]. The principal idea is to encrypt the outsourced data according to some priority or criteria and to avoid creating large data fragments that are expensive to encrypt or query.

More recently, De Capitani Di Vimercati et al. [7] worked on encryption policies for outsourced data. The policies are supported by hierarchical key management schemes [23, 24]. Each fragment of the data is encrypted with a separate key so that access authorizations can be distinguished quite easily. Additionally, these key management approaches aim to minimize the number of keys distributed in order to make security management an efficient process. Kayem et al. also proposed a solution to improving the cost of encryption to cope with security policy updates at the service provider's end [8]. Other relevant work includes that by Foresti [25] to minimize the number of fragments of data created and ensure maximal visibility of the data through querying and to minimize the number of fragments of data created while ensuring maximal visibility of the data through querying.

We note that the previous work focuses mainly on management of the data at the server-end and assumes that the data is primarily read-intensive which is well suited to scenarios in which the data stored at the service provider changes relatively infrequently. In the m-health care scenario however, situations can arise in which the data is updated fairly frequently resulting in frequent security policy updates which is expensive encryption/decryption-wise. Additionally, storage limitations on the mobile device imply that relying on manual mechanisms to outsource data from the mobile device to the cloud can result in privacy and/or security violations of the e-healthcare providers' policies. Therefore, a secure and efficient approach to managing the m-health data is needed to minimize the cost of updating and/or querying the data while ensuring that the data remains protected.

3 Fragmentation and Caching

Our framework for handling healthcare-as-a-mobile-service is illustrated in Fig. 1. In our framework, we assume that a user (patient) can collect data, via sensors on his/her body, in relation to an illness(es) he/she may have. This data is stored on the user's mobile device and can be made accessible via a mobile web service (MWS) to a caregiver. Mobile web services offer an interoperable interface that facilitates data sharing via ubiquitous protocols and data formats such as HTTP and XML that eliminate the need to worry about the implementation details of the service(s) being accessed [26]. In the mobile healthcare scenario this is an advantage because it implies that healthcare providers and patients are able to use a uniform communication platform irrespective of the device or operating system. For simplicity, we define a caregiver as anyone who administers a health related operation on the patient.

In order to handle data archiving, or cases in which all of the data cannot be stored on the mobile device, we envisage that the patient has access to a cloud storage service provider. Access to the data on the cloud storage provider is controlled by an authentication manager (AM) that ensures that the caregivers requesting access to the cloud data have the authorization to do so. For consistency with similar solutions in the literature [5–8], we consider that a cryptographic key management scheme is implemented to enforce the access control policies on the data that is stored on the

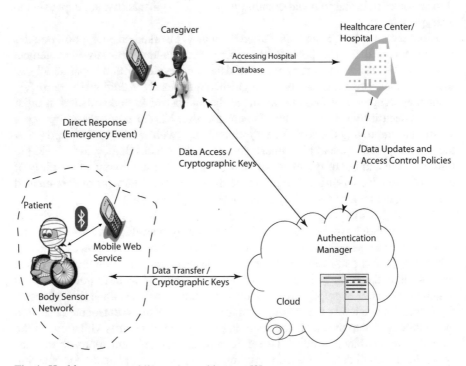

Fig. 1 Healthcare-as-a-mobile-service architecture [9]

cloud. As described in the scheme proposed by Di Vimercati De Capitani et al. [7], a data owner can protect the data that he/she transfers for storage and management to a service provider by using a cryptographic key management scheme to encrypt the data. The protection mechanism works by requiring that the data owner, in this case the hospital or healthcare center, encrypt the data before it is transferred to the service provider (or cloud). This is to ensure that the data remains secret even to the service provider. In order to enforce non-repudiation, the service provider imposes a second layer of encryption on the data and shares the key(s) used with the data owner. All users wanting to access the data receive two keys from the data owner, one that is used to decrypt the encryption layer imposed by the service provider and the other key, to decrypt the layer initially imposed by the data owner. The data at the cloud service can also be structured hierarchically as is the case in role-based access control models where the types of access granted to a user depend on the role or authorizations the user has in the system. As shown in Fig. 1, a caregiver or patient wanting to access the health data on the cloud receives two keys from the healthcare center or hospital. The first key decrypts the encryption layer imposed by the cloud service provider while the second key decrypts the encryption layer that the hospital imposed before the data was transferred to the cloud.

We address the problem of efficient data querying and privacy enforcement in the provision of healthcare-as-a-mobile-service, with two algorithms that are based on the concepts of data fragmentation and caching. The data caching algorithm employs a hierarchical tree-like structure to store data according to fragments that are formed on the basis of the privacy and temporal constraints. In the caching hierarchy, data fragments that are the most recent and most frequently accessed appear closer to the root node while the least accessed fragments are placed closer to the leaf nodes. The fragmentation algorithm, uses the concept of marginal gain to determine which fragments of the data contain attributes that permit an efficient execution of queries on the mobile device. Both algorithms are aimed at complimenting each other, the goal being to use the mobile storage capacity optimally to hold a copy of the data that contains information that is most relevant to the patient's present condition. By this we mean a copy of the data that can satisfy all or most of the queries without requiring access to the cloud data.

The advantage of this approach to healthcare-as-a-mobile-service is twofold, first, in cases of disconnections from the cloud service provider, we can guarantee a best-effort and reliable response to most queries; second, since the data is structured according to the security policies of the Electronic Healthcare Provider (EHP), we can ensure that the data is always handled in ways that comply with the security requirements of the EHP. In the following section, we specify the confidentiality constraints that are used to guide the data caching and fragmentation algorithms.

3.1 Conditions for Confidentiality

Dependent on the available mobile storage capacity, the fragmentation algorithm categorizes the data on the hospital database into small chunks or files based on the healthcare provider's confidentiality constraints and attribute affinity. This enables the caching algorithm to determine which fragments of the data to store locally on the hospital's database and which to outsource to the cloud storage provider. Depending on the storage constraints of the mobile device, the copy of the data stored on the hospital's database can be further fragmented to extract a more concise version of the data for storage on the mobile device. In order to ensure that the data is kept secure both on the cloud and on the mobile device it is important to model the privacy requirements of the healthcare provider by applying confidentiality constraints on the sets of attributes that make up the patient's data.

Basically what this means is that the confidentiality constraints indicate which attributes need to be kept together and protected from unauthorized access. For instance, {Name, DOB} are considered to be sensitive bits of data that are correlated and so must be kept in the same fragment of the data. Enforcing confidentiality constraints of this sort requires that the healthcare organization correctly specifies which attributes of the data need to be protected. In order to enforce confidentiality constraints we must ensure that the constraints over one set of attributes does not contain a constraint that belongs in another set of constraints. This is to avoid redundancies that might lead to violations of the security policy enforcing the confidentiality constraints. A well-defined confidentiality constraint can be defined formally as follows:

Definition 1 (*Well-Defined Confidentiality Constraints*) Given a set of attributes $A = \{a_0, ..., a_i, a_j, ..., a_n\}$, and a set of confidentiality constraints $C = \{c_0, ..., c_i, c_j, ..., c_m\}$ where m is the maximum number of confidentiality constraints that have been specified, and n the maximum number of attributes that have been specified, C is said to be well-defined if and only if $\forall c_i, c_j \in C, i \neq j$ $c_i \not\subseteq c_j$ and $c_i \subseteq A$.

Example 1 In Fig. 2 we illustrate an example of a patient's health record and some of the confidentiality constraints that can be applied to the data. In this case we have defined three constraints c_0, c_1, and c_2 that state that the association of a patient's name and ID with any other information is considered sensitive.

We note that $c_0 \not\subseteq c_1 \not\subseteq c_2 \not\subseteq c_3$ to avoid cases in which overlapping information can lead to security violations. If c_1 were a subset of c_2 or vice versa a user with the authorization to view information on a patient's name, address, and illness would also have access to the patient's date of birth. Essentially, the confidentiality constraints are aimed at specifying access control policies that minimize data exposure while guaranteeing efficiency and correctness in query responses.

In order to enforce the well-defined confidentiality constraints on the data, we assume that we are dealing with a storage system with a schema, such as a relational database management system. In the storage system we have a finite set of attributes that each take on a series of values within some defined domain.

ID Number	Name	DOB	Postal Code	Illness	Physician
1235641188389	A. Bossi	12/12/1982	7701	Diabetes	Saleem
1235641188389	A. Bossi	12/12/1982	7701	HIV	Coyne
1235641188389	A. Bossi	12/12/1982	7701	Hypertension	Ndaba
1235641188389	A. Bossi	12/12/1982	7701	Gastritis	Jackson

$c_0 = \{ID\}$

$c_1 = \{Name, DOB\}$

$c_2 = \{Name, Postal Code, Illness\}$

$c_3 = \{DOB, Postal Code, Physician\}$

Fig. 2 An example of a user's healthcare data and the associated well defined confidentiality constraints

3.2 Fragmentation Scheme

As mentioned before, the confidentiality constraints are used to model the privacy requirements that the healthcare provider applies to the data. The reason for this, is to ensure that the security and privacy policies of the healthcare provider are enforced on the data, even in cases where the data is outside of the security domain of the healthcare provider.

On the basis of the confidentiality constraints, the healthcare provider begins by fragmenting the data into a set of disjoint fragments F_i represented by a tree $(T, \prec)$, where $T = \{F_0, F_1, ..., F_{n-1}\}$. By definition, $F_i \prec F_j$ implies that F_j contains attributes that hold information that is more important or relevant to the patient's current condition than the attributes contained in F_i. Importance is derived from the specification of the confidentiality constraints that we defined in Sect. 3.1. For instance, from the confidentiality constraints specified in Example 1, $c_0 \equiv F_0$ and $c_1 \equiv F_1$. So in this case, since c_0 contains information that is more sensitive than that c_1 contains, it can be assumed that F_0 is more important than F_1.

In addition to creating fragments of the relation schema we need to ensure that the fragments created obey the confidentiality constraints in order to protect the information in the fragments from inference attacks. A fragment is considered to correctly enforce the confidentiality constraints if and only if the fragments are protected from linking attacks by requiring that fragments are disjoint so that no attribute appears in more than one fragment. This aspect is formally captured in the following definitions.

Definition 2 (*Distributed Databases*) A distributed database is a database that is managed by a software system (Distributed Database Management System) transparently (i.e. knowledge of the storage location of the data is not required for access). Distribution can be handled in terms of data fragmentation where a data fragment constitutes some subset of the original database.

Definition 3 (*Fragmentation*) Given a relation schema R. This is the partitioning of a distributed database into subsets of data. Three basic rules are required to support fragmentation and database consistency:

- **Completeness.** A database decomposed into fragments must be such that each data item found in the database appears in at least one fragment.
- **Reconstruction.** There exists at least one relational expression that will reconstruct the global database from the fragments.
- **Disjointness.** A data item appearing in a fragment, should not appear in any other fragment.

Definition 4 (*Horizontal Fragmentation*) Given a relation schema R, we say that a set of data fragments F is formed by subdividing R into subsets of tuples that adhere to a set of well-defined confidentiality constraints and that satisfy the following conditions:

- **Completeness.** The fragmentation is correct if each tuple of R is mapped unto at least one tuple of the fragments created.
- **Reconstruction.** A union operation can be used to obtain R from the set of fragments created.
- **Disjointness.** Each tuple of R is mapped unto exactly one tuple of one of the fragments to control duplication explicitly at the fragmentation level.

The resulting fragments have the same schema structure as R, but differ in the data they contain.

Definition 5 (*Fragment Correctness*) Let R be a relation schema in a distributed database, C a set of well-defined constraints over R, and F a set of fragments of R formed by horizontal fragmentation. We say that the fragments F are a result of applying some combination of the relational expressions E_1 and E_2 on R using some combination of the following rules:

- **Union.** $E_1 \cup E_2$: F is a result of merging data from E_1 and E_2
- **Difference.** $E_1 - E_2$: F is a result of eliminating unwanted tuples E_2 from the set of tuples E_1
- **Cartesian Product.** $E_1 \times E_2$: F is a result of combining each tuple in E_1 with every tuple in E_2
- **Selection.** $\sigma_p (E_1)$ for predicate p on attributes of E_1 : F is a result of extracting a set of tuples from R with respect to a given list of attributes of E_1.
- **Projection.** $\pi_s (E_1)$ where s is a subset of attributes of E_1 : F is a resulting set that is obtained when all tuples in R are restricted to the set of attributes in E_1
- **Duplication.** $\rho (Q (L), E_1)$ where Q is a new relation name and L is a list of (Old name $\rightarrow$ New name) mappings of attributes of E_1

In addition, we say that F correctly enforces C if and only if the following conditions are satisfied:

1. Each fragment belonging in F satisfies the confidentiality constraints C.
2. $\forall F_i, F_j \in F, i \neq j, F_i \nsubseteq F_j$.

The first condition satisfies the constraint given in Definition 1 where we describe the rules for ensuring that the data within a fragment remains confidential or inaccessible to unauthorized parties. While the second condition enforces the constraint in Definition 1, to prevent overlaps that could lead to redundancy.

Example 2 Combining the notions of confidentiality and fragmentation, from Fig. 2, we can create fragments based on the confidentiality constraints c_0, c_1, c_2, and c_3 to obtain fragments: $F_0 = c_0 = $ ID; $F_1 = c_1 = $ Name, DOB; $F_2 = c_2 = $ Name, Postal Code, Illness; $F_3 = c_3 = $ DOB, Postal Code, Physician.

In order to enforce the confidentiality constraints on the fragments of data the security administrator(SA) of the EHP selects a secret key K_0 that is used to generate keys, using a one-way function, that are used to encrypt each one of the data fragments F_i [8, 23, 24]. In the key management hierarchy, the keys are organized according to a partial order such that $K_i \prec K_j$ indicates that the key K_i is less important than the key K_j. One approach to enforcing hierarchical key generation is to have the SA select two large primes p and q, in addition to selecting the secret key K_0 and compute the required keys using an exponentiation function of the form:

$$K_i = K_0^{t_i} \bmod M \tag{1}$$

where $M = p \times q$ and t_i is a random integer value [23]. In this case, since the keys are inter-dependent, when the $K_i \prec K_j$ condition is enforced this implies that holders of the key K_j can derive the key K_i.

Each data fragment F_i can then be encrypted with a corresponding key K_i and transferred to the patient's mobile device. For simplicity, we assume that in order to structure the fragments of data to form a tree we will use an importance rating that we discuss in Sects. 3 and 4, and we use a hierarchical key management scheme to ensure that irrespective of the position of a data fragment in the tree hierarchy—access always happens according to the rules of the specified security policy. Example 3 provides an illustration of how the key management scheme works with data fragmentation.

Example 3 In the example illustrated in Fig. 3, we suppose that for each patient's medical data the EHP defines a hierarchy of keys to encrypt the fragments of data created from the patient's medical data. A "summarized" version of the data is then encrypted and transferred to the patient's mobile device. In order to access the data created, the patient and the caregiver(s) receive a cryptographic key from the EHP. This key is used to authenticate the user requesting access to the data.

In order to enforce the confidentiality constraints specified in Fig. 2, the EHP will create a hierarchy of keys, using Eq. 2, that are interconnected in ways that enforce the hospital's access control policy. For example, a hierarchy of keys K_0, K_1, K_2, and K_3 can be used to enforce the role-based access control (RBAC) policy where 0 indicates that no access is allowed, whereas 1 indicates that access is possible. In Fig. 3, we show a representation how the data fragments can be stored. In this case, we opted to use a complete binary tree to facilitate handling caching which we discuss in Sect. 3.4. For example as illustrated in Fig. 4, the hierarchy of keys K_0, K_1, K_2, and K_3 is used to enforce the role-based access control (RBAC) policy represented in the access control matrix.

Confidentiality constraints are useful in ensuring that the fragmented data obeys the security policies of the healthcare provider. However, since multiple security

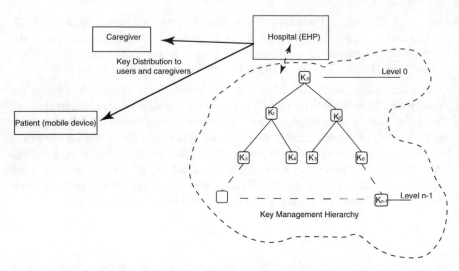

Fig. 3 Encrypting to protect mobile health data

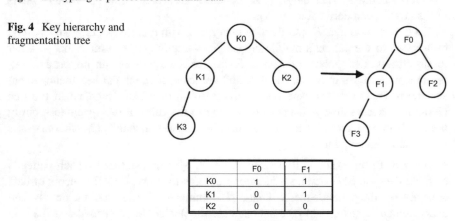

Fig. 4 Key hierarchy and
fragmentation tree

	F0	F1
K0	1	1
K1	0	1
K2	0	0

policies might exist, it follows that several confidentiality constraints might be specified to enforce these security policies. Creating fragments solely according to the
confidentiality constraints can result in several small but related fragments. Querying
small but disjoint fragments is expensive, particularly when these queries need to be
run on data that has been encrypted to enforce a security policy. Therefore, it makes
sense to aggregate fragments with similar confidentiality constraints to form bigger
fragments of data that guarantee data security and at the same time optimize query
execution. We are interested therefore in finding a fragment that contains all the relevant information required to make query execution on the mobile device efficient and
maximize the utility of mobile device resources such as battery power. This problem
can be formalized as follows.

Problem 1 (*Optimal Fragmentation*) Given a relation schema R, and a set C of well-defined constraints over R find a fragmentation F of R such that all of the following conditions hold.

1. F correctly enforces C
2. F maximizes attribute affinity
3. F is complete and minimal in that there is not a fragmentation F' that is a subset of F and that satisfies the first two conditions.

The optimal fragmentation problem is NP-Hard, as formally stated by the following theorem.

Theorem 1 *The optimal fragmentation problem is NP-Hard.*

Proof The proof is a reduction from the NP-Hard problem of computing a vertex cover of minimum size in a given undirected graph [27]. The vertex-cover problem can be formulated as follows: given an undirected graph $G = (V, E)$ the aim is to find a vertex-cover of minimum size in the undirected graph. So essentially the vertex cover of an undirected graph $G = (V, E)$ is a subset $V' \subseteq V$ such that if $(u, v) \in E$ then $u \in V'$ and/or $v \in V'$. Each vertex "covers" its incident edges and so the vertex cover for the graph G would be the set of vertices that include all the edges in E. If we restate the optimization problem as a decision problem then basically, we wish to determine whether a graph has a vertex cover of a given size k.

By analogy we can translate the vertex cover description to the case of a relational schema R and set C of well-defined constraints that affect the data that we aim to fragment optimally. We can then reformulate the optimal fragmentation problem as follows: A fragmentation of a relational schema R is a subset $R' \subseteq R$ such that if two attributes $a_i, a_j \in C$ then $a_i \in R'$ and/or $a_j \in R'$. So, essentially, any vertex v of the graph G corresponds to an attribute $a_i \in R$ and any edge e_i in G which connects $v_1, ..., v_n$ corresponds to a constraint $c_i = \{a_1, ..., a_n\} \in C$ where c_i is not a singleton constraint and is enforced in addition to attribute affinity.

A fragmentation F of a relational schema R that satisfies all the constraints in C corresponds to a solution S for the corresponding vertex cover problem. More specifically, S is composed of a set of size k attributes that are linked by confidentiality constraints in such a way that all the attributes contained in the fragments of F are also contained in the set S. As a consequence any algorithm that can find an optimal fragmentation for R can also be exploited to solve the vertex cover problem.

3.3   Heuristics for Enabling Efficient Fragmentation

Our heuristic to handle fragmentation optimally combines the notions of the confidentiality constraints and marginal gain. Confidentiality constraints are useful in forming data fragments and in deciding how to position the fragments in the data (tree) hierarchy while marginal gain is useful in determining fragment importance and in deciding when to merge data fragments as opposed to maintaining separate copies.

Marginal gain is computed by evaluating the number of hits (queries) involving the attributes of a fragment say F_i, in comparison to another fragment say, F_j. Marginal gain is computed as follows:

$$MG\left(F_i, F_j\right) = \text{Hits}\left(F_i\right) - \text{Hits}\left(F_j\right) \tag{2}$$

When $MG\left(F_i, F_j\right) > 0$ the implication is that F_i has a higher number of hits than F_j, and when $MG\left(F_i, F_j\right) < 0$ the implication is that F_j has a high number of hits than F_i. A value of 0 for $I\left(F_i, F_j\right)$ indicates that both fragments have the same number of hits.

In order to determine whether or not data fragments should be merged to minimize the number of data fragments created we use a distance measure to determine whether or not to merge data fragments. We compute the distance $\text{Dist}\left(F_i, F_j\right)$ between two fragments F_i and F_j as follows

$$\text{Dist}\left(F_i, F_j\right) = \sqrt{\sum_{h}^{k}\left(\text{Hits}\left(a_{i,h}\right) - \text{Hits}\left(a_{j,h}\right)\right)^2} \tag{3}$$

where $\text{Hits}\left(a_{i,h}\right)$ is the number of hits on attribute h in fragment F_i, $h \geq 1$, and $k \geq 1$ such that k represents the highest number of attributes in the largest fragment. A small distance value, that is lower than a predefined threshold value, indicates that merging both fragments of data is advantageous whereas a high distance value indicates the reverse. The assumption here is that the corresponding attributes in F_i and F_j are similar.

Periodically, the importance ratings of the fragments are compared to decide which ones should be moved higher up in the data hierarchy and which should be merged. We discuss how to assign importance ratings and how the promote/demote functions are applied in Sect. 3.4 when we describe the cache management algorithm.

Example 4 An example of how our fragmentation algorithm works is as follows. Each attribute that has been accessed is associated with a counter variable to log the number of times the attribute has been involved in responding to a query. If we suppose that we have two fragments F_1 and F_2. F_1 contains attributes a_2, and a_3 while F_2 contains attribute a_4. After a period say T an evaluation of the counters for attributes a_2, a_3, and a_4 reveals that the attributes have been accessed 10, 24, and 30 times respectively. In this case the $MG(F_1, F_2)$ is 4 so F_1 gets rated as having a higher number of hits than F_2.

In the next step we determine whether or not it is advantageous to merge F_1 and F_2. The distance between F_1 and F_2 is computed using Eq. 3, as follows:

$$\text{Dist}(F_1, F_2) = \sqrt{\sum_{h=1}^{k=2} \left(\text{Hits}(a_{i,h}) - \text{Hits}(a_{j,h})\right)^2}$$

$$= \sqrt{\left(\text{Hits}(a_{1,1}) - \text{Hits}(a_{2,1})\right)^2}$$

$$+ \sqrt{\left(\text{Hits}(a_{1,2}) - \text{Hits}(a_{2,2})\right)^2}$$

$$= (10 - 30)^2 + (24 - 0)^2$$

$$= 31.241$$

If we assume that the merging threshold value were preset to a value of 15 then in this case we wouldnot merge the fragments. Since the marginal gain between both fragments is not that high, and a_3 is getting significantly higher hits than the other two attributes in F_1 we can choose not to merge the two fragments. In this case, since F_1 still has a higher marginal gain than F_2 their relative importance ratings remain unchanged. However, if the rating of F_1 were to drop significantly so that the importance rating of F_2 is higher, or the threshold value for merging were raised, both fragments would be merged to form a new fragment $F_x = \{a_2, a_3, a_4\}$.

3.4 Caching to Support Fast Access to Data

Our approach to storage utilization is based on the architecture illustrated in Fig. 1. Consistent with how the confidentiality constraints are defined, all of the data collected from the sensors is stored in encrypted form on the user's mobile device or some portion of the data is transferred to the cloud for storage. The data is encrypted in order to ensure that it is protected from unauthorized access. As mentioned before, for simplicity and consistency with other proposals [5–8], we consider that the cloud service provider can impose additional layers of encryption on the data to provide stronger guarantees of data integrity and performance efficiency in handling updates to the data.

Caching to optimize storage utilization on the mobile device is guided by the following conditions. In determining which records to store on the mobile device and which ones need to be transferred to the cloud, we basically need to decide based on affinity between attributes and regularity (frequency) of access. Affinity between attributes is useful in handling queries efficiently. In placing attributes that are most likely to be needed to satisfy a query on the mobile device we ensure that the information can be made available to the caregiver in cases of emergency when access to the Internet is temporarily unavailable due to bandwidth limitations. Regularity of access on the other hand provides added information to use in deciding which attributes of the data are useful to store on the mobile device. Data consistency, both on the mobile device and the cloud, is maintained via updates to the medical data that are effected during periods when access to the network occurs. For instance, when a patient moves into an area with good Internet coverage.

Problem 2 (*Mobile Storage Management*) Given a relation schema R, and a set C of well-defined constraints over R compute a set of fragments F of R such that all of the following conditions hold.

1. F correctly enforces C
2. F maximizes the storage space S available on the mobile device
3. F is complete and minimal in that there is not F' that is a subset of F and that satisfies the first two conditions. In other words F represents the most 'useful' summarized version of the patient's healthcare data and ensures optimal query responses.

The mobile storage management problem is NP-Hard, as formally stated by the following theorem.

Theorem 2 *The mobile storage management problem is NP-Hard.*

Proof The proof is a reduction from the NP-Hard knapsack combinatorial optimization problem [10]. The knapsack problem can be formulated as follows: given a set of items, each with a mass and a value, determine the number of each item to include in a collection so that the total weight is less than or equal to a given limit and the total value is as large as possible. This is similar to the problem faced by a person who is constrained by a fixed size knapsack and needs to fill it with only the most important items. If we restate the knapsack optimization problem as a decision problem then the problem is that we wish to determine whether a value of at least V can be achieved without exceeding the weight W.

We can translate the knapsack problem's description to our relational schema R and set of constraints C on the set of data fragments F to be stored optimally on some limited storage space S' that is available on the mobile device. Reformulating our mobile storage management problem we say that: Given a set of data fragments $F_i, F_j \in F$ with marginal gain $MG\left(F_i, F_j\right) \geq 0$ we wish to assign F_i to S' such that all F_i fit into the space S'. Consequently, any algorithm that can find an optimal method of assigning the F_i to S' can be used to solve the knapsack problem.

We are now ready to discuss the storage management optimization problem in relation to storing medical data on a mobile device and to emphasize the need for storing a "summarized" version of the data on the mobile device as opposed to a complete copy. Since the storage management optimization problem is provably NP-hard we will use a heuristic approach to maximizing storage utilization on a mobile device.

3.5 A Heuristic Approach to Maximizing Storage Utilization

In this section we consider the case in which we would like to maximize storage utilization by prioritizing and caching the data. Prioritization is helpful in deciding which fragments of data are necessary for successful queries and caching is useful is maintaining a record of the fragments of data that are more frequently accessed. As indicated in Fig. 5, we assume that the data can be broken up into fragments that are structured hierarchically in terms of importance. We also assume that, as described in Definition 1, the fragments of data are formed based on confidentiality constraints that are specified by the EHP.

In the data hierarchy, there are n cache levels where $n \geq 1$, and the hierarchy is organized as a tree rooted at $cache_1$ which is attached to the storage partition on the mobile device's storage. In the cache hierarchy each node is a data fragment and the cache hierarchy sits at the cloud server. The cached data is handled in such a way as to have the most recent and frequently accessed information at the higher levels of the hierarchy while the least frequently accessed and, by comparison, older information is stored at the lower levels of the hierarchy. We define three operations for handling the cache hierarchy, namely:

- Buffer (y, temp): Move fragment y from the cache at level $\lfloor \frac{i}{2} \rfloor$ into a temporary buffer
- Promote (x, i): Move fragment x from the cache at level i to the cache at level $\lfloor \frac{i}{2} \rfloor$
- Demote (y, i): Move fragment y from the buffer to cache level i

In other words, a fragment gets moved to a lower priority cache level if it is not accessed frequently and gets promoted to a higher level, if it is frequently accessed or contains new material. We use the buffering operation to temporarily store the fragment that was previously at position $\lfloor \frac{i}{2} \rfloor$ in order to avoid data loss. Our algorithm for managing the fragments of data in the cache hierarchy works by assigning a priority to each fragment according to frequency of access. Initially a new fragment

Fig. 5 Caching hierarchy structure

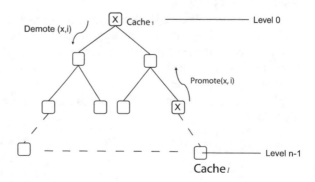

is assigned the highest priority and depending on the frequency of access on the fragment, get assigned a lower priority in which case the affected fragment gets demoted to a lower cache level that is representative of the importance level associated with the information. For simplicity, we assume that caches are stationary and that the data fragments are moved between them. In addition, each cache holds at most one data fragment at a time.

In our proposed storage management scheme caching priority is expressed as the importance I_R of a fragment. The importance of a data fragment basically indicates how recently it was updated and in addition how frequently it has been accessed. So, essentially we express importance of a data fragment as follows:

$$I_R = \frac{\text{Freq}_R}{U_R}$$

where U_R denotes the freshness of the data fragment and is calculated in terms of the number of time units (seconds, minutes, days...) for which the data has been on the system, and Freq_R expresses the frequency of access to a data fragment. A sliding window is used to evaluate frequency of access to fragments so that Freq_R is is a value between 0 and 1. Initially U_R is assigned a value of 1 time units and this grows proportionately to the 'age' of fragment in the hierarchy. The I_R of a fragment is evaluated with respect to Freq_R and U_R. Consequently, a constant Freq_R and a high U_R decreases the I_R of a fragment and similarly a low U_R and high Freq_R implies a high I_R. As expected, consistent proportionate increases or decreases in Freq_R and U_R result in a relatively constant I_R.

Periodically, as shown in Fig. 5 the data fragments in the caching hierarchy need to be evaluated and restructured to move the more frequently accessed fragments to the top of the hierarchy. We accomplish this with the Buffer (y, temp), Promote(x, i), and Demote(x, i) functions that we described earlier. In order to do this, we evaluate the importance of each fragment using the importance function described above to re-evaluate the value of I_R for each fragment and use this value to decide which fragments to move. For instance, as shown in Fig. 5 a data fragment associated with Cache$_1$ might get demoted to Cache$_2$ if the I_R value associated to the data in Cache$_1$ is less than that in Cache$_2$.

In order to formalize the implementation of our caching hierarchy we use a priority queue that is modeled as a binary heap. Initially, the caching hierarchy is populated on a first come first served basis so that the first data fragment to be processed is stored in Cache$_1$ and the last, in Cache$_l$, assuming that n is the maximum number of levels in the caching hierarchy and that $l = 2^n - 1$.

The caching hierarchy population algorithm can be summarized as shown in Algorithm 1.

1: **Input:** d_i, I_R /*A data fragment d_i of importance I_R*/
2: **Output:** H /*A hierarchy of n levels of data, with nodal degree 2*/
3: $H_1 = d_1$, I_R /*d_0: Initial fragment based on confidentiality constraints*/
4: **for** $i = 2$ to $2^n - 1$ **do**
5:   /*Assign fragment to all hierarchy levels based on prioritization*/
6:   $H_i = d_i$, I_R;
7: **end for**

Algorithm 1: Cache Hierarchy Population

In line 4 of Algorithm 1 we initialize the caching hierarchy with the first data fragment to arrive. Lines 5–10 basically populate the hierarchy from the leftmost node at each level to the rightmost node. This is in order to ensure that we form a complete binary tree to obey the heap structure property [10]. The ordering property is obeyed because all insertions have the same priority (importance) at the beginning [10].

As mentioned above, the caching hierarchy is evaluated periodically and the cache re-ordered in the form of a priority queue in which the most frequently accessed items get placed in the front of the queue. In order to do this, we use a version of the heap sort algorithm and order data fragments according to importance I_R. Our heap sorting algorithm works by iteratively evaluating each level of the cache hierarchy promoting and/or demoting data fragments in terms of importance. So the data fragment with the maximum importance is stored at the root of the caching hierarchy while the least import data fragment is stored at the lowest level of the caching hierarchy. Algorithm 2 summarizes the heap sort procedure:

1: **Input:** H /*An unsorted cache hierarchy of n levels of data, where each node can have a maximum of n children nodes*/
2: **Output:** H_S /*A sorted cache hierarchy of n levels of data, where each node can have a maximum of n children nodes*/
3: **for** $i = l$ to 1 **do**
4:   **if** $(H[i]) > \left(H\left[\left\lfloor\frac{i}{2}\right\rfloor\right]\right)$ **then**
5:     Buffer $\left(\left(\left(H\left[\left\lfloor\frac{i}{2}\right\rfloor\right]\right), \text{temp}\right)\right)$
6:     Promote $\left((H[i]), \left(H\left[\left\lfloor\frac{i}{2}\right\rfloor\right]\right)\right)$
7:     Demote $(\text{temp}, (H[i]))$
8:   **end if**
9: **end for**

Algorithm 2: Sorting Caching Hierarchy

Example 5 An example of how our cache hierarchy sorting algorithm works is given in Fig. 6. Basically the integer values in each node denote the current importance rating associated with a data fragment contained in the cache position of the hierarchy.

As shown in Fig. 6, Algorithm 2 is applied iteratively through all the levels of the hierarchy rearranging the data fragments to ensure that the fragment with the highest importance rating is located at the top of the binary tree that represents the priority queue. We note that in Step 3, the fragment with a rating of 9 has been progressively moved up to the root position.

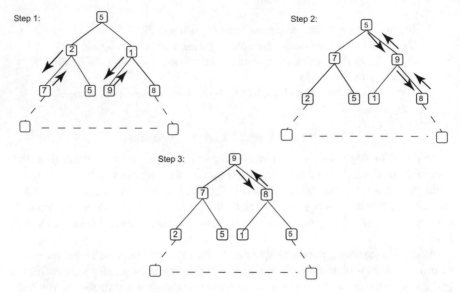

Fig. 6 Caching hierarchy: sorting example

3.6 Insertion and Deletion Schemes

Since the data fragment with the maximum importance is stored at the root of the caching hierarchy, when a new element is inserted what happens is that the new element is inserted in the last position of the caching hierarchy. We use a complete binary tree to implement this so the next empty slot would be the current size of the heap augmented by one (i.e. $H[max + 1]$). The tree can then be re-ordered using the cache sorting algorithm.

Example 6 Insertions happen as shown in Fig. 7 where the new data fragment and its associated importance rating are put in the last empty slot of the binary tree. Once this happens, the caching hierarchy sorting algorithm is applied to move the inserted data fragment to its "correct" position. As illustrated in Fig. 7, the insertion of a fragment with an importance rating of 8 will result in the fragment getting moved into the root position if we considered that the importance rating of all the other fragments of data that currently exist in the hierarchy is less than 8.

Finally, in order to decide which fragments of data to move off the mobile device to the cloud, we begin by using Algorithm 2 to sort the data fragments in order of maximum importance. A threshold value that is defined by the mobile device user, is used to determine how many of the fragments need to be transferred. Using this value the caching hierarchy management algorithm will use a delete operation to delete data fragments from the caching hierarchy and have these fragments transferred to the mobile device.

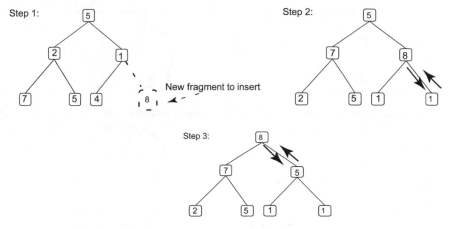

Fig. 7 Insertion of a new data fragment in the cache hierarchy

Example 7 Deletions of data fragments happen as illustrated in Fig. 8 where the caching hierarchy is reverse sorted to position the fragments with the least importance at the top of the hierarchy. Once the cache has been reverse sorted deletions happen by removing the fragment at the root of the binary tree (caching hierarchy). This creates, a sort of unbalanced tree, so in order to ensure that the binary tree structure obeys the structure and ordering properties of a complete binary tree, we need to reorder the data fragments till we find a suitable place for the fragment contained in $Cache_7$. In order to do this we compare the I_R values of $Cache_2$ and $Cache_3$ and move the lesser value up to the root position if this lesser value is less than the value in $Cache_7$. If the value in $Cache_7$ is smaller than both values, the value in $Cache_7$ is placed in the empty slot and the deletion process is terminated by reordering the hierarchy in terms of maximum first as shown in step 6. However, since this is not the case here, the result is that the empty slot that was at $Cache_1$ is moved to position $Cache_2$. In the next step we compare the I_R values of $Cache_4$ and $Cache_5$, since these are the children nodes of $Cache_2$. We find that the value in $Cache_5$ is lesser, so 5 gets moved into $Cache_2$ and the empty slot is now located at $Cache_5$. $Cache_5$ has no descendant nodes so the value in $Cache_7$ gets moved into it and the empty slot is now at $Cache_7$, thus retaining the complete binary tree structure of the caching hierarchy. This process is repeated until all the data fragments that were below the transfer threshold I_R have been moved off the mobile device. The caching hierarchy is then reordered to give a hierarchy such as the one in Step 6 of Fig. 8 which is the one we get after deleting 1 in Step 4.

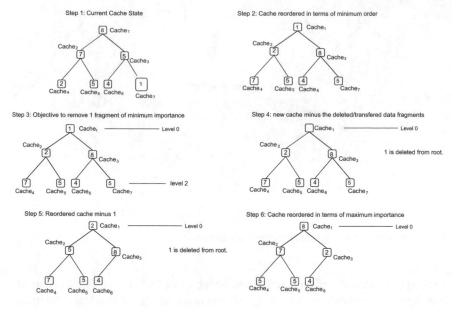

Fig. 8 Deletion and re-ordering in the caching hierarchy

4 Experimental Results

The heuristic algorithms that we presented in Sects. 3.5 and 3.3 have been implemented on a hybrid platform comprising a mobile device and a cloud server. We developed a prototype to evaluate the performance of our approach. The proposed fragmentation and caching algorithms are implemented in Python. Python comes with an embedded lightweight database engine, SQLite. The prototype is deployed on a Samsung Galaxy II I9100 smartphone (Dual-core 1.2 GHz Cortex-A9, 1 GB RAM) with a rooted Android 4.0.4 platform, connected to a WiFi network. This device consumes approximately 1.3 W per second to send data over the wireless link. The cloud server is represented by an Amazon EC2 virtual machine of the type 't1.large' with an EC2 pre-configured image (AMI) of Ubuntu Server 12.04 LTS, 64 bits.

We assess the behavior of the heuristic algorithms in terms of response (execution) time, quality of the returned response, storage capacity, and energy (battery) consumption. In order to compare the performance of our heuristic algorithms, we implemented two naive versions of the storage utilization and fragmentation algorithms. In the first case, the storage utilization algorithm randomly decides on a threshold limit to use in determining when to offload data from the mobile device to the cloud while the fragmentation algorithm uses frequency of access to randomly select the attributes to include in the fragment that is offloaded to the cloud. In the second case, we augment both naive algorithms to incorporate the confidentiality constraints that we implement by generating cryptographic keys that are used to encrypt the data.

For simplicity and consistency with Ciriani et al.'s work on combining fragmentation with encryption [22], we used a relation schema that is composed of 32 attributes from a database of medical information. We expressed 30 confidentiality constraints that are composed of 2–4 attributes. Singleton constraints were not considered since, as mentioned before, these cannot be used to determine what other attributes might be required to satisfy a query with the data available in the mobile device's storage M.

4.1 Performance Results

In these experiments, the patient's mobile device collects the vital signs and stores it locally. The device responds to queries requesting the status of specific health attributes. The objective is to enable the mobile device to efficiently manage storage utilization, where part of the patient's data is offloaded to the cloud while another part, that is frequently accessed, remains on the mobile device. The mobile device runs our fragmentation algorithm to partition the patient's data while satisfying the confidentiality constrains set by the healthcare provider or the patient. The mobile device also runs our proposed caching approach to secure data and allow efficient query handling. The device runs these algorithms once the size database reaches a pre-specified threshold based on available resources on the mobile side.

We have created a database to hold the patient information in multiple relations (tables). One table maintains the patient's biographical data and another one keeps the patient's vital signs. We also created a table to maintain the access frequency of each attribute. The vital signs are generated randomly within the reference range of each attribute. We test over a set of 32 attributes, such as Electrocardiography (ECG), Oxygen Saturation (SPO2), Temperature, Blood Sugar (Glucose), Forced Vital Capacity (FVC) and Functional Residual Capacity (FRC) that measure the lung function, etc. The fragments are generated based on confidentiality constraints, attribute affinity, and data freshness.

First, we split the data vertically based on confidentiality constraints and then calculate the marginal gain between the various data chunks based on the access frequency of their attributes. The marginal gain is used to decide which data chunks are better to stay together. Then, each data chunk is fragmented horizontally based on data freshness, which is captured by the record's date and time attribute. These fragments are generated using database views. Each view represents a fragment or a patient data chunk. Table 1 illustrates a summary of our experimental setup and parameters

Table 1 Summary of experimental setup and parameters

Parameter	Range of values
Number of attributes	32
Confidentiality constraints	30
Number of records	1000–1,000,000
Database size	100 KB–100 MB
Number of fragments	30–90

We query the database with a 1000 randomly generated queries. Each query requests a random set of attributes. The objective of these queries is to build a sufficient frequency access data set and therefore assess the attribute affinity. Figure 9 shows the execution time of our fragmentation algorithm versus a patient database of varying size. The small increase in the execution time as the database size gets bigger is due to creating the set of fragments. As the database size increases, either the number of fragments increases or the size of each fragment gets bigger. However, the complexity of our data fragmentation algorithm is relatively linear if one considers that our theoretical complexity analysis is in $O\,(n\log n)$ where n is the number of data fragments, with respect to the size of the database. We note also that the standard deviation error on fragmentation execution time is ±0.99s which is reasonable. It shows that caching improves the overall response time due to the positive impact of attribute affinity. We observe that data fragmentation improves the query response time by almost 25 %, while caching offers an additional 35 % relative improvement in both cases. Figure 10 shows the average query response time in both cases, when data is un-fragmented and fragmented based on our approach. Query response times on uncached data incur a standard deviation error of ±0.06s while the standard

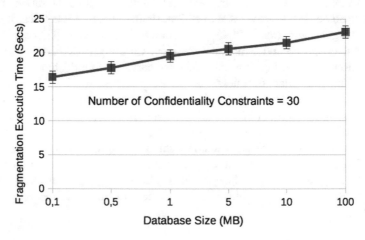

Fig. 9 Fragmentation time versus database size (confidentiality constraints)

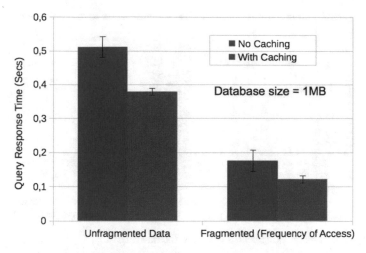

Fig. 10 Query response time: fragmented versus un-fragmented data (database size = 1 MB)

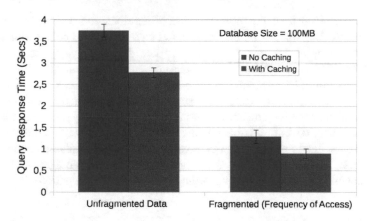

Fig. 11 Query response time: fragmented versus un-fragmented data (database size = 100 MB)

deviation error on cached data is ±0.027 s. Figure 11 shows results for a database of size 100MB and additionally that, the cost of responding to queries is scalable with increases in data sizes. The standard deviation error is also proportionate with the error on uncached data at ±0.48 s and ±0.19 s on cached data respectively. Figure 12 illustrates the query failure rate for un-fragmented data and fragmented data, where in the latter case we compare between our data fragmentation approach and a random data fragmentation. The query failure rate is defined by the number of failed queries to the total number of queries. The experimental results reveals that our approach significantly outperforms the random data fragmentation, where both the attributes and fragments that get to remain on the mobile device are chosen randomly. We note that our approach results in 13.2 % query failure rate in contrast to 58.7 % for the random fragmentation approach, resulting in more than a 4 times improvement. We observe that the un-fragmented data yields a 100 % query success rate by keeping

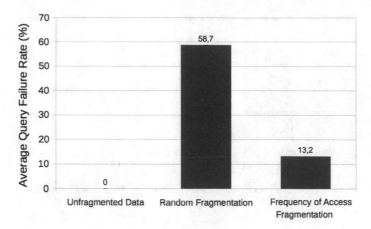

Fig. 12 Query failure rate: fragmented versus un-fragmented data

the entire patient's data on the mobile device. We attribute this difference in success rates to the fact that queries on the fragmented data experience delays in going out to the cloud copy of the data and hence are not able to provide a response in a time window that is complaint to the quality of service agreements. Other factors to which this might be attributed include low bandwidth or inaccessibility due to network failures. However, we need to strike a tradeoff balance between efficient utilization of the resource-constrained storage and better query handling. Figure 13 depicts the energy cost of transferring data to the cloud. The energy consumption is directly proportional to the amount of transferred data and available bandwidth on wireless link between the mobile device and the cloud data storage provider. The energy consumption is calculated according to the following equation:

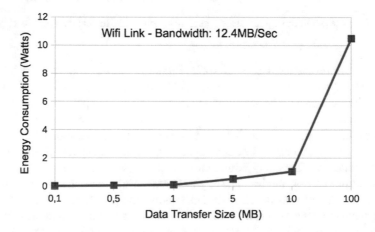

Fig. 13 Energy cost: data transfer—mobile device to cloud

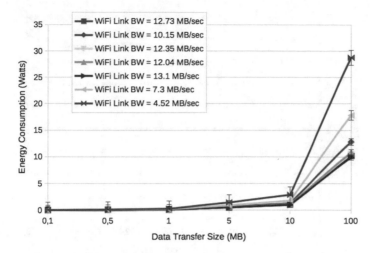

Fig. 14 Energy cost: data transfer—mobile device to cloud under varied bandwidth (BW)

$$E = \frac{D}{B} \times p_t$$

where E represents the total energy consumption in Watts, D indicates the size of transferred data, B represents the available bandwidth, and p_t is the energy consumption unit that the device consumes to transfer data over its network interface per second.

The rationale behind this equation is that a high bandwidth reduces the total energy consumption of the data transmission because more data can be transmitted over a short period. Hence this lowers the energy consumption per second. Whereas under low bandwidth conditions we want to avoid data transfers because the energy cost of transmission will be high and impact negatively on the limited battery power of the mobile device. We note additionally that the standard deviation error is ±1.69 W in terms of energy consumption with respect to varied data transfer sizes. As a further step we vary the bandwidth and note as shown in Fig. 14 that the cost (in terms of energy consumed is proportionate to the size of the data.

However, lower bandwidths result in a high energy consumption cost which is possibility due to cost of handling failed connections or re-initiating unsuccessful downloads. The error margin in this case is ±5 % W with respect to the bandwidth. Finally, we consider the average download speed with repeat to bandwidth and show our results in Fig. 15. As expected the cost of downloads grows linearly with an increase in bandwidth and as shown in Fig. 16, the cost of downloads in terms of speed is proportionate to the observations we made in Fig. 14. As well, considering that tests on user tolerance for bandwidth delays indicate that margins of 6–7 s are acceptable [28], we observe that our worst case download time at <2 s for a download rate 0.565 MB/s at a bandwidth of 4.52 MB/s is acceptable.

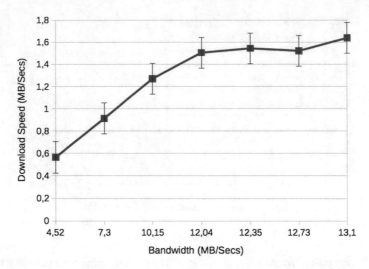

Fig. 15 Data transfer—mobile device to cloud: download speed with respect to bandwidth (BW)

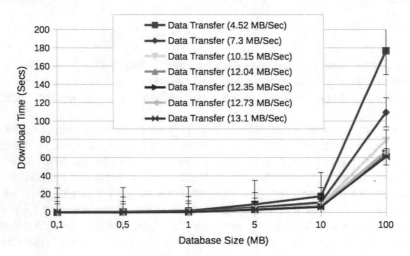

Fig. 16 Data transfer—mobile device to cloud: varied download speeds and varied bandwidth

5   Conclusion

In mobile health applications tasks such as maintaining data consistency and security are difficult to handle effectively in lossy networks where communications are negatively affected by factors such as low bandwidth and intermittent connections. Hence, to facilitate management of an mobile healthcare system, we have proposed a framework that enables suitable data management via a mobile web service architecture. In our proposed framework, a patient's vital signs are recorded periodically and stored locally on a mobile device such as a mobile device. Periodically, a subset of the mobile device data is transferred off to a third party storage provider for secure storage and management. In order to decide on how the data is transferred off the mobile device we employ two algorithms namely a fragmentation algorithm and caching algorithm. The fragmentation algorithm, segments the data using an importance rating metric while the caching algorithm uses an "aging" metric to determine which portions of the data to move off the mobile device to the cloud service provider. In order to preserve data privacy, we use confidentiality constraints supported by a cryptographic key management scheme to support data fragmentation.

Though the proposed solution was primarily designed to aid deciding which portions of the data to transfer to the cloud storage provider and which to maintain on the mobile device, it has all the features required for use in a variety of applications that involve deciding on how to transfer or outsource data management to a third party storage provider. Our experimental results indicate that data fragmentation improves query response time by almost 25 % while caching offers a 35 % performance improvement in query response time on various query workloads on fragmented data when compared to cases involving un-fragmented data that are solely cloud centric.

As a potential avenue for future work we plan to look at issues that emerge in composing data from a variety of sources with possibly conflicting security policies. In this case, a good negotiation model is needed to determine a set of minimum security requirements that satisfies the security policy constraints of all the services and/or domains participating in the composition. Another interesting topic would be to extend our approach to multimedia data from multiple sources (text, xrays, etc.) and to consider the media types in the storage management and how one might optimize existing storage with respect to the data source.

References

1. Goodman, S., Harris, A.: The coming African tsunami of information insecurity. Commun. ACM **53**(12), 24–27 (2010)
2. Greenhalgh, T., Potts, H.W., Wong, G., Bank, P., Swinglehurst, D.: Tensions and paradoxes in electronic patient record research: a systematic literature review using the meta-narrative method. Milbank Q. **87**(4), 729–788 (2009)
3. Mandl, K.D., Simons, W.W., Crawford, W., Abbett, J.M.: Indivo: a personally controlled health record for health information exchange and communication. BMC Med. Inform. Decis. Mak. **7**(1), 25 (2007)
4. Robison, J., Bai, L., Mastrogiannis, D.S., Tan, C.C., Wu, J.: A survey on phr technology. In: Proceedings of the 2012 IEEE 14th International Conference on e-Health Networking, Applications and Services, Healthcom '12, pp. 184–189. IEEE (2012)
5. De Capitani di Vimercati, S., Foresti, S., Jajodia, S., Paraboschi, S., Samarati, P.: Over-encryption: management of access control evolution on outsourced data. In: Proceedings of the 33rd International Conference on Very Large Databases, VLDB '07, pp. 123–134. VLDB Endowment (2007)

6. Samarati, P., De Capitani di Vimercati, S.: Data protection in outsourcing scenarios: issues and directions. In: Proceedings of the 5th ACM Symposium on Information, Computer and Communications Security, ASIACCS '10, New York, NY, USA, ACM, pp. 1–14 (2010)

7. De Capitani di Vimercati, S., Foresti, S., Jajodia, S., Paraboschi, S., Samarati, P.: Encryption policies for regulating access to outsourced data. ACM Trans. Database Syst. **35**(2), 12:1–12:46 (2010)

8. Kayem, A.V.D.M., Martin, P., Akl, S.G.: Effective cryptographic key management for outsourced dynamic data sharing environments. In: Proceedings of the 10th Annual Information Security Conference (ISSA 2011), 15–17 August 2011, Johannesburg, South Africa, pp. 1–8. IEEE (2011)

9. Kayem, A.V.D.M., Elgazzar, K., Martin, P.: Secure and efficient data placement in mobile healthcare services. In: Database and Expert Systems Applications—25th International Conference, DEXA 2014, Munich, Germany, 1–4 September 2014. Proceedings, Part I, pp. 352–361 (2014)

10. Cormen, T.H., Leiserson, C.E., Rivest, R.L., Stein, C.: Introduction to Algorithms, 3rd edn. MIT Press, Cambridge (2009)

11. Baker, D.B., Masys, D.R.: Pcasso: a design for secure communication of personal health information via the internet. Int. J. Med. Inform. **54**(2), 97–104 (1999)

12. Grimson, J.: Delivering the electronic healthcare record for the 21st century. Int. J. Med. Inform. **64**, 111–127 (2010)

13. Adida, B., Sanyal, A., Zabak, S., Kohane, I.S., Mandl, K.D.: Indivo x: developing a fully substitutable personally controlled health record platform. In: Proceedings of the AMIA Annual Symposium, Healthcom '12, p. 6. American Medical Informatics Association (2010)

14. Mandl, K.D., Mandel, J.C., Murphy, S.N., Bernstam, E.V., Ramoni, R.L., Kreda, D.A., McCoy, J.M., Adida, B., Kohane, I.S.: The smart platform: early experience enabling substitutable applications for electronic health records. J. Am. Med. Inf. Assoc. **19**(4), 597–603 (2012)

15. Win, K.T., Susilo, W., Mu, Y.: Personal health record systems and their security protection. J. Med. Syst. **30**(4), 309–315 (2006)

16. Damiani, E., De Captani Di Vimercati, S., Jajodia, S., Paraboschi, S., Samarati, P.: Balancing confidentiality and efficiency in untrusted relational dbmss. In: Proceedings of the 10th ACM Conference on Computer and Communications Security, CCS '03, New York, NY, USA, ACM, pp. 93–102 (2003)

17. Damiani, E., De Captani Di Vimercati, S., Finetti, M., Paraboschi, S., Samarati, P., Jajodia, S.: Implementation of a storage mechanism for untrusted dbmss. In: In Proceedings of the Second International IEEE Security in Storage Workshop, Washington DC, USA (2003)

18. Hacigümüş, H., Bala, I., Sharad, M.: Providing database as a service. In: Proceedings of the 18th International Conference on Data Engineering, San Jose, California, USA (2002)

19. Hacigümüş, H., Balakrishna, R.I., Sharad, M.: Ensuring the integrity of encrypted databases in the database-as-a-service model. In: Proceedings of the IFIP Conference on Data and Applications Security, Estes Park Colorado, USA, pp. 61–74 (2003)

20. Ciriani, V., De Capitani di Vimercati, S., Foresti, S., Jajodia, S., Paraboschi, S., Samarati, P.: Fragmentation and encryption to enforce privacy in data storage. In: Biskup, J., Lpez, J. (eds.) Computer Security ESORICS 2007. Volume 4734 of Lecture Notes in Computer Science, pp. 171–186. Springer, Berlin (2007)

21. Ciriani, V., De Capitani di Vimercati, S., Foresti, S., Jajodia, S., Paraboschi, S., Samarati, P.: Fragmentation design for efficient query execution over sensitive distributed databases. In: ICDCS 2009: IEEE International Conference on Distributed Computing Systems, pp. 32–39 (2009)

22. Ciriani, V., De Capitani di Vimercati, S., Foresti, S., Jajodia, S., Paraboschi, S., Samarati, P.: Combining fragmentation and encryption to protect privacy in data storage. ACM Trans. Inf. Syst. Secur. **13**(3), 22:1–22:33 (2010)

23. Akl, S.G., Taylor, P.D.: Cryptographic solution to a problem of access control in a hierarchy. ACM Trans. Comput. Syst. **1**(3), 239–248 (1983). August

24. Atallah, M.J., Blanton, M., Fazio, N., Frikken, K.B.: Dynamic and efficient key management for access hierarchies. ACM Trans. Inf. Syst. Secur. **12**(3), 18:1–18:43 (2009)
25. Foresti, S.: Preserving Privacy in Data Outsourcing. Volume 51 of Advances in Information Security. Springer, New York (2011)
26. Elgazzar, K., Martin, P., Hassanein, H.: Personalized mobile web service discovery. In: 2013 IEEE Ninth World Congress on Services (SERVICES), pp. 170–174. IEEE (2013)
27. Gavey, M.R., Johnson, D.S.: Computers and Intractability: A Guide to the Theory of NP-Completeness. W. H. Freeman (1979)
28. Nielsen, J.: User interface directions for the web. Commun. ACM **42**(1), 65–72 (1999)

On Vague Computers

Apostolos Syropoulos

Abstract Vagueness is something everyone is familiar with. In fact, most people think that vagueness is closely related to language and exists only there. However, vagueness is a property of the physical world. Quantum computers harness superposition and entanglement to perform their computational tasks. Both superposition and entanglement are vague processes. Thus quantum computers, which process exact data without "exploiting" vagueness, are actually vague computers.

1 Introduction

Vagueness is something we all are familiar with. A very rough definition of vagueness is this: the property of objects or entities that lack definite shape, form, or character. For many years *vagueness* was considered just a linguistic phenomenon. This simply means that vagueness is part of our everyday expression and not something real, which turned out not to be true. In the linguistic realm, a property of some object is vague when it is not clear to which group or, more generally, category the object belongs to. Thus when Garnet is 1.68 m tall, it is not obvious if she is tall or not tall. Similarly, if Jim is 1.90 m tall, he might be classified as tall, in general, but as short when his height is compared to the height of the average NBA player. Now, if John's height is 2.00 m and we are sure he is tall, then any other person whose height is slightly different (e.g., Mike, whose height is 1.98 m) is also considered tall, thus Mike is *similar* to John with respect to his height. But what exactly is slightly different? Obviously, this similarity degree is an indirect way to define vagueness. In particular, similarity between physical objects can be used to show that vagueness exists in the natural world. Although, elementary particles of the same kind (i.e., protons) are considered *indistinguishable* by most physicists, still there are some who argue that elementary particles are in fact distinguishable and also similar (e.g., see [5] for an overview). Provided this approach is valid, and I will say more on this later on, one can talk about vagueness in the physical reality.

A. Syropoulos (✉)
Xanthi, Greece
e-mail: asyropoulos@yahoo.com

© Springer International Publishing Switzerland 2017
A. Adamatzky (ed.), *Emergent Computation*, Emergence, Complexity and Computation 24, DOI 10.1007/978-3-319-46376-6_17

Sometimes people confuse vagueness with *ambiguity*. For example, the sentence "Garnet ate the cookies on the couch" is ambiguous because one can understand it in more than one way. In particular, did Garnet eat the cookies that were on the couch or did she bring cookies that she later ate on the couch? Now contrast the previous sentence with "the room was gray," which is vague because there are many shades of gray and it is not clear to which one is the color of the room. People also confuse *imprecision* with vagueness. For example, the sentence "bring me the cup" is not precise when there are many cups.

A modern computer is fed with exact data, processes them as such and delivers exact answers. Of course, this scheme is quite reasonable as we usually have specific problems and we want concrete answers to them, but in most, if not all cases, we do not care if the internal working involves *vague* data and operations as long as this does not affect the final result. Of course, it is widely assumed that the computational process does not involve any form of vagueness, yet there are error correction protocols because errors happen. But can we attribute these errors to vagueness? Furthermore, can we use vagueness constructively in the computational process? Or, in different words, is there room for vagueness in computation? As far as the second question is concerned, the answer is affirmative since there are realistic and not so realistic models of computation that employ vagueness [e.g., the fuzzy Turing machine is not a so realistic model, while fuzzy P systems and fuzzy chemical machines are realistic models of computation (see [15] for details)]. In addition, it seems that there is a connection between quantum mechanics and vague computing.

The pillars of modern physics are quantum mechanics and general relativity (special relativity explains only the special case when motion is uniform). One could say that general relativity is the physics of the macrocosm while quantum mechanics is the physics of the microcosm. In different words, one could say that quantum mechanics helps us understand the behavior of molecules atoms, and elementary particles while general relativity is the theory we use to explain phenomena near very massive objects, such as planets, stars, and galaxies (gravity weakens as we go away from massive objects). It is really weird that the two theories do not "mix". So far all efforts to quantize gravity have failed![1] Quantum mechanics started when Max Karl Ernst Ludwig Planck explained the problem of the radiation of a black body in 1900. Roughly, he proposed that energy can have only certain discrete values something that helped him to solve this problem (see [2] among others for a short description of the genesis of quantum mechanics).

Quantum computing is making use of the laws of quantum mechanics, and quantum mechanics is explained by, among others, statistical probabilities (i.e., a combination of statistics and probability theory). Quantum mechanics was formalized in 1926 while probability theory was formalized in 1930 [3]. Until that time, probability theory was considered a prediction tool, something that most people still believe. For example, today many people think they can use probabilities to make educated bets at a blackjack table and other games of chance. Naturally, they use statistical

---

[1] Speculations about higher dimensions, parallel universes, etc., will remain speculations until there is solid proof about their existence.

probabilities. Of course, probability theory is not about chance and games. In mathematics, (pure) probabilities are *ratios of the measure of subsets of a given set.* Here the word "measure" means "counting" in case one deals with finite sets. However, when one has to deal with sets that contain an infinite number of elements (e.g., the set of integer numbers), then one must employ a suitable measuring process to "count" elements. Thus when one knows how to count the elements of a set, then one can calculate probabilities. Obviously, this has nothing to do with chance or randomness.

In what follows I will explore the connection between (mathematical models of) vagueness and quantum computing. In particular, after a concise introduction to fuzzy set theory, I will introduce possibility theory. Then I will discuss vagueness at the quantum level and I will explain how possibilities can replace probabilities in quantum mechanics thus giving rise to real vague computers.

2 Fuzzy Set Theory: A Mathematical Model of Vagueness

Fuzzy set theory is a mathematical model of vagueness that was introduced by Lotfi Askar Zadeh [16]. Fuzzy sets are a natural extension of ordinary sets. Zadeh defined fuzzy sets by generalizing the membership relationship. In particular, given a universe X, he defined a fuzzy subset of X to be an object that is characterized by a function $A : X \rightarrow [0, 1]$. The value $A(x)$ specifies the degree to which an element x belongs to A. Thus if A denotes tallness and g is Garnet, then $A(g)$ is the degree to which Garnet is tall. A fuzzy set A for which there is an $x \in X$ such that $A(x) = 1$ is called normalized.

Most newcomers tend to take fuzzy set theory for an alternative formulation of probability theory, nevertheless, this is not the case. For instance, there are probability theorists that still believe that fuzziness is unnecessary since they argue that probability theory can be used to solve all problems that can be tackled by fuzzy set theory. Zadeh [17] has argued that the two theories are complementary, that is, they are different facets of vagueness. Kosko [10] and other researchers, including this author [15], have argued that fuzzy set theory is more fundamental than probability theory. However, I do not plan to say anything more on this matter (a very detailed discussion is included in [15]). Instead, let me now present the basic operations between fuzzy subsets.

Assume that $A, B : X \rightarrow [0, 1]$ are two fuzzy subsets of X. Then, their union and their intersection are defined as follows:

$$(A \cup B)(x) = \max\{A(x), B(x)\} \tag{1}$$

and

$$(A \cap B)(x) = \min\{A(x), B(x)\}. \tag{2}$$

Also, if $\bar{A}$ is the complement of the fuzzy subset A, then $\bar{A}(x) = 1 - A(x)$. More generally, it is quite possible to use functions other than min and max to define the intersection and the union of fuzzy subsets. These functions are known in the literature as *t-norms* and *t-conorms*, respectively. For more information on t-norms and t-conorms see [9] or any other textbook on fuzzy set theory.

In the years that followed the publication of Zadeh's paper, various researchers proposed and defined various fuzzy structures (e.g., fuzzy algebraic structures, fuzzy topologies, etc.). For instance, the concept of fuzzy languages was introduced by E.T. Lee and Zadeh [11]:

Definition 17.1 A fuzzy language λ over an alphabet S (i.e., an ordinary set of symbols) is a fuzzy subset of S^*.

If $s \in S^*$, then $\lambda(s)$ is the grade of membership that s is a member of the language.

Example 17.1 Consider the following set that includes all the sequences of zeros followed by ones:

$$L = \left\{ 0^i 1^j \mid i \neq j \text{ and } i, j > 0 \right\}.$$

Then, the following function

$$\lambda(0^i 1^j) = \begin{cases} j/i, & \text{if } i > j \\ i/j, & \text{otherwise} \end{cases}$$

defines a fuzzy language.

Ordinary set theory is built out of two predicates: membership and equality. This means that in a fuzzy theory of sets both the membership and the equality should be fuzzy. Unfortunately, and for unknown reasons, Zadeh *fuzzified* only the membership predicate whereas he left crisp the equality predicate, thus, making the resulting theory somehow incoherent. It is not difficult to fuzzify the equality predicate and Barr [1] has provided a solution to this problem. In addition, he showed how to construct categories of "fuzzy" sets that form a topos. Interestingly, a topos is a non-fuzzy mathematical universe, thus, he showed how to actually embed "fuzzy" sets in such a universe. Although a topos is an intuitionistic universe, that is, a universe that is strongly connected to recursion theory, still it is one that has no respect for vagueness! This implies that it is necessary to define fuzzy universes, whatever this may mean.

3 From Probabilities to Possibilities

Most textbooks on quantum mechanics introduce the reader to the *statistical interpretation* of the theory in the first pages of the book (e.g., Griffiths's [7] excellent textbook follows this convention). Of course, the reason is that the statistical interpretation plays a central rôle in quantum mechanics. Now, this interpretation is based on the pre-Kolmogorov probability theory and uses it for the estimation of the likelihood of various events. For example, if $\Psi(x, t)$ is the *wave function* of a particle that moves on a straight line and a and b are two points of this line, then

$$\int_a^b |\Psi(x, t)|^2 dx = \left\{ \begin{array}{l} \text{the probability of finding the par-} \\ \text{ticle between } a \text{ and } b, \text{ at time } t. \end{array} \right\} \tag{3}$$

Obviously, here we are talking about events that we cannot control and so they can be classified as random. This does not surprise anyone since nonspecialists perceive probabilities as a mathematical "measure" of how likely it is to see some event to happen. Statements like the following ones express exactly this view:

- it is quite probable that Bayern Munich will win the Champions League this season, or
- there is a 20% probability that it will rain tomorrow, or
- the probability of throwing two dice and obtaining two sixes is 1/36.

A rigorous and mathematically sound definition of probabilities have been given by Andrey Nikolaevich Kolmogorov in his *Analytical Methods of Probability Theory*, which was published in 1931. Kolmogorv's formulation appeared almost 6 years after the formalization of quantum mechanics (see [3]).

Kolmogorov employed measure theory in order to rigorously define probability theory. In particular, a probability measure is a function taking sets as arguments and assigns the number 0 to the empty set and a nonnegative number to any other set. Also, it has to be countably additive. Thus given a nonempty set X and a nonempty class $\mathbf{C}$ of subsets of X, and a function $\mu : \mathbf{C} \to [0, 1]$ such that

- $\mu(\emptyset) = 0$;
- $\mu\left(\bigcup_{i=1}^{\infty} E_i\right) = \sum_{i=1}^{\infty} \mu(E_i)$ for any disjoint sequence $\{E_n\}$ of sets in $\mathbf{C}$ whose union is also in $\mathbf{C}$;
- $\mu(X) = 1$;

then μ is a probabilistic measure on $\mathbf{C}$.

As was outlined above, quantum mechanics is using probability theory to explain and predict physical phenomena. But one could use *possibility theory* to give the same explanations and predictions in a more natural way. In particular, Kosko [10], a prominent fuzzy set theorist, argued in favor of the superiority of fuzzy set theory when compared to probability theory by saying that fuzziness "measures the degree to which an event occurs, not whether it occurs. Randomness describes the uncertainty of event occurrence." Thus if the particle lies between a and b, we need to know how likely it is for the particle to be at $a \leq c \leq b$ and not whether it is between a and b. Possibility theory is based on possibility measures, which are are based on fuzzy sets [18].

A possibility measure π is different from a probability measure in that

$$\pi\left(\bigcup_{i=1}^{\infty} E_i\right) = \sup_{i=1}^{\infty} \pi(E_i). \tag{4}$$

In simple words, the difference between the two approaches is that in probability theory one demands that the sum of probabilities for given *event* should be 1 whereas in possibility theory there should be at least one plausible event (i.e., one whose possibility is 1). And this is clearly closer to what actually happens. A particle that lies between a and b is definitely somewhere between them.

Starting from some measure one can define a corresponding integral. For example, when using a probabilistic measure one may define the Lebesgue integral. Similarly, using a possibility measure one can define the *Sugeno* integral. Assume that $(X, \mathbf{F})$ is measurable space, where X is some set and $\mathbf{F}$ is a σ-algbera,[2] $\mu : \mathbf{F} \to [0, +\infty]$ is continuous *monotone* measure,[3] and $\mathbf{G}$ is the class of all finite nonnegative measurable functions.[4] For any $f \in \mathbf{G}$, $F_\alpha = \{x \mid f(x) \geq \alpha\}$ and $F_{\alpha^+} = \{x \mid f(x) > \alpha\}$, where $\alpha \in [0, +\infty]$. Suppose that $A \in \mathbf{F}$ and $f \in \mathbf{G}$. Then the Sugeno integral of f on A with respect to μ is defined by

$$\fint_A f d\mu = \sup_{\alpha \in [0, +\infty]} \Big(\alpha \wedge \mu(A \cap F_\alpha) \Big), \tag{5}$$

When $A = X$, the Sugeno integral is also denoted by $\fint f d\mu$. This form of integration could be used instead of the Lebesgue integral in Eq. (3) to compute the possibility of finding the particle between a and b, at time t.

4   Vagueness in the Physical Reality

If vagueness is not just part of our everyday expression, then there should be vague objects. But are there such objects?[5] I will not give a "yes" or "no" answer but instead I would like to ponder about the length of the UK coastline. The British Cartographic Society does not give an exact answer on their web page. Instead, they give this answer: *The true answer is: it depends! It depends on the scale at which you measure it.* Mandelbrot [14] gave exactly this answer in 1967. So in a sense it is not exactly known what is inside the UK and what is outside. And of course it is quite possible that some objects may lie somewhere in the middle. Thus one could say that the UK is actually a vague object since its boundaries are rigid. Similarly, clouds are vague objects for exactly the same reasons. On the other hand, there are objects that appear to be genuine vague objects (e.g., think of heaps of grain or men with few hair), still most of them are classified as such because the terms that describe

---

[2]$\mathbf{F}$ has to be a subclass of the power set 2^X. Also, it must satisfy the following conditions: (a) $X \in \mathbf{F}$; (b) for all $E, F \in \mathbf{F}$, $E - F \in \mathbf{F}$; and (c) for all $E_i \in \mathbf{F}$, $i = 1, 2, \ldots, \bigcup_{i=1}^{+\infty} E_i \in \mathbf{F}$.

[3]μ is monotone if and only if $E, F \in \mathbf{F}$ and $E \subset F$ imply $\mu(E) \leq \mu(F)$.

[4]A function $f : X \to (-\infty, +\infty)$ on X is measurable if and only if $f^{-1}(B) = \{x \mid f(x) \in B\} \in \mathbf{F}$ for any Borel set $B \in \mathcal{B}$. Now, assume that X is the real line. Then, the class of all bounded, left closed, and right open intervals, denoted by $\mathcal{B}$, is the class of Borel sets.

[5]A mathematical response to this question has been recently given by Gerla [6].

them are vague. However, there is a third approach to the problem of finding vague objects in Nature. In quantum mechanics, the "standard" view is that elementary particles are indistinguishable, nevertheless, not everybody shares this view. More specifically, Lowe [12], has argued against this view thus showing that vagueness exists in the subatomic level:

> Suppose (to keep matters simple) that in an ionization chamber a free electron a is captured by a certain atom to form a negative ion which, a short time later, reverts to a neutral state by releasing an electron b. As I understand it, according to currently accepted quantum-mechanical principles there may simply be no objective fact of the matter as to whether or not a is identical with b. It should be emphasized that what is being proposed here is not merely that we may well have no way of telling whether or not a and b are identical,which would imply only an epistemic indeterminacy. It is well known that the sort of indeterminacy presupposed by orthodox interpretations of quantum theory is more than merely epistemic— it is ontic. The key feature of the example is that in such an interaction electron a and other electrons in the outer shell of the relevant atom enter an 'entangled' or 'superposed' state in which the number of electrons present is determinate but the identity of any one of them with a is not, thus rendering likewise indeterminate the identity of a with the released electron b.

The idea behind this example is that "identity statements represented by '$a = b$' are 'ontically' indeterminate in the quantum mechanical context" [4]. In different words, in the quantum mechanical context a is equal to b to some degree, which is one of the fundamental ideas behind fuzzy set theory. For a thorough discussion of the problem of identity in physics see [5].

5   Superposition and Entanglement Revisited

The well-known *Schrödinger's cat paradox* (see [7]) is about a cat that is placed inside a box along with a Geiger counter. The box contains a tiny amount of a radioactive substance whose atoms may or may not decay within an hour. If there is a decay, it triggers the Geiger counter which, in turn, triggers a hammer that breaks a glass that contains a poison capable to kill the cat. The obvious question is: What would happen to the cat after exactly 1 h? At the end of the hour the wave function of the cat would be

$$\psi = \frac{1}{\sqrt{2}}\psi_{\text{alive}} + \frac{1}{\sqrt{2}}\psi_{\text{dead}}. \tag{6}$$

This implies that the cat is neither dead nor alive! Schrödinger regarded this as patent nonsense, however, I tend to disagree. The reason of course is that there are many things that are not either black or white. After all, this is exactly the essence of vagueness. Thus, a patient who is in coma is not exactly alive and not exactly dead. Regardless of our objections, *superposition*, that is, the ability of particles to be in more than one state at the same time, is what makes quantum computing really interesting.

In "classical" computing a bit is either the digit 0 or the digit 1. In quantum computing a *qubit* is a quantum system (typically a polarized photon, a nuclear spin, etc.) in which the two digits are represented by two quantum states: $|0\rangle$ and $|1\rangle$. These states are represented by the following matrices:

$$|0\rangle = \begin{pmatrix} 1 \\ 0 \end{pmatrix} \text{ and } |1\rangle = \begin{pmatrix} 0 \\ 1 \end{pmatrix}. \tag{7}$$

Also, these two states are "basic" states (i.e., they form a basis of a Hilbert space) and any other state of the qubit can be written as a superposition $\alpha |0\rangle + \beta |1\rangle$, where α and β are complex numbers that are called normalization factors and they must obey the normalization condition $|\alpha|^2 + |\beta|^2 = 1$. For example, consider a photon that can be polarized in the x direction or in the y direction and assume that these states are represented by the vectors $|\uparrow\rangle$ and $|\rightarrow\rangle$, respectively, then one can use $|\uparrow\rangle$ for $|0\rangle$ and $|\rightarrow\rangle$ for $|1\rangle$.

The standard interpretation of $\alpha |0\rangle + \beta |1\rangle$ is that a particle is in states $|0\rangle$ or $|1\rangle$ with probability that depends on α and β. Of course, according to a layman's interpretation of probability theory, these two numbers express the change that a particle is in one of these states. A fuzzy theoretic interpretation of this state is that the particle is in both states but with some degree. In fact, one can define a fuzzy set as follows:

$$\Psi(|0\rangle) = |\alpha|^2 \tag{8}$$
$$\Psi(|1\rangle) = |\beta|^2 \tag{9}$$

However, here there is no reason to demand that $|\alpha|^2 + |\beta|^2 = 1$. In fact, there is no reason to impose any restriction other than $|\alpha|^2 \leq 1$ and $|\beta|^2 \leq 1$. One may argue that these two restrictions are not that different, however, the fuzzy theoretic approach assumes that the particle is in fact in a state that is partly $|0\rangle$ and partly $|1\rangle$. In different words, $\alpha |0\rangle + \beta |1\rangle$ is like a shade of gray, where, for instance, $|0\rangle$ is like black and $|1\rangle$ is like white.

Assume Ψ describes the state of quantum particle in superposition. Then, the superposition collapses upon a measurement, but the question is why this happens. Perhaps, the measurement forces a defuzzification of Ψ, that is, a process by which one gets bivalent data from multivalued data (in this case a vague state is transformed into a crisp one). But if defuzzification is possible, then one might expect that fuzzification is also possible. Indeed, the Hadamard gate is a mechanism that creates "vague" states as follows:

$$H |0\rangle = \frac{1}{\sqrt{2}} |0\rangle + \frac{1}{\sqrt{2}} |1\rangle \tag{10}$$
$$H |1\rangle = \frac{1}{\sqrt{2}} |0\rangle - \frac{1}{\sqrt{2}} |1\rangle \tag{11}$$

Thus superposition corresponds to the fuzzification of a quantum system by means of the H operator, while measurement is a "natural" defuzzification process.

Entanglement is another important quantum mechanical phenomenon. Consider a physical system with two degrees of freedom, A and B. The states of such a system belong to $\mathcal{E} = \mathcal{E}_A \otimes \mathcal{E}_B$. Some states can be expressed as

$$|\Psi\rangle = |\alpha\rangle \otimes |\beta\rangle. \tag{12}$$

However, there are states that cannot be *factorized* (i.e., they cannot be written as "products"). Such states are called *entangled states*. For example, the following is such a state:

$$|\Psi\rangle = \frac{1}{\sqrt{2}}\left(|\alpha_1\rangle \otimes |\beta_1\rangle + |\alpha_2\rangle \otimes |\beta_2\rangle\right). \tag{13}$$

First of all, there are two "special" forms of entanglement, namely entanglement of cost, E_C, and entanglement of distillation, E_D, that vague in a particular case (see [8] for details). More generally, Lowe [13] proposed a thought experiment that showed that entanglement is vague. Assume that there are two determinately distinct electrons. One of them (call it a) is determinately absorbed by an atom and then becomes entangled with a single electron (call it a^*) determinately already in the atom. Because these electrons exist in an entangled state inside the atom they are not determinately distinct but of course we know that there are two of them. At some moment one electron is emitted and so one electron is still inside the atom and one is outside the atom. Since these two electrons were in an entangled state, it is impossible to tell which electron left the atom. In a nutshell, this is the root of vagueness in entanglement.

Quantum computing is so attractive because it is harnessing both superposition and entanglement to achieve its exponential computational power. Since both superposition and entanglement are vague in their nature, this means that quantum computers operate on vague data using vague operations.

6   Conclusions

I have briefly explained why vagueness is not only a linguistic phenomenon but also a property of the physical world. Also, it is a fact that quantum computers harness quantum mechanical properties of matter to perform their computations. These properties of matter have been shown to be vague, thus quantum computers internally employ vagueness, which makes them automatically vague computers. Of course, these vague computers process non-vague data in a non-vague way, however, it would be really interesting to see if processing vague data vaguely would broaden our understanding of computation. This is certainly an open problem and I think a very interesting one.

Acknowledgements I thank Andromahi Spanou and Christos KK Loverdos for reading the man-uscript and helping me to imporve it.

References

1. Barr, M.: Fuzzy set theory and topos theory. Can. Math. Bull. **29**, 501–508 (1986)
2. Basdevant, J.L., Dalibard, J.: Quantum Mechanics. Springer, Berlin (2005)
3. Cook, D.B.: Probability and Schrödinger's Mechanics. World Scientific, Singapore (2002)
4. French, S., Krause, D.: Quantum Vagueness. Erkenntnis **59**, 97–124 (2003)
5. French, S., Krause, D.: Indentity in Physics: A Historical Philosophical and Formal Analysis. Oxford University Press, Oxford, UK (2008)
6. Gerla, G.: The existence of vague objects. Fuzzy Sets Syst. **276**(C), 59–73 (2015)
7. Griffiths, D.J.: Introduction to Quantum Mechanics. Pearson Prentice Hall (2004)
8. Hwang, W.Y., Matsumoto, K.: Irrversibility of entanglement manipulations: vagueness of the entanglement of cost and entanglement of distillation. Phys. Lett. A **310**(2–3), 119–122 (2003)
9. Klir, G.J., Yuan, B.: Fuzzy Sets and Fuzzy Logic : Theory and Applications. Prentice Hall (Sd) (1995)
10. Kosko, B.: Fuzziness vs Probability. Int. J. Gen. Syst. **17**(2), 211–240 (1990)
11. Lee, E., Zadeh, L.A.: Note on fuzzy languages. Inf. Sci. **1**, 421–434 (1969)
12. Lowe, E.J.: Vague identity and quantum indeterminacy. Analysis **54**(2), 110–114 (1994)
13. Lowe, E.J.: Vague identity and quantum indeterminacy: further reflections. Analysis **59**(4), 328–330 (1999)
14. Mandelbrot, B.: How long is the coast of Britain? Statistical self-similarity and fractional dimension. Science **156**(3775), 636–638 (1967)
15. Syropoulos, A.: Theory of Fuzzy Computation. No. 31 in IFSR International Series on Systems Science and Engineering. Springer, New York (2014)
16. Zadeh, L.A.: Fuzzy sets. Inf. Control **8**, 338–353 (1965)
17. Zadeh, L.A.: Discussion: probability theory and fuzzy logic are complementary rather than competitive. Technometrics **37**(3), 271–276 (1995)
18. Zadeh, L.A.: Fuzzy sets as a basis for a theory of possibility. Fuzzy Sets Syst. **100**(SUPPL. 1), 9–34 (1999)

Parallel Evolutionary Optimization
of Natural Convection Problem

Matjaž Depolli, Gregor Kosec and Roman Trobec

Abstract Computer simulations of complex natural phenomena become an approach of choice if experimental work is impractical or dangerous. Often, optimization approaches are used in a closed cycle with the simulation to obtain the desired performances. To test and validate such cases an optimization of a coupled thermo-fluid transport in a two dimensional cavity is elaborated. We seek for optimal positions and dimensions of obstacles in the cavity to minimize the heat flux through the domain. One can apply such an approach to maximize the insulation by using minimal amount of insulation material. The governing equations are solved with a meshless numerical method while the optimization is performed with differential evolution. The solution and optimization procedures are designed for execution on parallel computers. Incentive scalability and speed-up are demonstrated on the presented test case.

1 Introduction

High performance computer simulations are routinely used in the development of new technologies or in situations where experimental work is impossible, expensive or dangerous. For example, climate changes cannot be predicted by real experiments because setting up an experiment is practically impossible; a temperature distribution inside a beating human hearth during a surgical procedure cannot be obtained with measurements because of potential hazards for patients. Many practical problems are even more complex, because the required performances cannot be obtained with just a single development cycle, e.g., optimal shapes in hydraulic machinery or in specialized vehicles, optimal mixtures of ingredients in drug production, etc. Many trials could last too long for a practical usage or could become too expensive for a

M. Depolli (✉) · G. Kosec · R. Trobec
Jožef Stefan Institute, Jamova cesta 39, 1000 Ljubljana, Slovenia
e-mail: matjaz.depolli@ijs.si

G. Kosec
e-mail: gkosec@ijs.si

R. Trobec
e-mail: roman.trobec@ijs.si

© Springer International Publishing Switzerland 2017
A. Adamatzky (ed.), *Emergent Computation*, Emergence, Complexity
and Computation 24, DOI 10.1007/978-3-319-46376-6_18

reasonable investment. To cope with such situations, optimization approaches have to be used in a closed cycle with the simulation. Often, the real cases require high fidelity solutions which results in excessive amount of data and long computation times. Hence, the simulation and optimization methods have to be implemented using parallel high performance computers in order to solve the problems with required accuracy and in acceptable time frames.

To test and validate the described situations, an optimization of a coupled thermo-fluid transport in a closed two dimensional cavity is elaborated in more details. The two opposite edges of the cavity are set on different predefined temperatures. The cavity area is partially covered with non-permeable rectangles that obstruct the natural convection flow. We seek optimal positions and dimensions of the obstacles to minimize the heat flux through the domain, i.e., we maximize insulation. There are numerous practical examples where one would be interested in a similar optimization, e.g., designing windows or other insulating elements for buildings, optimizing heat storage systems, optimizing temperature distribution within rooms or warehouses, etc.

The energy and momentum transport is modelled by a set of Partial Differential Equations (PDEs), coupled with Boussinesq approximation. A momentum transport is modelled with the Navier-Stokes equation that is coupled with a mass continuity equation form the modelled fluid flow part, which is further coupled with heat transport, modelled with a diffusion-convection equation. There are many natural and technological problems that can be tackled with such diffusive-convective models, e.g., weather dynamics, aerodynamics, solidification, semiconductor simulations [9], and others.

The practical aspect of our test case is a minimization of energy loss in a non-uniformly heated air-filled square cavity [17] by obstructing the natural convection flow. Because the cavity is differentially heated on two opposite edges (left and right) and thermally isolated on the two remaining edges, the differences in air density due to the temperature gradients drive the fluid flow into pronounced natural convection flow patterns. The energy transport over the domain, i.e., the energy loss, is therefore not governed solely by a diffusion but also by a convection. In more realistic problems, the shape of the closed cavity and obstacles could also change and introduce supplementary optimization parameters. However, to keep the numerical solution methodology simple, and to present the methodology in an easy-to-grasp model, we omit any additional complications.

One of the important aspects of the solution procedure is its execution time. The simulation time depends on the calculation complexity of the simulation, and by selected output quality, determined through spatial and temporal resolutions. For reasonable results, thousands of discretization nodes and thousands of time steps are necessary, resulting in a simulation time in the range of minutes. In addition, stochastic optimization, as implemented by an evolutionary algorithm, typically requires vast number of iterations to converge, counted in thousands or millions. Soon, the computational cost becomes too high for practical use. Consequently, the efficiency of the computer implementation depends in its ability to yield reasonable results in a reasonable time frame.

The methodology applied in our test case is based on an efficient coupling of programs for computer simulation and optimization. The simulation and optimization procedures are designed for execution on parallel computers. The parallel simulator that runs on a shared-memory computer cooperates with an evolutionary optimizer executed on a cluster of interconnected multi-core processors. The obtained results are presented by temperature and velocity fields and convergence analyses. Incentive scalability and speed-ups are demonstrated on the selected test case, which confirms the relevance of the proposed methodology.

2 Test Case Definition

2.1 Numerical Model and Geometry

The natural convection is modelled by three coupled PDEs: diffusion equation for energy transport, Navier-Stokes equation for momentum transport, and mass continuity equation. The Boussinesq approximation is used for coupling the heat and momentum transport. The basic model is a well-know fluid flow benchmark test, in the literature usually referred to as de Vahl Davis test [17]. The model is defined by the following system of equations:

$$\nabla \cdot \mathbf{v} = 0,$$

$$\rho \frac{\partial \mathbf{v}}{\partial t} + \rho \nabla \cdot (\mathbf{v}\mathbf{v}) = -\nabla P + \nabla \cdot (\mu \nabla \mathbf{v}) + \mathbf{b},$$

$$\rho \frac{\partial (c_p T)}{\partial t} + \rho \nabla \cdot (c_p T \mathbf{v}) = \nabla \cdot (\lambda \nabla T),$$

$$\mathbf{b} = \rho \left[1 - \beta_T (T - T_{\text{ref}})\right] \mathbf{g},$$

where λ stands for thermal conductivity, $\mathbf{v}$ for velocity, t for time, c_p for specific heat, ρ for density, P for pressure, μ for viscosity, $\mathbf{b}$ for body force, T for temperature, β_T for thermal expansion coefficient, T_{ref} for reference temperature and $\mathbf{g}$ for gravitational acceleration. The problem is fully characterized by two dimensionless numbers: the Prandtl number ($\text{Pr} = \mu c_p / \lambda$) and the Rayleigh number ($\text{Ra} = |\mathbf{g}| \beta_T (T_H - T_C) \Omega^3 \rho^2 c_p / \lambda \mu$), where T_H and T_C are temperatures on the left end right edge, respectively.

A 2-D quadratic cavity $\Omega = \Omega_H \times \Omega_W$ filled with air at $\text{Pr} = 0.71$ and $\text{Ra} = 10^6$ is considered (Fig. 1). The cavity is covered by rectangular obstacles that both replace the air within the cavity and alter the air-flow streams. The number and shape of obstacles could be arbitrary, but we tested four cases with 1, 4, 9, and 16 rectangular obstacles. The obstacles are constrained to only have edges parallel

Fig. 1 Schematic representation of the test case domain

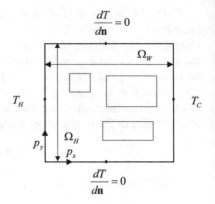

$$\frac{dT}{d\mathbf{n}} = 0$$

Ω_W

T_H T_C

p_y Ω_H

p_x

$$\frac{dT}{d\mathbf{n}} = 0$$

to the cavity edges. The obstacle material is non-permeable and significantly better thermal insulator than the fluid. The thermal conductivity of the obstacles is set to 25 % of the fluid. Each simulation starts with initial velocity and pressure set to 0 in the whole domain. The non-permeable and no-slip velocity boundary conditions are used, i.e., the velocity is zero on all boundaries. The left and right edges of the cavity have fixed temperatures T_H and T_C, while the remaining two edges are thermally isolated.

Since our goal is to construct an effective and scalable parallel implementation, a local numerical approach is preferred in order to minimize the communication between processors. We use the Meshless Local Strong Form Method (MLSM) [8, 10, 18], a variant of meshless methods [15, 16] for spatial discretization and solution of governing PDEs. The explicit time stepping is used to obtain the evolution of the solution in time. The artificial compressibility method is applied for treating the pressure velocity coupling [12]. The basic idea behind MLSM is to approximate the unknown field in each discretization node over a small subset of nearest nodes, named as support domains. The partial differential operators can be now expressed analytically from the locally approximated fields. Basically, the MLSM can be understood as a generalized Finite Difference Method (FDM), where instead of Taylor expression for spatial derivatives of unknown fields, a more general concept is used to approximate the derivatives. Like in other strong form methods, the continuous spatial variables and its derivatives from the PDE are approximated first in discretization nodes and then the treated PDEs are solved by collocation.

2.2 Optimization

The simulator is coupled with an optimizer that can handle optimization of functions, which cannot be defined analytically. Stochastic optimizers that are inspired by biological processes are often used for such cases, since they require very little knowledge of the function to be optimized. Swarm intelligence algorithms [2] and evolutionary algorithms [7] are the two largest classes of biologically inspired algorithms for stochastic optimization.

For the presented case, the simulator is coupled with an optimizer that implements the Asynchronous Master Slave Differential Evolution for Multi-objective Optimization (AMS-DEMO) algorithm [6], which is a parallel evolutionary algorithm for multi-objective optimization [1, 5, 19] of real-valued functions. As its name suggests, AMS-DEMO can perform multi-objective optimization but it is used in our case for single-objective optimization. AMS-DEMO performs stochastic optimization [3] and can solve problems for which the cost function can be evaluated for any given set of parameters even though the analytical form of the cost function is unknown. In the presented problem, the numerical simulation serves as the cost function evaluator.

With given input parameters, i.e., positions of the obstacles, the simulator computes steady-state temperature, pressure and velocity fields. The value of the cost function—the total heat flux through the domain with specified obstacles—can be easily determined from the computed fields. This value is then passed to the optimizer, which computes a new input parameter set for the simulator. The procedure iterates until the optimization convergence criterion is not met or the number of performed iterations grows too large.

For single objective problems, AMS-DEMO acts as a parallel version of Differential Evolution (DE) [13], which is an iterative algorithm operating on a set of solutions called *population*. In each iteration, every solution from the population acts as a parent $\mathbf{p}$ to a newly created trial solution $\mathbf{c}$ (also called candidate). To obtain trial solutions, parents are modified using evolutionary operators: differential mutation, uniform crossover, and selection. Differential mutation takes three or more members of the population $\mathbf{x}_1, \mathbf{x}_2, \mathbf{x}_3, [\mathbf{x}_4 \ldots \mathbf{x}_{n_p}] \in \mathbb{P}$, where n_p is the population size, to help construct a mutation vector $\mathbf{v}$ by vector addition and scalar multiplication. A common way of calculating the mutation vector, also used here, is using the formula $\mathbf{v} = \mathbf{x}_1 + F \cdot (\mathbf{x}_2 - \mathbf{x}_3)$, where $F \in \mathbb{R}$ is a constant scaling factor, most often from the interval $(0, 1]$. The mutation is followed by uniform crossover, which either takes the elements of the parent vector or the mutation vector, with a fixed probability, creating a trial solution:

$$\forall i \in \{1, 2, \ldots, n\} : \mathbf{c}_i = \begin{cases} \mathbf{p}_i \text{ with probability } 1 - P_c, \\ \mathbf{v}_i \text{ with probability } P_c. \end{cases}$$

Trial solutions are then tested against parents in a pair-wise manner, also called the selection. The losers, i.e., the solutions with higher cost, are discarded while the winners form the population of the next iteration.

The described operators are local in nature, since they operate on the parent and three other members of the population only, with the possible effect of locally optimizing the parent. Global optimization property emerges from the continuous application of evolutionary operators on individual solutions. By having the evolutionary operators applied to each population member in turn, they become structured; yet global optimization would be achieved even if they were applied in an unstructured manner. The structure is used only to keep track of the optimization progress and to make the algorithm less stochastic and, therefore, its results easily repeatable.

DE finishes either after a fixed number of performed cost function evaluations, after a solution of predefined quality or better is found, after the solutions have converged with a satisfyingly high confidence, or after a similar termination condition is met. In AMS-DEMO, the execution persists a bit after the selected termination condition is met, to allow all the evaluations that are in progress at that time to complete.

2.3 Optimizer Parameters

Although AMS-DEMO converges towards a good solutions using just about any but the most pathological parameter sets, this is not good enough in real life. Computational resources are valuable resources, and computational time determines which problems will get solved efficiently and which not. Therefore, selecting a good set of parameters to help achieve fast convergence towards optimal solutions is an important task. The best set of optimizer parameters is not known in advance and therefore cannot be determined prior to the actual optimization runs. Also, it is not clear how sensitive is the optimization on the small changes in parameters. To shed some light onto the problem of parameter selection, a scan over the parameter space is performed. A task of filling the cavity with four obstacles, described in Sect. 4, is selected for analysing the optimizer parameters.

The three most important parameters of the optimizer are taken into consideration for the scan: crossover probability P_c, scaling factor F, and population size n_p. For both P_c and F, the input set of values is set to $[0.1, 0.2, ..., 0.9]$, while for n_p, the input set of values is $[10, 20, ..., 50]$. Each of the parameters is varied across its input set values with five repetitions of the optimization run for each value. The best solutions after 8000 simulations are used to asses the impact of varying optimizer parameters.

Mean normalized heat fluxes with error bars that denote their standard deviations are shown in Fig. 2 for the P_c, F, and n_p. The optimizer works well enough with almost any combination of parameter values, except for the most extreme ones. The greatest influence on the solution quality comes from the crossover probability P_c. Low values lead to better results (lowest heat flux), with minimum being around 0.3. Only the highest two values, namely 0.8 and 0.9, seem to be bad choices. The scaling factor F is much less dynamic, which implies that the optimizer is quite robust to its variation. The results are clear enough to show that just about any value of F between 0.2 and 0.9 works well. Finally, the results of varying population size n_p are shown in the right part of Fig. 2. Population size is known to have a pronounced effect on the evolutionary algorithms, and AMS-DEMO is no exception. Low values will lead to fast convergence, but not to the best solutions, while higher values will lead to increasingly slow convergence towards increasingly better overall result. Since slow

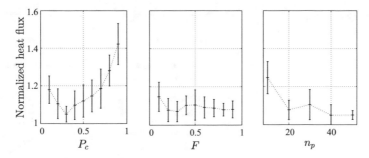

Fig. 2 Initial scan of optimizer parameters

convergence translates to long execution times, the population size is used to tweak the trade-off between the execution time and the quality of solutions found.

3 Parallel Implementation

The simulation-based optimization is parallelized on two levels. On the first level, each simulation exploits multi-core architecture of computing cluster nodes with shared-memory parallelism, using OpenMP [4]. On the second level, the optimization makes use of all interconnected computing cluster nodes by distributing separate simulations among the cluster nodes, using Message Passing Library (MPI) [14].

3.1 OpenMP

Since the simulator uses explicit temporal stepping and local spatial discretization, the solution in the next time step depends only on the previous known solutions in the nodes from local support domains. The spatial operations for discretization nodes can be executed independently, consequently, operations for all nodes in the spatial loops can be executed in parallel. The conceptual flow-chart of the simulator is shown in Fig. 3, where all parallel loops are marked. Note that most of the execution time is used for calculations in spatial loops.

In the C++ programming language, each spatial loop of the MLSM code is marked by a pre-processor directive #pragma omp parallel for, which divides the spatial loop into equally sized sub problems. An OpenMP enabled program divides the assigned task into disjunct parts and forks into several threads on request, with each thread then processing one or more task parts. After the execution of the parallelized code, the threads join back into the parent process and the program executes sequentially from there on. Fine control is possible over the task division by specifying the number of threads, the synchronization of threads, the scope of variables,

Fig. 3 Block diagram of shared-memory parallel simulation

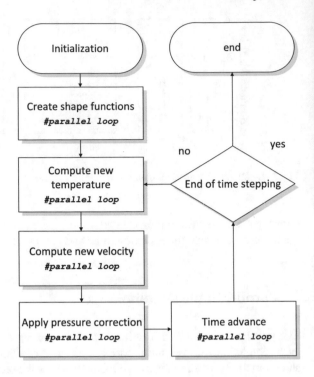

and so on. For effective parallel execution, the threads have to be bound to the cores by using the runtime environment, which helps to maximally exploit the local core caches. This can be done setting the KMP_AFFINITY environment variable before executing the program. The following setting has been used in our experiments: export KMP_AFFINITY="explicit,proclist=[0,1,2,3]".

Recently, a super-linear speed-up of such a simulation with a similar MLSM approach has been reported in [8]. The reason for super-linearity is in the accumulated L3 caches of multiple processors, which allows for significantly higher data bandwidth compared to the bandwidth on a single processor.

3.2 MPI

The AMS-DEMO algorithm supports parallel execution on distributed memory computers, e.g., computing clusters [6]. The master-slave methodology splits the algorithm into two parts: (i) master, usually the main algorithm logic that cannot be done in parallel and resides in the master process, and (ii) slave, comprising the tasks that can be done in parallel by the slave processes.

The evaluation of a cost function is often, and also in our case, the most time-demanding task within the optimization procedure. Furthermore, it is being repeated in the order of several thousands to million times, in the course of a single optimization. Fortunately, separate evaluations of cost function on different sets of parameters are independent of each other and can therefore reside in the slave part of AMS-DEMO. When comparing it to sequential DE, the advantage of AMS-DEMO is the parallel processing with no practical limits on the number of solutions that are processed in parallel, while the disadvantage is in slightly slower convergence of solutions. Nevertheless, AMS-DEMO provides a way for very efficient utilization of all the available processing nodes, even when faced with unpredictable and varying duration of the cost function evaluation.

MPI is used to implement the distributed job processing, by providing means of distribution of job input variables. The master process serves as job producer, where jobs can be defined on demand and in real-time. A job is defined as a predefined cost function evaluation that will be executed by the slave processes. Input parameters of cost function represent the input of the job and are encoded as a real-valued vector. The result of a job is encoded as a real-valued vector comprising the cost function value and optionally also some internal variables. The flow of jobs is designed around the processing node utilization efficiency.

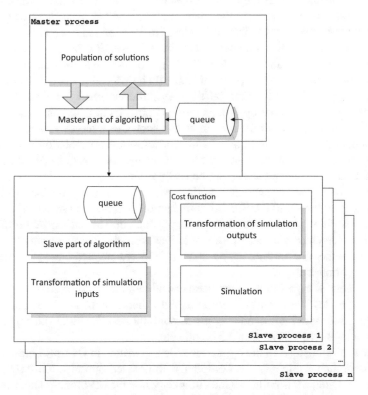

Fig. 4 Block diagram of MPI-based distributed optimization with AMS-DEMO

The master process starts the communication with the slave processes by generating a number of jobs and asynchronously distributing them among the slave processes. Slave processes then process their jobs in isolation and asynchronously return results to the master. To avoid the waiting time in between processing of consecutive jobs, slave processes maintain a queue of jobs, which the master process aims to keep full at all times (see [6] for more details). The scheme of the MPI-based parallel optimization is shown in Fig. 4.

4 Test Case Experiments

4.1 Experimental Setup

The optimizer is given a differentially heated square cavity and the option to place one or several obstacles in the domain. Obstacle positions are unconstrained, while their sizes are constrained upwards by the size of the cavity divided by the number of obstacles. Note that the orientation of the obstacles is also constrained in the sense that their edges are always parallel to the edges of the cavity. The obstacles are allowed to overlap, with the overlapping areas considered to have the same properties as regular obstacle. Overlapping thus provides no benefit to the solution, but it is enabled so that obstacles are not constrained by other obstacles, to ease the optimization procedure.

Since the obstacles are non-permeable and their material is a better insulator than the air, the optimal result—minimal flux—is achieved when the obstacles fill the whole cavity. Although the optimal solution is obvious to a human, that might not hold for the stochastic optimization logic, and such a task serves as a benchmark of the optimization procedure. In other words, a "closed form" solution is known and can be compared against the solutions obtained by the optimization procedure, hence the performance of the optimizer can be easily evaluated.

Four scenarios of the test case with varying difficulty are devised to test the performance of the simulation-based optimization. Within each scenario, the flux is normalized relative to the flux through an empty domain. The settings of the optimizer for all four scenarios are listed in Table 1. The scenarios are increasingly difficult, since the number of cost function parameters is increasing and these parameters are all equally important.

All numerical experiments have been executed on a homogeneous cluster of 20 computing nodes interconnected with Gigabit Ethernet network. The heart of each computing node is a single quad-core processor Intel Xeon E5520. In all experiments, the same clock frequency was used on all cores.

Both, the simulator and the optimizer code are written in C++ programming language, compiled with GCC 4.8 compiler with enabled optimization through -O3 switch. The OpenMP is built into the compiler while the MPI functionality is implemented with the Open MPI library. The optimizer and the simulator programs com-

Table 1 Settings of the optimizer for the test case scenarios

	Sub-cases			
Number of obstacles	1	4	9	16
Obstacle max size	1×1	$1/2 \times 1/2$	$1/3 \times 1/3$	$1/4 \times 1/4$
Number of cost function parameters	4	16	36	64
Population size	20	30	40	50
Number of simulations	2000	6000	12000	18000

municate via the file system and bash scripts, using the tools provided by Ubuntu 12.04.

All simulations were executed with the following parameters: 81×81 uniformly distributed discretization nodes on the domain (cavity and edges) with dimensionless size 1×1, MLSM time step $dt = 2.5 \cdot 10^{-5}$, maximal allowed dimensionless time $t_{max} = 20$ and steady state criteria $\mid T_{i+1} - T_i \mid < 10^{-7}$, where i is a time-step index.

4.2 Quantitative Results

For each scenario, the single optimum solution is to set obstacle sizes to their maximum and to arrange them in an orthogonal grid. The best solutions (out of 10 runs) obtained by the optimizer are plotted in Fig. 5. Obstacles are marked with numbers in their lower left corners. A single obstacle (scenario 1) is placed almost optimally. The optimizer performs very well also with four obstacles. For higher number of obstacles, i.e., 9 and 16, a slight drop in solution quality can be noticed. With increasing complexity of the optimization problem (with more obstacles), the optimizer finds less optimal solutions in the sense that the cavity is not fully covered, even though the scenarios with more obstacles are optimized with increased population size.

Because the holes between the obstacles become minimal after optimization, the temperature gradients and consequently thermal fluxes differ just slightly in all four scenarios. For example, in the scenarios with 4 obstacles, the temperature gradient is different because of four sizeable holes, which have pronounced impact on the convection. Although the optimization settings could be set to produce better results, the fine tuning of several parameters of optimization is beyond the scope of this work. It should also be noted that better results could be found on the account of running optimization for a longer time.

The convergence of solutions was measured by making 10 runs of the optimization for each scenario and extracting the best solution per scenario from these runs. The evolution of the best solution relative to the number of simulations is shown for each

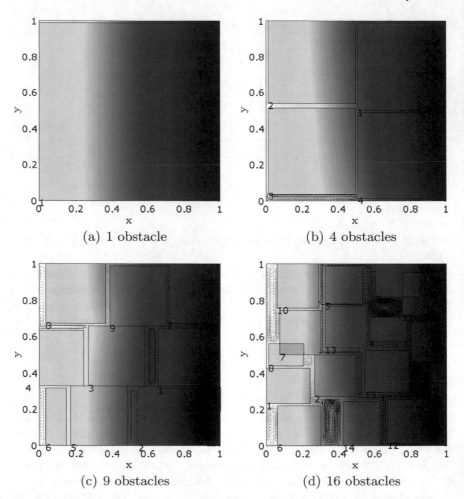

Fig. 5 Temperature contour plot and obstacle positions as set by the optimizer for the four scenarios

scenario in Fig. 6. On the figure, the heat flux is normalized with the true optimal solution heat flux value, therefore the minimal value of normalized heat flux is 1, and values above 1 represent sub-optimal solutions. We observe that the convergence is slower for the scenarios with more obstacles, which is an expected result. Larger populations would likely help in obtaining better results in terms of minimizing the heat flux, but for the price of even higher number of performed simulations for each optimization run.

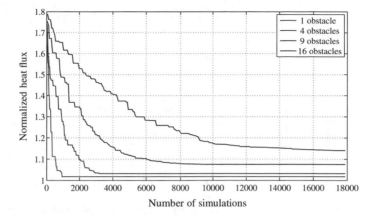

Fig. 6 Convergences of optimal solutions obtained for each optimization scenario

4.3 Parallel Speedup and Efficiency

The presented problem can easily become numerically so complex that exceeds the performances of a single computer, therefore, the parallel implementation is of high importance. With parallel implementation, great caution has to be paid on the parallelization speedup or efficiency evaluated as:

$$S = \frac{t_1}{t_N},$$

$$E = \frac{S}{n} = \frac{t_1}{n \cdot t_n},$$

where the symbols used are: speedup S, efficiency E, number of computing units n, execution time on a single computing unit t_1, and execution time on n computing units t_n. In the flowing, we analyze only the speedup.

As described, the parallelization is performed on two levels, which reflects in two-part speedup:

- speedup due to the execution of a single simulator run on multiple cores of a cluster node;
- speedup due to the execution of different simulation runs on distributed computing nodes.

An exact assessment of the speedup is difficult to obtain. While the simulations results are identical, either obtained from runs on one or more cores, the results of the optimization procedure depend on the number of computing nodes engaged in the optimization run [6]. The serial DE is deterministic with the same random generator seed, however, the parallel AMS-DEMO attempts to minimize the processor idle time by allowing a non-deterministic execution. Therefore, several repetitions of AMS-

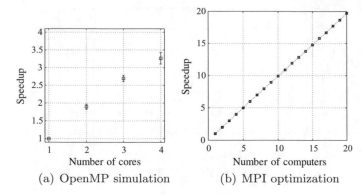

Fig. 7 Speedup of 100 simulations with different parameters (*left*) and speedup of optimization (*right*)

DEMO, even with the same random generator seed, are unlikely to produce the same results. Consequently, the run on n processing nodes cannot be quantitatively compared to the run on a single processing node, which induces difficulties in the calculation of speedup.

The correlation between the number of computers and the solution quality is not known, but it can be assumed that in most cases with more computers also more simulations are needed to produce the same solution quality. The detailed statistical analysis of the correlation on a large number of repetitions of AMS-DEMO has been performed in [6], and we will only rehash the main findings here. No statistically significant differences are found in the convergence rate of solutions produced by different runs of the AMS-DEMO as long as the number of computing nodes is lower than the population size, which is true in our test case. A justified simplification can thus be made to ease the speedup measurements. For each tested number of computing nodes, the AMS-DEMO is left to run until a predefined number of simulations is completed.

To avoid some exceptionally long optimization runs, the speedup is measured in parts—first for the simulator and then for the optimizer. The total speedup can be obtained as a product of both speedups. The execution times always include serial pre-processing, computation, post-processing, and input/output operations. The simulator runs vary in execution time length, depending on how fast they reach the steady state, which in turn depends on the input parameters—the placement of obstacles. To obtain a robust speedup measurement, one hundred parameter sets are randomly taken and used as input for separate simulations on 1–4 computing cores. Since the simulation times do not significantly differ between the scenarios, the joined results for the simulation speedup are shown in the left part of Fig. 7. Dots represent mean simulation speedups and their standard deviations, which are both satisfactory for a shared memory architecture.

To obtain the optimizer speedup, a separate test optimization is set up that stops after 1000 simulations. The test optimization time only minimally depends on the choice of problem scenario and other details of the simulation, and is therefore performed for one scenario only. The right part of Fig. 7 shows the optimization speedup, again incorporating all needed calculations and the overhead. To eliminate the noise introduced by the variation in simulation times from the graph, each optimization run time is normalized by the mean simulation time of that run. The result is a near-linear optimization speedup that scales well over the tested number of computers.

Finally, both speedups can be multiplied to get the total speedup. For example, on 10 and 20 computers, each using 4 cores, the total speedups are 32.1, and 63.9, respectively, which is satisfactory well, in particular, because the scalability of the parallelization is also near ideal.

5 Conclusions

The proposed methodology is focused on displaying the potency of the simulation-based optimization, taking into account the solution quality, the total execution time and the ability to further speedup the execution if more parallel processors are available. The presented case of obstructing air-flow with obstacles is an evidence of how the optimization achieves high quality solutions of complex problems. Although presented here in its simplest form, it is intuitively clear that given a more complex scenario, high quality obstacle placings would be hard to predict without the synergy between the numerical simulator and the multi-objective optimization. The test case could be extended in the optimization of various areas of design: structure of insulation material, large living and working spaces, air conditioning, heat storage and heat engines, and in many similar domains.

Both, the optimizer and the simulator, exploit the emergent properties of very simple and local operations, and are therefore able to make good use of parallelization. The optimization operates on the level of interconnected computing nodes, either on a homogeneous computing cluster, as presented in this work, or also on distributed systems built from computing nodes with diverse performances. The computer simulation, on the other hand, exploits the shared-memory model of a multi-core computer, and is also capable of switching to GPUs [11] and other computing accelerators that proliferate in the modern high performance computing hardware. The combination of the both approaches is shown to be efficient at utilizing hardware resources and providing an emergent tool for handling simulation-based optimization problems.

Acknowledgments We acknowledge the financial support from the Slovenian Research Agency under the programme group P2-0095.

References

1. Abraham, A., Jain, L., Goldberg, R. (eds.): Evolutionary Multiobjective Optimization. Springer, London (2005)
2. Bonabeau, E., Dorigo, M., Theraulaz, G.: Swarm Intelligence: From Natural to Artificial Systems. Oxford University Press (1999)
3. Burke, E.K., Kendall, G. (eds.): Introduction to Stochastic Search and Optimization. Wiley, Hoboken (2003)
4. Chandra, R., Dagum, L., Kohr, D., Maydan, D., McDonald, J., Menon, R.: Parallel Programming in OpenMP. Academic Press, San Diego (2001)
5. Coello, C.A.C., Lamont, G.B., Veldhuizen, D.A.V.: Evolutionary Algorithms for Solving Multi-Objective Problems (Genetic and Evolutionary Computation). Springer New York Inc, Secaucus, NJ, USA (2006)
6. Depolli, M., Trobec, R., Filipi, B.: Asynchronous master-slave parallelization of differential evolution for multiobjective optimization. Evolutionary Computation **21**(2), 261–291 (2013)
7. Eiben, A.E., Smith, J.E.: Introduction to Evolutionary Computing. Springer, Berlin (2003)
8. Kosec, G., Depolli, M., Rashkovska, A., Trobec, R.: Super linear speedup in a local parallel meshless solution of thermo-fluid problems. Comput. Struct. **133**, 30–38 (2014)
9. Kosec, G., Trobec, R.: Simulation of semiconductor devices with a local numerical approach. Eng. Anal. Boundary Elements **50**, 69–75 (2015)
10. Kosec, G., Šarler, B.: Solution of thermo-fluid problems by collocation with local pressure correction. Int. J. Numer. Methods Heat Fluid Flow **18**, 868–882 (2008)
11. Kosec, G., Zinterhof, P.: Local strong form meshless method on multiple graphics processing units. CMES. Comput. Model. Eng. Sci. **91**(5), 377–396 (2013)
12. Malan, A.G., Lewis, R.W.: An artificial compressibility cbs method for modelling heat transfer and fluid flow in heterogeneous porous materials. Int. J. Numer. Methods Eng. **87**, 412–423 (2011)
13. Price, K., Storn, R.M., Lampinen, J.A.: Differential Evolution: A Practical Approach to Global Optimization. Natural Computing Series. Springer, Berlin (2005)
14. Snir, M., Otto, S., Huss-Lederman, S., Walker, D., Dongarra, J.: MPI—The Complete Reference. The MIT Press, Cambridge (1996)
15. Šterk, M., Trobec, R.: Meshless solution of a diffusion equation with parameter optimization and error analysis. Eng. Anal. Boundary Elements **32**(7), 567–577 (2008)
16. Trobec, R., Kosec, G.: Parallel Scientific Computing: Theory, Algorithms, and Applications of Mesh Based and Meshless Methods. Springer (2015)
17. de Vahl Davis, : G.: Natural Convection of Air in a Square Cavity: A Bench Mark Numerical Solution. Int. J. Numer. Methods Fluids **3**, 249–264 (1983)
18. Wang, C.A., Sadat, H., Prax, C.: A new meshless approach for three dimensional fluid flow and related heat transfer problems. Comput. Fluids **69**, 136–146 (2012)
19. Zitzler, E., Thiele, L.: Multiobjective optimization using evolutionary algorithms—a comparative case study. In: Proceedings of the Fifth Conference on Parallel Problem Solving from Nature—PPSN V, pp. 292–301. Springer, Heidelberg (1998)

Theory and Practice of Discrete Interacting Agents Models

Adrian-Horia Dediu, Joana M. Matos and Carlos Martín-Vide

Abstract We review several distributed, discrete time, and probabilistic models with interacting multi-agents. We discuss the basic principles together with some variants of the frog model. We also present an experimental approach, talking about implementing and checking one of the less investigated variants of the model, where the frogs die if not meeting other frogs for some time. We follow the same lines for broadcasting and gossiping models, our experimental approach checks the validity of the broadcasting time for a wide range of the number of agents existing in the system. We also study the emergent behaviour of a multi-agent system whose agents follow only several simple rules; we performed our experiments with an implementation in StarLogo.

1 Introduction

Centralized versus *distributed* systems, *discrete* versus *continuous models*, *deterministic* versus *probabilistic* behaviour, these are only few of the decisions we have to take in complex design activities. Each choice offers various advantages and disadvantages. Kim et al. [17] show that we consider various aspects related to reliability, scalability, security, communication complexity, efficiency, etc. when choosing between distributed and centralized solutions. About using continuous time models, there are many opinions in favor, believing that there exist phenomena for which continuous models fit better and should be preferred (Jarrow and Protter [16]). There are also some people finding discrete models more intuitive and the simula-

A.-H. Dediu (✉) · C. Martín-Vide
Rovira i Virgili University, Avinguda Catalunya, 35, 43002 Tarragona, Spain
e-mail: adrian.dediu@urv.cat

C. Martín-Vide
e-mail: carlos.martin@urv.cat

J. M. Matos
CMA, Department of Mathematics, FCT—New University of Lisbon,
Quinta da Torre, 2829-516 Caparica, Portugal
e-mail: jmf.matos@fct.unl.pt

© Springer International Publishing Switzerland 2017
A. Adamatzky (ed.), *Emergent Computation*, Emergence, Complexity
and Computation 24, DOI 10.1007/978-3-319-46376-6_19

419

tion code easy to implement. There are also models where it is not possible to obtain analytical results, then numerical methods which involve discretizing time should be used, as showed by Gumel [14]. Deterministic versus probabilistic approaches reflect our level of knowledge about the system we work with, its components and the interactions among them (Kirchsteiger [18]). Without trying to argue in favor of one or another options, we focus our survey on distributed, discrete time, and probabilistic models. We mention only several books and reviewing articles relevant for this domain, as the literature dealing with these problems is very vast.

Easley and Kleinberg [9] dedicate an entire chapter of their book to epidemic models. They show that between epidemic disease and information spreading through social networks there is a striking similarity. The simplest model of contagion referred as a *branching process*, is actually a *tree*. An infected person meets other k persons and the disease is transmitted with probability p. The initially infected person is represented as the root of a tree and the contacted persons are vertices connected to the root (we can find a formal definition of trees in the next section). Assume that each infected person meets other k persons, we represent this as new layers in the contagion tree, and the process continues. Note that if the product $pk < 1$, then the disease dies out after some time, else if the product is greater than 1, then the disease is persistent, infecting more and more persons. If the product is approximatively 1, then the epidemic contagion is in a fragile equilibrium. We find out about two basic epidemic models, *SIS* (Susceptible-Infectious-Susceptible, these are the states an infected person passes through) and *SIR* (Susceptible-Infectious-Removed from the population, either getting immune, or dying). Variations of these basic models are discussed, for example *SIRS* represents a process where an infected person gets a temporary immunity (being removed from the population), after some time can contact again the disease, becoming susceptible once more.

In general, *percolation theory* describes the emergence of connected clusters of constituents in a multi-component system. The main phenomenon studied in percolation models is the emergence of a dramatic change in the qualitative behavior, triggered by an infinitesimal change in the parameters. For example, a small difference in the probability of the contamination can make the difference between a limited spreading disease and a global epidemic (Erez et al. [10]). Various mathematical models deal with percolation. In an infinite grid, choose randomly each edge to be open with probability p and closed with probability $1 - p$. We consider an open cluster C as the vertices that can be reached from the origin[1] through open edges. Let the *bond percolation* be the probability p_c^{bond} for which the cluster C becomes unbounded. If instead of edges we declare vertices to be open with probability p, we define the *site percolation* as the probability p_c^{site} for which the cluster C of open vertices becomes unbounded. For more details on this subject we recommend the book of Grimmett [13]. Other authors focus on *continuum percolation*, defining a random graph with vertices as the set of points randomly scattered over a region of space according to some probability distribution, and any two points separated by a distance less than a certain specified value r are connected by an edge. We denote by

---

[1]The vertex with the coordinates (0, 0).

$G(X; r)$ the undirected graph with vertex set X and with undirected edges connecting vertices whose distance is no longer than r. Percolation is studied as a function of r (Penrose [21]).

Random walks in graphs is a natural model well established to test graph connectivity. Cooper et al. [4, 5] study properties of multiple random walks in regular graphs. They consider particles making simple random walks and they give results for two cases, *oblivious particles* and *interacting particles*. The interacting particles are also categorised as:

- *Talkative particles*, when meeting, they exchange information;
- *Predator-Prey*, predators eat preys when meeting;
- *Annihilating particles*, which destroy each other (pairwise) on meeting;
- *Coalescent particles*, which coalesce on meeting.

In the models we discuss, we assume that particle interaction occurs only when meeting at a vertex not at edges. Chapter 14 from Aldous and Fill [2] gives formal definitions for the random variables associated to the mentioned various types of particles taking random walks. Also, Aldous [1] introduces *Finite Markov Information Exchange* (FMIE) processes, however, this model uses continuous time.

In this chapter, after the section containing preliminary notions and notations, we discuss together with detailed references about the *frog model*, we study two variants of information exchange models, namely the *broadcasting and gossiping models*, and we conclude with a practical section showing several characteristics of an *emergent behaviour*.

2 Preliminaries

We assume that the reader is familiar with the basic notions of graph theory, probability theory, and complexity theory. We briefly present an overview of the basic concepts we use in this paper. We denote by $\mathbb{N}$ the set of natural numbers, that is $\{0, 1, 2, \ldots\}$. We denote by $\mathbb{Z}$ the set of integers and by $\mathbb{R}$ the set of real numbers.

Let f, g be functions defined from $\mathbb{N}$ into $\mathbb{R}$, $n, n_0 \in \mathbb{N}$ and $c \in \mathbb{R}$ (we follow the notations from Rothlauf [27]). We define:

- $f \in O(g) \Leftrightarrow \exists c > 0, \exists n_0 > 0$ such that $|f(n)| \leq c \cdot |g(n)|, \forall n \geq n_0$ (*asymptotic upper bound*).
- $f \in o(g) \Leftrightarrow \forall c > 0, \exists n_0 > 0$ such that $|f(n)| < c \cdot |g(n)|, \forall n \geq n_0$ (*asymptotically negligible*).
- $f \in \Omega(g) \Leftrightarrow g \in O(f)$ (*asymptotic lower bound*).
- $f \in \Theta(g) \Leftrightarrow f \in O(g)$ and $g \in O(f)$ (*asymptotically tight bound*).

Some authors use the "soft-O" notation that ignores also the logarithmic factors, for example, $\tilde{\Theta}(f(m))$ represents $\Theta(f(m) \log^c m)$, for some constant c.

A *graph* $G = (V, E)$ consists of a set V of *vertices* (or *nodes*), and a set $E \subseteq V \times V$ of *edges*. A *path* in a graph is a sequence of vertices, $v_1 v_2 \ldots v_k$ for $k \geq 2$

such that every (v_i, v_{i+1}) is an edge in the graph, for $1 \leq i \leq k - 1$. A path is a *cycle* if the first and the last vertex are the same. The *distance* between two vertices v and w of a graph is defined as the number of edges in the shortest path connecting the vertices v and w, and we denote it by $D(v, w)$.

A graph $G = (V, E)$ is *connected* if for all $v, w \in V$ there exists a path between v and w. In this case, let $D_G = max_{v,w} D(v, w)$ be the *diameter* of the graph G.

A graph G is *undirected* if for all $v, w \in V$ we have $(v, w) \in E \Leftrightarrow (w, v) \in E$, otherwise G is *directed*. A graph G is called *acyclic* if there are no cycles in G. A *tree* is a directed graph that has no cycles and that has one distinct vertex, called the *root*, such that there is exactly one path from the root to every other vertex. A graph $G' = (V', E')$ is a *subgraph* of $G = (V, E)$ if $V' \subseteq V$ and $E' \subseteq E \cap (V' \times V')$. A *spanning tree* for a connected graph G is a tree subgraph of G including all the vertices of G.

The *2-dimensional grid with n^2 vertices* is the undirected and connected graph Z_n^2 with the set of vertices $V = \{(i, j) \mid 1 \leq i, j \leq n\}$ and the set of edges $E = \{((i, j), (i, j + 1)) \mid 1 \leq i \leq n, 1 \leq j \leq n - 1\} \cup \{((i, j), (i + 1, j)) \mid 1 \leq i \leq n - 1, 1 \leq j \leq n\}$. A discrete *torus* modulo n, denoted by T_n^2, is the finite grid Z_n^2 adding to E the following set of edges: $\{((1, j), (n, j)), ((j, 1), (j, n)) \mid 1 \leq j \leq n\}$. In a natural way, we extend the notion of a finite 2-dimensional grid to an *infinite grid* (with vertices from $\mathbb{Z} \times \mathbb{Z}$) and we denote it by Z^2.

On a graph $G = (V, E)$, we define a *random walk* $\overline{v}$ starting at the vertex v_1 for a given v_1, as a sequence of vertices $v_1 v_2 \ldots v_t \ldots$ from V for $t > 1$, where v_{t+1} is chosen uniformly at random from those vertices such that $(v_t, v_{t+1}) \in E$. For a graph $G = (V, E)$ and a random walk starting at $v \in V$, let C_v be the expected time to visit all the vertices of G. The *cover time* C_G is defined as $max_{v \in V} C_v$.

A remarkable result by Feige, gives the upper and lower bound for the cover time for any graph.

Theorem 1 (Feige [11, 12]) *For any undirected and connected graph G with $m = |V|$ vertices, the cover time C_G satisfies the relations:*

$$m \ln m + O(m \ln m) \leq C_G \leq \frac{4}{27} m^3 + O(m^{5/2}).$$

In particular, for large values of n, the cover time in a discrete torus T_n^2 satisfies the relation:

Theorem 2 (Dembo et al. [6])

$$\lim_{n \to \infty} \frac{C_{T_n^2}}{(n \log n)^2} = \frac{4}{\pi} \text{ in probability.}$$

Marking visited vertices/edges, neighborhood look-ahead, multiple random walks, are only several approaches for speeding up the cover times.

For a graph $G = (V, E)$ and k random walks starting at $v \in V$, let C_v^k be the expected time to visit all the vertices of G by at least one of the k random walks.

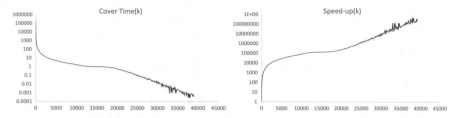

Fig. 1 Cover time and speed-up for k random walks with random starting points on Z_{50}^2

The *cover time of k random walks* C_G^k is defined as $max_{v \in V} C_v^k$. We also define the *speed-up* obtained by k parallel random walks as $S^k = \frac{C_G^k}{C_G}$.

When studding the properties of graphs and the speed-up given by multiple random walks, it makes sense to consider only multiple random walks starting at the same vertex. However, we are interested in the dissemination/propagation phenomena and for us the random walks do not start at the same vertex. We give the results of a simple experiment we made on Z_{50}^2. We represent the cover time of k parallel random walks with random initial points and the speed-up we get as functions of k in Fig. 1.

The proofs of the following two lemmas on random walks can be found in Pettarin et al. [24].

Lemma 1 (The meeting probability of a random walk and a vertex) *Given a random walk $\bar{v}$ on a graph G with m vertices, starting at the vertex v_0 at time 0, there exists a positive constant c such that for any vertex $v \neq v_0$ we have*

$$P(v \text{ is visited by } \bar{v} \text{ within } D(v_0, v)^2 \text{ steps}) \geq \frac{c}{max\{1, log\ D(v_0, v)\}}.$$

Lemma 2 (The meeting probability of two random walks) *Let $\bar{v}$ and $\bar{w}$ be two independent random walks on a graph G with m vertices, starting at time 0 at the vertices v_0 and w_0, respectively, with $v_0 \neq w_0$. Let v_t and w_t be the locations of the walks at time t and let $t_0 \geq D(v_0, w_0)^2$. Then, there exists a positive constant d such that*

$$P(\exists t \leq t_0 \text{ such that } v_t = w_t) \geq \frac{d}{max\{1, log\ D(v_0, w_0)\}}.$$

3 Frog Model

In this section we follow the lines of the surveying article about the frog model, a system of random walks on a graph, largely described by Popov [25]. The model works on a given graph G with one of the vertices designated as the *root*. Initially there is a random number of particles on each vertex of G, the ones in the root are active, the others are sleeping. The active particles perform random walks on G, obeying the following rules in each step:

1. Each active particle lives with probability p or die with probability $1 - p$.
2. A living particle moves to one of the neighbor vertices chosen with uniform probability.
3. When an active particle reaches a vertex with sleeping particles, it activates all the sleeping particles.

We can find a detailed description of the model as well as an example in Lebensz-tayn [19]. Let η be a random variable giving for each vertex x of G the initial number of particles in x. We denote by $FM(G, p, \eta)$ the frog model on the graph G with the living probability for particles p and the initial configuration given by η.

A frog model becomes *extinct* if there are no more active particles, otherwise it *survives*. A model $FM(G, p, \eta)$ is called *recurrent* if the probability p_r that the root is hit infinitely often is greater than 0, otherwise, the model is called *transient*. If we assume that active particles live some fixed time t, rather than having a random geometrically distributed lifetime, let T_G be the smallest value of t such that with probability at least 0.5 all vertices of G are visited by active particles.

Theorem 3 (Popov [25]) *There is a constant C such that $C \log n \leq T_{Z_n^2} \leq \log^2 n$.*

In Popov [25], we find also different versions of the shape theorem, conditions for recurrent and transience, extinction and survival for various frog models. There is also a list of open problems; unfortunately, this research line was not continued according to our knowledge. In the last part of the article we find several modifications introduced to frog models, we give here only several examples.

1. Instead of dying, the active particles become sleeping again.[2]
2. Let us introduce an integer parameter $M \geq 1$, also called the movement counter, such that any active particle which did not wake up anybody for M consecutive steps dies.[3]

We have implemented and run some tests with the second modified frog model; our results are similar with the ones presented by Popov [25], assuming that all the particles arriving to a site with sleeping particles reset their movement counter M. We considered $G = Z^2$. We also studied the case $\eta = 1$, that means there exists one particle on each vertex.

The following pictures (Fig. 2) show the results of our simulations, with red we represent the traces of the active particles, with black there are the remaining active particles.

We believe that many things can be improved when studying this model, for example, not saying only that a model survives or becomes extinct, also computing several expectations, like the surface covered by the active particles, the living time of a certain model, etc.

Dutta et al. [7] studied a frog model $FM(G, 1, k)$ where the meeting particles coalesce. This model is known as the *coalescing-branching random walk* (*cobra*

---

[2] Proposed by Hervé Guiol.
[3] Preliminary research by Fábio Machado and Lucas Meyer.

(a) $M = 1, p = 1$, after 200 steps, 836 active particles

(b) $M = 1, p = 0.96$, after 500 steps, 31 active particles

Fig. 2 Modified frog models

walk, for short). The models based on cobra walks are useful in understanding the SIS-type of epidemic processes in networks and can also be helpful in performing light-weight information dissemination in resource-constrained networks. The following lemma gives the cover time for 2-dimensional grids.

Lemma 3 (Dutta et al. [7]) *Let* $G = Z_n^2$. *The cover-time of a cobra walk with a branching factor 2 on G is $O(n^2 \log n^2)$ with high probability.*

4 Broadcasting and Gossiping Models

Information dissemination is a clear topic of interest in the literature, where its dynamics has been studied in several contexts and with different goals such as virus infection, rumor spreading and mobile or social networks (Chierichetti et al. [3], Hromkovic et al. [15]). This section is devoted to the work of Pettarin et al. [23, 24] on rumor spreading, under two possible models: the *broadcasting* and the *gossiping*.

Let us consider k mobile agents uniformly distributed at random among the n^2 vertices of the 2-dimensional grid Z_n^2. In the *broadcasting model*, at time 0, randomly one of the agents receives a rumor and starts a random walk. When an agent with the rumor meets an uninformed agent, the rumor is shared and the latest agent starts a new random walk. In the *gossiping model*, at time 0, each agent receives a distinct rumor and starts a random walk. When two agents meet, they share their rumors and continue their own random walks transporting the shared rumors.

Two natural questions arise from these scenarios, respectively. How long does it take until all agents of the broadcasting model share the rumor? How long does it take until all agents of the gossiping model are aware of all k rumors? An approach to these answers are given by Peres et al. [22] when the rumor transmission process is above the percolation point. Pettarin et al. [23, 24] complement their results for the case the rumor transmission process is below the percolation point.

We call *broadcasting time* to the first time T_B that all agents are sharing the rumor and *gossiping time* to the first time T_G that all agents are sharing all the k rumors. The time is a discrete variable with rumor informed agents performing synchronized movements.

In Pettarin et al. [23], the authors provide a tight characterization (up to logarithmic factors) of the rumor spreading time for both models. They prove, with high probability, that both broadcasting and gossiping times are $\tilde{\Theta}(n^2/\sqrt{k})$.

Theorem 4 (Upper and lower bounds for the broadcasting and gossiping times) *Let* $T \in \{T_B, T_G\}$. *With high probability,*

$$\Omega\left(\frac{n^2}{\sqrt{k}\log^2 n^2}\right) = T = \tilde{O}\left(\frac{n^2}{\sqrt{k}}\right).$$

Although the separated proofs presented by Pettarin et al. for each time T follow a similar line, due to time dependencies they can not be deduced from each other.

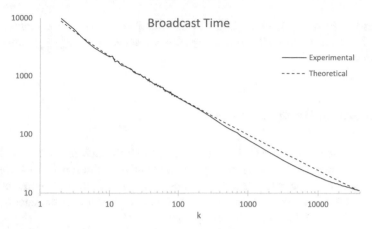

Fig. 3 Values of T_B depending on the number of agents k on Z_{50}^2

In fact, in the first model the random walks need to meet static uninformed agents, while in the second one random walks of rumor informed agents need to meet the random walks of the uninformed ones.

A direct corollary with a clear translation in the virus infection context can be obtained from these results.

Corollary 1 (Virus infection speed) *Static agents placed at random locations and dynamic agents moving in independent random walks are infected at about the same time.*

We run some tests on Z_{50}^2, studying the experimental values for T_B depending on the number of agents k. For each number of agents we tested the value of T_B for 200 different executions and we recorded the average values. We also compare our results with the theoretical values computed as $c_1 2500/(\sqrt{k} \log_2(c_2 k))$, where $c_1 = 15$ and $c_2 = 4$, the constants were obtained interpolating the experimental values. In Fig. 3 we represent graphically our results.

Panagiotou et al. [20] present a faster spreading rumors model. In each vertex of a connected graph with n vertices there is one agent. The so-called *Push* protocol starts with a single vertex that knows a rumor. In each round, informed agents make a random call to one of their neighbours, sending the rumor. The *Push-Pull* protocol works similarly, in addition, each uninformed agent calls a random neighbor and thus may also learn the rumor. Panagiotou et al. extend the known models allowing the number of calls of an agent to be chosen independently according to a probability distribution R. The main results of Panagiotou et al. give the number of rounds to inform all the nodes under certain assumptions.

Theorem 5 (Panagiotou et al. [20]) *Assume that R is a power law distribution with $2 < \beta < 3$. Then the Push-Pull protocol informs all nodes in $\Theta(\log \log n)$ rounds with probability $1 - o(1)$.*

Theorem 6 (Panagiotou et al. [20]) *Assume that R is a power law distribution with* $\beta = 3$. *Then the Push-Pull protocol informs all nodes in* $\Theta(\frac{\log n}{\log \log n})$ *rounds with probability* $1 - o(1)$.

5 Emergent Behaviour

In a general sense, *emergent* is associated with terms such prominent and unexpected. According to Dyson [8], emergent behaviour, by definition, is what's left after everything else has been explained. In multi-agent systems, the emergent behaviour represents the unexpected results arising from the interaction of the agents, rather than from the individual agent behaviour specification.

In this section we follow a model proposed by Resnik [26] about termites and wood-chips. In his article, Resnick, one of the parents of StarLogo, emphasizes the decentralized modeling and decentralized thinking aspects, required for understanding the new emergent paradigms related to several self organizing systems. StarLogo is a programming environment designed for simulations to understand complex systems; its current versions are known as *StarLogo TNG: The Next Generation*[4] and *StarLogo Nova*.[5]

Suppose that each termite executes only the following simple rules:

1. If you do not carry anything and you find a wood chip, then pick it up.
2. If you are carrying a wood chip and you bump into another wood chip, then put down the wood chip you are carrying.

It is almost incredible how starting from the following two simple rules, we get such an unexpected behaviour of the global system. Only from picking up and dropping wood chips, they group together in larger and larger piles, until eventually remaining a single pile.

We find a detailed description of a simulation based on these simple rules on a tutorial page about StarLogo.[6]

The instructions of each termite are coded into three simple procedures, that we present as Algorithm 1:

Algorithm 1: Termite Instructions

```
search-for-chip                              // First rule
find-new-pile      // Carrying a wood chip and finding a new pile,
find-empty-patch     // put down the wood chip in an empty place
                        // at the border of the pile
```

---

[4]http://education.mit.edu/portfolio_page/starlogo-tng/.

[5]http://www.slnova.org/.

[6]http://web.mit.edu/mitstep/starlogo/tutorial/tutorial.html.

The old version of StarLogo promoted a recursive technique, for example, the termite procedure `search-for-chip`, if not finding a wood chip, then after a random move, calls itself `search-for-chip` (see Algorithm 2).

Algorithm 2: search-for-chip

if *you find a wood chip* **then**
| take it
| move away
| **return**
end
else
| random walk one step
| search-for-chip
end

We have tried to update the study of this problem by making two implementations in StarLogo TNG, taking advantage of the facilities offered by the newer language.

In the first experiment, we only made some cosmetic improvements of the implementation. We preferred instead of three (recursive) procedures for a termite, a single one, coding the status of a termite with colors. We used black for a termite not carrying a chip, looking to find one, red for a termite with a chip, looking for a new pile, and grey for a termite looking for an empty place to put down its wood chip.

After running the experiment, we see the results in Fig. 4.

The first picture represents the initial configuration, with the wood chips (in red) distributed uniformly at random over all the working environment, *SpaceLand* (the green field). The following pictures show different stages of our simulation. We note in the first picture that the wood chips are embossed, and having also a height. This was leading us to a new variant of this problem. How about a 3-D version of the piles?

Let us see the new rules, actually only the second rule is a bit modified.

1. If you do not carry anything and you find a wood chip, then pick it up.
2. If you are carrying a wood chip and in front there is a wall or a cliff, then reduce the height difference.

The second rule could be better explained as follows:

2.a If in front there is a wall, then put down the wood chip you are carrying.
2.b If in front there is a cliff, move first, then put down the wood chip you are carrying.

For implementation, we used only two colors for termites, black for termites not carrying wood chips, and red for the ones with wood chips. For testing if a termite stands in front of a wall or a cliff, we use the height of the terrain ahead (StarLogo tile: `ph ahead`) compared to the current height (StarLogo tile: `patch height`). Walls and cliffs show a difference of height of at least two. The following pictures show different stages of our 3-D simulation (Fig. 5).

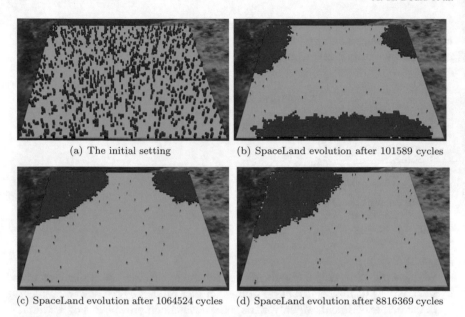

(a) The initial setting   (b) SpaceLand evolution after 101589 cycles

(c) SpaceLand evolution after 1064524 cycles   (d) SpaceLand evolution after 8816369 cycles

Fig. 4 Termites following only pick-up and drop instructions; they organize wood chips into a single pile

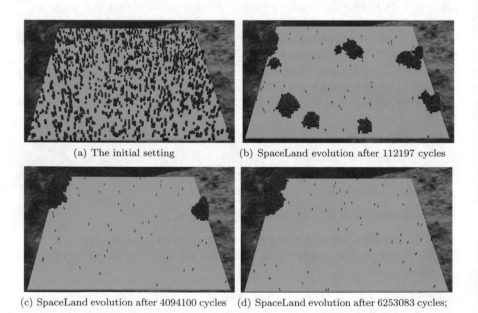

(a) The initial setting   (b) SpaceLand evolution after 112197 cycles

(c) SpaceLand evolution after 4094100 cycles   (d) SpaceLand evolution after 6253083 cycles;

Fig. 5 A simple modification in termites instructions leads to a different global behaviour

We conclude this section showing up several differences between StarLogo TNG and StarLogo Nova, with respect to this simulation problem. In StarLogo Nova there is no terrain height, therefore, we should use another method for simulations. There is anyway the altitude of the agents, thus the patches become active, they could execute movements by themselves. We could think to some self assembling constructions, not only pyramids like, as in our 3-D simulation, but also more complex buildings, adding rules for domes, tunnels, or bridges, etc.

An appropriate research direction can study several formal models helping us to understand better the relations between the individual rules and the global behaviour of the whole system. We also feel the lack of theoretical results in this field, computing several expectations, for example the number of cycles needed for k termites to move m patches into a single pile in a grid Z_n^2 would be highly appreciated.

6 Concluding Remarks

We studied several distributed, discrete time, and probabilistic models. For a better understanding, we illustrated the implementations of some of these models, showing the obtained results. We also pointed out several possible research directions.

The methods we discussed were inspired by reality, and the derived results have implications for the real world applications as well. For example, studies about cover times for various graphs are useful in search algorithms, when an exhaustive search is needed and there is no direct method to find faster the result. Studying about the broadcasting time of information in computer networks and social networks of our days, can have implications for world epidemic studies, helping us to prevent, fight against disease spreading, limiting the effects and risks we are exposed. Analysing the emergent behaviour and the relation between the simple individual rules and the global system response can be important for many fields, especially for robotics as a branch of the new domains of nanotechnology and biotechnology.

Acknowledgments This work was developed within the FCT Project UID/MAT/00297/2013 of CMA and of Departamento de Matematica da Faculdade de Cincias e Tecnologia da Universidade Nova de Lisboa.

References

1. Aldous, D.: Interacting particle systems as stochastic social dynamics. Bernoulli **19**(4), 1122–1149 (2013). doi:10.3150/12-BEJSP04
2. Aldous, D., Fill, J.A.: Reversible markov chains and random walks on graphs (2002). (Unfinished monograph, recompiled 2014, available at http://www.stat.berkeley.edu/~aldous/RWG/book.html)
3. Chierichetti, F., Lattanzi, S., Panconesi, A.: Almost tight bounds for rumour spreading with conductance. In: Schulman, L.J. (ed.) STOC, pp. 399–408. ACM (2010). doi:10.1145/1806689.1806745

4. Cooper, C., Frieze, A., Radzik, T.: Multiple random walks and interacting particle systems. In: Albers, S., Marchetti-Spaccamela, A., Matias, Y., Nikoletseas, S., Thomas, W. (eds.) Automata, Languages and Programming: 36th Internatilonal Collogquium, ICALP 2009, Rhodes, greece, July 5–12, 2009, Proceedings, Part II, pp. 399–410. Springer Berlin Heidelberg, Berlin, Heidelberg (2009). doi:10.1007/978-3-642-02930-1_33

5. Cooper, C., Frieze, A.M., Radzik, T.: Multiple random walks in random regular graphs. SIAM J. Discret. Math. **23**(4), 1738–1761 (2009). doi:10.1137/080729542

6. Dembo, A., Peres, Y., Rosen, J., Zeitouni, O.: Cover times for Brownian motion and random walks in two dimensions. Ann. Math. **160**, 433–464 (2004). doi:10.4007/annals.2004.160.433

7. Dutta, C., Pandurangan, G., Rajaraman, R., Roche, S.: Coalescing-branching random walks on graphs. In: Proceedings of the Twenty-fifth Annual ACM Symposium on Parallelism in Algorithms and Architectures, SPAA '13, pp. 176–185. ACM, New York, NY, USA (2013). doi:10.1145/2486159.2486197

8. Dyson, G.B.: Darwin Among the Machines: The Evolution of Global Intelligence. Addison-Wesley Longman Publishing Co., Inc, Boston, MA, USA (1997)

9. Easley, D., Kleinberg, J.: Networks, Crowds, and Markets: Reasoning About a Highly Connected World. Cambridge University Press, New York, NY, USA (2010)

10. Erez, T., Moldovan, S., Solomon, S.: Social anti-percolation, resistance and negative word-of-mouth. In: Sanchez, S., Lavigne, S. (eds.) Modeling an Artificial Stock Market: When Information Influence Market Dynamics, Handbook of Research on Nature Inspired Computing for Economics and Management. Idea Group (2006). http://www.idea-group.com/

11. Feige, U.: A tight lower bound on the cover time for random walks on graphs. Random Struct. Algorithms **6**(4), 433–438 (1995). doi:10.1002/rsa.3240060406

12. Feige, U.: A tight upper bound on the cover time for random walks on graphs. Random Struct. Algorithms **6**(1), 51–54 (1995). doi:10.1002/rsa.3240060106

13. Grimmett, G.: Percolation, 2nd edn. A Series of Comprehensive Studies in Mathematics, vol. 321. Springer (1999)

14. Gumel, A., Lenhart, S.: Modeling Paradigms and Analysis of Disease Transmission Models. DIMACS Series in Discrete Mathematics and Theoretical Computer Science. American Mathematical Society (2010). https://books.google.pt/books?id=oeQ-BAAAQBAJ

15. Hromkovic, J., Klasing, R., Pelc, A., Ruzicka, P., Unger, W.: Dissemination of Information in Communication Networks: Broadcasting, Gossiping, Leader Election, and Fault-Tolerance (Texts in Theoretical Computer Science. An EATCS Series). Springer New York, Inc., Secaucus, NJ, USA (2005)

16. Jarrow, R., Protter, P.: Discrete versus continuous time models: local martingales and singular processes in asset pricing theory. Financ. Res. Lett. **9**(2), 58–62 (2012). doi:10.1016/j.frl.2012. 03.002. http://www.sciencedirect.com/science/article/pii/S1544612312000177

17. Kim, Y., Perrig, A., Tsudik, G.: Simple and fault-tolerant key agreement for dynamic collaborative groups. In: Proceedings of the 7th ACM Conference on Computer and Communications Security, CCS '00, pp. 235–244. ACM, New York, NY, USA (2000). doi:10.1145/352600. 352638

18. Kirchsteiger, C.: On the use of probabilistic and deterministic methods in risk analysis. J. Loss Prev. Process Ind. **12**(5), 399–419 (1999)

19. Lebensztayn, É.: Um limitante superior para a probabilidade crtica do modelo dos sapos em rvores homogneas. Ph.D. thesis, Universidade de So Paulo (USP). Instituto de Matemtica e Estatstica, Brazil (2015). http://www.teses.usp.br/teses/disponiveis/45/45133/tde-24052013-125727/publico/Principal.pdf

20. Panagiotou, K., Pourmiri, A., Sauerwald, T.: Faster rumor spreading with multiple calls. Electr. J. Comb. **22**(1), P1.23 (2015). http://www.combinatorics.org/ojs/index.php/eljc/article/view/v22i1p23

21. Penrose, M.: Random Geometric Graphs. University Press, New York, Oxford (2003)

22. Peres, Y., Sinclair, A., Sousi, P., Stauffer, A.: Mobile geometric graphs: detection, coverage and percolation. In: Randall, D. (ed.) Proceedings of the Twenty-Second Annual ACM-SIAM Symposium on Discrete Algorithms, SODA 2011, San Francisco, California, USA, January 23–25, 2011, pp. 412–428. SIAM (2011). doi:10.1137/1.9781611973082.33

23. Pettarin, A., Pietracaprina, A., Pucci, G., Upfal, E.: Infectious random walks. CoRR abs/1007.1604 (2010). http://arxiv.org/abs/1007.1604
24. Pettarin, A., Pietracaprina, A., Pucci, G., Upfal, E.: Tight bounds on information dissemination in sparse mobile networks. In: Proceedings of the 30th Annual ACM SIGACT-SIGOPS Symposium on Principles of Distributed Computing, PODC '11, pp. 355–362. ACM, New York, NY, USA (2011). doi:10.1145/1993806.1993882
25. Popov, S.Y.: Frogs and some other interacting random walks models. In: Banderier, C., Krattenthaler, C. (eds.) DMTCS Proceedings, Discrete Random Walks, DRW'03, vol. AC, pp. 277–288. Discrete Mathematics and Theoretical Computer Science (2003). http://www.dmtcs.org/proceedings/html/dmAC0126.abs.html
26. Resnick, M.: Decentralized modeling and decentralized thinking. In: Modeling and Simulation in Precollege Science and Mathematics, pp. 114–137 (1999)
27. Rothlauf, F.: Design of Modern Heuristics: Principles and Application, 1st edn. Springer Publishing Company, Incorporated (2011)

Vehicular Clouds: Ubiquitous Computing on Wheels

Sherin Abdelhamid, Robert Benkoczi and Hossam S. Hassanein

Abstract Vehicular clouds have recently coalesced from two popular technologies, vehicular networks and cloud computing systems. In this chapter, we overview the vehicular cloud computing paradigm, discussing its unique features in relation with the conventional cloud computing models and highlighting several application domains where vehicular clouds can be useful. We summarize the main design elements of vehicular clouds and we discuss a broad range of research problems that are crucial for the performance, energy efficiency, and privacy of vehicular clouds. Our review is a gentle introduction for researchers and practitioners interested in learning about this promising new technology.

1 Introduction

Smart vehicles and the vehicular networking paradigm have received great attention due to the wide scope of benefits that can be unleashed through utilizing vehicles beyond just transportation. The ubiquity of vehicles along with a wide array of diversified in-vehicle sensing, computing, and communication resources have positioned smart vehicles in the heart of ubiquitous service provisioning. In addition to being key enablers for intelligent transport systems, smart vehicles have been engaged in a wider scope of services including safety, infotainment, and sensing services [2].

With the increased capability of the cloud computing paradigm, along with the aspirations of utilizing the powerful computing resources available in smart vehicles,

S. Abdelhamid (✉)
Faculty of Computer Science and Information, Ain Shams University, Cairo, Egypt
e-mail: shereen@cis.asu.edu.eg

R. Benkoczi
Department of Mathematics and Computer Science, University of Lethbridge,
Lethbridge, AB, Canada
e-mail: robert.benkoczi@uleth.ca

H.S. Hassanein
School of Computing, Queen's University, Kingston, ON, Canada
e-mail: hossam@cs.queensu.ca

© Springer International Publishing Switzerland 2017
A. Adamatzky (ed.), *Emergent Computation*, Emergence, Complexity
and Computation 24, DOI 10.1007/978-3-319-46376-6_20

a new paradigm has emerged from integrating both vehicular networking and cloud computing. This emerging paradigm is known as the Vehicular Cloud (VC). Due to the promising capabilities of VCs, their architectures, features, and applications have been hot topics of research and investigation in the past few years.

In this chapter, we present an overview of the VC paradigm delineating its unique features compared to the conventional cloud computing paradigm. We also discuss different architectures of VCs along with diversified application and service scopes. In addition, we shed light on some fundamental considerations that should be taken into account by the designers and researchers of VCs.

The chapter is organized as follows. In Sect. 2, we pave the way towards discussing the VC paradigm through an overview of the vehicular and cloud computing paradigms. In Sect. 3, we expound the VC paradigm and discuss its architectures and service scopes, as well as some potential applications. We touch upon some fundamental design considerations and approaches in Sect. 4. Finally, we summarize the discussion in Sect. 5.

2 Background

Vehicular clouds have emerged from the consolidation of the vehicular and cloud computing paradigms. In this section, we present an overview of the vehicular paradigm and smart vehicles. We also shed light on the cloud computing paradigm.

2.1 The Vehicular Paradigm and Smart Vehicles

Motivated by the urgent need to reduce on-road fatalities, and improve the driving experience, the Vehicular Ad-hoc Network (VANET) paradigm has emerged to connect vehicles on roads [3, 12, 15, 24, 40]. In a VANET, vehicles communicate with one another to exchange safety and navigation messages, and share traffic status, road conditions, and information regarding events on the road. A VANET can also involve some Road Side Units (RSUs) that are stationary nodes deployed to assist vehicles on roads. RSUs can provide vehicles with real-time traffic information, data of interest (e.g., digital maps), and drive-through Internet access.

A VANET is a category of the wireless multi-hop networks that depend on multi-hop communication over intermediate nodes working as relays to connect a source node to a destination node. A number of communication standards have been proposed for the use in VANETs [7, 30, 39]. The most dominant is the Wireless Access for Vehicular Environment (WAVE) standard that makes use of Dedicated Short-Range Communication (DSRC) [36]. The WAVE standard consists of two sub-standards: the IEEE 802.11p standard [14] for managing the lower layers and the IEEE 1609.x family of standards that manage the upper layers. Three types of communication can be found in a VANET:

1. Communication between vehicles, known as vehicle-to-vehicle (V2V) communication
2. Communication between vehicles and infrastructure (e.g., RSUs), known as vehicle-to-infrastructure (V2I) and infrastructure-to-vehicle (I2V) communication
3. Communication between a vehicle and any neighbouring object, known as vehicle-to-any (V2X) communication

An example of a VANET with its basic entities engaged in the three aforementioned types of communication is shown in Fig. 1.

Connected vehicles in a VANET are known as "smart vehicles". A smart vehicle is equipped with components that add computing capabilities. A typical smart vehicle has an abundant number of sensors of different types that monitor the interior and exterior surroundings of the vehicle for diagnostic purposes and for detecting hazards/events on roads [1]. In addition, a smart vehicle has wireless communication capabilities that support communication with its neighboring vehicles and RSUs. Some vehicles have broadband connectivity on-board to support communication with remote entities. A main component of a smart vehicle is the on-board unit (OBU), which is known as the in-vehicle PC. An OBU works as the interface that provides the driver with information/alerts about events that are either detected by the in-vehicle sensors or received through the communication module. It is also the means of receiving input from the driver, when needed. OBUs are now get-

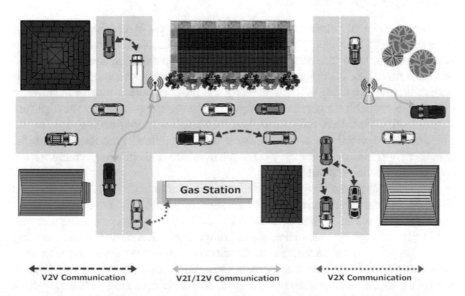

Fig. 1 Smart vehicles and RSUs forming a VANET and involved in V2V, V2I, and I2V communications. A vehicle is communicating with a gas station showing an example of V2X type of communication

ting as powerful as personal computers in terms of their processing and storage capabilities [35].

With these components available on-board, a smart vehicle can be considered a mobile resource for diverse service scopes including sensing, storage, computing, relaying, infotainment, and localization [2]. Such an abundance of resources along with the ubiquity of vehicles bring smart vehicles to the forefront in service provisioning, compared to other mobile resource providers, such as smart-phones, that suffer from resource scarcity and unpredictable mobility. Resources of moving and/or parked vehicles can be augmented to form a powerful resource and service provider that can benefit a wide scope of users.

2.2 Cloud Computing

Cloud computing emerged as a disruptive technology providing a cost effective alternative to the traditional, in-house, information technology solutions that players in the private and public sector depend on [37]. According to Mell and Grance from the National Institute of Standards and Technology [22],

> Cloud computing is a model for enabling ubiquitous, convenient, on-demand network access to a shared pool of configurable computing resources (e.g., networks, servers, storage, applications, and services) that can be rapidly provisioned and released with minimal management effort or service provider interaction.

Three classes of service are usually offered by cloud computing providers: infrastructure as a service (IaaS), platform as a service (PaaS) and software as a service (SaaS). With IaaS, customers have access to computing, storage, network, and other fundamental computing resources on which they can deploy arbitrary software, including operating systems, and applications. PaaS offers a range of software development tools to support the development of custom applications to be deployed in the cloud. Clients have full control over the applications they develop but they do not control the underlying cloud infrastructure. SaaS allows consumers to use the provider's applications, commonly via a web interface, and offer very limited control over the application capabilities.

The vision of recognizing computing as a utility after water, gas, electricity, and telephony, is not new. During the inauguration of ARPANET in 1969, Kleinrock contemplated the rise of "computer utilities" enabled by advances in computer networks [16]. Forty-five years later, cloud computing comes very close to realizing this vision [6]. The key advantages of cloud computing—low cost of entry for compute intensive business analytics, increased scalability, capability to support new and innovative applications that are not supported in a traditional IT environment—have pushed an increasing number of businesses to adopt "the cloud" [21] and have consolidated the position of several established providers like Microsoft, IBM, Google, Cisco, or AT&T. However, with virtualization, customers have little control on the physical devices that execute their applications and store their data, and thus concerns regarding privacy and compliance with local laws have been raised. In response to these concerns, the US Federal Government has created the *FedRAMP* program to evaluate and monitor the security and suitability of cloud solutions for use by government agencies, and a number of cloud providers have obtained certification [10]. In Europe, the European Union Agency for Network and Information Security (ANISA) published a guide assisting the member nations with implementing cloud systems for government agencies [9], while in Canada, Public Works and Government Services has issued a request for information to determine the appropriate cloud adoption strategy for Canadian government agencies [32].

Beyond government agencies, consumers of cloud services are faced with similar concerns. For some applications, private clouds may address issues of jurisdiction, and in this respect, vehicular clouds which are localized services by nature, may provide a cost effective alternative to the computing services offered by dedicated providers. Certainly, privacy and security are important considerations for vehicular clouds, and we briefly discuss these in Sect. 4.5. In the following paragraphs, we provide a broad review of vehicular clouds, highlighting the current state of research and pointing out some of the design considerations.

3 Vehicular Clouds—Definition, Architecture, and Applications

Motivated by the abundant storage, processing, and communication resources of smart vehicles, a futuristic vision of "taking vehicles to the cloud" has been attracting many researchers in the past few years resulting in the emerging paradigm of "vehicular clouds". In this section, we shed light on the definition, architectures, and potential applications of this paradigm.

3.1 What Is a Vehicular Cloud?

With the plethora of vehicular resources, researchers have predicted that such resources would be underutilized if their use was limited to intelligent transport system (ITS) applications [29, 38]. That was the main motivation for coming up with the vehicular cloud paradigm to allocate physical resources of smart vehicles to users for utilizing them in computing tasks. Researchers argued that the benefit of accessing such vehicular resources would be maximized when resources of multiple vehicles are combined. In essence, we sum up our definition of a vehicular cloud as:

A vehicular cloud is a pool of vehicular computing resources that can be coordinated dynamically and on-demand for performing a computing task.

The VC paradigm shares similarities with the conventional fixed clouds. Generally speaking, a VC follows the general concept of renting out computing resources that are not in use to authorized users on-demand based on a rental model. A VC brings unique advantages compared to fixed clouds. Due to the mobility of vehicles, their computing resources can be utilized in areas with restricted access to the Internet, and consequently to fixed clouds. Another advantage is the possibility of autonomous formation of VCs. Neighbouring vehicles can automatically collaborate to form a VC to handle instantaneous services (e.g., collecting traffic information for managing congested areas). More details about autonomous VCs are presented later in this section. One more advantage is the ability to survive and operate in emergency and natural disasters where infrastructure may be broken down and conventional clouds cannot be reached. A detailed comparative study of vehicular and conventional clouds is presented in [38].

3.2 Vehicular Cloud Architectures

We distinguish between two types of VCs: centralized and autonomous. The building architecture of the VC is the main distinction between these two types. Below, we discuss these two architectures and delineate the differences between them.

3.2.1 Centralized Vehicular Clouds

A typical centralized VC architecture can follow the general architecture of the open-source Eucalyptus Cloud Computing system [28]. The architecture consists of a central cloud controller interacting with node controllers. Based on the size of the VC, the architecture may also include some cluster controllers. The functionalities of such entities are discussed next.

Cloud Controller
The cloud controller is the central entity that manages the whole operation of the VC. It communicates with the participating vehicles to discover and manage their resources, assign computing tasks to them, and handle any necessary exchange of control/data messages. To perform such functionalities, the cloud controller includes three underlying components: a broker, a resource manager, and a task scheduler.

A centralized VC follows a client/server model where clients can reach the cloud controller with access requests to computing resources managed by this controller.

The **broker** component of the controller is the entity responsible for receiving client requests and negotiating the terms of service/access on behalf of the VC.

The **resource manager** is a crucial entity of the cloud controller managing many functionalities. The resource manager:

- discovers and monitors the availability of resources,
- predicts the availability of resources based on a use history and presence patterns,
- decides on the sufficiency of the available resources to match a computing task requirements,
- allocates resources to computing tasks in cooperation with the task scheduler,
- migrates running tasks from vehicles leaving the VC to other available vehicles,
- ensures that recruitment requirements and fairness of use are met.

The **task scheduler** works in cooperation with the resource manager to create an access schedule to the resources allocated to perform released computing tasks. The schedule is built based on the availability span of the resources, the task temporal span, and the dependency among the tasks/sub-tasks.

Node Controller
Each vehicle interested in participating in a VC sets up a node controller on-board. The node controller works as the interface between the cloud controller and the on-board resources of the vehicle managing all the interactions between them. It can be considered as a local resource controller that reports the resource availability and controls the access to the resources.

Cluster Controller
When a VC is large in size (defined by the number of participating vehicles), the vehicles can be divided into clusters with each cluster being managed by a cluster controller. In such a case, each cluster controller would work between the central cloud controller and the node controllers of the vehicles in its cluster. The purpose of such fragmentation is to reduce the management load of the cloud controller through offloading a part of its functionalities to the cluster controllers (e.g., having each cluster controller monitor and manage the vehicular resources of its cluster).

The centralized VC architecture is depicted in Fig. 2a through a VC formed at a parking garage.

3.2.2 Autonomous Vehicular Clouds

In some scenarios, a VC might be only needed for a spontaneous computing service associated with unplanned event. For example, a VC can be formed among vehicles in a traffic jam to alert and re-route vehicles thereby mitigating the congestion. In such a case, an autonomous VC can be temporarily formed in a self-organizing fashion without the aid/control of a central entity. Vehicles in the vicinity of the event cooperate to pool their computing resources and utilize them for handling the required task.

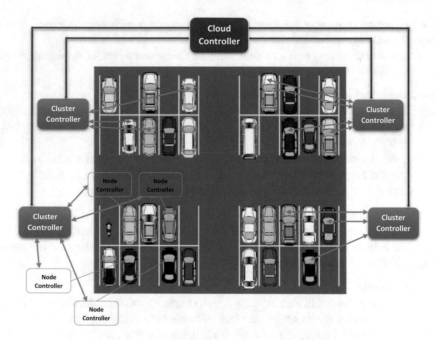

(a) The typical architecture of a centralized VC.

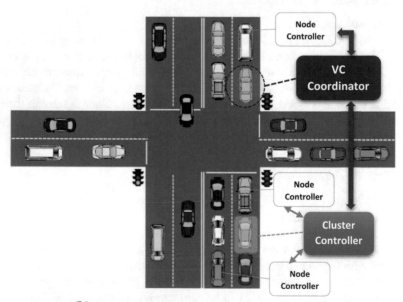

(b) The typical architecture of an autonomous VC.

Fig. 2 The different architectures of vehicular clouds

Due to their short life-span compared to the relatively long-lasting centralized VCs, the autonomous VCs do not involve a central cloud controller, moving the control to the node controllers. Since the computing requirements of autonomous VCs are also relatively low compared to the well-planned centralized VCs, many of the functionalities of the cloud controller are not needed for the operation of the autonomous VCs. The client-server model adopted in the centralized VCs does not conform to the nature of the ad-hoc autonomous VCs therefore, the broker functionality is not needed. Limited resource management and task scheduling functionalities however might be needed in such VCs. To handle these situations, a vehicle can be elected by the vehicles forming the VC to serve as a VC coordinator and manage such functionalities.

Similar to their use in the centralized VCs, for easier management of large VCs, vehicles can be grouped into clusters, each is managed by a cluster controller. The cluster controllers take care of managing the vehicles in their clusters through communicating with the corresponding node controllers. In autonomous VCs, cluster controllers link between the node controllers and the elected VC coordinator, when needed.

The typical architecture of an autonomous VC is depicted in Fig. 2b.

3.3 Services and Applications

3.3.1 Scopes of Services

The types of services that can be provided by a VC can be categorized into four different scopes: Processing as a Service (PRaaS), Storage as a Service (STaaS), Network as a Service (NaaS), and Information as a Service (INaaS).

Processing as a Service (PRaaS)
With processing capabilities as high as the capabilities of personal computers, smart vehicles can be utilized for handling processing tasks. Their idle processing resources can be rented out to authorized users for either accessing software/applications already available on the on-board OBUs, or for running their own virtual machines after migrating them to the OBUs. Resources of multiple vehicles can be combined to handle tasks with high processing requirements that exceed a single vehicle's capabilities. In such a case, efficient task scheduling mechanisms would be needed to manage task partitioning, offloading, and migration.

Storage as a Service (STaaS) Smart vehicles are anticipated to have Terabytes of storage. Those vehicles that would have unutilized storage capabilities can share their resources with others in need of storage. For example, other vehicles with limited on-board storage can rent resources from neighbouring vehicles to store their generated/downloaded data for later retrieval. In addition, aggregated vehicular storage resources can be utilized as a dynamic data center to be rented out to third parties with interest as is the case with the storage services offered by conventional clouds. As an advantage over conventional cloud storage, mobility of vehicles can support their use as data mules carrying data between a pair of nodes in need of data delivery but do not have direct communication capabilities between each other.

Network as a Service (NaaS)

Although some vehicles would have the capability of connecting to the Internet while moving, some would not have such a capability; especially in the early stages of deployment and in areas with low numbers of connected vehicles. The networking capabilities of such connected vehicles can be offered to unconnected vehicles in a 'Drive-by Internet Access' model working as mobile hot spots. Such a service involves advertisements by the drivers of connected vehicles who are interested in sharing their network resources.

Information as a Service (INaaS)

Smart vehicles are considered major resources of information that can be generated by them utilizing their in-vehicle sensors, obtained from other vehicles/RSUs on roads, or downloaded from the Internet. Examples include information about road and traffic conditions, events on roads, news of interest, and store/restaurant offers. Such information can be offered to other users who need it.

3.3.2 Potential Applications

Under the different service scopes discussed above, many potential applications can be proposed utilizing the vehicular cloud paradigm. Examples of these applications are highlighted next.

Computing Engines at a Company Parking Lot

During a workday, the vehicles of a company's employees are parked idle with ample computing resources unutilized. These untapped resources can be utilized as computing engines that can carry out computing tasks offloaded to them from the company's IT department in lieu of renting/outsourcing a computing infrastructure. The owners of the utilized vehicles can be compensated for the use of their resources so both the company and employees benefit. Such an application can be an example of PRaaS and/or STaaS VC services.

Data Center at an Airport

A similar scenario to the previous one can be applied to vehicles left by travelers at an airport. These vehicles are left idle for days resulting in massive unutilized vehicular resources. With proper management and scheduling, such resources can turn an airport parking lot into a data center with a capability to rent out its collective resources, as an instance of the STaaS service scope. The airport can build an access schedule to the vehicles based on the travellers' plans that they share with the airport. Facilitating access to such parked vehicles requires plugging them into a power supply and providing them with an Ethernet connection. The same use setup applies to all applications utilizing resources of parked vehicles.

Dynamic Traffic Management

An autonomous VC of vehicles at an intersection can be formed to manage the traffic dynamically at this intersection and alleviate congestion. Computational resources of vehicles at an intersection can be consolidated to come up with a dynamic schedule to adjust the traffic lights according to the current status of the intersecting roads. The vehicles can elect a vehicle to work as the VC coordinator. This coordinator can send the computed schedule to the authority that manages the traffic light controller to put it in action.

Autonomous Congestion Alleviation

Vehicles in the vicinity of a congested road can form a VC to handle the situation and alleviate the congestion. Vehicles on different neighbouring road segments can share real-time traffic information with one another and cooperate to compute detouring routes to help alleviate the congestion. A coordinator would be elected to manage the VC and dissemination/announcement of the computed routes. This application and the traffic management application described earlier are examples of PRaaS, STaaS, and INaaS services.

Vehicular Public Sensing

As highlighted in Sect. 2.1, vehicles are equipped with an abundance of sensors that enable them to work as mobile sensors. A vehicle's sensing and computational resources can be utilized by a service provider interested in collecting road information for provisioning an information service. Collective vehicular resources of multiple vehicles can be pooled to provide correlated sensing information of an event with high levels of information fidelity and integrity.

Mobile Laboratory

A vehicle's computing resources coupled with its sensing capabilities can enable a vehicle to be a mobile laboratory. Phenomena of interest (e.g., ambient pressure, temperature, and pollution) can be monitored by a vehicle, and its on-board computing resources can be utilized to handle experimental analyses of such data while the vehicle is on the go covering broad areas of interest. Such an autonomous application can enable scientific experiments in areas with limited/restricted access to remote facilities. This would be a definitive example of a benefit of a VC versus a conventional cloud.

Mobile Marketplace

Vehicles' owners can utilize the resources of the vehicles for mobile business purposes. For example, owners interested in selling/trading products while on the go can make use of a vehicle's communication capabilities to advertise such products and receive purchase/trading orders. The computational resources of the vehicle can be exploited to carry out the corresponding transactions and to maintain a database of the products, orders, and their details.

4   Vehicular Clouds—Fundamental Considerations

Some design considerations need to be taken into account when designing/deploying VCs. In this section, we highlight these considerations and discuss different approaches to handle them.

4.1   Resource Discovery

Resource discovery is the first step when it comes to building a VC. Appropriate resources that match the task requirements should be searched for and passed to the resource manager to start the task allocation process. Many factors make resource discovery in VCs a challenge, including the huge number of potential resources, intermittent resource availability, distributed ownership, and diversified communication interfaces to reach vehicular resources.

Fortunately, resource discovery is a well-studied topic with many mechanisms available in the literature for use by other related paradigms such as grid computing [11, 27] and peer-to-peer networks [23, 26]. Such mechanisms can be adapted for use in VCs in a means that handles the unique features of the VC paradigm and the factors mentioned so far. The resource discovery techniques can be classified into three categories: *centralized*, *decentralized*, and *hierarchical*.

Centralized Approach
This approach follows a client-server architecture where a designated controller/set of controllers take care of discovering the resources. The controller(s) stores information about all the candidate resources. Such information includes the resource features and availability span. When an entity is interested in accessing/renting resources, it sends a request to the central controller defining the access/resource requirements. From its registered resource list, the controller then finds a set of appropriate resources matching the requirements defined in the received request. Although this approach is the fastest in terms of the search time, it suffers from a scalability concern since the controller has to keep information about all the candidate resources.

Decentralized Approach
In this approach, the need for a central controller is avoided to solve the scalability concern faced in the centralized approach. Nodes carrying the resources cooperate together to handle the discovery process in a distributed fashion. Discovery requests are shared among the nodes until a resource is found or the requests expire. Although this approach is more suitable for large-scale systems, it suffers from overhead resulting from the wide exchange of the discovery requests among the nodes.

Hierarchical Approach
In this approach, nodes are divided into a hierarchical structure assembling nodes in the same layer into clusters. Each cluster is assigned a controller that manages the communication and discovery of the resources of its cluster members. By dividing

the discovery burden on multiple controllers, the scalability issue of the centralized approach is mitigated. In addition, by limiting the communication/query exchange handled by each node only within its cluster, the request dissemination overhead of the decentralized approach is reduced. Although this approach succeeds in handling the concerns of the two aforementioned approaches, it risks not reaching some nodes in case their cluster controller fails, moves, or gets overloaded. For this reason, dynamic clustering should be adopted where clusters are periodically rearranged.

We remark that the hierarchical approach is the most suitable for the centralized VC architecture. Each cluster controller would handle resource discovery of the node controllers of its cluster and work as a gateway between them and the resource manager in the cloud controller. For the autonomous VC architecture, the decentralized approach is the candidate for resource discovery due to the distributed nature of this architecture. The dissemination overhead of the decentralized approach would not be a severe issue in autonomous VCs as the dissemination of discovery requests would be usually limited within a region of interest.

4.2 Task Allocation and Scheduling

Once the resources available in the VC are identified, the system must assign or schedule the customers' requests to the appropriate cloud resources. We discuss in this section the research problems relevant to scheduling two of the services mentioned in Sect. 3.3, processing and storage.

There is a large body of literature on scheduling processing tasks in domains such as machine scheduling [18], distributed, and grid computing [8]. The scheduling of processes in grid systems is closest, from the point of view of objectives and constraints, to scheduling in vehicular clouds, however grid systems serve applications that are primarily computational intensive and exhibit a high degree of parallelism. Vehicular clouds, on the other hand, may serve a large number of customers by supporting information rich applications rather than scientific computation.

One task allocation model for vehicular clouds consists of a set of jobs, each composed of a set of tasks. Tasks may be independent of each other, or may exhibit precedence constraints. Optionally, tasks may request storage service for data persistence. Memory required for running applications is not viewed as a separate STaaS service but is part of PRaaS. As in the case of grid systems, vehicular clouds do not have control on the availability of the computational and storage resources. In vehicular clouds, however, the system must be able to respond in real time to events triggered by resources becoming unavailable as vehicles move away from the area that defines the VC.

Two objectives can be considered when scheduling tasks in a vehicular system, minimizing makespan and minimizing cost. The makespan for a processing job represents the total length of time that a job takes from the moment one of its tasks begins execution until its last task completes execution. The cost objective is related to the incentive mechanisms mentioned in Sect. 4.4. Both objectives may also be considered, for example, by imposing a deadline for the completion of a job and minimizing the cost. Sets of pareto-optimal solutions can be generated by imposing several different constraints for the deadline.

Depending on the application supported by the vehicular cloud, makespan might not be an appropriate measure for scheduling jobs. Consider an application where customers rent access to the cloud for a given period of time during which they interact with various stored databases (see STaaS applications from Sect. 3.3.2). In this case the objective of the scheduler is to minimize inter-vehicle communication cost which amounts to solving a *facility location* type of problem: assign storage requests and databases to vehicles and assign customer jobs to vehicles in such a way that the amount of data transferred to and from local resources is maximized over all jobs. An objective of this type has been considered in the data placement problem [5].

These scheduling problems are NP-hard in general, and so researchers have focused on approximations, exact heuristics, and meta-heuristic algorithms without performance guarantees [5, 8, 18]. For autonomous vehicular clouds, designing distributed or localized algorithms is of great interest. If the availability of the resources is not known ahead of time or if it cannot be estimated, we are faced with solving on-line versions of these scheduling problems. In such a setting, the resource availability is unknown ahead of time and the scheduling needs to react to changes in resources. Another possibility is that resource availability is uncertain and it is known only to lie within some interval of values. In such cases, we need to solve robust optimization problems, where the objective is to compute a solution that minimizes regret. Regret is defined by the maximum absolute difference between the cost of the solution under a scenario where the uncertainties are within the prescribed limits and the cost of the optimal solution had the scenario been known from the beginning.

4.3 Task Migration and Offloading

As inferred from the previous section, once a resource is about to become unavailable, the vehicular cloud has little time to migrate the tasks assigned to it. The overhead of migrating pure processing tasks may be significantly reduced if the task schedule and resource availability is known in advance. In this case, the system could transfer the machine code of the tasks on the resources scheduled to run the tasks the moment these resources become available which allows to transfer only the process state when migration is performed. Similarly, several instances of STaaS objects can be spawned on the resources in anticipation of migration so that during migration, the data objects are only synchronized rather than fully transferred. Although task migration and offloading can be performed efficiently in this way, task scheduling may seek to also minimize migration operations.

4.4 Incentive Mechanisms

In vehicular clouds, incentives are needed to encourage vehicle owners to offer their resources to other parties. Some of the incentive mechanisms proposed for other paradigms, such as P2P file sharing [4, 33], can be borrowed by VCs. Generally, incentives can be classified into three types: (1) getting nothing in return, (2) getting service in return, and (3) getting monetary rewards in return. Among the three types, mechanisms using monetary rewards appear to be the most effective.

The use of monetary incentives requires adopting a pricing model for calculating the monetary reward paid to each participant. Two distinct pricing approaches can be utilized: *identical pricing* and *dynamic pricing*.

Identical Pricing Approach

In this approach, all participants get the same monetary reward per a defined rental period regardless of the difference in the capabilities of their vehicular resources. Although this approach is easy to implement, it lacks fairness with respect to differentiating various levels of resource quality.

Dynamic Pricing Approach

To avert the perceived unfairness of the identical pricing approach, pricing models in the dynamic pricing approach calculate a participant's incentive in proportion to the features and quality of the rented resources and their access period. In that way, vehicles with high-end resources get high rewards compared to those with less powerful resources given that they are utilized for the same period.

Dynamic pricing models can be classified into two different types based on the entity that holds the pricing model and administers the incentive computation. These two types are the *controller-based* and *owner-based* models. In the controller-based models, a dedicated controller manages the pricing model and computes the incentives of all participants. Such a controller can be the main cloud controller in centralized VCs, and the VC coordinator in autonomous VCs. In the owner-based models, the vehicle/resource owners set their own rewarding value and announce it to the interested parties before tapping into their resources. Reverse auction techniques are currently popular with the owner-based pricing models [17, 19].

Irrespective of the pricing model used, the incentive can be computed as a number of tokens that can be translated to any form of monetary value such as cash, vouchers, and passes. A unique incentive for parked vehicles can offer free parking while the vehicle's resources are utilized.

4.5 Privacy

The open access feature of VCs brings concerns regarding privacy to the forefront. Since vehicle resources would be shared with others, privacy mechanisms are needed to ensure data protection for all users renting the same resources. Furthermore, privacy should be guaranteed for vehicle owners themselves so that they are not deterred from joining such an open-access paradigm.

Privacy in VCs can be supported using multiple techniques [13, 34]. Virtualization and scheduling techniques can be used to coordinate the multi-access nature of VCs. Other techniques can be used to grant exclusive access to the data's owner. An example of such techniques is the use of data vaults [25] as individually controlled data repositories.

Some participants would prefer to hide their identity while offering their vehicular resources. Pseudonymity [31] is one of the popular anonymity techniques that can be deployed to ensure that the real identity of participants is not exposed nor tracked.

4.6 Powering and Connecting Parked Vehicles

In the model of VCs utilizing parked vehicles, such vehicles would be connected to a power supply and a data port offering Internet access. Despite being power-supplied, parked vehicles cannot have their on-board computers and resources on all the time waiting for task assignments. Techniques are needed to power up partici-pating vehicles only when required. We categorize such techniques into *on-demand* and *pre-scheduled* techniques.

On-Demand Techniques
With this approach, vehicle PCs are powered up when a task is assigned to their hosting vehicles, then turned off after the completion of the task. An example of such techniques is proposed in [20]. This technique utilizes the Controller Area Network (CAN) communication bus of the vehicle connecting the in-vehicle PC and control units for powering up the needed resources. The CAN-connected units can operate on a sleep mode to consume minimal energy while a vehicle is not utilized. Once a parked vehicle is assigned a task, its CAN-connected computing resources can be remotely powered up.

Pre-Scheduled Techniques
These techniques require the use of scheduling mechanisms to schedule the power-up times of vehicles a priori. Schedules would be sent to vehicles whenever they are on. Mobility prediction mechanisms can be utilized to anticipate a vehicle's parking times and plan the power-up times accordingly.

5   Summary

Vehicular networks and cloud computing have received a lot of attention over the past decade. Motivated by the considerable benefits and wide scope of applications of these two paradigms, a new computing paradigm, the Vehicular Cloud (VC), emerged. This chapter presents an overview of the VC paradigm to familiarize interested researchers and developers. We delineate the unique features of VCs and compare them to the conventional cloud computing systems. Architecture, service scope, and potential applications of VCs are discussed. In addition, we highlight several fundamental considerations in the design of VCs. We anticipate that VCs will revolutionize service provisioning and pervasive computing.

References

1. Abdelhamid, S., Hassanein, H.S., Takahara, G.: Vehicle as a mobile sensor. In: International Conference on Future Networks and Communications (FNC '14), Elsevier Procedia Computer Science, vol. 34, pp. 286–295 (2014)
2. Abdelhamid, S., Hassanein, H.S., Takahara, G.: Vehicle as a resource (VaaR). IEEE Netw. Mag. **29**(1), 12–17 (2015)
3. Al-Sultan, S., Al-Doori, M.M., Al-Bayatti, A.H., Zedan, H.: A comprehensive survey on vehicular ad hoc network. J. Netw. Comput. Appl. **37**, 380–392 (2014)
4. Antoniadis, P., Courcoubetis, C., Mason, R.: Comparing economic incentives in peer-to-peer networks. Comput. Netw. **46**(1), 133–146 (2004)
5. Baev, I., Rajaraman, R., Swamy, C.: Approximation algorithms for data placement problems. SIAM J. Comput. **38**(4), 1411–1429 (2008)
6. Buyya, R., Yeo, C.S., Venugopal, S., Broberg, J., Brandic, I.: Cloud computing and emerging IT platforms: vision, hype, and reality for delivering computing as the 5th utility. Future Gener. Comput. Syst. **25**(6), 599–616 (2009)
7. Dar, K., Bakhouya, M., Gaber, J., Wack, M., Lorenz, P.: Wireless communication technologies for ITS applications [Topics in Automotive Networking]. IEEE Commun. Mag. **48**(5), 156–162 (2010)
8. Dong, F., Akl, S.G.: Scheduling algorithms for grid computing: state of the art and open problems. Techniacl report, Queen's University (2006)
9. European Union Agency for Network and Information Security: Security framework for governmental clouds (2015)
10. FedRAMP compliant systems. http://www.fedramp.gov/marketplace/compliant-systems/. Accessed 3 April 2016
11. Hameurlain, A., Cokuslu, D., Erciyes, K.: Resource discovery in grid systems: a survey. Int. J. Metadata Seman. Ontol. **5**(3), 251–263 (2010)
12. Hartenstein, H., Laberteaux, K.P.: A tutorial survey on vehicular ad hoc networks. IEEE Commun. Mag. **46**(6), 164–171 (2008)
13. Heurix, J., Zimmermann, P., Neubauer, T., Fenz, S.: A taxonomy for privacy enhancing technologies. Comput. Secur. **53**, 1–17 (2015)
14. Jiang, D., Delgrossi, L.: IEEE 802.11p: towards an international standard for wireless access in vehicular environments. In: IEEE Vehicular Technology Conference (VTC-Spring '08), pp. 2036–2040 (2008)
15. Karagiannis, G., Altintas, O., Ekici, E., Heijenk, G., Jarupan, B., Lin, K., Weil, T.: Vehicular networking: a survey and tutorial on requirements, architectures, challenges, standards and solutions. IEEE Commun. Surv. Tut. **13**(4), 584–616 (2011)
16. Kleinrock, L.: An Internet vision: the invisible global infrastructure. Ad Hoc Netw. **1**(1), 3–11 (2003)
17. Krontiris, I., Albers, A.: Monetary incentives in participatory sensing using multi-attributive auctions. Int. J. Parallel Emergent Distrib. Syst. **27**(4), 317–336 (2012)
18. Lawler, E.L., Lenstra, J.K., Kan, A.H.R., Shmoys, D.B.: Sequencing and scheduling: algorithms and complexity. Handbooks Oper. Res. Manage. Sci. **4**, 445–522 (1993)

19. Luo, T., Tham, C.K.: Fairness and social welfare in incentivizing participatory sensing. In: 9th IEEE Communications Society Conference on Sensor, Mesh and Ad Hoc Communications and Networks (SECON '12), pp. 425–433 (2012)
20. Mahfoud, M., Al-Holou, N., Baroody, R.: Next generation vehicle network: web enabled. In: 3rd IEEE International Conference on Information and Communication Technologies: From Theory to Applications (ICTTA '08), pp. 1–7 (2008)
21. Marston, S., Li, Z., Bandyopadhyay, S., Zhang, J., Ghalsasi, A.: Cloud computing the business perspective. Decis. Support Syst. **51**(1), 176–189 (2011). doi:10.1016/j.dss.2010.12.006. http://www.sciencedirect.com/science/article/pii/S0167923610002393
22. Mell, P., Grance, T.: The NIST definition of cloud computing. Technical report, Computer Security Division, Information Technology Laboratory, National Institute of Standards and Technology Gaithersburg (2011)
23. Meshkova, E., Riihijärvi, J., Petrova, M., Mähönen, P.: A survey on resource discovery mechanisms, peer-to-peer and service discovery frameworks. Comput. Netw. **52**(11), 2097–2128 (2008)
24. Moustafa, H., Zhang, Y.: Vehicular Networks: Techniques, Standards, and Applications. Auerbach Publications (2009)
25. Mun, M., Hao, S., Mishra, N., Shilton, K., Burke, J., Estrin, D., Hansen, M., Govindan, R.: Personal data vaults: a locus of control for personal data streams. In: 6th ACM International Conference on Emerging Networking Experiments and Technologies (CoNEXT '10), p. 17 (2010)
26. Navimipour, N.J., Milani, F.S.: A comprehensive study of the resource discovery techniques in peer-to-peer networks. Peer-to-Peer Netw. Appl. **8**(3), 474–492 (2014)
27. Navimipour, N.J., Rahmani, A.M., Navin, A.H., Hosseinzadeh, M.: Resource discovery mechanisms in grid systems: a survey. J. Netw. Comput. Appl. **41**, 389–410 (2014)
28. Nurmi, D., Wolski, R., Grzegorczyk, C., Obertelli, G., Soman, S., Youseff, L., Zagorodnov, D.: The Eucalyptus open-source cloud-computing system. In: 9th IEEE/ACM International Symposium on Cluster Computing and the Grid (CCGRID '09), pp. 124–131 (2009)
29. Olariu, S., Eltoweissy, M., Younis, M.: Towards autonomous vehicular clouds. ICST Trans. Mobile Commun. Appl. **11**(7–9), 1–11 (2011)
30. Papadimitratos, P., La Fortelle, A., Evenssen, K., Brignolo, R., Cosenza, S.: Vehicular communication systems: enabling technologies, applications, and future outlook on intelligent transportation. IEEE Commun. Mag. **47**(11), 84–95 (2009)
31. Petit, J., Schaub, F., Feiri, M., Kargl, F.: Pseudonym schemes in vehicular networks: a survey. IEEE Commun. Surv. Tut. **17**(1), 228–255 (2015)
32. Public Works and Government Services Canada: Request for information cloud computing solutions (2014). Document number EN578-151297/B
33. Salek, M., Shayandeh, S., Kempe, D.: You share, I share: network effects and economic incentives in P2P file-sharing systems. Lect. Notes Comput. Sci. **6484**, 354–365 (2010)
34. Serna-Olvera, J.M., Pacheco, R.A.M., Parra-Arnau, J., Rebollo-Monedero, D., Forné, J.: Privacy in vehicular ad hoc networks. In: Privacy in a Digital, Networked World, pp. 167–187. Springer (2015)
35. SINTRONES: In-vehicle computing. http://www.sintrones.com/products/invehiclecomputing.php (2015). Accessed 18 March 2016
36. Uzcategui, R., Acosta-Marum, G.: WAVE: a tutorial. IEEE Commun. Mag. **47**(5), 126–133 (2009)
37. Weiss, A.: Computing in the clouds. netWorker **11**(4), 16–25 (2007). doi:10.1145/1327512.1327513
38. Whaiduzzaman, M., Sookhak, M., Gani, A., Buyya, R.: A survey on vehicular cloud computing. J. Netw. Comput. Appl. **40**, 325–344 (2014)
39. Willke, T.L., Tientrakool, P., Maxemchuk, N.F.: A survey of inter-vehicle communication protocols and their applications. IEEE Commun. Surv. Tut. **11**(2), 3–20 (2009)
40. Zeadally, S., Hunt, R., Chen, Y.S., Irwin, A., Hassan, A.: Vehicular ad hoc networks (VANETs): status, results, and challenges. Telecommun. Syst. **50**(4), 217–241 (2012)

Computational Approaches to Epigenetic Drug Discovery

Emese E. Somogyvari, Selim G. Akl and Louise M. Winn

Abstract The misregulation of epigenetic mechanisms has been linked to disease. Current drugs that treat these dysfunctions have had some success, however many have variable potency, instability in vivo and lack target specificity. This may be due to the limited knowledge on epigenetic mechanisms, especially at the molecular level, which restricts the development and discovery of novel therapeutics and the optimization of existing drugs. Computational approaches, specifically in molecular modeling, have begun to address these issues by complementing phases of drug discovery and development. Here is presented a review of current computational efforts in drug discovery and development, with a focus on molecular modeling approaches including virtual screening, molecular dynamics, molecular docking, homology modeling and pharmacophore modeling.

1 Introduction

The term epigenetics describes the regulation of genomic functions leading to heritable changes in gene expression that are outside of the DNA sequence, and is thought to be the link between environmental factors and gene expression. Epigenetic modifications include DNA methylation, histone modifications (acetylation, methylation, phosphorylation) and ATP-dependent chromatin remodeling (Fig. 1). The misregulation of these components, regardless of the DNA sequence, has been shown to lead to an increase incidence in several diseases including Type II diabetes, cancer and Alzheimers [22]. The development of drugs treating disorders of the epigenome has

E.E. Somogyvari (✉) · S.G. Akl
School of Computing, Queen's University, Kingston, ON, Canada
e-mail: somogyva@cs.queensu.ca

S.G. Akl
e-mail: akl@queensu.ca

L.M. Winn
Department of Biomedical and Molecular Sciences, School of Environmental Studies
and School of Computing, Queen's University, Kingston, ON, Canada
e-mail: winnl@queensu.ca

© Springer International Publishing Switzerland 2017
A. Adamatzky (ed.), *Emergent Computation*, Emergence, Complexity
and Computation 24, DOI 10.1007/978-3-319-46376-6_21

been gaining interest in recent years due to the potential to reverse inappropriate epigenetic modifications [9]. Knowledge about the underlying mechanisms of the epigenome is growing, however the development of specific drugs has been difficult due to issues with instability in vivo and lack of specificity [8]. A potential solution may lie in computational approaches to complement phases of the drug discovery process, which may facilitate the development of epigenetic drugs.

1.1 Epigenetic Mechanisms

DNA can be compacted into chromatin by wrapping around an octamer of histone proteins into a nucleosome (see Fig. 1). The interaction of multiple nucleosomes is what makes up chromatin. Histone modifications, such as acetylation, which is the addition of an acetyl group, can change the conformational state of chromatin. When histones become acetylated, they lose their positive charge, which decreases their interaction with DNA. This leads to a more open, active state of chromatin, known as euchromatin, which allows DNA to be accessible to transcription factors leading to increased gene expression. Histone acetyltransferases (HATs) are among the many enzymes that catalyze the acetylation of histone proteins. Histone deacetylases (HDACs) on the other hand have the opposite effect and lead to heterochromatin, a closed, inactive chromatin structure in which DNA is inaccessible thus leading to decreased gene expression [26]. Methylation of DNA, which is the addition of methyl groups to nucleic acids, is carried out by DNA methyltransferases (DNMTs) and primarily occurs within cytosine/guanine (CpG) dinucleotides. CpG dinucleotides are located preferentially in the genome, and can be found in gene promoter regions as clusters known as CpG islands [8]. DNMTs promote the addition of a methyl group from a methyl donor such as S-Adenosyl methionine (SAM) to cytosine [18]. The methylation status of an organism's genome or DNA of a particular cell or tissue is known as its methylome. DNA methylation may reduce gene expression via two mechanisms. First, methylated DNA may physically impede the ability of a transcription factors to bind to genes. Second, methyl-CpG-binding domain proteins (MBDs) may bind to methylated DNA. MBD proteins then recruit other proteins, such as HDACs that act to change the conformation of chromatin into the transcriptionally inaccessible state, heterochromatin [3].

1.2 Epigenetic Drugs

Currently, there exists two primary classes of epigenetic drugs; DNA methylation inhibitors and histone deacetylases inhibitors. DNA methyltransferase inhibitors can work in two ways. First, the DNA methyltransferase inhibitors may be nucleoside analogues. When these nucleoside-like inhibitors are phosphorylated into nucleotides, they are incorporated into DNA. There they can prevent methylation

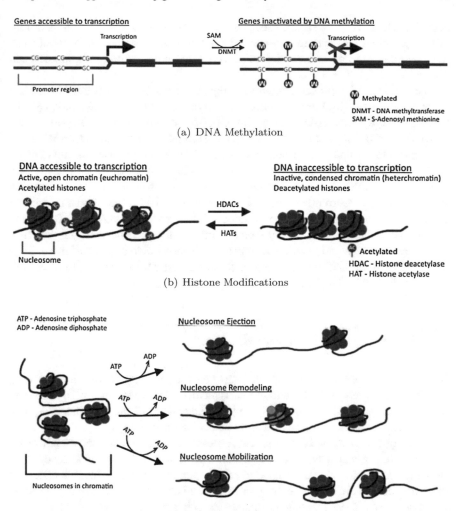

(a) DNA Methylation

(b) Histone Modifications

(c) ATP-dependent Chromatin Remodeling

Fig. 1 **a** Methylation primarily occurs in CpG rich promoter regions of DNA. Carried out by DNMTs and supplied by SAM, methylation alters accessibility of DNA to transcription [8, 18]. Modified from [20]. **b** Histone modifications, such as acetylation, alter the accessibility of DNA to transcription. Acetylation is mediated by HATs and HDACs [26]. Modified from [14] (**c**). The energy from ATP hydrolysis alters the accessibility of nucleosomal DNA through nucleosome ejection, restructuring, or mobilization [30]. Modified from [30]

by trapping any DNMTs that attempt to methylate them. Second, DNA methyltransferase inhibitors that are non-nucleoside analogues can inhibit methylation by reversibly binding to the active sites of DNMTs, preventing them from binding to DNA. HDAC inhibitors are much more numerous and diverse and can act through a variety of mechanisms to modify gene expression and other cellular processes.

Blocking the activity of HDACs leads to the acetylation of both histone and non-histone proteins. This may alter transcription either directly or indirectly, affect DNA replication and repair, and can influence cell differentiation and programmed cell death [28].

The majority of approved epigenetic drugs, or drugs that are in clinical trials are either HDAC or DNMT inhibitors [8]. These drugs include 5-azacytidine (AZA), 5-aza-2'deoxycytidine (decitabine [DAC]), Suberoylanilide hydroxamic acid (SAHA), valproic acid and entinostat [5]. Developing drugs that target DNMTs is of particular interest due to their known association with certain diseases and because of the complex and less understood effects of HDAC inhibitors. Additionally, targeting specific DNMT isoforms may also reduce the off-target effects that existing drugs suffer from. Currently, there is a need to develop DNMT inhibitors that do not incorporate into DNA. DNA-incorporating inhibitors such as AZA and DAC have been found to have a lack of DNA incorporation at high concentrations, are limited by cytotoxicity, and have variable potency. The variable potency of these drugs may be due to the ability of diseased cells, specifically tumour cells, to limit their incorporation into DNA [8].

1.3 Drug Discovery and Development

Drug discovery and development is a long and costly process. In the United States, bringing a new therapeutic drug to market typically takes an average of 10 years, and costs an average of $2.6 billion (Fig. 2) [21]. Key stages of the drug discovery and development process are briefly outlined below (see Fig. 3) [11].

The drug discovery and development process begins with research, often in academia, which results in a hypothesis regarding a protein or pathway and its association with a disease state that can be used to select a target. The target identification and validation phases are crucial in drug discovery and development since the inhibition or activation of the target should ultimately result in a therapeutic effect. Targets may

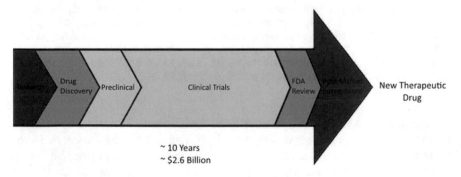

Fig. 2 Bringing a new drug to market in the United States. Modified from [21]

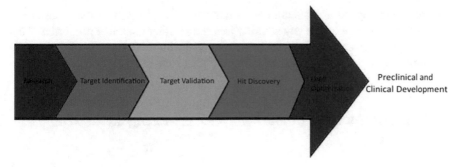

Fig. 3 Phases of drug discovery and development including research, target identification and validation, hit discovery and lead optimization. Modified from [11]

include proteins, genes, and RNA, and a good target is efficacious, safe and "drug-gable", meaning that a potential drug can bind to it and induce a biological response. Data mining has significantly contributed to the identification of drug targets. Data sources include publications and patent information, transgenic phenotyping, compound profiling data, and genome wide association data. Alternatively, phenotypic screening has been used, which involves identifying a target that is found to alter the phenotype of a cell or organism.

The selected target must then undergo a thorough validation process. This usually involves the use of in vitro and in vivo models such as cells and whole animals respectively. There are many tools that are used in target validation. For example, antisense technology, which involves the use of RNA-like molecules to prevent the synthesis of an encoded protein, is often used. It allows researchers to study the role of a target in a given disease by preventing its synthesis and observing the effects. Transgenic animals are also frequently used since they allow the observation of the phenotype of whole animals that have been genetically manipulated, which gives insight on the potential functions of the gene.

Next is the hit discovery process that involves developing assays to identify a hit molecule. A hit may be defined as a compound that has a desired and confirmed activity resulting from a compound screen. There are many screening techniques that exist, among which is high throughput screening (HTS). HTS is often an automated process that involves screening a compound library against the drug target or a cell-based assay and looking for a target induced response. The intent of these screens is to identify compounds that interact with the target, improve the potency, selectivity, and physiochemical properties of the compound and to verify the initial hypothesis that interaction with the target will elicit a desired biological response.

Prior to preclinical and clinical trials, the lead molecule selected from the hit discovery process enters the lead optimization phase. Here, the goal is to maintain and improve the desirable properties of the lead compound. The drug candidate is observed in various in vitro and in vivo models to ensure that it does not induce any genetic mutations or undesirable behavioural or physiological functions. Various pharmacological studies are also conducted to establish how the candidate is metabolised and to explore any stability issues and other chemical properties. This information is assembled along with control considerations to create a target candidate profile in order to be considered for preclinical and clinical trials.

2 Current Computational Approaches in Epigenetic Drug Discovery

Computational methods have been shown to be powerful tools in drug discovery and development [19]. The benefits of these approaches lie in their ability to represent real-time events in a fraction of the time, quickly analyse mass amounts of data, and find complex patterns. In epigenetic drug discovery, computational methods provide a means for gaining a better understanding of epigenetic mechanisms and identifying potential drugs, and drug targets. Several computational approaches have been proposed to advance epigenetic drug discovery. Many of these approaches use molecular modeling techniques such as molecular dynamics simulations, molecular docking, homology modeling and pharmacophore modeling, or virtual screening, although several other methods have been proposed.

2.1 Virtual Screening

Also known as computational or in silico screening, virtual screening involves computationally searching databases such as small molecule libraries for specific structures of interest [32]. The criteria for these structures are often determined using information from X-ray crystallography or nuclear magnetic resonance and molecular modeling, with the intent of selecting a small number of compounds that are likely to be active. Virtual screening is an attractive method to guide hit identification and lead optimization, and has been successfully used to identify potential epigenetic drugs [2, 16, 18, 32, 37].

In an early application, virtual screening was used to identify novel DNMT inhibitors. An initial set of 1990 compounds obtained from the Diversity Set available from the National Cancer Institute (NCI) was screened using molecular modeling and the top 2 ranking compounds were validated in vitro and in vivo [24]. Later, a larger subset of the NCI database consisting of 260,000 compounds was screened, out of which 65,000 compounds were selected. The application of several molecular modeling techniques created a set of 24 compounds out of which 13 continued for experimental testing. Seven of these compounds were found to have detectable DNMT inhibitory activity and at least 6 of these compounds were selective for a specific DNMT isoform [32].

In some cases, drugs have been found to treat diseases and disorders other than the ones they were originally approved for. For example, valproic acid, which is generally used for its anticonvulsant properties in the treatment of epilepsy, has also been found to exhibit HDAC inhibition [27]. This idea was explored in a study which aimed to re-purpose bioactive food compounds. In this study, a database of 4600 bioactive food compounds was screened using 32 approved antidepressant drugs. On the basis of chemical similarity, the 10 compounds found to be the most similar to the antidepressant drugs were experimentally screened against an HDAC. Interestingly, these 10 compounds were most similar to valproic acid. Out of the 10 compounds, 2 showed HDAC inhibition equivalent to valproic acid [15].

2.2 Molecular Dynamics

Molecular dynamics simulations provide information on the dynamic behaviour of atoms and molecules. Although computationally expensive, molecular dynamic simulations offer many advantages, such as detailed structural data, the microscopic interactions between molecules, and time-dependent responses to perturbations, which complement traditional experiments [1]. Simulations may validate whether theoretical models predict empirical information and can give insight on details not available in experiments. For example, molecular dynamics simulations reveal information on protein dynamics at the atomic-level which may help improve experimental and predicted protein structures [13, 27]. Specifically, molecular dynamics simulations consist of algorithms which evaluate many mathematical physics equations, such as equations of motion [1]. Molecular dynamics provides more detail than other molecular modeling approaches and has applications in enhancing conformational sampling and calculating free-energy changes upon ligand binding [27].

In 1988, hydralazine which is normally used as a potent arterial vasodilator, was found to exhibit DNA methylation inhibition, and in 2008 showed antitumour effects when combined with valproic acid during clinical trials [6, 32]. In order to gain a better understanding of the underlying molecular mechanisms of the methylation inhibitory activity of hydralazine, molecular modeling techniques, which included molecular dynamics, were used to model the binding mode of a DNMT isoform [32]. These simulations revealed that hydralazine shares similar binding behaviour as nucleoside analogs, which are known to be important in DNA methylation mechanisms [16].

With the intent of gaining a better understanding of DNMTs, molecular dynamics simulations were used to model the catalytic domains of DNMTs upon binding to SAM. Crystal structures and other molecular modeling techniques were used represent the different DNMT isoforms. However, on a nanosecond scale, no significant conformational changes were found upon binding of SAM. Nevertheless, the study provided insight on the the protein dynamics of DNMTs when binding to this cofactor [7].

2.3 Molecular Docking

Molecular docking often uses experimental data, such as the crystal structures of compounds, in parallel with molecular dynamics and other molecular modeling techniques, to predict how a molecule might fit into a specific binding site, such as the catalytic binding site of a DNMT [16, 36]. Each docking pose is scored in order to find the best position and orientation of the molecule within the binding site [32, 37]. Applications of this method benefit from the ability of scoring a large number of compounds in a small amount of time and generally produce meaningful results [18]. Docking has been used to study protein-ligand interactions of known DNMT inhibitors and has been shown to be important in the drug discovery process, as it can lead to a better understanding of the molecular interactions involved with potential drugs and can also be used to improve existing epigenetic drugs [19, 37].

With the intent of gaining a better understanding of the different binding poses of DNMTs, 14 compounds with different structural classes, which included nucleoside and non-nucleoside inhibitors, were used for docking. Because the study was conducted prior to the availability of the crystallographic structure of the DNMT, a molecular model of the catalytic domain was used. A comparison between the docking score and experimental data was not possible, however docking revealed similar binding interactions among the different compounds with the binding site that are thought to be crucial in DNA methylation [37].

In a later study, the crystal structure of a DNMT bound to DNA containing unmethylated CpG sites was used to dock known DNMT inhibitors. First, molecular dynamics was used to model the catalytic binding site of the crystal structure, as it was in an inactive state, into an active conformation. The binding poses of the inhibitors were found to share common interactions with the catalytic domain of the DNMT that are involved with the proposed mechanisms of DNA methylation. To further the study, compounds that were recently identified through high-throughput screening were also docked in an attempt to understand their binding modes. These docking models were then used in virtual screening with the goal of finding other inhibitors in large databases. The compounds that were identified were found to have favourable docking scores and included approved drugs ideal for drug re-purposing [17].

Similar docking studies have been conducted to explore the binding of SGI-1027, a known DNMT inhibitor, and propose mechanisms for its inhibitory activity [18, 34].

2.4 Homology Modeling

Homology modeling is among the top three three-dimensional (3D) structure prediction techniques and is often used as an alternative in the absence of experimental data or 3D structural information of a molecule [12, 27]. Homology modeling involves constructing a 3D model of a protein using its amino acid sequence and an homologous protein as a template. This approach is based on the observation that the amino acid sequence of a protein determines its structure and that related sequences will fold into similar structures [12]. The success of many homology models can be attributed to the well conserved catalytic domain of DNMTs [32]. Before the availability of crystal structures of DNMTs, many structure-based design studies, such as docking, relied on homology models that used the crystal structures of bacterial DNMTs [18, 37]. Homology modeling of this type was essential for the identification of novel DNMT inhibitors [18].

RG108 was the first DNMT inhibitor identified using a homology model of a human DNMT in combination with virtual screening [18, 24]. Later, two more DNMT inhibitors were discovered using the same homology model and have been used in the optimization of novel DNMT inhibitors. Although the crystal structures of many human DNMTs have since increased, homology modeling still provides an excellent starting point in many molecular modeling studies.

Homology modeling was used in one of the first contributions of molecular modeling to the research of DNMT inhibitors. In 2003, the catalytic domain of a human DNMT isoform was modeled using the crystal structure of a related human DNMT isoform. This model was used to develop N4-fluoroacetyl-5-azacytidine, which was found to successfully inhibit DNA methylation in human tumour cell lines [23]. This homology model was also combined with docking and molecular dynamics to develop a binding mode of hydralazine [25]. Later, using the crystal structures of bacterial DNMTs, two homology models of the catalytic domain of a human DNMT were constructed. Although the two models were created using different homologous templates, both homology models shared common key interactions in the catalytic site. The models were later validated by superimposing them on their recently published crystal structure, and it was found that the homology model was in agreement with proposed mechanisms of DNA methylation [32].

2.5 Pharmacophore Modeling

The concept of a pharmacophore has existed for over a century [31]. Although the basic idea of a pharmacophore hasn't changed, recently, the International Union of Pure and Applied Chemistry (IUPAC) has formally defined a pharmacophore as

> an ensemble of steric and electronic features that is necessary to ensure the optimal supramolecular interactions with a specific biological target and to trigger (or block) its biological response

[29]. More simply, a pharmacophore model can be defined as a structure-based model that describes features necessary for molecular recognition [31]. It is often used in virtual screening, docking, lead optimization and is one of the major tools in drug discovery [31, 37].

In 2011, Yoo and Medina-Franco proposed a computational method that could be used in parallel with experimental data and molecular dynamics, to develop a pharmacophore. An active conformation of the catalytic binding site of a DNMT inhibitor was modeled using its crystal structure and molecular dynamics. Molecular docking was then conducted by using known active compounds in the catalytic site. Comparison of the molecular docking results with previous research confirmed the model. They proposed that the information gained from this model could be used to determine binding behavior of DNMTs, which may lead to the development of effective non-nucleoside inhibitors [33, 36]. This method has been shown to be important in the drug discovery process, as it can lead to a more thorough understanding of the molecular interactions involved with potential drugs and their possible structures. This technique may also be used in toxicity screening and to improve existing epigenetic drugs [19].

Pharmacophore modeling is frequently performed alongside virtual screening and molecular docking, and can use either homology models or crystal structures [17, 33, 36, 37]. This ensemble of techniques allows researchers to explore ligand-binding interactions of DNMTs, giving insight into possible inhibitors [17]. For example, Yoo et al. developed pharmacophore models for 16 known DNMT inhibitors. Using the best scoring docking poses and a homology model of the catalytic binding site, researchers found that many of the inhibitors matched the pharmacophore features of the model [35]. Results led to the identification of aurintricarboxylic acid (ATA) as a novel DNMT inhibitor.

2.6 Additional Methods

Identifying epigenetic targets has recently become an area of particular interest for successful drug research. However, current research is limited by a lack of thorough understanding of the underlying biology of many epigenetic targets [4]. In this regard, researchers developed a method to gain a better understanding of genome functionality through database analysis to determine epigenetic targets. A 2015 study utilised several databases such as The Encyclopedia of DNA Elements Consortium (ENCODE), the University of California Santa Cruz (UCSC) genome browser and The National Centre for Biotechnology Information (NCBI) to collect information that may help drive future in vivo studies. The goal of the ENCODE project is to reveal and characterize functional elements of the human genome. This data can then be passed to the UCSC genome browser to identify epigenetic modifications, which can be further superimposed with functional genomic studies from NCBI. Through this process, researchers may gain a better understanding of relevant systems and may use this information to guide in vivo epigenetic target studies [10].

Computational approaches are also used to gather more information about epigenetic systems. Epigenetic protein targets and drug molecules are difficult to identify because they are highly dynamic. In 2012, Baron and Vellore approached this problem by using high-performance computing technology to dynamically visualize the conformational changes of different molecules. They performed a molecular dynamics computer simulation on X-ray crystal structures of a protein involved in epigenetic changes. This involved taking several static images and creating a dynamic visualization of the mechanisms of the protein. Visualization of these dynamic changes helped explain the ability of the protein to affect a variety of molecules involved in epigenetic regulation, and gave researchers insight into designing more targeted epigenetic drugs [2].

3 Conclusions and Prospects

Epigenetics and its relationship to disease is a complex area of research with many aspects that still remain to be explored. Because of the limited, although growing knowledge on epigenetic mechanisms, the availability of epigenetic drugs and thes discovery of novel therapeutics has been restricted. Combined, epigenetics and computational methods have been shown to have enormous potential in drug discovery and development. Specifically, molecular modeling approaches, including virtual screening, molecular dynamics, molecular docking, homology modeling and pharmacophore modeling, as well as other techniques, have led to a better understanding of the dynamic behaviour of DNMTs, the discovery and design of novel inhibitors as well as the optimization of existing therapeutics, and can ultimately be used to guide future theoretical and experimental studies.

These computational methods play a key role in the multidisciplinary effort to develop epigenetic drugs for the treatment of various diseases. But disease is a combination of epigenetic changes and genetic mutation. Although these studies have contributed many successes to the field of epigenetic drug discovery, few efforts in drug development exist that combine both genetic and epigenetic data. Such studies may elucidate new epigenetic targets and have become an area of particular interest because they may provide novel scaffolds for the treatment of disease. Additionally, although crystal structure availability of many proteins involved in epigenetic mechanisms has increased, and homology modeling has been found to be a successful alternative, many molecular modeling techniques still depend on the availability of crystal structures. Several of these approaches can be computationally expensive, and with the diverse field of computer science, many existing computational approaches, other than molecular modeling, have yet to be explored. Nevertheless, with the emergence of new technologies, modeling approaches, and epigenetic research, future prospects in epigenetic drug discovery look promising.

References

1. Allen, M.P.: Introduction to molecular simulation. In: Attig, N., Binder, K., Grubmüller, H., Kremer, K. (eds.) Computational soft matter: From synthetic polymers to proteins, Lecture notes. NIC Series, Jülich (2004)
2. Baron, R., Vellore, N.A.: LSD1/CoREST reversible opening-closing dynamics: discovery of a nanoscale clamp for chromatin and protein binding. Biochemistry **51**(15), 3151–3153 (2012)
3. Choy, M.K., Movassagh, M., Goh, H.G., Bennett, M.R., Down, T.A., Foo, R.S.: Genome-wide conserved consensus transcription factor binding motifs are hyper-methylated. BMC Genomics. **11**, 519 (2010)
4. Comley, J.: Epigenetic targets: on the verge of becoming a major new category for successful drug research. Drug Discovery World. http://www.ddw-online.com/summer-15/p303686-epigenetic-targets-:-on-the-verge-of-becoming-a-major-new-category-for-successful-drug-research.html. Accessed 15 Oct 2015
5. Cramer, S.A., Adjei, I.M., Labhasetwar, V.: Advancements in the delivery of epigenetic drugs. Expert Opin. Drug Deliv. **12**(9), 1501–1512 (2015)
6. Dueñas-González, A., García-López, P., Herrera, L.A., Medina-Franco, J.L., González-Fierro, A., Canderlaria, M.: The prince and the pauper. A tale of anticancer targeted agents. Mol. Cancer **7**, 33 (2008)
7. Evans, D.A., Bronowska, A.K.: Implications of fast-time scale dynamics of human DNA/RNA cytosine methyltransferases (DNMTs) for protein function. Theoret. Chem. Acc. **125**, 407–418 (2010)
8. Foulks, J.M., Parnell, K.M., Nix, R.N., Chau, S., Swierczek, K., Saunders, M., Kanner SB.: Epigenetic drug discovery: targeting DNA methyltransferases. J. Biomol. Screen **17**(1), 2–17 (2012)
9. Francis, R.C.: Epigenetics: How Environment Shapes our Genes. W.W. Norton and Company, New York (2012)
10. Hay, E.A., Cowie. P., MacKenzie. A.: Determining epigenetic targets: A beginner's guide to identifying genome functionality through database analysis. Methods Mol. Biol. 1–17 (2015)
11. Hughes, J.P., Rees, S., Kalindjian, S.B., Philpott, K.L.: Principles of early drug discovery. Br. J. Pharmacol. **162**(6), 1239–1249 (2011)
12. Krieger, E., Nabuurs, S.B., Vriend, G.: Homology modeling. In: Bourne, P.E., Weissig, H. (eds.) Structural Bioinformatics. Wiley-Liss Inc, Hoboken (2003)
13. Lindahl, E.R.: Molecular dynamics simulations. Methods Mol. Biol. **443**, 3–23 (2008)
14. Luong, L.D.: Basic principles of genetics (2009). http://cnx.org/contents/41c4c77e-a44c-431f-bbc0-32eb72726630@1/Basic-Principles-of-Genetics
15. Martinez-Mayorga, K., Peppard, T.L., López-Vallejo, F., Yongye, A.B., Medina-Franco, J.L.: Systematic mining of generally approved safe (GRAS) flavor chemicals for bioactive compounds. J. Agric Food Chem. **61**(31), 7507–7514 (2013)
16. Medina-Franco, J.L., Caulfield, T.: Advances in the computational development of DNA methyltransferase inhibitors. Drug Discovery Today **16**(9–10), 418–425 (2011)
17. Medina-Franco, J.L., Yoo, J.: Docking of a novel DNA methyltransferase inhibitor identified from high-throughput screening: insights to unveil inhibitors in chemical databases. Mol. Diversity **17**, 337–344 (2013)
18. Medina-Franco, J.L., Méndez-Lucio, O., Dueñas-González, Yoo J.: Discovery and development of DNA methyltransferase inhibitors using in silico approaches. Drug Discovery Today **20**(5), 569–577 (2015)
19. Mishra, N.K.: Computational modeling of P450s for toxicity prediction. Expert Opin. Drug Metab. Toxicol. **7**(10), 1211–1231 (2011)
20. National Cancer Center Research Institute: DNA methylation (2010). http://www.ncc.go.jp/en/nccri/divisions/14carc/14carc01_1.html
21. Pharmaceutical Research and Manufacturers of America: Biopharmaceutical research and development: The process behind new medicines. PhRMA (2015)

22. Ptak, C., Petronis, A.: Epigenetics and complex disease: from etiology to new therapeutics. Annu. Rev. Pharmacol. Toxicol. **48**, 257–276 (2008)
23. Siedlecki, P., Garcia Boy, R., Comagic, S., Schirrmacher, R., Wiessler, M., Zielenkiewicz, P., Lyko, F.: Establishment and functional validation of a structural homology model for human DNA methyltransferase 1. Biochem. Biophys. Res. Commun. **306**(2), 558–563 (2003)
24. Siedlecki, P., Boy, R.G., Musch, T., Brueckner, B., Suhai, S., Lyko, F., Zielenkiewicz, P.: Discovery of two novel, small-molecule inhibitors of DNA methylation. J. Med. Chem. **49**, 678–683 (2006)
25. Singh, N., Dueñas-González, A., Lyko, F., Medina-Franco, J.L.: Molecular modeling and molecular dynamics studies of hydralazine with human DNA methyltransferase 1. Chem. Med. Chem. **4**(5), 792–799 (2009)
26. Turner, B.M.: Histone acetylation and an epigenetic code. Bioessays **22**(9), 836–845 (2000)
27. Vellore, N.A., Baron, R.: Epigenetic molecular recognition: a biomolecular modeling perspective. Chem. Med. Chem. **9**(3), 484–494 (2014)
28. Ververis, K., Hiong, A., Karagiannis, T.C., Licciardi, P.V.: Histone deacetylase inhibitors (HDACIs): multitargeted anticancer agents. Biologics **7**, 47–60 (2013)
29. Wermuth, C.G., Ganellin, R.C., Lindberg, P., Mitscher, L.A.: Chapter 36—Glossary of terms used in medical chemistry (IUPAC recommendations 1997). Anuu. Rep. Med. Chem. **33**, 385–395 (1998)
30. Xu, Y.Z., Kanagaratham, C., Radzioch, D.: Chromatin remodelling druing host-bacterial pathogen interaction. In: Radzioch, D. (ed.) Chromating remodelling (2013). http://www.intechopen.com/books/chromatin-remodelling/chromatin-remodelling-during-host-bacterial-pathogen-interaction
31. Yang, S.Y.: Pharmacophore modeling and applications in drug discovery: challenges and recent advances. Drug Discovery Today **15**, 444–450 (2010)
32. Yoo, J., Medina-Franco, J.L.: Discovery and optimization of inhibitors of DNA methyltransferase as novel drugs for cancer therapy. In: Rundfeldt, C. (ed.) Drug development—a case study based insight into modern strategies, InTech. http://www.intechopen.com/books/drug-development-a-case-study-based-insight-into-modernstrategies/discovery-and-optimization-of-inhibitors-of-dna-methyltransferase-as-novel-drugs-for-cancertherapy
33. Yoo, J., Medina-Franco, J.L.: Homology modeling, docking and structure-based pharmacophore of inhibitors of DNA methyltransferase. J. Comput. Aided Mol. Des. **25**(6), 555–567 (2011)
34. Yoo, J., Choi, S., Medina-Franco, J.L.: Molecular modeling studies of the novel inhibitors of DNA methyltransferases SGI-1027 and CBC12: implications for the mechanism of inhibition of DNMTs. PLoS One **8**(4), e62152 (2012)
35. Yoo, J., Kim, J.H., Robertson, K.D., Medina-Franco, J.L.: Molecular modeling of inhibitors of human DNA methyltransferase with a crystal structure: discovery of a novel DNMT1 inhibitor. Adv. Protein Chem. Struct. Biol. **87**, 219–247 (2012)
36. Yoo, J., Medina-Franco, J.L.: Computer-guided discovery of epigenetics drugs: molecular modeling and identification of inhibitors of DNMT1. J. Cheminform. **4**, 25 (2012)
37. Yoo, J., Medina-Franco, J.L.: Inhibitors of DNA methyltransferases: insights from computational studies. Curr. Med. Chem. **19**(21), 3475–3487 (2012)

Dimensionality Reduction for Intrusion Detection Systems in Multi-data Streams—A Review and Proposal of Unsupervised Feature Selection Scheme

Naif Y. Almusallam, Zahir Tari, Peter Bertok and Albert Y. Zomaya

Abstract An Intrusion Detection System (IDS) is a security mechanism that is intended to dynamically inspect traffic in order to detect any suspicious behaviour or launched attacks. However, it is a challenging task to apply IDS for large and high dimensional data streams. Data streams have characteristics that are quite distinct from those of statistical databases, which greatly impact on the performance of the anomaly-based ID algorithms used in the detection process. These characteristics include, but are not limited to, the processing of large data as they arrive (real-time), the dynamic nature of data streams, the curse of dimensionality, limited memory capacity and high complexity. Therefore, the main challenge in this area of research is to design efficient data-driven ID systems that are capable of efficiently dealing with data streams by considering these specific traffic characteristics. This chapter provides an overview of some of the relevant work carried out in three major fields related to the topic, namely *feature selections* (FS), *intrusion detection systems* (IDS) and *anomaly detection* in multi data streams. This overview is intended to provide the reader with a better understanding of the major recent works in the area. By critically investigating and combining those three fields, researchers and practitioners will be better able to develop efficient and robust IDS for data streams. At the end of this chapter, we provide two basic models: an Unsupervised Feature Selection to Improve Detection Accuracy for Anomaly Detection (UFSAD) and its extension (UFSAD-MS) for multi streams, that could reduce the volume and the dimensionality of the big data resulting from the streams. The reduction is based on the selection of only the relevant features and removing irrelevant and redundant ones. The last section of the chapter provides an example of the developed UFSAD model, followed by some

N.Y. Almusallam (✉) · Z. Tari · P. Bertok
RMIT University, Melbourne, Australia
e-mail: naif.almusallam@rmit.edu.au

Z. Tari
e-mail: zahir.tari@rmit.edu.au

P. Bertok
e-mail: peter.bertok@rmit.edu.au

A.Y. Zomaya
University of Sydney, Sydney, Australia
e-mail: albert.zomaya@sydney.edu.au

© Springer International Publishing Switzerland 2017
A. Adamatzky (ed.), *Emergent Computation*, Emergence, Complexity
and Computation 24, DOI 10.1007/978-3-319-46376-6_22

467

experimental results. UFSAD-MS is provided as a conceptual model as it is in the implementation phase.

1 Existing Work

This section reviews the progress that has been made in the field of Feature Selections (FS), Intrusion Detection Systems (IDS) as well as anomaly detection in data streams. A review of these three areas will hopefully provide the reader with a comprehensive understanding of existing approaches, challenges and solutions that could help properly investigating the solutions in the selected area.

1.1 Feature Selection (FS) Techniques

Extensive research has been carried out in the field of feature selections aimed at reducing the high dimensionality of high volume data. Methods for Dimensionality reduction can be categorised into feature selection and feature extraction [60]. FS is an efficient data pre-processing technique that improves accuracy and reduces computational complexity by eliminating redundant and irrelevant features while maintaining the original features of the data [29]. FS can be carried out either by ranking the features based on particular criteria and adopting the top N features or by selecting the minimum subset of features without weakening the learning performance. The former can automatically set the number of features whereas the latter depends on a pre-defined threshold to determine the number of features [55]. On the other hand, feature extraction transforms the original feature space into a new reduced dimensional feature space. However, it seems unable to solve the problem of redundancy as redundant features might be included in the transformation phase [42].

There are various *challenges* resulting from the existence of irrelevant and redundant features in the datasets. Firstly, they diminish the accuracy of the mining algorithms by misleading the classification or clustering process [53]. Also, the existence of the redundant and irrelevant features would negatively affect the performance of the algorithms due to the large volume of data [38]. Moreover, they increase the processing time of the mining algorithms, which would result in very expensive complexity [20]. Furthermore, a large storage capacity is required for the storing of the large volume of data [39]. Finally, the curse of dimensionality is a challenge for feature selection algorithms due to the sparseness of the data, which would deceive the mining algorithms by looking equally in terms of distance between each other [11]. Consequently, various researchers have proposed feature selection as an efficient technique, which would help to address the aforementioned challenges.

The FS process comprises (i) subset generation, (ii) subset evaluation, (iii) stopping criterion and (iv) result validation [55]. This process is illustrated in Fig. 1.

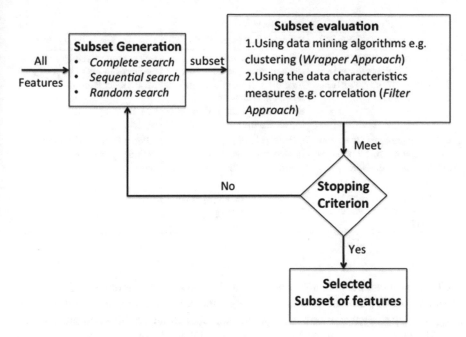

Fig. 1 Feature selection process

Subset generation searches for a set of features based on a particular strategy to be ready for the evaluation at the next step. The three main types of search strategy, in addition to their strengths and weaknesses, are illustrated in Table 1. Subset evaluation is the second step of the FS process, where every generated candidate features is evaluated for its quality based on a specific evaluation criterion [97]. Evaluation criterion is broadly classified into *filter* and *wrapper* approaches whether the mining algorithms are to be applied in the evaluation of the selected features or not [82]. The filter approach [23, 62, 78] relies on the general characteristics of the data to evaluate the quality of the generated candidate features without involving any mining algorithm. This includes, but is not limited to distance, information and consistency measures. Filter-based algorithms have faster processing time than wrapper-based algorithms, as they do not include any data mining algorithm [43]. Conversely, the wrapper-based algorithms [41, 71, 96] require the use of specific data mining algorithms such as clustering in the evaluation process of the generated candidate features by exploiting their specific performance requirements [32]. Despite the fact that the wrapper approach can discover better quality candidate features than does the filter approach, this incurs high computational overheads [39].

Subset generation and evaluation of the feature selection process is iteratively repeated until they meet the requirement of the stopping criterion. The stopping criterion is activated by the completeness of the search, by a maximum iteration times or when the classification error rate is less than the pre-set threshold [54]. Then, the

Table 1 Search strategies for subset generation

Complete search [16, 52, 89]	Sequential search [56, 69, 75]	Random search [8, 22, 77]
• Starts with an empty feature set, and adds the features for the purpose of the evaluation and vice versa	• Starts with an empty feature set, and adds one feature at a time till reaching the stage when the features do not enhance the quality of the subset features	• Starts the search by selecting random subsets to be produced for the evaluation
• **Pros**: guarantees the search for the optimal result based on the adopted evaluation criterion	• **Pros**: it is simple to be implemented and fast in getting the results	• **Pros**: Ensure the global optimality of the selected subset
• **Cons**: exhaustive search, which induces performance overheads	• **Cons**: It does not produce "optimal" features set	

selected best candidate features are validated by conducting before and after experiment testing of different aspects such as classification error rate, number of selected features, the existence of redundant/irrelevant features and the time complexity [55].

Although there are many FS schemes for reducing the data dimensionality, they are not capable of efficiently working in data streams. This is because they were designed to select the relevant features and remove the redundant ones from statistical databases. In fact, we believe that an FS scheme must take into account the following properties in order to work efficiently in data streams. It should be restricted to read the data only once as it is impossible to store the entire stream. Also, it should take into account that many stream applications stream the features one-by-one and do not assume the existence of the entire feature space in advance (called dynamic feature space). An FS scheme has to incrementally measure and update the relevance of the features, as one feature might be relevant in a time t but not in $t + 1$ (concept drift). Furthermore, it is not enough to reduce the feature space from the stream; the instances must be reduced as well because they usually contain great amounts of noise, redundancy and irrelevance. An FS scheme should not be limited to data class labels; instead, it should be (Unsupervised), as the data class labels are not available for most applications. Finally, an FS scheme should also be able to select the relevant features from multiple streams in order to measure the relevance of the features accurately.

There are very few FS schemes that work in data stream applications. Every scheme consists of some properties but not all of them. OSFS [86] handles a stream of features one by one as they arrive. However, it requires the data to be labeled; it removes irrelevant/redundant features but not instances and only works for a single data stream. By contrast, Kankanhalli et al. [44] selects a subset of relevant features from multiple streams based on the Markovian decision problem. However, it requires the full feature space to be known in advance and the data to be labeled, and removes irrelevant/redundant features but not instances. Toshniwal et al. [80] developed an un-supervised FS that does not require the data labels in order to select the relevant features. It is designed primarily for the purpose of outlier detection. However, it does not handle stream features one by one as they arrive; it removes irrelevant/redundant features but not instances, and works only for a single data stream stream. Finally, the Zhang et al. [93] approach incrementally measures and updates the relevance of the features in order to accurately evaluates their relevance. On the other hand, it requires the full feature space to be known in advance and is designed to work only in a single data stream.

1.2 Intrusion Detection Systems (IDS)

Feature selection is a pre-processing step that helps to optimise the performance of security mechanisms (e.g. firewalls, cryptography or access controls), which have been mainly designed to protect computer or information systems from malicious attacks. In addition to those security mechanisms, ID systems have been developed as a second-line defence to discover attacks after they have been successfully launched [49]. IDS can be host-based (e.g. to monitor the logs), network-based (e.g. to monitor the networks traffic flow) or data-driven (e.g. to detect any deviations from the normal pattern of the data), which is the focus of our interest. Broadly, IDS is classified in term detecting intrusions into signature based and anomaly based [66].

The signature-based ID approach [6, 21, 30] discovers suspicious behaviours by comparing them with pre-defined signatures. Signatures are patterns associated with attacks, which are verified in advance by the human experts and used to trace any suspicious patterns. If the suspicious patterns and the signatures match, an alarm is activated to warn the administrators or to take a pre-defined actions in response to the alarm [51]. The algorithms that are signature-based ID are efficient in detecting *known* attacks with low false alarms and are reasonably quick to do so. Despite the fact that most existing commercial IDs are signature-based, most of them cannot detect new types of attacks (also called *un-known attacks*), as their signatures are new and not known in advance [85].

Unlike the signature-based ID algorithms, anomaly-based ID algorithms [17, 18, 88] can identify new attacks because they "appropriately" model the 'normal' behaviour of a non-attacked system. They can therefore identify serious deviations from the normal profile to be considered as *anomalies* (also called *outliers*) [4]. Anomalies can emerge as a result of fraudulent behaviour, mechanical faults or attacks [46]. Figure 2 illustrates how the majority of the data points (triangle points) take a particular distribution while the circle points have a significant deviation from the rest. The circle points are considered as *outliers*.

Based on the form of the input data they are using, anomaly-based ID techniques generally can be categorised under three approaches: *supervised anomaly detection* [28, 34], *semi-supervised anomaly detection* [48, 87] and *unsupervised anomaly detection* [76, 79]. Supervised-based anomaly detection techniques require training data in advance along with their class labels for both normal and abnormal

Fig. 2 Deviation of circle
points from the normal
triangle ones [35]

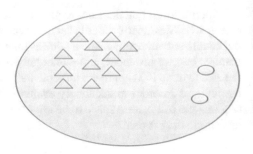

data, so as to accurately detect anomalies. The model is then trained with both classes and applied on un-labeled data to determine the class to which it belongs. Although there are plenty of classification methods that could be applied in this category, the classes of the data are un-balanced because the "normal class" is much bigger than the "anomaly class", which therefore negatively affects the detection recall. Additionally, it is challenging to find accurate and representative data class labels, particularly for the anomalies, as they emerged periodically and they are uncountable [35]. On the other hand, semi-supervised based anomaly detection techniques require only one class label, which is either *normal* or *outlier*. The corresponding model could be trained with the normal class only, and then any instance that does not belong to that class would be classified as an outlier. These techniques are more much applicable than supervised ones because they do not require the specification of anomalous behaviour. In addition, as the models for semi-supervised techniques could also be trained with anomaly class only, this provides substantial limitations because it is difficult to recognise all anomalies for the training of the data [14].

Both of the aforementioned approaches are limited as they rely on the availability of the labeled data. As a result, they are restricted for specific applications such as spacecraft fault detection and therefore they are not generic. On the other hand, the unsupervised anomaly detection approach is generic and widely applicable as it does not need the data to be labeled [70]. This approach assumes that the normal data has a pattern that is significantly different from the pattern of the outliers. For instance, the normal data should form groups with instances that are very similar to each other and dissimilar to the outliers. Although this approach is widely applicable, the related techniques experience a high rate of false alarms [67].

Anomaly-based ID can mainly be categorised into classification methods, statistical methods, proximity-based methods and clustering methods. Classification methods [15, 25] are supervised by nature, and they are applicable only if there are class labels in the training data. The classifier is trained with the labeled data and then applied for the testing of unlabeled data. The test data is then classified as an outlier if it is not classified as normal by the classifier. Classification methods seem to provide good accuracy in distinguishing between data and their related classes. Although such methods demonstrate good performance during the testing phase in comparison to the other methods, their detection accuracy depends on the accuracy of the labeled data [66].

Statistical methods [72, 74] are another type of approach, which observe the activity of the data so as to create profiles representing acceptable behaviour. There are two kinds of profiles: current and stored profiles. The former logs and updates regularly the distribution of the data as long as the data is processed. Additionally, the data is assigned with an anomaly score by comparing them with the stored profile. If any anomaly score exceeds a pre-defined threshold, it is labeled as an outlier. Statistical methods do not need knowledge about labeled data or attacks patterns in advance. Hence, they seem to be efficient in detecting recent attacks. On the other hand, it is difficult to establish a threshold that balances the occurrence of false positives and false negatives [91].

Proximity-based methods use distance metrics to calculate the similarity between data. It assumes that the proximity between an outlier and its nearest neighbour is different from its proximity to the remaining data. Such methods can be distance-based or density based. Distance-based methods [1, 12] search for a minimum pre-defined number of neighbours of a data point within a specific range in order to decide its normality. The point is labeled as an outlier if the neighbours within the range are less than the pre-defined threshold. On the other hand, density-based methods [64, 94] compare the density of data with its neighbours densities so to decide about its normality. The point is labeled as an outlier if its density is considerably less than the density of its neighbours. Generally, the effectiveness of proximity-based methods varies based on the adopted measure as it is challenging to ensure effectiveness in particular situations. Furthermore, proximity-based methods seem to be inefficient in detecting outliers that form groups and are close to each other.

Lastly, clustering methods [57, 83] work in unsupervised mode to recognise patterns of un-labeled data by grouping similar instances into groups. They cluster data by examining their relationships with other clusters. Indeed, normal data are those data that belong to clusters that are dense as well as large. On the other hand, outliers can be identified based on the three assumptions [46]: (1) outliers are objects, which have not been allocated to any cluster. In fact, the initial goal of clustering is to find clusters in particular and not the outliers; (2) outliers are objects that are far in term of measured distance measure to their closest cluster centroids. Indeed, every object is given a score based on its distance to its closest cluster centroid and it should not exceed a pre-defined distance in order to be considered as normal. The limitation of this assumption is that outliers cannot be found if they have already formed a cluster. The aforementioned assumptions have a common limitation in that they seem to detect only individual outliers but not groups of outliers, which form clusters by themselves [35]. To overcome this limitation, (3) the last assumption defines the outliers as objects, which have been allocated to sparse or small clusters.

Generally, clustering methods do not require the data to be labeled so it can handle zero-day attacks. Also, it can adapt to cluster "complex objects" by adopting existing clustering algorithms that can handle those particular types of objects. Furthermore, clustering methods are fast in the testing phase because every object is compared with the clusters only, which are relatively small in comparison with all the objects. On the other hand, the efficiency of clustering methods depends on the clustering algorithms in establishing the normal behaviour of the objects. Also, clustering methods work

Table 2 Characteristics of clustering methods

Methods	Characteristics
Partitioning [13, 19, 63, 65]	• Use mostly a distance-based, where the dataset is partitioned into n parts, each representing a cluster with minimum data points • Each object is allocated to only one cluster • Does not maintain any hierarchal structure • Adopts iterative relocation mechanism in the partitioning to produce "optimal" results • Works efficiently with small to medium size datasets • k-means is an example clustering algorithm used as a partitioning method
Hierarchical [40, 45, 58, 73]	• Clustering is maintained based on hierarchal decomposition of the dataset • It is either agglomerative or divisive decomposition • Uses either distance-based or density-based • Clusters cannot be corrected when they have been merged or split
Density-based [2, 5, 36, 47]	• Has been defined under proximity-based methods • Has good accuracy in detecting outliers • Capable of discovering clusters with arbitrary shape as it is based on density, not distance • DBSCAN clustering algorithm used as density-based algorithm
Grid-based [3, 37, 51, 84]	• The feature space is divided into a limited number of cells to form the grid • Clustering operations are performed inside the cells • Has fast processing time, as complexity depends on the number of grid cells and not the number of instances

efficiently when the outliers are individuals but not when they form groups of clusters. Finally, clustering methods are still computationally expensive even with some recent work attempting to resolve the performance problem [24].

Clustering methods can be broadly categorised into partitioning methods, density-based methods, hierarchal methods and grid-based methods. Table 2 below provides a classification of these methods as well as their characteristics.

1.3 Anomaly Detection for Multiple Data Streams

Anomaly detection is no longer limited to statistical databases due to the the emergence of very large data (**Big Data**) with specific characteristics: Volume, Variety, Velocity (3V). Volume relates to the huge amount of data generated. Such data can be found in different formats such as videos, music and large images. Velocity refers to the high speed at which data is generated, captured, and shared. Variety refers to the proliferation of new data types. The real world has produced big data in many different formats, posing a challenge that needs to be addressed. A data stream is an ideal example of big data because: (a) a huge (Volume) of data is gathered from different sources (i.e. sensors) to extract knowledge by mining and analysing the collected big data; (b) a data stream arrives in a timely manner at different speed rates (Velocity); (c) sensors can stream different data types (Variety).

Although anomaly detection for data streams has been investigated intensively, most of the recent research has focused only on *single data stream*. Therefore, we believe it is crucial to investigate how to detect anomalies or launched attacks arriving from *multiple data streams*. In fact, attacks like Denial of Service (DoS) might cause severe damage to the systems if they have been flooded through multiple streams. Therefore, anomaly detection algorithms need to be improved and adapted to multiple data streams. A *data stream* could be defined in [81] as a set of infinite data points that consist of attribute values along with an implicit or explicit timestamp. Anomaly-based ID methods are applied to detect outliers from not only a single stream but also from various data streams. This is often carried out by mining the relationships between those multiple streams, either by: (a) computing the correlations between multiple data streams and identifying points who have a high correlation; or (b) computing the similarity by querying multiple data streams to figure out high similarity points; or (c) utilizing clustering methods to discover the relationship between the streams in order to filter the outliers [90].

Existing anomaly-based algorithms, which have been covered in Sect. 1.2, might not be able to mine the data points in data streams for the following reasons. *Firstly*, data arrives in the form of streams and should be tested for outlier-ness as long as they arrive which could result in wrong decisions due to the dynamic nature of the data streams [33]. *Secondly*, data streams produce a very high volume of data, which would be too expensive to store. In fact, it has been suggested in [50] that data stream algorithms should be executed in the main memory and not requisite secondary storage. *Thirdly*, unlike traditional methods for anomaly detection that assume the existence of the entire datasets in advance, the mining of data streams requires the consumption of a minimum amount of memory [10]. Therefore, the model should have only a single scan to access the data points in the storage for the purpose of detection.

In addition to the above-mentioned characteristics, it is challenging to determine whether or not the data streaming points are outliers as the characteristics of the data streams may change over time. This phenomena is called *concept evolution* [92], and it takes place when new class emerges from streaming data overtime. Therefore, clustering techniques in particular should adapt to the concept evolution in order to reflect the real characteristics of data points. Additionally, data streams do not form a unified distribution of the data points, which seems to increase the complexity of detecting outliers [59]. High dimensionality is also a characteristic of data streams due to the sparseness of the data, which could degrade the efficiency of detecting outliers, as high dimensional data appears to be equal in terms of distance between the data points due to the sparse data [31]. Moreover, in some situations, different data streams with different data types need to be mined, such as categorical or numerical;

hence, it becomes challenging to finding the relationship between them [27]. Finally, most data mining algorithms have high computational complexity when applied to data streams [9]. As a result, new algorithms should be designed, or improved from existing algorithms, to meet the requirements as well as the characteristics of multi-data streams so they can mine patterns efficiently and accurately.

There are a few existing solutions that specifically apply to anomaly detection in multi-data streams. The algorithm proposed in [26] attempts to solve the problem of judging the stream data points for outlier-ness as soon as they arrive due to limited memory capacity, which could result in wrong decisions. This is carried out by partitioning the data streams into chunks and later clustering each one by applying the k-means algorithm. Then, every point that deviates significantly from its clusters centroid would be saved temporarily as a candidate outliers, and the normal points are discarded after computing their mean values in order to free the memory. To decide whether or not they are outliers, the mean value of the candidates clusters is then compared with the mean value of a pre-set L number of previous chunks. Although this algorithm seems to be computationally efficient because it does not rely on distance measures, it has low detection accuracy. Additionally, several parameters need to be properly defined (e.g. number of clusters and L number of chunks to compare the mean value and the chunk size as well), which makes the algorithm less attractive for multi-stream data.

Another clustering-based approach is proposed in [68] to detect anomalies for multi-data streams. It partitions a stream into windows or chunks, each of which is clustered and associated with a reference. Then, the numbers of adjacent clusters, along with representation degree references, are computed to find outlier references that contain potential anomalies. This model is believed to have better scalability and accuracy.

Koupaie et al. [46] proposed an incremental clustering algorithm that has two main phases to detect outliers in multi-data streams. In the online phase, the data in the windows is clustered using the k-mean algorithm, where clusters that are relatively small or quite far from other clusters are considered to be online outliers and therefore need further investigation. During the offline phase, the outlier from previous windows is added to the current window to be re-clustered by the k-mean algorithm. With higher confidence, it guarantees that any small or far clusters are real outliers as they have been given a survival chance. The work claims that the proposed algorithm is more accurate than existing techniques in discovering outliers; however, no evaluation results have been provided. Similarly to other algorithms, many of its parameters need to be adjusted.

2 Issues and Methodology

Redundant and irrelevant features are two factors that result in large volume and high dimensional data, which obviously degrade performance when outlier detection algorithms are used for data streams. Some of the aspects that should be carefully looked at are storage, detection accuracy and time complexity [20]. Many researchers have adopted feature selection as a pre-processing step to help the detection models to overcome these drawbacks or at least improve their performance by removing redundant and irrelevant features. However, there have been few proposals to date on feature selection that are mainly designed for outlier detection. After investigating an important body of work, we found that the only feature selection model, which takes outlier detection into consideration as an end goal, is the one detailed in [7]. However, their model works in *supervised* mode, and hence it assumes the availability of labeled data. Unfortunately, in most of situations, labelled data is not available.

Online FS selection is an emerging field where the relevant features are selected from data streams. Existing schemes for FS try to solve individual problems to cope with the data stream characteristics, although no single scheme has met the required properties (see FS in Sect. 1.1) for working efficiently in data streams.

In summary, the fundamental research considerations that need to be tackled are (Q1) and (Q2) below.

(Q1) How to identify representative features from un-labeled data? High dimensional and large size datasets have negative consequences on any data-mining algorithms because of their curse of dimensionality, high computational complexity, low detection accuracy and large memory consumption. Redundant and irrelevant features are commonly believed to increase dimensionality and the size of the datasets. Therefore, the inclusion of redundant and irrelevant features is a dilemma that can degrade the performance of the applied anomaly detection algorithms. Also, it can mislead the clustering algorithms during their detection, which could result in a high false positive rate.

(Q2) How to simultaneously identify representative subset of features and instances from data streams? In addition to reducing the feature space, we believe that it is essential to select relevant instances, and remove redundant ones, from a big data stream rather than analysing the entire stream. This is because thousands of instances usually contain irrelevant, noisy and redundant data, which consequently do not help the applied anomaly-based algorithms in their detection target as they will add extra processing time, consume extra memory space and increase the false alarm rate of these algorithms. Therefore, it is very important to exclude those irrelevant or noisy instances and to include only the relevant instances in the reduced sample. Although sampling would reduce time complexity and enhance detection accuracy, the main challenge is how to find a reduced sample from a data stream while maintaining the original characteristics of that stream. For example, it is impossible to have the entire data in advance for sampling and alternatively sampling done on windows of the last n elements or a specific period of time of the data stream. The second challenge is how to simultaneously reduce the dimensionality of both features and instances, taking into account the data stream characteristics.

The following is an overview of the methodology used to address the issues listed above, with the ultimate aim of designing an efficient intrusion detection system for multi-data streams.

- To answer (Q1), we proposed a filter-based approach, called UFSAD (Unsupervised Feature Selection to Improving Detection Accuracy for Anomaly Detection), that is primarily intended for outlier detection, and produces a reduced, relevant and non-redundant feature set. The reason for adopting a filter-based approach feature selection is its ability to provide fast processing as it does not involve any data mining algorithm to evaluate the generated feature subset, which seems to work efficiently for large and high dimensional datasets. The k-mean algorithm is a well-known unsupervised clustering technique for clustering features as well as finding the relevant subset of features. However, this algorithm needs to be extended to integrate more similarity measures so as to "properly" cluster data, as the use of a single similarity measure will be biased towards specific models, thereby not guaranteeing the accurate detection of outliers. We have used three similarity measures, namely PCC—Pearson Correlation Coefficient, LSRE—Least Square Regression Error and MICI—Maximal Information Compression Index, and these cover most possible linear dependent correlations between features. Therefore the integration of these metrics will guarantee real accuracy during the detection.
- To answer (Q2), UFSAD has been improved to work in data streams applications. The proposed Unsupervised Feature Selection to Improving Detection Accuracy for Anomaly Detection in Multi Streams (UFSAD-MS) is designed taking into account the entire properties of a FS scheme to work in data streams (see FS in Sect. 1.1). UFSAD-MS reads the data only once due to the limited memory. It does not assume the existence of the whole feature space but consider the applications where the features in a stream arrive one by one. It incrementally measures and updates the relevance of the features, as one feature might be relevant in a time t but not in $t + 1$. It reduces the dimensionality of the data streams simultaneously in terms of both features and instances. UFSAD-MS is an unsupervised learning approach that does not require the data stream classes to be known. Finally, it can work with both single and multiple data streams.

The following software is required in order to perform the experiments such as building and evaluating various algorithms.

- The Weka tool can be used to evaluate our developed Algorithms. It offers variety of classification, clustering and regression algorithms. Weka is free software that is provided by the University of Waikato in New Zealand. This can be downloaded from http://www.cs.waikato.ac.nz/ml/weka/downloading.html
- Matlab is a high level language which helps us to develop our algorithm. It can be easily integrated with Weka for evaluation purposes and all the results can be displayed in the Matlab platform. http://au.mathworks.com/academia/student_version/

Each **dataset** included in this research has a class label even though our target is to work in unsupervised model. The aim is to conduct before (labels included) and after (labels not included) evaluation to determine the effectiveness of the model. Some datasets that can be used for our experiments are:

- DARPA 2000 for details refer to http://www.ll.mit.edu/ideval/data/2000data.html

- Water Treatment Plant for details refer to https://archive.ics.uci.edu/ml/datasets/Water+Treatment+Plant
- KDD Cup 1999 for details refer to http://kdd.ics.uci.edu/databases/kddcup99/kddcup99.html
- Spambase for details refer to https://archive.ics.uci.edu/ml/datasets/Spambase
- PAMAP2 for details refer to https://archive.ics.uci.edu/ml/datasets/PAMAP2+Physical+Activity+Monitoring

The **evaluation metrics** that can be used to test the performance of the various algorithms are listed below:

- *True Positive TP*: the number of anomalies that have been detected correctly.
- *False Negative FN*: the number of anomalies that have not been detected correctly as outliers.
- *False Positive FP*: the number of normal points that have been incorrectly labeled as outliers.
- *True Negative TN*: the number of anomalies that have been correctly labeled as outliers.
- *Detection rate or Recall*: the proportion of anomalies that have been detected correctly to the actual size of the entire existing anomalies on the tested dataset.
- *False Positive Rate FPR*: the proportion of the normal points that are correctly flagged as anomalies to the actual size of the entire existing normal points on the tested dataset.
- *Precision*: used to test the robustness of IDS by minimising the FPR.
- *F-measure*: is the harmonic mean of precision and recall, which precisely demonstrates the accuracy of the evaluated ID algorithm.

Here we outline the progress of our work with regards (Q1) and (Q2), namely the development of an unsupervised feature selection algorithm and its extension to work efficiently in data streams. The proposed UFSAD and UFSAD-MS schemes are pre-processing steps to generate reduced, relevant and non-redundant samples as input for anomaly detection algorithms. They remove the redundant and irrelevant features in order to improve detection accuracy, reduce computational complexity and conserve memory capacity. The reduced feature set can represent the entire data stream with lower dimensionality and smaller data size. We aimed for an unsupervised learning approach because it might be impossible to obtain in advance the data class labels for the data streams.

UFSAD comprises two main stages: **partitioning** the features space and **selecting** the relevant features. The following steps broadly demonstrate the methodology of the proposed UFSAD algorithm in selecting a reduced, relevant and non-redundant subset of features:

- Firstly, UFSAD partitions the feature space by applying k-mean into k clusters using every similarity measure, namely: MCI—Maximal Information Compression Index, PCC—Pearson Correlation Coefficient and LSRE—Least Square Regression Error. Each similarity measure is computed individually.

- Secondly, the centroids are initialised to be the first features vectors from the feature space based on the k value. For example, if $k = 10$, then the first ten features vectors are the initial cluster's centroids.
- Thirdly, we assign every feature to a cluster. To do so, the similarity between every centroid and all the features in the feature space is computed. Every feature is therefore assigned to its relevant cluster centroid. This process is repeated until the re-assigning of features does not change the centroids, meaning that you have stable cluster centroids (i.e. the means of all clusters do not change).
- Fourthly, we find the relevant feature of every cluster. A feature of a cluster that has the highest similarity (i.e. highest PCC or lowest LSRE and MICI) to its centroid (mean) is selected as the relevant feature for the cluster.
- Lastly, UFSAD ignores all the remaining features of every cluster (and therefore keeps only the relevant features). This guarantees the removal of both redundant and irrelevant features, and produces a set of all relevant features.

For the purpose of evaluation, the UFSAD algorithm was compared with two other well-known algorithms [61, 95]. All the reduced features sets, which are generated by these three algorithms, are evaluated with three classifiers, namely Naive Bayes, Lazy Nearest Neighbour (also called IB1) and J48 decision tree. Also, the performance metrics used to evaluate the detection accuracy are FPR, Precision and F-measure. Figure 3 shows an experimental comparison of the proposed UFSAD with the algorithms proposed in [61, 95].

Figure 3 shows that UFSAD generated a reduced feature set that achieved the best accuracy according to the evaluation metrics for the Water Treatment Plant dataset. It has the lowest FPR (False Positive Rate), the highest precision and F-measure in comparison with SPEC and scheme in [61], whether the evaluation model is Naive Bayes, Lazy Nearest Neighbour or J48 Decision Tree. The strength of the UFSAD scheme compared to existing solutions is clearly the improvement in detection accuracy, and this is for the following reasons. Firstly, the methodology in selecting the relevant features contributes to improving the detection accuracy by guarantee the relevance of the selected features. Although all the features grouped in a given cluster are relevant, they are not enough to properly measure the relevance of the features. By contrast, UFSAD strongly limits the relevance of features to only those features that have the highest similarity to the cluster centroids, and this for every similarity measure. This methodology of selecting relevant features assists UFSAD to achieve better detection accuracy by guaranteeing the relevance of the selected features. Secondly, all the features in a cluster other than relevant features are discarded to ensure the removal of any redundant features. As a result, the scheme ensures the generation of a reduced, relevant and non-redundant feature set that helps the classifiers to detect outliers accurately.

UFSAD was later extended to consider the specific characteristics of data streams, called UFSAD-MS. Below is a summary of the different steps of UFSAD-MS:

(a) *Initialising the clusters*

FS scheme	Similarity Measure	FPR	Precision	F-measure
Scheme in [93]	PCC	0.0214	0.4500	0.5294
	LSRE	0.0390	0.2593	0.3415
	MICI	0.0117	0.6471	0.7097
Proposed scheme	PCC	0.0175	0.5500	0.6471
	LSRE	0.0195	0.5455	0.6667
	MICI	0.0058	0.7692	0.7407
SPEC	RBF Kernal	0.0312	0.4286	0.5714

Comparison of detection accuracy metrics of different schemes in Water Treatment Plant dataset with Naive Bayes evaluation classifier

FS scheme	Similarity Measure	FPR	Precision	F-measure
Scheme in [93]	PCC	0.0058	0.6250	0.4545
	LSRE	0.0039	0.7778	0.6087
	MICI	0.0039	0.7143	0.4762
Proposed scheme	PCC	0.0019	0.8889	0.6957
	LSRE	0.0019	0.8750	0.6764
	MICI	0.0019	0.8899	0.6957
SPEC	RBF Kernal	0.0058	0.7273	0.6400

Comparison of detection accuracy metrics of different schemes in Water Treatment Plant dataset with IB1 evaluation classifier

FS scheme	Similarity Measure	FPR	Precision	F-measure
Scheme in [93]	PCC	0.0117	0.60	0.6207
	LSRE	0.0078	0.6923	0.6667
	MICI	0.0058	0.7692	0.7407
Proposed scheme	PCC	0.0019	0.90	0.75
	LSRE	0.0039	0.8333	0.7692
	MICI	0.0039	0.8462	0.8148
SPEC	RBF Kernal	0.0078	0.7143	0.7143

Comparison of detection accuracy metrics of different schemes in Water Treatment Plant dataset with J48 decision tree evaluation classifier

Fig. 3 Experimental comparison of UFSAD with algorithms proposed in [61, 95]

- Assign the features that are part of the current window, as an initial mean of k clusters.
- For every initialised cluster, assign k = relevant feature. If k > number of features of the current window, waits for the arrival of features from the second window.

(b) *Partitioning the stream of features in a window*

- When a window of features arrives, the following steps are carried out:
  - compute the similarity (PCC, LSRE or MICI) between every feature in the window to every cluster.
  - assign every feature in a window to its most similar cluster (using the highest value PCC, Lowest LSRE or MICI).
  - re-compute the mean of every cluster.

(c) *Finding a relevant feature from every cluster*

- Compare the similarity of a highest similar feature and the relevant feature to the cluster mean.
- Set the one with highest similarity to the mean as the relevant feature.
- retain the mean, number of features so far, and the relevant feature. Otherwise, drop all other features to free the memory.

As illustrated in Figs. 4 and 5, this work is followed by two versions for multiple streams: the centralised and distributed versions of UFSAD-MS. These two versions are evaluated so as to select the best approach in terms of performance. Figure 4 shows the centralised version where UFSAD-MS waits for windows of features from every stream, and later clusters them to find the relevant features. Conversely, Fig. 5 depicts the distributed version where UFSAD-MS clusters every stream individually and then aggregates their selected relevant features. Experimental results are not shown as the two versions of UFSAD-MS are currently being implemented. This implementation covers all the required properties for FS schemes to work in multi-data streams.

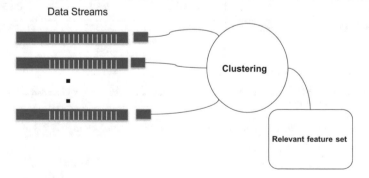

Fig. 4 UFSAD-MS centralised version

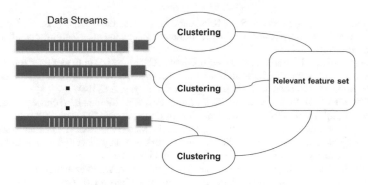

Fig. 5 UFSAD-MS distributed version

References

1. Aldahdooh, R.T., Ashour, W.: DIMK-means distance-based initialization method for K-means clustering algorithm. Int. J. Intell. Syst. Appl. (IJISA) **5**, 41 (2013)
2. Amini, A.: An adaptive density-based method for clustering evolving data streams. University of Malaya (2014)
3. Amini, A., Wah, T.Y., Saybani, M.R., Yazdi, S.R.A.S.: A study of density-grid based clustering algorithms on data streams. Eighth Int. Conf. Fuzzy Syst. Knowl. Disc. (FSKD) **2011**, 1652–1656 (2011)
4. Angiulli, F., Basta, S., Lodi, S., Sartori, C.: Distributed strategies for mining outliers in large data sets. IEEE Trans. Knowl. Data Eng. **25**, 1520–1532 (2013)
5. Ankerst, M., Breunig, M.M., Kriegel, H.-P., Sander, J.: OPTICS: ordering points to identify the clustering structure. In: ACM SIGMOD Record, pp. 49–60 (1999)
6. Au, M.H., Liu, J.K., Susilo, W., Yuen, T.H.: Secure ID-based linkable and revocable-iff-linked ring signature with constant-size construction. Theor. Comput. Sci. **469**, 1–14 (2013)
7. Azmandian, F., Yilmazer, A., Dy, J.G., Aslam, J.A., Kaeli, D.R.: Harnessing the power of GPUs to speed up feature selection for outlier detection. J. Comput. Sci. Technol. **29**, 408–422 (2014)
8. Banati, H., Bajaj, M.: Fire fly based feature selection approach. Int. J. Comput. Sci. Issues (IJCSI) **8** (2011)
9. Bifet, A., Morales, G.D.F.: Big data stream learning with SAMOA. In: IEEE International Conference on Data Mining Workshop (ICDMW), pp. 1199–1202 (2014)
10. Bifet, A., Holmes, G., Kirkby, R., Pfahringer, B.: MOA: massive online analysis. J. Mach. Learn. Res. **11**, 1601–1604 (2010)
11. Boratto, L., Carta, S.: Using collaborative filtering to overcome the curse of dimensionality when clustering users in a group recommender system. In: Proceedings of 16th International Conference on Enterprise Information Systems (ICEIS), pp. 564–572 (2014)
12. Ceberio, J., Irurozki, E., Mendiburu, A., Lozano, J.A.: A distance-based ranking model estimation of distribution algorithm for the flowshop scheduling problem. IEEE Trans. Evol. Comput. **18**, 286–300 (2014)
13. Celebi, M.E., Kingravi, H.A., Vela, P.A.: A comparative study of efficient initialization methods for the k-means clustering algorithm. Expert Syst. Appl. **40**, 200–210 (2013)
14. Chandola, V., Banerjee, A., Kumar, V.: Anomaly detection: A survey. ACM Comput. Surv. **41**, 1–58 (2009)
15. Choi, L., Liu, Z., Matthews, C.E., Buchowski, M.S.: Validation of accelerometer wear and nonwear time classification algorithm. Med. Sci. Sports Exerc. **43**, 357 (2011)
16. Dai, J., Wang, W., Tian, H., Liu, L.: Attribute selection based on a new conditional entropy for incomplete decision systems. Knowl.-Based Syst. **39**, 207–213, 2 (2013)

17. Damopoulos, D., Menesidou, S.A., Kambourakis, G., Papadaki, M., Clarke, N., Gritzalis, S.: Evaluation of anomaly-based IDS for mobile devices using machine learning classifiers. Secur. Commun. Netw. **5**, 3–14 (2012)

18. Damopoulos, D., Kambourakis, G., Portokalidis, G.: The best of both worlds: a framework for the synergistic operation of host and cloud anomaly-based ids for smartphones. In: Proceedings of the Seventh European Workshop on System Security, p. 6 (2014)

19. De Carvalho, F.D.A., Lechevallier, Y., De Melo, F.M.: Partitioning hard clustering algorithms based on multiple dissimilarity matrices. Pattern Recognit. **45**, 447–464 (2012)

20. de la Hoz, E., de la Hoz, E., Ortiz, A., Ortega, J., Martnez-lvarez, A.: Feature selection by multi-objective optimisation: application to network anomaly detection by hierarchical self-organising maps. Knowl.-Based Syst. **71**, 322–338 (2014)

21. Debiao, H., Jianhua, C., Jin, H.: An ID-based proxy signature schemes without bilinear pairings. annals of telecommunications-annales des tlcommunications **66**, 657–662 (2011)

22. Diao, R., Shen, Q.: Fcature selection with harmony search. IEEE Trans. Syst. Man Cybern. B Cybern. **42**, 1509–1523 (2012)

23. Doquire, G., Verleysen, M.: Feature selection with missing data using mutual information estimators. Neurocomputing **90**, 3–11 (2012)

24. Dua, S., Du, X.: Data mining and machine learning in cyber-security. CRC press (2011)

25. Dukart, J., Mueller, K., Barthel, H., Villringer, A., Sabri, O., Schroeter, M.L., et al.: Meta-analysis based SVM classification enables accurate detection of Alzheimer's disease across different clinical centers using FDG-PET and MRI. Psychiatry Res. Neuroimaging **212**, 230–236 (2013)

26. Elahi, M., Li, K., Nisar, W., Lv, X., Wang, H.: Efficient clustering-based outlier detection algorithm for dynamic data stream. pp. 298–304 (2008)

27. Elahi, M., Li, K., Nisar, W., Lv, X., Wang, H.: Detection of local outlier over dynamic data streams using efficient partitioning method, pp. 76–81 (2009)

28. Eskin, E., Arnold, A., Prerau, M., Portnoy, L., Stolfo, S.J.: Methods of Unsupervised Anomaly Detection Using a Geometric Framework. Google Patents (2013)

29. Fahad, A., Tari, Z., Khalil, I., Habib, I., Alnuweiri, H.: Toward an efficient and scalable feature selection approach for internet traffic classification. Comput. Netw. **57**, 2040–2057 (2013)

30. Fan, X., Gong, G.: Accelerating signature-based broadcast authentication for wireless sensor networks. Ad Hoc Netw. **10**, 723–736 (2012)

31. Feng, L., Liu, S., Xiao, Y., Wang, J.: Subspace detection on concept drifting data stream. In: Proceedings of ELM-2014, vol. 1. Springer, pp. 51–59 (2015)

32. Freeman, C., Kuli, D., Basir, O.: An evaluation of classifier-specific filter measure performance for feature selection. Pattern Recognit. (2014)

33. Golab, L., Zsu, M.T.: Issues in data stream management. ACM SIGMOD Rec. 32, pp. 5–14 (2003)

34. Grnitz, N., Kloft, M.M., Rieck, K., Brefeld, U.: Toward supervised anomaly detection. J. Artif. Intell. Res. (2013)

35. Han, J., Kamber, M., Pei, J.: Data Mining: Concepts and Techniques, 3rd edn. http://RMIT. eblib.com.au/patron/FullRecord.aspx?p=729031

36. Hinneburg, A., Keim, D.A.: An efficient approach to clustering in large multimedia databases with noise. In: KDD, pp. 58–65 (1998)

37. Hinneburg, A., Keim, D.A.: Optimal grid-clustering: Towards breaking the curse of dimensionality in high-dimensional clustering (1999)

38. Hong, Y., Kwong, S., Chang, Y., Ren, Q.: Unsupervised feature selection using clustering ensembles and population based incremental learning algorithm. Pattern Recognit. **41**, 2742–2756, 9 (2008)

39. Hong, Y., Kwong, S., Chang, Y., Ren, Q.: Consensus unsupervised feature ranking from multiple views. Pattern Recognit. Lett. **29**, 595–602 (2008)

40. Horng, S.-J., Su, M.-Y., Chen, Y.-H., Kao, T.-W., Chen, R.-J., Lai, J.-L., et al.: A novel intrusion detection system based on hierarchical clustering and support vector machines. Expert Syst. Appl. **38**, 306–313 (2011)

41. Hsu, C.-N., Huang, H.-J., Dietrich, S.: The ANNIGMA-wrapper approach to fast feature selection for neural nets. IEEE Trans. Syst. Man Cybern. B Cybern. **32**, 207–212 (2002)
42. Hua-Liang, W., Billings, S.A.: Feature subset selection and ranking for data dimensionality reduction. IEEE Trans. Pattern Anal. Mach. Intell. **29**, 162–166 (2007)
43. Jiang, S., Wang, L.: Unsupervised feature selection based on clustering. In: BIC-TA, pp. 263–270 (2010)
44. Kankanhalli, Mohan S., Wang, Jun, Jain, Ramesh: Experiential sampling in multimedia systems. IEEE Trans. Multimedia **8**, 937–946 (2006)
45. Karypis, G., Han, E.-H., Kumar, V.: Chameleon: hierarchical clustering using dynamic modeling. Computer **32**, 68–75 (1999)
46. Koupaie, H.M., Ibrahim, S., Hosseinkhani, J.: Outlier detection in stream data by clustering method. Int. J. Adv. Comput. Sci. Inf. Technol. **2**, 25–34 (2013)
47. Kriegel, H.P., Krger, P., Sander, J., Zimek, A.: Density-based clustering. Wiley Interdisc. Rev. Data Min. Knowl. Discov. **1**, 231–240 (2011)
48. Kuusela, M., Vatanen, T., Malmi, E., Raiko, T., Aaltonen, T., Nagai, Y.: Semi-supervised anomaly detectiontowards model-independent searches of new physics. In: Journal of Physics: Conference Series, p. 012032 (2012)
49. Law, K.H.: IDS false alarm filtering using KNN classifier. In: Information Security Applications. Springer, pp. 114–121 (2005)
50. Leskovec, J., Rajaraman, A., Ullman, J.D.: Mining of Massive Datasets. Cambridge University Press (2014)
51. Leung, K., Leckie, C.: Unsupervised anomaly detection in network intrusion detection using clusters. In: Proceedings of the Twenty-Eighth Australasian Conference on Computer Science, vol. 38, pp. 333–342 (2005)
52. Li, S., Wu, H., Wan, D., Zhu, J.: An effective feature selection method for hyperspectral image classification based on genetic algorithm and support vector machine. Knowl.-Based Syst. **24**, 40–48 (2011)
53. Lian, C., Ruan, S., Denux, T.: An evidential classifier based on feature selection and two-step classification strategy. Pattern Recognit
54. Liu, H., Yu, L.: Toward integrating feature selection algorithms for classification and clustering. IEEE Trans. Knowl. Data Eng. **17**, 491–502 (2005)
55. Liu, H., Motoda, H.: Computational Methods of Feature Selection. CRC Press (2007)
56. Luukka, P.: Feature selection using fuzzy entropy measures with similarity classifier. Expert Syst. Appl. **38**, 4600–4607 (2011)
57. Ma, Y.P., Ma, B., Jiang, T.H.: Applying improved clustering algorithm into EC environment data mining. In: Applied Mechanics and Materials, pp. 951–959 (2014)
58. Mahmood, A.N., Leckie, C., Udaya, P.: An efficient clustering scheme to exploit hierarchical data in network traffic analysis. IEEE Trans. Knowl. Data Eng. **20**, 752–767 (2008)
59. Mierswa, I., Wurst, M., Klinkenberg, R., Scholz, M., Euler, T.: Yale: rapid prototyping for complex data mining tasks. In: Proceedings of the 12th ACM SIGKDD International Conference on Knowledge Discovery and Data Mining, pp. 935–940 (2006)
60. Min, F., Hu, Q., Zhu, W.: Feature selection with test cost constraint. International Journal of Approximate Reasoning **55**, 167–179, 1 (2014)
61. Mitra, P., Murthy, C., Pal, S.K.: Unsupervised feature selection using feature similarity. IEEE Trans. Pattern Anal. Mach. Intell. **24**, 301–312 (2002)
62. Nandi, G.: An enhanced approach to Las Vegas Filter (LVF) feature selection algorithm. In: 2nd National Conference on Emerging Trends and Applications in Computer Science (NCETACS), pp. 1–3 (2011)
63. Ng, R.T., Jiawei, H.: CLARANS: a method for clustering objects for spatial data mining. IEEE Trans. Knowl. Data Eng. **14**, 1003–1016 (2002)
64. Parimala, M., Lopez, D., Senthilkumar, N.: A survey on density based clustering algorithms for mining large spatial databases. Int. J. Adv. Sci. Technol. **31** (2011)
65. Park, H.-S., Jun, C.-H.: A simple and fast algorithm for K-medoids clustering. Expert Syst. Appl. **36**, 3336–3341 (2009)

66. Patcha, A., Park, J.-M.: An overview of anomaly detection techniques: existing solutions and latest technological trends. Comput. Netw. **51**, 3448–3470 (2007)

67. Portnoy, L.: Intrusion detection with un-labeled data using clustering (2000)

68. Ren, J., Wu, Q., Zhang, J., Hu, C.: Efficient Outlier Detection Algorithm for Heterogeneous Data Streams, pp. 259–264 (2009)

69. Ruiz, R., Riquelme, J.C., Aguilar-Ruiz, J.S., Garca-Torres, M.: Fast feature selection aimed at high-dimensional data via hybrid-sequential-ranked searches. Expert Syst. Appl. **39**, 11094–11102 (2012)

70. Sadik, S., Gruenwald, L.: Research issues in outlier detection for data streams. SIGKDD Explor. Newsl. **15**, 33–40 (2014)

71. Sainin, M.S., Alfred, R.: A genetic based wrapper feature selection approach using nearest neighbour distance matrix. In: 3rd Conference on Data Mining and Optimization (DMO), pp. 237–242 (2011)

72. Saligrama, V., Chen, Z.: Video anomaly detection based on local statistical aggregates. In: IEEE Conference on Computer Vision and Pattern Recognition (CVPR) in 2012, pp. 2112–2119

73. Saunders, D.G., Win, J., Cano, L.M., Szabo, L.J., Kamoun, S., Raffaele, S.: Using hierarchical clustering of secreted protein families to classify and rank candidate effectors of rust fungi. PLoS One **7**, e29847 (2012)

74. Simmross-Wattenberg, F., et al.: Anomaly detection in network traffic based on statistical inference and alpha-stable modeling. IEEE Trans. Dependable Secure Comput. **8**, 494–509 (2011)

75. Singh, V., Pathak, S.: Feature selection using classifier in high dimensional data. arXiv preprint arXiv:1401.0898 (2014)

76. Skudlarek, S.J., Yamamoto, H.: Unsupervised anomaly detection within non?numerical sequence data by average index difference, with application to masquerade detection. Appl. Stoch. Models Bus. Ind. **30**, 632–656 (2014)

77. Srivastava, M.S., Joshi, M.N., Gaur, M.: A review paper on feature selection methodologies and their applications. Int. J. Comput. Sci. Netw. Secur. (IJCSNS) **14**, 78 (2014)

78. Suri, N.N.R.R., Murty, M.N., Athithan, G.: Unsupervised feature selection for outlier detection in categorical data using mutual information. In: 12th International Conference on Hybrid Intelligent Systems (HIS), pp. 253–258 (2012)

79. Tang, A., Sethumadhavan, S., Stolfo, S.J.: Unsupervised anomaly-based malware detection using hardware features. In: Research in Attacks, Intrusions and Defenses, Springer, pp. 109–129 (2014)

80. Toshniwal, Durga: A framework for outlier detection in evolving data streams by weighting attributes in clustering. Procedia Technol. **6**, 214–222 (2012)

81. Tu, L., Cui, P.: Clustering over uncertain data stream. Future Comput. Inf. Technol. **86**, 291 (2014)

82. Wald, R., Khoshgoftaar, T.M., Napolitano, A.: How the Choice of Wrapper Learner and Performance Metric Affects Subset Evaluation, pp. 426–432 (2013)

83. Wang, X.: A fast exact k-nearest neighbors algorithm for high dimensional search using k-means clustering and triangle inequality. In: The International Joint Conference on Neural Networks (IJCNN), pp. 1293–1299 (2011)

84. Wang, W., Yang, J., Muntz, R.: STING: A statistical information grid approach to spatial data mining. In: VLDB, pp. 186–195 (1997)

85. Wang, W., Guan, X., Zhang, X.: Processing of massive audit data streams for real-time anomaly intrusion detection. Comput. Commun. **31**, 58–72 (2008)

86. Wu, X., et al.: Online feature selection with streaming features. In: IEEE Trans. Pattern Anal. Mach. Intell. **35**, 1178–1192 (2013)

87. Wulsin, D., Gupta, J., Mani, R., Blanco, J., Litt, B.: Modeling electroencephalography waveforms with semi-supervised deep belief nets: fast classification and anomaly measurement. J. Neural Eng. **8**, 036015 (2011)

88. Xie, M., Hu, J., Han, S., Chen, H.-H.: Scalable hypergrid k-NN-based online anomaly detection in wireless sensor networks. IEEE Trans. Parallel Distrib. Syst. **24**, 1661–1670 (2013)

89. Xue, B., Cervante, L., Shang, L., Browne, W.N., Zhang, M.: A multi-objective particle swarm optimisation for filter-based feature selection in classification problems. Connect. Sci. **24**, 91–116 (2012)
90. Yeh, M.-Y., Dai, B.-R., Chen, M.-S.: Clustering over multiple evolving streams by events and correlations. IEEE Trans. Knowl. Data Eng. **19**, 1349–1362 (2007)
91. Yu, J.: A nonlinear kernel Gaussian mixture model based inferential monitoring approach for fault detection and diagnosis of chemical processes. Chem. Eng. Sci. **68**, 506–519 (2012)
92. Zang, W., Zhang, P., Zhou, C., Guo, L.: Comparative study between incremental and ensemble learning on data streams: case study. J. Big Data **1**, 5 (2014)
93. Zhang, C., Ruan, J., Tan, Y.: An incremental feature subset selection algorithm based on boolean matrix in decision system. J. Converg. Inf. Technol. **6** (2011)
94. Zhang, X., Shen, Q., Gao, H., Zhao, Z., Ci, S.: A density-based method for initializing the k-means clustering algorithm. In: Proceedings of International Conference on Network and Computational Intelligence (ICNCI 2012), IPCSIT, pp. 46–53 (2012)
95. Zhao, Z., Liu, H.: Spectral feature selection for supervised and unsupervised learning. In: Proceedings of the 24th International Conference on Machine learning, pp. 1151–1157 (2007)
96. Zhou, H., Wu, J., Wang, Y., Tian, M.: Wrapper approach for feature subset selection using GA. In: International Symposium on Intelligent Signal Processing and Communication Systems, 2007. ISPACS 2007, pp. 188–191 (2007)
97. Zhu, W., Si, G., Zhang, Y., Wang, J.: Neighbourhood effective information ratio for hybrid feature subset evaluation and selection. Neurocomputing **99**, 25–37 (2013)

Physical Maze Solvers. All Twelve Prototypes Implement 1961 Lee Algorithm

Andrew Adamatzky

Abstract We overview experimental laboratory prototypes of maze solvers. We speculate that all maze solvers implement Lee algorithm by first developing a gradient of values showing a distance from any site of the maze to the destination site and then tracing a path from a given source site to the destination site. All prototypes approximate a set of many-source-one-destination paths using resistance, chemical and temporal gradients. They trace a path from a given source site to the destination site using electrical current, fluidic, growth of slime mould, Marangoni flow, crawling of epithelial cells, excitation waves in chemical medium, propagating crystallisation patterns. Some of the prototypes visualise the path using a stream of dye, thermal camera or glow discharge; others require a computer to extract the path from time lapse images of the tracing. We discuss the prototypes in terms of speed, costs and durability of the path visualisation.

Gradient:

...a continuous increase or decrease in the magnitude of any quantity or property along a line from one point to another; also, the rate of this change, expressed as the change in magnitude per unit change in distance.

Oxford English Dictionary

1 Introduction

To solve a maze is to find a route from the source site to the destination site. If there is just a single path to the destination the maze is called a labyrinth. To solve a labyrinth one must just avoid dead ends. In a maze there are at least two paths leading from the entrance to the exit. To solve a maze one must find a shortest path. Not rarely concepts of 'maze', 'labyrinth' and 'collision-free shortest path' are mixed in experimental laboratory papers. We will not differentiate either. All algorithms and

A. Adamatzky (✉)
University of the West of England, Bristol, UK
e-mail: andrew.adamatzky@uwe.ac.uk

© Springer International Publishing Switzerland 2017
A. Adamatzky (ed.), *Emergent Computation*, Emergence, Complexity
and Computation 24, DOI 10.1007/978-3-319-46376-6_23

physical prototypes that solve shortest collision-free path on a planar graph solve mazes [14]. All algorithms and prototypes that solve mazes solve labyrinths.

There are two scenarios of the maze problem: the solver does not know the whole structure of the maze and the solver knows the structure of the maze.

The first scenario—we are inside the maze—is the original one. This is how Theseus, the Shannon's maze solving mechanical mouse, was born [49].[1] The mouse per se was a magnet with copper whiskers. The mechanism was hidden under the maze. A circuit with hundred of relays and mechanical drives grabbed the mouse from below the floor and moved to a randomly chosen direction. When the mouse detected an obstacle with its whiskers the underfloor mechanism moved the mouse away from the obstacle and other direction of movement has been selected. The task was complemented when the destination site was found. Being placed at any site of the maze the mouse was able to find a path towards the destination site. Several electro-mechanical devices have been built in 1950–1970, including well know Wallace's maze solving computer [57]. The Theseus also inspired a range of robotic mice competitions [61]. The algorithms of a maze traversing agent, see overviews in [12, 58] include random walk [10]; the Dead Reckoning (the mouse travels straight, when it encounters a junction it turns randomly, when it finds itself in the dead end it turns around), Dead End Learning (the agent remembers dead ends and places a virtual wall in the corridor leading to each dead end); Flood Fill (the agent assigns a distance, as crow flies, to each site of the maze and then travel in the maze and updates the distance values with realistic numbers); and, Pledge algorithm where the maze traversing agent is equipped with a compass, which allows to maintain a predetermined direction of motion (e.g. always north); the intersections between the corridors oriented north-south and walls are treated as graph vertices [1, 14]. Hybrids of the Wall Follow and the Pledge algorithms are used in industrial robotics and space explorations: a robot knows coordinates of the destination site, has a compass, turns on a fixed angle and counts turns [33, 34]. There are also genetic programming and artificial neural networks for maze solving [27].

The second scenario of maze solving—we are above the maze—is the one we study here. In 1961 Lee proposed the algorithm [31, 47] which became one of the most famous, reused and rediscovered algorithms in last century. We start at the destination site. We label neighbours, first order neighbours, of the site with '1'. Then we label second order their neighbours with '2'. Being at the site labelled i we label its non-yet-labelled neighbours with $i + 1$. Sites occupied by obstacles, or the maze walls, are not labelled. When all accessible sites are labelled the exploration task is completed. Eventually each site gets a label showing a number of steps someone must make to reach the site from the destination site.

To extract the path from any given site of the maze to the destination site we start at the source site. Then we select a neighbour of the source site with lowest value of its label. We add this neighbour to the list. We jump at this neighbour. Then we select its neighbour with lowest value of the label. We add this neighbour to the list. We jump at this neighbour. We continue like that till we get at the destination site. Thus

---

[1] See http://cyberneticzoo.com/mazesolvers/.

the algorithm computes one-destination-many-sources shortest path. A set of shortest paths starting from each site of the maze gives us a spanning tree which nodes are sites of the maze and a root is the site where wave pattern of labelling started to grow. In robotics the Lee algorithm was transformed into a potential method pioneered in [39] and further developed in [25, 59]. The destination is assigned an infinite potential. Gradient is calculated locally. Streamlines from the source site to the destination site are calculated at each site by selecting locally maximum gradient [17]. Also, some algorithms assume that the destination has an 'attracting' potential and obstacles are 'repellents' [6, 8, 28].

All experimental laboratory prototype of maze solvers implement the Lee algorithm. The gradients developed are resistance (Sect. 2), chemical (Sect. 2), temporal (Sect. 2) and thermal (Sect. 2). The paths are traced along the gradients by electrical current (Sect. 2.1), fluids (Sect. 2.2), cellular cytoplasm (Sect. 2.3), Marangoni flow (Sect. 3.1), living cells (Sect. 3.2), excitation waves (Sect. 5.2) and crystallisation (Sect. 5.3). We present brief descriptions of known experimental laboratory prototypes of maze solvers and analyse them comparatively.

2 Resistance Gradient

Imagine a maze filled with hard balls. The entrance and the exist are open. We put our hand in the entrance and push the balls. Balls in the dead ends have nowhere to move. The pressure is eventually transferred to the balls nearby exit. These balls start falling out. We add more balls thought entrance and push again. Balls fall out through the exit. Thus a movement of balls is established. The balls are moving along the shortest path between the entrance and the exit. The balls explore the maze in parallel and 'calculate' the path from the exit to the entrance. In this section we discuss prototypes which employe electrical and hydrodynamic resistances.

2.1 Electrical Current

Approximation of a collision-free path with a network of resistors is proposed in [53, 54]. A space is discretised as a resistor network. The resistors representing obstacles are insulators or current thinks. Other resistors have the same initial resistance. An electoral power source is connected to the destination and the source sites. The destination site is the electrical current source [15]. Current flows in the grid. Current does not flow into obstacles. To trace the path one must follow a current streamline by performing gradient descent in electrical potential. That is for each node a next move is selected by measuring the voltage difference between current node and each of its neighbours, and moving to the neighbours which shows maximum voltage. We are not aware of any large-scale prototype of such path solver. Two VLSI processors have been manufactured [50]. They feature 16×16 and 18×18 cells, $2\,\mu m$ nwell SCMOS technology, $3960\,\mu \times 4240\,\mu$ and $4560\,\mu \times 4560\,\mu$ frame size, 16-bit asynchronous data bus. For gradient descent the source is $5\,V$ and the destination is $0\,V$, and vice verse for gradient ascent.

There are two ways to represent walls of a maze [15]: Dirichlet boundary conditions and Neumann boundary conditions. When the Dirichlet boundary conditions is adopted the walls, or obstacles, have zero electrical potential and act as sinks of the electrical current [17]. Then the current lines are perpendicular to the walls and an agent, e.g. a robot, travelling along the current lines stays away from the walls [15]. In case of the Neumann boundary conditions the walls are insulators [53]. The walls are not 'felt' by the electrical current. Then the current lines are parallel to the walls. This gives the travelling agent less clearance. The original approach of [53] has been extended to networks of memristors (resistors with memory) [40], which could allow for computation of a path in directed planar graph.

A shortest path can be visualised, though not digitally recorded, without discretisation of the space. A maze is filled with a continuous conductive material. Corridors are conductors, walls are insulators. An electrical potential difference is applied between the source and the destination sites. The electrical current 'explores' all possible pathways in the maze. As proposed in [11] the electrons, driven by the applied electrical field, move along the conductive corridors in a maze until they encounter dead end or the destination site. When the electrons reach dead end they are cancelled inside the conductor. The electric field inside the dead ends becomes zero. The flow of electrons on the conductive pathways leading the destination does not stop. Thus the electrical flow calculates the shortest path. This path is detected via glow-discharge or thermo-visualisation.

2.1.1   GLOW: Glow-Discharge Visualisation

This maze solver is proposed in [44]. A drawing of a maze is transferred on a glass wafer and channels are etched in the glass. The channels are c. $250\,\mu m$ wide and c. $100\,\mu m$ deep. Electrodes are inserted in the source and the destination sites. The glass maze is covered tightly and filled with helium at $500\,Torr$. A voltage of up to $30\,kV$, above the breakdown voltage, is applied. Luminescence of the discharge shows the shortest path in the maze. Shortest path is visualised in $500\,ms$.

A maze solver using much less pressure and much lower voltage is proposed in [18]. The maze is made of a plexiglas disk, diameter $287\,mm$, $50\,mm$ height. Channels $25\,mm$ wide and $40\,mm$ deep are cut in the disk. Hole for the anode is made in the center of the disk. Cathode is placed in the destination site. The maze is filled with air, pressure $0.110\,Torr$. A gas-discharge chamber is made of polyamide in $450\,mm$ diameter and $50\,mm$ height, copper cathode in the form of rectangular plate of $158\,mm^2$ size is fitted on its side surface in a cathode holder. The maze is placed in the gas-discharge chamber. Stainless steel rod anode of $10\,mm$ diameter is placed at the center of the chamber. Voltage of $2\,kV$ is applied between the electrodes. The path is visualised by the glow of ionised air in the maze's channels. Experiments [18] also demonstrate propagation of striations, the ionisation waves [29], along the path.

2.1.2   ASSEMBLY: Assembly of Nano Particles

The maze is solved with nano particles in [35]. A maze is made of polydimethylsilicoxane and filled with silicon oil. A drop of a dispersion of conductive nano particles: spherical copper nano particles $10\,\mu$m diameter and metallic carbon nanotubes $10\,\mu$m long—is added at the source site. An electrical potential 1–5 kV is applied between the source and the destination. The particles diffusing from the source site become polarised. The polarised particles experience dipole interactions. The dipole interactions make the particles to form a chain along line of the electric field with maximal strength. The chain is formed to maximise the electrical current and to minimise the potential drop. The chain of the particles forms a conductive bridge between the electrodes at the source site and the destination site which represents the shortest path.

2.1.3   THERMO: Thermo-visualisation

A maze solver using electrical current is proposed in [11]. The prototype employs thermal visualisation of the electrical current. A maze 10×10 cm is made from copper tracks on a printed board. Copper tracks represent corridors. The electrical current 2.4 A is applied between the source and the destination sites. The flow of electrons heats the conductor, due to Joule heating. A local temperature of the conductor is proportional to intensity of the flow. In experiments [11] the temperature along the shortest path increased by c. 10 °C. The heating is visualised with the infrared camera. The shortest path is represented by the brightest loci on the thermographic image.

2.2   FLUIDIC: *Fluidic*

In a fluidic maze solver developed in [19] a maze is the network of micro-channels. The network is sealed. Only the source site (inlet) and the destination site (outlet) are open. The maze is filled with a high-viscosity fluid. A low-viscosity coloured fluid is pumped under pressure into the maze, via the inlet. Due to a pressure drop between the inlet and the outlet liquids start leaving the maze via the outlet. A velocity of fluid in a channel is inversely proportional to the length of the channel. High-viscosity fluid in the channels leading to dead ends prevents the coloured low-viscosity fluid from entering the channels. There is no pressure drop between the inlet and any of the dead ends. Portions of the 'filler' liquid leave the maze. They are gradually displaced by the colour liquid. The colour liquid travels along maximum gradient of the pressure drop, which is along a shortest path from the inlet to outlet. When the colour liquid fills in the path the viscosity along the path decreases. This leads to increase of the liquid velocity along the path. The shortest path—least hydrodynamic resistance path—from the inlet to the outlet is represented by channels filled with coloured fluid.

In the experiments [19] channel width varied from c. 90–200 μm, maze size c. 40×50 mm. The maze is filed with ethanol-based solution of bromophenoll. A dark coloured solution of a food dye in mix of water and ethylene glycol is injected in the maze at a constant flow, velocity c 5–10 mm/s. A drop of pressure between the inlet and the outlet is c. 0.75–2.25 Torr. The channels along the shortest path become coloured. The path is visualised in half-a-minute.

2.3 PHYSARUM I: *Slime Mould*

The prototype based on reconfiguration of protoplasmic network of acellular slime mould *Physarum polycephalum* is proposed in [36]. The slime mould is inoculated everywhere in a maze. The slime mould develops a network of protoplasmic tubes spanning all channels of the maze. Oat flakes are placed in the source and the destination site. A tube lying along the shortest (or near shortest) path between sites with nutrients develop increased flow of cytoplasm. This tube becomes thicker. Tubes branching to sites without nutrients become smaller due to lack of cytoplasm flow. They eventually collapse. The sickest tube represents the path between the sources of nutrients, and therefore, the path between the source and the destination sites. The selection of the shortest protoplasmic tube is implemented via interaction of propagating bio-chemical, electric potential and contractile waves in the plasmodium's body, see mathematical model in [55].

3 Diffusion Gradient

A source of a diffusing substance is placed at the destination site. After the substance propagates all over the maze a concentration of the substance develops. The concentration gradient is steepest towards the source of the diffusion. Thus starting at any site of the maze and following the steepest gradient one can reach the source of the diffusion. The diffusing substance represents one-destination-many-sources shortest paths. To trace a shortest path from any site, we place a chemotactic agent at the site and record its movement towards the destination site.

3.1   MARANGONI: *Marangoni Flow*

A diffusion gradient determines a surface tension gradient. A flow of liquid runs from the place of low surface tension to the place of high surface tension. This flow transports droplets. A maze solver proposed in [30] is as follows. A maze is made of polydimethylsiloxane, size c. 16×16 mm, channels have width 1.4 mm, and walls are 1 mm high. The maze is filled with a solution of potassium hydroxide. A surfactant is added to reduce the liquid's surface tension. An agarose block soaked in a hydrochloric acid is placed at the destination site. In c. 40 s a pH gradient establishes in the maze. Then a 1 μL droplet of a mineral oil or dichloromethane mixed with 2-hexyldecanoic acid is placed at the source site. The droplet does not mix with the liquid filling the maze. The droplet moves along the steepest gradient of the potassium hydroxide. The steepest gradient is along a shortest path. Exact mechanics of the droplet's motion is explained in [30] as follows. Potassium hydroxide, which fills the maze, is a deprotonating agent. Molecule of the potassium hydroxide removes protons from molecules of 2-hexyldecanoic acid diffusing from the droplet. A degree of protonation is proportional to concentration of hydrochloric acid, diffusing form the destination site. Protonated 2-hexyldecanoic acid at the liquid surface determines the surface tension. The gradient of the protonated acid determines a gradient of the surface tension. The surface tension decreases towards the destination site. A flow of liquid—the Marangoni flow—is established from the site of low surface tension to the site of high surface tension, i.e. from the start to the destination site. The droplet is moved by the flow [30]; see also discussion on mobility of surface in [41] and more details on pH dependent motion of self-propelled droplets in [13].

In the prototype [30] a path from the start site to the destination site is traced by a droplet but not visualised. To visualise the path fully one must record a trajectory of the droplet. A visualisation is implemented in [32]. A dye powder, Phenol Rd, is placed at the start site. The Marangoni flow transports the dye form the start to the destination. The coloured channels represent a path connecting the source site and the destination.

Another prototype of a droplet maze solver is demonstrated in [16]. The maze c. 45×75 mm in size, with channels c. 10 mm wide, is filled with water solution of a sodium decanoate. A nitrobenzene droplet loaded with sodium chloride grains is placed at the destination site. A 5 μL decanol droplet is placed at the source site. The sodium chloride diffuses from its host nitrobenzene droplet at the destination site. A gradient of salt is established. The gradient is steepest along a shortest path leading from any site of the maze to the destination site. A decanol droplet moves along the steepest gradient till the droplet reaches the nitrobenzene with salt droplet at the destination site.

3.2 Living Cells

A source of a chemo-attractant is placed at the destination site. The chemo-attractant diffuses along the channels of the maze. It reaches the destination site eventually. The maximum gradient is along the shortest path from any given site of a maze to the destination site. A living cell is placed at the source site. The cell follows the maximum gradients thus moving along the shortest path towards the destination site.

3.2.1 PHYSARUM II: Slime Mould

The slime mould maze solver based on chemo-attraction is proposed in [4]. An oat flake is placed in the destination site. The slime mould *Physarum polycephalum* is inoculated in the source site. The oat flakes, or rather bacterias colonising the flake, release a chemoattractant. The chemo-attractant diffuses along the channels. The Physarum explores its vicinity by branching protoplasmic tubes into openings of nearby channels. When a wave-front of diffusing attractants reaches Physarum, the Physarum halts the lateral exploration. Instead it develops an active growing zone propagating along the gradient of the attractant's diffusion. The problem is solved when Physarum reaches the source site. The sickest tube represents the shortest path between the destination site and the source site. Not only nutrients can be placed at the destination site but any volatile substances that attract the slime mould, e.g. roots of the medicinal plant *Valeriana officinalis* [46].

3.2.2 EPITHELIUM: Epithelial Cells

Experimental maze solver with epithelial cells is proposed in [48]. Epithelial cells move towards sites with highest concentration of the epidermal growth factor (EGF). An epithelial cell uptakes EGF. Thus the cell depletes EGF's concentration in the cell's vicinity. A $400\,\mu m \times 400\,\mu m$ maze is made of orthogonal channels c. $10\,\mu m$ wide [48]. The channels are filled with epithelium culture medium. There is a uniform distribution of the EGF inside the maze at the beginning of an experiment. The maze is placed in the medium with 'unlimited' supply of EGF. A cell is placed at the entrance channel. The cell enters the maze and crawls along its first channel. The cell consumes EGF and decreases EGF concentration in its own neighbourhood. EGF from all channels, accessible from the current position of the cell, diffuses towards the site with low concentration. Supply of EFG in channels ending with dead ends is limited. Unlimited supply of EGF into the maze is provided via exit channel. An EGF diffusion gradient from the exit through the maze to dead ends and the entrance is established. The cell follows the diffusion gradient. The gradient is maximum along the shortest path. The cell moves along the shortest path towards the exit.

4   Temperature Gradient: TEMPERATURE

A Marangoni flow is a mass-transfer of a liquid from a region with low surface tension to a region with high surface tension [32]. The mass transfer can move droplets, or any other objects, or dyes. Any methods of establishing a surface tension gradient is OK for tracing a shortest path with Marangoni flow. In [32] a temperature gradient is used. A maze is made from polydimethylsiloxane with channels 1.4 mm wide and 1 mm deep. The maze is filled with hot, c. 99 °C, aqueous solution of sodium hydroxide with hexyldecanoic acid. A steel sphere, diameter 4 mm, is cooled with dry ice and placed at the destination site in the maze. A phenol red dye powder is placed on the surface of the liquid at the start site. The cold sphere creates temperature gradient. The temperature gradient creates a surface tension gradient. The Marangoni flow is established along a shortest path from any site of the maze to the destination site. The dye powder applied at the source site is dragged by the flow towards the destination site. The trace of the dye represents the shortest path.

5   Temporal Gradient

A wave front advances for a fixed distance per unit of time in a direction normal to the front. A wave generated at the source site of a maze reaches the destination site along the shortest path. The wave 'finds' the exit. We just need to record the path of the wave.

5.1   VLSI: *VLSI Array Processor*

An array processor solving shortest path over terrain with elevations is reported in [26]. This is a digital processor of 24×25 cells, manufactured with $2 \, \mu m$ CMOS. A cell size is $296 \, \mu m \times 330 \, \mu m$, the processor's size is $7.9 \, mm \times 9.2 \, mm$. An elevations map is encoded to 255 levels of grey and loaded into the processor array. A signal is originated at the destination site. Wave-front of the signal propagates on the array. Each cell delays a signal by time proportional to the 'elevation' value loaded into the cells. Then the cell broadcasts the value to its neighbours. When a processor receives signal, the incoming signal direction is stored and further inputs to the cell are ignored. Starting from each cell we can follow the fastest path towards the destination site.

5.2   WAVE: *Excitation Waves*

In [51] a labyrinth solution in excitable chemical medium is proposed. A c. 3×3 cm labyrinth is made of vinyl-acrylic membrane. The membrane is saturated with Belousov-Zhabotinsky (BZ) mixture. Impenetrable walls are made by cutting away parts of the membrane. The channels are excitable. The walls are non-excitable. Excitation waves are initiated at the source site by touching the membrane with a silver wire. The wave-fronts propagate with a speed of c. 2 mm/min. Dynamics of the waves is recorded with 50 s intervals. The locations of the wave-fronts are colour-mapped: the colour depends on the time of recording. A shortest path from the source site to the destination site is extracted from the time lapse colour maps. In this setup excitation waves explore the labyrinth but the path is extracted by a computer.

The approach is slightly improved in [9]. By recording time lapse images of excitation wave fronts propagating in a two-dimensional medium we can construct a set of isochrones: lines which points are at the same distance from the site of the wave origination. By extracting intersection sites of isochrones of waves propagating from the source site to the destination site with isochrones of waves propagating from the destination site to the source site we can extract the shortest path. Experiments report in [9] deal not with a maze but a space with two obstacles. The approach would work in the maze as well. The BZ medium does not do any computation. The results are obtained on a computer by analysing dynamics of the excitation wave fronts. Something is better than nothing: the approach is successfully used in unconventional robotics [6, 7].

Another BZ based maze solver is proposed in [43]: it exploits light-sensitive BZ reaction. Extraction of the path requires an extensive image analysis of the excitation dynamics, therefore this prototype is not worthy of discussion here.

5.3   CRYSTALL: *Crystallization*

In [2] we proposed that a propagating excitation wave-front sets up pointers at each site of the medium. The pointers indicate a direction from where the wave-fronts came from. A shortest path towards the the source of the excitation can be recovered by following the pointers. This approach is experimentally implemented in [3] where crystals play a role of pointers. A set of one-source-many-destinations shortest path in a room with obstacles is constructed using crystallisation of sodium acetate tri-hydrate [3]. The room with obstacles is technically a maze with irregularly shaped walls. Obstacles are made of an adhesive resin attached to a bottom of a Petri dish. A super-saturated clear solution of sodium acetate trihydrate is poured into the Petri dish. The solution is cooled down. A crystallisation is induced by briefly immersing an aluminium wire powdered with fine crystals of the sodium acetate into the target site in the solution. Crystals growing from nucleation sites bear distinctive elongated shapes expanding towards their proximal ends. Not only a crystal's overall shape but also the orientation of saw-tooth edges indicate the direction of the crystal's growth. The crystallisation patterns compete for space. A crystallisation pattern following longer than shortest path is unable to reach the source site because the space available is already occupied by the crystallisation pattern following the shortest path. Thus the direction of crystal growth can be detected by conventional image processing techniques, e.g. edge detection procedures, or by a complementary method of detecting directional uniformity of image domains. A configuration of local vectors, which indicate direction of crystallisation propagation, is calculated. A vector at each point indicates the direction from where the wave of crystallisation came from.

6 Analysis

With regards to time it takes the prototypes to solve the maze, i.e. to produce a traced path from the source to the destination, we can split the solvers into three groups:

- solve instantly (milliseconds or less): GLOW, THERMO, VLSI, CRYSTALL
- solve in minutes: ASSEMBLY, FLUIDIC, MARANGONI, TEMPERATURE, WAVE
- solve in hours: PHYSARUM I, PHYSARUM II, EPITHELIUM

The classification is very rough. We estimate the solution times based on reported sizes of mazes and expected propagation time of substances or perturbations in the maze solving substrates.

Only CRYSTALL visualises a gradient. Other prototypes do not. In CRYSTALL maze solver the temporal gradient can be seen by an unaided eye as a pattern of crystal needles.

With regards to path tracing two groups can be specified:

- path is visible as a state of the solver, no computer is necessary: GLOW, THERMO, ASSEMBLY, FLUIDIC, MARANGONI, TEMPERATURE, PHYSARUM I, PHYSARUM II,
- a computer is necessary to extract the path from the states of the solver: VLSI, CRYSTALL, WAVE, EPITHELIUM

In MARANGONI no computer assistance is required if the path is traced using dyes. If the path is traced by a mobile droplet then we must record time lapse images of the travelling droplet and extract the path form the images. In EPITHELIUM, if a single cells is used to trace the path then time lapse recording is necessary. However, if we send a procession of epithelial cells, following one another, the path will become visible as a chain of cells. Because CRYSTALL and WAVE definitely require an external computing device to visualise the path we exclude them from further analyse. We also exclude VLSI because it a conventional hardware.

Several maze solvers are 'too demanding' in their requirements. Thus, GLOW requires up to 0.1–500 Torr pressure and 2–30 kV electrical potential; ASSEMBLY requires 1 kV electrical potential; FLUIDIC requires 0.7–2 Torr pressure, and TEMPERATURE requires fluid to be scalding hot.

Let us check five remaining prototypes: THERMO, MARANGONI, PHYSARUM I, PHYSARUM II, EPITHELIUM. The prototypes MARANGONI and EPITHELIUM are not ideal because they require some kind of a lab equipment not just a maze. The prototype PHYSARUM I is computationally inefficient because it requires the slime mould to be placed in all channels of the maze.

The remaining 'contestants' are THERMO and PHYSARUM II. Both prototypes are easy to make without a specialised lab equipment, and the whole setup of an experiment can be implemented in few minutes. It takes hours for PHYSARUM II to trace and to visualise the path. The prototype THERMO calculates the path instantly.

To conclude, the maze solver THERMO is the fastest and the easiest to implement physical maze solver. However, the THERMO does not visualise the path. We need a thermal camera to see the path. The PHYSARUM II visualises the path by morphology of the slime mould cell. However, the PHYSARUM II 'computes' the maze in many hours.

7 Discussion

The experimental laboratory prototypes of maze solvers might look differently but they all implement the same two tasks. First, they explore mazes in parallel and develop resistance, chemical, thermal or temporal gradients. Thus they approximate a one-destination-many-sources set of shortest paths. Second, they trace a path from a given source site to the destination site using fluid flows, electricity or living cells. Paths traced with electricity or flows are visualised with glow-charge gas, thermal camera, droplets, dyes. Paths traced with slime mould do not require any additional visualisation. Paths approximated with excitation wave-fronts or crystallisation patterns are visualised on computers.

A comparative plot of the maze solvers is shown in Fig. 1. We estimated speeds of maze solving from the experimental laboratory results reported, scaling all mazes to the same size. We estimated 'costs of prototyping' based on descriptions of experimental setups. The 'costs' are not in monetary terms but include fuzzy estimates of efforts necessary to prepare a maze, experimental setup, auxiliary equipment, and an inconvenience of using extreme physical parameters. For example, the glow-discharge maze solver requires a high pressure of gas filling the maze and very high electrical potential; the maze solver exploiting temperature gradients demands filling fluid to be scalding hot.

We have not discussed optical maze solving because we found reports only of theoretical designs [23, 24, 38]. In [24] the symmetrical properties of optical systems are utilised, the light diffraction is calculated and the path is extracted based on minimisation of the optical interferences. A detailed design of a two-dimensional nonlinear Fabry-Perot interferometer for maze solving is proposed in [38]. The optical solutions are interesting but impractical because they require a maze to be equipped with mirrors, so the light explores all paths in parallel; and, every junction of the

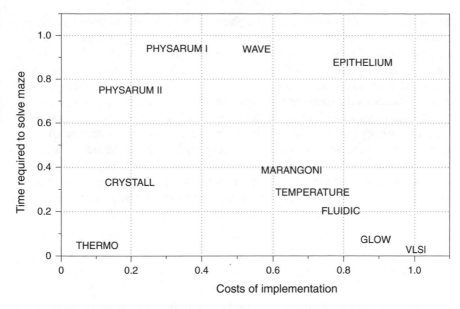

Fig. 1 Scatter plot of physical maze solvers based on estimated efforts of their prototyping and speed of maze solving

maze channels must be checked with interferometer. Also, to trace the path one must analyse all data on a computer.

We have not included any results on solving mazes with living creatures but the slime mould. Most papers published on experimental laboratory 'maze solving' by living creatures do actually mean animals learning to make a left turn or a right turn in 'T'- or 'Y'-shaped junctions, as reported for ants [52], honey bees [62], nematodes [42]. Ants can approximate shortest paths [20, 21, 37, 56]. Chances are high the ants can solve complex (not just 'T'-junctions) mazes. Some of the laboratory results, e.g. chemotactic behaviour and learning of nematodes [42, 45], are enriched by simulations of a bit more complex than just 'Y'-junction mazes. Whatever case, most living creatures will not display the whole path in a maze. Slime mould does. In principle, plant roots can do as well.

With regards to plants, apexes of plant roots show chemotactic behaviour. The roots can make a choice between direction of motion in Y-shape junctions and select a route towards the attractant [60]. Experiments on plant root maze solving, where gravity is the only guiding force, were not successful because the roots often stuck in a dead end [5]. This is not surprising thought cause the gravity does not provide a length-dependent tracing of a path. More experiments should be done in future with plant roots.

Do humans employ gradient developing and subsequent path tracing techniques when they solve maze visually? Given a full structure of a maze, i.e. being above the maze, humans solve the maze by scanning the maze, memorising critical cues and then tracing the path visually [63]. The phases of the maze exploration and the path tracing are clearly distinct. These phases can be seen as roughly corresponding to the gradient developing phase and the path tracing, or visualisation, phase. There is a chance that a topological model of a maze is 'physically' mapped onto a neuronal activity of a human brain cortex, where some neurons or their ensembles play a role of obstacles, other neurons are responsible for developing a gradient, and some neurons are responsible for tracing the path. In [22] a neuronal activity in superior parietal lobule of human subjects solving the maze is recorded with a whole-body magnetic resonance imaging and spectroscopy system. A mental traversing of the maze path is reflected in directional tuning of the volume unit. As [22] suggest, this might indicate an existence of a directionally selective synaptic activity of spatially close neuronal ensembles. Micro-volumes of the parental lobe showing absence of tuning might be corresponding to obstacles of the maze.

References

1. Abelson, H., Di Sessa, A.: Turtle Geometry: The Computer as a Medium for Exploring Mathematics. MIT Press (1986)
2. Adamatzky, A.: Cellular automaton labyrinths and solution finding. Comput. Graph. **21**(4), 519–522 (1997)
3. Adamatzky, A.: Hot ice computer. Phys. Lett. A **374**(2), 264–271 (2009)
4. Adamatzky, A.: Slime mold solves maze in one pass, assisted by gradient of chemo-attractants. IEEE Trans. NanoBiosci. **11**(2), 131–134 (2012)
5. Adamatzky, A.: Towards plant wires. Biosystems **122**, 1–6 (2014)
6. Adamatzky, A., de Lacy Costello, B.: Collision-free path planning in the Belousov-Zhabotinsky medium assisted by a cellular automaton. Naturwissenschaften **89**(10), 474–478 (2002)
7. Adamatzky, A., de Lacy Costello, B., Melhuish, C., Ratcliffe, N.: Experimental implementation of mobile robot taxis with onboard Belousov-Zhabotinsky chemical medium. Mater. Sci. Eng. C **24**(4), 541–548 (2004)
8. Adamatzky, A., Teuscher, C. (eds.): From Utopian to Genuine Unconventional Computers. Luniver Press (2006)
9. Agladze, K., Magome, N., Aliev, R., Yamaguchi, T., Yoshikawa, K.: Finding the optimal path with the aid of chemical wave. Physica D **106**(3), 247–254 (1997)
10. Aleliunas, R., Karp, R.M., Lipton, R.J., Lovasz, L., Rackoff, C.: Random walks, universal traversal sequences, and the complexity of maze problems. In: FOCS, vol. 79, pp. 218–223 (1979)
11. Ayrinhac, S.: Electric current solves mazes. Phys. Educ. **49**(4), 443–446 (2014)
12. Babula, M.: Simulated maze solving algorithms through unknown mazes. In: Organizing and Program Committee, p. 13 (2009)
13. Ban, T., Yamagami, T., Nakata, H., Okano, Y.: pH-dependent motion of self-propelled droplets due to Marangoni effect at neutral pH. Langmuir **29**(8), 2554–2561 (2013)
14. Blum, M., Kozen, D.: On the power of the compass. In: Proceedings of 19th Annual Symposium on Foundations of Computer Science, pp. 132–142 (1978)
15. Bugmann, G., Taylor, J.G., Denham, M.: Route finding by neural nets. Neural Networks 217–230 (1995)
16. Čejkova, J., Novak, M., Stepanek, F., Hanczyc, M.: Dynamics of chemotactic droplets in salt concentration gradients. Langmuir **30**(40), 11937–11944 (2014)
17. Connolly, C.I., Burns, J.B., Weiss, R.: Path planning using Laplace's equation. In: IEEE International Conference on Robotics and Automation. Proceedings, pp. 2102–2106. IEEE (1990)

18. Dubinov, A.E., Maksimov, A.N., Mironenko, M.S., Pylayev, N.A., Selemir, V.D.: Glow discharge based device for solving mazes. Phys. Plasmas (1994–present) **21**(9), 093503 (2014)
19. Fuerstman, M.J., Deschatelets, P., Kane, R., Schwartz, A., Kenis, P.J.A., Deutch, J.M., Whitesides, G.M.: Solving mazes using microfluidic networks. Langmuir **19**(11), 4714–4722 (2003)
20. Goss, S., Aron, S., Deneubourg, J.-L., Pasteels, J.M.: Self-organized shortcuts in the Argentine ant. Naturwissenschaften **76**(12), 579–581 (1989)
21. Goss, S., Beckers, R., Deneubourg, J.-L., Aron, S., Pasteels, J.M.: How trail laying and trail following can solve foraging problems for ant colonies. In: Behavioural Mechanisms of Food Selection, pp. 661–678. Springer (1990)
22. Gourtzelidis, P., Tzagarakis, C., Lewis, S.M., Crowe, D.A., Auerbach, E., Jerde, T.A., Uğurbil, K., Georgopoulos, A.P.: Mental maze solving: directional fMRI tuning and population coding in the superior parietal lobule. Exp. Brain Res. **165**(3), 273–282 (2005)
23. Haist, T., Osten, W.: Wave-optical computing based on white-light interferometry. DGaO Proc. (2008)
24. Huang, L.-D., Wong, M.D.F.: Optical proximity correction: friendly maze routing. In: Proceedings of the 41st Annual Design Automation Conference, pp. 186–191. ACM (2004)
25. Hwang, Y.K., Ahuja, N.: A potential field approach to path planning. IEEE Trans. Robot. Autom. **8**(1), 23–32 (1992)
26. Kemeny, S.E., Shaw, T.J., Nixon, R.H., Fossum, E.R.: Parallel processor array for high speed path planning. In: Proceedings of IEEE Custom Integrated Circuits Conference, vol. 6, pp. 1–5 (1992)
27. Khan, G.M., Miller, J.F.: Solving mazes using an artificial developmental neuron. In: ALIFE, pp. 241–248 (2010)
28. Khatib, O.: Real-time obstacle avoidance for manipulators and mobile robots. Int. J. Robot. Res. **5**(1), 90–98 (1986)
29. Kolobov, V.I.: Advances in electron kinetics and theory of gas discharges. Phys. Plasmas (1994–present) **20**(10), 101610 (2013)
30. Lagzi, I., Soh, S., Wesson, P.J., Browne, K.P., Grzybowski, B.A.: Maze solving by chemotactic droplets. J. Am. Chem. Soc. **132**(4), 1198–1199 (2010)
31. Lee, C.Y.: An algorithm for path connections and its applications. IRE Trans. Electron. Comput. **3**, 346–365 (1961)
32. Lovass, P., Branicki, M., Tóth, R., Braun, A., Suzuno, K., Ueyama, D., Lagzi, I.: Maze solving using temperature-induced Marangoni flow. RSC Adv. **5**(60), 48563–48568 (2015)
33. Lumelsky, V.J.: A comparative study on the path length performance of maze-searching and robot motion planning algorithms. IEEE Trans. Robot. Autom. **7**(1), 57–66 (1991)
34. Lumelsky, V.J., Stepanov, A.A.: Path-planning strategies for a point mobile automaton moving amidst unknown obstacles of arbitrary shape. Algorithmica **2**(1–4), 403–430 (1987)
35. Nair, A., Raghunandan, K., Yaswant, V., Pillai, S.S., Sambandan, S.: Maze solving automatons for self-healing of open interconnects: modular add-on for circuit boards. Appl. Phys. Lett. **106**(12), 123103 (2015)
36. Nakagaki, T., Yamada, H., Toth, A.: Path finding by tube morphogenesis in an amoeboid organism. Biophys. Chem. **92**(1), 47–52 (2001)
37. Narendra, A.: Homing strategies of the Australian desert ant Melophorus bagoti I. Proportional path-integration takes the ant half-way home. J. Exp. Biol. **210**(10), 1798–1803 (2007)
38. Okabayashi, Y., Isoshima, T., Nameda, E., Kim, S.-J., Hara, M.: Two-dimensional nonlinear Fabry-Perot interferometer: an unconventional computing substrate for maze exploration and logic gate operation. Int. J. Nanotechnol. Mol. Comput. (IJNMC) **3**(1), 13–23 (2011)
39. Pavlov, V.V., Voronin, A.N.: The method of potential functions for coding constraints of the external space in an intelligent mobile robot. Sov. Autom. Control **17**(6), 45–51 (1984)
40. Pershin, Y.V., Di Ventra, M.: Solving mazes with memristors: a massively parallel approach. Phys. Rev. E **84**(4), 046703 (2011)
41. Pimienta, V., Antoine, C.: Self-propulsion on liquid surfaces. Curr. Opin. Colloid Interface Sci. **19**(4), 290–299 (2014)

42. Qin, J., Wheller, A.R.: Maze exploration and learning in C. elegans. Lab Chip **7**(2), 186–192 (2007)
43. Rambidi, N.G., Yakovenchuk, D.: Chemical reaction-diffusion implementation of finding the shortest paths in a labyrinth. Phys. Rev. E **63**(2), 026607 (2001)
44. Reyes, D.R., Ghanem, M.M., Whitesides, G.M., Manz, A.: Glow discharge in microfluidic chips for visible analog computing. Lab Chip **2**(2), 113–116 (2002)
45. Reynolds, A.M., Dutta, T.K., Curtis, R.H.C., Powers, S.J., Gaur, H.S., Kerry, B.R.: Chemotaxis can take plant-parasitic nematodes to the source of a chemo-attractant via the shortest possible routes. J. R. Soc. Interface **8**(57), 568–577 (2011)
46. Ricigliano, V., Chitaman, J., Tong, J., Adamatzky, A., Howarth, D.G.: Plant hairy root cultures as plasmodium modulators of the slime mold emergent computing substrate Physarum polycephalum. Front. Microbiol. 6 (2015)
47. Rubin, F.: The Lee path connection algorithm. IEEE Trans. Comput. **100**(9), 907–914 (1974)
48. Scherber, C., Aranyosi, A.J., Kulemann, B., Thayer, S.P., Toner, M., Iliopoulos, O., Irimia, D.: Epithelial cell guidance by self-generated EGF gradients. Integr. Biol. **4**(3), 259–269 (2012)
49. Shannon, C.E.: Presentation of a maze-solving machine. In: 8th International Conference of the Josiah Macy Jr. Found. (Cybernetics), pp. 173–180 (1951)
50. Stan, M.R., Burleson, W.P., Connolly, C.I., Grupen, R.A.: Analog VLSI for robot path planning. Analog Integr. Circ. Sig. Process. **6**(1), 61–73 (1994)
51. Steinbock, O., Tóth, Á., Showalter, K.: Navigating complex labyrinths: optimal paths from chemical waves. Science-New York Then Washington-, 868–868 (1995)
52. Stratton, L.O., Coleman, W.P.: Maze learning and orientation in the fire ant (Solenopsis saevissima). J. Comp. Physiol. Psychol. **83**(1), 7 (1973)
53. Tarassenko, L., Blake, A.: Analogue computation of collision-free paths. In: IEEE International Conference on Robotics and Automation, Proceedings, pp. 540–545. IEEE (1991)
54. Tarassenko, L., Brownlow, M., Marshall, G., Tombs, J., Murray, A.: Real-time autonomous robot navigation using VLSI neural networks. In: Advances in Neural Information Processing Systems, pp. 422–428 (1991)
55. Tero, A., Kobayashi, R., Nakagaki, T.: Physarum solver: a biologically inspired method of road-network navigation. Physica A **363**(1), 115–119 (2006)
56. Turner, C.H.: The homing of ants: an experimental study of ant behavior. J. Comp. Neurol. Psychol. **17**(5), 367–434 (1907)
57. Wallace, R.A.: The maze solving computer. In: Proceedings of the 1952 ACM National Meeting (Pittsburgh), pp. 119–125. ACM (1952)
58. Willardson, D.M.: Analysis of micromouse maze solving algorithm. In: Learning from Data (2001)
59. Wyard-Scott, L., Meng, Q.-H.M.: A potential maze solving algorithm for a micromouse robot. In: IEEE Pacific Rim Conference on Communications, Computers, and Signal Processing, 1995. Proceedings, pp. 614–618. IEEE (1995)
60. Yokawa, K., Derrien-Maze, N., Mancuso, S., Baluška, F.: Binary decisions in maize root behavior: Y-maze system as tool for unconventional computation in plants. Int. J. Unconv. Comput. 10 (2014)
61. Zhang, H.M., Peh, L.S., Wang, Y.H.: Study on flood-fill algorithm used in micromouse solving maze. In: Applied Mechanics and Materials, vol. 599, pp. 1981–1984. Trans Tech Publications (2014)
62. Zhang, S., Mizutani, A., Srinivasan, M.V.: Maze navigation by honeybees: learning path regularity. Learn. Mem. **7**(6), 363–374 (2000)
63. Zhao, M., Marquez, A.G.: Understanding humans' strategies in maze solving. arXiv:1307.5713 (2013)

Computer Chess Endgame Play With Pawns: Then and Now

M. Newborn and R. Hyatt

> *In the past Grandmasters came to our computer tournaments to laugh. Today they come to watch. Soon they will come to learn.*
> Monty Newborn 1977. [12]

Abstract Forty years ago, PEASANT, a program designed to play chess king and pawn endgames, was tested on a set of sixteen positions. The results reflected the weak state of play by computers during this stage of the game. To examine the progress made since then, CRAFTY was recently tested on the same set, and this paper reports on its far stronger results. A new more difficult set is then proposed in this paper, tested on CRAFTY, and the results presented herein. The set is intended to serve others wishing to measure and compare their program's performance.

1 Introduction

Forty years ago, in the mid-1970s, computer chess programs were beginning to be recognized as capable chess players. However, while their middle-game tactical play was on a par with (USCF) Expert-level chess players, their end-game play was

The article COMPUTER CHESS ENDGAME PLAY WITH PAWNS: THEN AND NOW appears here in the slightly adapted version from the original publication which appeared in the Journal of the International Computer Games Association, Vol. 37, No. 4, pp. 195–205 (2015). We thank Jaap van den Herik, Editor-in-Chief of the journal, for allowing us to reprint the article.

M. Newborn (✉)
School of Computer Science, McGill University, Montreal, QC, Canada
e-mail: newborn@cs.mcgill.ca

R. Hyatt
Department of Computer and Information Sciences, University of Alabama at Birmingham, Birmingham, AL, USA
e-mail: hyatt@uab.edu

significantly weaker, maybe at most at the level of (USCF) B-rated players. It was often argued that it would be many years before one could boast of their play during this final stage of the game. Now it is forty years later, and it is time to examine the progress made.

It was back then that a set of sixteen positions, taken from Reuben Fine's *Basic Chess Endings* [5], was used to test the capabilities of a chess program named PEAS-ANT on king and pawn endgames. PEASANT was designed to specifically handle this phase of a game. The program, designed by this paper's first author, with help from two students, Israel Gold and Leon Piasetski, an International Master, was tested on the Fine positions, and the results were reported in Peter Frey's 1977 book *Chess Skill in Man and Machine* [17]. In essence, the program probably did better than expected at the time, but nevertheless its results were unimpressive. To observe the progress since then we asked the classic chess program CRAFTY, designed by this paper's second author, to attempt the same positions.

In the next section, the sixteen test positions chosen from Fine are presented (called Test Set 1 and hereafter abbreviated as TS1). The computing system, hardware and software, used by PEASANT in 1975 for its test are described in Sect. 3. PEASANT's results are summarized in Sect. 4. Following these two sections, two similar sections are presented for CRAFTY. Based on the performance of CRAFTY on TS1, a new more difficult test set (called Test Set 2 and hereafter abbreviated by TS2) was proposed and constitutes Sect. 7. CRAFTY's efforts with TS2 are given in Sect. 8. This set can serve to compare other current and future chess engines with CRAFTY's current capabilities. Some remarks along with an acknowledgement and a number of references end the paper.

2 TS1

TS1 is presented below in Fig. 1. The FEN notation for each position is given in Table 6 in the Appendix. In general, the positions get increasingly more complex with increasing Fine number. The first three positions have only three pawns on the board, while the last four have ten or more. Exclamation characters under a position indicate the move shown is the only winning move.

3 PEASANT's Hardware and Software

Work on PEASANT began in 1973 when programming still meant working with a deck of punched cards! The program was written to run on the IBM 360/370 series computers. It consisted of about 2000 FORTRAN instructions and executed in about 25 K words of memory. Data structures occupied about 3 K of the 25 K words. It examined approximately 300 terminal positions per second when tested on an IBM 360/158 located in the McGill University Computing Centre. The IBM 370/158,

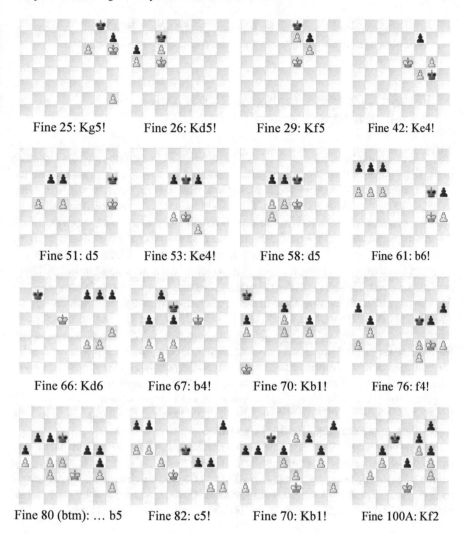

Fine 25: Kg5! Fine 26: Kd5! Fine 29: Kf5 Fine 42: Ke4!

Fine 51: d5 Fine 53: Ke4! Fine 58: d5 Fine 61: b6!

Fine 66: Kd6 Fine 67: b4! Fine 70: Kb1! Fine 76: f4!

Fine 80 (btm): … b5 Fine 82: c5! Fine 70: Kb1! Fine 100A: Kf2

Fig. 1 TS1

debuting in 1972, was given credit for executing one million instructions/second [9]; the size of its main memory was 4 MB.

PEASANT was implemented as a conventional Shannon–Turing program using the alpha-beta algorithm along with an evaluation function tailored to king and pawn endgames and pruning heuristics to narrow the search. The killer heuristic was used to further improve the effectiveness of the search. There were no hash tables and no endgame tables. Two subroutines called ONEPAWN and TWOPAWNS, described in Piasetski's M.Sc. Thesis [20], were called to handle positions in the search tree with a single pawn or two pawns by one side on the board. PEASANT searched to depths of about ten plies when there were three or four pawns on the board.

4 PEASANT's Performance on TS1 at Two Minutes/Position

PEASANT attempted TS1 with a limit of 2 min of CPU time per position. (In those days, IBM systems time-shared program execution on a single CPU, and the actual time a program ran was often considerably less than the real clock time.) According to the write-up, "PEASANT selected the correct move for eleven positions, although the correct move was selected for reasons somewhat secondary to the main theme. In the other five positions, PEASANT had no understanding of the real issues involved in solving the position." The time taken on any particular position was not given.

From the write-up in Frey, it is not clear which of the eleven positions PEASANT correctly solved. It seems that the program saw a win in only three of them, Fine 29, 51, and 61, although it came up with the correct move in six others, Fine 26, 42, 80, 82, 90, and 100A. It failed to solve Fine 25, 53, 67, 70, and 76. The write-up leaves unclear whether it solved Fine 58, as the write-up says it "vacillated" between two moves, one the correct move and the other an incorrect move. No move was indicated for Fine 29, but it was implied PEASANT solved the position and credit for it is given here. In summary, of the sixteen positions, PEASANT played the wrong move in six, the correct move in nine, and the write-up left unclear its selection in one. Of the sixteen positions, PEASANT saw that it could reach a winning position in only three. Table 1 summarizes PEASANT's results.

When speculating on the effort necessary for PEASANT to understand that several of the positions were won, some predictions were made in the paper. It was felt that Fine 25, 53, and 58 would each require 1000 h of computing time and a 25-ply search to find a win. Similarly Fine 70 would require 25,000 h and a 30-ply search, while Fine 90 and 100A would require over 300,000 years and a 40-ply search.

5 CRAFTY's Hardware and Software

For this test and the new one presented in Sect. 7, CRAFTY ran on a 3.2 GHz, four-core Intel I7 processor with 16 GB of memory and with a 64-bit word size. The system cost less than $2,000 even though it included a 512 GB solid-state drive (SSD) and a 21-in. color high definition monitor.

CRAFTY [7] is an outgrowth of the former world champion program CRAY BLITZ [8]. It came into existence in the middle 1990s when the capabilities of the Cray-series supercomputers were being challenged by much simpler and far less expensive sys-

Table 1 Summary of PEASANT's results on TS1 at 2 min/position

#	Position, to move	Initial material	Correct move/ PEASANT's move		Observations
1	Fine 25, w	KPPKP	Kg5/Kh5	×	Played incorrectly
2	Fine 26, w	KPPKP	Kd5/Kd5	√	Played correctly. Other moves lead quickly to negative consequences
3	Fine 29, w	KPPKP	Kf5/Kf5	√	Played correctly
4	Fine 42, w	KPPKP	Ke4/Ke4	√	Played correctly. Other moves lead quickly to negative consequences
5	Fine 51, w	KPPKPP	d5/d5	√	Played correctly with a shallow 5-ply search, and saw a win
6	Fine 53, w	KPPKPP	Ke4/Kd4	×	Played incorrectly
7	Fine 58, w	KP(3)KPP	d5+/??	?	Vacillated between c5 and d5; selection not clear. Did not see win
8	Fine 61, w	KP(4)KP(4)	b6/b6	√	Played correctly and saw a win
9	Fine 66, w	KP(3)KP(3)	Kd6/??	?	Move selected not given, though said to be wrong
10	Fine 67, w	KP(3)KP(3)	b4/Kf4	×	Played incorrectly
11	Fine 70, w	KP(4)KP(3)	Kb1/Kb2	×	Played incorrectly
12	Fine 76, w	KP(5)KP(4)	f4/h4	×	Played incorrectly
13	Fine 80, b	KP(6)KP(6)	b5/b5	√	Played correct move but did not see win
14	Fine 82, w	KP(5)KP(5)	c5/c5	√	Played correct move but did not see win
15	Fine 90, w	KP(6)KP(6)	e4/e4	√	Played correct move but did not see win
16	Fine100A, w	KP(5)KP(5)	Kf2/Kf2	√	Played correct move only because all others were seen to lose

tems. Tord Romstad, the developer of STOCKFISH, described CRAFTY as "arguably the most important and influential chess program ever" [21].

CRAFTY uses the traditional ideas such as iterative deepening, alpha–beta pruning, transposition/refutation tables and similar things. It also uses more recent ideas like Late Move Reductions (LMR) and Late Move Pruning (LMP). The net result is an effective branching factor well below 2.0 for the endgame positions tested. It is designed to take advantage of hardware with multiple cores with its parallel search, and easily searches 30M nodes per second and beyond even on the inexpensive hardware platform used for the tests reported here. Finally, it also uses Nalimov endgame tables [16] for three through five piece endgames and plays them perfectly. It should be repeated out that PEASANT did not use transposition tables or endgame tables.

6 CRAFTY's Performance on TS1 at Three Minutes/Position

CRAFTY's output from running TS1 is summarized in Table 2. Column 3 of the table gives the initial material on the board. Column 4 gives the material on the board at the end of the final continuation provided by the program. Column 5 indicates the best move in the position followed by CRAFTY's selection. Column 6 gives the time taken by CRAFTY to find the move that it stuck with thereafter. That is, often, for a very short period of time, CRAFTY would bounce around from one move to another, and then settle with one move for the rest of the time; column 6 indicates that time. Column 7 indicates the score assigned by CRAFTY to the selected move. Column 8 gives the approximate number of nodes searched and Column 9 gives the approximate number of nodes per second searched. Lastly, Column 10 gives the length of the continuation found by CRAFTY. For a continuation leading to a position in an endgame table, the length of the continuation is supplemented by a parenthetical note that indicates the length of the continuation leading to the endgame table. (For example, for Fine 53, the entry "47 (11 + EGTB)" means that a continuation of length 11 ended at a position in an endgame table, and that the table found a mate-in-36 at that point, resulting in a 47 ply continuation to mate.)

First and foremost, CRAFTY found a winning move for the side to move in all sixteen positions. The first eleven positions were assigned mates, while the last five were assigned scores that called for a resignation by its opponent. Although CRAFTY was given 3 min to search each of the sixteen positions, it needed no more than 0.03 s to find a winning move in each position, thus solving the entire suite in less than half a second. For Fine 80, however, while the move selected by CRAFTY leads to a win and was the move Fine said was the one to play, when given more time to search the position (as discussed later in this section), another move was found best. A question mark appears in Column 5 for Fine's move b6 (also questioned in Table 3).

Table 2 Summary of CRAFTY's results on TS1 at 3 min/position

#	Position, to move	Initial material	Final material	Fine's move/ CRAFTY's move		Time to final selection (s)	Final score	Time to 'mate!'	Nodes searched	Nodes/ second	Length of continuation
1	Fine 25, w	KPPKP	KPPKP	Kg5/Kg5	✓	0.01	M18	0.01	856	85.6 K	28 (1 + EGTB)
2	Fine 26, w	KPPKP	KPPKP	Kd5/Kd5	✓	0.00	M17	0.00	1426	142.6 K	33 (1 + EGTB)
3	Fine 29, w	KPPKP	KPPKP	Kf5/Kf5	✓	0.00	M13	0.01	895	89.5 K	25 (1 + EGTB)
4	Fine 42, w	KPPKP	KPPKP	Ke4/Ke4	✓	0.00	M17	0.01	1.5 K	145.3 K	33 (1 + EGTB)
5	Fine 51, w	KPPKPP	KPPPKP	d5/d5	✓	0.00	M19	0.01	169.9 K	772.1 K	37 (2 + EGTB)
6	Fine 53, w	KPPKPP	KPPKPP	Ke4/Ke4	✓	0.03	M24	0.12	6.9 M	16.7 M	47 (11 + EGTB)
7	Fine 58, w	KP(3)KPP	KPPKP	d5+/d5+	✓	0.00	M24	0.07	50.2 M	25.8 M	47 (3 + EGTB)
8	Fine 61, w	KP(4)KP(4)	KQQKQPP	b6/b6	✓	0.01	M20	43.35	7.7 B	42.7 M	39
9	Fine 66, w	KP(3)KP(3)	KPPKP	Kd6/Kd6	✓	0.03	M21	6.31	7.7 B	42.7 M	41 (16 + EGTB)
10	Fine 67, w	KP(3)KP(3)	KP(3)KP	b4/b4	✓	0.01	M27	7.87	6.8 B	37.5 M	53 (37 + EGTB)
11	Fine 70, w	KP(4)KP(3)	KQQPPKP	Kb1/Kb1	✓	0.00	M32	11.44	7.6 B	41.9 M	63
12	Fine 76, w	KP(5)KP(4)	KQPKP	f4/f4	✓	0.00	15.65		5.2 B	28.8 M	42
13	Fine 80, b	KP(6)KP(6)	KQPKP	b5??/b5	✓	0.03	−13.93		4.7 B	26.0 M	36
14	Fine 82, w	KP(5)KP(5)	KP(4)KP	c5/c5	✓	0.00	>7.27		5.4 B	29.9 M	28
15	Fine 90, w	KP(6)KP(6)	KQPPKPP	e4/e4	✓	0.01	>8.07		5.0 B	27.5 M	38
16	Fine100A, w	KP(5)KP(5)	KQQPPKP	Kf2/Kf2	✓	0.00	>29.20		6.9 B	38.3 M	47

Table 3 Summary of CRAFTY's results on the last five TS1 positions at 1 h/position

#	Position, to move	Initial material	Final material	Fine's move/ CRAFTY's move	Time to final selection (s)	Final score	Time to 'mate!'	Nodes searched	Nodes/ second	Length of continuation
12	Fine 76, w	KP(5)KP(4)	KQPKP	f4/f4 √	0.01	M32	52 min 28 s	147.1 B	40.9 M	63 (45 + EGTB)
13	Fine 80, b	KP(6)KP(6)	KQPKP	b5?!/Kc7 √	0.03	−23.45		127.7 B	35.5 M	39
14	Fine 82, w	KP(5)KP(5)	KP(4)KP	c5/c5 √	0.00	>13.31		108.9 B	30.2 M	32
15	Fine 90, w	KP(6)KP(6)	KQPPKPP	e4/e4 √	0.01	>12.26		90.7 B	25.2 M	44
16	Fine100A, w	KP(5)KP(5)	KQQPPKP	Kf2/Kf2 √	0.00	M24	6 min 32 s	148.9 B	41.4 M	47 (33 + EGTB)

CRAFTY's five-piece endgame tables came into play in finding mates to the first eleven of the sixteen positions. It did not find mates for positions that started with nine or more pawns on the board, as was the case for the final five positions. The principal continuations found for the first two of these positions, Fine 76 and Fine 80, ended with only five pieces on the board and one might think that CRAFTY's endgame tables would have assigned win, loss, or draw values to these positions. However, CRAFTY tries to limit the I/O overhead caused by endgame table probes to a file kept on its SSD, and is programmed not to probe positions at the last six plies of any search branch. Because of this, some of the continuations ended in 5-piece positions without a corresponding mate or draw assigned. Using a modern SSD, as opposed to the outdated rotating disk drive used, would reduce the overhead in making these probes and allow a more aggressive probing. The principal continuations for the final three positions, Fine 82, Fine 90, and Fine 100A, each ended with seven pieces on the board. For all these five positions, CRAFTY's continuations led to scores that were clearly wins for the side on the move. The most difficult position for CRAFTY was Fine 82, ending nevertheless in a position calling for a resignation by its opponent as shown in (Fig. 2).

While CRAFTY used a five-piece endgame table for this test, endgame tables exist for all seven-piece endgames though they require far more memory than available for CRAFTY [6]. Using a 7-piece endgame table that was accessible at all positions in the search tree would have allowed CRAFTY to find mates in all 16 positions [13]. Six-piece endgame tables require on the order of a terabyte of memory [15].

When examining CRAFTY's results on the 3-min test, the authors became curious how CRAFTY would do if given more time on the five positions where mate was not found. Thus a second run was carried out with an hour of computing time allotted to each of these positions. Table 3 summarizes the results. CRAFTY found a mate-in-32 for Fine 76 after computing for 52 min; it also found a mate-in-32 for Fine 100A after computing for 6.32 min, subsequently reducing the length of the mate to mate-in-24 after computing for 46:19 min. It also found the analysis in Fine's book for Fine 80 to be incorrect. In fact the best move in this position is 1 ... Kc7, winning for Black and winning more quickly than Fine's choice of 1 ... b5. Though CRAFTY was not able to find a mate here when given an hour, it did find a line leading to a much-improved score of 23.45 compared to a score of 13.93 when searching for only 3 min. For Fine 82 and Fine 90, much improved scores were also obtained on this second run.

One especially illuminating example of the great improvement in endgame play is shown by efforts to solve Fine 70. In PEASANT's write-up, it was estimated that PEASANT would require 25,000 h, almost three years, to solve the position using its hardware. It is now solved in a millisecond by CRAFTY, approximately 100 trillion (10^{14}) times faster than estimated. Hardware speeds have improved, but nowhere near that much. As another example in this regard, in 1975 a CDC Cyber 176 (one of the most powerful computers of its time running one of the strongest programs of its time) searched 2,600 nodes per second, about 10 times as many as PEASANT. Today CRAFTY easily hits 26,000,000 nodes per second on a 4-core Apple iMac. If, for round numbers, we assume 300,000,000 nodes per second can be searched on the best hardware currently available, then we have a performance gain of approximately

Fig. 2 CRAFTY's 28-ply
continuation 1. c5 Ke6 2. c6
b6 3. axb6 axb6 4. Ke4 g3
5. h4 f3 6. Kxf3 Kd6 7. Kxg3
Kc7 8. Kf4 Kc8 9. Kg5 Kc7
10. Kh6 Kd6 11. g4 Kc7
12. g5 Kd8 13. Kxh7 Ke7
14. g6 Ke6

Fine 82: c5!   After CRAFTY's 14. ... Ke6

10^6. And as a third approach to compare then and now, CHESS 4.5 (importantly, using transposition tables, unlike PEASANT) was reported [4] to have solved Fine 70 in 26 min, or 1,560 s. CRAFTY solves it in 0.001 s, more than a factor of 10^6 faster, and very close to our previous comparison. There is a difference of 10^8 between the first estimate of a speedup of 10^{14} and the second and third estimates of about 10^6. This difference must be attributed to two factors: (1) an overestimate of 25,000 h to solve Fine 70, and (2) not taking into account software improvements. Given that CHESS 4.5 solved Fine 70 in 26 min, or about 50,000 times or 5×10^4 times as fast as estimated that PEASANT would take, this would imply software improvements accounted for a speedup by a factor of 2000. These numbers are probably easy to dispute.

From another perspective of the performance gain, CRAFTY searched approximately one hundred thousand times as many nodes per second as searched by PEASANT. If the same progress had been made with rocket ships that traveled to the moon 240,000 miles from Earth at say an average speed of 4000 miles per hour in 1977, and thus taking 60 h to travel to the moon, the spaceship of today would be flying at 400,000,000 miles per hour and would take about 2 s for the trip, racing there at near the speed of light's approximately 670,000,000 miles per hour!

7   A New Test Set 'TS2'

Initially the objective of this paper was to report on the great improvement in endgame play made over the years by testing CRAFTY on the TS1 in 2015. But upon finishing the testing, it became clear that a more difficult test set was in order to measure future performance improvements as CRAFTY neared perfection on the TS1. A new set of sixteen miscellaneous positions was thus put together and called Test Set Two (TS2). Table 3 presents the 16 positions. They are given in FEN notation in Table 7 in the Appendix. The source of the positions is given in Table 4. The first seven (Positions 1–7) are from Averbakh and Maiselis's book *Comprehensive Chess Endings 4, Pawn Endings* [1]. Peter Mackenzie suggested them as a test set for computers in a 2000 Computer Chess Club posting [14]. Vincent Lejeune suggested another six test positions at that time, although only four are included here (Positions 8–11) [11]. Lejeune's positions had appeared earlier in John Dunn and Frederic Friedel's two ICCA Journal articles [18, 19], where they point out that Position 9 was previous published by Grigoriev in 1938, and Position 10 was previously published by J. Krivcun in 1961. Position 12 dates back to the 1919 British Chess Magazine, where it was named "The Christmas Cracker" [3]. The final four positions are from Averbakh and Maiselis's *Pawn Endings* [2] (Fig. 3).

Table 4 Sources of TS2 positions

Position	Source of position
CCE4 479	Grigoriev, 1932. CCE, V.4., pg 166.
CCE4 491a	Grigoriev, 1934. CCE, V. 4, pp. 169–170.
CCE4 530	Grigoriev 1933. CCE, V. 4, pg. 184.
CCE4 608	Gheorghiu—Gligoric, Hastings 1964. CCE, V. 4, pg. 217.
CCE4 679	Eliskases—Skalicka, Podebrady 1936. CCE, V. 4, pg 354.
CCE4 680	Kryukovsky—Bishard, Leningrad 1974. CCE, V. 4, pp. 254–255.
CCE4 765	Zinar, 1983. CCE, V. 4, pg. 290.
Lejeune 2	J. Nunn, in Nunn and Friedel, Brains of the Earth Challenge, *ICCA Journal*, V. 22, No. 3, pp. 188–189.
Lejeune 4	Grigoriev, 1938. in Dunn and Friedel (1999). The "Brains of Earth" Challenge, *ICCA Journal*, Vol. 22, No. 4, pp. 259–263. Also in CCE, V. 4, pg. 183.
Lejenue 5	Krivcun, 1961, in Dunn and Friedel (1999). The "Brains of Earth" Challenge, *ICCA Journal*, Vol. 22, No. 4, pp. 259–263.
Lejeune 6	J. Nunn, in Nunn and Friedel, The "Brains of Earth" Challenge, ICCA Journal, V. 22, No. 3, pp. 188–189.
Christmas Cracker	A. Baker, British Chess Magazine 1916.
Pillsbury1895	Pillsbury–Gunsberg, Hastings 1895. PE, pg. 256.
Capablanca1919	Capablanca–Condé, Hastings 1919. PE, pg. 248.
Botvinnik1944	Botvinnik–Flohr, Moscow 1944. PE, pg. 285.
Botvinnik1958	Botvinnik–Smyslov, 20th World Championship 1958. PE, pg. 270.

Note CCE = *Comprehensive Chess Endings*; PE = *Pawn Endings*

8 CRAFTY's Performance on TS2 at One Hour/Position

To begin, CRAFTY found the recommended move to all but two positions, CCE4 765 and Capablanca1919, when given an hour of computing time per position; mates were found to eight of the sixteen. Winning continuations were found to fourteen, while two positions were correctly understood as drawn. While selecting a different move to CCE4 765, CRAFTY's choice of 1. Kd3 led to a draw as did the recommended move

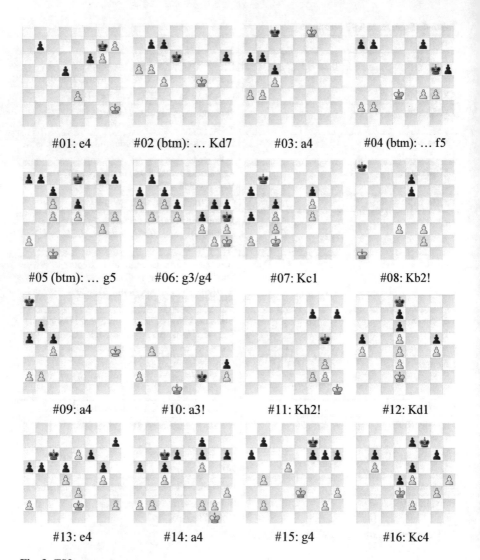

#01: e4   #02 (btm): ... Kd7   #03: a4   #04 (btm): ... f5

#05 (btm): ... g5   #06: g3/g4   #07: Kc1   #08: Kb2!

#09: a4   #10: a3!   #11: Kh2!   #12: Kd1

#13: e4   #14: a4   #15: g4   #16: Kc4

Fig. 3 TS2

1. Kc1. For the position Capablanca1919, CRAFTY selected 1. g4, a move different than recommended, i.e., 1. a4, though both moves led to clear wins.

For historical purposes, the final continuation found for each position is given here along with some information on the search. In retrospect, it would have been good if more information were given in the paper on PEASANT regarding its efforts on each of the Fine positions. This would have made our comparison between PEASANT and CRAFTY more thorough and relevant. By doing so here with CRAFTY's results, future comparisons between CRAFTY and some future engine will not suffer from this shortcoming.

CCE4 479/wtm.... CRAFTY selected 1. e4 leading to a mate-in-24 after calculating for 21.73 s. The endgame table assigned a mate to the 15-ply continuation: 1. e4 dxe4 2. Kg3 e3 3. Kf3 e2 4. Kxe2 b5 5. Ke3 b4 6. Kd4 b3 7. Kc3 f5 8. Kxb3 {EGTB}. CRAFTY liked 1. Kg3 prior to settling on 1. e4.

CCE4 491a/btm.... CRAFTY selected 1. ... Kd7 after searching for 0.04 s, and it stuck with this selection throughout the remainder of the 1-h search. It realized after 0.79 s that the move led to a draw, assigning it a score of −0.01. The final 27-ply continuation found when search ended in one of its five-piece endgame tables was: 1. ...Kd7 2. Kg4 Kc8 3. Kh5 Kb8 4. Kh4 Ka7 5. Kg4 c6 6. Kf4 cxb5 7. cxb5 Kb8 8. Kf5 Kc8 9. Ke5 Kc7 10. Ke6 h5 11. Ke5 h4 12. Kf4 Kd6 13. Kg4 Kc5 14. a6 bxa6 {EGTB}.

CCE4 530/wtm.... CRAFTY selected 1. a4 and stuck with it after 2.30 s. It initially found the move yields mate-in-46 after 3 min and 59 s, and later found it led to a shorter mate-in-42 after 4 min and 15 s. The final 49-ply continuation found was: 1. a4 Kd7 2. a5 Kd6 3. Kf7 Kd7 4. Kf6 Kd6 5. Kf5 Kc7 6. Ke6 Kc6 7. Ke7 Kc7 8. Ke8 Kc6 9. Kd8 bxa5 10. Ke7 Kc7 11. Ke6 Kc6 12. Ke5 Kc7 13. Kd5 Kb6 14. Kd6 Kb7 15. Kxc5 Kc7 16. Kd5 Kd7 17. c5 Kc7 18. c6 Kb8 19. Kd4 Kc8 20. Kc4 Kd8 21. Kd5 Ke7 22. Kc5 Ke6 23. c7 Kd7 24. Kb6 Kd6 25. Kxa5 {EGTB}.

CCE4 608/btm .... CRAFTY selected 1 ... f5 after 0.01 s and stuck with this move for the remainder of the search, ending with a continuation having a score of −33.26. The final 35-ply continuation was: 1. ...f5 2. Ke2 f4 3. Kf2 fxg3+ 4. Kxg3 h4+ 5. Kh3 b5 6. Kg2 Kf4 7. a3 Ke3 8. f4 Kxf4 9. Kf2 a5 10. Ke2 Ke4 11. Kd2 h3 12. a4 bxa4 13. Kc3 a3 14. b3 h2 15. b4 h1=Q 16. bxa5 a2 17. Kb4 a1=Q 18. a6 Qxa6, though CRAFTY's score was failing high when search terminated.

CCE4 679/btm .... CRAFTY selected 1 ... g5 after 0.05 s and then stuck with it thereafter. The final score was −16.25 and the final 62-ply continuation was: 1. ...g5 2. Kd2 Kf7 3. Ke3 Kg6 4. Kf3 Kh5 5. hxg5 Kxg5 6. Kg2 Kg4 7. Kf2 Kh3 8. Kf3 a6 9. a3 a5 10. a4 h5 11. Kf2 Kh2 12. Kf3 Kg1 13. g4 h4 14. g5 h3 15. Kg4 h2 16. Kf5 h1=Q 17. Kxe5 Qh2+ 18. Ke6 Qg3 19. Kf5 Qf3+ 20. Kg6 Qxe4+ 21. Kh5 Qe2+ 22. Kh6 Qxc4 23. g6 Qh4+ 24. Kg7 Qxa4 25. Kg8 Qb3+ 26. Kh7 Qh3+ 27. Kg7 a4 28. Kg8 a3 29. g7 a2 30. Kf8 Qc8+ 31. Kf7.

CCE4 680/wtm .... CRAFTY selected 1. g4, leading to mate which it found after searching for 0.57 s. The final 43-ply continuation was: 1. g4 fxg3+ 2. Kg2 g4 3. hxg4 Kg5 4. gxh5 Kxh5 5. Kxg3 Kg5 6. f4+ Kh5 7. Kh3 Kh6 8. Kg4 Kg6 9. f5+ Kf6 10. Kf4 Kf7 11. Kg5 Kg7 12. f6+ Kh7 13. Kh5 Kg8 14. Kg6 Kf8 15. f7 Ke7

16. Kg7 Ke6 17. f8=Q Kd7 18. Qb8 b5 19. axb6 Ke6 20. Qe5+ Kd7 21. Kf8 Kd8 22. Qe8#

CCE4 765/wtm .... CRAFTY selected 1. Kd3. Crafty's final 35-ply continuation was: 1. Kd3 Kc7 2. Ke4 Kd6 3. Kf3 Ke7 4. Kg2 Kf7 5. Kg3 Kf8 6. Kg4 Kf7 7. Kh5 Kg7 8. a3 a5 9. Kh4 Kf7 10. Kg3 Ke8 11. Kg2 Kd7 12. Kf3 Ke7 13. Ke3 Kd7 14. Kf2 Kd8 15. Ke2 Ke8 16. Kf1 Kf7 17. Kg2 Ke8 18. Kf1 led to a draw. Both 1. Kc1 and 1. Kd3 lead to draws.

Lejeune 2/wtm .... CRAFTY selected 1. Kb2 after 0.01 s and stuck with it for the remainder of the search. It saw a mate-in-33 after 0.48 s. The final 14-ply continuation was: 1. Kb2 e5 2. Kc3 Kb7 3. d4 Kc6 4. dxe5 Kd5 5. f4 Ke4 6. Kc4 Kf5 7. Kc5 Kxf4 {EGTB}.

Lejeune 4/wtm .... CRAFTY selected 1. a4 after 0.04 s and stuck with it for the remainder of the search. It saw a mate-in-32 after 19.03 s, and then a mate-in-31 after 41.99 s. The final 40-ply continuation was: 1. a4 Kb7 2. Kh5 Kc6 3. Kg5 Kc7 4. Kg6 Kc6 5. b3 Kc7 6. Kg7 Kc6 7. Kf8 Kb7 8. Kf7 Kb8 9. Ke6 Kc7 10. Ke7 Kc6 11. Kd8 Kb7 12. Kd7 Kb8 13. Kc6 Ka7 14. Kc7 Ka6 15. Kb8 b5 16. axb5+ Kb6 17. Kc8 a4 18. bxa4 Ka5 19. Kd7 Kb4 20. b6 Kxc4 {EGTB}.

Lejeune 5/wtm .... CRAFTY selected 1. a3 after 0.05 s and stuck with it thereafter, finding a mate-in-31 after 2.07 s, and a shorter mate-in-25 after 2.58 s. The final 19-ply continuation was: 1. a3 Kf1 2. Kd2 Kf2 3. Kd3 Kf3 4. Kd4 Kf2 5. Ke4 Kg2 6. Ke3 Kg1 7. Kf3 Kxh2 8. Kf2 Kh1 9. Kg3 Kg1 10. Kxh3 {EGTB}.

Lejeune 6/wtm .... CRAFTY selected 1. Kh2 after 0.41 s and stuck with it thereafter. It saw a mate-in-42 after 1 min and 14 s, and then after 1 min and 35 s, it found a mate-in-39. The final 33-ply continuation was: 1. Kh2 f6 2. Kh1 f5 3. Kg1 Kf6 4. Kf1 Ke5 5. Ke1 Kd5 6. Kd1 Ke5 7. Kc2 Kd4 8. Kd2 Ke4 9. Ke2 h6 10. Kf1 Kd4 11. Kg1 Ke5 12. Kh2 Kf6 13. Kh3 h5 14. Kh4 Kg6 15. f3 Kh6 16. g4 hxg4 17. fxg4 {EGTB}.

Christmas Cracker/wtm .... CRAFTY selected 1, Kd1, though initially finding mate-in-34 with 1. Kc2 after 6.52 s. It was after 8.49 s that CRAFTY preferred 1. Kd1, finding mate-in-29, then improving on that result, finding mate-in-28 after 14.15 s. The 55-ply mating continuation was: 1. Kd1 Kc7 2. Ke2 Kd8 3. Kd2 Ke7 4. Kc3 Ke8 5. Kc4 Kd8 6. Kb5 Kc7 7. Kxa5 Kb7 8. Kb5 Ka7 9. a5 Kb7 10. a6+ Ka7 11. Ka5 Ka8 12. Kb6 Kb8 13. a7+ Ka8 14. Kc7 Kxa7 15. Kxd7 Kb6 16. Kxd6 Kb7 17. Ke6 Ka6 18. d6 Ka5 19. d7 Kb4 20. d8=Q Kc3 21. d5 Kd4 22. d6 Ke3 23. Qf6 Kxd3 24. d7 Kc2 25. d8=Q Kb3 26. Qb6+ Kc2 27. Qfb2+ Kd1 28. Qg1#.

Pillsbury1895/wtm .... CRAFTY selected 1.e4 after 0.01 s and stuck with it thereafter. The final 44-ply continuation was: 1. e4 dxe4 2. Ke3 Kd6 3. d5 b4 4. Kxe4 a4 5. Kd3 Ke7 6. Kc4 b3 7. axb3 a3 8. Kc3 f5 9. gxf5 h5 10. b4 g4 11. b5 h4 12. b6 a2 13. Kb2 g3 14. d6+ Kxd6 15. b7 Kc7 16. e7 g2 17. b8=Q+ Kxb8 18. e8=Q+ Kc7 19. Qe7+ Kc6 20. Qe4+ Kd6 21. Qxg2 a1=Q+ 22. Kxa1 Ke5. The continuation was assigned a score of 11.94, though CRAFTY's search had failed high just before time ran out, with a final score of at least 12.10.

Capablanca1919/wtm .... CRAFTY selected 1. g4 after calculating for 0.00 s. It stuck with this move throughout its calculation. The final 41-ply continuation was: 1. g4 d5 2. b3 d4 3. Kf1 Kd6 4. f4 Kc6 5. h4 Kd7 6. Ke2 Ke7 7. Kd3 Ke8 8. h5 Ke7 9. a4 Ke8 10. b4 axb4 11. a5 Kd7 12. Kc2 b3+ 13. Kxb3 d3 14. Kc3 d2 15. Kxd2

Kc7 16. g5 fxg5 17. fxg5 hxg5 18. h6 Kb7 19. h7 Ka6 20. h8=Q Kxa5 21. Ke3. It failed high when search ended with a score of at least 10.90.

Botvinnik1944/wtm .... CRAFTY selected 1. g4 after calculating for 0.54 s. It stuck with this more thereafter leading to the 42-ply continuation: 1. g4 Ke7 2. h4 Kd6 3. Ke4 Kc7 4. h5 gxh5 5. gxh5 Kd6 6. Kf5 Kxd5 7. Kxf6 Ke4 8. Kg6 Ke5 9. Kxh6 Kf6 10. b5 a5 11. b6 a4 12. Kh7 Kf7 13. h6 Kf8 14. Kg6 Kg8 15. Kf6 Kh7 16. Ke6 Kxh6 17. Kd6 Kg5 18. Kc7 Kf4 19. Kxb7 a3 20. bxa3 Ke4 21. Kc7 Kd5 and a final score of 9.25.

Botvinnik1958/wtm .... CRAFTY selected 1. Kc4 after calculating for 0.56 s. It stuck with this move thereafter leading to the 54-ply continuation: 1. Kc4 e6 2. Kd3 Kg7 3. Kd2 g5 4. hxg5 Kg6 5. Ke1 Kxg5 6. Kf2 Kg4 7. Kg2 Kh5 8. Kh3 Kg5 9. g4 Kg6 10. Kh4 Kh6 11. g5+ Kg6 12. Kg4 Kf7 13. Kh5 Kg7 14. g6 Kg8 15. Kg4 Kh8 16. Kh4 Kg8 17. Kh5 Kg7 18. Kg5 Kg8 19. Kf6 Kf8 20. g7+ Kg8 21. Kxe5 Kxg7 22. Kxe6 Kg6 23. e5 Kg5 24. Kf7 Kf4 25. e6 Ke3 26. e7 Kxe2 27. e8=Q+ Kd3, and a score of 9.06 after calculating for 56 min and 52 s. It failed high when search ended with a score of at least 9.38.

9 Remarks

The logs of CRAFTY's output from the three tests summarized in Tables 2, 3 and 5 can be found at: http://www.cis.uab.edu/hyatt/crafty/ICGA/x where x = {FTS3minutes.log, FTS1hour.log, MTS1hour.log}. It might be pointed out that the FEN notation for the test positions in the appendix can be used as input if and when others wish to carry out these tests on their chess engines.

There is considerable room for improvement on TS2 by current and future chess engines. While CRAFTY found mates to eight of the positions, future chess engines can be expected to find mates to these eight positions much faster. And for those for which mates were not found and for which CRAFTY only found winning lines (six of the positions; two of the other eight were draws), chess engines with 6-piece endgame tables should find mates to four of them. CRAFTY's lines for CCE679 and Casablanca1919 terminated in positions with too many pieces on the board for the endgame tables of the current crop of chess engines to be of use.

Table 5 Summary of CRAFTY's results on TS2 at 1 h/position

#	Position, to move	Initial material	Final material	Recommended move/ CRAFTY's MOVE		Time to final selection (s)	Final score	Time to 'mate!'	Nodes searched	Nodes/ second	Length of continuation
1	CCE4 479/w	KP(3)KP(3)	KPPKP	e4/e4	√	21.7	M24	21.73 s	147.3 B	40.9 M	47 (15 + EGTB)
2	CCE4 491a/b	KP(3)KP(3)	KPKPP	Kd7/Kd7	√	0.04	−0.01	–	83.4 B	23.2 M	27
3	CCE4 530/w	KP(3)KP(3)	KPPKP	a4/a4	√	2.30	M42	2.30 s	128.8 B	35.8 M	81 (49 + EGTB)
4	CCE4 608/b	KP(4)KP(4)	KKQQ	f5/f5	√	0.01	−33.26	–	151.1 B	42.0 M	35
5	CCE4 679/b	KP(6)KP(6)	KPPKQP(3)	g5/g5	√	0.05	−16.25	–	139.2 B	38.7 M	62
6	CCE4 680/w	KP(7)KP(7)	KQP(4)KP(3)	g3, g4/g4	√	0.00	M22	57.11 s	158.3 B	44.0 M	43
7	CCE4 765/w	KP(5)KP(4)	KP(5)KP(4)	Kc1/Kd3	≠	0.06	0.01	–	90.8 B	25.2 M	35
8	Lejeune 2/w	KP(3)KPP	KPPKP	Kb2/Kb2	√	0.01	M33	0.48 s	138.6 B	38.5 M	65 (14 + EGTB)
9	Lejeune 4/w	KP(3)KP(3)	KPPKP	a4/a4	√	0.05	M31	19.03 s	147.3 B	40.9 M	61 (40 + EGTB)
10	Lejeune 5/w	KP(3)KPP	KPPKP	a3/a3	√	0.05	M25	0.05 s	1.2 B	37.1 M	49 (19 + EGTB)
11	Lejeune 6/w	KP(3)KPP	KPPKP	Kh2/Kh2	√	0.41	M39	1 min 14 s	130.3 B	36.2 M	77 (33 + EGTB)
12	Xmas Cracker/w	KP(5)KP(4)	KQQPKP	Kd1/Kd1	√	7.00	M28	8.49 s	3.5 B	45.4 M	55
13	Pillsbury 1895/w	KP(6)KP(6)	KQPPKP	e4/e4	√	0.01	>12.10	–	92.5 B	25.7 M	44
14	Capablanca 1919/w	KP(7)KP(6)	KQPPKP(3)	a4/g4	≠	0.00	>10.90	–	97.6 B	27.1 M	41
15	Botvinnik 1844/w	KP(5)KP(5)	KPPK	g4/g4	√	0.54	9.25	–	101.4 B	28.2 M	42
16	Botvinnik 1958/w	KP(5)KP(5)	KQPKPP	Kc4/Kc4	√	0.56	>9.38	–	95.6 B	26.5 M	54

While the positions considered in this paper have a narrow focus, more generally it is pretty amazing that positions once thought to require exceptional brilliance and to be beyond the capabilities of computers for years and years to come are now routine for computers as was the case for many of the positions considered here.

The strongest chess engines are super-grandmasters today, and they will continue to improve. Their endgame play certainly can no longer be considered a weakness. Many programs today are in the FIDE 3400 territory, which is well more than 400 rating points higher than the highest human rating ever achieved. In the future, multiprocessing systems with far more processors than currently exist will be used for increasing the strength of the programs. Memory sizes, too, will continue to grow thus strengthening programs. On the software side, endgame databases will continue to be developed as will better search heuristics.

More specifically for CRAFTY, there are several ways to improve its overall performance as well as its performance on the positions considered herein. Moving to Intel's new 18-core processor [10] from the current 4-core system would be one way and this is under consideration. In addition, recoding the C language instructions in CRAFTY with machine instructions would be a second way. Thirdly, with increasing memory sizes, incorporating some 6-piece endgame tables would be another way to further strengthen the program's endgame play. Lastly, attaching a faster SSD to CRAFTY will permit more frequent probes into the endgame tables and consequently strengthen its endgame play.

Whether one can contend that computers have artificial intelligence, they certainly play chess king and pawn endgames at a level well beyond the most brilliant human minds. And please do your best to relax and enjoy it when you see your first (electric) car driving itself down the road! They will be better drivers than us too before very long!

The quotation that began this paper might be updated to, "Once the grandmasters came to laugh, later they came to watch, and then they came to learn. Now, in 2015, they come to watch in awe and bewilderment while learning as well."

Acknowledgments The authors would like to thank the anonymous reviewers who led us to the source of many of the TS2 positions and who made many important specific and general suggestions for improvements to the submission.

Appendix

Table 6 TS1 in FEN notation

#	Position	Material	FEN	Best move
01	Fine 25	KPPKP	6k1/7p/5P1K/8/8/8/7P/8 w	Kg5
02	Fine 26	KPPKP	8/2k5/p1P5/P K5/8/8/8/8 w	Kd5
03	Fine 29	KPPKP	4p3/4Pp2/5P2/4K3/8/8/8/8 w	Kf5
04	Fine 42	KPPKP	8/5p2/8/4K1P1/5Pk1/8/8/8 w	Ke4
05	Fine 51	KPPKPP	8/8/2pp3k/8/1P1P3K/8/8/8 w	d5
06	Fine 53	KPPKPP	8/8/3pkp2/8/8/3PK3/5P2/8 w	Ke4
07	Fine 58	KP(3)KPP	8/8/2ppk3/8/2PPK3/2P5/8/8 w	d5+
08	Fine 61	KP(4)KP(4)	8/ppp5/8/PPP3kp/8/6KP/8/8 w	b6
09	Fine 66	KP(3)KP(3)	8/1k3ppp/8/3K4/7P/5PP1/8/8 w	Kd6
10	Fine 67	KP(3)KP(3)	8/2p5/3k4/1p1p1K2/8/1P1P4/2P5/8 w	b4
11	Fine 70	KP(4)KP(3)	8/k7/3p4/p2P1p2/P2P1P2/8/8/K7/w	Kb1
12	Fine 76	KP(5)KP(4)	8/8/p6p/1p3kp1/1P6/P4PKP/5P2/8 w	f4
13	Fine 80	KP(6)KP(6)	8/8/1ppk4/p4pp1/P1PP2p1/2P1K1P1/7P/8 b	b5
14	Fine 82	KP(5)KP(5)	8/pp5p/8/PP2k3/2P2pp1/3K4/6PP/8 w	c5
15	Fine 90	KP(6)KP(6)	8/7p/2k1Pp2/pp1p2p1/3P2P1/4P3/P3K2P/8 w	e4
16	Fine 100A	KP(5)KP(5)	8/6p1/3k1p2/2p2Pp1/2P1p1P1/1P4P1/4K3/8 w	Kf2

Table 7 TS2 in FEN notation

#	Position	Material	FEN	Best move
01	CCE4 479	KP(3)KP(3)	8/1p4kP/5pP1/3p4/8/4P3/7K/8 w	e4
02	CCE4 491a	KP(3)KP(3)	8/1pp5/3k3p/PP6/2P2K2/8/8/8 b	Kd7
03	CCE4 530	KP(3)KP(3)	2k2K2/8/pp6/2p5/2P5/PP6/8/8 w	a4
04	CCE4 608	KP(4)KP(4)	8/pp3p2/8/6kp/8/3K1PP1/PP6/8 b	f5
05	CCE4 679	KP(6)KP(6)	8/pp2k1pp/2p5/2P1p3/2P1P2P/6P1/P7/2K5 b	g5
06	CCE4 680	KP(7)KP(7)	8/1p6/p1p5/P1Pp2pp/1P1P1p1k/5P1P/6PK/8 w	g3/g4
07	CCE4 765	KP(5)KP(4)	8/1k6/p4p2/2p2P2/p1P2P2/2P5/P1K5/8 w	Kc1
08	Lejeune 2	KP(3)KPP	k7/4p3/4p3/8/8/3P1P2/5P2/K7 w	Kb2
09	Lejeune 4	KP(3)KP(3)	k7/8/1p6/p1p5/2P4K/8/PP6/8 w	a4
10	Lejenue 5	KP(3)KPP	8/8/p7/8/1P6/7p/P4k1P/3K4 w	a3
11	Lejeune 6	KP(3)KPP	8/5p1p/8/6k1/8/6P1/5PP1/7K w	Kh2
12	Christmas Cracker	KP(5)KP(4)	3k4/3p4/3p4/p2P2p1/P2P2P1/3P4/3K4/8 w	Kd1
13	Pillsbury 1895	KP(6)KP(6)	8/7p/2k1Pp2/pp1p2p1/3P2P1/4P3/P3K2P/8 w	e4
14	Capablanca 1919	KP(7)KP(6)	8/5p2/2kp1p1p/p1p2P2/2P5/7P/PP3PP1/6K1 w	a4
15	Botvinnik 1944	KP(5)KP(5)	8/1p3k2/p4ppp/3P4/1P6/4K2P/1P4P1/8 w	g3/g4
16	Botvinnik 1958	KP(5)KP(5)	8//4pk2/1p4p1/1P2p3/3pP2P/3K2P1/4P3/8 w	Kc4

References

1. Averbakh, Y., Maiselis, I.: Comprehensive Chess Endings 4. Ishi Press, Pawn Endings (2012)
2. Averbakh, Y., Maiselis, I.: Pawn Endings. Harper Collins Distribution Services, 1st edn. (1974). ISBN 978-0-7134-2797-4
3. Baker, A.: The Christmas Cracker. The British Chess Magazine (1919)
4. Church, K.: Co-ordinate squares: a solution to many chess pawn endgames (abbreviated version of B.S. Thesis). In: International Joint Conference on Artificial Intelligence, pp. 149–154. Tokyo, Japan (1979)
5. Fine, R.: Basic Chess Endings. David McKay (1941). ISBN 978-0-8129-3493-9
6. Haworth, G.: Chess endgame news. ICGA J. 36(1), 41–43 (2013)
7. Hyatt, R.: www.cis.uab.edu/hyatt/crafty
8. Hyatt, R.H., Nelson, H.L., Gower, A.E.: Cray Blitz, Computers, Chess, and Cognition, pp. 111–130. Springer (1990)
9. Instructions Per Second: http://en.wikipedia.org/wiki/Instructions_per_second
10. Intel® Xeon® Processor E5-2699 v3 (45M Cache, 2.30 GHz): http://ark.intel.com/products/81061/Intel-Xeon-Processor-E5-2699-v3-45M-Cache-2_30-GHz
11. Lejeune, V.: Re: really very difficult pawn endgames PGN's+solutions+diagrams (2000). http://www.stmintz.com/ccc/index.php?id=131929
12. Levy, D.: A welcome from the president (2014). http://ilk.uvt.nl/icga/organisation/welcome.php
13. Lomonosov Endgame Tablebases: http://chessok.com/?page_id=27966
14. Mackenzie, P.: Pawn endgame test suite (2000). http://www.stmintz.com/ccc/index.php?id=131885
15. Meyer-Kahlen, S.: All 6 men chess endgame databases available online, shredder computer chess. http://www.shredderchess.com/chess-news/shredder-news/all-6-men-chess-endgame-databases.html
16. Nalimov, E.V., Haworth, G.M., Heinz, E.A.: Space-efficient Indexing of chess endgame tables. ICGA J. 23(3), 148–162 (2000)
17. Newborn, M.: PEASANT: an endgame program for kings and pawns. In: Frey, P.W. (ed.) Chess Skill in Man and Machine, pp. 119–130 (1977). ISBN 978-1-4612-5515-4
18. Nunn, J., Friedel, F.: The "brains of earth" challenge. ICCA J. 22(3), 259–263 (1999a)
19. Nunn, J., Friedel, F.: Brains of earth: solutions of the pawn endings. ICCA J. 22(3), 188–189 (1999b)
20. Piasetski, L.: An evaluation function for simple king and pawn endings. Master's Thesis, McGill University, School of Computer Science (1977)
21. Quisinsky, F.: Interview with Tord Romstad (Norway), Joona Kiiski (Finland) and Marco Costalba (Italy) Programmers of Stockfish. Schachwelt (2010)

Parallel Algorithms to Align Multiple Strings in the Context of Web Data Extraction

Christine Gfrerer, Marián Vajteršic and Rade Kutil

Abstract The alignment of multiple strings generated from web pages represents a crucial problem in the processing of daily increasing amounts of data in the Internet. The complexity of this problem grows exponentially with the number of strings. Since it is not possible to achieve practically acceptable results on serial computers even with efficient heuristic approaches, parallel processing seems to be an inevitable option. There already exist emerging parallel solutions for the alignment of multiple strings in areas such as bioinformatics and genome applications. However, to our knowledge, no parallel solution has been published so far for a problem which arises in the context of web data extraction. In this work, we present two algorithms for a parallel solution of this problem, where input web data records are represented as a two-dimensional array of symbols. The algorithms differ in the assignment of the array data to the parallel processes. In the first one a distribution according to symbols is considered, whereas the second one operates by partitioning its columns. Communication among processes is handled via message passing in both cases. The algorithms are analyzed with respect to time and space complexity. We implemented both algorithms and have studied their properties by running them on a multiprocessor system. For the version with distributed columns, we observed that its speedup significantly suffers from the communication overhead. However, the results for the version with data distribution by symbols are convincing. In this case, reasonable performance has been obtained.

C. Gfrerer · M. Vajteršic · R. Kutil (✉)
Department of Computer Sciences, University of Salzburg, Salzburg, Austria
e-mail: rkutil@cosy.sbg.ac.at

C. Gfrerer
e-mail: christine.gfrerer@gmail.com

M. Vajteršic
Mathematical Institute, Department of Informatics,
Slovak Academy of Sciences, Bratislava, Slovakia
e-mail: marian@cosy.sbg.ac.at

© Springer International Publishing Switzerland 2017
A. Adamatzky (ed.), *Emergent Computation*, Emergence, Complexity
and Computation 24, DOI 10.1007/978-3-319-46376-6_25

1 Introduction

The alignment of multiple strings arises in many areas where data has to be extracted from similarly structured sources. However, this problem belongs to a class of NP-hard problems, and, hence, there is a pressing need for algorithmic solutions which are able to scale with the permanently growing size of processed datasets. Until now, most alignment algorithms are proposed for serial computers.

Recently, however, a remarkable increase of parallel approaches can be observed, particularly in bioinformatics and genomics [9]. One of the first parallel systems in this field was ParAlign [22], which parallelizes the Smith-Waterman algorithm and is designed for a SIMD (Single Instruction Multiple Data) parallel system. Most efficient algorithms in this research field are based on the so called progressive strategy, which performs consecutively pairwise alignments on the most similar sequences until all sequences become aligned. The first parallelization of these methods for a shared-memory system was presented in [20]. Approaches based on the MPI (Message Passing Interface) implementation for distributed-memory systems are published e.g. in [16, 21]. A recent implementation on GPUs (Graphics Processing Units) is presented in [12]. Algorithms for the alignment of multiple sequences in large-scale databases on a massively parallel supercomputer system are presented for the IBM BlueGene/P in [6]. Even the PVM (Parallel Virtual Machine) computing paradigm was considered for a design of an efficient parallel method for this problem [18].

In this work, our focus is on the problem area of information extraction from structured data obtained from web pages. When data records from databases are displayed on web pages, the raw information is enriched with visual and structural formatting elements. While this is helpful for humans to visually identify data elements, machines have to use heuristics for analyzing document structures in order to separate formatting templates from actual data items. This can be done by transforming web pages into a sequence of symbols representing HTML elements and aligning the resulting symbol strings so that aligned elements can be identified as template elements.

Existing algorithms for web data extraction are all sequential. A comprehensive survey and a comparison of existing systems is given e.g. in [3]. The underlying methods often exceed polynomial complexity, even for heuristics, because they involve comparisons of code and text regions that grow in number and size with the size of web data. Moreover, data sets accessible through web interfaces are increasing in size.

Processing of such large data sets imposes a big computational demand, which justifies efforts to employ parallel computing for solving this task in order to achieve a reasonable reduction of computing times and improvement of performance. In our work, we present two parallel algorithms, which could be seen as a first practical attempt for the alignment of strings originating from web pages.

This chapter is organized as follows. In Sect. 2 we introduce the web data extraction problem. We give a detailed description of how the input data is generated, and we define parameters for the estimation of the quality of the output. We also present there a motivation and definitions which help to formalize the problem.

The alignment system for this application and an overview of the respective modules is presented in Sect. 3. It is also shown there how the token sequences are generated and it encompasses a definition of parameters needed for the alignment algorithms. In terms of these parameters, a sequential algorithm is formulated, which is a modification of the adopted heuristic alignment strategy from [14].

The core of our contribution is Sect. 4, where we formulate and illustrate both parallel alignment algorithms. Section 5 reports about experiments for real web data collections and discusses obtained results. The final section is devoted to conclusions and outlooks.

2 Model

In order to introduce our web data extraction system, we first need to lay out the basic notions and components of web data extraction, the string alignment problem, and their interrelationship.

2.1 Web Data Extraction

The World Wide Web has grown enormously after it was made public in the early 1990s. Meanwhile, there are countless web pages, and with every minute more and more content is produced. The number of users also increases steadily. With the increasing availability of documents on the web, the need for analysis of the included data has emerged.

Relevant web data for our studies comes from the so-called deep web. The deep web (or hidden web) is associated with public available databases, which are connected to the Internet and which store the data of interest. Such databases are queried via search forms on web pages after a user enters some key words. In most cases the query results are presented in a list of semi-structured entries, which are denoted as data records. Hence, the web pages containing the query results are generated dynamically. What motivates us to accomplish data extraction is the fact that conventional search engines are not able to catch the data from the deep web.

The format of the generated web content is denoted as semi-structured. This means that the structured data from the database query is mixed up with HTML code to present data on the web. The purpose of the extraction is to separate the data items from the web pages, where the items are embedded. The extraction of the data goes hand in hand with finding the generation model that formulates a possible embedding of the data. Designated programs, which perform the latter task, are denoted as wrappers.

When we speak about web data extraction, the term information extraction has to be explained. "Information extraction refers to the automatic extraction of structured information such as entities, relationships between entities, and attributes describing the entities from unstructured sources" [23].

Web data extraction is a kind of information extraction. For the web data extraction, the input is restricted to content available on the World Wide Web.

2.2 System Classification

In the surveys of Laender et al. [15] and Chang et al. [3], the authors present their own taxonomies, which lead to classify a wrapper induction system in various dimensions. Inspired by both surveys, we classify our system by the degree of automation, the stages of the extraction, the utilized features and the applied techniques.

2.2.1 Degree of Automation

We distinguish manual, semi-automatic and fully automatic systems for web data extraction, as it is done in [3]. The degree of automation corresponds to the degree of user interaction. Moreover, the distinction corresponds to the historic development of tools for web data extraction.

Early systems required users to be programmers who have to write a program to extract data from a certain web site. Although special programming languages reduce the effort of wrapper generation, such systems are still expensive. However, they produce the desired output precisely.

Semi-automatic wrappers work with annotations from users. Such annotations can be labels in training examples that provide relevant information to build the template. For example, OLERA [4] needs a user to label a data record of interest and the user improves the segmentation of attributes. Contrary, in IEPAD [5] the user has to interact with the produced output. The system determines some possible extraction patterns of the records. Thus the possible patterns are presented to a user, who selects the most suitable target.

Most recent wrappers are classified as automatic. Automatic wrappers do not need any human involvement during processing. Therefore, they are faster. There must be a predefined schema that guides the template generation. Most of our studied systems are based on heuristics like e.g. [2, 7, 13, 14, 17, 24, 31, 35, 36]. The method in [37] uses statistical features. Ontologies support the extraction task in [26]. In general, the experiments show that automated approaches have a fairly high degree of accuracy. Moreover, such tools are robust against changes in the template. This means, if for example the layout of a web page is updated and the template is altered, then the extraction program can still be utilized.

Fig. 1 Assumed structure of an example web page

We perform automatic data extraction. Our implementation is based on a fully automatic approach that uses heuristics. Moreover, we want to improve overall system performance.

2.2.2   Stages in the Extraction

Before a classification in the direction of the stage is given, let us picture the general arrangement of web pages, which are relevant for our studies. A web page[1] has generally a centrally located data region. Within the region, one or multiple data records are found. Further, each data record contains several data items. Figure 1 shows an example of a web page, which has the assumed structure.

---

[1] Web page excerpt from http://www.barnesandnoble.com/.

From the arrangement we derive different input/output scenarios. To our knowledge, all preceding work takes one, two or multiple web pages as input. The targets are either data records or data items. In either case the identification of a data region is a possible intermediate stage, e.g. like in [35]. The next stage is then the distillation of data records. Systems with this objective concentrate on the detection of record boundaries and/or the similarity of the HTML structure covered by the records. The extraction of data items follows optionally. In contrast, FiVaTech [14], ExAlg [2] and RoadRunner [8] focus on the template detection of the web pages and thus data items emerge directly. In the survey of Chang et al. [3], this latter type is referred to as page-level extraction.

Our contribution differs from all others because we manually attain the data records from the web pages. These web data records build the input for our system.

2.2.3 Utilized Features

The web pages as input provide several valuable features. Usually systems utilize text content and structure of the input web pages. Newer approaches also incorporate visual information.

Early tools were mainly text-based. String based computations require to split up the input into a sequence of tokens. HTML tags offer themselves as tokens. Text between tags can be generalized to an artificial text token, as e.g. in [24] or [31]. Consequently the text content disappears for the processing. In [2], text is separated into the words to associate different roles with text components. In DEPTA [35], similar text values indicate matching nodes to support a tree matching algorithm.

The nested structure of HTML elements leads to a representation as a rooted tree, the DOM (Document Object Model) tree. Zhai and Liu introduced data extraction based on partial tree alignment (DEPTA) [35]. In [19], Miao et al. find repetitions of tag paths and, thus, data regions and records. The work in [14] and [13] partition the computation problem upon the tree nodes, where the tree nodes are processed together if they are direct descendants of equal parent nodes from different page representations.

Simon and Lausen introduced visual perception based extraction of records (ViPER) [24]. They use visual information to identify the data region out of a range of computed candidate regions. In [24] each candidate data region is analyzed for its containing bounding boxes and structural similarity of the text that appears on the web page. Liu et al. [17] propose some hypothesis about the web pages containing data from the deep web. It is assumed that data records are similar in their appearance. This includes the used fonts, the size of images and the position of the data items inside the records.

Our system is based on the textual representation of the input. Further, we utilize structural information from the tree representation, particularly the level of the tree nodes. Additionally, we incorporate style related attributes. We assume that different values of `class` attributes correspond to different layouts in the style sheet.

2.2.4   Extraction Methods

One of the tasks in web data extraction is to distinguish tokens that belong to the template from those that are data. In the past, several methods have been studied. Arasu and Garcia-Molina [2] implemented a method that is based on building equivalence classes of tokens. It counts the occurrences and evaluates the name and parse the tree path of a token in web pages. Tokens with the same occurrence vector form an equivalence class.

A given input HTML code is partitioned into a string of tokens. Next, the different strings are aligned. Alignment means placing equal tokens opposed each other. Our work uses string alignment. A formal description is given in Sect. 2.5.

Tree alignment incorporates the tree structure of the HTML input. The tokens of different trees are aligned level by level subsequently. Therefore, the tokens under a common parent node are arranged with a string alignment technique.

The construction of a suffix tree requires a sequence of tokens. If a suffix tree is built from a sequence of multiple records, then the template results from tree paths, which cover multiple suffixes. Suffix trees enable to detect nested structures in the template [31]. The system called DeLa [31] requires alignment of the data after extracting them with a wrapper, which is generated with a suffix tree.

2.3   Description of the Input

Input documents are given in the HyperText Markup Language (HTML). It is defined by the W3C [30]. The latest specification is HTML 5.0 [29], released in October 2014. However, our input data is mostly based on the prior HTML 4.01 [28].

HTML is structured in the form of the DOM [27]. HTML pages contain data records that conform to a common schema [2]. A template supplements the data with structural and presentation components. Figure 2 gives an example for the structure of an HTML document. In the first line the declaration of a Document Type Definition (DTD) is given. The declaration defines which schema the document has to apply to. The root element of every HTML document is called html. Within it, the elements head and body are stated. Within the head section, meta information is given. The text of the mandatory title element is displayed on the top of the browser window. Additionally, it is possible to include a reference to a Cascading Style Sheet (CSS) that contains the separated style commands. This is realized in line 7. The body section contains everything that will be published on the web page.

HTML documents have the structure of a rooted tree according to their specification. The interface of the DOM enables access to this tree structure. We utilize the generated DOM tree after parsing a data record by traversing the created DOM tree in depth-first order to acquire the content of each input document. Figure 3 shows an example of a DOM tree which correspond to the HTML example in Fig. 2.

```
 1  <!DOCTYPE html PUBLIC "-//W3C//DTD HTML 4.01 Transitional//EN" "
        http://www.w3.org/TR/html4/loose.dtd">
 2  <html>
 3    <head>
 4      <title>
 5        Title of the web page
 6      </title>
 7      <link rel="stylesheet" type="text/css" href="style.css">
 8    </head>
 9    <body>
10      <div id="mainsection">
11      <h1>
12        Hello world!
13      </h1>
14      <p class="par">
15        This is a paragraph.
16      </p>
17      <ul type="circle">
18        <li class="listitem">
19          First point
20        </li>
21        <li class="listitem">
22          Second point
23        </li>
24      </ul>
25      <p>
26        Another paragraph with a picture
27        <img src="pic.jpg" alt="picture description">
28      </p>
29      </div>
30    </body>
31  </html>
```

Fig. 2 HTML document example

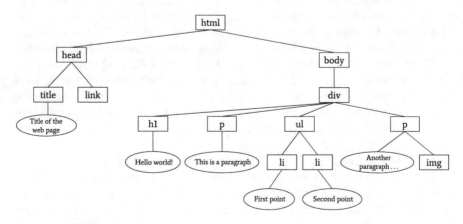

Fig. 3 An example DOM tree

Based on the work of Arasu and Garcia-Molina [2], Kayed and Chang [14] and Wang and Lochovsky [31], we define the composition of data items. The composition is defined recursively by the following types:

- β is denoted as basic type. An instance of β is a data item, derived after querying the database.
- $(\tau_1 \mid \tau_2 \mid \ldots \mid \tau_n)$ is denoted as disjunctive type. For instantiation of this type, there is exactly one τ_i, $1 \leq i \leq n$, selected. The optional type $(\tau_1 \mid \epsilon)$ can be expressed also as (τ)?, where ϵ is the type for the empty string.
- $\langle \tau_1 \tau_2 \ldots \tau_n \rangle$ is denoted as tuple type. The order of the types needs to be preserved.
- $\{\tau\}$ is denoted as set type. There are any number of instances.

A schema S is a type τ that describes the nested structure of data items for any data record in the input set. The schema only consists of basic types. Thus the recursive relationships of types must be dissolved.

Figure 4 contains the HTML code for two different data records. It is a possible code for the examples shown in Fig. 5. The actual values from the database are bold in the listings.

The common schema of the data records in Fig. 4 can be expressed by $S = \langle \beta \ (\beta)? \ (\beta)? \ \{\langle \beta \ \beta \rangle\} \rangle$. Each data record covers four different data items, namely the title of the album, the artist, the format and the price. Conforming to the schema requires at least the album title. The artist name is declared optionally. Next there follow several tuples of format and price.

```
1  <div id="record1">
2    <img src="br_cover.jpg">
3    <strong>Bankrupt!</strong>
4    <span class="ed">[Deluxe
         Edition]</span>
5    <br>
6    by Phoenix <br>
7    <table>
8      <tr class="head_row">
9        <td>Format </td>
10       <td>Price </td>
11     </tr>
12     <tr class="content_row">
13       <td>CD </td>
14       <td>$12.58 </td>
15     </tr>
16   </table>
17 </div>
```

```
1  <div id="record2">
2    <img src="sc_cover.jpg">
3    <strong>Sound City: Real to
         Reel</strong>
4    <br>
5    <table>
6      <tr class="head_row">
7        <td>Format </td>
8        <td>Price </td>
9      </tr>
10     <tr class="content_row">
11       <td>CD </td>
12       <td>$10.79 </td>
13     </tr>
14     <tr class="content_row">
15       <td>Vinyl LP </td>
16       <td>$28.49 </td>
17     </tr>
18   </table>
19 </div>
```

Fig. 4 HTML code extract

Fig. 5 Two examples of data records

The template itself consists of several template strings. Template strings are usually HTML tags to structure and to present the data. Template strings may also be extended with text content placed before or after a data item. For example, consider "5" is a data item, then a user hardly knows what the number means. If the text "articles are left" is placed after the number, then the meaning is clear.

Therefore, we define a template $\mathcal{T}_S$ for a given schema $\mathcal{S}$. The template $\mathcal{T}_S$ describes the arrangement of template strings for a given schema $\mathcal{S}$. Basically it has the same structure as the schema. Based on [2, 37], we define the composition of the template as follows:

- A template string is a sequence of template tokens (HTML tags or text). A template string may also be an empty string.
- There is exactly one pair of template strings associated with each instance of a type τ. We denote tl as the left template string and tr as the right template string.
- If tl contains HTML opening tags, that are not closed in tl, then the corresponding HTML closing tags must occur in tr of that pair.
- A template string has to be disjoint from the next occurring template string in the template.

A template of a particular schema $\mathcal{T}_S$ consists of all template strings to produce a data record from the data of an arbitrary instance of schema $\mathcal{S}$.

Consider the example schema $\mathcal{S}$ from above. The corresponding template will be $\mathcal{T}_S = tl_1 \langle tl_2 * tr_2 \ tl_3 \ (tl_4 * tr_4)? \ tr_3 \ tl_5 \ (tl_6 * tr_6)? \ tr_5 \ tl_7 \ \{tl_8 \langle tl_9 * tr_9 \ tl_{10} * tr_{10} \rangle tr_8 \} tl_7 \rangle tr_1$. The symbol $*$ is a wild card for a data item and actually does not belong to the template.

In Table 1, the values of the template variables are given for our example. The template strings are listed pairwise in the table, which differs from their appearance in the template $\mathcal{T}_S$. This illustrates better the partition of the tags. For instance, the opening tag of the HTML element `strong` is covered in tl_2 and its closing tag is assigned to tr_2. The template string tl_7 contains two text instances to emphasize the meaning of the respective data items. tr_7 only contains the ending tag of the table element, because it is left unclosed from template string tl_7. tl_4 and tr_4 belong to an optional type and thus these strings only occur, if an instance of this type exists. Template strings belonging to a set type, like tl_8, tr_8, tl_9, tr_9, tl_{10}, tr_{10}, occur as many times as instances of the set type occur.

Table 1 Values for the template strings of the example in Fig. 4

Template String	Value
tl_1	`<div><img>`
tr_1	`</div>`
tl_2	`<strong>`
tr_2	`</strong>`
tl_3	ϵ
tr_3	` `
tl_4	`<span>[`
tr_4	`]</span>`
tl_5	ϵ
tr_5	ϵ
tl_6	`by`
tr_6	` `
tl_7	`<table><tr><td>Format</td><td>Price</td></tr>`
tr_7	`</table>`
tl_8	`<tr>`
tr_8	`</tr>`
tl_9	`<td>`
tr_9	`</td>`
tl_{10}	`<td>$`
tr_{10}	`</td>`

2.4 Output Quality

Unless wrappers are tailor-made to handle input pages from a certain source, programs have its limitations. Generally extraction tools take different inputs as long as they fulfill some constraints. To cover a range of possible input, as a consequence the produced output is not perfect.

Errors in the data extraction process means the wrong interpretation of data as template or vice versa. Data can wrongly be assumed to belong to the template, or template tokens can be identified as data. The reasons for this can be schema ambiguities, text tokens that can serve partly or fully as template, decorative tags that are part of the data, multiple data items that are aggregated into a single text string with only delimiter characters to separate them, and DOM tree level crossings due to tags within data records.

The quality of the output is measured based on the extracted data. The standard metrics are precision and recall. For this, we need three values:

- the number of actual data items a, which are found in the input,
- the number of extracted data items e, which correspond to the output,
- the number of correctly extracted data items c.

The precision P is calculated as $P = \frac{c}{e}$, and the recall R is calculated as $R = \frac{c}{a}$.

2.5 The Problem of String Alignment

In the following we state the motivation for introducing the alignment problem for the extraction of data items out of data records from web pages. Next we describe the composition of the strings over an alphabet. This is followed by definitions for the string alignment problem.

2.5.1 Motivation

Multiple string alignment or Multiple Sequence Alignment (MSA) is a computational problem that has its origin in biology. Its significance lies in finding highly conserved sub-patterns within a set of biological sequences [10, 32]. Another important reason for its application is to infer the evolutionary history of species from their associated sequences [10, 32]. The adjustment of multiple sequences is well studied for DNA, RNA or for sequences from amino acids [10]. The same idea is applicable also in other disciplines, for example in computer science for web data extraction.

The objective for our system is to get the data values for each distinct attribute from the outgoing database query. For example, we want to place together all artist names of music records or we want to get all album titles of the data records. Further we want to determine the common template of the data records.

Unless the schema for record generation is straightforward, there are misalignments in the input. A straightforward schema consists of basic types only. If all data records conform to such a schema, then it makes the alignment a simple task. In this case, the template is defined by the patterns that are found within every data record and the data is mostly different from one record to another.

Usually, the schema includes optional, disjunctive and repetitive types. As a consequence, the strings derived from the data records are misaligned. Misalignment of data items means for example that some album titles occur wrongly in the group of artist names. The objective of the alignment procedure is to move characters to the right position. Characters from different strings should occupy the same position within each respective string, when they represent the same type of information.

2.5.2 Alignment of Two Strings

Based on the work of Gusfield [11] and of Wang and Jiang [32], we describe formally the problem of two-string alignment and the related functions.

Fig. 6 An example for the alignment of two strings

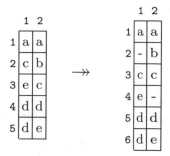

Definition 1 Let X, Y be two strings. X is of length k and Y is of length l. The strings consist of characters from an alphabet Σ. The alignment is achieved by inserting space characters $(-)$ at the beginning, at the end, or in between of each string X and Y, such that the resulting strings X' and Y' have equal length m. Resulting strings are placed next to each other, so that a character of the first string X' is opposed a character of the second string Y'. Thus the characters of X' and of Y' are from the extended alphabet $\Sigma' = \Sigma \cup \{-\}$.

Example 1 Let $\Sigma = \{a, b, c, d, e\}$, $X = acedd$, $Y = abcde$. A possible alignment $\mathcal{A}$ consists of the strings $X' = a - cedd$ and $Y' = abc - de$. Figure 6 visualizes the input strings on the left side and the alignment is placed on the right side. Strings are arranged column-wise.

This general definition only ensures that the resulting strings X' and Y' have equal length m. Any two opposing characters in the alignment form either a match, a mismatch or a character is opposite a space. A mismatch refers to the case when the two characters are from the alphabet Σ and they are dissimilar. This is the case for the last two characters in the example.

The alignment is a product [11]. It results as output of a procedure that requires the two strings as input. There are several ways how an alignment is achieved. We state some metrics for the alignment to be able to compare the achieved results later. The score function and the value of the alignment are defined as follows.

Definition 2 Let Σ' be the alphabet for the obtained strings X' and Y'. Let x, y be any two characters from Σ'. The score function s assigns a numerical value to a pair of characters, $s : \Sigma' \times \Sigma' \to \mathbb{N}$.

Definition 3 Let X' and Y' be the strings of an alignment $\mathcal{A}$, where X', Y' are strings over the alphabet Σ', each of length m. Let respectively x_i, y_i be the ith character in X' and in Y'. The value of the alignment $V(\mathcal{A})$ is the sum of the scores of each opposing pair of characters, $V(\mathcal{A}) = \sum_{i=1}^{m} s(x_i, y_i)$.

Example 2 Let be $\Sigma = \{a, b, c, d, e\}$ and $\Sigma' = \{\Sigma \cup \{-\}\}$. The score s of two characters $x, y \in \Sigma'$ be

$$s(x, y) = \begin{cases} 0 & \text{if } x = y, \quad x, y \in \Sigma' \\ 1 & \text{if } x = -, \quad y \in \Sigma \\ 1 & \text{if } x \in \Sigma, \quad y = - \\ 2 & \text{if } x \neq y, \quad x, y \in \Sigma. \end{cases}$$

Consequently, the example alignment $\mathcal{A}$ with the strings $X' = a - cedd$ and $Y' = abc - de$ has a value $V(\mathcal{A}) = 0 + 1 + 0 + 1 + 0 + 2 = 4$.

Different alignment procedures result in different alignments. Further, each alignment procedure has an impact on the resulting value of the alignment. In other words, the resulting value $V(\mathcal{A})$ states the goodness of an alignment. The question arises what the best possible alignment is.

Definition 4 The optimal alignment is an alignment $\mathcal{A}$, which has the minimum value for $V(\mathcal{A})$.

According to Gusfield [11], the optimal alignment is closely related to the (weighted) edit distance problem between the two strings.

2.5.3 Alignment of Multiple Strings

Based on the work of Gusfield [11] and of Wang and Jiang [32], we describe formally the problem of the alignment of multiple strings and a related function.

Definition 5 For multiple string alignment, the input is a set of strings $\mathcal{X} = \{X_1, X_2, ...X_n\}, n > 2$ over the alphabet Σ. Each string is extended by chosen spaces at the beginning, at the end or in-between any two characters, such that all resulting strings have the same length, denoted to be m. The resulting strings are positioned into a two-dimensional array, which has m rows and n columns. Each column contains a string. Each row contains characters or spaces from the n different strings.

The computation of an optimal alignment of n strings, each with a string length of k, has a time complexity of $\mathcal{O}(k^n)$ [1, 5, 24]. This is feasible for a very small number of strings [11]. Hence the computation cannot be handled practically if n becomes large. Consequently, methods were developed that are faster in computation, but they cause some errors. Such methods are denoted as approximation algorithms, as is the algorithm presented in Sect. 3.4.

In order to measure the quality of alignment for multiple strings we use the sum-of-pairs, which is defined as follows:

Definition 6 Let $\mathcal{X}' = \{X_1', X_2', ...X_n'\}, n > 2$, be the set of strings belonging to an alignment $\mathcal{A}$. Then the sum-of-pairs measure of the alignment $\mathcal{A}$ is the sum of the score values for all pairs of strings from $\mathcal{X}'$.

Then, the sum-of-pairs alignment problem is to find the alignment of multiple strings with minimum sum-of-pairs value [11]. Wang and Jiang proved that this problem is NP-complete [32].

Generally, the value of the alignment is a theoretical measure. Gusfield [11] pointed out that the goodness of alignment should be measured by the biological meaning. The same applies to web data extraction: It is more important to evaluate the number of correctly determined data items than to measure the sum-of-pairs.

3   Design of the System for Web Data Extraction

In the following we describe our system for web data extraction. It starts with an overview of the developed system to get the global overview of the processing steps. Next, the required pre-processing is explained. The pre-processing covers the preparation of the input to acquire a matrix for subsequent alignment. This matrix has certain properties depending on the input, i.e. the data records from the web. The algorithm described in Sect. 3.4.2 gives an idea how to align a matrix sequentially. The intention for its presentation is to simplify the understanding of the parallel algorithms for alignment, which will be given in Sect. 4. Finally, we describe optional post-processing of the output matrix.

3.1   Overview of the System Modules

Our developed system consists of five modules as illustrated in Fig. 7. The arrangement of the modules is inspired by the work of Kayed and Chang [14].

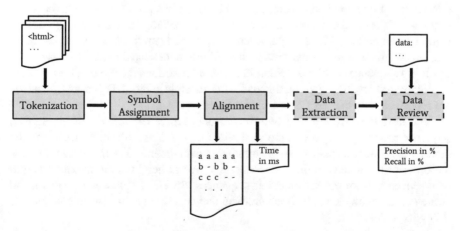

Fig. 7 Processing modules including I/O

A set of HTML documents serves as input, where each of them contains the HTML code for a record. The input files are parsed and a sequence of tokens is created. The next module produces a matrix based on the created tokens. Moreover it assigns a symbol to each token, respective matrix element. This concludes the pre-processing.

The actual core module consists of the parallel alignment, where we focus on fast and accurate computation. There are two versions of the (parallel) adjustment, which we want to compare by their execution time. Both adjustments lead to an aligned matrix as output.

The data extraction and data review complete our system for web data extraction. The post-processing was relevant during the development of the system to ensure accurate output of the alignment. Since in our experiments we focus on minimizing the execution times of the alignment, we may stop at this point. Hence, the post-processing modules are optional.

3.2 Creation of Token Sequences

After removing unneeded content, such as comment tags, line breaks, and non-text spaces, every HTML code snippet is parsed and stored intermediately in a DOM tree. The tree representation simplifies the partitioning procedure. Moreover, the DOM tree nodes contain relevant information, which is required for the symbol assignment. Further it is relevant for the alignment at which level a token corresponds to a node in the DOM tree.

Each DOM tree is traversed in pre-order. During the traversal we create one or two tokens per tree node. A single token results from an empty tag like <img> or <br>. Empty tags are HTML tags, which do not have any content between their opening and closing tag pair. Therefore, the closing tag is usually omitted. Also, one token results from the text content of a tree node. Two tokens derive from tree nodes that represent opening and closing tags. In between two produced tokens, the token sequence of the child tree nodes is inserted. The procedure starts with the pair of tags from the root node and it puts all sequences deriving from the sub-trees in between. The procedure continues recursively until all tree nodes have been processed. The result is a sequence of tokens, which in principle partitions the input HTML string.

Figure 8a shows the resulting tokens for the example in Fig. 4. Different text values are represented by the term text in the figure.

As a next step, we create a two-dimensional array. The tokens belonging to a sequence are mapped to a column in the matrix. So the width of the matrix is the number of sequences, respective the number of data records. The mapping preserves the order of the tokens in the sequences. The matrix height is the maximal length of a sequence from the set of token sequences. Shorter sequences do not fill the entire column with elements. Each element that corresponds to a token contains the following fields:

Fig. 8 **a** The tokenization of the data records from the example in Fig. 4. **b** The assigned symbols to the tokens from (**a**)

(a)	1	2
1	`<div>`	`<div>`
2	`<img>`	`<img>`
3	`<strong>`	`<strong>`
4	text	text
5	`</strong>`	`</strong>`
6	`<span>`	` `
7	text	`<table>`
8	`</span>`	`<tr>`
9	` `	`<td>`
10	text	text
11	` `	`</td>`
12	`<table>`	`<td>`
13	`<tr>`	text
14	`<td>`	`</td>`
15	text	`</tr>`
16	`</td>`	`<tr>`
17	`<td>`	`<td>`
18	text	text
19	`</td>`	`</td>`
20	`</tr>`	`<td>`
21	`<tr>`	text
22	`<td>`	`</td>`
23	text	`</tr>`
24	`</td>`	`<tr>`
25	`<td>`	`<td>`
26	text	text
27	`</td>`	`</td>`
28	`</tr>`	`<td>`
29	`</table>`	text
30	`</div>`	`</td>`
31		`</tr>`
32		`</table>`
33		`</div>`

(b)	1	2
1	b	b
2	d	d
3	g	g
4	o	o
5	h	h
6	e	a
7	r	i
8	f	t
9	a	k
10	s	p
11	a	m
12	i	k
13	t	p
14	k	m
15	p	v
16	m	u
17	k	l
18	p	q
19	m	n
20	v	l
21	u	q
22	l	n
23	q	w
24	n	u
25	l	l
26	q	q
27	n	n
28	w	l
29	j	q
30	c	n
31		w
32		j
33		c

- The position in the matrix, given by the column index and the row index.
- The level, derived from the corresponding DOM tree node.
- The symbol, which is the primary indicator for alignment.

3.3 Symbol Assignment

For the symbol assignment we try to preserve structural features. The structure information derives from the DOM tree nodes. Such features give us valuable hints for the subsequent alignment. Each DOM tree node is associated with a token. Consequently, the similarity of elements or tokens is referred to the similarity of the associated tree nodes/tags. Tags are similar if they have the same name and type. Furthermore, it is evaluated whether their class attributes are equal or the path in the DOM tree is equal. We want to achieve that tokens that most probably belong to the same part of the template, are represented by the same symbol. Symbols are implemented by consecutive natural numbers.

In case of text tokens, only the tags which encapsulate the text are concerned, since the text belongs to data and not to the template structure. If these surrounding template tokens have the same symbol, then the text in between similar template tokens very likely belong to the same data item.

In principle, all tokens are compared with each other. The same symbol is allocated if the tokens have at least the same tag name and the same tag type. These criteria reduce the number of necessary comparisons of tokens. Hence the tokens are partitioned into groups by tag name and tag type. In the implementation all tokens belonging to a group are compared, since tokens from other groups certainly have different symbols.

Further conditions are evaluated only for tokens in a group. Such a condition tests whether each of the both associated tags has an attribute named class. If the class attributes are equal, then the same symbol is assigned.

The class attribute of tags is used to connect them with certain style information. The style instructions, such as font shapes or positions on the web page, are stated within a Cascading Style Sheet (CSS) document. In the CSS file the value of the class attribute is referred as selector. The advantage of the class selector is that it can be assigned arbitrarily often in contrast to the id selector. However, tags are most likely to concern the same template token if the tags have equal name and type and, further, if the tags have the same value of the class attribute. If the same style is used, then tags probably have the same meaning. In our case, the same meaning applies to the common part of the template.

The evaluation of the class attribute fails if it is absent in one or both cases. If one attribute value is missing, it means the same as in case different attribute values occur, so different symbols are assigned. If both tags do not have any class attribute, then the paths are compared. A path consists of successive tags in the DOM tree, which starts at the root and end at the desired node. Here also the equality leads to the same symbol.

The symbol assignment is a sequential task, such that symbol assignment depends on previous determined similarities between tokens. We use transitive relations to abbreviate the necessary comparisons between tokens. If there are 3 tokens and 2 pairs of them are already identified as similar, then the third pair combination needs not to be computed. The transitivity holds for both, the condition for equal class attributes and the condition for the equal path from the root, but only without considering the node positions. In addition, the initial segmentation of tokens by their name and type also reduces the computational demand. We summarize the computing steps for symbol assignment in Algorithm 1.

Algorithm 1 SymbolAssignment(T, M)

---

computes the symbols for all generated tokens. The input for the symbol assignment is a data structure T that keeps the sequences of tokens. The output is given by the symbol values for the elements in the matrix M. Symbols are represented by consecutive natural numbers.

(1) Get the list of distinct combinations of tags and their respective type from T. The list contains e.g. <a>, </a>, <br>, <div>, </div>, etc.
(2) For each list entry from step (1) get all tokens with same name and type, e.g. all tokens with the tag <a>. Execute steps (3) and (4) for each token set.
(3) For each ordered pair of tokens (t_i, t_j), where $i < j$, and tokens belong to the same set after step (2), if at least one of the tokens does not already have a symbol:

    (a) If class attributes of t_i and t_j are equal, assign the tokens the same symbol. If t_i has a symbol, then assign this symbol to t_j and vice versa. If both do not have a symbol, then give both of them a new symbol.

    (b) If both tokens have no class attribute, then the nodes on the path are compared, which lead to the next ancestor node with class attribute or the root nodes. If paths are equal and the ancestor nodes have the same class attribute, assign them the same symbol. Also, if the paths lead up to the root nodes and the entire path is equal without considering the nodes' positions, then assign the tokens t_i and t_j the same symbol.

(4) If a token t_i has no symbol assigned after step (3), then a new symbol is attached to t_i. The same applies to t_j.

Figure 8b shows the assignment of the symbols in case of the example in Fig. 4. For better understanding, symbols are displayed as letters. Mostly tags with the same name and type have equal symbols. Exceptions are characters for tags <tr>, </tr>, <td>, </td> and characters for the different strings denoted as text. The <tr> tags have different symbols because the values of their class attributes differ (head_row versus content_row). The same argument holds for the corresponding tags </tr>. The tag pairs for td have different values just like the tokens of the table row, because the equality of class attributes of the parent nodes are checked as well. Different characters for the text depend on the ancestor nodes of the trees.

Comparison with the Related Procedure of FiVaTech Our symbol assignment module has the same purpose as the peer node recognition of the FiVaTech system in [14]. Conceptually we also compare each token with each other, but the algorithms are completely different. Kayed and Chang [14] utilize a tree matching algorithm to compute a similarity score for each pair of two tags, respective their DOM tree nodes. Their algorithm accepts different text values in the leaf nodes as a match. Furthermore, repetitive nodes in the trees are considered. The computed score is normalized such that it can be compared with other scores. Basically similar structures of the sub-trees induce the assignment of the same symbol. So the score is calculated depending on descendant tree nodes. In contrast, we use the structural features directly connected with nodes or the structure information of ancestor nodes.

3.4 Matrix Alignment

Before we describe the sequential algorithm for matrix alignment, we introduce some definitions for the computation. It gives a view about the functionality of the alignment in general. Also, it gives an idea about which parts have the potential for parallel implementation.

3.4.1 Alignment Preliminaries

An *aligned row* contains elements that all have the same symbol. Additionally, an aligned row can contain empty elements, which correspond to spaces defined in Sect. 2.5. Our algorithm does not allow any misalignments. This means that, after the alignment of a row, there can only be one symbol. Hence, other elements are moved, such that they do not longer occupy the aligned row.

A *shift operation* corresponds to the insertion of spaces between two elements in the matrix, and these elements are above each other. All applied shifts together transform the unaligned matrix into the aligned matrix. Obviously, with the shifts the height of the matrix increases.

The alignment algorithm computes the necessary shifts. Relevant for shifts are the row r, the column c, and the length l. Thus, the shift operation is performed for all elements in the referred column c and in or below the row r by a certain length l. The length l is equal or greater to 1 (negative values are not allowed, 0 indicates that no shift needs to be performed).

In case elements are shifted by a length of 1, the foremost aligned element is considered for the adjustment of the next row again. If k is the height of the input matrix, then, in the worst case, k elements are shifted down in a column.

The alignment algorithm is mainly based on the positions of the symbols in the matrix. These positions are characterized by *position parameters* r_{down}, r_{up} and *span*, which will be defined below. Basically, the definitions are similar to the definitions made by Kayed and Chang for FiVaTech [14].

Definition 7 If a symbol s occurs in the current processed row r, then r_{down} is the nearest row below r, $r_{down} > r$, where s also occurs. In addition, the symbol s has to be in a different column in r_{down}. That means if s is found at possible multiple positions (r, c_i) and (r_{down}, c_j), then $c_i \neq c_j, \forall i, j$.

Definition 8 If r is the current processed row, then r_{up} defines the nearest row above r, $r > r_{up}$, where the symbol s is also found. The parameter r_{up} is related to a row that is already aligned, due to the fact that the matrix is aligned row by row starting at the top.

Definition 9 The *span* of a symbol s is the maximal distance of two subsequent occurrences of s in a column, taken over all matrix columns. The distance is defined as the difference of row numbers, where s occurs in a column.

These parameters are computed per symbol and for each alignment step. An alignment step means an iteration of a matrix row. An adjustment of a matrix row may cause the shift of elements in the matrix if different symbols occur in a row. When it comes to the shift of elements, these elements (symbols) change their row positions. With an update of the row position the previous calculated parameters become invalid.

The definition of *span* differs here from the source definition in FiVaTech, such that we do not consider repetitions of symbols between two occurrences of s. One reason is the subsequent parallel algorithm with symbol distribution. There, it is impractical to determine possible repetitive symbols between two symbol occurrences in a column, because such symbols may be processed by another instance. The second reason is the meaning of the symbols. In the original definition, symbols of the matrix represent tree nodes of the same level and it is useful to consider repetitions, whereas in our system the matrix consists of elements concerning all tree levels. Additionally, tree nodes are split up in start tag and end tag as far as they exist. Hence, the *span* measures the distance between two symbols of the same starting tag or the same end tag. The *span* value gives the maximal number of produced tokens from the underlying sub-trees plus eventually the number of some other tokens until the symbol appears again. The sub-trees of nodes with the same symbol may differ in size. Consequently, the number of symbols is different, because they cover all subordinate tree levels.

Example 3 Figure 9 shows a partially aligned matrix. The arrow indicates that row 4 is the current row to adjust. For the symbol a, r_{down} is 6, because a in row 5 cannot be considered for r_{down}. For symbol a, r_{up} is equal to 1 and the *span* value is 5.

3.4.2   The Algorithm for Sequential Matrix Alignment

As already pointed out, the sequential algorithm is the basis for our parallel versions for matrix alignment. Although not implemented, it is formulated and presented by Algorithm 2 solely to understand the entire system.

Algorithm 2 SequentialMatrixAlignment(M)

---

aligns a two-dimensional array M of size $k \times n$. The output is the matrix M' of size $m \times n$. Practically, the algorithm updates row indices of elements in M, which correspond to shift down of elements. The shifts of elements cause spaces in the affected matrix columns. Empty elements correspond to these inserted spaces and they need not to be stored. The aligned elements are put in the output matrix M', where each element has the same symbol as the other elements in the row.

(1) Creation of necessary data structures.
    The list $sr\_objects$ keeps sr_obj tuples of a row. Each tuple of type sr_obj merges the information concerning a symbol in a row. A sr_obj contains the values for r_{down}, r_{up} and $span$, the $level$, the count of elements with the actual symbol, and a list of the columns where these elements are found. The array $shifts\_col$ keeps the shift lengths per column for a row iteration.

Steps (2)–(6) are performed iteratively row by row.

(2) Collect elements of the current row r into the set E_r, where E_r does not contain empty elements.
(3) Determine from E_r the distinct symbols, denoted as set S_r.
    If $|S_r| = 1$, then the row is aligned. Go to step (2) for the next row $r + 1$.
(4) For each symbol s_i in S_r, determine necessary values for alignment as given in steps (4a)–(4e), and store them in a structure sr_obj per symbol.

   (a) Count the occurrences of s_i in E_r.
   (b) Store at which columns s_i occurs in E_r.
   (c) Determine the maximal $level$ of s_i from E_r.
   (d) Compute r_{down} for s_i from M.
   (e) Compute r_{up} and $span$ for s_i from M.

(5) Compute the alignment of row r.
    Determine the symbol that remains in row r from a list $sr\_objects$ of all sr_obj. Details for this (sub-)algorithm are given in Sect. 3.5. It returns the shift length for each matrix column in an array $shifts\_col$.
(6) Apply shift operations.
    Update row values of elements for each affected column in row r and below. Continue with the next row to align at step (2), except no more rows are left.
(7) Transform M to M'.
    Create a new matrix, where elements are positioned due to their alignment.

Fig. 9 A partially aligned matrix to demonstrate position parameters

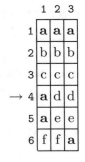

3.4.3 Time Complexity

For Algorithm 2, we now state the theoretical time complexity considering the worst case. The time for sequential matrix alignment T_{mat} is evaluated with respect to the matrix width n, the height k of the input matrix, and the height m of the output matrix. We evaluate T_{mat} as a sum of times for each of the steps of the algorithm.

Since the construction of temporal data structures requires a constant number of operations, the complexity of step (1) is

$$T_1(n) = \mathcal{O}(1).$$

In the worst case there are n different elements in step (2) for one iteration. The number of iterations is given by the height of the output matrix. Therefore,

$$T_2(m, n) = \mathcal{O}(mn).$$

When each element carries a different symbol, the loop in step (3) needs, again, n computational steps for each of m iterations, so

$$T_3(m, n) = \mathcal{O}(mn).$$

Step (4) is computed per symbol in each row, which means mn times. Steps (4a)–(4c) are linear with respect to the different symbols in the row, so a maximum of n computing steps per row are required. The position parameter r_{down} is computed in step (4d), which takes a maximum of $(k - 1)(n - 1)$ operations per symbol and per row. For the computation of r_{down}, only symbols in the rows below the current row must be considered. Computing r_{up} in step (4e) takes a maximum of $(k - 1)n$ comparisons per symbol and row. It is sufficient to consider only the rows above the current one. The calculation of the *span* value in (4e) requires looking at each matrix element, which results in kn computational steps per symbol and row. Altogether, the complexity is

$$T_4(k, m, n) = \mathcal{O}(mn^2 + kmn^2).$$

The time complexity for step (5) is $\mathcal{O}(n)$ for each unaligned row. For details see Sect. 3.5. Hence,

$$T_5(m, n) = \mathcal{O}(mn).$$

Applying shifts requires a maximum of $k(n - 1)$ steps per row, because rows at or below the current one are affected, giving

$$T_6(k, m, n) = \mathcal{O}(kmn).$$

The transformation of the matrix requires kn computational steps, thus

$$T_7(k, n) = \mathcal{O}(kn).$$

The computation of the position parameters dominates the theoretical time requirements as given in the time estimation for step (4). Hence, the sequential matrix alignment requires $T_{mat} = \mathcal{O}(kmn^2)$ time.

In [14], a time complexity of $\mathcal{O}(k^2n^2)$ is given. This is because of the computation of the *span* value, which influences to the greatest extent the worst case estimation of the time. If an element is shifted, then its symbol occurs again in a row below and we need to compute *span* and the other position parameters again. Due to the executed shifts, the parameters may change. The repetitive computation of parameters affects more rows than k. Precisely, these parameters need to be determined maximal m times for maximal n different symbols. Thus, the *span* value is computed in kn steps and this explains our worst case estimation.

3.4.4 Space Complexity

Space requirements are dominated by the size of the input matrix, which is $k \times n$. All other necessary data structures are linear to the number of the columns n. The structures are reused in each row iteration. These are the list for elements E_r, the array *shifts_col* and the list *sr_objects*. An entry of *sr_objects* is of type *sr_obj*. The structure *sr_obj* needs 5 integers to store the position parameters, the level and the count. Theoretically a *sr_obj* tuple has additionally a variable component. The field for the columns, where the designated symbol is found, can be realized through a list. Such a list has between 1 and n entries. In fact, the number of all list entries is bounded by n. For the computation of the alignment decision for a row, $\mathcal{O}(n)$ space is required. Summing up, in total $\mathcal{O}(kn)$ space is required.

3.4.5 Comparison with Matrix Alignment of FiVaTech

Algorithm 2 is an adaption of the matrix alignment algorithm presented by Kayed and Chang [14, Fig. 8]. In the following, we figure out the distinctions.

- Our matrix contains symbols related to any token from the data records. The matrix in [14] is related to tree nodes on the same level under a common parent node from the trees that represent the records. Furthermore, there is not a single matrix to align. Indeed the matrix alignment procedure is called recursively to cover all nodes in the trees.
- In [14] a row is aligned, if either the same symbol occurs in the non-empty elements or the row consists of symbols corresponding to variant leaf nodes. Variant leaf nodes are text nodes with different text values which result in symbols that occur only once in the matrix from their symbol assignment. Further img tags are candidates for variant leaf nodes, if the src attributes differ. In our system design, we do not distinguish text tokens by their values and img tags are treated like other tags. Consequently here are no variations to consider. In our system a row is aligned, if and only if the symbol is the same in every non-empty column.

- In [14], the columns are processed iteratively for row alignment. That means the computation of the shifts depends on the order of the elements. We changed the decision about shifting to be based on the symbols, because elements with the same symbol in the same row should be shifted the same length.
- The alignment algorithm in [14] returns a list of the symbols corresponding to the aligned matrix. In our system, we return the aligned matrix. The difference is associated with the next processing step in which the systems require different return types. They perform pattern recognition on the returned list of symbols, whereas we extract data items from the obtained matrix.

3.4.6   Potential for Parallel Algorithm Design

While iterating the matrix rows, the current processed row gets aligned. Due to the shift operations, the alignment of a row depends on all previously aligned rows. It is crucial for the design of parallel algorithms. In fact, it restricts the concurrent processing of rows. Therefore, it would be ineffective to compute in parallel the alignment of the rows.

What indeed can be parallelized is the inner loop at step (4). Parameters for different symbols can be computed individually. Also step (6) contains potential for parallelism. Here, the shift operations can be executed per element individually. Subsequently we identified two possibilities to decompose the elements, once by symbols and once by columns. As a consequence, steps (2) and (3) are adapted for the chosen decomposition. The elements and the symbols can be determined in parallel for a given row.

For the alignment procedure at step (5), we noticed that this is a sequential task. Thus it can be either performed by each process or one process computes the decision and communicates it. We applied the latter approach by including an additional process. We note that processing the alignment decision that way minimizes communication cost.

3.5   Alignment of a Matrix Row

Our algorithm for row alignment is similar to the function `getShiftedColumn()` presented in the source paper [14]. Like the original algorithm, we implemented this function based on rules.

3.5.1   Details of the Algorithm

Basically, the alignment algorithm (Algorithm 3) requires the elements of the row and the positions of all related elements anywhere else in the matrix. The related elements are the ones with a symbol that is also found in the current row. The

---

Algorithm 3 AlignElementsOfTheRow(sr_objects, r)

---

computes which symbol should be kept in the row r and shift lengths for all other symbols.
The input is given in $sr\_objects$, which is a list of sr_obj tuples. A tuple of type sr_obj summarizes
elements in r with the same symbol. Thus it stores the *count* of the symbol s in r and it stores the
columns, where s is found in r. Furthermore, it contains the previously computed values r_{down}, r_{up},
span and *level*.

(1) Create necessary data structures $shifts\_sym$ and $shifts\_col$.
    The array $shifts\_sym$ stores the shift lengths for each symbol. The array $shifts\_col$ is
    designated to store the shift lengths per column.
(2) Determine a symbol to stay.

    (a) Select a symbol to stay, if r_{up} and *span* exist for it and if the condition $r - r_{up} = span$ is
    fulfilled for that symbol. If multiple symbols fulfill this condition, then choose the symbol
    to stay that has the most associated elements in the row.

    (b) If a symbol is determined to stay according to step (2a), then set for all other symbols
    the shift lengths in the array $shifts\_sym$. The shift length is 1 if r_{up} and *span* exist for
    a symbol and $r - r_{up} < span$. Otherwise, the shift length is $(r_{down} - r)$ if r_{down} exists
    for a symbol. In all other cases, the shift length is 1 for a symbol.

(3) Determine shifts upon r_{up} and *span* if at least two symbols are remaining after step (2).

    (a) If r_{up} and *span* exist for a remaining symbol and $r - r_{up} < span$, then set its shift length
    to 1.

    (b) If all remaining symbols are supposed to be shifted, then no symbol would stay in the
    current row r. In this case, select the symbol with the highest *level* value to stay in r. If
    it cannot be decided by the *level* value, then choose the symbol with the lowest symbol
    number to stay. The shift length is reset to 0 in $shifts\_sym$ for this symbol.

(4) Determine shifts upon r_{down} if at least 2 symbols are left after step (3).

    (a) If r_{down} exists for a symbol, then set its shift length to $(r_{down} - r)$.

    (b) If all remaining symbols are supposed to be shifted at step (4), then determine the symbol
    to stay like in step (3b).

(5) If there are at least 2 different symbols left in the row after steps (3) and (4), determine which
    symbol will stay in the row.

    (a) Select the elements with the most occurrences of a symbol in r to stay. In case of ambiguity,
    select the symbol with the lowest value to stay.

    (b) Set the shift length to 1 for remaining symbols that have not already assigned a shift
    length.

(6) Transfer shifts per symbol to shifts per column.
    Determine per sr_obj the columns of a symbol and then put the shift length concerning that
    symbol into $shifts\_col$ for each relevant column.
(7) Return $shifts\_col$.

---

position parameters r_{down}, r_{up} and *span* depend on the symbol and the current row. The elements which are associated with a certain symbol, are subordinate for the alignment. Consequently, we choose a structure that contains data per symbol as input. Another reason concerns the performance: The number of symbols in a row is less or equals the number of elements in a row. So we need to evaluate rules on fewer entries.

We apply heuristics for the distribution of symbols in the matrix. Step (2) covers the case when the *span* value is reached for at least one symbol. That means such a symbol occurs repeatedly in a column and the maximal distance between two occurrences is found for at least one symbol. Here the designated symbol has to be kept in the current row. The example in Fig. 10 shows the application of the rule in step (2). Therefore, row 3 needs to be aligned and symbol a is forced to stay.

The rule in step (3) applies to repetitive occurrences of a symbol in a column including optional symbols in between. The *span* value gives therefore the maximal distance between two consecutive elements with the same symbol in the same column. If the condition $r - r_{up} < span$ holds, then it indicates, whether the symbol is within the *span*. If the symbol is within the *span*, then the concerning elements need to be shifted to get a consistent alignment. Figure 11 shows how this rule is applied to symbol a.

Step (4) of the algorithm forces elements to be shifted down to reach r_{down}. Therefore, the number of elements with the same symbol in row r_{down} increases. Remember that the parameter r_{down} states the nearest row below the current row r, where an element with the same symbol occurs. Additionally, the column of this

Fig. 10 Demonstration of the rule, when $r - r_{up} = span$

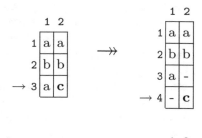

Fig. 11 Demonstration of the rule, when $r - r_{up} < span$

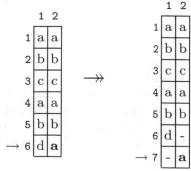

Fig. 12 Demonstration of
the rule to align symbols by
r_{down}

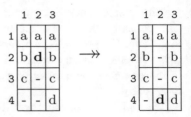

element differs from all columns where the symbol is found in r. This rule has its application if symbols occur optionally. From a global point of view, this means there is some optional data in the data records. Figure 12 gives an example for this case. Here, the alignment of the second row causes the symbol d to be moved to row 4.

Step (5) of the algorithm contains commands for the default case. It is applied to symbols without any other occurrences in the matrix. Such symbols usually derive from leaf nodes from the DOM tree. A suitable criterion is the count of the symbol for the current row. If the remaining symbols occur equally often, then the symbol with the lowest value is kept in the row r.

At some points we utilize the *level* value to decide secondly which symbol will stay in the current row. The *level* value indicates how deep in the DOM tree the node has been found. A high value of *level* means that the element derives from a tree node near the leaf. Tags concerning the structure of the HTML from the data record are situated near the root node in the DOM tree. Those elements have a low *level* value. On the other hand, elements that are related to data items have a higher *level*. Data is usually represented as text and text is positioned in the leaf nodes of a DOM tree. The *level* is given per element, so the maximal value is calculated, when elements with the same symbol are subsumed.

For example, let us assume that there is a conflict between two symbols as it is illustrated in Fig. 13. The first symbol of the conflict concerns an opening tag of a HTML hyperlink <a>. The second symbol concerns a closing tag of a HTML list item </li>. Both symbols have a value for r_{down}. The hyperlink is nested within a list item, so the symbol f for the tag <a> has a higher *level*. As a consequence, the second symbol d needs to be moved down. In such a case, there are some missing tags in a data record. This causes the premature end of the list item concerning the second symbol.

By design it is ensured that the algorithm terminates. After each step one or more symbols are left for the current row. If only one symbol is left for the row, then the subsequent steps are not evaluated. Otherwise, the next rules are applied.

Finally, the shift lengths per symbol are transferred to another array. In the target array are the shift lengths laid out for the row, i.e. each entry corresponds to the shift length for a certain column. Due to the fact that elements below the current row are affected from being shifted down, it is necessary to return a column-based array. A shift length of a symbol s is assigned to all columns, where the elements occur with the symbol s.

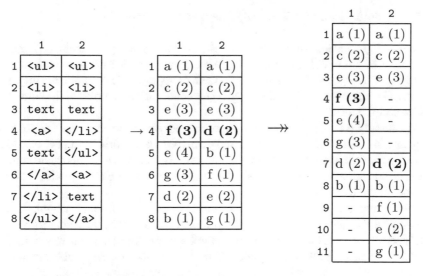

Fig. 13 Demonstrates the shift depending on the *level* of each symbol

3.5.2 Time Complexity

The input of the algorithm is the number of columns, denoted as n. In the worst case, all elements have a different symbol and in each column an element exists. In average, some empty elements are found in a row through shifts and these entries need not be considered for the alignment.

Step (1) is carried out in constant time. The evaluation of the rules (steps (2)–(5)) is linear to the input list. It maximally takes two times to go through the list. The evaluation of which rule is applicable to which symbol requires n computing steps. If the symbol parameters are all equal, then the refinement of the decision is carried out in maximal n further steps. Thus, the rules computation concludes in $\mathcal{O}(n)$. Step (6) can be computed in linear time to the length of the output array, which is n. Altogether, computation of the alignment for a row takes $\mathcal{O}(n)$ time.

3.5.3 Space Complexity

For the required space we neglect some counter variables. We need two arrays for the shift lengths, which makes $2n$ in worst case. An additional array helps to note which rule applies to which symbol, again this requires n memory positions in worst case. In total, the space requirement is $\mathcal{O}(n)$.

3.5.4 Comparison with Row Alignment of FiVaTech

Our row alignment algorithm is based on the function called `getShiftColumn()` presented in the FiVaTech paper [14]. We carefully studied the presented rules to select symbols to be shifted. The situation leaves opportunities to introduce improvements. The algorithm for row alignment is essential for the outcome of the alignment of the entire matrix.

As already pointed out, our algorithm decides about the shifts for the different symbols in contrast to an element based decision. The algorithm in [14] considers the rules for the elements. Additionally, the authors mentioned to abbreviate computations and take advantage of already determined results for elements with the same symbol in the row.

In our case, the alignment of a row results in elements with equal symbol. In [14], the aligned row may contain different symbols because of the definition for variant leaf nodes. For FiVaTech, a row is also aligned if the symbols in the row belong to text nodes from the DOM tree with different text values or the symbols belong to nodes with `<img>` tag and different values for the `src` attribute. We already handle this situation with the different kind of symbol assignment.

Considering the original presented rules within function `getShiftColumn()`, we reverse the order of the three rules.

Step (2) considers rule R3 in [14]. The original formulation leaves some space for interpretation. We decided to choose a symbol to stay in the row, when the condition $r - r_{up} = span$ is fulfilled. This condition prevents empty rows in the aligned matrix.

The rule applied in step (3) is based on the presented rule R1 in [14]. The only distinction here is, that we do not consider the order of the columns. Columns may be permuted, because their order depends on the input order of the data records. In case all symbols in a row could be shifted due to R1, the last column stays according to the original definition. This leads to a non-deterministic behavior, which we avoid by extending this rule with other criteria and also in the other cases.

Step (4) shifts elements according to r_{down}, as with rule R2 in [14]. This rule implies a higher shift value and so it produces more empty elements, which need not be considered for alignment in the subsequent rows. So it reduces computation effort. In case, we consider a symbol that is the first of a repetition, which means $span$ exists, but r_{up} does not exist, it makes sense to shift elements according to step (4). Another case is if the symbol occurs once in every column, i.e. $span$ and r_{up} do not exist.

In step (5), we introduced another rule as the default case. It selects a symbol to stay, if symbols are left. Remaining symbols are shifted down to the next row.

3.6 Data Extraction

Relevant data can be extracted from the aligned matrix. The output of this module is a table with the assumed data items. A table column ideally contains the data for a data item from different data records. Due to the fact that the output of the alignment is error-prone, the assumption of a data item per column may fail.

We consider text values of text tokens as possible data items. For text tokens it is necessary that the values are distinct in a row. Otherwise, text tokens are considered to belong to the template. For example, an aligned row contains text elements that all have the content "Price:".

The challenge of decorative tags is not covered in this post-processing module, because our focus is on the parallel alignment module. As a consequence, decorative tags from the input forces a data item to be separated into several parts in the output. For example, if individual words are marked as important in a continuous text block, then the necessary formatting tags `<strong>` and `</strong>` cause the partition of the text block into multiple tokens. This tokenization schema results in multiple data items, although there should be one item.

From the point of view of the input data types, we are able to recognize optional data items. Optional data is indicated by empty elements in an aligned row, whereas the other elements in such a row contain text values. The recognition of repetitive and disjunctive data is left for further improvement in this work.

3.7 Review Extracted Data

In the last module we provide a procedure to measure the quality of the web data extraction. Concretely, the extracted data is compared with the actual data. The extracted data corresponds to the produced output from the data extraction module. The actual data is the desired outcome of the system. A user must define which data is actually denoted as correct. However, we provide a procedure that takes a `.txt` file as input for the actual values.

The quality is measured in terms of precision and recall. That means the actual data items and the extracted data items are counted. Furthermore, the values are compared and matches count as correctly identified data items. From this three count variables the precision and the recall are determined.

4 Parallel Algorithms

Based on the sequential algorithm of Sect. 3.4, we develop parallel algorithms using the message passing paradigm. Message passing is chosen instead of shared memory programming due to reasons of hardware independence and better cache coherence when accessing local copies of data.

There are two approaches how to apply data parallelism to the algorithm, one being a decomposition of data along different symbols, as presented in Sect. 4.1, the other one a decomposition along columns of the matrix, as presented in Sect. 4.2.

We utilize a hybrid approach that is a combination of the data-parallel model and the master-worker model. The master process performs all computations before and after the matrix alignment. During the row iterations, the worker processes prepare the data and then the master process makes a decision about the alignment of each row. Handling the decision by only one process reduces the number of necessary messages by avoiding all-to-all communication when each worker process would have to collect global information to make decisions redundantly. Also, contention is reduced when I/O is delivered from and to one process.

4.1 Parallel Matrix Alignment with Element Distribution by Symbols

In this version, matrix elements are partitioned by their symbols. The communication path is always between a worker process and the master process to keep the communication at a minimum. The alignment parameters r_{down}, r_{up} and $span$ can be computed for every symbol individually. These parameters depend only on positions of other matrix entries with the same symbol. Additionally, the shifts on matrix elements can be applied independently from each other. The resulting parallel algorithm is presented as Algorithm 4.

4.1.1 Details of the Algorithm

The number of employed worker processes is limited by the number of different symbols. However, it turned out that the symbols outnumber the processes in our experiments. We made the load distribution dependent on the chosen number of processes. This design gives us flexibility to vary the number of processes. Hence, the element sets per symbol are accumulated to larger sets. Each worker instance handles such a portion of elements. Therefore, the per-process sets are built according to balance the working load.

The distribution works as follows: The occurrences of each symbol are counted in a single pass of the matrix. Next, the symbols are sorted in descending order by their occurrences. Next, the elements for a symbol are assigned to a process in a round-robin fashion. The first process gets the elements with the symbol that has the most occurrences. The second process gets the elements, where the symbols have the second most occurring number. This continues until all elements are distributed.

The workers compute their symbol's parameters and make them available to the master. The master process computes the different shift lengths per column and it broadcasts the shift lengths. The worker processes apply the shifts according to the received messages on all relevant elements. As mentioned before, shifts also affect elements in rows below the current one. Hence a worker needs the shift lengths for

---

Algorithm 4 ParallelMatrixAlignmentWithDistributedSymbols(M)

---

computes the alignment of the input matrix M of size $k \times n$. The output is an aligned matrix M' of size $m \times n$. The computing steps are performed by a master process and p worker processes, $p \geq 1$. At each matrix row, the workers prepare the relevant information in parallel and sends it to the master, which then decides which elements remain in the current row. After that, shifts are performed by the workers simultaneously.

(1) Creation of necessary data structures.

    (a) Each worker process creates a tuple sr_obj to store the position parameters r_{down}, r_{up}, $span$, the *level* of the symbol, the number of elements with the symbol as count value, and the column numbers of the elements with the concerning symbol.

    (b) The master process creates an array of tuples $sr\_objects$ to keep the same values of all symbols in a row, an array of integers $shifts\_col$ to keep the intermediate shift values for a row, and an array of integers $lengths$ to store the intermediate row lengths for each column.

(2) The master determines the distribution of elements of the matrix M. Due to the matrix preparation, the master process holds the complete matrix.

    (a) Partition the set of matrix elements by the symbols. Elements with the same symbol s_i form a set E_{s_i}, $E_{s_i} \cap E_{s_j} = \emptyset \; \forall s_i \neq s_j$.

    (b) Distribute sets E_{s_i} among the worker processes. Elements from different sets E_{s_i} are merged to a larger set of elements E_w, which are sent to the worker process w.

Steps (3)–(5) are performed iteratively row by row.

(3) Each worker process prepares data for the alignment of the current row r.

    (a) Find from E_w those elements where the row index is r and build the set S_{wr} of distinct symbols in the current row.

    (b) For each symbol $s_j \in S_{wr}$, compute the position parameters r_{down}, r_{up} and $span$, determine the maximum *level* of elements with symbol s_j, count the occurrences of s_j in r, and determine the columns where the symbol s_j occurs. Collect these values in the structure sr_obj and send it to the master.

(4) The master determines the alignment of r. If the master receives just one message, then r is already aligned. In this case, the master broadcasts that r is aligned. All processes skip step (5) and continue with step (3) for the next row $r + 1$.

Otherwise, the master determines which symbol remains in r. The received tuples of type sr_obj are put together in the array $sr\_objects$. The algorithm, described in Sect. 3.5 is employed. The resulting shift lengths are stored in $shifts\_col$ and broadcast to all workers.

(5) Each worker performs shifts of elements in parallel. Shifts are applied on each element with the position (r', c), where the row $r' >= r$ and c is a column where $shifts\_col[c] > 0$.

(6) The master updates the values within $lengths$. When for each column c the condition $lengths[c] = r$ holds, the master sends a broadcast to the workers in order to finish the alignment procedure.

(7) Each worker sends aligned elements to the master, which forms the complete aligned matrix M'.

---

all columns, because a shift operation may affect any element in its processing set. The master does not know explicitly, where the elements reside during the iterations.

Theoretically, in a row iteration the symbols are distributed among the processes. Thus, the parameters are computed in parallel. In the worst case scenario, however, parameter computation could be executed by one worker process per iteration, which is equivalent to the sequential case. Only the shift operations are indeed distributed.

The master process initiates the termination of the repeated row adjustment. It does so by updating the number of elements per column, when an unaligned row is processed. If the maximum length of all updated sequences is reached, then no more elements are to be processed and the procedure can be terminated.

4.1.2   Illustration of the Algorithm

Figure 14 gives a small example. The matrix M is created by the master process. The master instance computes the partitioning of the symbols. Next the master transfers the two distinct parts, E_1 and E_2. The parts are illustrated as matrices, although they are stored efficiently in the implementation. In this example, the row 1 is already aligned, whereas for row 2 the master has to decide which symbol to keep. After the shift operations, both E_1 and E_2 are updated, because both worker processes handle elements in column 3.

4.1.3   Time Complexity

Now we state the theoretical time complexity of Algorithm 4 considering the worst case. We denote the required time for theoretical analysis of the parallel algorithm with symbol distribution as T_{sym}. The time T_{sym} is evaluated with respect to the matrix width n, the height k of the input matrix, the height m of the output matrix, the number of worker processes p, the maximum occurrence number of a symbol as o and d as the number of different symbols in the matrix. We evaluate T_{sym} as a sum of times for each of the steps of the algorithm. The time complexity for computation and communication are considered together.

As in the sequential algorithm, the construction of temporal data structures has a constant complexity

$$T_1(n) = \mathcal{O}(1).$$

The time complexity of the distribution task depends on the size of the input matrix. In a single pass through the matrix the sets of elements are created. The merging of the sets into larger groups of elements requires sorting the sets by their size. This takes $\mathcal{O}(d \log d)$ time depending on the different symbols in M. The distribution of the matrix elements requires p messages. Altogether,

$$T_2(k, n, p, d) = \mathcal{O}(kn + d \log d + p).$$

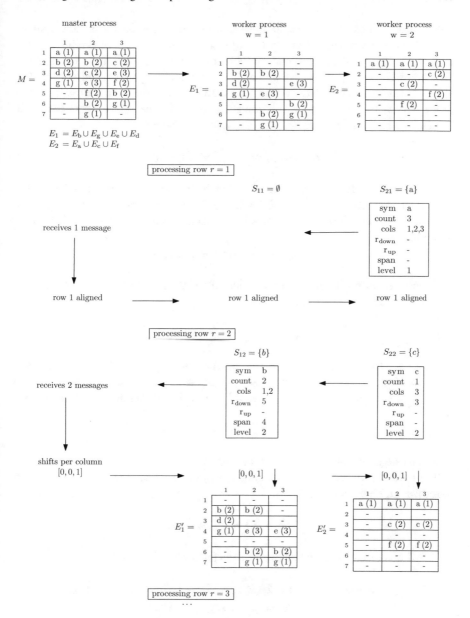

Fig. 14 Illustration of the parallel algorithm with symbol distribution

Since each worker process handles about kn/p elements, the determination of which elements are in the current row needs to iterate over its elements. The number of different elements that a process may work on is bounded by the number of columns n. Assuming elements of the row have all different symbols and these elements occur accidentally at the same worker instance, the number of different symbols is also at most n. For the parameter computation, we clearly benefit from the symbol distribution. It takes at most $\mathcal{O}(o)$ steps for every different symbol because for a symbol s_i only the corresponding subset of elements E_{s_i} within E_w is examined. This holds for each position parameter. The computations are carried out for each row of the output matrix. Transferring each required sr_obj from a worker process to the master requires $\mathcal{O}(n)$ communication steps, because the master receives at most n messages for each row of the output matrix. Hence,

$$T_3(k, m, n, p, o) = \mathcal{O}(kmn/p + mno).$$

As described in Sect. 3.5, finding the shift lengths is linear with respect to the number of different symbols in a row, which takes at most $\mathcal{O}(n)$ time. The row alignment has to be computed at most m times. A broadcast is executed in $\log(p + 1)$ steps per iteration, so

$$T_4(m, n, p) = \mathcal{O}(mn + m \log p).$$

Updating the row positions needs to consider all elements residing at a worker process for each row of the output matrix, thus

$$T_5(k, m, n, p) = \mathcal{O}(kmn/p).$$

The master process may need to check the maximum possible row length in each iteration. The designated structure $lengths$ of size n is iterated for each row. To broadcast the termination messages takes $\log(p + 1)$ communication steps, giving

$$T_6(m, n) = \mathcal{O}(mn + \log p).$$

In the last step, the master receives p messages containing the aligned elements. The collection of the matrix elements is bounded by $\mathcal{O}(kn)$ since the master process has to receive all elements. Therefore,

$$T_7(k, n, p) = \mathcal{O}(kn + p).$$

The time complexities are summed up as follows:

$$\begin{aligned}
T_{sym}(k, m, n, p, d, o) &= T_1(n) + T_2(k, n, p, d) + T_3(k, m, n, p, o) + T_4(m, n, p) \\
&\quad + T_5(k, m, n, p) + T_6(m, n) + T_7(k, n, p) \\
&= \mathcal{O}(1) + \mathcal{O}(kn + d \log d + p) + \mathcal{O}(kmn/p + mno) \\
&\quad + \mathcal{O}(mn + m \log p) + \mathcal{O}(kmn/p) + \mathcal{O}(mn + \log p) \\
&\quad + \mathcal{O}(kn + p) \\
&= \mathcal{O}(kn + d \log d + kmn/p + mno + mn + m \log p + p).
\end{aligned}$$

Since the term mn is covered by mno, $k \le m$, and $p \le kn$, this results in

$$T_{sym}(k, m, n, p, d, o) = \mathcal{O}(kmn/p + m \log p + mno + d \log d).$$

Any further reductions of the terms in T_{sym} would require additional assumptions. Analyzing this result, the two last terms in this complexity formula depend on the symbols, which are characterized by o and d. The term kmn/p shows a decreasing number of computational steps with the growth of processes. In contrast, $m \log p$ grows logarithmically with the number of processes.

4.1.4 Space Complexity

The master process holds the input matrix of size kn. The output matrix can occupy the space for the input matrix after the alignment and so it does not demand more memory for that. During the iterations over the rows, it requires at most n structures sr_obj, kept in $sr\_objects$, and an array $shifts\_col$ of size n. The array called $lengths$ has size n. For the row alignment, the memory requirement is $\mathcal{O}(n)$. In total, the master process needs $\mathcal{O}(kn)$ space.

Each worker processes about kn/p elements, which makes the most part of the required memory. The other factors concern the row iterations, so they can be reused with every new processed row. A worker process requires at most n times the structure sr_obj and it requires the array $shifts\_col$ of linear size to the matrix width n. So each worker process requires $\mathcal{O}(kn/p)$ space. In total, the required space is $\mathcal{O}(kn) + p \, \mathcal{O}(kn/p) = \mathcal{O}(kn)$.

4.2 Parallel Matrix Alignment with Element Distribution by Columns

From the point of view of the computation, this version has a similar concept as the previous one. It implements the master-worker model. Worker processes prepare data for the alignment of each row and the master process decides about the alignment. Matrix elements are distributed by the columns. An entire column is assigned to a single process. Thus, the position parameters are evaluated per column, which gives intermediate results. The shift operations can be performed in parallel per column. This process is formulated by Algorithm 5.

4.2.1 Details of the Algorithm

Data decomposition by columns provokes that elements with same symbols are distributed among the processes. This means that the structural relations are partially broken up and a process does not have the entire information for a symbol to compute the position parameters. This circumstance demands an extra communication and

---

Algorithm 5 ParallelMatrixAlignmentWithDistributedColumns(M)

---

computes the alignment of the input matrix M of size $k \times n$. The output is an aligned matrix M' of size $m \times n$. The computing steps are performed by a master process and p worker processes, $1 \le p \le n$. Parameters for the alignment of a row are computed by the worker processes with a reduce operation. The master decides, which elements stay in the current row, and workers apply the shifts in parallel.

(1) Creation of necessary data structures.

    (a) The master process creates an array of tuples $sr\_objects$ and an array of integers $shifts\_col$ to keep intermediate parameters and shift values, as well as an array of integers $lengths$ to store the intermediate column lengths.

    (b) Each worker process creates an array of tuples called $params$, which is used to keep intermediate results for the position parameters r_{down}, r_{up}, $span$, and the $level$ values for each possible symbol in a row.

(2) The master process determines which columns are handled by each worker and transfers them to the workers, where they are stored in a set E_q for worker process q. Due to the matrix preparation, the master process holds the complete matrix.

Steps (3)–(7) are performed iteratively row by row.

(3) Each worker determines the elements in the current row r from the set E_q and sends the symbol of each element to the master.

(4) From the received elements, the master extracts the set of all distinct symbols of the current row r, denoted as S_r. The master broadcasts the set of symbols S_r. If $|S_r| = 1$, then the master process continues with the next row $r + 1$. Otherwise, it reduces the implicit data provided by the elements for each symbol. It determines for each symbol s_j the count of elements and the columns, where s_j occurs.

(5) Each worker process receives the set of symbols S_r. If $|S_r| = 1$, then each worker continues with the next row $r + 1$ at step (3). Otherwise, it computes intermediate values of r_{down}, r_{up}, $span$, and $level$ for each symbol $s_j \in S_r$. Intermediate means that each process can only compute values on its available set of elements E_q.

(6) The workers reduce the parameters to $\min(r_{down})$, $\max(r_{up})$, $\max(span)$, and $\max(level)$ for each symbol $s_j \in S_r$, and send them to the master.

(7) To compute the row alignment of r, the master process performs the algorithm described in Sect. 3.5. The resulting shift array $shifts\_col$ is distributed to the workers, which perform the shifts in parallel.

(8) The master updates the values within $lengths$. When for each column c the condition $lengths[c] = r$ holds, the master sends a broadcast to the workers in order to finish the alignment procedure.

(9) Each worker sends aligned elements to the master, which forms the complete aligned matrix M'.

---

computation step by determining the position parameters for each different symbol in the row.

The advantage of this distribution is that elements are stored per column in a sorted form. This situation simplifies the iteration procedure from one row to another. Indices are used to remember the position of the elements within a column, which makes the determination of the elements at the start of every loop easier.

The number of processes is variable, although it is bounded by the number of columns. The maximum number of employed processes is therefore the number of columns plus 1 for the master process. The work load is distributed, such that each worker gets a strip of columns. If there are n columns in total, then $\lfloor n/p \rfloor$ is the minimum number of columns per process. The first n mod p worker processes get an additional column.

Each worker instance carries out the computation of the parameters on its available portion of columns. As a consequence, the worker processes compute intermediate values, which need to be collected and evaluated by the master. The master reduces the received information to determine the final parameters per symbol. While the workers compute the parameters, the master process counts the elements for each symbol. This means that step (4d) and step (5b) are executed in parallel.

4.2.2 Illustration of the Algorithm

Figure 15 illustrates the algorithm for column distribution. The input matrix M is the same as in Fig. 14. The master process partitions M by the columns into the sets E_1 and E_2. Row 1 is already aligned, which is indicated by the number of symbols that are broadcast by the master. Row 2 is unaligned. After all processes know the current set of symbols S_2, they compute in parallel the parameters, while the master instance determines the count and the columns for the symbols. The diagonal arrow indicates a reduction, which is executed to determine the final parameters. The decision for the shifts affects only process $q = 1$ in this example (compare Fig. 14).

4.2.3 Time Complexity

We define T_{col} as the time required for the alignment with element distribution by columns. The computation and communication steps are described for the width n and height k of the input matrix, the height m of the output matrix, the number of worker processes p, and c as the maximal number of columns for a worker. Again, it requires constant time to allocate temporal data structures:

$$T_1(n) = \mathcal{O}(1).$$

The distribution of elements at the start of the parallel alignment procedure depends on the size of the input matrix and requires p messages, so

$$T_2(k, n, p) = \mathcal{O}(kn + p).$$

The current row elements are determined in every row iteration. For each row, at most c columns are considered. The use of an array of indices alleviates this task. The elements are transmitted in p communication steps at m rows. Therefore,

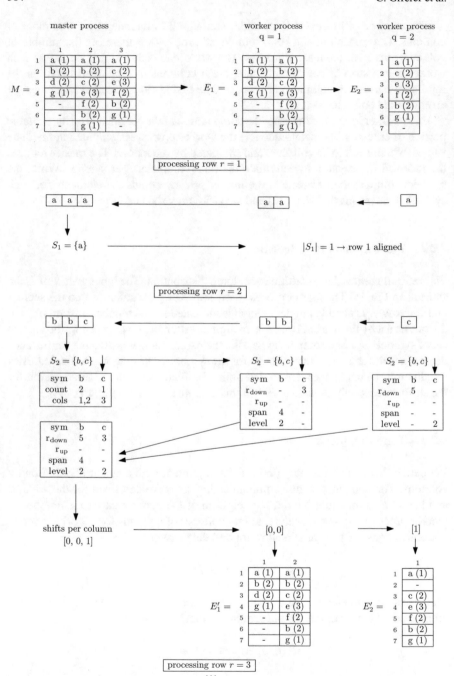

Fig. 15 Illustration of the parallel algorithm, column distribution

$$T_3(m, p, c) = \mathcal{O}(mc + mp).$$

At the master process, the reduction to the set of different symbols requires n steps and this is done in m rows. Subsequent computations in case of an unaligned row also require at most n computational steps in m rows. To broadcast the symbols, $\log(p + 1)$ communication steps are required for m rows, giving

$$T_4(m, n, p) = \mathcal{O}(mn + m \log p).$$

In the worst case, a row is not aligned and the position parameters must be calculated. There are m row iterations and at most n different symbols for each row are considered. Since each of the position parameters is computed in ck time,

$$T_5(k, m, n, c) = \mathcal{O}(kmnc).$$

Note that this estimation is based on the maximum number of processed elements per worker.

The parameters of a symbol can be reduced in $\log p$ computational steps, when p processes are assumed. This is necessary in at most m rows and at most n symbols per row. Hence,

$$T_6(m, n, p) = \mathcal{O}(mn \log p).$$

The computation of the shift lengths is linear with respect to the maximum number of different symbols n. Sending the shift information takes a maximum of mp communication steps. In the worst case, all elements in a row must be shifted except one, where the element (or the symbol) stays in place. Consequently, this task affects c columns and maximal k elements per column, giving

$$T_7(k, m, n, p, c) = \mathcal{O}(mn + kmc + mp).$$

In the worst case, the termination requires the update of the length value per column for each of the m rows. The broadcast message requires $\log(p + 1)$ time, so

$$T_8(m, n, p) = \mathcal{O}(mn + \log p).$$

The master receives p messages at the end of the procedure. After collecting the elements, the processing of the aligned elements takes kn computing steps:

$$T_9(k, n, p) = \mathcal{O}(kn + p).$$

This results in the total time complexity

$$
\begin{aligned}
T_{col}(k, m, n, p, c) &= T_1(n) + T_2(k, n, p) + T_3(m, p, c) + T_4(m, n, p) \\
&\quad + T_5(k, m, n, c) + T_6(m, n, p) + T_7(k, m, n, p, c) \\
&\quad + T_8(m, n, p) + T_9(k, n, p) \\
&= \mathcal{O}(1) + \mathcal{O}(kn + p) + \mathcal{O}(mc + mp) + \mathcal{O}(mn + m \log p) \\
&\quad + \mathcal{O}(kmnc) + \mathcal{O}(mn \log p) + \mathcal{O}(mn + kmc + mp) \\
&\quad + \mathcal{O}(mn + \log p) + \mathcal{O}(kn + p) \\
&= \mathcal{O}(kn + mc + mp + mn + m \log p + kmnc + mn \log p + kmc) \\
&= \mathcal{O}(kmnc + mn \log p) .
\end{aligned}
$$

Since the number of columns c depends on the number of processes, we can replace it with $c = \lceil n/p \rceil$. So the resulting time complexity is

$$
T_{col}(k, m, n, p, c) = \mathcal{O}(kmn^2/p + mn \log p) .
$$

From the received time estimation, some conclusions can be drawn. The first term kmn^2/p dominates the required computing and communication steps. Further, it shows that the computational effort decreases with the growth of the number of processes. In contrast, the second term $mn \log p$ shows a growth of computing steps with increasing p. Thus, for reasonable speedups the inequality $kn/p > \log p$ should be fulfilled.

4.2.4 Space Complexity

The master process holds the matrix elements, at most kn. For the alignment of a row, the master process needs memory for n elements. Additionally, it requires an array for the different symbols of maximal size n. Next, it needs an array $sr\_objects$ with size n in the worst case, and an array $shifts\_col$ of size n. Computing the decision needs $\mathcal{O}(n)$ space. In total we have a count $\mathcal{O}(kn)$ because the required space is dominated by the size of the input matrix.

A worker carries out each row loop on at most kc matrix elements. The other data structures are reused during the row iterations. Here each instance needs memory for c elements. For the computation of the parameters for all different symbols in the row it requires an array $params$ with maximal size n. The shift lengths need space for c integers. An array of indices is necessary to remember which element in the column is currently processed. This takes again c space elements. Summarized, all space requirements, each worker needs $\mathcal{O}(kn/p)$ space, because $c = \lceil n/p \rceil$. In total, the space requirements are $\mathcal{O}(kn) + p\mathcal{O}(kn/p) = \mathcal{O}(kn)$.

4.3   Comparison of Complexities

Both T_{sym} and T_{col} have their own variables, which depend on the corresponding version. Common are the variables k, m, n concerning the matrix, and the number of processes p. Further, the time requirement for the symbol version depends on properties depending on the symbol distribution. These properties are the number of different symbols d and the maximum of occurrences of a symbol o. For comparison, we assume 3 different cases for the symbol dependent variables d and o.

1. Each element of the matrix has the same symbol, i.e. $d = 1$, $o = kn$. This gives us $T_{sym}(k, m, n, p) = \mathcal{O}(kmn^2 + kmn/p + m \log p)$. Compared with T_{col}, it requires less time to compute the alignment.
2. Each element of the matrix has another symbol, i.e. $d = kn$, $o = 1$. This leads to $T_{sym}(k, m, n, p) = \mathcal{O}(kmn/p + kn \log kn + m \log p + mn)$. If the inequality $\log(kn) \leq mn/p$ holds, then the symbol version is better. This can be achieved easily by the proper choice of p.
3. In the matrix are m different symbols and each of them occurs at most n times, $o = n$, $d = m$. This is close to the real case in the context of web data extraction. This results in $T_{sym}(k, m, n, p) = \mathcal{O}(mn^2 + kmn/p + m \log m + m \log p)$. If $k/p > 1$ and $\log m < n \log p$, then the symbol version has the lower time complexity. If $k/p < 1$ and $\log m > n \log p$, then the column version is better.

The spatial complexity is equal for both parallel algorithms.

5   Experiments

In this chapter, we report about the performance of our developed algorithms. First, we describe the corpus of the used test data. Afterwards, we give details of the parallel shared memory system, on which the algorithms were running. Finally, sequential and parallel execution times are measured and discussed.

5.1   Test Data

There exist a couple of test suites for the purpose of web data extraction. A widely used suite is the test bed for the deep web, described by Yamada et al. [34]. It contains 51 different sets of web pages. Each set consists of 5 web pages, which are search result pages from a web site. These 51 sets were randomly chosen out of over 100,000 sites from the World Wide Web. The provided test bed additionally contains the actual data items of the first data record found on each web page. The TBDW test bed has been used in many research projects in the field of web data extraction, e.g. in [13, 14, 19, 24]. However, we found that the test sets from the established TBDW test

Table 2 Properties of the generated matrices for the test sets

Cat.	Set name	Symbols	Columns	Rows in	Rows out	+% of rows
1	Autos	128	999	177	188	+6
	Music	141	1,000	138	152	+13
	Rentals	61	1,000	73	73	+0
2	Books	197	500	195	488	+150
	e-books	145	500	125	224	+79
3	Games	414	30	2,015	6,325	+214
	Properties	530	30	1,883	2,005	+6
	Sports	525	30	3,524	6,221	+77

bed result in matrices of small size. Approximately 100 data records exist for each test set in the test bed. Additionally, many sets have fairly small data records. For parallel execution it is desirable to have large matrices with numerous entries. Thus, we used the TBDW test bed only during the development of the program, especially to test the alignment algorithm.

Sleiman and Corchuelo provide a test bed within their CEDAR framework [25], which is available online.[2] They proposed the framework to introduce a more comparable study of different web data extraction systems. We used three sets of the available test collection, which match our needs for resulting large matrices.

Another extensive test collection is presented by Álvarez et al. [1], which is also available online.[3] This collection covers web pages that are acquired after searching for a certain term via the web form on the respective web site. We utilized two of the proposed web sites, but we used different search terms to gain other search result records.

For our experiments we collected eight different test sets. The properties of the resulting matrices for the alignment are given in Table 2. Our test suite covers variations in size, source domain and in the template structure. The sets are divided into three different categories depending on the width and height of the resulting matrices. Each set consists of data records, respective HTML code snippets that are from a certain web site. The columns number is the number of data records. The number of input rows is the maximal length of a string of symbols within a set. The number of symbols depends on the string lengths and the similarity of the produced tokens from the input HTML code. The number of output rows is greater or equal than the number of input rows. A plus of a few per cent means most rows are already aligned. The differences between input and output rows in per cent also indicate the similarity of the structure of the data records. A significant growing of matrix rows may correspond to a template consisting of many variations like optional, disjunctive or repetitive tokens.

---

[2]http://www.tdg-seville.info/Hassan/CEDAR.

[3]http://www.tic.udc.es/~mad/resources/projects/dataextraction/testcollection_0507.htm.

Category 1 contains three data sets, where the number of data records prevails the string length attained from a data record. In other words, the matrices are wide and short. Category 1 is also characterized by sets of relative homogeneous data records, which means the alignment increases slightly the number of input rows.

Data records from category 2 have a higher variety in the template, respective in the sequences resulting from the data records. Here the accretion of rows during the alignment is proportionally greater in comparison to category 1. Concerning the size, sets in category 2 are denoted as "middle" both in width and height of the matrices.

The sets from category 3 have comparable few columns, but large number of rows. Thus the constituted matrices are "thin and long". Category 3 contains detailed data records, where each of them occupies an entire web page. The given test sets provides 30 columns each. In this case we execute a program run with maximal 31 processes.

The test sets are chosen to obtain a high number of matrix elements. Thus we are able to sufficiently investigate the performance of the parallel alignment. For the categories 1 and 2 we gained an extensive high number of data records to produce many columns. Xia et al. [33] report for their tree alignment method, that the resulting tree template starts to converge at 30 input pages. That means an alignment usually requires a fraction of the sets in the first two categories. The category 3 sets are not typical due to the size of each record. Usually, the alignment is applied to data records, where multiple of them occur on a web page. It means, the row numbers in category 3 outweighs.

The sources of the test sets are the following:

autos: Extracted from the AutoTrader web site[4] through a search for cars within 10 miles of zip code 90210 from private sellers only, and selecting 999 of the acquired search result records.

music: Extracted from the web site of Barnes & Noble[5] for the category Brit-pop offers from the music department.

rentals: Extracted from the overview page for flats to rent in London, provided on the web site from Homes & Property.[6]

books: Extracted from Amazon[7] through a search for the key word "parrot" in the books department.

e-books: Extracted from Kobo[8] for offers from the category Art and Architecture.

games, properties, sports: Taken from the CEDAR framework.

Each web page is transformed into a valid XHTML page by using the program .tidy[9] The main purpose of tidy is to repair malformed HTML code. The additional transformation into XHTML is required for parsing via the libxml++ library.

---

[4]http://msn.autotrader.com/cars-for-sale/.

[5]http://productsearch.barnesandnoble.com/search/results.aspx?CAT=1000652\&STORE=music.

[6]http://zoopla.homesandproperty.co.uk/to-rent/property/london/.

[7]http://www.amazon.com/.

[8]http://www.kobobooks.com/browse/Art__Architecture/LjUHGEwvR0KbkEbC9oZP_Q-2.html.

[9]http://www.w3.org/People/Raggett/tidy/.

Data records from category 1 and 2 need to be manually extracted, because multiple records are embedded in a web page.

5.2 Hardware and Software Infrastructure

The source code is written in C++. Within the program we used the C bindings of the OpenMPI library to enable multiple processes and communication with each other. We utilize features of the Standard Template Library (STL), especially the provided data structures. For parsing the input we use the libxml++ bindings, which is the C++ interface for the libxml2 library. Consequently the original HTML documents are transformed into XHTML files, which are parsed with the XML parser of the library.

We compiled the sources by invoking `mpic++`, specifically the version 1.4.3 of the OpenMPI C++ wrapper compiler. Compiling the whole project, the wrapper compiler utilizes the g++ compiler, version 4.4.6. Compilation is carried out with option -O3 for optimizations.

We conducted the experiments on the Doppler cluster of the University of Salzburg. The available computing power enables research experiments in various high performance computing areas.

Our experiments are run on a single cluster node, namely doppler23. The utilized node consists of four AMD Opteron processors, model 6274. They are placed together on a G34 socket. Each Opteron processor has 16 cores with a clock rate of 2.2 GHz, and 504 GB RAM memory. The installed operating system is CentOS, version 6.2. Program runs are executed starting with 1 process up to 60 processes, so there is a one-to-one mapping between processes and cores.

Because of variations in execution time, we run the program five times for each setting and select the best time for the performance comparisons.

5.3 Evaluation

Primarily, we investigate the run-time of the alignment procedure for each of the two versions of element decomposition. Computations are sequential for pre- and for post-processing, so only the alignment module is considered here. We aim to answer the following questions:

- Which implemented version performs better for the sequential alignment?
- Which implemented version of the parallel alignment performs better?
- Which speedup, if any, do we achieve in each of our versions of parallel alignment?
- Do we discover differences in the performance between the test sets?

5.3.1   Sequential Execution Times

The results are given in Fig. 16. The column version performs better for data sets of category 1 and the data set 'e-books'. Element distribution by columns has the advantage, that indices are kept to iterate from one row to another. Additionally the shift operations are executed faster, because shifts affect only elements in certain columns and elements are stored column-wise. In the symbol version, the computation of the position parameters can be evaluated faster because elements are sorted by the symbols. This effect is more prominent when the matrices contain plenty of symbols. Data sets 'books', 'games', 'properties' and 'sports' lead to such matrices for the alignment. In category 3 the times are significantly higher because of the high number of output rows.

5.3.2   Parallel Execution Times

Figure 17, 18, 19, 20, 21, 22, 23 and 24 present the results of the alignment times for parallel execution. The parallel performance is evaluated from 1 to 60 worker processes for the data sets in category 1 and 2. Sets in category 3 have only 30 columns, thus maximally 30 worker processes are employed.

   The version with the symbol decomposition shows decreasing execution times for increasing number of processes up to a point where the limited number of symbols per row poses problems for efficient parallelization. Above this point, execution times may increase due to increased communication overhead. Matrices are aligned

Fig. 16 Execution times of the sequential alignment

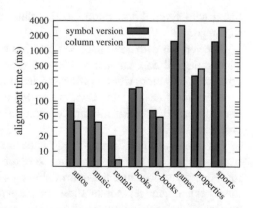

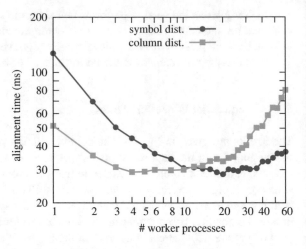

Fig. 17 Performance of the parallel alignment for the test set 'autos'

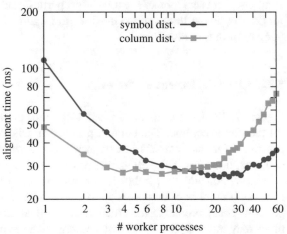

Fig. 18 Performance of the parallel alignment for the test set 'music'

faster with symbol distribution for the data set 'books' and the data sets belonging to category 3.

The version with column decomposition has benefits for lower process numbers. The additional communication steps of this version lead to severe communication cost for 5, 6 or more worker processes, leading to a growth in the required alignment time. For the data sets from category 1 and the data set 'e-books', the symbol version becomes faster than the column distribution at a point which lies between 8 and 14 worker instances.

For the sets 'autos' and 'music', the communication dominates the run-time because of wide matrices. The data set 'rentals' does not profit from the parallel implementation because the matrix elements are practically aligned for this set. On the other hand, the set 'books' shows better parallel performance because a significant portion of the matrix rows needs to be aligned.

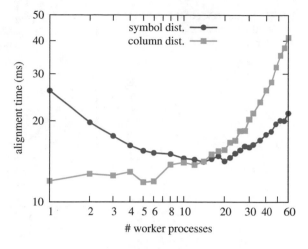

Fig. 19 Performance of the parallel alignment for the test set 'rentals'

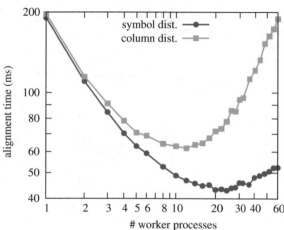

Fig. 20 Performance of the parallel alignment for the test set 'books'

All sets in category 3 have a decreasing run-time of the alignment for symbol decomposition. The reasons are the reduced number of processes and the higher total alignment times compared to the other categories. This also explains a relatively moderate increase for the column distribution.

Basically, the results show two different pictures. For one of them, the column version is initially better and then its performance decreases. The curve of the symbol version starts high, falls down and then it remains steady for a high number of processes. This picture is captured for data sets in category 1 and data set 'e-books'. All other sets show curves, where the symbol version always performs better than the column version regardless of the amount of processes. This second picture is covered by data sets of category 3 and data set 'books'. Considering the properties of the input matrices, it seems that the number of symbols influences the different performance behaviours.

Fig. 21 Performance of the parallel alignment for the test set 'e-books'

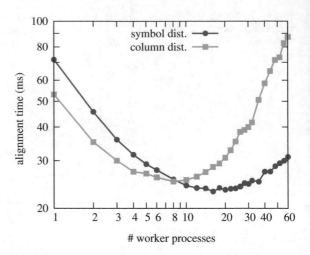

Fig. 22 Performance of the parallel alignment for the test set 'games'

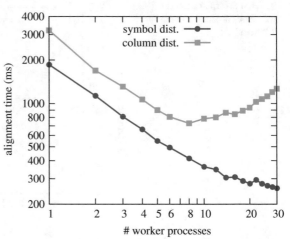

The best run-times indicate which configuration should be chosen for a given data set. Table 3 states them for each data set. The gained results provide a clear recommendation. If computing power allows, the alignment should be computed with the symbol version and between 16 and 30 processes, depending on the number of symbols. In case of the data set 'rentals', the sequential execution for column distribution performs best.

5.3.3 Quality of the Alignment

The quality of the matrix alignment algorithm is presented in Table 4. For the actual data items, we omitted the template text. The extracted data items are the items,

Fig. 23 Performance of the parallel alignment for the test set 'properties'

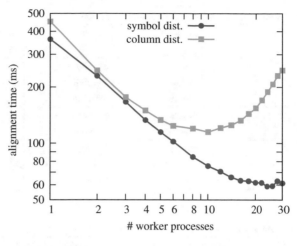

Fig. 24 Performance of the parallel alignment for the test set 'sports'

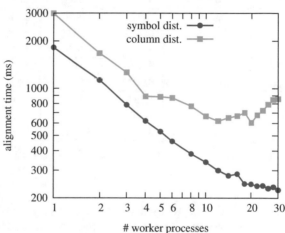

which we compute with the algorithm. The correct data items are determined as matches between actual and extracted items as described in Sect. 3.6.

As the last row of Table 4 shows, overall results for recall and precision are rather good. Particularly, data sets from the category 1 show similar results compared to other web data extraction systems. This means that a high number of data records does not bring down the output quality significantly. Data sets in category 2 show satisfactory results. Recall and precision for the sets 'games' and 'sports' indicate that the algorithm is not suitable here. These two sets from category 3 produce input sequences with maximal length in our test data corpus. It shows that, the longer sequences are, the more inaccurate the alignment becomes. The applied heuristics of the row alignment tend to fail with an extensive height of the output matrix. This is the case when the data set consists of web page covering records. The utilized heuristics

Table 3 Best run-times for parallel execution

Cat.	Set name	Version	Worker processes	Time (ms)
1	autos	Symbol	20	28.37
	music	Symbol	22	26.33
	rentals	Column	5	11.83
2	books	Symbol	24	42.93
	e-books	Symbol	16	23.07
3	games	Symbol	30	257.76
	properties	Symbol	24	58.89
	sports	Symbol	30	225.16

Table 4 Precision and recall of the test data

Cat.	Set name	Actual	Extracted	Correct	Recall (%)	Precision (%)
1	autos	8,683	9,525	8,336	96	88
	music	10,393	13,514	9,389	90	69
	rentals	8,990	8,990	8,990	100	100
2	books	7,605	7,804	6,246	82	80
	e-books	6,386	8,752	6,273	98	72
3	games	4,737	7,876	390	8	5
	properties	1,067	2,301	906	85	39
	sports	3,054	3,158	1,355	44	43
	Total	50,915	61,920	41,885	82	68

are intended for fairly smaller matrices in the original context. The heuristics for row alignment have been introduced to align nodes of DOM trees, which are on the same level.

6 Conclusion

This work presents the design of parallel algorithms for solving the alignment problem in the area of web data extraction. First, we adapted and improved an existing sequential alignment algorithm that works with heuristics. Based on that, we identified two suitable strategies for the parallel alignment. Following those, we developed two parallel algorithms, investigated their complexities, and evaluated their implementation in a parallel computing environment.

The results of the experiments showed that these parallel implementations are adequate for today's general-purpose computers with multi-core processors. We demonstrated that parallelism is able to improve the performance of solving the alignment problem when a sufficient number of processes is utilized.

It was shown that the best performance is achieved with the algorithm that decomposes the matrix by groups of symbols when the matrix has a high number of rows and there is a sufficient number of matrix rows that need to be aligned. However, with many rows already aligned, the computational demand decreases and communication becomes more dominant, reducing parallel efficiency.

The work done so far leaves space for further research. From the point of view of the algorithm development, the two versions may be combined. We expect that such a hybrid algorithm has a potential for performance improvements. Moreover, it may be worth investigating whether utilizing the shared memory programming paradigm instead of message passing could improve the performance. Another research question for the future is whether other heuristic methods could also be considered for an efficient parallelization.

References

1. Álvarez, M., Pan, A., Raposo, J., Bellas, F., Cacheda, F.: Finding and extracting data records from web pages. In: Proceedings of the 2007 International Conference on Embedded and Ubiquitous Computing, pp. 466–478 (2007)
2. Arasu, A., Garcia-Molina, H.: Extracting structured data from web pages. In: Proceedings of the 2003 ACM SIGMOD International Conference on Management of Data, pp. 337–348 (2003)
3. Chang, C.H., Kayed, M., Girgis, M.R., Shaalan, K.F.: A survey of web information extraction systems. IEEE Trans. Knowl. Data Eng. **18**(10), 1411–1428 (2006)
4. Chang, C.H., Kuo, S.C.: OLERA: Semisupervised web-data extraction with visual support. IEEE Intell. Syst. **19**(6), 56–64 (2004)
5. Chang, C.H., Lui, S.C.: IEPAD: Information extraction based on pattern discovery. In: Proceedings of the 10th International Conference on World Wide Web, pp. 681–688 (2001)
6. Church, P.C., Goscinski, A., Holt, K., Inouye, M., Ghoting, A., Makarychev, K., Reumann, M.: Design of multiple sequence alignment algorithms on parallel, distributed memory supercomputers. In: 33rd Annual International Conference of the IEEE Engineering in Medicine and Biology Society, pp. 924–927. IEEE (2011). doi:10.1109/IEMBS.2011.6090208
7. Crescenzi, V., Mecca, G.: Automatic information extraction from large websites. J. ACM **51**(5), 731–779 (2004)
8. Crescenzi, V., Mecca, G., Merialdo, P.: RoadRunner: Towards automatic data extraction from large web sites. In: Proceedings of the 27th International Conference on Very Large Data Bases, pp. 109–118 (2001)
9. Daugelaite, J., Driscoll, A.O., Sleator, R.D.: An overview of multiple sequence alignments and cloud computing in bioinformatics. ISRN Biomath. **2013**, 1–14 (2013). doi:10.1155/2013/615630
10. Gusfield, D.: Efficient methods for multiple sequence alignment with guaranteed error bounds. Bull. Math. Biol. **55**(1), 141–154 (1993)
11. Gusfield, D.: Algorithms on Strings, Trees, and Sequences, 1st edn. Cambridge University Press, New York (1997)
12. Hunga, C.L., Linb, Y.S., Linc, C.Y., Chungb, Y.C., Chungc, Y.F.: CUDA ClustalW: An efficient parallel algorithm for progressive multiple sequence alignment on multi-GPUs. Comput. Biol. Chem. **58**, 62–68 (2015)

13. Kayed, M.: Peer matrix alignment: A new algorithm. In: Proceedings of the 16th Pacific-Asia Conference on Advances in Knowledge Discovery and Data Mining, pp. 268–279 (2012)
14. Kayed, M., Chang, C.H.: FiVaTech: Page-level web data extraction from template pages. IEEE Trans. Knowl. Data Eng. **22**(2), 249–263 (2010)
15. Laender, A.H.F., Ribeiro-Neto, B.A., da Silva, A.S., Teixeira, J.S.: A brief survey of web data extraction tools. SIGMOD Rec. **31**(2), 84–93 (2002)
16. Li, K.B.: ClustalW-MPI: ClustalW analysis using distributed and parallel computing. Bioinformatics **19**(12), 1585–1586 (2003)
17. Liu, W., Meng, X., Meng, W.: ViDE: A vision-based approach for deep web data extraction. IEEE Trans. Knowl. Data Eng. **22**(3), 447–460 (2010)
18. Lopes, H.S., Lima, C.R.E., Moritz, G.L.: A parallel algorithm for large-scale multiple sequence alignment. Comput. Inform. **29**(6), 1233–1250 (2010)
19. Miao, G., Tatemura, J., Hsiung, W.P., Sawires, A., Moser, L.E.: Extracting data records from the web using tag path clustering. In: Proceedings of the 18th International Conference on World Wide Web, pp. 981–990 (2009)
20. Mikhailov, D., Cofer, H., Gomperts, R.: Performance optimization of ClustalW: Parallel ClustalW. HT Clustal and MultiClustal, SGI ChemBio (2001)
21. Nguyen, L.T., Schmidt, H.A., von Haeseler, A., Minh, B.Q.: IQ-TREE: A fast and effective stochastic algorithm for estimating maximum-likelihood phylogenies. Mol. Biol. Evol. **32**(1), 268–274 (2015). doi:10.1093/molbev/msu300
22. Rognes, T.: ParAlign: A parallel sequence alignment algorithm for rapid and sensitive database searches. Nucl. Acids Res. **29**(7), 1647–1652 (2001)
23. Sarawagi, S.: Information extraction. Found. Trends Databases **1**(3), 261–377 (2008)
24. Simon, K., Lausen, G.: ViPER: Augmenting automatic information extraction with visual perceptions. In: Proceedings of the 2005 ACM CIKM International Conference on Information and Knowledge Management, pp. 381–388 (2005)
25. Sleiman, H.A., Corchuelo, R.: An architecture for web information agents. In: 11th International Conference on Intelligent Systems Design and Applications, pp. 18–23 (2011)
26. Su, W., Wang, J., Lochovsky, F.H.: ODE: Ontology-assisted data extraction. ACM Trans. Database Syst. **34**(2), 12:1–12:35 (2009)
27. W3C: Document Object Model (DOM). http://www.w3.org/DOM/. Accessed 6 Jan 2009
28. W3C: HTML 4.01 specification. http://www.w3.org/TR/html401/. Accessed 24 Dec 1999
29. W3C: HTML 5. http://www.w3.org/TR/2014/REC-html5-20141028/. Accessed 27 Dec 2015
30. W3C: World Wide Web Consortium. http://www.w3.org/. Accessed 18 April 2013
31. Wang, J., Lochovsky, F.H.: Data extraction and label assignment for web databases. In: Proceedings of the 12th International Conference on World Wide Web, pp. 187–196 (2003)
32. Wang, L., Jiang, T.: On the complexity of multiple sequence alignment. J. Comput. Biol. **1**(4), 337–348 (1994)
33. Xia, Y., Yu, H., Zhang, S.: Automatic web data extraction using tree alignment. In: Proceedings of the 18th ACM Conference on Information and Knowledge Management, pp. 1645–1648 (2009)
34. Yamada, Y., Craswell, N., Nakatoh, T., Hirokawa, S.: Testbed for information extraction from deep web. In: Proceedings of the 13th International World Wide Web Conference, pp. 346–347 (2004)
35. Zhai, Y., Liu, B.: Structured data extraction from the web based on partial tree alignment. IEEE Trans. Knowl. Data Eng. **18**(12), 1614–1628 (2006)
36. Zhao, H., Meng, W., Wu, Z., Raghavan, V., Yu, C.: Fully automatic wrapper generation for search engines. In: Proceedings of the 14th International Conference on World Wide Web, pp. 66–75 (2005)
37. Zhao, H., Meng, W., Yu, C.: Mining templates from search result records of search engines. In: Proceedings of the 13th ACM SIGKDD International Conference on Knowledge Discovery and Data Mining, pp. 884–893 (2007)

Community Detection Using Synthetic Coordinates and Flow Propagation

Paraskevi Fragopoulou, Harris Papadakis and Costas Panagiotakis

Abstract Various applications like finding web communities, detecting the structure of social networks, or even analyzing a graph's structure to uncover Internet attacks are just some of the applications for which community detection is important. In this paper, we propose an algorithm that finds the entire community structure of a network, based on local interactions between neighboring nodes and on an unsupervised distributed hierarchical clustering algorithm. In this paper, we describe two novel community detection algorithms, one for full graph communities detection and one for single community detection. The novelty of the first proposed approach, named SCCD (to stand for Synthetic Coordinate Community Detection), is the fact that the algorithm is based on the use of Vivaldi synthetic network coordinates computed by a distributed algorithm. We also present an extended version of said algorithm, modified to deal efficiently with community detection on dynamic graphs. Finally, we present a new algorithm which partially analyzes a graph to detect the community of a single node. The current paper not only presents two efficient community finding algorithms, but also demonstrates that synthetic network coordinates could be used to derive efficient solutions to a variety of problems. Experimental results and comparisons with other methods from the literature are presented for a variety of benchmark graphs with known community structure, derived by varying a number of graph parameters and real dataset graphs. The experimental results and comparisons

P. Fragopoulou (✉) · H. Papadakis
Department of Informatics Engineering,
Technological Educational Institute of Crete, 71004 Heraklion, Crete, Greece
e-mail: fragopou@ics.forth.gr

H. Papadakis
e-mail: adanar@staff.teicrete.gr

P. Fragopoulou · C. Panagiotakis
Foundation for Research and Technology-Hellas, Institute of Computer Science,
70013 Heraklion, Crete, Greece
e-mail: cpanag@staff.teicrete.gr

C. Panagiotakis
Department of Business Administration, Technological Educational Institute of Crete,
72100 Agios Nikolaos, Crete, Greece

© Springer International Publishing Switzerland 2017
A. Adamatzky (ed.), *Emergent Computation*, Emergence, Complexity
and Computation 24, DOI 10.1007/978-3-319-46376-6_26

579

to existing methods with similar computation cost on real and synthetic data sets demonstrate the high performance and robustness of the proposed scheme.

1 Introduction

Networks in various application domains present an internal structure, where nodes form groups of tightly connected components which are more loosely connected to the rest of the network. These components are mostly known as *communities*, *clusters*, *groups*, or *modules*, the first two terms interchangeably used in the rest of this paper. Uncovering the community structure of a network is a fundamental problem in complex networks which presents many variations. With the advent of Web 2.0 technology, came along the emerging need to analyze network structures like web communities, social network relations, and in general user's collective activities. The newly emerging applications came along with a different set of parameters and demands due to the enormous data size, rendering prohibitive the static manipulation of data and raising the demand for flexible solutions.

Several attempts have been made to provide a formal definition to the generally described "community finding" concept, providing different approaches. Some of them aim at detecting the so-called, *strong communities*, groups of nodes for which each node has more edges to nodes of the same community than to nodes outside the community [1]. Others aim at detecting *weak communities*, which is defined as a subgraph in which the sum of all node degrees within the community is larger than the sum of all node degrees towards the rest of the graph [2]. Variations also appear in the method used to identify communities: Some algorithms follow an iterative approach starting by characterizing either the entire network, or each individual node as community, and splitting [3–5] or merging [2] communities respectively, producing a hierarchical tree of nested communities, called *dendrogram*. Several researchers aim to find the entire hierarchical community dendrogram [3, 4] while others wish to identify only the optimal community partition [1]. More recently used approaches aim to identify the community surrounding one or more seed nodes [6]. Some researchers aim at discovering distinct (non-overlapping) communities, while others allow for overlaps between communities [7]. Several proposals [8, 9] have been made to tackle the issue of efficient community detection in dynamic graphs (i.e., graphs whose structure changes over time). Even though community detection on dynamic graphs can be solved by running a static (i.e.: non-dynamic) algorithm on each snapshot of the graph, most dynamic algorithms try to capitalize on the facts that (i) graph changes between consecutive snapshots are usually small so (ii) the communities detected in the previous snapshot can help the algorithm speed up the detection process in the current snapshot. Finally, another main distinction between community detection algorithms lies in the ability to detect only a single community. Most of the aforementioned literature is comprised of examples of algorithms which analyze the entirety of the graph and locate all communities. However, several algorithms, such as [7] can be employed to detect a single community around a node, only partly analyzing the graph in question.

For the reminder of this paper, we shall present our contributions in the topic of community detection both for static and dynamic graphs, as well as single and total community detection.

2 Distributed Community Detection in Complex Dynamic Networks Using Synthetic Coordinates

In this part we propose SCCD (to stand for Synthetic Coordinate Community Detection), an algorithm that identifies the entire community structure of a network based on interactions between neighboring nodes. In the core of our proposal lies the spring metaphor which inspired the Vivaldi synthetic network coordinate algorithm [10]. The algorithm comprises two main phases. First, each node selects a "local" set containing mostly nodes of the same community, and a "foreign" set containing mostly nodes of different communities. As the algorithm evolves, and the springs connecting local and foreign nodes are tightened and relaxed, nodes of the same community pull each other close together, while nodes of different communities push each other further away. Given that the initial selection of local and foreign sets is "mostly" correct, nodes of the same community eventually gravitate to the same area in space, while nodes of different communities are placed further away. In other words nodes belonging to the same community will form natural clusters in space. In the second phase of the algorithm, a distributed hierarchical clustering algorithm has been proposed to automatically identify the natural communities formed in space. Extensive experiments on several benchmark graphs with known community structure indicate that our algorithm is highly accurate in identifying community membership of nodes. A first version of our algorithm was presented in [11]. The algorithm presented in this paper is a heavily modified and improved version. A new simpler and more accurate algorithm termination mechanism has been introduced. More importantly, the algorithm can now dynamically make an effort to correct the "foreign" and "local" sets as we shall see later on, increasing the obtained accuracy. Furthermore, we added an optional third phase in the algorithm, which allows for a user defined value on the number of communities requested. As far as the experimental evaluation is concerned, we performed experiments on real world graphs, in addition to the benchmark graphs. In particular, we performed experiments using new benchmark graphs of diverse community sizes and node degrees. We then compared our algorithm based on a new accuracy metric, for a total of two metrics to evaluate performance on benchmark graphs. Finally, we present a modification of the algorithm (DSCCD) to allow for fast, efficient community detection on dynamic graphs.

Apart from its accurate detection of communities, the main reason for choosing this algorithm is the fact that it is based on Vivaldi's spring metaphor [10]. As we mentioned, the first phase of the algorithm is terminated when the nodes' positions have stabilized. Given that any subsequent changes in the graph (node and edge insertions and departures) do not change the entire graph into a new one, it will take much less time for the nodes to stabilize to new positions, in consecutive runs of the algorithm.

The remaining of the paper is organized as follows: Sect. 2.1 presents an overview of some of the methods developed over the years for community detection in networks. Our distributed community detection algorithm is presented and analyzed in Sect. 3.3. Section 3.4 describes the experimental framework and comparison results with other known algorithms on a number of benchmark graphs, whereas, Sect. 2.4 describes experimental results of the static algorithm on Real World graphs. Section 2.5 describes the experimental results of DSCCD. Finally, we conclude in Sect. 3.5 with some directions for future research.

2.1 Related Work

Below we review some of the known methods for community detection and give insight on the approach they follow. For the interested reader, two comprehensive and relatively recent surveys covering the latest developments in the field can be found in [12, 13]. While the first algorithms for the problem used the agglomerative approach trying to derive an optimal community partition by merging or splitting other communities, recent efforts concentrate on the derivation of algorithms based exclusively on local interaction between nodes. A community surrounding a seed node is identified by progressively adding nodes and expanding a small community.

One of the most known community finding algorithms was developed by Girvan and Newman [3, 4]. This algorithm iteratively removes edges participating in many shortest paths between nodes (indicating bridges), connecting nodes in different communities. By gradually removing edges, the graph is split and its hierarchical community structure is revealed. The algorithm is computationally intensive because following the removal of an edge, the shortest paths between all pairs of nodes have to be recalculated. However, it reveals not only individual communities, but the entire hierarchical community dendrogram of the graph. In [5], a centralized method for decomposing a social network into an optimal number of hierarchical subgroups has been proposed. With a perfect hierarchical subgroup defined as one in which every member is automorphically equivalent to each other, the method uses the REGGE algorithm to measure the similarities among nodes and applies the k-means method to group the nodes that have congruent profiles of dissimilarities with other nodes into various numbers of hierarchical subgroups. The best number of clusters is determined by minimizing the intra-cluster variance of dissimilarity subject to the constraint that the improvement in going to more clusters is better than a network whose n nodes are maximally dispersed in the n-dimensional space would achieve.

In a different approach, the algorithm presented in [2], named *CiBC*, starts by assuming that each node is a different community, and merges closely connected communities. This algorithm is less intensive computationally since it starts by manipulating individual nodes rather than the entire graph.

The authors of [6] introduce a local methodology for community detection, named *Bridge Bounding*. The algorithm can identify individual communities starting at seed nodes. It initiates community detection from a seed node and progressively expands a community trying to identify *bridges*. An edge is characterized as a bridge by computing a function related to the *edge clustering coefficient*. The edge clustering coefficient is calculated for each edge, looking at the edge's neighborhood, and edges are characterized as bridges depending on wether their clustering coefficient exceeds a threshold. The method is local, has low complexity and allows the flexibility to detect individual communities, albeit less accurately. Additionally, the entire community structure of a network can be uncovered starting the algorithms at various unassigned seed nodes, till all nodes have been assigned to a community.

In [14], a local partitioning algorithm using a variation of PageRank with a specified starting distribution, which allows to find such a cut in time proportional to its size. A PageRank vector is a weighted sum of the probability distributions obtained by taking a sequence of random walk steps starting from a specified initial distribution. The cut can be found by performing a sweep over the PageRank vector, which involves examining the vertices of the graph in an order determined by the PageRank vector, and computing the conductance of each set produced by this order. In [15], three distributed community detection approaches based on Simple, K-Clique, and Modularity metrics, that can approximate their corresponding centralized methods up to 90 % accuracy.

Other community finding methods of interest involve [1] in which the problem is regarded as a maximum flow problem and edges of maximum flow are identified to separate communities from the rest of the graph. In clique percolation [16, 17] a complete subgraph of k nodes (k-*clique*) is rolled over the network through other cliques with $k - 1$ common nodes. This way a set of nodes can be reached, which is identified as a community. A method based on voltage drops across networks and the physics kirchhoff equations is presented in [18]. A mathematical Markov stochastic flow formulation method known as *MCL* is presented [19], and a local community finding method in [20], just to mention a few.

We will now describe the four state-of-the-art algorithms that we compare our approach with, in the experimental evaluation section. An exceptionally interesting method for community detection was developed by Lancichinetti et al. and appears in [7]. Although most previous approaches identify distinct (non-overlapping) communities, this algorithm is developed based on the observation that network communities may have overlaps, and thus, algorithms should allow for the identification of overlapping communities. Based on this principle, a local algorithm is devised developing a community from a starting node and expanding around it based on a *fitness* measure. This fitness function depends on the number of inter- and intra-community edges and a tunable parameter α. Starting at a node, at each iteration, the community is either expanded by a neighboring node that increases the community fitness, or shrinks by omitting a previously included node, if this action results in higher fitness for the resulting community. The algorithm stops when the insertion of any neighboring node would lower the fitness of the community. This algorithm is local, and able to identify individual communities. The entire overlapping and hierarchical structure of complex networks can be found by initiating the algorithm at various unassigned nodes.

Another efficient algorithm is the one described by Chen et al. in [21]. The algorithm follows a top down approach where the process starts with the entire graph and sequentially removes inter-community links (bridges) until either the graph is partitioned or its density exceeds a certain desired threshold. If a graph is partitioned, the process is continued recursively on its two parts. In each step, the algorithm removes the link between two nodes with the smallest number of common neighbors. The density of a graph is defined as the number of edges in the graph divided by the number of edges of a complete graph with the same number of nodes.

The algorithm described by Blondel et al. in [22] follows a bottom-up approach. Each node in the graph comprises a singleton community. Two communities are merged into one if the resulting community has larger modularity value [23] than both the initial ones. This is a rapid and accurate algorithm which detects all communities in the graph. In suffers however, in the sense, from the fact that during its execution, it constantly requires the knowledge of some global information of the graph, namely the number of its edges (which changes during the execution since the algorithm modifies the graph), limiting, to a certain extend, its distributed nature.

Finally, we compare our algorithm with the one described in [24], called Infomap. This algorithm transforms the problem of community detection into efficiently compressing the structure of the graph, so that one can recover almost the entire structure from the compressed form. This is achieved by minimizing a function that expresses the tradeoff between compression factor and loss of information (difference between the original graph and the reconstructed graph).

Most of the approaches found it the literature are centralized, heuristic without a global optimality criterion. On the contrary, in this paper, we have proposed a fully distributed method that solves the community detection problem. In addition, another strong point of the proposed method is that according to the experimental results and comparisons to existing methods on real and synthetic data sets, the proposed method clearly outperforms the other methods.

We will now present some representative works in the dynamic graph community detection literature. In [25], the Louvain static algorithm [22] is used to detect the communities of the first snapshot. The communities of the following snapshots are calculated based on certain rules and the observed changes in the graph. Four events are defined, namely node insertion, node departure, edge insertion and edge departure. In the case of node insertion, the new node is added to a community which maximizes the new modularity after checking likewise whether any of its neighbors should change community too. In the case of node departure, they use the clique percolation method [26] in order to find out whether the previous community should be split into two new ones. In case of edge insertion, if the new edge connects two nodes in the same community, the community structure is left unchanged. If it connects nodes to different communities, it is iteratively checked whether one of the nodes should change community along with its neighbors and so forth, in a similar manner as new node insertion. In the case of edge removal, a similar method using clique percolation is used, as in node departure, to check whether the previous community should be split.

LabelRankT [8] is another state of the art algorithm for fast, dynamic, overlapping community detection. It belongs to the second category of dynamic community detection algorithms where the same algorithm is used for the first and subsequent snapshots. It is based on Label propagation, where a set of random nodes is initialized with different labels. Iteratively, each node adopts the label of the majority of its neighbors. By allowing for more than one label per node, the algorithm is able to detect overlapping communities. During subsequent snapshots, this algorithm only needs to update nodes which have been modified compared to the previous snapshot, or nodes whose neighbors' label has changed during this procedure, making its execution faster.

The D-GT algorithm presented in [9] is based on a game theoretic approach to community detection. Each node of the underlying graph is a selfish agent who periodically takes certain actions (join, switch, leave, or no operation) to maximize its total utility. The utility function tries to connect each node to "similar" nodes. In the algorithm, the metric of similarity used is neighborhood similarity [27]. The squared complexity of calculating those similarities makes this algorithm the slowest by far of the three algorithms compared in this paper (DynamicSCCD, LabelRankT and D-GT).

2.2 SCCD Community Finding

The proposed local community finding algorithm comprises the following steps:

- The position estimation algorithm, which is a distributed algorithm inspired by Vivaldi [10].
- The community detection algorithm using hierarchical clustering.

2.2.1 Vivaldi Synthetic Coordinates

Network coordinate systems predict latencies between network nodes, without the need of explicit measurements using probe queries. These algorithms assign synthetic coordinates to nodes, so that the distance between two nodes' coordinates provides an accurate latency prediction between them. This technique provides to applications the ability to predict round trip time with less measurement overhead than probing.

Vivaldi is a fully decentralized, light-weight, adaptive network coordinate algorithm that was initially developed to predict Internet latencies with low error. Vivaldi uses the Euclidean coordinate system (in *n-dimensional* space, where *n* is a parameter) and the associated distance function. Conceptually, Vivaldi simulates a network of physical springs, placing imaginary springs between pairs of network nodes.

Let $G = (V, E)$ denote the given graph comprising a set V of nodes together with a set E of edges. Each node $x \in V$ participating in Vivaldi maintains its own coordinates $p(x) \in \Re^n$ (the position of node x that is a point in the *n-dimensional* space). The Vivaldi method consists of the following steps:

- Initially, all node coordinates are set at the origin.
- Periodically, each node communicates with another node (randomly selected among a small set nodes of nodes known to it). Each time a node communicates with another node, it measures its latency and learns that node's coordinates. Subsequently, the node allows itself to be moved a little by the corresponding imaginary spring connecting them (i.e. the positions change a little so as the Euclidean distance of the nodes to better match the latency distance).
- When Vivaldi converges, any two nodes' Euclidean distance will match their latency distance, even though those nodes may never had any communication.

Unlike other centralized network coordinate approaches, in Vivaldi each node only maintains knowledge for a handful of other nodes, making it completely distributed. Each node computes and continuously adjusts its coordinates based on measured latencies to a handful of other nodes. Finally, Vivaldi does not require any fixed infrastructure as for example landmark nodes.

2.2.2 The Position Estimation Algorithm

As we mentioned, in the core of our proposal lies the spring metaphor which inspired the Vivaldi algorithm. Vivaldi uses the spring relaxation metaphor to position the nodes in a virtual space (the *n*-dimensional Euclidean space), so as the Euclidean distance of any two node positions approximates the actual distance between those nodes. In the original application of Vivaldi, the actual distances were the latencies between Internet hosts. Our algorithm is based on the idea that by providing our own, appropriate, definition of distance between nodes, we can use Vivaldi to position the nodes in a way as to reflect community membership, i.e. nodes in the same community will be placed closer in space than nodes of different communities. In other words nodes belonging to the same community will form natural clusters in space.

Let $C(x), C(y)$ denote the communities' sets of two nodes $x, y \in V$, respectively, of a given graph. Since two nodes either belong to the same community ($C(x) = C(y)$) or not, we define the initial node distance between two nodes x and y as $d(x, y)$:

$$d(x, y) = \begin{cases} 0, & C(x) = C(y) \\ 1, & C(x) \neq C(y) \end{cases} \qquad (1)$$

When $C(x) \neq C(y)$, we have set $d(x, y) = 1$ in order to normalize the distances in range between 0 and 1. Given this definition of distance, we can employ the core part of the Vivaldi algorithm to position the nodes appropriately in the n-dimensional Euclidean space ($\Re^n$). As one can expect from those dual distances, Vivaldi will position nodes in the same community close-by in space, while place nodes of different communities away from each other. This is the reason for the dual nature of the distance function, otherwise all nodes, regardless of community membership, would gravitate to the same point in space.

In addition, Vivaldi requires a selection of nodes to probe. Each node calculates a "local" set containing nodes of the same community, and a "foreign" set containing nodes of different communities. The size of the local set as well as the size of the foreign set of a node equals the degree of the node. The perfect construction of these sets depends on the apriori knowledge of node community membership, which is the actual problem we are trying to solve. However, even though we do not know the community each node belongs to, there are two facts we can exploit to make Vivaldi work without this knowledge:

- The first is the fact that, by definition, the number of *intra-community links* of a node exceeds the number of its *inter-community links*. This means that, if we assume that all of a node's neighbors belong to the same community, this assumption will be, *mostly*, correct, which in turn means that even though some times the node may move to the wrong direction, most of the time it will move to the right direction and thus, will eventually acquire an appropriate position in space. Thus, we let the local set $L(x)$ of a node $x \in V$, be its "neighbor set".

$$L(x) = \{y \in V : x \sim y\} \qquad (2)$$

The distance from node x to nodes in $L(x)$ is set to 1 according to Eq. (1).

- The second fact we exploit concerns the *foreign links*. Since we consider all a node's links as local links, we need to find some nodes which most likely do not belong to the same community as that node, and therefor will be considered as foreign nodes. This can simply be done by randomly selecting a small number of nodes from the entire graph. Assuming that the number of communities in the graph is at least three, the majority of the nodes in this set will belong to a different community than the node itself. These nodes will comprise the "foreign set" $F(x)$ of node $x \in V$:

$$F(x) \subset \{y \in V : x \nsim y\}. \tag{3}$$

The distance from node x to the nodes in $F(x)$ is set to 1 according to Eq. (1).

The pseudo-code of the position estimation algorithm is given in Algorithm 1 and it is described hereafter. The function $getRandomNumber(0, 1)$ returns a random number in $[0, 1]$. Initially, each node is placed at a random position in $\Re^n$. Iteratively, each node $x \in V$ randomly selects a node from either its $L(x)$ or its $F(x)$ set (see lines 8, 11 of Algorithm 1). It then uses Vivaldi to update its current position using the appropriate distance (i.e. 0 or 1) to the selected node (see lines 9, 12 of Algorithm 1).

Each node continues this process until it deems its position to have stabilized as much as possible (see line 36 of Algorithm 1). This is done by calculating the sum of the distances between each two consecutive positions of the node between 40 iterations (corresponding to 40 position updates, experiments showed a larger number only slows down the algorithm without adding to efficiency). Each node also calculates the distance, in a straight line, between the two positions before and after the 40 updates. Should this value be less than half the actual traveled distance (the aforementioned sum) for 20 consecutive times, the node declares itself to have stabilized. Each node continues, however, to execute the algorithm until at least 90 % of its "foreign" and "local" sets have also stabilized. If the algorithm has not stabilized yet, the 20 oldest distances are removed and the stabilization check will be performed again after another 20 position updates.

As we have already mentioned, both the "local" and the "foreign" sets of a node will initially contain erroneous nodes. One of the most important augments of our algorithm is its ability to dynamically correct those sets (see lines 23–32 of Algorithm 1). This is based on the fact that, as the algorithm progresses, those nodes in the "local" set of a node x which do not actually belong in the same community as x, will be located a long distance away from X. As a result, even though the distances between a node and the nodes in its "local" set will initially be uniformly distributed, after a while we will notice the nodes of the local set to be divided into two groups of smaller and larger distance values. This is a good indication that we can separate the wheat from the chaff, which is implemented in the following fashion:

Let $ite(x)$ denotes the number of updates of node x (see lines 3, 14 of Algorithm 1). After several number of updates ($5 \cdot (|L(x)| + |F(x)|$, where $|L(x)|$, $|F(x)|$ denote the number of elements of the sets $L(x)$, $F(x)$), the node x as will have been updated from most of the nodes of $L(x)$, $F(x)$. This means that x is able to check its "confidence level" in identifying the erroneous nodes of its local-foreign sets (see line 15 of Algorithm 1).

```
     input  : L(x), F(x), ∀x ∈ V.
     output: p(x), ∀x ∈ V.
 1  foreach x ∈ V do
 2       p(x) = random position in ℜⁿ
 3       ite(x) = 0
 4  end
 5  repeat
 6       foreach x ∈ V do
 7            if getRandomNumber(0,1) < 0.5 then
 8                 Let v be a random vertex from L(x)
 9                 p(x) = Vivaldi(p(x), p(v), 0)
10            else
11                 Let v be a random vertex from F(x)
12                 p(x) = Vivaldi(p(x), p(v), 1)
13            end
14            ite(x) = ite(x) + 1
15            if ite(x) > 5 · (|L(x)| + |F(x)|) then
16                 ite(x) = 0
17                 maxD = max_{y∈L(x)}(||p(x) − p(y)||)
18                 minD = min_{y∈L(x)}(||p(x) − p(y)||)
19                 T₂ = minD + max(maxD−minD/3, 1/3)
20                 μ = minD+maxD/2
21                 σₙ = √(F_{y∈L(x)}[(||p(x)−p(y)||−μ)²]) / T₂
22                 if σₙ > 0.6 then
23                      foreach y ∈ L(x) do
24                           if ||p(x) − p(y)|| > T₂ then
25                                L(x) = L(x) − {y}
26                           end
27                      end
28                      foreach y ∈ F(x) do
29                           if ||p(x) − p(y)|| ≤ T₂ then
30                                F(x) = F(x) − {y}
31                           end
32                      end
33                 end
34            end
35       end
36  until ∀x ∈ V  p(x) is stable
```

Algorithm 1: The position estimation algorithm.

Let $minD$ and $maxD$ denote the minimum and maximum Euclidean distance values from x to all the nodes in its "local" set, respectively (see lines 17–18 of Algorithm 1). Let μ be the average of $minD$ and $maxD$ (see line 19 of Algorithm 1). Next, we calculate the normalized standard deviation σ_n of its distances to the nodes in its "local" set, normalizing σ_n based on its distance of the closest and furthest away node in the local set (the same formula used in Eq. 5):

$$\sigma_n = \frac{\sqrt{E_{y\in L(x)}[(||p(x) - p(y)|| - \mu)^2]}}{T_2} \tag{4}$$

In addition, σ_n is computed using the value μ instead of the mean value of distances. This is because our "confidence" should be high when most neighbor distances are located in the extreme ends (close to minD and maxD). Should σ_n exceed a certain threshold $T_1 = 0.6$, the node iterates between the nodes in both its "local" and "foreign" sets, removing any inappropriate nodes (see line 22 of Algorithm 1). In order to identify a node as "local" or "foreign" based on its distance, another threshold is required. This threshold T_2 is calculated as follows:

$$T_2 = minD + max\left(\frac{maxD - minD}{3}, \frac{1}{3}\right),$$ (5)

The idea behind this formula is that the distance of a "foreign" node is both related to the distance values of the rest of the nodes but also has a fixed minimum value. Overall, this dynamic set correction gave us on the benchmark graph tests an increase of about 6 % on average.

Figure 1 shows a small time-line of the execution of our algorithm on a graph with 1024 nodes, degree 20, and a known community structure comprising four communities. We have used different colors for the nodes of each different community. Initially, nodes were randomly placed in $\mathfrak{R}^2$. As we can see, in the beginning all colors are dispersed on the entire space. As the algorithm progresses, we see that nodes of the same color, belonging to the same community, gradually gravitate to the same area, forming distinct clusters in space.

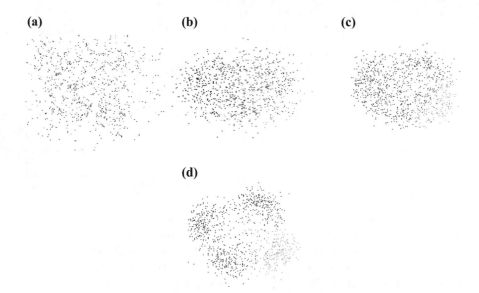

(a) **(b)** **(c)**

(d)

Fig. 1 Snapshots of the execution of the first phase of our algorithm for a graph with known community structure. **a** Initialization, **b** after 150 iterations, **c** after 400 iterations

2.2.3 Hierarchical Clustering

After each node has converged to a point in space, we use a hierarchical clustering algorithm to perform the actual grouping of nodes into clusters. The main advantages of the hierarchical clustering algorithms is that the number of clusters need not be specified a priory, and problems due to initialization and local minima do not arise [28]. The pseudo-code of the proposed hierarchical clustering method is given in Algorithm 2 and it is described hereafter. Let $c(x)$ denote the cluster id of node x. The function $getClosest(x)$ returns the closest neighboring cluster of x.

Firstly, each node is considered as a (singleton) cluster (see lines 2–5 of Algorithm 2). In addition, to make the procedure completely distributed, each node-cluster is aware of the location only of its neighboring node-clusters. Then, the following loop is executed repeatedly, until no appropriate pair of clusters can be located: Given a pair of neighboring clusters x and y, if both of them are each other's closest neighbor and the distance between the two clusters is less than a threshold $T_3 = \frac{1}{2}$, then those two clusters are merged in the following fashion (see line 10 of Algorithm 2) (Fig. 2):

- The merged cluster contains the union of the neighbors of A and B (see lines 14–15 of Algorithm 2).
- Its position is calculated as the weighted based on the population of nodes ($|x|$ and $|y|$) in each cluster x and y average of the positions of x and y, $p(x)$ and $p(y)$ (see line 12 of Algorithm 2):

$$p(x) = \frac{|x| \cdot p(x) + |y| \cdot p(y)}{|x| + |y|} \tag{6}$$

---

input : $p(x), \forall x \in V$.
output: $c(x), \forall x \in V$.

1 $i = 1$
2 **foreach** $x \in V$ **do**
3      $c(x) = i$
4      $i = i + 1$
5 **end**
6 **repeat**
7      $S = 0$
8   **foreach** $x \in V$ **do**
9          $y = getClosest(x)$
10   **if** $getClosest(y) = x \wedge |p(x) - p(y)|_2 < \frac{1}{2}$ **then**
11             $S = S + 1$
12             $p(x) = \frac{|x| \cdot p(x) + |y| \cdot p(y)}{|x| + |y|}$
13             $c(y) = c(x)$
14             $x = x \cup y$
15             $V = V - y$
16   **end**
17   **end**
18 **until** $S = 0$

Algorithm 2: The Hierarchical clustering algorithm.

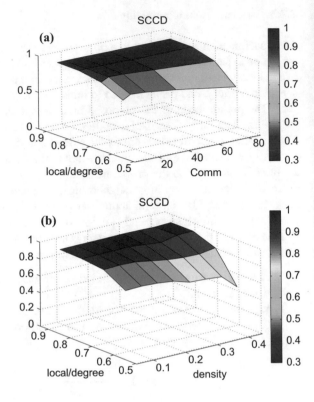

Fig. 2 The mean value of accuracy under **a** different ratios of total degree to local links (*local/degree*) and number of communities (*Comm*) and **b** total degree to local links (*local/degree*) and densities for our algorithm

2.2.4   Communication Load and Computational Complexity

SSCD can be implemented as a fully distributed system, since both of the two main parts of the proposed method are distributed (Fig. 3).

- The first part concerns the position estimation algorithm that uses the Vivaldi synthetic network coordinates [10] (see Sects. 2.2.1 and 2.2.2). This part can be computed by a distributed algorithm. It holds that each node only maintains knowledge for a handful of other nodes, computing and continuously adjusting its coordinates based on the coordinates of the nodes that belong on its local and foreign sets, making it completely distributed.
- The second part concerns the hierarchical clustering that can be computed by a distributed algorithm. In order to make this procedure distributed, each node-cluster is aware of the location only of its neighboring node-clusters (see Sect. 2.2.3).

Hereafter, we provide an analysis of the communication load and computational complexity (Figs. 4 and 5).

- Concerning the position estimation algorithm, it holds that during the update process each node communicates with a node of its local or its foreign set. So, the communication load depends on the time of convergence. In order to measure the

Fig. 3 The mean value of accuracy under **a** different ratios of total degree to local links (*local/degree*) and number of communities (*Comm*) and **b** total degree to local links (*local/degree*) and densities for the Lancichinetti algorithm

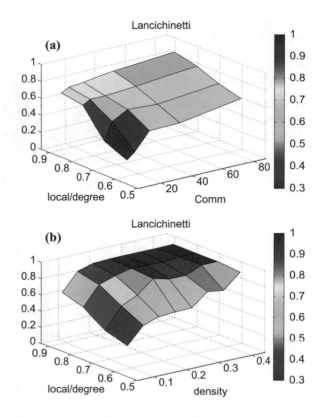

dependance of the number of messages to the size of the graphs, we performed experiments on graphs with identical parameters but varying sizes. Namely, we used 9 graphs with 2000, 3000, …, 10000 nodes, with 500 nodes per community, 10 degree per node and a 0.75 ratio of local to foreign links per node. Figure 7 shows that the average number of update messages *per node* required by Vivaldi in order to stabilize is approximately the same, regardless of the size of the graph. This means that the convergence time (computational complexity) *per node* is also independent of the size of the graph.

- Concerning the hierarchical clustering algorithm, it holds that during the merging process each node communicates with the nodes of its neighborhood in order to find the closest. In a distributed implementation, the initial communication load is $O(degree)$ for the first merging. Next, each new cluster sends its updated position to its neighborhood that needs $O(degree)$ messages. In the second level of merging, each new cluster sends its updated position to its neighborhood that have size $O(2^1 \cdot degree)$ (worst case). In the last level of merging ($l = log(\frac{N}{Comm})$), when the hierarchical clustering tree is balanced, each new cluster sends its updated position to its neighborhood that have size $O(2^1 \cdot degree) = O(\frac{N}{Comm} \cdot degree)$ (worst case). The total communication load is $O(N \cdot degree + \frac{N}{2} \cdot 2^1 \cdot degree +$

Fig. 4 The mean value of
accuracy under **a** different
ratios of total degree to local
links (*local/degree*) and
number of communities
(*Comm*) and **b** total degree
to local links (*local/degree*)
and densities for the Chen et
al. algorithm

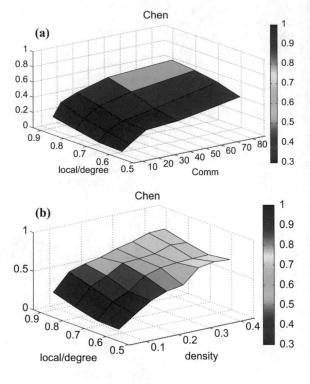

$$\cdots + \frac{N}{2^l} \cdot 2^l \cdot degree) = O(l \cdot N \cdot degree) = O(log(\frac{N}{Comm}) \cdot N \cdot degree). \text{The com-}$$
putation cost per node is $O(l \cdot degree) = O(log(\frac{N}{Comm}) \cdot degree)$.

The communication load as well as the computational complexity of the proposed
distributed framework make possible the execution of SCCD on graphs of very large
scale (e.g. 50 millions of nodes with a billion of links) (Fig. 6).

2.2.5 DynamicSCCD

As mentioned before, SCCD is comprised of two main phases.

- The position estimation algorithm, which is a distributed algorithm inspired by
  Vivaldi [10].
- The community detection algorithm using hierarchical clustering.

In the first phase of the SCCD algorithm, the Vivaldi spring metaphor is used.
On each iteration, each node slightly updates its position in order to either move
closer to one of its neighbors in the graph (which it considers as belonging in the
same graph) or further away from a random node (which is considers to belong to a
different community). The aforementioned assumptions are true for the majority of
the neighbors and the random nodes respectively (SCCD detects strong communities,

Fig. 5 The mean value of accuracy under **a** different ratios of total degree to local links (*local/degree*) and number of communities (*Comm*) and **b** total degree to local links (*local/degree*) and densities for the Blondel algorithm

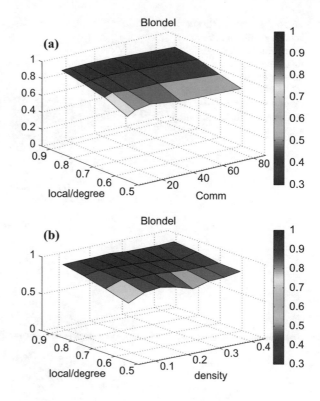

meaning that it holds for each member-node of the community that the majority of its neighbors belong to the same community as itself). This way, SCCD positions the nodes in a way as to reflect community membership, i.e. nodes in the same community will be placed closer in space than nodes of different communities. In other words nodes belonging to the same community will form natural clusters in space. The position updates stop based on a new termination threshold described below. The hierarchical clustering algorithm then comes into play, which we describe below. For more information on the SCCD algorithm we refer the reader to [29] (Fig. 7).

In the first snapshot of the graph (as per its static version) each node position is initialized to a random point in space. In the following snapshots, the remaining nodes (i.e.: the majority of the graph) retain their previous positions. It is easy to realize that, given two (relatively similar) consecutive snapshots of the dynamic graph, the first phase will terminate much faster than in the first snapshot. The fact that the majority of the graph nodes are already positioned in appropriate points in space, regarding their community membership, will help any new (or pre-existing, with modified community membership) nodes to oscillate to their own proper positions much faster.

Fig. 6 The mean value of
accuracy under **a** different
ratios of total degree to local
links (*local/degree*) and
number of communities
(*Comm*) and **b** total degree
to local links (*local/degree*)
and densities for the Infomap
algorithm

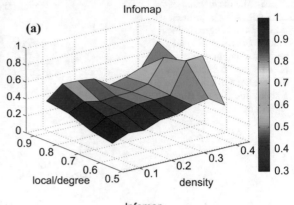

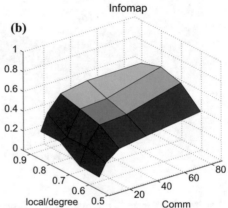

Fig. 7 Average number of
Vivaldi update messages per
node, per graph size

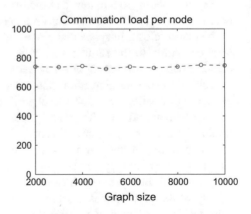

2.2.6 Vivaldi Termination Threshold

We implemented a new termination threshold for the Vivaldi phase of our algorithm which yielded better execution times while maintaining the same accuracy levels. A good termination criterion is essential for dynamic community detection algorithms since the main idea behind them is to exploit previous results in order to speed-up community detection of later snapshots. We periodically monitor two metrics during the execution of the first phase, as the nodes continuously update their positions. The first is the average distance between any two communities. The second is the Jaccard similarity between two corresponding communities (as indicated by the Hungarian algorithm [30]) between two consecutive periods. When the value of both those metrics does not change for 20 consecutive periods, the algorithm stops. The first metric is necessary due to the fact that, in complicated graphs, community memberships do not change during the fist period of the execution, where all nodes are still far from their appropriate positions. However, what does change is the average distance between pairs of community centers, as nodes try to distance themselves from other nodes which are considered to belong to different communities. The second metric is also necessary in order for the algorithm to realise that the memberships of the detected communities have stabilized.

2.2.7 Hierarchical Clustering

We modified the original hierarchical clustering algorithm to allow it to exploit previous clustering information to speed up its execution. In the static version of the algorithm, the hierarchical algorithm is initialized with a number of clusters equal to the nodes in the graph (singleton communities). It then iteratively merges neighboring close-by clusters. In order for two clusters to be merged several criteria have to be met.

- The two clusters have to contain at least a pair of neighboring nodes
- Cluster A's center has to be the closest cluster center to cluster's B center and vice versa.
- The distance between the centers of the two clusters has to be below a certain threshold.

When two clusters merge, a new center is calculated for the new cluster based on the positions of the membership nodes. When no more merges can be done, the algorithm terminates. For more information, again we refer the reader to [29].

The dynamic version of this algorithm takes into consideration the clusters detected in the previous snapshot. Any new nodes in the graph which appeared in the current snapshot are added in the previous partitioning as singleton communities-clusters. In addition, we iterate over the previous nodes. If any of those node's current distance from its cluster's center is over a certain threshold (one fourth of the threshold we use to merge clusters) then this node is removed from the cluster and instead also forms a singleton cluster. This threshold is quite strict, however if the node still belongs to the same community in the current snapshot, it will be re-merged by the normal execution of the algorithm which operates on the modified previous partition instead of being initialized with only singleton clusters.

2.3 Benchmark Graph Experiments

2.3.1 Benchmark Graphs

We have created a variety of benchmark graphs with known community structure to test the accuracy of our algorithm. Benchmark graphs are essential in the testing of a community detection algorithm since there is an apriori knowledge of the structure of the graph and thus one is able to accurately ascertain the accuracy of the algorithm. Since there is no consensus on the definition of a community, using a real-world graph makes it more difficult to assess the accuracy of a community partition (Fig. 8).

Our benchmark graphs were generated randomly given the following set of parameters:

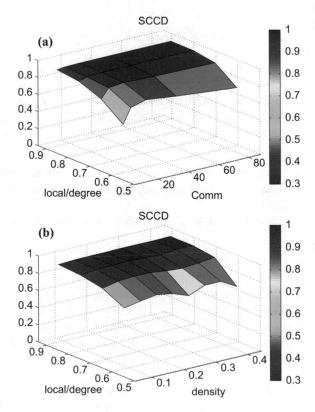

Fig. 8 The mean value of normalized mutual information under **a** different ratios of total degree to local links (*local/degree*) and number of communities (*Comm*) and **b** total degree to local links (*local/degree*) and densities for our algorithm

Fig. 9 The mean value of normalized mutual information under **a** different ratios of total degree to local links (*local/degree*) and number of communities (*Comm*) and **b** total degree to local links (*local/degree*) and densities for the Lancichinetti algorithm

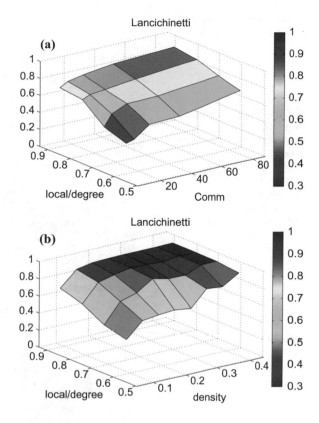

- The number of nodes N of the graph.
- The number of communities *Comm* of the graph.
- The ratio of local links to node degree *local/degree*.
- The (average) degree of nodes *degree* (Fig. 9).

Notice that even though the number of the nodes, the number of the communities and the degree of the nodes are parameters of the construction of the graph, the degree of each node as well as the number of nodes in a single community varies based on a pareto distribution. This enables us to create graphs of community sizes and individual degrees varying up to an order of magnitude.

The parameters used by the algorithm and their corresponding values are shown in Table 1. In total, we created a number of 208 benchmark graphs (Fig. 10).

A demonstration of the propose method is given in,[1] that contains the benchmark graphs, related articles and an executable of the proposed method.

---

[1] http://www.csd.uoc.gr/~cpanag/DEMOS/commDetection.htm.

Table 1 The different values
for the used parameters

N	1000, 5000, 10,000
$Comm$	5, 10, 20, 40, 80
$local/degree$	0.55, 0.65, 0.75, 0.85
$degree$	10, 20, 30, 40

Fig. 10 The mean value of
normalized mutual
information under **a** different
ratios of total degree to local
links (*local/degree*) and
number of communities
(*Comm*) and **b** total degree
to local links (*local/degree*)
and densities for the Chen et
al. algorithm

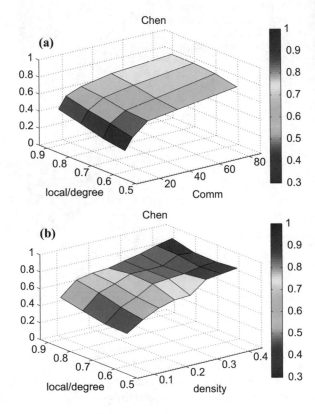

2.3.2 Comparison Metrics

We compared the five algorithms using two accuracy-related metrics found mostly
used in the literature. The first is a simple accuracy metric paired however with
the Hungarian algorithm. The simple accuracy is defined as follows: Let S_i, $i \in \{1, ..., Comm\}$ be the estimated and $\hat{S}_i$, $i \in \{1, ..., Comm\}$ the corresponding actual
communities. The accuracy (acc) is given by the average (of all communities) of the
number of nodes that belong to the intersection of $S_i \cap \hat{S}_i$ divided by the number of
nodes that belongs to the union $S_i \cup \hat{S}_i$ (Fig. 11).

$$acc = \frac{1}{Comm} \cdot \sum_{i=1}^{Comm} \frac{|S_i \cap \hat{S}_i|}{|S_i \cup \hat{S}_i|} \quad (7)$$

Fig. 11 The mean value of normalized mutual information under **a** different ratios of total degree to local links (*local/degree*) and number of communities (*Comm*) and **b** total degree to local links (*local/degree*) and densities for the Blondel algorithm

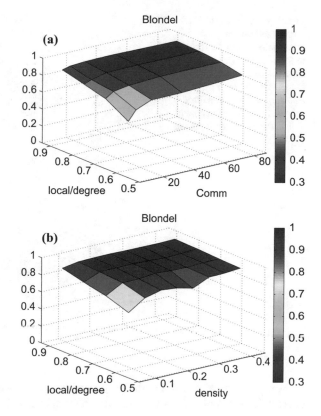

It holds that $acc \in [0, 1]$, the higher the accuracy the better the results. When $acc = 1$ the community detection algorithm gives perfect results. The Hungarian algorithm [30] is used to better match the estimated communities with the actual communities, in order to calculate and average the accuracies over all communities.

We also used the Normalized Mutual Information (NMI) metric to evaluate the correctness of the detected communities [31]. NMI is calculated using the following formula:

$$NMI = \frac{\sum_{i=1}^{k} \sum_{j=1}^{l} n_{i,j} log(\frac{n \cdot n_{i,j}}{n_i \cdot n_j})}{\sqrt{(\sum_{i=1}^{k} n_i log(\frac{n_i}{n}))(\sum_{j=1}^{l} n_j log(\frac{n_j}{n}))}} \quad (8)$$

where i iterates through the population of the "correct" communities, j iterates through the communities detected by the algorithm, $n_{i,j}$ is the size of the union of the nodes of the ith and jth communities, n_i is the size of the ith community and n

Fig. 12 The mean value of
normalized mutual
information under **a** different
ratios of total degree to local
links (*local/degree*) and
number of communities
(*Comm*) and **b** total degree
to local links (*local/degree*)
and densities for the Infomap
algorithm

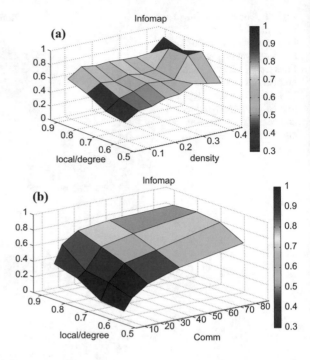

is the total number of nodes in the graph. In both cases, a value close to one indicates
correct detection of the communities of the graph (Fig. 12).

2.3.3 Results on Benchmark Graphs

The performance of four different algorithms, SCCD presented in this paper, the
Lancichinetti [7], Chen [21], and Blondel [22] algorithms are compared on Bench-
mark graphs. A brief description of these algorithms has been provided in the Related
Work section of the paper.

Each graph shows the performance of a single algorithm using either the first
or the second metric, on all 208 benchmark graphs. Since there are four types of
parameters which describe the graphs of the experiments, we decided to use the two
most important factors which affect the algorithms' performances, in order to plot
the accuracies in 3D graphs. The first of these factors is always the local links to
node degree value, which dictates how strong the clear the communities in the graph
are. The second factor in some cases is the number of communities in the graph,
while in other cases is the (average) density of these communities. Thus, we decided
to plot, for each algorithm and accuracy metric, two 3D graphs using, in one case,
the number of communities as the values on one the axes and the average density
on the other. Each accuracy value in the graphs is the average of the accuracies

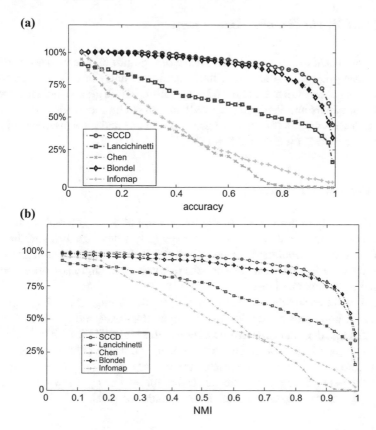

Fig. 13 Percentage of "successful" experiments, given an accuracy success threshold (x axis) for **a** the accuracy metric and **b** the NMI metric

of all the experiments based on the benchmark graphs with the same value on the aforementioned factors (Fig. 13).

We can see from these 3D graphs how our algorithm greatly outperforms almost all other algorithms, with the exception of the Blondel algorithm. Compared with Blondel, we see that the performance is comparable. However overall, our algorithm still has, on average over all experiments, a 4 % higher accuracy in both metrics. Chen and Infomap obtained very low results. Apart from SCCD and Blondel, only Lancichinetti has produced some respectable results, however the figures show that it fails to work on less dense graphs. Lancichinetti has the advantage of being able to detect just one community (whereas SCCD and Blondel only produce all communities). In order to do so, however, it relies on the existence of triangles in the graph, which is the case only in more dense graphs, hence the observed results.

Finally, Fig. 29 shows the percentage of "successful" experiments, given an accuracy threshold to define "success". One can see the better performance of our algorithm, especially in "tougher" cases.

2.4   Real World Graphs Experiments

We also conducted experiments on five real world graphs of diverse sizes. Due to the size of those graphs, the only algorithms capable of analyzing them in reasonable time were our algorithm and the Blondel algorithm. These graphs include a network of scientific papers and their citations [32], an email communication network from Enron, two web graphs (of Stanford.edu and nd.edu) and an Amazon product co-purchasing network, all obtained from [33].

2.4.1   Comparison Metrics

Three different metrics are used for the comparison of results on real world graphs, namely modularity, conductance and coverage [34]. These are different than those used in the case of benchmark graphs, since the real decomposition of graphs into communities is not known and as such the resulted community structure cannot be compared against the real one.

One of the most popular validation metrics for topological clustering, *modularity* states that a good cluster should have a bigger than expected number of internal edges and a smaller than expected number of inter-cluster edges when compared to a random graph with similar characteristics. The modularity Q for a clustering given below, where $e \in \Re^{k,k}$ is a symmetric matrix whose element e_{ij} is the fraction of all edges in the network that link vertices in communities i and j, and $Tr(e)$ is the trace of matrix e, i.e., the sum of elements from its main diagonal.

$$Q = Tr(e) - \sum_{i=1}^{k} \left(\sum_{j=1}^{k} e_{ij} \right)^2 \tag{9}$$

The modularity Q often presents values between 0 and 1, with 1 representing a clustering with very strong community characteristics.

The *conductance* of a cut is a metric that compares the size of a cut (i.e., the number of edges cut) and the number of edges in either of the two subgraphs induced by that cut. The conductance $\phi(G)$ of a graph is the minimum conductance value between all its clusters.

Consider a cut that divides $G = (V, E)$ into k non-overlapping clusters $C_1, C_2, ..., C_k$. The conductance of any given cluster $\phi(C_i)$ is given by the following ratio:

$$\phi(C_i) = \frac{\sum_{(u,v)\in E \wedge u\in C_i \wedge v\notin C_i} 1}{\min\left(\alpha(C_i), \alpha(V \setminus C_i)\right)} \tag{10}$$

where

$$\alpha(C_i) = \sum_{(u,v)\in E \wedge u\in C_i \wedge v\in V} 1 \tag{11}$$

Essentially, $a(C_i)$ is the number of edges with at least one endpoint in C_i. This $\phi(C_i)$ value represents the cost of one cut that bisects G into two vertex sets C_i and $V \setminus C_i$ (the complement of C_i). Since we want to find a number k of clusters, we will need $k - 1$ cuts to achieve that number. The conductance for the whole clustering is the average value of those $k - 1$ ϕ cuts, as follows:

$$\phi(G) = avg(\phi(C_i)), \forall C_i \subseteq V \qquad (12)$$

The final metric used to assess the performance of clustering algorithms on real world graphs is called *Coverage*. The coverage of a clustering C (where $C = C_1, C_2, ..., C_k$) is given as the fraction of the intra-cluster edges (E_C) with respect to all edges (E_G) in the whole graph G, $coverage(C) = \frac{E_C}{E_G}$. Coverage values usually range from 0 to 1. Higher values of coverage mean that there are more edges inside the clusters than edges linking different clusters, which translates to a better clustering.

2.4.2 Results on Real World Graphs

In Table 2, we present the results of running SCCD and Blondel [22] on these graphs, using three metrics for comparison, namely modularity, conductance and coverage. An explanatory survey of those metrics can be found in [23]. Both a high modularity and a high coverage indicate a better partitioning of the graph whereas in the case of conductance, a lower value is better. Although we included three metrics to get a better understanding of how the two algorithms behave in real world graphs, it is widely accepted that the modularity is the metric which better captures all the characteristics of a good partitioning of the graph into clear communities.

Two observations are quickly apparent from the results. One is that although the two algorithms locate a completely different number of communities for each graph, in most cases, the respective modularities are very comparable (in the order of 0.01). This shows that a "good" partitioning cannot be achieved in one way only. Coverage values are also comparable. This is not the case for conductance values where Blondel

Table 2 Results on real world graphs without merging

Graph	Nodes	Nr of coms		Modularity		Conductance		Coverage	
		SCCD	Blondel	SCCD	Blondel	SCCD	Blondel	SCCD	Blondel
Citations	27,771	375	171	0.58	0.59	0.25	0.05	0.7	0.74
Enron email	36,693	1590	1247	0.5	0.51	0.09	0.03	0.73	0.73
Amazon	262,111	8943	177	0.87	0.89	0.3	0.1	0.88	0.92
stanford. edu	281,904	3997	793	0.91	0.91	0.12	0.01	0.96	0.98
nd.edu	325,730	6825	475	0.91	0.93	0.2	0.04	0.93	0.96

outperforms our algorithm. This is because conductance favors algorithms which "produce" a smaller number of communities. The main reason our algorithm prefers many, denser communities is the fact that it tries to locate strong communities. This stems from the algorithm assumption that the short links *per node* are more in number than the long links.

```
    input  : C = {c(x), ∀x ∈ V}.
    output: c′(x), ∀x ∈ V.
 1  foreach x ∈ V do
 2      c′(x) = c(x)
 3  end
 4  repeat
 5      S = 0
 6      for i = 1 to |C| do
 7          A = {x : c(x) = Cᵢ}
 8          foreach B ~ A do
 9              if getDM(A, B) > 0 then
10                  c′(B) = c(A)
11                  A = A ∪ B
12                  V = V − B
13                  S = S + 1
14              end
15          end
16      end
17  until S = 0
```

Algorithm 3: The Correction Clustering algorithm.

In order to verify this, we modified our algorithm by adding a third phase which iteratively merges the communities found, if this results in an increase of the modularity. The pseudo-code of this face is given in Algorithm 3 and it is described hereafter. Let $c'(x)$ denotes of updated cluster id of node x. Although the modularity is calculated on the entire graph, the change in the modularity value dm can be computed only using information related to the communities (i, j) which are to be merged, since the subtraction eliminates the values in the modularity table of communities not participating in the merge. This is implemented by function $getDM(., .)$ (see line 9 of algorithm 3).

$$dm = \dot{e}_{ii} - \dot{a}_i^2 - (e_{ii} + e_{jj} - a_i^2 - a_j^2),\qquad(13)$$

where $\dot{e}$ is the modularity table before the merge and e is after the merge. Thus e_{ii} and $\dot{e}_{ii}$ are the fraction of all edges in the graph with both ends connected to nodes in community i, before and after the merge, and $\dot{a}_i$ and a_i are the fractions of edges with at least one end vertex in community i before the merge and after the merge, respectively. Note that community i in the merged graph corresponds to the community that results from merging communities i and j before the merge.

The only global information required in this phase is the number of edges in the initial graph, in order to calculate the aforementioned fractions. Although this somehow limits the completely distributed nature of our algorithm, it is an information

Table 3 Results on real world graphs with merging

Graph	Nodes	Nr of coms		Modularity		Conductance		Coverage	
		SCCD	Blondel	SCCD	Blondel	SCCD	Blondel	SCCD	Blondel
Citations	27,771	181	171	0.58	0.59	0.07	0.05	0.78	0.74
Enron email	36,693	1300	1247	0.5	0.51	0.04	0.03	0.74	0.73
Amazon	262,111	594	177	0.88	0.89	0.1	0.1	0.93	0.92
stanford. edu	281,904	1122	793	0.93	0.91	0.02	0.01	0.98	0.98
nd.edu	325,730	995	475	0.94	0.93	0.08	0.04	0.97	0.96

which is easily obtained and furthermore, in contrast to the Blondel algorithm, does not change as communities are merged. Table 3 shows the new results after we apply the merging step. We can see now that both the number of communities and the conductance value are also comparable in the two algorithms. This shows that there is a range in the number of communities that equally produce partitions of quality (modularity values).

It is also noteworthy that using this final step, our algorithm is able to produce any desired (user-defined) number of communities from the range of quality partitioning (highest modularity value)

2.5   Dynamic Graph Experiments

We have chosen two other state of the art algorithms to compare DSCCD with, namely LabelRankT [8] and D-GT [9]. We have also used both real and synthetic datasets to provide a better understanding of the efficiency of the algorithms.

2.5.1   Synthetic Datasets

We used the dynamic graph generator used in [35] to generate a more challenging synthetic graph dataset than the one used in that paper. This generator is based on the well-known static graph generator of [36]. We used several parameters values in all combinations, which led to the generation of 72 graphs of distinct communities (no overlap) of 10 snapshots each. Table 4 summarizes the parameter values used.

The evolutionary event types are the following:

- Birth-death: a number of pre-existing communities disappear in their entirety in the next snapshot, while other entirely new communities appear.

Table 4 Parameter values of the synthetic benchmark graphs

Number of nodes	10,000
Average degree	10, 20
Nr of communities	50, 100, 200
Ratio of inter-community neighbors	0.4
probability of a node switching community membership between time steps	0.1
Community evolution method (events)	Birth-death, expand, hide, merge-split
Ratio of communities per time step which experience evolution events	0.05, 0.1, 0.2

- Expand: Several nodes appear and disappear from the previous snapshot leading to the expansion and contraction of the existing communities in the next snapshot.
- Hide: Entire communities disappear for a certain number of snapshots, after which they appear again.
- Merge-split: Entire communities are either merged or splitted (by adding and removing edges accordingly) in the next snapshot.

Apart from those evolutionary events, for each graph, regardless of time, there is a 10 % chance that any node in the previous snapshot will change its community membership in the new snapshot.

We have used the same two metrics, as before, to evaluate the three algorithms on the synthetic datasets. The first is the simple accuracy metric paired however with the Hungarian algorithm described in (7). The second is the NMI metric described in (8)

2.5.2   Real World Datasets

We also used three real-world datasets. Two Autonomous Systems graph datasets (AS-Internet Routers Graph [37] and AS-Oregon Graph [38]) as well as HEP citation graph dataset used in the 2003 KDD Cup [39]. Table 5 summarizes the datasets' characteristics

In order to evaluate the algorithms in real world datasets, we used the well-known modularity metric [23]. However, even though out algorithm only detects non-overlapping communities, both LabelRankT and DG-T detect overlapping communities. In order to be fair in out comparison, we used an extended version [40] of

Table 5 Real-world datasets information

	AS-Inet Routers	AS-Oregon	KDDCup-HEP
Nodes	6474	10,670	27,769
Edges	13,895	22,002	250,262
Snapshots	733	9	12

modularity which works on overlapping partitions, since the original metric also only works for non-overlapping communities. The formula for the extended modularity is

$$Q_{Ov}^{E} = \frac{1}{2m} \sum_{c} \sum_{i,j\varepsilon c} \left[A_{ij} - \frac{k_i k_j}{2m} \right] \frac{1}{O_i O_j} \qquad (14)$$

where m is the number of links in the graph, c the set of detected communities, array A is the adjacency matrix of the graph nodes, k_i the number of neighbors of node i and O_i the number of communities node i belongs to. It is easy to see that for all $O_i = 1$ (as is the case with non-overlapping communities) this formula is identical to the one of the classic modularity. This metric was used for all three algorithms.

2.5.3   Results

In this section we present the experimental results of our work. Figures 14 through 16 show the performance of all three algorithms, over the temporal snapshots of the three real-world datasets. One can see how DynamicSCCD outperforms the other two algorithms in all three cases. An interesting observation can be made in the AS-Internet Routers dataset of Fig. 16 where we can see a sudden drop in the modularity values for all three algorithms. This is due to the fact that at that point, the new snapshot is vastly different that the previous one, confusing the algorithms which rely on previous information. The KDDCup-Hep dataset appears to be the most complicated of the three, as seen by the steady drop of modularity values in all three algorithms, with DynamicSCCD still having higher values. Figures 17 through 19 show the Jaccard similarity in time between two consecutive snapshots of each dataset. Each point in the y axis is calculated using the following equation (Figs. 15 and 18):

$$change = 1 - \frac{|S_t \cap S_{t-1}|}{|S_t \cup S_{t-1}|} \qquad (15)$$

I.e.: The higher the value, the higher the difference between two consecutive snapshots. One can easily, for instance, notice the aforementioned peak, around the

Fig. 14 Modularity values in time for the AS-Oregon dataset

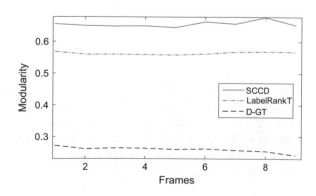

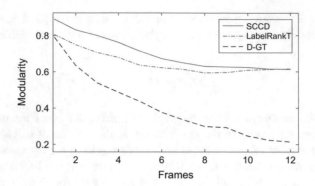

Fig. 15 Modularity values in time for the KDDCup dataset

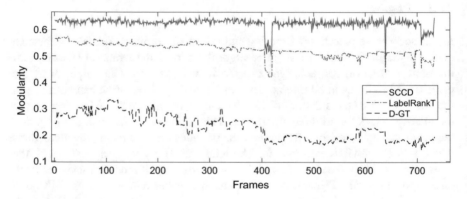

Fig. 16 Modularity values in time for the AS-Internet Routers dataset

Fig. 17 Change between two consecutive snapshots in time for the AS-Oregon dataset

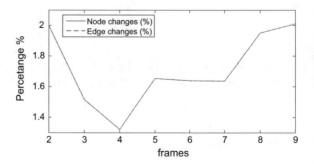

400th snapshot in the AS-Internet Routers dataset. One can see that the changes are very small in the AS-Oregon dataset, which explains the fact that all algorithms exhibit a stable modularity through time, whereas in the KDDCup-HEP dataset the majority of the graph changes during the the first snapshots and the change is still substantial even in the later snapshots. This can explain the decreasing modularity values we see in Fig. 15.

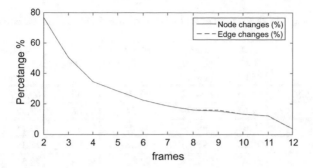

Fig. 18 Change between two consecutive snapshots in time for the KDDCup dataset

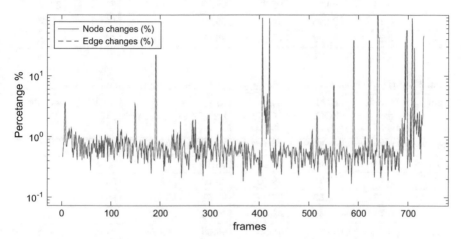

Fig. 19 Change between two consecutive snapshots in time for the AS-Internet Routers dataset (Y axis is log scale)

We now present the results of the synthetic, benchmark graph experiments. Tables 6 through 9 provide an thorough view on the accuracy and NMI values obtained for each algorithm. We have grouped the results first of all based on the graph evolution event types (namely, birth-death, hide, merge-split and expand). In addition, we provide the two most important characteristics of the graphs, which mostly affect its complexity and thus the efficiency of the algorithms, namely the number of communities in the graph as well as the average degree of its nodes. According to these Tables, the average value of NMI of DynamicSCCD, LabelRankT and D-GT is 99.3, 99.2 and 75.7 %, respectively. The average value of ACC of DynamicSCCD, LabelRankT and D-GT is 97.2, 97.2 and 58.3 %, respectively. So, one can see in these tables that DynamicSCCD and LabelRankT exhibit comparable, high values of both NMI and Accuracy, while D-GT exhibits much lesser performance. We have emphasized the corresponding values in the tables where one algorithm mostly outperforms the other (Tables 7 and 8).

Table 6 NMI and Accuracy values of all three algorithms for birth-death evolution events based graphs

coms	events freq	NMI SCCD (%)	Acc SCCD (%)	NMI LBRT (%)	Acc LBRT (%)	NMI DGT (%)	Acc DGT (%)
50	0.05	99.03	94.52	**99.79**	**99.11**	63.31	44.42
50	0.1	**99.65**	**98.99**	98.77	94.68	57.77	42.77
50	0.2	**99.63**	**99.02**	98.70	95.42	45.51	41.41
100	0.05	**99.25**	**97.32**	99.22	94.66	78.32	60.19
100	0.1	**99.43**	97.54	99.42	**98.28**	73.57	60.40
100	0.2	99.36	97.03	**99.60**	**98.68**	63.60	61.37
200	0.05	99.11	95.70	**99.59**	**98.88**	87.63	80.46
200	0.1	98.96	94.07	**99.69**	**99.03**	69.03	62.08
200	0.2	98.90	93.82	**99.65**	**98.85**	67.52	67.24

Table 7 NMI and Accuracy values of all three algorithms for expand evolution events based graphs

coms	events freq	NMI SCCD (%)	Acc SCCD (%)	NMI LBRT (%)	Acc LBRT (%)	NMI DGT (%)	Acc DGT (%)
50	0.05	99.00	95.11	**99.79**	**98.87**	68.31	42.98
50	0.1	**99.65**	**99.25**	98.38	94.14	66.96	41.75
50	0.2	**99.70**	**99.40**	98.10	93.92	64.95	39.74
100	0.05	**99.65**	**99.25**	98.61	94.99	83.63	62.77
100	0.1	99.49	98.03	**99.70**	**98.57**	82.24	61.12
100	0.2	99.54	98.36	**99.68**	**98.54**	81.02	59.55
200	0.05	99.36	98.07	**99.53**	**98.15**	93.44	83.93
200	0.1	99.03	94.71	**99.77**	**98.62**	92.96	83.73
200	0.2	98.98	94.60	**99.79**	**98.81**	92.20	81.72

Figures 20 and 21 show a histogram of the Accuracy and NMI values of the two more well-performing algorithms (namely DynamicSCCD and LabelRankT) over the entire synthetic graph dataset. It holds that the NMI of DynamicSCCD, LabelRankT and D-GT belongs in the ranges [96.9 %, 99.9 %], [98.5 %, 99.9 %] and [43.8 %, 97.1 %] respectively. The ACC of DynamicSCCD, LabelRankT and D-GT belongs in the ranges [92.2 %, 99.8 %], [90.5 %, 99.8 %] and [11.8 %, 99.4 %] respectively. Conclusively, it holds that while the performances of those two algorithms in the synthetic datasets are comparable, it is clear that DynamicSCCD outperforms LabelRankT in the real-world datasets.

Finally, Fig. 22 shows the reduction in execution time accomplished by DynamicSCCD over all the snapshots following the first one, compared to the execution time of the algorithm on the first snapshot. One can see that for the majority of the

Table 8 NMI and Accuracy values of all three algorithms for hide evolution events based graphs

coms	events freq	NMI SCCD (%)	Acc SCCD (%)	NMI LBRT (%)	Acc LBRT (%)	NMI DGT (%)	Acc DGT (%)
50	0.05	99.13	95.19	**99.80**	**98.90**	67.30	41.81
50	0.1	**99.69**	**99.29**	98.77	95.98	65.22	41.56
50	0.2	**99.68**	**99.27**	98.75	95.96	62.09	41.71
100	0.05	**99.69**	**99.40**	97.84	93.06	81.97	63.79
100	0.1	99.49	97.94	**99.63**	**98.22**	80.14	63.22
100	0.2	99.48	97.91	**99.70**	**98.51**	76.22	62.40
200	0.05	99.59	**98.46**	**99.64**	98.43	91.87	85.13
200	0.1	98.98	94.40	**99.81**	**98.91**	89.66	84.59
200	0.2	98.98	94.31	**99.79**	**98.78**	85.66	83.43

Table 9 NMI and Accuracy values of all three algorithms for merge-split evolution events based graphs

coms	events freq	NMI SCCD (%)	Acc SCCD (%)	NMI LBRT (%)	Acc LBRT (%)	NMI DGT (%)	Acc DGT (%)
50	0.05	98.86	95.93	**99.71**	**98.72**	67.09	41.08
50	0.1	**99.65**	**99.25**	97.64	92.06	62.63	31.95
50	0.2	**99.72**	**99.55**	98.06	93.89	61.70	29.18
100	0.05	**99.57**	**99.08**	97.84	92.98	81.18	55.41
100	0.1	99.43	97.88	**99.60**	**98.26**	79.11	49.92
100	0.2	99.32	97.54	**99.66**	**98.62**	75.67	44.59
200	0.05	**99.19**	**97.40**	99.13	96.94	91.43	75.47
200	0.1	98.95	94.86	**99.74**	**98.68**	89.40	68.46
200	0.2	98.91	95.41	**99.74**	**98.73**	85.01	57.51

experiments on the synthetic graphs, DynamicSCCD achieved a 5 to 10 times less execution time compared to the execution time of the corresponding first snaphot. In about 10 % cases, this reduction is higher than 30 times. This mainly appears when the graph structure changes are smooth, since it holds that the smoother graph changes, the higher the reduction in execution time, an expected conclusion. This shows how DynamicSCCD takes advantage of the available community information of the previous snapshot, as compared to running the static algorithm for every single snapshot from scratch.

Fig. 20 Histogram of the accuracy values of DynamicSCCD and LabelRankT

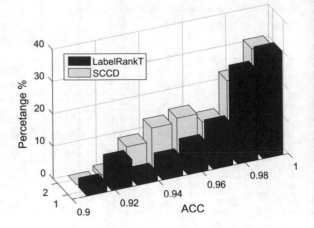

Fig. 21 Histogram of the NMI values of DynamicSCCD and LabelRankT

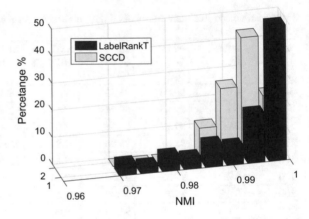

Fig. 22 DynamicSCCD speedup compared to the first snaphot

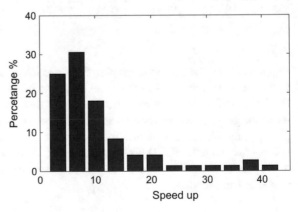

2.6 Conclusions and Future Work

We presented a community finding algorithm which is based on a custom-tailored version of the Vivaldi network coordinate system. The proposed algorithm has been tested on a large number of benchmark graphs with known community structure comparing it with several state-of-the-art algorithms, proving its effectiveness against all other algorithms. In addition we performed experiments on large, real world datasets and compared it with the next most efficient algorithm resulting in very comparable effectiveness. Moreover, our algorithm can employ a simple third step to allow it to provide a wider range of similarly optimal results.

We plan to expand the algorithm, in order to enable the detection of also overlapping communities. In addition, another goal is the modification of the algorithm in order to locate only a single community. Both of these problems are of great interest to the field of social networks, since overlaps can be appeared between communities. In addition, it an is important to provide the single community detection (per node) when each node of the social network ask for its community instead of entire community detection. Another possible extension of the proposed scheme is the application in weighted networks that can measure the strength of social relationships in social networks [41]. In iterative process of position estimation algorithm, this extension can be done by setting the probability of edge selection according to the edge weight, so that the strong edges would have higher selection probability corresponding to high-tension springs.

3 Local Community Detection via a Flow Propagation Method

We propose a flow propagation algorithm (FlowPro) that finds the community surrounding a node in a complex network. In each iteration of the main process of FlowPro, the initial node propagates a flow that is shared among its neighbors. Each node is able to store, propagate to its neighbors, and return, part of the flow it receives to the initial node. When the algorithm converges, the flow stored in the nodes that belong to the community of the initial node, is generally higher than the flow stored in the rest of the graph nodes, thus the requested community emerges. The novelty of the proposed approach lies in the fact that Flow-Pro is local, allows to visualize the community and does not require the knowledge of the entire graph as most of the existing methods found in the literature. This makes possible the application of FlowPro in extremely large graphs or in cases where the entire graph is unknown like in most social networks. Experimental results on real and synthetic data sets demonstrate the high performance and robustness of the proposed scheme.

3.1 Related Work

We will now describe some state-of-the-art algorithms that solve the local community detection problem that are able to detect a community starting at a given node without the knowledge of the entire graph [42]. In [43], the L-Shell algorithm is proposed that expands a community, stopping the expansion whenever the network structure does not allow any further expansion, i.e. the bridges are reached. This expansion is controlled the given threshold α. Disadvantages of this method are its quadratic computation cost and the fact that it is possible to "spill over" the community when the starting node is close to some non-community vertex. In [44], a random walks based local clustering algorithm is proposed called Nibble, that runs in time proportional to the size of the cluster it outputs. The cluster that Nibble produces consists of nodes that are among the most favored destinations of random walks starting from initial node. However, Nibble may not find a local cluster for some input vertices and is sensitive on the two given parameters φ, b. The Lancichinetti et al. algorithm [7] is developed based on the observation that network communities may have overlaps, and thus, algorithms should allow for the identification of overlapping communities. Based on this principle, a local algorithm is devised developing a community from a starting node and expanding around it based on a *fitness* measure. This fitness function depends on the number of inter- and intra-community edges and a tunable parameter α. Starting at a node, at each iteration, the community is either expanded by a neighboring node that increases the community fitness, or shrinks by omitting a previously included node, if this action results in higher fitness for the resulting partitioning of the graph. The algorithm terminates when the insertion of any neighboring node would lower the fitness of the community. This algorithm is local, and able to identify individual communities. Similarly with Lancichinetti et al. algorithm, in [21], the authors define a density quality measure to be optimized and then recursively merge clusters if this move produces an increase in the quality function. In the experimental evaluation section, we compare our approach with the Lancichinetti et al. algorithm, that is the most recent method solving the local community detection problem without difficult to be tuned parameters.

3.2 Problem Formulation

This section, presents the local community detection problem and studies some issues that have been taken into account in the proposed algorithm. Let $G = (V, W)$ denote the given graph comprising a set V of nodes together with a set W of weighted edge. In order to simplify the problem formulation, we suppose that the graph is undirected and all edges' weights are equal to one. So, if there exists an edge from node $i \in V$ to node $j \in V$, then the edge weight is given by $W(i, j) = 1$, otherwise $W(i, j) = 0$. The proposed problem formulation as well as the proposed method can be extended to undirected weighted graphs.

According to the problem definition of local community detection, the initial node ($s \in V$) is given and the goal is to find the set of nodes $C(s)$ that belong to the community of s, with $C(s) \supseteq \{s\}$. This means that there exist a high number of edges between the nodes of $C(s)$ compared to the number of edges that connect nodes of $C(s)$ to the rest of the graph. In addition, we have assumed that the given graph mainly has non-overlapping community structure. In the case of graphs with overlapping communities [40], the proposed method can be also applied but we get lower performance results (see Sect. 3.4.1). This is due to the fact that our problem formulation is proposed mainly for non-overlapping networks, e.g. Eq. (16) is not true for nodes that belong to several communities, since it yields low values for them.

Let $p(x), x \in V$ denote the probability that node x belongs to $C(s)$. Let $local(x)$ be the number of edges of node x that belong to $SG(C(s))$, where $SG(C(s))$ denotes the subgraph that is defined by the set of nodes $C(s)$. If community $C(s)$ is known, then $p(x/C(s))$ is defined by the ratio of $local(x)$ to the degree of x ($degree(x)$):

$$p(x/C(s)) = \frac{local(x)}{degree(x)} \tag{16}$$

Let $d(x), x \in V$ denote the shortest path distance between nodes x and s. Let $\{e_1, e_2, \cdots, e_{d(x)}\}$ be the set of edges of a shortest path between nodes s and x. Then, an estimation of $p(x)$ is given by the probability that $e_1 \in SG(C(s)) \wedge e_2 \in SG(C(s)) \cdots \wedge e_{d(x)} \in SG(C(s))$. Let ρ be the average ratio of local links to node degree value, which dictates how strong and clear the communities of the graph are. Thus, if we ignore the graph structure then the probability of an edge e_1 to belong to $SG(C(s))$ is given by ρ. Therefore, a simple estimation of $p(x)$ can be given by Eq. 17 under the assumption that the possibilities $e_i \in SG(C(s))$ and $e_{i+1} \in SG(C(s))$ are independent.

$$p(x) = \rho^{-d(x)} \tag{17}$$

This estimation ignores the graph structure and it takes into account only $d(x)$. In addition, we have assumed that Eq. 18 is true for $p(x)$.

$$p(x) \leq \frac{\sum_{y \in n(x)} p(y)}{|n(x)|} \tag{18}$$

where $n(x)$ denotes the set of neighbors of node x and $|n(x)|$ denotes the number of neighbors of node x. We have used inequality instead of equality in order to take into account the case of a bridge, meaning that it is possible that x does not belong to $C(s)$ even if x is connected to s. The proposed algorithm is based on Eqs. 17 and 18 in order to estimate a quantity $S(x)$, that we call stored flow, which is analogous to $p(x/C(s))$. According to the problem formulation of FlowPro, the probability of a node x to belong to the community of node s is analogous to its stored flow.

3.3 Flow Propagation Algorithm

In this paper, we propose a local algorithm that can be implemented as a fully distributed method (FlowPro). The main goal of this work is not to analyze the structure
of the whole social network, but to provide a useful community detection tool for a
simple user of a social network, which is impossible to know the entire graph structure. This makes possible the application of FlowPro in extremely large graphs or in
cases where the entire graph is unknown like in most social networks. In addition,
FlowPro can be applied in weighted networks that can measure the strength of social
relationships in social networks [41]. According to the problem formulation of Flow-
Pro, the probability of a node belonging to the requested community is analogous to
its stored flow. So, the stored flow can be used as a belief (rating) of a node belonging
to that (the seed node's) community, and to answer various questions like finding
the k-nearest neighbors of a community that are considered the most important for
several applications, such as social networks. The CoViFlowPro, described in [45],
successfully applies the stored flow of FlowPro on the visualization of a community on the Archimedean spiral. Another novelty of the proposed method, is that we
optionally give the community size that is quite important in social networks where
the communities are not well separated, which cannot be done by other local methods
of literature. In such social networks, a community finding algorithm that automatically estimates a community would probably fail. In addition, compared to other
local methods, the proposed method has the extra ability of automatically removing
(e.g. bridges) and adding edges to the initial (seed) node, decreasing the diameter of
the community. These changes on the graph structure increase the converge and the
performance of FlowPro.

In each iteration of the main process of FlowPro, the initial node propagates a
flow that is shared among its neighbors. Each node is able to store, propagate to
its neighbors, and return part of the flow it receives to the initial node. When the
algorithm converges, the flow stored in the nodes that belong to the community of
the initial node is generally higher than the flow stored in the rest of the graph nodes,
thus the requested community emerges. A flow propagation algorithm that has been
successfully used for the point clustering problem [46] is the affinity propagation (AP)
method [47]. AP takes as input measures of similarity between pairs of data points
using negative squared error, thus solving the entire community detection problem.
Real-valued messages are exchanged between data points until a high-quality set of
exemplars and corresponding clusters gradually emerges. The number of clusters is
automatically estimated by the AP, influenced by the values of the input preferences,
but also emerges from the message-passing procedure. A detailed and more formal
description of our algorithm follows:

FlowPro requires as input the initial node ($s \in V$) that searches for its community.
This node has an initial flow for propagation $T(s) = |n(s)|$. In each iteration of the
main process of FlowPro, the initial node propagates a flow that is shared to its
neighbors. Each node x is able to store, propagate to its neighbors, and return part of
this flow to the initial node. Finally, when the algorithm converges, the flow stored

at the nodes that belong to the community of the initial node is generally higher than the flow stored in the rest of the graph nodes, since the stored flow of a node will be analogous to $p(x)$, resulting to the requested community.

The proposed method is similar to the belief propagation algorithms [48] that are normally presented as message update equations on a factor graph, involving messages between variable nodes and their neighboring factor nodes and vice versa. Considering messages between regions in a graph is one way of generalizing a belief propagation algorithm [48]. However, based on the estimation of $S(x)$ FlowPro has the extra ability of removing and adding edges to s in order to increase $d(x)$ for nodes x that do not belong to $C(s)$ (e.g. removing bridges) and to decrease $d(x)$ for nodes x that belong to $C(s)$, respectively. This property increases the convergence and the performance of FlowPro. Hereafter, we give a detailed description of the FlowPro algorithm which comprises the following steps:

1. At each iteration of the main process, the initial node s propagates a flow that is shared to its neighbors according to their edge weights. Each node that receives a flow stores half of it, and sends the other half of the flow to its neighbors only if the flow is greater than a threshold T_0 (e.g. in our experiments $T_0 = 0.001$) in order to allow the process to terminate. In Sect. 3.4.3, we describe how threshold T_0 affects the performance of the method and we show that the selection of $T_0 = 0.001$ suffices to get high performance results while keeping the total number of messages low. So, when the graph is undirected and without edge weights, the flow is equally distributed to the neighbors since the edges are equivalent and node x considers the probability of each of the neighbors belonging to $C(s)$ equal. Since $s \in C(s)$, the flow propagation is executed under the assumption that node s does not receive/store flow meaning that we set the edge weight $W(x, s) = 0$, $\forall x \in V$ (see line 5 of Algorithm 4).

   Let $T(x)$ be the quantity of flow that node $x \in V$ is going to transmit. The fact that each node stores half of the receiving flow can be considered as a physical way to reduce the flows and to terminate the process. So, this process will terminate when there does not exist any flow to be sent (see lines 8–18 of Algorithm 4).

   This process is based on Eq. 17 in the sense that the nodes that are close to s will have high $S(x)$. In addition, it takes into account the graph structure in the sense that a node x that has a lot of connections with nodes of high stored flow, will receive high quantity of flow. Therefore, when two nodes have the same distance from s, the node with more "important" connections[2] will have higher stored flow. For example, in Fig. 24a, nodes x and y have the same distance from s, but $S(x) > S(y)$, due to the connections of x with nodes a and b. Thus, node x belongs to $C(s)$ with higher probability than node y.

2. Based on $S(.)$, in the next step, the proposed method removes or adds edges to s in order to remove bridges, thus decreasing $d(x)$ for nodes that belong to $C(s)$. We sort vector S in descending order getting the node indices S_{ind} (see line 19 of Algorithm 4). If there exists a neighbor v of node s, that does not belong to the

---

[2]Connections with nodes that have high stored flow.

first $|n(s)|$ nodes of S_{ind} ($S_{ind}(1 : |n(s)|)$), then v is removed from the neighbors of s, and the stored as well as the transmitted quantity of v is transferred to node s,

$$T(s) = T(s) + T(v) + S(v), \tag{19}$$

since, with high probability it holds that v does not belong to the community of s (see lines 22–29 of Algorithm 4). In the next main step of the algorithm, we check if u is the last point of $S_{ind}(1 : |n(s)|)$ and then we remove it from the neighbors of s (see lines 31–33 of Algorithm 4). Otherwise, we add node $u = S_{ind}(|n(s)| + 1)$ to node's s neighbors (see lines 35–38 of Algorithm 4). The new edge weight is given by the average of edge weights between node s and its neighbors (see lines 39–43 of Algorithm 4). The goal of the removal and the expansion of neighbors of s is

- to decrease the shortest paths between the nodes that belong to the community of s and s in order to be able to increase their stored flow in the next iterations,
- to gradually keep most of the flow to nodes of the community by removing bridges and
- to keep $|n(s)|$ balanced.

In order to measure the importance of this step, we have executed FlowPro without this step on the Benchmark graphs (see Sect. 3.4). Without this step, the average acc (that measures the performance, see Sect. 3.4) is only 58.7 % over all Benchmark graphs, while the original FlowPro yields $acc = 81.7\%$ over all Benchmark graphs.

3. The probability of a node x belonging to the community of s is equal or less than the average of the corresponding probabilities of node's x neighbors (see Eq. 18). Thus, based on this inequality, in case $S(x)$ is greater than $E(S(n(x)))$, we set it to the mean stored flow for the nodes $n(x)$, $E(S(n(x)))$, where $E(.)$ is the symbol for the mean value. Quantity $S(x) - E(S(n(x)))$ is added to $T(s)$ (see lines 44–50 of Algorithm 4) in the next main step of the algorithm.

   This step will significantly reduce the stored flow of nodes that do not belong to the community. For example, in Fig. 24b, there exists a bridge (edge $s \sim d$) and due to this step the reduction of $S(d)$ will be high, since node d is only connected to the almost fully connected subgraph G2. Without this step, $S(d)$ would be less but close to $S(a)$, $S(b)$ and $S(c)$.

4. Finally, if the size of a community is not given we sort vector S in descending order and we compute the differences between adjacent elements of the sorted vector (DS). Let K be the position of the global minimum[3] of DS. The community of node s is defined by the first K nodes with the highest $S(x)$. This trivial procedure is implemented by function $getCluster$ (see line 51 of Algorithm 4).

---

[3]In order to speed up the algorithm, we search for the global minimum in the range $[|n(s)|, \frac{|\{x \in V : S(x) > 0\}|}{2}]$.

Fig. 23 Quantity $S(x)$ for a synthetic graph. Blue and red colors are used for the nodes that belong to $C(s)$ and those that do not belong to $C(s)$

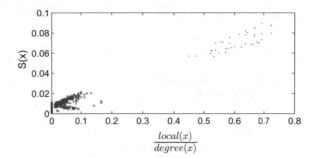

If the size of the community is given, the algorithm is called FlowPro-S. Let K_0 be the number of community nodes, then the procedure *getCluster* yields the first $K_0 - 1$ nodes with highest $S(x)$.

5. The main process ends when

- the community finding algorithm converges to a solution (e.g. the last 10 iterations we receive the same community) or
- quantity $\frac{T(s)}{\sum_{x \in V} S(x)}$ is lower than a threshold meaning that $S(.)$ has converged (see line 57 of Algorithm 4).

In Fig. 23, we plot quantity $S(x)$ for a synthetic graph $N = 1000$, $Comm = 20$, $degree = 20$ and $local/degree = 0.65$ (see Sect. 3.4.1) using blue and red colors for the nodes that belong to $C(s)$ and those that do not belong to $C(s)$, respectively. It holds that quantity $S(x)$ is generally higher for the nodes that belong to $C(s)$, since the stored flow of a node $S(x)$ is almost analogous to $p(x/C(s))$. In these example, the Pearson product-moment correlation coefficient r between $S(x)$ and $p(x/C(s))$ is 0.92 showing a strong linear dependence between the two variables (Fig. 24).

(a)   **(b)**

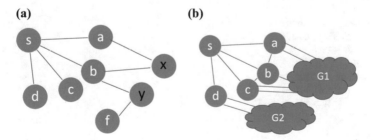

Fig. 24 a A graph with 8 nodes. **b** A graph with two almost fully connected subgraphs G1 and G2

```
    input  : V, W(i, j), i, j ∈ V, s ∈ V.
    output: C
1   C = C_prev = {s}
2   T(x) = S(x) = 0, ∀x ∈ V
3   T(s) = |n(s)|
4   lastNode = doExit = iter = 0
5   W(x, s) = 0, ∀x ∈ V
6   repeat
7       iter = iter + 1
8       repeat
9           SET = {x ∈ V : T(x) > T_0}
10          foreach x ∈ SET do
11              foreach y ∈ n(x) do
12                  ds = T(x)/|n(x)|
13                      S(y) = S(y) + 0.5 · ds
14                      T(y) = T(y) + 0.5 · ds
15              end
16              T(x) = 0
17          end
18      until SET = ∅
19      S_ind = sort(S)
20      SET = S_ind(1 : |n(s)|)
21      change = 0
22      foreach v ∈ n(s) do
23          if v ∉ SET then
24              W(s, v) = 0
25              T(s) = T(s) + S(v) + T(v)
26              T(v) = V(v) = 0
27              change = 1
28          end
29      end
30      u = S_ind(|n(s)| + 1)
31      if S_ind(|n(s)|) = lastNode then
32          W(s, lastNode) = 0
33      end
34      lastNode = 0
35      if change = 0 then
36          SET = SET ∪ u
37          lastNode = u
38      end
39      foreach x ∈ SET do
40          if x ∉ n(s) then
41              W(s, x) = 1/|n(s)| ∑_{y ∈ n(s)} W(s, y)
42          end
43      end
44      SET = {x ∈ V : S(x) > E(S(n(x)))}
45      S_prev = S
46      foreach x ∈ SET do
47          DS = S(x) − E(S_prev(n(x)))
```

48	$T(s) = T(s) + DS$
49	$S(x) = S_{prev}(x) - DS$
50	**end**
51	$C = getCluster(S) \cup \{s\}$
52	**if** $C = C_{prev}$ **then**
53	$doExit = doExit + 1$
54	**else**
55	$doExit = 0$
56	**end**
57	$C_{prev} = C$
58	**until** $doExit \geq 10 \vee \frac{T(s)}{\sum_{x \in V} S(x)} < 0.02$

Algorithm 4: The proposed FlowPro algorithm.

3.4 Experimental Results

We have created a variety of benchmark graphs with known community structure to test the accuracy of our algorithm. Benchmark graphs are essential in the testing of a community detection algorithm since there is an apriori knowledge of the structure of the graph and thus one is able to accurately ascertain the accuracy of the algorithm. Since there is no consensus on the definition of a community, using a real-world graph makes it more difficult to assess the accuracy of a community partition. Our benchmark graphs were generated randomly given the following set of parameters: The number of nodes N of the graph, the number of communities $Comm$ of the graph, the (average) degree of nodes $degree$ and the ratio of local links (intra-community links) to node degree $local/degree$. The parameters used by the algorithm and their corresponding values are shown in Table 10. In addition, we have used 14 low overlapping density LFR [12, 40] networks. We created a total of 208 benchmark graphs. Notice that even though the number of nodes, the number of communities and the node degree are construction parameters of the graphs, the degree of each node as well as the number of nodes in a single community varies based on a pareto distribution. This enables us to create graphs of community sizes and individual degrees varying up to an order of magnitude.

A demonstration of the proposed method is given in,[4] that contain the benchmark graphs, related articles and an executable of the proposed method.

We have used an accuracy-related metric found in the literature mostly to measure the performance of community detection. The simple accuracy is defined as follows: Let A be the estimated and $\hat{A}$ the corresponding actual community. The accuracy (acc) is given by the fraction of the number of nodes that belong to the intersection of $A \cap \hat{A}$ divided by the number of nodes that belongs to the union $A \cup \hat{A}$, $acc = \frac{|A \cap \hat{A}|}{|A \cup \hat{A}|}$. It holds that $acc \in [0, 1]$, the higher the accuracy the better the results. When $acc = 100\%$ the community detection algorithm gives perfect results.

---

[4] goo.gl/867M4z.

Table 10 The different
values for the parameters

N	1000, 5000, 10,000
Comm	5, 10, 20, 40, 80
local/degree	0.55, 0.65, 0.75, 0.85
degree	10, 20, 30, 40

Figures 25, 26 and 27 show an example of the evolution of the FlowPro for the synthetic graph with $N = 1000$, $Comm = 10$, $degree = 20$ and $local/degree = 0.75$. In Fig. 25a the variation of the stored flow ($\sum_{x \in V} S(x)$) and $T(s)$ during the execution of the main process are illustrated. It holds that $\sum_{x \in V} S(x)$ is increasing and $T(s)$ is decreasing during the execution of FlowPro. In Fig. 25b the variation of acc during the execution of the main process is plotted. The acc of the method increases during the execution of FlowPro.

In Figs. 26 and 27 the stored flow and community detection results (see black line) are depicted for the first, 30, 59 and 88 iterations of the main process, demonstrating the evolution of Flow-Pro. In Fig. 26, for each node x, the x-axis shows the shortest distance between x and initial node s ($d(x)$) and the y-axis shows the stored flow of x ($S(x)$). In Fig. 27, the nodes are sorted in descending order of their stored flow. It holds that during the execution of the main process the shortest distance from the initial node s decreases for the nodes that belong to the community and increases for the rest of the nodes due to the fact that FlowPro removes and adds edges to s (see third step of Algorithm FlowPro in Sect. 3.3). According to the $getCluster$ procedure (see Sect. 3.3), the nodes are sorted by their stored flow (S). The nodes that really belong to the community of s are plotted with blue spots while the rest of the nodes are plotted with red circles. In this example FlowPro terminates, since it converges to a solution. It holds that the acc after the first, 30 and 59 and 88 iterations is 23.1, 94.8, 96.7 and 100 %, respectively.

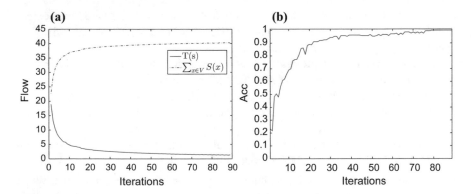

Fig. 25 An example of the variation of the **a** stored flow and $T(s)$ and **b** the acc during execution of the main process of FlowPro

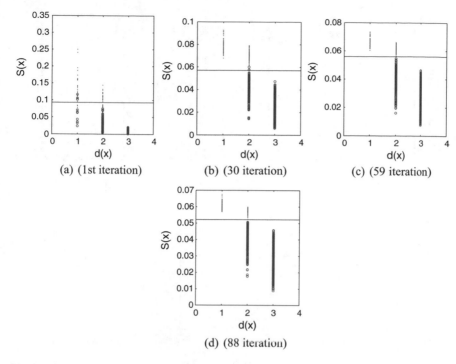

(a) (1st iteration)   (b) (30 iteration)   (c) (59 iteration)

(d) (88 iteration)

Fig. 26 The FlowPro community detection results (above the *black line*) after the first, 30, 59 and 88 iteration of the main process

In Fig. 28, we illustrate an example of the community visualization CoViFlowPro method [45] on a synthetic graph with $N = 5000$, $Comm = 40$, $degree = 10$ and $local/degree = 0.85$ has been used. In this example the community size is 40 nodes. The initial node ($s = 26$) has 13 neighbors. Two of them are detected as bridges. According of CoViFlowPro, it seems that the only one out of first five most important nodes of the community of s are also neighbors of s, while the node with the highest belief to belong on the $C(s)$ has label 14. Concerning the structure of the $C(s)$, we can group it into three categories, the strong community nodes: the first six nodes (14, 12, 4, 30, 33 and 13), the medium community nodes: the next 32 nodes and the weak community nodes: the last node (34).

3.4.1 Experiments on Benchmark Graphs

The performance of FlowPro-S, FlowPro and the Lancichinetti [7] are compared on Benchmark graphs. Over all benchmark graphs, it holds that the mean accuracy of FlowPro-S, FlowPro and Lancichinetti is 91.4, 81.7 and 34.1 %, respectively. In order to estimate the mean accuracy of FlowPro-S, FlowPro and Lancichinetti, we have tested each algorithm on 10 different nodes yielding the average *acc* per graph.

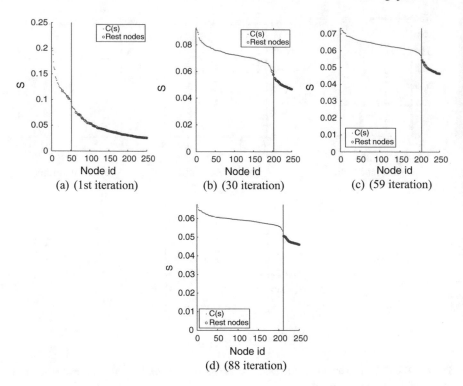

Fig. 27 The FlowPro community detection result (before the *black line*) in the first, 30, 59 and 88 iteration of the main process

Figure 29 shows the percentage of "successful" experiments, given an accuracy threshold to define "success". One can see the high performance and the stability of FlowPro-S and FlowPro. Each 3D graph in Figs. 30 and 31 shows the performance of a single algorithm (FlowPro-S, FlowPro and Lancichinetti) using either the first or the second metric, on all 208 benchmark graphs. Since there are four types of parameters which describe the graphs of the experiments, we decided to use the two most important factors which affect the algorithms' performance, in order to plot the accuracies in 3D graphs. The first of these factors is always the local links to node degree value, which dictates how strong and clear the communities in the graph are. The second factor in some cases is the number of communities in the graph, while in other cases (depending on the algorithm) is the (average) density of these communities. Thus, we decided to plot, for each algorithm two 3D graphs using, in one case, the number of communities as the values on one of the axes and the average density on the other. Each accuracy value in the graphs is the average of the accuracies of all the experiments based on the benchmark graphs with the same values of the aforementioned factors.

We can see from Fig. 29 and from the 3D graphs how FlowPro outperforms the Lancichinetti algorithm. The performance of FlowPro-S is about 10 % higher than

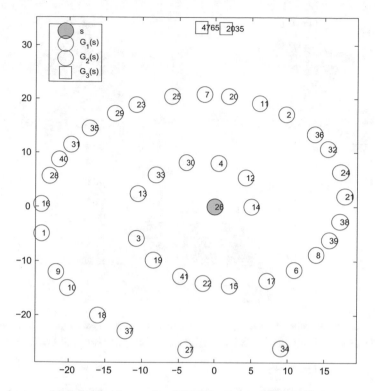

Fig. 28 An example of visualization results of CoViFlowPro method [45] on a synthetic graph

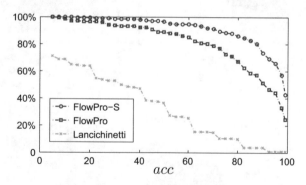

Fig. 29 Percentage of "successful" experiments, given an accuracy success threshold (x axis) for the accuracy metric

FlowPro, since the size of the community is given. Hereafter, we have performed several experiments to show how the parameters *size of the graph, degree, Comm* and *local/degree* of benchmark graphs affect the accuracy of FlowPro and Lancichinetti algorithms. For each parameter, we have removed the most difficult cases in order to measure its influence to the method performance.

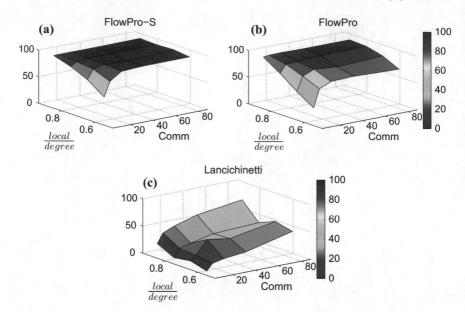

Fig. 30 The mean value of accuracy under different ratios of ($local/degree$) and number of communities ($Comm$) for the **a** FlowPro-S and **b** FlowPro and **c** Lancichinetti algorithms

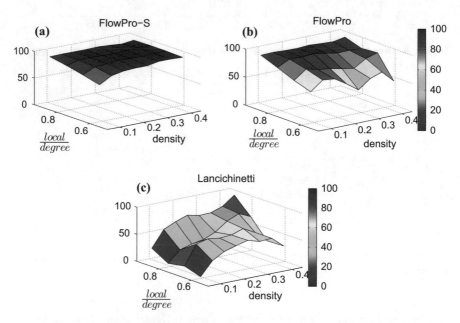

Fig. 31 The mean value of accuracy under different ratios of $local/degree$ and $density$ for the **a** FlowPro-S and **b** FlowPro and **c** Lancichinetti algorithms

- If we ignore the cases of benchmark graphs that have N equal to 10,000, then the mean accuracy of FlowPro and Lancichinetti are 82.1 and 35.1 %, respectively.
- If we ignore the cases of benchmark graphs that have *degree* equal to 10, then the mean accuracy of FlowPro and Lancichinetti are 85.5 and 38.4 %, respectively.
- If we ignore the cases of benchmark graphs that have *Comm* equal to 5, then the mean accuracy of FlowPro and Lancichinetti are 86.9 and 39.8 %, respectively.
- If we ignore the cases of benchmark graphs that have *local/degree* equal to 0.55, then the mean accuracy of FlowPro and Lancichinetti are 87.9 and 37.8 %, respectively.
- If we ignore the cases of benchmark graphs that have *local/degree* equal to 0.55 or *degree* equal to 10 or *Comm* equal to 5, then the mean accuracy of FlowPro and Lancichinetti are 94.2 and 48.5 %, respectively.

According to these experiments it seems that the parameters *degree*, *Comm* and *local/degree* affect the performance of FlowPro and Lancichinetti since these parameters are clearly related to the community density, while the size of the graph doesn't really affect the performance of FlowPro.

In addition, the performance of FlowPro and Lancichinetti methods has been tested on 14 LFR overlapping networks [12, 40]. Even though the problem formulation of both of the methods is mainly proposed for non-overlapping communities, we have evaluate their performance on low overlapping density LFR networks of 1000 nodes, 10 % of them are overlapping nodes. Similarly with [40], we allow the number of communities to which each overlapping node belongs (Om) to vary from 2 to 8 indicating the overlapping diversity of overlapping nodes. By increasing the value of Om, harder detection tasks are created. So, we have performed two tests on $LFR_{\mu=0.1}$ (7 datasets) and $LFR_{\mu=0.3}$ (7 datasets), where the mixing parameter μ denotes the expected fraction of links through which a node connects to other nodes in the same community. Figure 32 depicts the mean value of accuracy under the **(a)** $LFR_{\mu=0.1}$ and **(b)** $LFR_{\mu=0.3}$ for the FlowPro-S and FlowPro and Lancichinetti algorithms. On $LFR_{\mu=0.1}$ dataset, the average *acc* of FlowPro and Lancichinetti is 65.03 and 52.09 %, respectively. On $LFR_{\mu=0.3}$ dataset, the average *acc* of FlowPro and

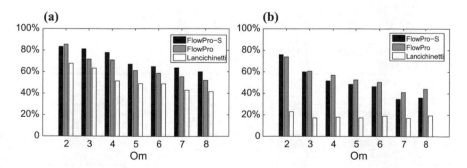

Fig. 32 The mean value of accuracy under the **a** $LFR_{\mu=0.1}$ and **b** $LFR_{\mu=0.3}$ for the FlowPro-S and FlowPro and Lancichinetti algorithms

Lancichinetti is 54.26 and 18.73 %, respectively. It holds that under any case, Flow-Pro clearly outperforms Lancichinetti algorithm. FlowPro gives high performance detection results when $Om = 2$ yielding 85.58 and 73.87 % under $LFR_{\mu=0.1}$ and $LFR_{\mu=0.3}$ datasets, respectively.

3.4.2 Experiments on Real World Graphs

We also conducted experiments on four real world graphs of diverse size (see Table 11). These graphs are obtained from [33] and from [49]. The *conductance* metric [23] is used for the comparison of the results on real world graphs.

The *conductance* of a cut is a metric that compares the size of a cut (i.e., the number of cut edges) and the number of edges in either of the two subgraphs induced by that cut. Thus, it is a metric that can be used to measure the performance of a single community detection. Consider a single community detection algorithm that divides a given graph G into two non-overlapping clusters $C(s)$ and $V \setminus C(s)$. The conductance $\phi(C(s))$ of cluster $C(s)$ can be obtained as shown in Eq. 20, where $a(C(s)) = \sum_{i \in C(s)} \sum_{j \in V} W(i, j)$ is the sum of the weights of all edges with at least one endpoint in $C(s)$. The $\phi(C(s))$ value represents the cost of one cut that bisects G into two vertex sets $C(s)$ and $V \setminus C(s)$.

$$\phi(C(s)) = \frac{\sum_{i \in C(s)} \sum_{j \notin C(s)} W(i, j)}{min(a(C(s)), a(V \setminus C(s)))} \tag{20}$$

Table 11 The Coverage of FlowPro and Lancichinetti on real world graphs

Graph	FlowPro	Lancichinetti
ca-GrQc (5.242 nodes, 28.980 edges)	0.264	0.996
ca-HepTh (9.877 nodes, 51.971 edges)	0.307	0.999
ego-Facebook (747 nodes, 30.025 edges)	0.135	0.239
Wiki-Vote (7.115 nodes, 103.689 Edges)	0.678	0.717
Enron email (36.692 nodes, 183.831 Edges)	0.398	–
Epinions (75.879 nodes, 508.837 Edges)	0.343	–
Slashdot (82.168 nodes, 948.464 Edges)	0.252	–
WWW (325.730 nodes, 1.497.135 Edges)	0.443	–

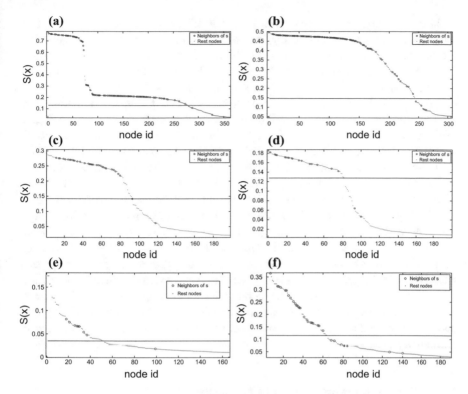

Fig. 33 The FlowPro community detection results (above the black line) for **a–d** four nodes of the ego-Facebook graph and **e, f** two nodes of the CA-HepTh graph

These metrics are different than the ones used in benchmark graphs, since the real decomposition of a real graph into communities is not known. We have tested each algorithm on 50 different nodes yielding the average conductance. The lower the conductance the better the community detection. In Table 11, we present the results for the FlowPro and the Lancichinetti algorithms on real graphs using the conductance metric. Due to the high computational cost of Lancichinetti, it was not possible to get its results on large graphs. According to these results FlowPro clearly outperforms Lancichinetti, since a lower conductance value is better.

Figure 33 depicts the FlowPro community detection results (above the black line) for four nodes of the ego-Facebook graph ((a),(b),(c),(d)) and two nodes of the CA-HepTh graph ((e), (f)). The nodes are sorted in descending order according to their stored flow. In cases, where the communities are not well discriminated like in several real social networks examples, we can use these graphs of sorted flow to analyze the communities, since the stored flow is highly correlated to the *local/degree* ratio. More specifically, in Fig. 33a the degree of the initial node is 110. According to this graph, it seems that this community can be separated into two sub-communities: the first 80 nodes strongly belong to the community and the rest 200 nodes weakly belong to the community. The detected community has 280 nodes, due to the constraint that the community size should be higher that $|n(s)|$, otherwise FlowPro would yield the first 80 nodes (strong community). In Fig. 33b the degree of the initial node is 121. The detected community has 270 nodes. However, according to their stored flow the first 150 nodes are strong community members. In Fig. 33c, d the degrees of the initial nodes are 34 and 19, respectively. The two detected communities have about 90 nodes with similar distribution of stored flow. In Fig. 33e, f the degrees of the initial nodes are 13 and 37, respectively. In these examples, according to the stored flow the communities are not well discriminated since the stored flow rapidly decreases, however we can use it to extract possible solutions for the community.

In all these examples, with high probability it holds that the neighbors of s (red circles) are nodes of the community. However, some of the neighbors of s are detected as bridges getting low stored flow. Moreover, it seems that some nodes of high stored flow may not belong to the initial neighbors of s, since the stored flow mainly depends on ratio $local/degree$. This also means that FlowPro is able to predict friendships in a social network. The following experiment also proves that FlowPro predicts friendships. In the ego-Facebook social network, we get the first ten neighbors $n_{10}(x)$ according to their stored flow that belong to the community of a node s. We remove the edges between them and s getting a graph G'. Then, we execute FlowPro on G' and we measure that on average 94.9 % of the nodes of $n_{10}(x)$ belong again to the community of s.

3.4.3 Communication Load and Complexity

FlowPro can be implemented as a fully distributed system without any knowledge of the entire graph. The communication load as well as the computational complexity of the proposed distributed framework make possible the execution of FlowPro on graphs of unlimited size (e.g. more than 100 millions of nodes with billions of links). The communication load and computational cost is not affected by the graph size, since the entire graph is not given and it holds that in the worst case the flow of messages is possible to reach at most the graph nodes that have distance less than $-log(T_0) + 1$ from of initial node s, where T_0 is a parameter of FlowPro defined in Sect. 3.3.

Hereafter, we provide an analysis of the communication load and computational cost. We measure the communication load by the total number of messages (mes) that are exchanged during the execution of FlowPro. The computation cost can be measured by the total number of iterations (ite) that are needed for the algorithm convergence.

Figure 34 illustrates the mean value of **(a)** mes and **(b)** ite under different ratios of ($local/degree$) and $density$ under all benchmarks. Similarly to Sect. 3.4.1, in 3D graphs we use $local/degree$ and $density$ to plot the mes and ite. It holds that mes as well as ite get low values when the communities are well discriminated, e.g. high density graphs and when $local/degree$ is high. When the communities are not

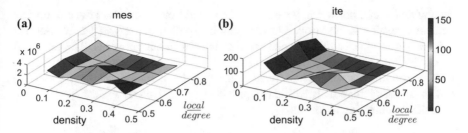

Fig. 34 The mean value of **a** *mes* and **b** *ite* under different ratios of ($local/degree$) and $density$

Table 12 The average *acc*, *mes* and *ite* for different values of T_0 over benchmark graphs

T_0	acc (%)	mes	ite
0.0005	81.9	$2.81 \cdot 10^6$	95
0.001	81.7	$1.75 \cdot 10^6$	100
0.05	77.2	$4.98 \cdot 10^5$	149.4

well discriminated, e.g. for low density graphs and when $local/degree$ is low, the algorithm is more difficult to converge and it needs more messages and iterations.

In addition, we have measured the stability of FlowPro for different values of T_0 over benchmark graphs. In Table 12, we present the average values for *acc*, *mes* and *ite* for different values of T_0 over benchmark graphs. For each benchmark graph we have tested each configuration on 10 different nodes getting the average values. It holds that, when $T_0 = 0.001$ we get high performance results keeping low the total number of messages and the number of iterations, so in our experimental results we have used $T_0 = 0.001$. When $T_0 = 0.0005$, we get only 0.2 % higher *acc* than the *acc* of $T_0 = 0.001$, however the average *mes* is very high. When $T_0 = 0.05$, the average *acc* is only 77.2 % and average *ite* is very high (149.4).

3.5 Conclusions

We presented a local community finding algorithm which is based on a flow propagation method. The stored flow can be used as a belief that a node belongs to the community which has been used on related problems such as finding the k-nearest neighbors of the community of a node that are important on several applications of social networks and on community visualization problem. The novelty of the proposed approach is the fact that FlowPro is local, fully distributed, it detects a single community and it does not require the knowledge of the entire graph as most of the existing methods in the literature. Thus, the application of FlowPro is possible in extremely large graphs or in cases where the entire graph is unknown like in most social networks. In cases, where the communities are not well discriminated like in several real social networks, we can use the sorted flows to analyze the community detection results. In addition, FlowPro can be applied in weighted networks that can measure the strength of social relationships in social networks. The proposed algorithm has been tested on a large number of benchmark graphs with known community structure and in graphs derived from real social networks. We have compared it with the Lancichinetti algorithm [7], proving its effectiveness against another local community detection algorithm.

Acknowledgments This research has been partially co-financed by the European Union (European Social Fund—ESF) and Greek national funds through the Operational Program "Education and Lifelong Learning" of the National Strategic Reference Framework (NSRF)—Research Funding Programs: ARCHIMEDE III-TEI-Crete-P2PCOORD.

References

1. Flake, G.W., Lawrence, S., Lee Giles, C., Coetzee, F.M.: Self-organization and identification of web communities. IEEE. Comput. **35**, 66–71 (2002)
2. Katsaros, Dimitrios, Pallis, George, Stamos, Konstantinos, Vakali, Athena, Sidiropoulos, Antonis, Manolopoulos, Yannis: Cdns content outsourcing via generalized communities. IEEE Trans. Knowl. Data Eng. **21**, 137–151 (2009)
3. Girvan, M., Newman, M.E.J.: Community structure in social and biological networks. Proc. Natl. Acad. Sci. U.S.A. **99**(12), 7821–7826 (2002)
4. Newman, M.E.J., Girvan, M.: Finding and evaluating community structure in networks. Phys. Rev. E **69**(2), 026113 (2004)
5. Hsieh, M.-H., Magee, C.L.: A new method for finding hierarchical subgroups from networks. Soc. Netw. **32**(3), 234–244 (2010)
6. Papadopoulos, S., Skusa, A., Vakali, A., Kompatsiaris, Y., Wagner, N.: Bridge bounding: a local approach for efficient community discovery in complex networks. Technical Report. arXiv:0902.0871 (2009)
7. Lancichinetti, A., Fortunato, S., Kertész, J.: Detecting the overlapping and hierarchical community structure in complex networks. New J. Phys. **11**(3):033015+ (2009)
8. Xie, J., Chen, M., Szymanski, B.K.: LabelRankT: Incremental Community Detection in Dynamic Networks via Label Propagation, at SIGMOD. arXiv:1305.2006 (2013). Comments: DyNetMM 2013, New York, USA (conjunction with SIGMOD/PODS 2013)
9. Alvari, H., Hajibagheri, A., Reese Sukthankar, G.: Community detection in dynamic social networks: a game-theoretic approach. In: 2014 IEEE/ACM International Conference on Advances in Social Networks Analysis and Mining, ASONAM 2014, Beijing, China, August 17–20, 2014, pp. 101–107 (2014)
10. Dabek, F., Cox, R., Kaashoek, F., Morris, R.: Vivaldi: a decentralized network coordinate system. In: Proceedings of the ACM SIGCOMM '04 Conference (2004)
11. Papadakis, H., Fragopoulou, P., Panagiotakis, C.: Distributed community detection: finding neighborhoods in a complex world using synthetic coordinates. In: ISCC'11, pp. 1145–1150 (2011)
12. Lancichinetti, Andrea, Fortunato, Santo: Community detection algorithms: a comparative analysis. Phys. Rev. E **80**(5 Pt 2), 056117 (2009)
13. Schaeffer, S.E.: Graph clustering. Comput. Sci. Rev. **1**(1), 27–64 (2007)
14. Andersen, R., Chung, F., Lang, K.: Local graph partitioning using pagerank vectors. In: 47th Annual IEEE Symposium on Foundations of Computer Science, 2006, FOCS'06, pp. 475–486. IEEE (2006)

15. Hui, P., Yoneki, E., Yan Chan, S., Crowcroft, J.: Distributed community detection in delay tolerant networks. In: Proceedings of 2nd ACM/IEEE International Workshop on Mobility in the Evolving Internet Architecture, p. 7. ACM (2007)
16. Derényi, Imre, Palla, Gergely, Vicsek, Tamás: Clique Percolation in random networks. Phys. Rev. Lett. **94**(16), 160–202 (2005)
17. Palla, G., Derenyi, I., Farkas, I., Vicsek, T.: Uncovering the overlapping community structure of complex networks in nature and society (2005)
18. Wu, F., Huberman, B.A.: Finding communities in linear time: a physics approach. Eur. Phys. J. B—Condens. Matter Complex Syst. **38**(2), 331–338 (2004)
19. Van Dongen, Stijn: Graph clustering via a discrete uncoupling process. SIAM J. Matrix Anal. Appl. **30**, 121–141 (2008)
20. Bagrow, James P., Bollt, Erik M.: Local method for detecting communities. Phys. Rev. E **72**(4), 46–108 (2005)
21. Chen, Jie, Saad, Yousef: Dense subgraph extraction with application to community detection. IEEE Trans. Knowl. Data Eng. **24**, 1216–1230 (2012)
22. Blondel, V.D., Guillaume, J.L., Lambiotte, R., Mech, E.L.J.S.: Fast unfolding of communities in large networks. J. Stat. Mech P10008 (2008)
23. Almeida, H., Guedes, D., Meira, W., Zaki, M.J.: Is there a best quality metric for graph clusters? In: Proceedings of the 2011 European Conference on Machine Learning and Knowledge Discovery in Databases—Volume Part I, ECML PKDD'11, pp.s 44–59. Springer, Berlin (2011)
24. Rosvall, M., Bergstrom, C.T.: An information-theoretic framework for resolving community structure in complex networks. Proc. Natl. Acad. Sci. **104**(18), 7327 (2007)
25. Nguyen, N.P., Dinh, T.N., Shen, Y., Thai, M.T.: Dynamic social community detection and its applications. Dynamic social community detection and its applications. PLoS ONE **9**(4), e91431 (2014)
26. Palla, G., Pollner, P., Barabási, A.-L., Vicsek, T.: Social group dynamics in networks. In: Adaptive Networks, pp. 11–38. Springer, Berlin (2009)
27. Alvari, H., Hashemi, S., Hamzeh, A.: Detecting overlapping communities in social networks by game theory and structural equivalence concept. In: Deng, H., Miao, D., Lei, J., Wang, F.L. (eds.) AICI (2), volume 7003 of Lecture Notes in Computer Science, pp. 620–630. Springer (2011)
28. Frigui, H., Krishnapuram, R.: A robust competitive clustering algorithm with applications in computer vision. IEEE Trans. Pattern Anal. Mach. Intell. **21**(5), 450–465 (1999)
29. Papadakis, Harris, Panagiotakis, Costas, Fragopoulou, Paraskevi: Distributed community detection in complex networks using synthetic coordinates. J. Stat. Mech: Theory Exp. **2014**(3), P03013 (2014)
30. Papadimitriou, Christos H., Steiglitz, Kenneth: Combinatorial Optimization: Algorithms and Complexity. Prentice-Hall Inc., Upper Saddle River (1982)
31. Strehl, Alexander, Ghosh, Joydeep, Cardie, Claire: Cluster ensembles—a knowledge reuse framework for combining multiple partitions. J. Mach. Learn. Res. **3**, 583–617 (2002)
32. Cornell kdd cup
33. Stanford large network dataset collection. http://snap.stanford.edu/data/
34. Almeida, H., Guedes, D., Meira, W., Zaki, M.: Is there a best quality metric for graph clusters? In: Machine Learning and Knowledge Discovery in Databases, pp. 44–59 (2011)
35. Greene, D., Doyle, D., Cunningham, P.: Tracking the evolution of communities in dynamic social networks. In: Proceedings of the 2010 International Conference on Advances in Social Networks Analysis and Mining, ASONAM '10, pp. 176–183, Washington, DC, USA. IEEE Computer Society (2010)
36. Lancichinetti, Andrea, Fortunato, Santo: Benchmarks for testing community detection algorithms on directed and weighted graphs with overlapping communities. Phys. Rev. E **80**(1), 016118 (2009)
37. Leskovec, J., Kleinberg, J., Faloutsos, C.: Graphs over time: densification laws, shrinking diameters and possible explanations. In: Proceedings of the Eleventh ACM SIGKDD International Conference on Knowledge Discovery in Data Mining, KDD '05, pp. 177–187, New York, NY, USA. ACM (2005)

38. Leskovec, J., Krevl, A.: SNAP Datasets: Stanford large network dataset collection. http://snap. stanford.edu/data (2014)

39. KDD 2003. Kddcup dataset. http://www.cs.cornell.edu/projects/kddcup/datasets.html (2003)

40. Xie, J., Kelley, S., Szymanski, B.K.: Overlapping community detection in networks: the state-of-the-art and comparative study. ACM Comput. Surv. **45**(4), 43:1–43:35 (2013)

41. Opsahl, Tore, Panzarasa, Pietro: Clustering in weighted networks. Soc. Netw. **31**(2), 155–163 (2009)

42. Coscia, Michele, Giannotti, Fosca, Pedreschi, Dino: A classification for community discovery methods in complex networks. Stat. Anal. Data Mining: ASA Data Sci. J. **4**(5), 512–546 (2011)

43. Bagrow, J.P., Bollt, E.M.: Local method for detecting communities. Phys. Rev. E **72**(4), 046108 (2005)

44. Spielman, D.A., Teng, S.-H.: A local clustering algorithm for massive graphs and its application to nearly-linear time graph partitioning. arXiv preprint arXiv:0809.3232 (2008)

45. Panagiotakis, C., Papadakis, H., Fragopoulou. P.: Coviflowpro: a community visualization method based on a flow propagation algorithm. In: International Conference on Bio-inspired Information and Communications Technologies (2014)

46. Panagiotakis, C., Fragopoulou, P.: Voting clustering and key points selection. In: International Conference on Computer Analysis of Images and Patterns (2013)

47. Frey, B.J., Dueck, D.: Clustering by passing messages between data points. Science **315**(5814), 972–976 (2007)

48. Yedidia, J.S., Freeman, W.T., Weiss, Y.: Constructing free-energy approximations and generalized belief propagation algorithms. IEEE Trans. Inform. Theory **51**(7), 2282–2312 (2005)

49. Network databases—University of Notre Dame. http://www3.nd.edu/

Index

© Springer International Publishing Switzerland 2017

A. Adamatzky (ed.), *Emergent Computation*, Emergence, Complexity
and Computation 24, DOI 10.1007/978-3-319-46376-6

Algorithm 2. Abstract uDir smart contract.

```
contract uDir {
    constructor (address Adjudicator);
    modifier (onlyAuthServ, onlyAdjAuth, onlyBA);
    function registerUser (string pseudonym, string pk, string certificate) onlyAuthServ public
    function deleteUser(string pk) onlyAdjAuth public
    function getUsersInfo(string pk) public
    function selfDestruct() onlyBA public }
```

Algorithm 3. Abstract oDir smart contract.

```
contract oDir {
    constructor (address Adjudicator);
    modifier (onlyObjectOwner, onlyOwnerAdj, onlyBA);
    function registerResource (bytes32 Oid, string pid, string pk, string Desc, address
    ACC_addr, string ACC_abi, address PR_addr, string PR_abi) public
    function updateResource(bytes32 Oid, string desc) onlyObjectOwner public
    function deleteResource(bytes32 Oid) onlyOwnerAdj public
    function getContractInfo(bytes32 Oid) public
    function getAdvertiseInfo(bytes32 Oid) public
    function selfDestruct() onlyBA public }
```

Algorithm 4. Abstract objPropRep smart contract.

```
contract objPropRep {
    constructor (address objACC);
    modifier (onlyOwner, onlyOwnerAdj);
    function setPropertyInfo (bytes32 Oid, string properties, string certificate) onlyOwner
    public
    function updatePropertyInfo(bytes32 Oid, string properties, string certificate) onlyOwner
    public
    function deletePropertyInfo(bytes32 Oid) onlyOwnerAdj public
    function getPropertyInfo(bytes32 Oid) public
    function selfDestruct() onlyOwner public }
```

Algorithm 5. Abstract objACC smart contract.

```
contract objACC {
    constructor ();
    modifier (onlyOwner, onlyOwnerAdj);
    function addAccessInfo (bytes32 Oid, string CM, string[] policy) onlyOwner public
    function updateAccessInfo(bytes32 Oid, string CM, string[] policy) onlyOwner public
    function deleteAccessInfo(bytes32 Oid) onlyOwnerAdj public
    function getAccessInfo(bytes32 Oid) public
    function setContractAddress(address oPropRep, address oDir) onlyOwner public
    function getRequestHistory(bytes32 Oid) public
    function selfDestruct() onlyOwner public }
```

Algorithm 6. Abstract Adj smart contract.

```
contract Adj {
    constructor ();
    modifier (onlyBA, onlyVerifier);
    function registerVerifier (address verifier) onlyBA public
    function reportMisbehavior(bytes32 Oid, string pk, string misbehaviour, uint time) public
    function setMisbehaviorState(string state, bytes32 Oid, string pk) onlyVerifier public
    function getLatestMisbehavior(string pk) public
    function selfDestruct() onlyBA public }
```

References

1. Airbnb, Inc.: Vacation rental company (2020). https://www.airbnb.com
2. Uber Technologies, Inc.: Transport company (2020). https://www.uber.com
3. Muni Venkateswarlu, K., Avizheh, S., Safavi-Naini, R.: A blockchain based approach to resource sharing in smart neighbourhoods. In: Bernhard, M., et al. (eds.) FC 2020. LNCS, vol. 12063, pp. 550–567. Springer, Cham (2020). https://doi.org/10.1007/978-3-030-54455-3_39
4. Bethencourt, J., Sahai, A., Waters, B.: Ciphertext-policy attribute-based encryption. In: 2007 IEEE Symposium on Security and Privacy, SP 2007, pp. 321–334. IEEE (2007)
5. Rivest, R.L., Shamir, A., Tauman, Y.: How to leak a secret. In: Boyd, C. (ed.) ASIACRYPT 2001. LNCS, vol. 2248, pp. 552–565. Springer, Heidelberg (2001). https://doi.org/10.1007/3-540-45682-1_32
6. Ganache: Ganache one click blockchain (2019). https://www.trufflesuite.com/ganache
7. Remix: Remix-solidity ide (2019). https://remix.ethereum.org
8. Daemen, J., Rijmen, V.: Announcing the advanced encryption standard (AES). In: Federal Information Processing Standards Publication 197, pp. 1–51, 3 (2001)
9. Baiardi, F., Falleni, A., Granchi, R., Martinelli, F., Petrocchi, M., Vaccarelli, A.: SEAS, a secure e-voting protocol: design and implementation. Comput. Secur. **24**(8), 642–652 (2005)
10. Chaum, D.L.: Untraceable electronic mail, return addresses, and digital pseudonyms. Commun. ACM **24**(2), 84–90 (1981)
11. Xie, Q., Hengartner, U.: Privacy-preserving matchmaking for mobile social networking secure against malicious users. In: 2011 9th Annual International Conference on Privacy, Security and Trust, pp. 252–259. IEEE (2011)
12. Sandhu, R.S., Samarati, P.: Access control: principle and practice. IEEE Commun. Mag. **32**(9), 40–48 (1994)
13. Sandhu, R.S.: Role-based access control. In: Advances in Computers, vol. 46, pp. 237–286. Elsevier (1998)
14. Hu, V.C., et al.: Guide to attribute based access control (ABAC) definition and considerations (draft). NIST Special Publication 800-162 (2013)
15. Ahmed, T., Sandhu, R., Park, J.: Classifying and comparing attribute-based and relationship-based access control. In: Proceedings of the 7th ACM on Conference on Data and Application Security and Privacy, pp. 59–70 (2017)
16. Sahai, A., Waters, B.: Fuzzy identity-based encryption. In: Cramer, R. (ed.) EUROCRYPT 2005. LNCS, vol. 3494, pp. 457–473. Springer, Heidelberg (2005). https://doi.org/10.1007/11426639_27
17. Goyal, V., Pandey, O., Sahai, A., Waters, B.: Attribute-based encryption for fine-grained access control of encrypted data. In: Proceedings of the 13th ACM Conference on Computer and Communications Security, pp. 89–98 (2006)
18. Boneh, D., Boyen, X., Goh, E.-J.: Hierarchical identity based encryption with constant size ciphertext. In: Cramer, R. (ed.) EUROCRYPT 2005. LNCS, vol. 3494, pp. 440–456. Springer, Heidelberg (2005). https://doi.org/10.1007/11426639_26
19. Shoup, V.: Lower bounds for discrete logarithms and related problems. In: Fumy, W. (ed.) EUROCRYPT 1997. LNCS, vol. 1233, pp. 256–266. Springer, Heidelberg (1997). https://doi.org/10.1007/3-540-69053-0_18
20. Narayan, S., Gagné, M., Safavi-Naini, R.: Privacy preserving EHR system using attribute-based infrastructure. In: Proceedings of the 2010 ACM Workshop on Cloud Computing Security Workshop, pp. 47–52 (2010)

21. Wan, Z., Deng, R.H., et al.: HASBE: a hierarchical attribute-based solution for flexible and scalable access control in cloud computing. IEEE Trans. Inf. Forensics Secur. **7**(2), 743–754 (2011)
22. Jung, T., Li, X.-Y., Wan, Z., Wan, M.: Privacy preserving cloud data access with multi-authorities. In: 2013 Proceedings IEEE INFOCOM, pp. 2625–2633. IEEE (2013)
23. Belguith, S., Kaaniche, N., Jemai, A., Laurent, M., Attia, R.: PAbAC: a privacy preserving attribute based framework for fine grained access control in clouds. In: 13th International Conference on Security and Cryptography, SECRYPT 2016, vol. 4, pp. 133–146. SciTePress (2016)
24. Waters, B.: Ciphertext-policy attribute-based encryption: an expressive, efficient, and provably secure realization. In: Catalano, D., Fazio, N., Gennaro, R., Nicolosi, A. (eds.) PKC 2011. LNCS, vol. 6571, pp. 53–70. Springer, Heidelberg (2011). https://doi.org/10.1007/978-3-642-19379-8_4
25. Zyskind, G., Nathan, O., et al.: Decentralizing privacy: using blockchain to protect personal data. In: 2015 IEEE Security and Privacy Workshops, pp. 180–184. IEEE (2015)
26. Di Francesco Maesa, D., Mori, P., Ricci, L.: Blockchain based access control. In: Chen, L.Y., Reiser, H.P. (eds.) DAIS 2017. LNCS, vol. 10320, pp. 206–220. Springer, Cham (2017). https://doi.org/10.1007/978-3-319-59665-5_15
27. Ouaddah, A., Elkalam, A.A., Ouahman, A.A.: FairAccess: a new blockchain-based access control framework for the internet of things. Secur. Commun. Netw. **9**(18), 5943–5964 (2016)
28. Zhang, Y., Kasahara, S., Shen, Y., Jiang, X., Wan, J.: Smart contract-based access control for the internet of things. IEEE IoT J. **6**(2), 1594–1605 (2018)
29. Cruz, J.P., Kaji, Y., Yanai, N.: RBAC-SC: role-based access control using smart contract. IEEE Access **6**, 12240–12251 (2018)
30. Xu, R., Chen, Y., Blasch, E., Chen, G.: BlendCAC: a blockchain-enabled decentralized capability-based access control for IoTs. In: 2018 IEEE International Conference on Internet of Things (iThings) and IEEE Green Computing and Communications (GreenCom) and IEEE Cyber, Physical and Social Computing (CPSCom) and IEEE Smart Data (SmartData), pp. 1027–1034. IEEE (2018)
31. Guo, H., Meamari, E., Shen, C.-C.: Multi-authority attribute-based access control with smart contract. In: Proceedings of the 2019 International Conference on Blockchain Technology, pp. 6–11 (2019)
32. Maesa, D.D.F., Mori, P., Ricci, L.: Blockchain based access control services. In: 2018 IEEE International Conference on Internet of Things (iThings) and IEEE Green Computing and Communications (GreenCom) and IEEE Cyber, Physical and Social Computing (CPSCom) and IEEE Smart Data (SmartData), pp. 1379–1386. IEEE (2018)
33. Maesa, D.D.F., Mori, P., Ricci, L.: A blockchain based approach for the definition of auditable access control systems. Comput. Secur. **84**, 93–119 (2019)
34. Tapas, N., Longo, F., Merlino, G., Puliafito, A.: Experimenting with smart contracts for access control and delegation in IoT. Fut. Gener. Comput. Syst. **111**, 324–338 (2020)
35. Raikwar, M., Gligoroski, D., Kralevska, K.: SoK of used cryptography in blockchain. IEEE Access **7**, 148550–148575 (2019)
36. Buterin, V., et al.: A next-generation smart contract and decentralized application platform, white paper (2014)
37. Wood, G., et al.: Ethereum: A secure decentralised generalised transaction ledger. Ethereum project yellow paper, vol. 151, pp. 1–32 (2014)

38. Bonneau, J., Clark, J., Goldfeder, S.: On bitcoin as a public randomness source. IACR Cryptol. ePrint Arch. **2015**, 1015 (2015)
39. Kelsey, J., Brandão, L.T., Peralta, R., Booth, H.: A reference for randomness beacons: format and protocol version 2, Technical report, National Institute of Standards and Technology (2019)
40. Code: Privacy-preserving resource sharing (2021). https://github.com/mnr-18/Privacy-preserving-Resource-Sharing
41. The OpenSSL Project: OpenSSL: The open source toolkit for SSL/TLS. www.openssl.org. Accessed April 2003
42. Bethencourt, J.: Advanced Crypto Software Collection. http://acsc.cs.utexas.edu/cpabe/. Accessed December 2006
43. Chase, M.: Multi-authority attribute based encryption. In: Vadhan, S.P. (ed.) TCC 2007. LNCS, vol. 4392, pp. 515–534. Springer, Heidelberg (2007). https://doi.org/10.1007/978-3-540-70936-7_28
44. Chase, M., Chow, S.S.: Improving privacy and security in multi-authority attribute-based encryption. In: Proceedings of the 16th ACM Conference on Computer and Communications Security, pp. 121–130 (2009)
45. Bender, A., Katz, J., Morselli, R.: Ring signatures: stronger definitions, and constructions without Random Oracles. In: Halevi, S., Rabin, T. (eds.) TCC 2006. LNCS, vol. 3876, pp. 60–79. Springer, Heidelberg (2006). https://doi.org/10.1007/11681878_4

WWTSC – Attacks' Analysis

SoK: Algorithmic Incentive Manipulation Attacks on Permissionless PoW Cryptocurrencies

Aljosha Judmayer[1,2]([✉]), Nicholas Stifter[1,2], Alexei Zamyatin[3], Itay Tsabary[4], Ittay Eyal[4], Peter Gaži[5], Sarah Meiklejohn[6], and Edgar Weippl[2]

[1] SBA Research, Vienna, Austria
{ajudmayer,nstifter}@sba-research.org
[2] Uni Wien, Vienna, Austria
edgar.weippl@univie.ac.at
[3] Imperial College London, London, England
a.zamyatin@imperial.ac.uk
[4] Technion and IC3, Haifa, Israel
Ittay@technion.ac.il
[5] IOHK, Singapore, Singapore
peter.gazi@iohk.io
[6] University College London, London, England
s.meiklejohn@ucl.ac.uk

Abstract. A long standing question in the context of cryptocurrencies based on Nakamoto consensus is whether such constructions are incentive compatible, i.e., the intended properties of the system emerge from the appropriate utility model for participants. Bribing and other related attacks, such as front-running or Goldfinger attacks, aim to directly influence the incentives of actors within (or outside) of the targeted cryptocurrency system. The theoretical possibility of bribing attacks on cryptocurrencies was discussed early on in the cryptocurrency community and various different techniques and approaches have since been proposed. Some of these attacks are designed to gain in-band profits, while others intend to break the mechanism design and render the cryptocurrency worthless. In this paper, we systematically expose the large but scattered body of research in this area which has accumulated over the years. We summarize these bribing attacks and similar techniques that leverage on programmatic execution and verification under the term *algorithmic incentive manipulation* (AIM) attacks, and show that the problem space is not yet fully explored. Based on our analysis we present several research gaps and opportunities that warrant further investigation. In particular, we highlight *no- and near-fork* attacks as a powerful, yet largely underestimated, AIM category that raises serious security concerns not only for smart contract platforms.

Keywords: Algorithmic incentive manipulation · Cryptocurrencies · Bribing · Goldfinger · Front-running

© International Financial Cryptography Association 2021
M. Bernhard et al. (Eds.): FC 2021 Workshops, LNCS 12676, pp. 507–532, 2021.
https://doi.org/10.1007/978-3-662-63958-0_38

1   Introduction

Bitcoin, and most of its cryptocurrency descendants, is based on what is termed Nakamoto consensus [16,47,55]. In a nutshell, Nakamoto consensus enables *any-one* to initiate valid state transitions to a replicated state machine if they solve a cryptographic puzzle of sufficient hardness that depends on the prior state. This is usually implemented by appending a block of ordered transactions and an appropriate proof-of-work [21] to a directed rooted and cryptographically linked tree of other blocks. The path to the leaf with the highest depth (resp. difficulty) is called the *longest (heaviest) chain* and thus the current state of the system.

A crucial part of this so-called *permissionless* [60] consensus concept is the utilization of incentives in the protocol design to provide a motivation for miners to participate. A long-standing question in this regard is whether or not this con-struction is incentive compatible, i.e., that the intended properties of the system emerge from the appropriate utility model for miners [14,15]. As Nakamoto did not provide a formal description of the protocol in [47], several attempts towards formalization[1] have been made to prove certain security properties of the pro-tocol. Thereby, most approaches, such as [13,26,28,48], do assume a sufficient *honest* majority of miners without considering incentives, or like [12] explicitly do not consider *bribing attacks* to manipulate incentives of participants.

Bribing attacks target incentive compatibility and assume that at least some miners accept bribes to maximize their profit. Hereby, bribing not necessarily refers to illegal activity, but merely that a payment is made in exchange for a certain action [15]. If the attacker, together with all bribable miners, can gain a sizable portion of the computational power even for a short period of time, attacks are likely to succeed. To the best of our knowledge, the first discussions of *bribery* attacks on Bitcoin date back to a bitcointalk forum post from 2012 by a user called *cunicula* [18]. Since then attacks on incentives in cryptocurrencies have been sporadically discussed in the cryptocurrency community [11,40], with the first peer reviewed paper on the subject presented in 2016 by Bonneau [14]. Over the years several different techniques and approaches of bribing attacks have been proposed [11,14,18,31,32,37,40,42–44,56,58,59,61]. These proposals vary regarding their system models, technical methods and evaluation criteria, which makes comparing them a challenging task. What all this approaches have in common is that they are targeted to manipulate the incentives of actors in the cryptocurrency ecosystem.

All these attacks, as well as meta arguments [14,15,17,25,36] and recent research [19,35] have fueled the debate around incentives in Nakamoto consen-sus. A key observation hereby noted in Bonneau [14] and Budish [17] is, that the security guarantees of Nakamoto consensus against bribing attacks to facil-itate double-spending, are *linear* in the number of blocks and the expenditure on mining power to produce them (in terms of financial resources required). In contrast, the achievable security guarantees of many other investments in IT security, like for example traditional usage of cryptography, are designed to

---

[1] For a summary see [55].

"yield convex returns" [17]. Large scale temporary majority attacks, in which an attacker overtakes a cryptocurrency for a short period of time, have gained further practical importance as they have been observed more frequently in recent history [5,6,8,9].

Lately, also other attempts to manipulate incentives targeted to influence the *order* of transactions within a not yet mined block [19,22,32] (front-running), or to *exclude* transactions [32,61] have received increased attention, as some less sophisticated variants of them have been observed in the wild already [4,19]. These attacks show that the security properties of Nakamoto consensus under real-world incentives are still not fully understood, not only by cryptocurrency users and smart contract developers, but also in the research community.

This paper aims to systematize the landscape of research on attacks that target the incentives of actors within – and through the use of – cryptocurrencies. To systematically expose the large body of research on bribing-, front-running- Goldfinger- and other related attacks, we jointly consider them under the general term *algorithmic incentive manipulation attacks* (AIM). Thereby, we want to distinguish programmatic ways to setup and execute incentive attacks on cryptocurrencies using cryptocurrencies, from "classical" bribing attacks, like for example using a suitcase full of cash to bribe miners, as the latter does not require technical means, but at the same time lacks technical enforcement [14]. The classification of AIM attacks in this paper forms a necessary prerequisite and basis for the comparison and discussion of work in this field – being it attacks or meta arguments. In summary, our contributions are:

1. A definition of algorithmic incentive manipulation (AIM) providing a unified view of different approaches targeting the incentives of actors.
2. A generalized attack model for AIM.
3. A classification framework for AIM that is applicable to describe a broad class of attacks.
4. A classification of existing AIM approaches and discussion of main observations and gaps.

1.1 Structure of This Work

We start by giving a definition of AIM in Sect. 2. Next, in Sect. 3, we provide a generalized attack model that can be readily applied to most presented attacks by adjusting the provided parameters. We then present the main classification criteria for AIM in Sect. 4. The classification and analysis of existing attacks is provided in Sect. 5, by comparing them property by property. In Sect. 6 we highlight the challenges when comparing costs and profits of AIM attacks. We conclude by discussing the relation of AIM to other ways of gaining capacity in Nakamoto consensus and present directions for future work in Sect. 7.

2 Algorithmic Incentive Manipulation

To meaningfully partake in a Nakamoto consensus protocol, a certain capacity of a scarce resource is required. In case of Proof-of-Work (PoW) these resources

are mining hardware and electricity to solve cryptographic puzzles. In [15] the different ways to gain capacity in Nakamoto consensus are grouped into four different strategies: *rent, bribe, build* and *buy out*. It is well known, that an actor who builds a new datacenter running specialized mining hardware, or rents GPU clusters, or buys existing mining hardware from current miners can increase his influence on the targeted cryptocurrency and thereby (depending on his resources) potentially launch attacks [15,17]. This permissionlessness [60] which allows such kinds of attacks is a desired property of Nakamoto consensus based on PoW[2].

In this paper we want to focus on methods of *algorithmic incentive manipulation* (AIM) to gain capacity in permissionless PoW based cryptocurrencies, as all existing attacks which fall into this category – and are classified in this paper – explicitly target PoW systems. Algorithmic or "virtual" methods of gaining capacity rely on the usage of game theory and cryptocurrencies to perform payments which are cryptographically secured and dependent on certain conditions. This ability of cryptocurrencies to tie payments to the fulfillment of certain conditions, like for example the existence of prior transactions, or the successful execution of smart contract invocations, are a way to promise *credible but conditioned* payments.

Utilizing such techniques, AIM methods do not involve the physical transfer of resources, like buying, or building and maintaining hardware. Instead, these methods assume that at least some fraction of actors within, or outside of the system behaves rationally in the sense that they want to maximize their short-term profits[3]. Some approaches for AIM have been referred to as *bribing*, but AIM goes beyond of what is currently viewed as a bribing attack in the literature, as they should incorporate Goldfinger [38] and front-running [19,22] attacks as well. Therefore our broader definition is as follows:

Definition 1. *Algorithmic Incentive Manipulation (AIM) utilizes either credible threats, or conditioned rewards denominated in cryptocurrency units, to incentivize certain actions, within a targeted cryptocurrency system, to be taken by capable actors.*

Hereby, the definition of *capable actor* depends on the requirements of the concrete attack, as well as the targeted system. For most attacks, the timely creation and submission of valid PoW solutions – complying to the required difficulty – is necessary to qualify as a capable actor. If the target would be a Proof-of-Stake (PoS) cryptocurrency for example, a capable actor would be required to control voting stake.

---

[2] In comparison, in proof-of-stake (PoS) cryptocurrencies it would not be possible to rent or build *new* capacity, as all stake eligible for voting has to exist in the system already [15].

[3] For a discussion on rationality in this context see, Sect. 7.

3   Generalized Attack Model for AIM

For all analyzed AIM attacks we describe the following generalized attack model, which can readily be applied in most cases.[4] If an analyzed attack deviates from this model, it is highlighted in detail when the attack is described.

As most bribing and related attacks in this area are designed to target Bitcoin, Ethereum, or other derived cryptocurrencies, we also focus in our model on AIM in *permissionless* [60] proof-of-work (PoW) cryptocurrencies. That is, we assume protocols adhering to the design principles of Bitcoin, or its backbone protocol [27,47,48], which is sometimes referred to as Nakamoto consensus [20,55]. Within the attacked cryptocurrency we differentiate between *miners*, who participate in the consensus protocol and attempt to solve PoW-puzzles, and *clients*, who do not engage in such activities. As in previous work on bribing attacks [14,42,44,56], the set of miners is assumed to be fixed, as well as their respective computational power within the network is assumed to remain constant.

To abstract from currency details, we use the term *value* as a universal denomination for the purchasing power of a certain amount of cryptocurrency units or any other out-of-band funds such as fiat currency. Miners and clients may own cryptocurrency units and are able to transfer them (i.e., their value) by creating and broadcasting valid transactions within the network. Moreover, in most prior work [42,44,57] the simplifying assumption is made that exchange rates remain constant over the duration of the attack.

Actors. Our generalized attack model splits participating miners into three groups and their roles remain static for the attack duration. Categories follow the *BAR (Byzantine, Altruistic, Rational)* [10,41] rational behavior model. Additionally, we define the *victim(s)* as another group or individual without hashrate.

- **Byzantine miners or attacker (s):** The attacker B wants to execute an AIM attack on a *target cryptocurrency*. B is in control of bribing funds $f_B > 0$ that can be in-band or out-of-band, depending on the attack scenario. He has some or no hashrate $p_B \geq 0$ in the target cryptocurrency. The attacker may deviate arbitrarily from the protocol rules.
- **Altruistic or honest miner (s):** Altruistic miners A are honest and always follow the protocol rules, hence they will not accept bribes to mine on a different chain-state or deviate from the rules, even if it would offer larger profit. Miners A control some or no hashrate $p_A \geq 0$ in the target cryptocurrency.
- **Rational or bribable miner (s):** Rational miners R control hashrate $p_R > 0$ in the target cryptocurrency. They aim to maximize their short term profits in terms of *value*. We consider such miners "bribable", i.e., they follow

---

[4] Only the Proof-of-Stale blocks [43,59] attack, as well as Fomo 3D [4] are fundamentally different: The former is targeted to attack mining pools, while the latter is designed as an exit scam, but can also lead to scenarios resembling an attack.

strategies that deviate from the protocol rules as long as they are expected to yield higher profits than being honest. For our analyses we assume that rational miners do not concurrently engage in other rational strategies such as selfish mining [24].

- **Victim (s):** The set of victims or a single victim, which loses value if the bribing attack is to be successful. The victims control zero hashrate, and therefore can be viewed as a client.

It holds that $p_B + p_A + p_R = 1$. The assumption that the victim of a AIM attack has no hashrate is plausible, as the majority of transaction in Bitcoin or Ethereum are made by clients which do not have any hashrate in the system they are using. Nonetheless, this assumption is often left implicit (e.g. [42]).

Some bribing attacks (e.g. [56]) implicitly model victims (in this case the betrayed collaborators of the double-spending attack) as honest, i.e., as strictly following the protocol. We emphasise that this is not necessarily the case, especially if economically rational counter-attacks by the victim should be considered. This distinction between rational and honest victims is more important if V is in possession of some hashrate, but even in a setting where V has no hashrate, he can use his funds (f_V) for counter attacks.

Whenever we refer to an attack as *trustless*, we imply that no trusted third party is needed between the briber and the bribee to ensure correct payments are performed for the desired actions. It is clearly desirable from the attacker's perspective to design AIM attacks in a way that the attacker (s) as well as the collaborating miners have no incentive to betray each other if they are economically rational.

Communication and Timing. As previous AIM attacks, we assume that all miners in the target cryptocurrency have *perfect knowledge* about the attack once it has started, if not stated differently. Miners with imperfect information can be modelled by adding their respective hashrate to the hashrate of altruistic miners (p_A). All participants communicate through message passing over a peer-to-peer gossip network, which we assume implements a reliable broadcast functionality. This does not mean, that every transmitted transaction will make it into the next block, as the block size is bounded by the underlying blockchain protocol. Analogous to [27], we model the adversary as "rushing", meaning that he gets to see all other players' messages before he decides his on strategy.

If more than one cryptocurrency is involved in the considered scenario, for example when out-of-band payments should be performed in another cryptocurrency, an additional *funding cryptocurrency* is assumed. While the attack is performed on a *target cryptocurrency*, the funding cryptocurrency is used to orchestrate and fund it. In such a case, we also assume that the difficulty and thus the mean block interval of the funding chain is fixed for the duration of the attack. Further, no additional attacks are concurrently being launched against either of the cryptocurrencies.

4   Classification Framework for AIM

We first introduce a general classification along four main dimensions: the *state of the targeted transaction(s)*; the *intended impact* on these transactions; the *required interference with consensus*, i.e., the depth of blockchain reorganizations caused by forks for the attack to be successful; and finally the *used payment methods*. Besides these main distinguishing properties, there are also other characteristics which are introduced when they become relevant during the classification of existing AIM attacks in Sect. 5. To get a feel for our classification framework and the herein introduced dimensions, see Sect. A for an example usage.

4.1   State of Targeted Transactions

A core goal for permissionless PoW cryptocurrencies is to achieve an (eventually) consistent and totally ordered log of transactions that define the global state of the shared ledger. Therefore, our classification uses a transaction-centric viewpoint to systematize different attacks and their relation to the underlying consensus. We differentiate between three *states* a transaction can be in from the perspective of a participant (miner or client):

- **unconfirmed**[5], the transaction has been broadcasted in the respective P2P network;
- **confirmed**, the transaction has been confirmed by at least one block, i.e., has been included in a block;
- **settled**, the transaction has been confirmed by at least k blocks, where k is defined by the recipient of the transaction. We denote $k_{participant}$ to refer to the confirmation policy of a participant if it is not clear from the context e.g., k_V denotes the confirmation policy of the victim.

4.2   Intended Impact/Influence on Transactions

We further separate between the following four main types of how AIM can have an influence on transactions and their ordering:

- **transaction revision**, change a previously proposed, possibly confirmed or settled transaction;
- **transaction exclusion/censorship**, exclude a specific transaction, or set of transactions, from the log of transactions for a bounded amount of time i.e., the transaction remains unconfirmed.
- **transaction ordering**, change either the proposed, confirmed or already settled upon order of transactions;
- **transaction triggering**, incentivize the creation of one or multiple transactions by a specific actor or group of actors, e.g., trigger spending transactions for *anyone-can-spend* outputs.

---

[5] Sometimes also referred to as *proposed*, or *published* in related literature.

The design paradigms of the underlying cryptocurrency have to be considered to assess the impact and effects of the mentioned manipulation methods. For example, the impact of transaction (re)ordering in a smart contract capable cryptocurrency, is greater than for a cryptocurrency platform which does not support smart contracts. Conversely, the censorship of undesired transactions is easier to define programmatically in an UTXO based model, as there can only be one transaction spending a certain unspent output, compared to a smart contract capable cryptocurrency where a transaction to a smart contract can be routed through several layers of contract invocations. Therefore, influence methods such as transaction ordering and exclusion have variable impact depending on the targeted platform. Similarly, the ability to invalidate a transaction can result from successfully performing one or more of the above transaction manipulation types. Thereby, the definition of "invalid" depends on the underlying cryptocurrency and is different for UTXO and account-based models. For example, to invalidate a transaction to a smart contract in Ethereum two approaches exists: Either a transaction is not accepted because a transaction with the same nonce was already included in Ethereum, or the transaction throws an exception during execution because it operates on a (unexpectedly) changed state. The first would be a result of transaction revision, while the later can happen because of a change in the order of invoked transactions or the exclusion of a previous transaction.

Some AIM attacks may allow multiple types of transaction manipulation at the same time, while others are specifically constructed to support only one method (see Table 1). Depending on the state of the targeted transaction(s) (proposed, confirmed, settled) the attack might vary in cost and in the required level of interference with consensus.

4.3   Required Interference with Consensus

While the previous classification of transaction manipulation attacks describes the intended impact, here we consider the *required interference* with consensus by which they can be achieved. Specifically, we introduce three different fork requirements:

- **Deep-fork required**, where a fork with depth of at least ℓ exceeding a security parameter k_V is necessary (i.e., $\ell > k_V$). The victim defines k_V [26,54] and it refers to its required number of confirmation blocks for accepting transactions[6]. In other words, the victim indirectly defines the required minimum fork length ℓ by his choice of k_V.
- **Near-fork required**, where the required fork depth is *not* dependent on k_V, but forks might be required. In other words, the attacker defines the gap k_{gap} (which can be smaller than k_V) he wants to overcome.[7]

---

[6] We emphasize that each transaction has a recipient (and thus a potential victim with an individual k_V), in practice there is no global security parameter k which holds for all transactions.

[7] The length of k_{gap} also depends on the attacker's resources and willingness to succeed (e.g., to exclude a certain block).

- **No-fork required**, where no blockchain reorganization is necessary at all (i.e., $\ell = 0$).

The required interference with consensus specifies the chain reorganization needed. A classical double-spending attack scenario [49, 54] can be considered as a *transaction revision* attempt in which a single attacker aims at producing a longer chain (possibly in secret [24, 52]) than the main chain to revert one (or possibly many) of *his own* transactions. Therefore, this attack requires *deep forks* ($\ell > k$) to reorganize the chain. Since the classic case attacker is assumed to have full control over the required hashrate to perform the attack, he can also arbitrarily order and exclude transactions from the longest chain. Clearly, an attacker with more than 50% of the hashrate is able to eventually produce a longer chain with probability one and thus can revert/undo any transaction and permanently perform all four kinds of transaction manipulation attacks by providing a longer chain.[8]

No-fork attacks distinguish themselves from the other two categories by aiming to manipulate miner's *block proposals* rather than (preliminary) consensus *decisions*, i.e., already mined blocks. In the context of PoW cryptocurrencies, manipulating a miner's block proposal means influencing the input block used for finding and adding a valid PoW. Deep- and near-fork attacks seek to undo state-updates to the ledger that are already confirmed by subsequent PoW.

4.4 Used Payment Method

AIM attacks either pay for compliant behaviour, or they penalize for non-compliant actions. How this mechanism is set up depends on the attack in question, but there are three general methods that differ in which currency is used for the payment.

- **In-band payment:** The payment is performed in the target cryptocurrency. Most early bribing attacks where designed to gain in-band profits, like for example checklocktime bribes [14], whale transactions [42] or history revision contracts in Ethereum [44].
- **Out-of-band payment:** The payment is performed in another currency, the so-called funding cryptocurrency. Some AIM attacks which utilize out-of-band funding where designed as Goldfinger attacks, like for example Goldfinger-Con [44] and Pitchforks [31]. Others can be executed as Goldfinger attack, or with the goal to gain in-band profits, like for example [56], or the out-of-band variants of P2W attacks [32]. This highlights that AIM attacks which are intended to destroy a Cryptocurrency, i.e., perform a Goldfinger attack, inherently requite methods of out-of-band funding.
- **Threat:** No direct payment is performed, but a credible threat is constructed that non-compliant behaviour could lead to losses [11, 46].

---

[8] Actually the heaviest chain by PoW, e.g., in Bitcoin measured in difficulty periods.

5   Classification of Existing AIM Approaches

Equipped with our generalized attack model and the classification by state of and intended impact on transactions as well as the resulting required interference with consensus, we now inspect and compare existing AIM attacks within this section. Table 1 presents an overview of our systematization of existing proposals. Each row represents a different attack (in chronological order of their release) and columns outline respective properties.

5.1   Impact on Transactions

The different ways of how AIM attacks can have an impact on transactions are outlined in Sect. 4.2.

Tx Revision: In the first bribing attack, proposed by Bonneau [14], the use of *lock time transactions* is suggested, which are only valid on the attacker's chain, but there they can be claimed by anyone (anyone-can-spend outputs). Miners are hence expected to be incentivized to mine blocks on the attacker's chain to collect these bribes as inputs in new transactions included in their new blocks. As a by-product one transaction per new block is triggered to claim the anyone-can-spend output. Therefore, transaction triggering is technically achieved, but set into parenthesis as it is not the main intent of the attack. A variation of the check-locktime bribes which does not trigger additional transactions was proposed by Liao and Katz [42] and uses high fee transactions (*whale transactions*) to provide incentives for miners to join the attack. In [44] they proposed a smart contract (`HistoryRevisionCon`) which pays additional in-band rewards to miners of the attacker's desired Ethereum chain branch, iff the effects of the double-spending transaction have occurred on this branch. Strictly speaking, this attack also triggers transactions as the promised rewards have to be claimed by the bribees from the smart contract. The mentioned attacks ([14,42,44]) rely on in-band payments and are designed to replace or *revise* a specific transaction, i.e., perform a single double-spend. As a consequence, they do not consider defining the order or exclusion of arbitrary transactions. Except for the double-spending transaction itself, the block content of subsequent blocks can freely be defined by the bribed miners. Thus – if not explicitly considered – also the blocks produced by the bribed miners will not be fully under control of the adversary. Therefore, it would be possible for such miners to also perform a double-spend of one of their transactions for free, by piggybacking on the attack financed by the original attacker.

Table 1. Comparison of existing AIM approaches on cryptocurrencies in *chronological order* according to their appearance. A property is marked with ✓ if it is achieved and with ✗ otherwise, — is used if a property does not apply. If the symbol is within brackets, e.g., (✓), this means that this property is achieved (or can be augmented), but this was initially not discussed or considered by the authors. ~ means that the property cannot be clearly mapped to any of the previously defined categories without further details or discussion which is given in the textual description. ⋆ means that this attack aims against mining pools and hence is not intended to manipulate the content of the blockchain. † means that the paper does not explicitly specify the out-of-band payment method but assumes its correctness.

	Tx rev.	Tx ord.	Tx excl.	Tx trig.	Required interference with consensus	Attacker hashrate p_S	Rational hashrate p_R	Distracts hashrate	Requires smart contract	Payment	Trustless for attacker	Trustless for collaborator	Subsidy	Compensates if attack fails
Bribery [18]	✓	✗	✗	(✓)	Deep fork	≈ $(0, \frac{1}{2})$	≈ $[\frac{1}{2}, 1)$	✗	✗	in-band	✓	?	✗	✗
Dark side attack [40]	✓	(✓)	(✓)	✗	Deep fork	≈ $(0, \frac{1}{2})$	≈ $[0, 1)$	✗	✗	in-band	✗	✓	✗	✗
Feather-forks [11]	✗	✓	✓	✗	Near-/No forks	≈ $(0, \frac{1}{2})$	≈ $[\frac{1}{2} - 1)$	✗	✗	threat	—	—	—	—
Checklocktime bribes [14]	✓	✗	✗	(✓)	Deep fork	✗	≈ $[\frac{1}{2}, 1]$	✗	✗	in-band	✓	?	✗	✗
Negative fee miningpool [14]	(✓)	(✓)	✓	✗	Near-/No-/Deep forks	✗	≈ $[\frac{1}{2}, 1]$	✗	✗	out-of-band	✗	✗	✓	✓
Script puzzle double-spend [56]	✓	(✓)	✓	(✓)	Deep fork	$(0, \frac{1}{2})$	$1 - p_S$	✗	✗	in-band	?	✗	✗	?
Script puzzle 38.2% attack [56]	✗	(✓)	✓	?†	Near-/No forks	$[0.382, \frac{1}{2})$	$1 - p_S$	✗	?†	out-of-band	?†	?†	✓	✓
Whale transactions [42]	✓	✗	✗	✗	Deep fork	$(0, \frac{1}{2})$	$1 - p_S$	✗	✗	in-band	✓	?	✗	✗
Proof-of-stale blocks [43,59]	—⋆	—⋆	—⋆	(✓)	—⋆	✗	$[0,1]$	✓	✓	out-of-band	?	?	✓	?
Fomo3D game [4]	—	✓	✓	✓	No fork	✗	$[0,1]$	✓	✓	in-band	✓	✓	✗	✗
CensorshipCon [44]	✗	(✓)	✓	(✓)	Near-/No forks	$[\frac{1}{3}, \frac{1}{2})$	$(\frac{1}{3}, \frac{2}{3})$	✓	✓	in-band	?	✓	✓	✗
HistoryRevisionCon [44]	✓	✗	✗	(✓)	Deep fork	✗	≈ $[\frac{1}{2}, 1]$	✗	✓	in-band	✓	?	✓	✗
GoldfingerCon [44]	—	—	✓all	(✓)	No fork	✗	≈ $[\frac{1}{2}, 1]$	✗	✓	out-of-band	✓	✓	✓	✓
Race to the door [15]	—	—	—	—	No fork	✗	$[0,1]$	✗	✓	o.o.-band/threat	?	?	✗	?
Pitchforks [31]	—	—	✓all	✗	No fork	✗	$(\frac{1}{3}, 1]$	?	✗	out-of-band	✓	✓	✓	✗
Front-running [19,22]	✓	✓	✗	(✓)	Near-/No fork	✗	$(0,1]$	✗	✗	in-band	✗	✓	✓	✓
Pay per miner censorship [61]	✗	✓	✓	—	No fork	✗	1	✗	✓	in-band	?	?	✓	✓
Pay per block censorship [61]	✗	✓	✓	—	No fork	✗	1	✗	✓	in-band	?	?	✓	✓
Pay per commit censorship [61]	✗	✓	✓	—	Near-/No fork	✗	1	✗	✓	in-band	?	?	✗	✗
P2W Tx Excl. & Ord [32]	✓	✓	✓	(✓)	Near-/No fork	✗	$[\frac{1}{2}, 1]$	✗	✓	out-of-band	✓	✓	✓	✓
P2W Tx Rev. & Excl. & Ord. [32]	✓	✓	✓	(✓)	Deep fork	✗	$[\frac{1}{2}, 1]$	✗	✓	out-of-band	✓	✓	✓	✓
P2W Tx Ord. (in-band) [32]	✗	✓	✗	(✓)	No fork	✗	$(0,1]$	✗	✓	out-of-band	✓	✓	✓	✓
P2W Tx Excl. (in-band) [32]	✗	✓	(✓)	(✓)	Near-/No fork	✗	$[\frac{1}{2}, 1]$	✗	✓	in-band	✓	✓	✗	✗
BDos [46]	—	✓all	✓all	✗	Near-/No fork	≈ $[0.21, \frac{1}{2})$ ($for\ BTC$)	1	✗/✓	✗	threat	✓	—	—	—
HTLC bribing [37,58]	✗	✓	✗	✗	Near-/No fork	✗	1	✗	✗	in-band	✓	?	✗	✗

Tx Exclusion: There is one notable exception which was specifically designed to exclude transactions: `CensorshipCon` [44] rewards mining uncle blocks to distract the hashrate of bribable miners, which in turn enables the attacker to overtake the Ethereum blockchain s.t., blocks exclusively come from the attacker. Since this attack is in-band, it only works in Ethereum and relies on the uncle block reward scheme of Ethereum to subsidise the attack, i.e., reduce the value of the required bribes. To succeed, it requires that the hashrate of the attacker is larger than $\frac{1}{3}$ and the hashrate of the bribable miners to be between $[\frac{1}{3}, \frac{2}{3})$. If the attack is successful it allows for arbitrary transaction ordering as well and thus also for arbitrary transaction exclusion, as all blocks appended to the main chain during the attack come from the attacker.

`GoldfingerCon` [44] can be seen as a special case of the transaction exclusion attack which rewards Bitcoin miners for mining empty blocks with the help of an Ethereum smart contract. In this case, all transactions are excluded to reduce the utility of the respective cryptocurrency for all its users. So called *Goldfinger attacks* have been first described by Kroll et al. [38], but Goldfinger-Con was the first practical instantiation. The name is derived from the James Bond movie villain Goldfinger, who seeks to destroy the gold reserves stored in Fort Knox to increase the value of his own holdings. An important aspect of Goldfinger attacks is that the payments have to be performed out-of-band since, if successful, the value of the targeted cryptocurrency is intended to drop. Similarly, *Pitchforks* [31] leverage merged mining [33] to subsidize the creation of empty (or specially crafted) blocks in the attacked parent chain [31]. As with all Goldfinger-style attacks, the attacker is required to achieve utility outside of the cryptocurrency economy he wants to attack [38]. In case of the Pitchfork attack, the external utility comes from a hard-fork, which creates a new cryptocurrency. In this new cryptocurrency, the merge-mined PoW consists of blocks which attack the forked parent cryptocurrency, e.g., are empty. As the hashrate is repurposed in this case, it is technically not directed anywhere else i.e., not distracted.

Distracted hashrate is redirected from the valid tip(s) of the attacked blockchain to some other form of puzzle, or alternative branch, that does not contribute to state transitions of the targeted cryptocurrency. The *Script puzzle* 38.2% [56] and `CensorshipCon` attack [44] distract hashrate of bribable miners to gain an advantage over the remaining honest miners. The former redirects the hashrate from the main chain towards puzzles which promise more rewards than honest mining, the later rewards uncle block mining in Ethereum. The goal of both attacks is that the attacker gains the majority of the hashrate in the respective main chain, and he can hence arbitrarily exclude, or order transactions. Although, the attack does not explicitly aim to allow the specific ordering of certain transactions, this capability is achieved as a by-product. Neither attack is reverting blocks to change history, which is a different scenario and requires further analysis in this context, as reverting blocks would change the incentives of miners which have produced them.

Also in *Script Puzzle double-spend* [56] PoW like puzzles, offering in-band rewards, are published within the respective cryptocurrency with the intent to distract the hashrate of rational miners. Using the gained advantage to overtake the main chain requires attacker hashrate. Again, transaction ordering comes as a by-product and was not an explicit design goal, but theoretically this is the only existing attack utilizing *in-band* payments, which can achieve the three properties: revision, ordering and exclusion. Although, upon successful execution rational miners are deprived of their bribes as the previously hidden attack chain becomes the longest chain and does not pay the promised puzzle rewards. This renders the attack non-repeatable against rational miners.

Tx rev./ord./excl.: There are only two proposed attack methods which achieve these three properties in an out-of-band payment scenario: *negative-fee mining pools* [14] and P2W Tx Rev. & Excl. & Ord. [32]. A negative-fee mining pool is like a classic mining pool, except that it pays out an above-market return. *"Because such a pool would lose money on expectation, no honest pool should be able to match this reward"* [14]. As with most classic mining pools[9] the pool operator can define the content of a block proposal and hence forge arbitrary attack blocks. Even if miners are rational and hence willing to actively participate in such operations, this approach has at least two major limitations: First, miners would still have to trust the pool owner to pay out the promised rewards. Second, miners could report only solutions which are below the current difficulty target (shares) to prove that they are working for the pool, but withhold blocks which actually match the difficulty target. Thereby, they would potentially gain profits by pretending to participate in the attack/pool without actually doing so. This *miner's dilemma* is a general problem for mining pools [23].

The smart contract design presented in [32] resolves the limitations of negative-fee mining pools by automating the payment of bribes to complacent miners without requiring any further interaction of the attacker. Thereby, the attacker publishes block templates to the smart contract and offers a bribe for the first miner who can provide a valid PoW solution for such a template. As only payments for valid PoW solutions are provided by the smart contract, it is ensured that the actions of bribees are specifically targeted to aid the attacker. If the attacker deems that the ongoing attack is not likely to succeed, he can stop the investment of further funds by not publishing any further block templates.

Tx Triggering: The are only two existing AIM techniques, which are intended to trigger transactions: The *Fomo3D game* [4] and the *race to the door* Goldfinger attack sketched by Bonneau [15]. In an race to the door, the attacker *"credibly commits"* to buy out half of all funds present in the targeted cryptocurrency, to utilize them for destroying the system. Therefore, the price the attacker has to pay for those funds is likely to drop the more users decide to sell, increasing the likelihood of the attack to succeed. This creates a vicious cycle, resulting in a race to the door. The idea was not presented in great detail and mainly discussed in

---

[9] In P2Pool for example, there is no single operator which can define the content of a block proposal.

context of overtaking PoS/PoW cryptocurrencies, but of course such an attack would also trigger sell transactions. Moreover, there are plenty of ways to attack the value of a cryptocurrency while holding substantial amounts of it that are left unexplored.

There are multiple variants of Fomo3D, but roughly the rules are as follows. In this game, which is open for everybody, the last account which has purchased a ticket wins when a timer goes to zero and every purchase again increases the timer by 30 s. This leads to the situation that transactions are triggered by rational players as soon as the timer gets close to zero. It was conjectured that the game would never end, but in august 2018 the first round of the game ended and the winner collected $10,469$ Ether ($\approx$ \$2.1M USD at that time)[10]. It can be argued that a single instance of this game does not qualify as an "attack", but the same concept of presumably "free money" available to grab from a smart contract can also be used as an attack method (see our discussion in Sect. 7). The interesting aspect about these tx triggering attacks is, that they have effects for any hashrate of rational miners as long as there are rational clients. Even if $p_{\mathcal{R}} = 0$, rational clients in the network will issue transactions.

Tx Ordering: Dedicated ordering attacks, like front-running [19,22], P2W Tx Excl.& Ord., or P2W Tx Ord. (in-band) [32], target unconfirmed or confirmed transactions and therefore are cheaper as their interference with consensus is less severe.

5.2 Required Interference with Consensus

The concept of required interference with consensus is outlined in Sect. 4.3 and classifies if an attack can be realized without, with a near- or with a deep-fork. Depending on the scenario and the desired attack outcome, e.g., if only ordering is relevant, deep forks are not necessarily required. For example if, the victim accepts unconfirmed transactions, transaction revision can happen without any fork by simply updating the transaction. Bitcoin [2] as well as Ethereum allow something like *replace-by-fee* i.e., if there is a transaction signed by the same sender with the same nonce but a significantly higher gas value [3], the transaction with the higher gas value replaces the original one in certain clients. This circumstance is also used in the context of *front-running* [19,22]. But front-running is only a subset of possible (re-)ordering attacks, as it might be desirable to place a transactions more accurately in between two other transaction, e.g., as required for exploiting the `BlockKing` contract [53].

Prior to 2018, ordering attacks on smart contract cryptocurrencies have not been intensively studied [34,53]. This has recently changed as order fairness has been exposed as a fundamental issue in leader based consensus protocols [19,35,39]. In context of Nakamoto consensus, every miner that is capable of producing blocks can define the order of the transactions in his blocks. This

---

[10] The winner flooded the network with unrelated high gas transactions to custom smart contracts which congested the network blocking other "last" payments to the game.

circumstance alone can be used to gain an advantage in certain scenarios e.g., where transactions race against each other to collect something that is claimable by everybody like an *anyone-can-spend* transaction, the reward of a puzzle, or arbitrage[11]. But when rational actors are assumed, there are also scenarios where the ordering of transactions can be manipulated by attackers which are not necessarily miners themself, but have funds at their disposal to launch incentive attacks. In classical front-running miners are incentivized to prioritize transactions because they carry a larger fee. This however is not a consensus rule and thus lacks enforcement, as transaction with the highest fee can still be included at the end of a block, resulting in an all-pay auction [19]. In [32] an in-band as well as an out-of-band AIM attack is proposed, which allow arbitrary transaction ordering while only paying if the desired ordering is observed. Both attacks can be executed without any hashrate assuming rational miners.

5.3   Required Hashrate

Required attacker hashrate. p_B specifies how much hashrate is required to be under direct control of the attack (without considering the effects of AIM) for the attack to be successful. As observable in Table 1 there are three attacks which require $p_B > 0$. The Script Puzzle 38.2% attack is designed to overtake the blockchain entirely by offering alternative script puzzles with higher rewards to distract the hashrate of rational miners. This allows an adversary with appropriate hashrate to establish a computational majority and gain a net profit without considering double-spending attacks. In Script Puzzle double-spend the adversary has no explicit minimum hashrate requirement, however low hashrate has to be compensated with more puzzle funds. Moreover, it is designed as a single-shot double-spending attack that, if successful, deprives rational miners of their bribes. `CensorshipCon` uses a smart contract to offer in-band bribes for mining uncle blocks to distract hashrate. Thus, it requires attacker hashrate to include uncle blocks from rational miners in the main chain. Since it has to include all mined uncle blocks, it requires the hashrate of the attacker to be larger than $\frac{1}{3}$ and the hashrate of the bribable miners to be between $[\frac{1}{3}, \frac{2}{3})$.

It makes sense to bound the attacker hashrate below $\frac{1}{2}$ since otherwise the attacker has no need to perform bribing attacks as he could overtake the chain single handedly.

Required rational hashrate. p_R specifies how much hashrate is requited to be under control of rational miners for the attack to have a chance to succeed as described and evaluated in the respective paper. Generally, all bribing attacks have to assume that at least some of the miners are rational and hence bribable. Generally, it makes sense to assume that more than half of the miners are rational s.t. attacks have at realistic change to win longer block races. Both Script Puzzle

---

[11] Interestingly the problem of racing transaction was known very early on in the cryptocurrency community, which lead to the first fork of Bitcoin, i.e., Namecoin [1, 33], which introduced a commit reveal scheme to prevent races while registering domain names on the blockchain.

attacks require all miners to be rational, i.e., $p_{\mathcal{B}} + p_{\mathcal{R}} = 1$, as well as the Pay per ... attacks ($p_{\mathcal{R}} = 1$).

However, the attacks observed in practise provide no guarantees for the attacker that the desired ordering is achieved even if the highest transaction fee has been paid as the resulting game is an all pay auction [19].

5.4   Payment Method

This specifies where the payments to the bribees are performed (see 4.4). It can be argued that miners will try not to harm the value of their own cryptocurrency holdings by accepting in-band bribes, hence out-of-band AIM are of particular interest. **Subsidy** means that the attack leverages some characteristic of the cryptocurrency, or the environment to become cheaper. In case of Censorship-Con the rewards from uncle blocks are used to subsidize the attack, whereas in Pitchforks the additional income from merged mining is used as an incentive.

Compensates if attack fails refers to the property that at least a portion of the bribe is paid irrespective of the outcome. To successfully engage rational miners, attacks such as Checklocktime bribes [14], Whale Transactions [42] and HistoryRevisionCon [44], must pay high rewards in case of success to compensate the financial risk faced by bribees if the attack fails despite of their participation. So far the only attack which facilitates transaction revision that achieves this property is [32]. Script Puzzle double-spend defrauds the bribed miners if successful and hence actually only pays out rewards if it fails. In front-running attacks, high transaction fees are usually incurred even if the desired ordering effect is not achieved. Thus, in this case it is also an undesirable property for the attacker. The same holds true for negative-fee mining pools as rewards have to be paid for performed work even if no attack block fulfilling the difficulty target has been submitted by a miner.

5.5   Trustlessness

Trustless for attacker specifies if the attack itself can be exploited by allowing collaborating/bribed miners to profit without adhering to the attack. For example, Script Puzzle attacks require some form of freshness guarantee to prevent bribees from intentionally waiting until the attack fails before computing puzzle solutions to obtain rewards. It is also possible to claim rewards for stale honest blocks that are later on submitted as uncles to the CensorshipCon. Also in naive front-running attacks the attacker has no guarantee that the desired ordering will be achieved by paying a high fee. The Pay per ... attacks are only modelled theoretically without providing concrete instantiation. Therefore, it cannot be evaluated in this regard.

Trustless for collaborator specifies if bribees have to trust the attacker that they will receive their payments, if they adhere to the attack. In Checklocktime bribes a lock time on individual transaction outputs intends to ensure that they cannot be spent before a particular block height, even by the creator. This

ensures that at each height a locked output is released and split into a anyone-can-spend and another locked output. However, the holder of the associated private key can cheat, by creating a conflicting/racing transaction, which also becomes valid after the intended lock time has passed. This conflicting transaction, transfers the whole output back to the owner without an additional anyone-can-spend output. However, this attempt is only possible if the attacker is under control of some hashrate $p_B > 0$, as a miner would never prefer this transaction before the other. The same holds true for Whale Transactions, or HTLC bribes since the attacker has to provide new high fee transactions for each block on the attack chain at each step of the attack. While HistoryRevisionCon does not explicitly consider trustlessness for collaborating miners, an augmentation is possible[12], CensorshipCon requires that the attacker includes blocks produced by collaborating miners as uncle blocks and thus is not trustless. The Script Puzzle double-spend attack is designed as a one-shot attack that defrauds collaborators. The Script Puzzle 38.2% attack does not specify how payments are performed and assumes a working trustless out-of-band payment method.

6 Costs, Profits and Extractable Value

In this section we want to highlight the challenges of comparing existing AIM attacks with respect to their costs and potential profits.

First of all, the presented attacks differ significantly with respect to their system- and attack models, which have diverse goals regarding their intended influence on transactions (revision, ordering, exclusion, triggering), as well as varying assumptions regarding the capabilities of the attacker, e.g., hashrate and funds.

Second, not all existing proposals have analyzed the involved costs and gains in a comparable way. Attacks such as the Script Puzzle double-spend or CensorshipCon express the required funds in terms of the hashrate which is also required to successfully execute it [44,56]. For transaction revision using Whale Transaction or P2W attacks [32,42] concrete values are provided while at the same time no hashrate is required. In GoldfingerCon [44] only the costs of invoking the smart contract are provided.

Costs: What stands out in the comparison of costs is that: i) Attacks which compensate collaborating rational miners even if the attack fails are cheaper. The reason for this is that such attacks do not have to provide high bribes to account for the risks faced by bribees if the attack is unsuccessful [32,61]. ii) Attacks which exclusively focus on transaction exclusion or (re)ordering of unconfirmed transactions are substantially cheaper as they only compete with the fee, i.e., extractable value, of the transaction(s) in conflict [19,32,37,58,61].

---

[12] The issue stems from the fact that the bribing contract checks the balance of the Ethereum account which should receive the bribing funds before issuing any bribes, but without any additional locking constraints these funds can be moved by the attacker once received.

Profit: To calculate the profit of the attack it is important to estimate the costs as well as the extractable value. In this context, the term *miner extractable value* [19] has been coined to describe the value which can be extracted by a miner by including a certain transaction in terms of fees or guaranteed profits through token arbitrage. In relation to other AIM attacks surveyed in this paper, this leads to an interesting observation: We argue that the extractable value of a transaction for a certain party can not readily be determined by exclusively looking at the cryptocurrency system in which this transaction is to be performed. The reason is that there might be additional protocols like colored coins [50] or out-of-band payments from AIM attacks at play, which can influence the (miner) extractable value of a given transaction. This is an instantiation of a more general observation that game-theoretic analysis is not composable.

The question whether AIM is profitable can be summarized by comparing the extractable value as well as the costs of the attack and the behaviour intended by the protocol designer. The following simplified equation was adapted from Böhme [7].

$$\text{EV}(\text{attack}) - costs_{\text{attack}} > \text{EV}(\text{follow protocol}) - costs_{\text{follow protocol}}$$

Let's assume two unconfirmed, but conflicting Bitcoin transactions (tx_1, tx_2) are competing for a place in the next block. If the extractable fee of one transaction is greater than for the other $\text{Fee}(tx_1) > \text{Fee}(tx_2)$, it would be rational from the miner to include tx_1, since $\text{EV}(tx_1) - \text{Fee}(tx_1)$. But if there is a side payment, due to an AIM attack on a different funding cryptocurrency (e.g., Ethereuem) for including tx_2, which leads to the situation that $\text{EV}(tx_2) > \text{EV}(tx_1)$, then the situation for the rational miner changes. In this case, the reason for the change is not directly visible in Bitcoin.

The question whether it is possible to upper-bound the extractable value was also touched by Budish [17] in a different setting and from the perspective of double-spending attacks only. Under a simplified model, the extractable value of a double-spend is the transferred value of coins[13]. To calculate the required rewards and fees for making double-spending attacks economically unattractive, the author assumed that in the worst case every transaction in a block is potentially up for double-spending and highlights that the relation between reward and fees, compared to the value transferred in Bitcoin makes such attacks economically feasible in theory. An instantiation of an attack in which every transaction of a block can theoretically become a target for double-spending, has been proposed in [32], where a crowdfunded attack is described, utilizing smart-contracts. The goal is to distribute the costs of multiple double-spend attempts in the same block to the set of transacting entities.

By these examples, we see that it is hard or even impossible to accurately bound the extractable value of transactions (and thus blocks) in a multi cryptocurrency ecosystem by solely looking at data from one cryptocurrency. A related meta argument was presented in [25].

---

[13] The dependency between transaction value and confirmation time k_V, is also discussed in [54].

7   Discussion

We finally discuss the relation of AIM to other ways of gaining capacity in Nakamoto consensus, as well as highlight open questions and directions for future work in this area.

Relationship of AIM to Other Ways of Gaining Capacity: In the paper [15] an excellent classification of different methods on how to gain capacity in Nakamoto consensus is provided. These methods are separated into: *rent*, *build*, *bribe* and *buy out*. Hereby, rent, buy out as well as build refer to classical methods of renting hardware, buying cryptocurrency units at exchanges, or building new datacenters for mining. We augment this classification and argue that AIM can be used to construct algorithmic ways for all these methods. Table 2 depicts an augmented version from [15] showing the different methods of how to obtain capacity in Nakamoto consensus.

According to the original classification in [15], bribing is a *temporary* attack, which utilizes *existing* resources of miners. If the terms *new* and *existing*, in the context of PoW capacity, are to be interpreted from the perspective of the targeted system, then some existing attacks which rely on out-of-band payments would also classify as *rent*. The reason for this is: They are also able to attract new capacity currently bound in other cryptocurrencies which utilize the same PoW algorithm, like [31,32,56]. Capacity which is present in a different cryptocurrency is also new to the targeted cryptocurrency if miners decide to switch for supporting an attack.

We further argue that *buy out* attacks can theoretically be done algorithmically using cross-chain atomic swaps [30] (or any other blockchain interlinking protocol). A *race to the door* style attack [15] in combination with cross-chain atomic swaps can be imagined to perform Goldfinger style attacks on smart contract capable PoW cryptocurrencies. Hereby, out-of-band payments are used to buy out cryptocurrency units, through a smart contract, which is going to use these previously bought cryptocurrency assets to perform a denial of service attack by dumping the previously bought crypto assets on the market as freely available for anyone to claim after a certain timeout. If there is a limit for what is claimable per transaction, as well as the requirement of a high fee, this on-chain faucet construction will trigger a flood of transactions as soon as the timeout is reached. In this case, existing funds are bought and permanently redistributed with the intent to perform a denial-of-service attack and at the same time collapse the market due to increased supply.

It remains to be shown that it is theoretically possible to build *permanent* AIM attacks. Arguably, any Goldfinger attack, such as GoldfingerCon [44], which creates enough external utility to refuel the attack, can in theory be constructed in a way to run permanently. Although, it is unlikely that a Goldfinger attack has to be continued infinitely long if the intended effects have already occurred. An attack which also discusses its perpetuity is the Script puzzle 38.2% attack. In this case the attack can also theoretically be used to permanently overtake the chain by supplying puzzles that provide out-of-band reward and thereby overtake

the original blockchain with 38.2% of the total hashrate. Also, the pitchfork [31], in which the additional revenue stream to sustain the attack comes from a fork of the targeted cryptocurrency and not from a previously determined bribing fund, can in theory be sustained infinitely long. Whether the attack can be sustained depends on the value of the newly generated cryptocurrency. An interesting analogy exists between any permanent AIM attack and a cryptocurrency itself. From the perspective of a miner who exclusively mines on puzzles for any of these three permanent attacks, there is no difference to mining on any other PoW based cryptocurrency other than the format of the associated PoW.

Mitigation and Counter Attacks: The presented systematization has a very attack centric view on the issue at hand. This is due to the selection of papers, which almost all have a very attack-focused viewpoint. Therefore, counter measures and counter attacks are often omitted in these papers, or not discussed to a great extent.

Nevertheless, for the victim(s) counter bribing might be a viable strategy against AIM. The difficulty of successfully executing counter bribing highly depends on the respective scenario. In the end, counter bribing can also be countered by counter-counter bribing and so forth. Therefore, as soon as this route is taken, the result becomes a bidding game. Against transaction exclusion attacks, counter bribing can be performed by increasing the fee of the transaction to be excluded such that it surpasses the value promised for not including the transaction. If defenders have imperfect information, they may not be able to immediately respond with counter bribes. In this case some of the attack chain blocks may have already been mined, or even take the lead, before they are recognized by defenders. Counter bribing then necessitates a fork, and thus a more expensive transaction revision attack, leading to asymmetric costs in the bidding game. This illustrates an important aspect of AIM, namely their visibility. On the one hand, sufficiently many rational miners of the targeted cryptocurrency have to recognize that an attack is occurring, otherwise they won't join in and the attack is likely to fail. On the other hand, if the victims of the attack recognize its existence, they can initiate and coordinate a counter bribing attack. So the optimal conditions for AIM arise if all rational miners have been informed directly about the attack, while all victims/merchants do not monitor the chain to check if an attack is going on and are not miners themselves. If the payments are made out-of-band, they are rendered more stealthy to victims who only monitor the targeted cryptocurrency. It can hence be argued that counter attacks by victims are harder to execute as they are not immediately aware of the bribing value that is being bet against them on a different funding cryptocurrency. We also follow the argument in [14] that requiring clients to monitor the chain and actively engage in counter bribing is undesirable, and out-of-band attacks further amplify this problem as clients would have to concurrently monitor a variety of cryptocurrencies.

To prevent repercussions, participating miners can make use of the fact that the PoW mining process itself does not require any strong identity by using different payout addresses. Of course their received rewards can be traced, but

available privacy techniques could be used to camouflage the real recipient of the funds, e.g., [29,45,51].

Situation in PoS: Since all considered attacks target PoW cryptocurrencies, the applicability of AIM on PoS cryptocurrencies is not sufficiently understood yet. It remains to be understood which techniques are transferable to PoS cryptocurrencies, and which additional mitigations (e.g., providing collateral, slashing) can increase the induced costs of attacks in this setting.

Rationality and Practicality of Attacks: All AIM attacks assume some form of rational behaviour of participants. In practise although, it is hard to define rational behaviour in a general way, as also the individual investments and the long term interests of miners play an important role. Although, there may be scenarios where miners are capable of providing PoW for a targeted cryptocurrency, but at the same time do not have any long-term interest in the well-being of the target. Consider the real-world example of Bitcoin and BitcoinCash which utilize the same form of PoW and can be considered rivals. Thus, the question if the proposed attacks are possible in practice is difficult to answer scientifically. There is already empirical evidence from previous large scale attacks by miners, especially on smaller cryptocurrencies as well as AIM attacks [5,6,8,9,19]. These cases demonstrate that large scale attacks happen and that the topic of incentives in cryptocurrencies is an area deserves further study.

Acknowledgements. We would like to thank the participants of the Dagstuhl Seminar 18152 (*Blockchains, Smart Contracts and Future Applications*), especially Samuel Christie and Sebastian Faust, as well as the participants of the Dagstuhl Seminar 18461 (*Blockchain Security at Scale*) for all the frutiful discussions.

This paper is based upon work partially supported by (1) the Christian-Doppler-Laboratory for Security and Quality Improvement in the Production System Lifecycle; The financial support by the Austrian Federal Ministry for Digital and Economic Affairs, the Nation Foundation for Research, Technology and Development and University of Vienna, Faculty of Computer Science, Security & Privacy Group is gratefully acknowledged; (2) SBA Research; the competence center SBA Research (SBA-K1) funded within the framework of COMET Competence Centers for Excellent Technologies by BMVIT, BMDW, and the federal state of Vienna, managed by the FFG; (3) the FFG Bridge 1 project 864738 PR4DLT. (4) the Israel Science Foundation (5) the Israel Cyber Bureau (6) the Technion Hiroshi Fujiwara cyber-security research center

A   Example Use of Our Classification Framework

Whether an attack is executable with or without a fork depends on the intended impact on transactions as well as on the state of the targeted transaction. For example, transaction revision where the victim accepts $k_V = 0$ (zero confirmations) may be executable as no-fork attacks. Other attacks, such as performing a *double spend* where the victim has been carefully chosen k_V [54], may require deep-forks because they need to substantially affect consensus and violate the

security assumption that the common prefix of the blockchain remains stable. Transaction exclusion (*censorship*) may require near-forks to exclude the latest blocks which include the respective transaction.

With our classification framework, we can map *front-running* [19,22,32] as an attack which aims to influence transaction ordering, while targeting unconfirmed transactions (state of targeted transactions). Compared to that, the so called *time-bandit attack* [19] also aims to influence transaction ordering, but targets confirmed or even agreed transactions. Note that strictly speaking a time-bandit attack is not AIM, as it does not incentivize other participants to aid the attack, but instead relies on "classic" methods like performing a rental attack to temporarily hold the majority of the hashrate.

B   Ways to gain capacity in Nakamoto Consensus

Table 2. Strategies to gain capacity in Nakamoto consensus according to [15], augmented with AIM strategies (colored background).

Source			Duration of control	
			Temporary	Permanent
	PoW	New	Rent	Build
			AIM	AIM
	PoW & PoS	Existing	Bribe	Buy out
			AIM	AIM

References

1. Namecoin: https://www.namecoin.org/. Accessed 15 Sept 2020
2. Replace by fee in bitcoin: https://en.bitcoin.it/wiki/Replace_by_fee. Accessed 23 Dec 2020
3. Replace by fee in openethereum: https://openethereum.github.io/Transactions-Queue.html. Accessed 23 Dec 2020
4. How the winner got Fomo3D prize - a detailed explanation. medium (2018). https://medium.com/coinmonks/how-the-winner-got-fomo3d-prize-a-detailed-explanation-b30a69b7813f. Accessed 15 Sept 2020
5. Bitcoin cash miners undo attacker's transactions with 51% attack'. coindesk (2019). https://www.coindesk.com/bitcoin-cash-miners-undo-attackers-transactions-with-51-attack. Accessed 15 Sept 2020
6. Ethereum classic 51% attack – the reality of proof-of-work. cointelegraph (2019). https://cointelegraph.com/news/ethereum-classic-51-attack-the-reality-of-proof-of-work. Accessed 15 Sep 2020
7. Talk: A primer on economics for cryptocurrencies. School of Blocks, Blockchain summer school at TU Wien (2019). https://bdlt.school/files/slides/talk-rainer-b%C3%B6hme-a-primer-on-economics-for-cryptocurrencies.pdf. Accessed 15 Sept 2020

8. Bitcoin gold (btg) was 51% attacked. github (2020), https://gist.github.com/metalicjames/71321570a105940529e709651d0a9765. Accessed 15 Sept 2020

9. Ethereum classic suffers second 51% attack in a week. coindesk (2020). https://www.coindesk.com/ethereum-classic-suffers-second-51-attack-in-a-week. Accessed 15 Sept 2020

10. Aiyer, A.S., Alvisi, L., Clement, A., Dahlin, M., Martin, J.P., Porth, C.: Bar fault tolerance for cooperative services. In: ACM SIGOPS Operating Systems Review, vol. 39, pp. 45–58. ACM (2005). http://www.dcc.fc.up.pt/~Ines/aulas/1314/SDM/papers/BAR%20Fault%20Tolerance%20for%20Cooperative%20Services%20-%20UIUC.pdf

11. socrates1024 (Andrew Miller): Feather-forks: enforcing a blacklist with sub-50

12. Badertscher, C., Garay, J., Maurer, U., Tschudi, D., Zikas, V.: But why does it work? A rational protocol design treatment of bitcoin. In: Nielsen, J.B., Rijmen, V. (eds.) EUROCRYPT 2018. LNCS, vol. 10821, pp. 34–65. Springer, Cham (2018). https://doi.org/10.1007/978-3-319-78375-8_2, https://eprint.iacr.org/2018/138.pdf

13. Badertscher, C., Maurer, U., Tschudi, D., Zikas, V.: Bitcoin as a transaction ledger: a composable treatment. In: Katz, J., Shacham, H. (eds.) CRYPTO 2017. LNCS, vol. 10401, pp. 324–356. Springer, Cham (2017). https://doi.org/10.1007/978-3-319-63688-7_11, https://eprint.iacr.org/2017/149.pdf

14. Bonneau, J.: Why buy when you can rent? Bribery attacks on bitcoin consensus. In: BITCOIN 2016: Proceedings of the 3rd Workshop on Bitcoin and Blockchain Research, February 2016. http://fc16.ifca.ai/bitcoin/papers/Bon16b.pdf

15. Bonneau, Joseph: Hostile Blockchain Takeovers (Short Paper). In: Zohar, A., et al. (eds.) FC 2018. LNCS, vol. 10958, pp. 92–100. Springer, Heidelberg (2019). https://doi.org/10.1007/978-3-662-58820-8_7, http://fc18.ifca.ai/bitcoin/papers/bitcoin18-final17.pdf

16. Bonneau, J., Miller, A., Clark, J., Narayanan, A., Kroll, J.A., Felten, E.W.: SoK: research perspectives and challenges for bitcoin and cryptocurrencies. In: IEEE Symposium on Security and Privacy (2015). http://www.ieee-security.org/TC/SP2015/papers-archived/6949a104.pdf

17. Budish, E.: The economic limits of bitcoin and the blockchain. Technical report, National Bureau of Economic Research (2018). https://faculty.chicagobooth.edu/eric.budish/research/Economic-Limits-Bitcoin-Blockchain.pdf

18. Cunicula: Bribery: The double double spend. Bitcoin Forum. https://bitcointalk.org/index.php?topic=122291. Accessed 31 Jan 2021

19. Daian, P., et al: Flash boys 2.0: frontrunning in decentralized exchanges, miner extractable value, and consensus instability. In: 2020 IEEE Symposium on Security and Privacy, SP 2020, San Francisco, CA, USA, 18–21 May 2020, pp. 910–927. IEEE (2020). https://doi.org/10.1109/SP40000.2020.00040, https://arxiv.org/pdf/1904.05234.pdf

20. Dembo, A., et al.: Everything is a race and nakamoto always wins (2020)

21. Dwork, C., Naor, M.: Pricing via processing or combatting junk mail. In: Brickell, E.F. (ed.) CRYPTO 1992. LNCS, vol. 740, pp. 139–147. Springer, Heidelberg (1993). https://doi.org/10.1007/3-540-48071-4_10, https://web.cs.dal.ca/~abrodsky/7301/readings/DwNa93.pdf

22. Eskandari, S., Moosavi, S., Clark, J.: Sok: transparent dishonesty: front-running attacks on blockchain. In: Bracciali, A., Clark, J., Pintore, F., Rønne, P.B., Sala, M. (eds.) Financial Cryptography and Data Security - FC 2019 International Workshops, VOTING and WTSC, St. Kitts, St. Kitts and Nevis, 18–22 February 2019, Revised Selected Papers. Lecture Notes in Computer Science, vol. 11599, pp. 170–189. Springer (2019). https://doi.org/10.1007/978-3-030-43725-1_13, https://arxiv.org/pdf/1902.05164.pdf

23. Eyal, I.: The miner's dilemma. In: 2015 IEEE Symposium on Security and Privacy (SP), pp. 89–103. IEEE (2015). http://arxiv.org/pdf/1411.7099

24. Eyal, I., Sirer, E.G.: Majority Is Not Enough: Bitcoin Mining Is Vulnerable. In: Christin, N., Safavi-Naini, R. (eds.) FC 2014. LNCS, vol. 8437, pp. 436–454. Springer, Heidelberg (2014). https://doi.org/10.1007/978-3-662-45472-5_28, http://arxiv.org/pdf/1311.0243

25. Ford, B., Böhme, R.: Rationality is Self-Defeating in Permissionless Systems (2019). https://arxiv.org/pdf/1910.08820.pdf, _eprint: arXiv:1910.08820

26. Garay, J.A., Kiayias, A., Leonardos, N.: The bitcoin backbone protocol: analysis and applications. In: Oswald, E., Fischlin, M. (eds.) Advances in Cryptology - EUROCRYPT 2015–34th Annual International Conference on the Theory and Applications of Cryptographic Techniques, 26–30 April 2015, Sofia, Bulgaria, Proceedings, Part II. Lecture Notes in Computer Science, vol. 9057, pp. 281–310. Springer (2015). https://doi.org/10.1007/978-3-662-46803-6_10, https://eprint.iacr.org/2014/765.pdf

27. Garay, J.A., Kiayias, A., Leonardos, N.: The bitcoin backbone protocol with chains of variable difficulty (2016). http://eprint.iacr.org/2016/1048.pdf. Accessed 6 Feb 2017

28. Gaži, P., Kiayias, A., Russell, A.: Tight consistency bounds for bitcoin. Cryptology ePrint Archive, Report 2020/661 (2020). https://eprint.iacr.org/2020/661

29. Heilman, E., Baldimtsi, F., Goldberg, S.: Blindly signed contracts: anonymous on-blockchain and off-blockchain bitcoin transactions. Cryptology ePrint Archive, Report 2016/056 (2016). https://eprint.iacr.org/2016/056.pdf. Accessed 3 Oct 2017

30. Herlihy, M.: Atomic cross-chain swaps. In: Newport, C., Keidar, I. (eds.) Proceedings of the 2018 ACM Symposium on Principles of Distributed Computing, PODC 2018, Egham, United Kingdom, 23–27 July 2018, pp. 245–254. ACM (2018). https://arxiv.org/pdf/1801.09515.pdf

31. Judmayer, A., Stifter, N., Schindler, P., Weippl, E.: Pitchforks in cryptocurrencies: enforcing rule changes through offensive forking- and consensus techniques (short paper). In: CBT 2018: Proceedings of the International Workshop on Cryptocurrencies and Blockchain Technology, September 2018. https://www.sba-research.org/wp-content/uploads/2018/09/judmayer2018pitchfork_2018-09-05.pdf

32. Judmayer, A., et al.: Pay to win: cheap, crowdfundable, cross-chain algorithmic incentive manipulation attacks on pow cryptocurrencies. Cryptology ePrint Archive, Report 2019/775 (2019). https://eprint.iacr.org/2019/775

33. Judmayer, A., Zamyatin, A., Stifter, N., Voyiatzis, A.G., Weippl, E.: Merged mining: Curse or cure? In: CBT 2017: Proceedings of the International Workshop on Cryptocurrencies and Blockchain Technology, September 2017. https://eprint.iacr.org/2017/791.pdf

34. Kalra, S., Goel, S., Dhawan, M., Sharma, S.: ZEUS: analyzing safety of smart contracts. In: 25th Annual Network and Distributed System Security Symposium, NDSS 2018, 18–21 February 2018, San Diego, California, USA. The Internet Society (2018). http://wp.internetsociety.org/ndss/wp-content/uploads/sites/25/2018/02/ndss2018_09-1_Kalra_paper.pdf

35. Kelkar, M., Zhang, F., Goldfeder, S., Juels, A.: Order-fairness for byzantine consensus. In: Micciancio, D., Ristenpart, T. (eds.) Advances in Cryptology - CRYPTO 2020–40th Annual International Cryptology Conference, CRYPTO 2020, Santa Barbara, 17–21 August 2020, CA, USA, Proceedings, Part III. Lecture Notes in Computer Science, vol. 12172, pp. 451–480. Springer (2020). https://doi.org/10.1007/978-3-030-56877-1_16, https://eprint.iacr.org/2020/269

36. Ketsdever, S., Fischer, M.J.: Incentives don't solve blockchain's problems (2019). https://arxiv.org/pdf/1905.04792.pdf

37. Khabbazian, M., Nadahalli, T., Wattenhofer, R.: Timelocked bribes. Cryptology ePrint Archive, Report 2020/774 (2020). https://eprint.iacr.org/2020/774

38. Kroll, J.A., Davey, I.C., Felten, E.W.: The economics of bitcoin mining, or bitcoin in the presence of adversaries. In: Proceedings of WEIS, vol. 2013, p. 11 (2013). https://pdfs.semanticscholar.org/c55a/6c95b869938b817ed3fe3ea482bc65a7206b.pdf

39. Kursawe, K.: Wendy, the good little fairness widget. IACR Cryptol. ePrint Arch. 2020, 885 (2020). https://eprint.iacr.org/2020/885

40. Lerner, S.D.: The bitcoin eternal choice for the dark side attack (ECDSA). https://bitslog.com/2013/06/26/the-bitcoin-eternal-choice-for-the-dark-side-attack-ecdsa/. Accessed 31 Jan 2021

41. Li, H.C., Clement, A., Wong, E.L., Napper, J., Roy, I., Alvisi, L., Dahlin, M.: Bar gossip. In: Proceedings of the 7th symposium on Operating systems design and implementation. pp. 191–204. USENIX Association (2006), http://www.cs.utexas.edu/users/dahlin/papers/bar-gossip-apr-2006.pdf

42. Liao, K., Katz, J.: Incentivizing blockchain forks via whale transactions. In: Brenner, M., et al. (eds.) FC 2017. LNCS, vol. 10323, pp. 264–279. Springer, Cham (2017). https://doi.org/10.1007/978-3-319-70278-0_17

43. Luu, L., Velner, Y., Teutsch, J., Saxena, P.: SmartPool: practical decentralized pooled mining. In: Kirda, E., Ristenpart, T. (eds.) 26th USENIX Security Symposium, USENIX Security 2017, Vancouver, BC, Canada, 16–18 August 2017, pp. 1409–1426. USENIX Association (2017). http://eprint.iacr.org/2017/019.pdf

44. McCorry, P., Hicks, A., Meiklejohn, S.: Smart contracts for bribing miners. In: Zohar, A., et al. (eds.) FC 2018. LNCS, vol. 10958, pp. 3–18. Springer, Heidelberg (2019). https://doi.org/10.1007/978-3-662-58820-8_1, http://fc18.ifca.ai/bitcoin/papers/bitcoin18-final14.pdf

45. Meiklejohn, S., Mercer, R.: Möbius: trustless tumbling for transaction privacy. Proc. Priv. Enhancing Technol. 2018(2), 105–121 (2018). https://doi.org/10.1515/popets-2018-0015, http://eprint.iacr.org/2017/881.pdf

46. Mirkin, M., Ji, Y., Pang, J., Klages-Mundt, A., Eyal, I., Juels, A.: BDoS: blockchain denial-of-service. In: Proceedings of the 2020 ACM SIGSAC conference on Computer and Communications Security, pp. 601–619 (2020)

47. Nakamoto, S.: Bitcoin: A peer-to-peer electronic cash system, December 2008. https://bitcoin.org/bitcoin.pdf. Accessed 1 Jul 2015

48. Pass, R., Seeman, L., Shelat, A.: Analysis of the blockchain protocol in asynchronous networks. In: Coron, J., Nielsen, J.B. (eds.) Advances in Cryptology - EUROCRYPT 2017–36th Annual International Conference on the Theory and Applications of Cryptographic Techniques, 30 April–4 May 2017, Paris, France, Proceedings, Part II. Lecture Notes in Computer Science, vol. 10211, pp. 643–673 (2017). https://doi.org/10.1007/978-3-319-56614-6_22, https://doi.org/10.1007/978-3-319-56614-6_22

49. Rosenfeld, M.: Analysis of hashrate-based double spending (2014). https://arxiv.org/pdf/1402.2009.pdf. Accessed 9 Mar 2016

50. Rosenfeld, M.: Overview of colored coins (2012). https://bitcoil.co.il/BitcoinX.pdf. Accessed 9 Mar 2016
51. Ruffing, Tim, Moreno-Sanchez, Pedro, Kate, Aniket: CoinShuffle: practical decentralized coin mixing for bitcoin. In: Kutyłowski, Mirosław, Vaidya, Jaideep (eds.) ESORICS 2014. LNCS, vol. 8713, pp. 345–364. Springer, Cham (2014). https://doi.org/10.1007/978-3-319-11212-1_20, http://crypsys.mmci.uni-saarland.de/projects/CoinShuffle/coinshuffle.pdf
52. Sapirshtein, A., Sompolinsky, Y., Zohar, A.: Optimal selfish mining strategies in bitcoin. In: Grossklags, J., Preneel, B. (eds.) Financial Cryptography and Data Security - 20th International Conference, FC 2016, 22–26 February 2016, Christ Church, Barbados, Revised Selected Papers. Lecture Notes in Computer Science, vol. 9603, pp. 515–532. Springer (2016). https://doi.org/10.1007/978-3-662-54970-4_30, http://arxiv.org/pdf/1507.06183.pdf
53. Sergey, I., Kumar, A., Hobor, A.: Temporal properties of smart contracts. In: Leveraging Applications of Formal Methods, Verification and Validation. Industrial Practice - 8th International Symposium, ISoLA 2018, 5–9 November 2018, Limassol, Cyprus, Proceedings, Part IV, pp. 323–338 (2018). https://ilyasergey.net/papers/temporal-isola18.pdf
54. Sompolinsky, Y., Zohar, A.: Bitcoin's security model revisited (2016). http://arxiv.org/pdf/1605.09193.pdf. Accessed 4 July 2016
55. Stifter, N., Judmayer, A., Schindler, P., Zamyatin, A., Weippl, E.: Agreement with satoshi - on the formalization of nakamoto consensus. Cryptology ePrint Archive, Report 2018/400 (2018). https://eprint.iacr.org/2018/400.pdf
56. Teutsch, J., Jain, S., Saxena, P.: When cryptocurrencies mine their own business. In: Financial Cryptography and Data Security (FC 2016), February 2016. https://www.comp.nus.edu.sg/~prateeks/papers/38Attack.pdf
57. Tsabary, I., Eyal, I.: The gap game. In: Proceedings of the 2018 ACM SIGSAC Conference on Computer and Communications Security, pp. 713–728. ACM (2018). https://arxiv.org/pdf/1805.05288.pdf
58. Tsabary, I., Yechieli, M., Eyal, I.: MAD-HTLC: because HTLC is crazy-cheap to attack. CoRR abs/2006.12031 (2020). https://arxiv.org/abs/2006.12031
59. Velner, Y., Teutsch, J., Luu, L.: Smart contracts make bitcoin mining pools vulnerable. In: Brenner, M., et al. (eds.) Financial Cryptography and Data Security - FC 2017 International Workshops, WAHC, BITCOIN, VOTING, WTSC, and TA, 7 April 2017, Sliema, Malta, Revised Selected Papers. Lecture Notes in Computer Science, vol. 10323, pp. 298–316. Springer (2017). https://doi.org/10.1007/978-3-319-70278-0_19, http://fc18.ifca.ai/bitcoin/papers/bitcoin18-final14.pdf
60. Vukolić, M.: The quest for scalable blockchain fabric: proof-of-work vs. BFT replication. In: Camenisch, J., Kesdoğan, D. (eds.) iNetSec 2015. LNCS, vol. 9591, pp. 112–125. Springer, Cham (2016). https://doi.org/10.1007/978-3-319-39028-4_9, http://vukolic.com/iNetSec_2015.pdf
61. Winzer, F., Herd, B., Faust, S.: Temporary censorship attacks in the presence of rational miners. In: 2019 IEEE European Symposium on Security and Privacy Workshops, EuroS&P Workshops 2019, 17–19 June 2019, Stockholm, Sweden, pp. 357–366. IEEE (2019). https://doi.org/10.1109/EuroSPW.2019.00046, https://eprint.iacr.org/2019/748

Pay to Win: Cheap, Cross-Chain Bribing Attacks on PoW Cryptocurrencies

Aljosha Judmayer[1,2]($\boxtimes$), Nicholas Stifter[1,2], Alexei Zamyatin[3], Itay Tsabary[4], Ittay Eyal[4], Peter Gaži[5], Sarah Meiklejohn[6], and Edgar Weippl[2]

[1] SBA Research, Vienna, Austria
{ajudmayer,nstifter}@sba-research.org
[2] Uni Wien, Vienna, Austria
edgar.weippl@univie.ac.at
[3] Imperial College London, London, UK
a.zamyatin@imperial.ac.uk
[4] Technion and IC3, Haifa, Israel
Ittay@technion.ac.il
[5] IOHK, Kowloon, Hong Kong
peter.gazi@iohk.io
[6] University College London, London, UK
s.meiklejohn@ucl.ac.uk

Abstract. In this paper we extend the attack landscape of bribing attacks on cryptocurrencies by presenting a new method, which we call *Pay-To-Win* (P2W). To the best of our knowledge, it is the first approach capable of facilitating double-spend collusion across different blockchains. Moreover, our technique can also be used to specifically incentivize transaction exclusion or (re)ordering. For our construction we rely on smart contracts to render the payment and receipt of bribes trustless for the briber as well as the bribee. Attacks using our approach are operated and financed *out-of-band* i.e., on a funding cryptocurrency, while the consequences are induced in a different target cryptocurrency. Hereby, the main requirement is that smart contracts on the funding cryptocurrency are able to verify consensus rules of the target. For a concrete instantiation of our P2W method, we choose Bitcoin as a target and Ethereum as a funding cryptocurrency. Our P2W method is designed in a way that reimburses collaborators even in the case of an unsuccessful attack. Interestingly, this actually renders our approach approximately one order of magnitude cheaper than comparable bribing techniques (e.g., the whale attack). We demonstrate the technical feasibility of P2W attacks through publishing all relevant artifacts of this paper, ranging from calculations of success probabilities to a fully functional proof-of-concept implementation, consisting of an Ethereum smart contract and a Python client.

Keywords: Algorithmic incentive manipulation · Bribing · Smart contracts · Ethereum · Bitcoin

© International Financial Cryptography Association 2021
M. Bernhard et al. (Eds.): FC 2021 Workshops, LNCS 12676, pp. 533–549, 2021.
https://doi.org/10.1007/978-3-662-63958-0_39

1   Introduction

*"The system is secure as long as **honest** nodes collectively control more CPU power than any cooperating group of attacker nodes."* Satoshi Nakamoto [19].

Despite an ever growing body of research in the field of cryptocurrencies, it is an open question if Bitcoin, and thus Nakamoto consensus, is incentive compatible under practical conditions, i.e., that the intended properties of the system emerge from the appropriate utility model [3,4]. *Bribing attacks*, in particular, target incentive compatibility and assume that at least some of the miners act **rationally**, i.e., they accept bribes to maximize their profit. If the attacker, together with all bribable miners, can gain a sizable portion of the computational power, even for a short period of time, attacks are likely to succeed.

Since the first descriptions of bribing attacks [4,6], various attack approaches, which tamper with the incentives of protocol participants, have been presented for different scenarios and models. As bribing [15,17,22,24], front-running [7,8,12] Goldfinger [3,11,13] and other related attacks, all intend to manipulate the incentives of rational actors in the system, we jointly consider them under the general term *algorithmic incentive manipulation* (AIM). So far, most proposed AIM attack strategies focus on optimizing a player's (miner's) utility by accepting *in-band* bribes, i.e., payments in the respective cryptocurrency [4,15,17,24] Thus, a common argument against the practicality of such attacks is that miners have little incentive to participate, as they would put the economic value of their respective cryptocurrency at risk, harming their own income stream. Another common counter argument against in-band bribing attacks is that they are considered expensive for an adversary (e.g., costs of several hundred bitcoins for one successful attack [15]), or require substantial amounts of computing power by the attacker.

In this paper, we present an AIM attack method called *Pay-To-Win* (P2W), which generalizes the construction of different AIM attacks on PoW Cryptocurrencies by leveraging smart contract platforms. Our attack requires no attacker hashrate, and an order of magnitude less funds than comparable attacks (i.e., the whale attack). To highlight the technical and economical feasibility of our approach, we provide a concrete instantiations of our P2W design, representing a new bribing attack. It uses a smart contract capable funding cryptocurrency (Ethereum) to finance and operate an attack on a (different) target cryptocurrency (Bitcoin). All bribes are paid in the funding cryptocurrency, i.e., out-of-band. Prior to our attacks, out-of-band payments have only been used in the context of Goldfinger-attacks, where the goal of an attacker is to destroy a competing cryptocurrency to gain some undefined external utility [13]. The attacks we present in this paper can be performed based on either strategy, using in-band profit, or as out-of-band Goldfinger-style attacks to destroy the value of the targeted cryptocurrency. In a multi-cryptocurrency world, P2W attacks demonstrate that utilizing out-of-band payments can pose an even greater threat to cryptocurrencies, as the argument that miners won't harm their own income

stream must be critically examined in this context. Consider as an example two PoW cryptocurrencies that share the same PoW algorithm and have competing interests, for example Bitcoin and Bitcoin Cash. If rational Bitcoin miners face the opportunity of earning Ether for performing attacks on Bitcoin Cash, they may be willing to redirect their hashrate for this purpose, especially if they are guaranteed to receive the promised out-of-band rewards/bribes.

We show that such sophisticated trustless *out-of-band* attacks on Bitcoin-like protocols can readily be constructed, given any state-of-the-art smart contract platform capable for verifying the consensus rules of the target for the duration of the attack. Moreover, we show that the cost for an attacker can be considerably reduced by guaranteeing that participating bribees are reimbursed. Furthermore, cross-chain transaction ordering attacks can also be executed as targeted bribing attacks using our method. This possibility for rational miners to (trustlessly) auction the contents of their block proposals (i.e., votes) to the highest bidder raises fundamental questions on the security and purported guarantees of most permissionless blockchains.

Contribution: We propose a new design pattern, called *Pay-To-Win* (P2W), for out-of-band algorithmic incentive manipulation (AIM) attacks. To highlight the concept behind our design approach, we provide a *new out-of-band AIM attack* to incentivize double-spend collusion (Sect. 3.1).[1] On the technical level, we introduce *ephemeral mining relays*, as an underlying construction which is required to execute our trustless, time-bounded, cross-chain attack method. Moreover we describe *guaranteed payment* of bribed miners even if the attack fails, which actually reduces the costs of such attacks. All artifacts reaching from calculations, simulations, a PoC and scripts used to derive the operational costs are available online.[2]

2 Model

We focus on *permissionless* proof-of-work (PoW) cryptocurrencies, as the majority of related bribing attacks target Bitcoin, Ethereum, and systems with a similar design. That is, we assume protocols adhering to the design principles of Bitcoin [19], generally referred to as Nakamoto consensus, or Bitcoin backbone protocol [10,20]. Within the attacked cryptocurrency we differentiate between *miners*, who participate in the consensus protocol and attempt to solve PoW-puzzles, and *clients*, who do not engage in such activities. Following the models of related work [4,15,17,22], we assume the set of miners to be fixed, and their respective computational power within the network to remain constant.

To abstract from currency details, we use the term *value* as a universal denomination for the purchasing power of cryptocurrency units, or any other

---

[1] Three other new attacks which we also described, as well as an in-depth analysis of the herein proposed attack, can be found in the extended version of the paper.

[2] Link to repository blinded for review.

out-of-band funds such as fiat currency. Miners and clients may own cryptocurrency units and are able to transfer them (i.e., their value) by creating and broadcasting valid transactions within the network. Moreover, as in prior work [15,17], we likewise make the simplifying assumption that exchange rates are constant over the duration of the attack.

In this work we follow the established *BAR-model* [14] and split participating miners into three groups and their roles remain static for the duration of the attack. Additionally, we define the *victim(s)* as another group or individual without hashrate.

Byzantine Miners or Attacker(s) (Blofeld): The attacker B wants to execute an incentive attack on a *target cryptocurrency*. B is in control of bribing funds $f_B > 0$ and has some, or no hashrate ($p_B \geq 0$) in the target cryptocurrency. B may deviate arbitrarily from the protocol rules.

Altruistic or Honest Miner(s) (Alice): Altruistic miners A are honest and always follow the protocol rules, hence they will not accept bribes to mine on a different chain-state or deviate from the rules, even if it would offer larger profit. Miners A control some or no hashrate $p_A \geq 0$ in the target cryptocurrency.

Rational or Bribable Miner(s) (Rachel): Rational miners R controlling hashrate $p_R > 0$ in the target cryptocurrency They aim to maximize their short term profits in terms of *value*. We consider such miners "bribable", i.e., they follow strategies that deviate from the protocol rules as long as they are expected to yield higher profits than being honest. For our analyses we assume rational miners do not concurrently engage in other rational strategies.

Victim(s) (Vincent): The set of victims, or a single victim, which loses value if the bribing attack is to be successful. The victims control zero hashrate, and therefore can be viewed as a client.

It holds that $p_B + p_A + p_R = 1$. The assumption that the victim of an AIM attack has no hashrate is plausible, as the majority of transactions in Bitcoin or Ethereum are made by clients which do not have any hashrate in the system they are using.

Whenever we refer to an attack as *trustless*, we imply that no trusted third party is needed between briber and bribee to ensure correct payments are performed for the desired actions. Thus the goal is to design AIM in a way that the attacker(s), as well as the collaborating miners, have no incentive to betray each other if they are economically rational.

Communication and Timing: Participants communicate through message passing over a peer-to-peer gossip network, which we assume implements a reliable broadcast functionality. As previous bribing attacks, we further assume that all miners in the target cryptocurrency have *perfect knowledge* about the attack once it has started. Analogous to [10], we model the adversary Blofeld as *rushing*, meaning that he gets to see all other players messages before he decides his strategy, e.g., executes his attack. While the attack is performed on a *target cryptocurrency*, the distinct *funding cryptocurrency* is used to orchestrate and

fund it. We also assume that the difficulty and the mean block interval of the funding chain is fixed for the duration of the attack, and that no additional attacks are concurrently being launched against either cryptocurrency.

3 P2W Attack Method

In this section, we introduce a new approach for algorithmic incentive manipulation attacks, which we call *Pay-To-Win* (P2W). Our approach relies on smart contracts and the specification of *block templates* by the attacker. These templates define the desired block structure for which Blofeld is willing to provide rewards in form of bribes. We consider *out-of-band* attacks to be technically more challenging, as well as more powerful regarding their capabilities (see below), therefore we focus on out-of-band attacks in this paper.[3] As the payment is performed *out-of-band*, we differentiate between a *target cryptocurrency*, where the attack is to be executed, and a *funding cryptocurrency*, where the attack is coordinated and funded. While the funding cryptocurrency must support sufficiently expressive smart contracts, there are no such requirements for the target cryptocurrency. For presentation purposes, we choose Bitcoin as target and Ethereum as the funding cryptocurrency to instantiate and describe our attacks. Theoretically, the attack can be funded on *any* smart contract-capable funding cryptocurrency, which is able to verify the PoW of the target. This advantage of being fund- and operable on any appropriate smart contract capable cryptocurrency renders these P2W attacks arguably more difficult to detect and protect against, as the victim(s) would have to monitor multiple, if not all, possible funding blockchains. Moreover, our attacks can also use additional privacy preserving techniques available on the funding cryptocurrency (e.g., [18]) to hinder the traceability of funds and transactions of involved parties. Another advantage of out-of-band payments is, that they are not bound to the exchange value of the targeted cryptocurrency and thus can also be used for Goldfinger style attacks [3,13], as the assumption that miners of the target cryptocurrency would not harm their own revenue channel does not necessarily hold true anymore. This is an even more compelling argument in a world where multiple cryptocurrencies either share the same PoW algorithm, or hardware can be effectively used for mining other forms of PoW.

Our construction requires a combination of a smart contract based mining pool [16,23] and a temporary chain relay.[4] We call this underlying construction an *ephemeral mining relay* (EMR).[5] Chain relays are smart contracts which allow

---

[3] We also describe and evaluate three new in-band attacks targeting transaction ordering and transaction exclusion in the extended version of this paper. A in-band transaction exclusion attack was also described and analyzed in concurrent work by Winzer et al. [24], but no concrete instantiation was given.

[4] cf. https://github.com/ethereum/btcrelay.

[5] We use the term "ephemeral" as the mining relay is instantiated only temporarily and does not require verification of the entire blockchain, but only the few blocks relevant for the attack.

to verify the state of other blockchains, however, a naive chain relay implementation only allows to verify that a certain block (or transaction) was included in a chain with the most accumulated proof-of-work (i.e., heaviest chain). It does not allow to verify whether the blocks and transactions included in this heaviest chain are indeed *valid*, i.e., adhere to the consensus rules of the corresponding blockchain. In contrast to previous proposals, our EMR needs to be capable of validating if blocks adhere to the consensus rules of the target cryptocurrency. This is achieved by sufficiently restricting the allowed block structure. In our case the set of transactions within blocks generated by collaborating miners is specified by the block template provided by the adversary. As Blofeld wants to submit collected PoW solutions to Bitcoin, it is in his best interest to provide only templates including valid transactions. Conversely, collaborating rational miners do not care if the block template they mine on is actually valid in Bitcoin, since the rewards they receive for solutions are guaranteed to be paid out by the smart contract in Ethereum.

Summarizing, our EMR takes care that the promised rewards are only paid to complacent bribees which have actively contributed to the attack. Therefore, the introduced attack can be considered *trustless*, both for the attacker as well as the collaborating bribed miners. Moreover, the attack does not require the adversary to control any hashrate, i.e., we assume $p_B = 0$. To demonstrate the feasibility of our approach and the described attack, we implemented a fully functional prototype of our attack and evaluated its costs in Ethereum. The source code and all other artifacts of the evaluation are available on Github.[6]

3.1 Transaction Revision, Exclusion and Ordering Attack

To illustrate all underlying concepts, we present them within the context of a concrete attack. While we focus on transaction revision in our description, the presented attack also bears the possibility for arbitrary transaction exclusion and ordering. To execute our attack, Blofeld must construct a smart contract which temporarily rewards the creation of attacker-defined blocks on the target cryptocurrency. After its initialization, the smart contract can be used by him as well as by other collaborating miners/attackers/bribees to coordinate the attack and manage the investment and payout of funds.

Initialization Phase (deploy, init): First the attacker (Blofeld) creates the uninitialized attack contract and publishes it on the Ethereum blockchain. This is done with a *deploy* transaction included in some Ethereum block e_0 from an Ethereum account controlled by the attacker.[7] Then, Blofeld creates a conflicting pair of Bitcoin transactions. The spending transaction tx_B is published on

---

[6] Link to repository blinded for review.

[7] It is also possible to deploy and initialize the attack contract at the same time (e_1), but publishing an uninitialized attack contract upfront ensures that potential collaborators can audit it and familiarize themselves with the procedure. In any case, it is important that the double-spend transaction tx'_B is disclosed after block b_{k_V} on the main chain, as otherwise Vincent may recognize the double-spending attack and refuse to release the goods.

the main chain in Bitcoin immediately, and the double-spending transaction tx'_B is kept secret. After the confirmation period of k_V blocks (defined by the victim V) has passed on the Bitcoin main chain, Blofeld releases an initialization transaction, which defines the conditions of the attack in the smart contract on the Ethereum chain. The block e_1 represents the first block on the Ethereum chain after the Bitcoin block b_{k_V} has been published.

In e_1 the contract is initialized with $k_V + 1$ new Bitcoin block templates, each carrying the transactions from the original chain to collect their fees, but instead of tx_B the conflicting transaction tx'_B is included. Collaborating miners are now free to mine these new block templates. For the first template they are only allowed to change the nonce and the coinbase field to find a valid PoW and include their payout Ethereum address in the coinbase. This prevents front running of solutions (see Sect. 3.1). Once a solution has been found, it has to be submitted by the respective miner to the attack contract, which verifies the correctness of the PoW and that only allowed fields (nonce and coinbase) have been changed. After the first block (b'_1) in the sequence, also the previous block hashes of subsequent blocks ($\{b'_2, \dots\}$) have to be adjusted by collaborating miners. If a submitted solution is valid, the contract knows which previous block hash it must use to verify the next solution and so forth.

As soon as Blofeld becomes aware that a valid solution was broadcasted in the Ethereum P2P network, he uses the PoW solution to complete the whole block and submits it to the Bitcoin P2P network. Blofeld and the collaborating miners have an incentive to submit solutions timely. The collaborating miners want to collect an additional bribe ϵ in case the attack succeeds, and the attacker wants his blocks included in the Bitcoin main chain to receive the Bitcoin block rewards to his Bitcoin address, and in the best case, perform a successful double-spend.

Attack Phase (update): Bribed miners now proceed to mine $k_V + 1$ blocks on the attack chain. If additional blocks are found on the main chain, the attacker can update the attack contract with new block templates for blocks $k_V + 2$ to N, where N is the maximum number of attack blocks that can be funded by the adversary. Note that N is not necessarily known by Vincent, Rachel or any other observer.

Payout Phase (pay): The payout phase starts as soon as the attack phase has ended. This happens when k_B blocks have been mined on top of the last block for which a block template has been provided to the smart contract. In the best case, this happens at block $T = k_V + 1 + k_B$, but in our example one `update` with an additional block template was required, leading to $T = k_V + 2 + k_B$. The delta of k_B is a security parameter defined by the attacker, which should ensure that every participant had enough time to submit information about the longest Bitcoin chain to the contract and that the sequence of blocks relevant for the attack has received sufficient confirmations.[8]

---

[8] Ideally k_B is specified as an acceptance policy logarithmic in the chain's length as described in [21].

The attack terminates as soon as the first block of height T is committed to the contract. This can be a block of the main chain, or the attack chain. After the attack has terminated, the contract unlocks the payment of compensations and rewards for the miners of the associated blocks. Now all miners who joined the attack and contributed blocks can collect their compensations and/or bribes from the contract.

To accurately pay out funds, the contract on Ethereum has to determine which chain in Bitcoin has won the race and is now the longest chain. Thereby, the contract has to distinguish between two possible outcomes:

Attack Failed (Main chain wins): In this case the contract must compensate the bribed miners for their contributed blocks to the attack chain, which are now stale. These are at most $\{b'_1, \ldots, b'_N\}$, Every collaborating miner who mined and successfully submitted a block to the attack contract receives the original Bitcoin reward (in Ether) for that block, without an additional ϵ.

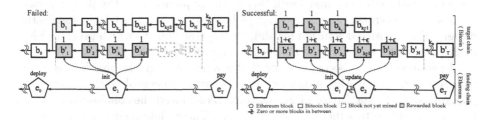

Fig. 1. Example blockchain structure and resulting payouts of a failed, and a successful attack. The colored blocks are rewarded by the attack contract, either with their original value (reward + fee normalized to 1) or with an additional ϵ if the attack was successful. The numbers above colored blocks indicate those normalized rewards.

Attack Succeeded (Attack chain wins): If the attack chain wins, then the contract executes the following actions: 1) Fully compensate the miners of k_V main chain blocks starting from b_1, which are now stale. This is necessary to provide an incentive also for those miners to switch and contribute to the attack chain, as they otherwise would lose their rewards from blocks they contributed to the main chain if the attack is successful. 2) Pay the miner of every attack chain block, b'_1 to b'_{k_V+2} in our example (max. till b'_N), the full block reward plus an additional ϵ as a bribe in Ether.

Figure 1 shows the different stages of the attack on the funding cryptocurrency, as well as two different outcomes (failed and successful attack) on the target cryptocurrency. The paid out compensations (block rewards normalized to 1) and bribes (ϵ) are given above the respective blocks. Upon being invoked with a miner's cash-out transaction, the contract checks if the attack has already finished, i.e., a valid chain up to block height T is known, and which chain has won the race. Then the contract pays out accordingly.

Incentives to Submit Blocks: Since collaborating miners are competing for mined attack chain blocks and want the attack to be successful to receive the

additional bribes, they have an incentive to timely submit their attack chain blocks to the contract. In any case, for every (valid) submitted Bitcoin attack chain block, the full Bitcoin reward is paid in Ether by the contract. In case of a successful attack an additional ϵ is paid. Therefore, Blofeld who initialized the contract and provided the funds has an incentive to submit the relevant part of the main chain, if a conflicting longer chain ($\{b_1, \ldots, b_T\}$) exists, since he would pay an additional ϵ for every block otherwise. Moreover, Blofeld has an incentive to submit every completed Bitcoin block (with the PoW provided by the bribees) to the Bitcoin network, because he is the one who receives the full Bitcoin block rewards as specified in his Bitcoin block template. **Ethereum Payout Address Derivation:** To determine the correct Ethereum payout addresses of collaborating miners, the following approaches are feasible: As soon as bribed miners start participating in the attack, they directly provide their Ethereum address as additional data in the coinbase field of every submitted Bitcoin block on the attack chain. As miners of the blocks b_1 to b_{k_V}, may not have disclosed their Ethereum address in the coinbase field already, another technique has to be used. For blocks where miners were not yet aware of the attack, they must prove to the contract that they indeed mined the respective block(s). If *Pay-to-Pubkey* outputs have been used in the respective coinbase transactions, the Bitcoin address public key can be used to derive the corresponding Ethereum address, as described and implemented in the Goldfinger attack example in [17]. This can also be achieved by providing the ECDSA public keys corresponding to *Pay-to-PubKey-Hash* payouts from the respective coinbase outputs to the smart contract. Thereby, the contract can verify if the keys correspond to the respective Bitcoin addresses and also derive the corresponding Ethereum addresses, as they rely on the same signature scheme.

3.2 Evaluation with Solely Rational Miners ($p_{\mathcal{R}} = 1$)

As rational miners will participate in the attack as long as it is expected to yield more profit than honest mining, the remaining question is, what budget in Ether is required by Blofeld (f_B) for the attack to succeed. As the Bitcoin block rewards and bribes have to be payed out in Ether, we assume a fixed exchange rate between cryptocurrencies to derive our lower bound in terms of BTC required. Blofeld has to lock funds in the attack contract for each submitted block template, to ensure complacent miners can be certain to receive their rewards if they submit blocks and thus are incentivized to join the attack. Therefore, the duration of the attack is the main driver for the required budget. As the duration is dependent on the security parameter k_V chosen by Vincent, $N > k_V$ has to hold for an attack to be feasible.

Necessary Attack Budget: For Bitcoin, a common choice is $k_V = 6$ requiring N to be at least 7. The budget of the attack contract must cover all rewards which could potentially be paid out by the contract. For the most expensive case, which is a successful attack, this encompasses: The bribes (ϵ) as well as

Bitcoin block rewards including fees[9] (r_b), which we previously normalized to 1 in Fig. 1. Assuming the current block reward (6.25 BTC), average fees ($\approx$2 BTC), operational costs ($c_{operational} = 0.5$ BTC), as well as a bribe of $\epsilon = 1$ BTC, this leads to a budget of 114.75 BTC which has to be provided to the attack contract in Ether upfront s.t., $f_B = k_V \cdot r_b + N \cdot (r_b + \epsilon) + c_{operational}$. As Blofeld receives the Bitcoin block rewards in case of a successful attack, the actual costs of the attack are *much smaller* than the required budget Blofeld has to lock in the contract.

Costs and Profitability of a Successful Attack: If the attack is successful, then Blofeld earns the block rewards on the main chain in BTC which compensate his payouts to bribed miners in Ether. The costs for a successful attack are thus reduced by $N \cdot r_b$ main chain blocks, whereas rewards must be paid for $N \cdot (r_b + \epsilon)$ block templates. The remaining costs of a successful attack stem from the $k_V \cdot r_b$ main chain blocks that have to be compensated on the attack chain s.t., $c_{success} = k_V \cdot r_b + N \cdot \epsilon + c_{operational}$. The initial k_V compensations are necessary to provide the same incentive for *all* miners that have already produced blocks on the main chain to switch to the attack chain. Since we assume rational miners, the attack in this scenario is always successful if $N > k_V$ and $\epsilon > 0$ hold. For Bitcoin, this means that the costs of a successful double spend with $k_V = 6$ and $r_b = 8.25$ and $\epsilon = 1$ are $c_{success} = 57$ BTC. For a successful attack to be profitable, the value of the double-spend (v_d) has to be greater than this value. In Bitcoin, transactions carrying more than 57 BTC are observed regularly.[10] For comparison, in its cheapest configuration, the whale attack costs approximately 770 BTC [15], but it was simulated for a previous Bitcoin reward epoch, where block rewards have been higher. Even if we assume $r_b = 12.5$ BTC, our attack would cost 94.5 BTC, which is considerably lower than the whale attack. The remaining difference to our approach is that the whale attack does not assume all miners to be rational. In Sect. 3.3 we also extend our evaluation to this model by introducing altruistic miners.

Costs of a Failed Attack: Although the attack cannot fail in a model where all miners are rational and the attacker has enough budget, it is relevant for a scenario where $p_{\mathcal{R}} < 1$ to determine the worst case cost for an unsuccessful attack. In the worst case, the attack duration is N and not a single block produced by complacent miners (according to a published block template) made it into the main chain. Then the costs are determined by the duration N and the block rewards including fees (r_b) s.t., $c_{fail} = N \cdot r_b + c_{operational}$. Setting the same values for r_b and N amounts to approximately $c_{fail} = 58.25$ BTC in our example.

---

[9] In a concrete attack of course r_b is not constant, but given by the coinbase output values of every submitted block.

[10] cf.   https://blockchair.com/bitcoin/outputs?s=value(desc),time(desc)&q=time (2020- 10),value(6000000000..)\#.

3.3   Evaluation with Altruistic Miners ($p_A > 0 \land p_R + p_A = 1$)

We now discuss a more realistic scenario where not all miners switch to the attack chain immediately, i.e., some of them act altruistically. Altruistic miners follow the protocol rules and only switch to the attack chain if it becomes the longest chain in the network – but do not attempt to optimize their revenue, contrary to economically rational, i.e., bribable, miners.[11]

We derive the probability of the attack chain to win a race against altruistic miners, based on the budget of the attacker and the initial gap between those chains which has to be overcome k_{gap} where k_{gap} is initially set to $k_{gap} = k_V$. The difference between k_V and k_{gap} is that k_{gap} can increase when altruistic miners find a new block, while k_V is static. In other words, the attack chain must find $k_{gap} + 1$ more blocks than the altruistic main chain – but must achieve this within the upper bound of N blocks (maximum funded attack duration). Each new block is appended to the main chain with probability p_A, and to the attack chain with probability p_R respectively ($p_A + p_R = 1$). We therefore seek all possible series of blocks being appended to either chain, and calculate the sum of the probabilities of the series which lead to a successful attack. In a successful series $i \in \mathbb{N}$ blocks are added to the main chain and $k_{gap} + i + 1$ blocks are added to the attack chain. The probability for such a series is $p_R^{k_{gap}+i+1} \cdot p_A^i$.

For any prefix strictly shorter than the whole series, the number of appended blocks to the attack chain is smaller than $k_{gap} + 1$, as otherwise the attack would have ended sooner. It follows that the last block in a successful series is always appended to the attack chain. The number of combinations for such a series is derived similarly to Bertrand's ballot theorem, with a difference of k_{gap} for the starting point. Assuming the attacker can only fund up to N blocks on the attack chain, the probability of a successful attack is hence given by:

$$\sum_{i=0}^{i \leq N - k_{gap} - 1} \left(\binom{k_{gap} + 2i}{i} - \binom{k_{gap} + 2i}{i - 1} \right) \cdot p_R^{k_{gap}+i+1} \cdot p_A^i \qquad (1)$$

Using formula 1 we can calculate the success probability of the attack. Clearly, the attack requires $N > k_V$ to have a chance of being successful. As with the classical 51% attacks, the attack eventually succeeds once the bribable hash rate is above the 50% threshold and the number of payable blocks N grows. In other words, assuming more than $p_R > 0.5$ rational hashrate, bribing attacks are eventually successful if they can be funded long enough. The relevant question is how expensive it is to sustain the attack for a long enough period s.t., the attack is expected to be successful.

---

[11] Another explanation can be that some miners have imperfect information, which might be the case in practice.

Table 1. Comparison of attack costs for $k_V = 6$, all costs given in BTC. The costs for the whale attack are the average from 10^6 simulation results provided in [15]. For comparision different Bitcoin block reward epochs (12.5 and 6.25 BTC) are provided for our P2W attack, all with $c_{operational} = 0.5$ BTC, and average fee per block of 2 BTC and a bribe $\epsilon = 1$ BTC.

Rational hashrate $p_{\mathcal{R}}$	Average whale attack costs epoch reward 12.5 c_{whale} in BTC	P2W epoch reward 12.5 $c_{expected}$ in BTC	P2W cost compared to whale	P2W N average	P2W epoch reward 6.25 $c_{expected}$ in BTC
0.532	293e+23	196.50	≈0.00%	109	159.00
0.670	999.79	108.50	10.85%	21	71.00
0.764	768.09	101.50	13.21%	14	64.00
0.828	1265.14	98.50	7.79%	11	61.00
0.887	1205.00	96.50	8.01%	9	59.00
0.931	1806.67	96.50	5.34%	9	59.00
0.968	2178.58	95.50	4.38%	8	58.00
0.999	2598.64	95.50	3.67%	8	58.00

Table 1 shows a comparison between the expected costs of a successful P2W attack, against the average costs of 10^6 simulations of the whale attack as presented in [15]. At a first glance, given that the attacker must pay collaborating miners regardless of the outcome of the attack, one may assume that the costs faced by the attacker are high compared to other bribing schemes. However, this is not the case. In our attack miners face *no risk* from participation – requiring only a *low bribe value* to incentivize sufficient participation for a successful attack, contrary to existing bribing attacks like the whale attack.

It can be observed that, in contrast to the whale attack, our attack becomes cheaper when $p_{\mathcal{R}}$ grows large since the race is won faster and therefore fewer bribes have to be paid. Moreover, the whale attack has to pay substantially more funds to account for the risk rational miners face if the attack fails. Our approach is hence approximately ≈87% to ≈96% cheaper than the whale attack. For $p_{\mathcal{R}} = 0.532$ the difference is so large, that the costs of our P2W attack are insignificant compared to the whale attack. The switch to a new Bitcoin block reward epoch has further reduced the costs of the attack s.t., the costs of a successful double-spending attack ($k_V = 6$) using our technique are around 60 BTC. In October 2020 alone, there where around 60 thousand Bitcoin transactions with outputs greater than 60 BTC.[12]

3.4   Evalution of the Operational Costs

We implement a fully functional attack contract including the EMR on Ethereum, which is capable of verifying the state of the Bitcoin blockchain.[13]

---

[12] c.f.   https://blockchair.com/bitcoin/outputs?s=value(desc),time(desc)&q=time (2020-10),value(6000000000..)\#.

[13] Blinded for review.

We use Solidity v0.6.2 and a local Ganache instance for cost analysis, with a current gas price of 45 Gwei and an exchange rate 500 USD/ETH. Submitting a block template for a Bitcoin block amounts to 302,228 Gas ($ 6.80 USD). The costs for submitting and verifying a new Bitcoin block are 468,273 Gas ($ 10.54 USD) in the worst case. In total the costs of an example attack on Bitcoin with $k_V = 6$ and $k_B = 6$ are about $ 355.24 USD. This confirms that the costs for maintaining an attack contract including an EMR are marginal when compared to the potential scale of incentive attacks described in this paper. For comparison, the reward for a single Bitcoin block (*excluding* transaction fees) at the time of writing is approximately $ 120 000 USD.

4   Discussion and Mitigations

Our AIM attack highlights the security dependency between transaction value and confirmation time k_V, as also stated in [21]. As with the negative-fee mining pools presented by Bonneau in [4], there exists an interesting analogy between such an incentive manipulation attack and a mining pool. At an abstract level, the presented attack relies on a construction comparable to a mining pool, where the pool owner/attack operator defines specific rules for block creation for the targeted cryptocurrency within a smart contract. Moreover, every participant must be able to claim their promised rewards in a trustless fashion, based on the submitted blocks and state of the targeted cryptocurrency. The construction of an *ephemeral mining relay*, presented within this paper, provides exactly this functionality. Luu et al. [16] also proposes a mining pool (Smart pool) which itself is governed by a smart contract. However, its design and intended application scenarios did not consider use-cases with malicious intent. Smart pool does not enforce any properties regarding the content and validity of submitted blocks beyond a valid PoW, as an intrinsic incentive among participants is assumed to earn mining rewards in the target cryptocurrency, which is only possible if valid blocks have been created.

Practical Possibility: The focus of this paper is to improve upon existing attacks and demonstrate the technical feasibility of advanced bribing attacks, as well as to evaluate the associated costs. Hereby, the long term interests of miners of course also play an important role. There may be scenarios where miners are capable of providing PoW for a target blockchain, but at the same time do not have any long-term interest in the well-being of the target. Consider the real-world example of Bitcoin and Bitcoin Cash which utilize the same form of PoW and can be considered competitors. Thus, the question if the proposed attacks are possible in practice is difficult to answer scientifically. There is already empirical evidence from previous large scale attacks by miners, e.g., recent 51% attacks on Ethereum Classic and Bitcoin Gold, as well as incentive manipulation attacks and front-running [7]. To the best of our knowledge, none of the observed attacks has been as sophisticated as the new technique proposed in this paper, but of course, they can get better over time. Nevertheless, these cases demonstrate that large scale attacks happen, and that the topic of incentives in cryptocurrencies

is an area which deserves further study. We see our paper as another important contribution in this direction.

Counter Attacks: Counter bribing refers to the technique of countering bribing attacks with other bribing attacks [3,4]. For the victim(s), counter bribing is a viable strategy against AIM. The difficulty of successfully executing counter bribing highly depends on the respective scenario. In the end, counter bribing can also be countered by counter-counter bribing and so forth. Therefore, as soon as this route is taken, the result becomes a bidding game. If defenders have imperfect information, they may not be able to immediately respond with counter bribes. This illustrates an important aspect of AIM, namely their visibility. On the one hand, sufficiently many rational miners of the target cryptocurrency have to recognize that an attack is occurring, otherwise they won't join in and the attack is likely to fail. On the other hand, if the victims of the attack recognize its existence, they can initiate and coordinate a counter bribing attack. So the optimal conditions for AIM arise if all rational miners have been informed directly about the attack, while all victims/merchants do not monitor the chain to check if an attack is going on and are not miners themselves.

The great benefit of the herein described attacks is that bribes are paid out-of-band. Hereby, our attacks are rendered more stealthy to victims, who only monitor the target cryptocurrency. Of course their received rewards can be traced in the funding cryptocurrency, but available privacy techniques may be used to camouflage the real recipient of the funds e.g., [18]. It can hence be argued that counter attacks by victims are harder to execute as they are not immediately aware of the bribing value that is being bet against them on a different funding cryptocurrency. We also follow the argument in [4] that requiring clients to monitor the chain and actively engage in counter bribing is undesirable, and our out-of-band attacks further amplifies this problem as clients would have to concurrently monitor a variety of cryptocurrencies.

Cross-chain Verifiability: One crucial aspect of our attacks is that a smart contract within the funding cryptocurrency must be able to validate core protocol and consensus rules of the target chain, in particular it must be able to determine the validity of blocks. If this is not possible, the attack cannot be executed trustlessly. For example, it is currently not possible to execute an AIM against Litecoin using Ethereum as a funding cryptocurrency in a fully trustless manner, as it is economically unfeasible to verify the Scrypt hash function within a smart contract. However, it is generally beyond the reach of an individual cryptocurrency to dictate or enforce what other cryptocurrencies support in future versions of their smart contract languages. Thus, any such defensive decision of the target cryptocurrency may be mitigated by future changes in another cryptocurrency. Hence, such measures can not guarantee lasting protection.

5   Implications and Future Work

In this paper we introduced a new AIM attack method called Pay-To-Win (P2W) and showed that attacks utilizing the described techniques can

readily be constructed given current smart contract platforms. The implications of our proposed method (and related AIM/bribing attacks) regarding the security guarantees of PoW cryptocurrencies are not yet conclusive and topic of future work. On the theoretical side, embedding and modeling incentive attacks in formalisms of Nakamoto style cryptocurrencies is non-trivial, as prevalent approaches do not consider rational participants [2,9,20], or explicitly exclude bribing [1]. Furthermore, no agreed upon game theoretic analysis technique for (PoW) cryptocurrencies currently exits, and it remains an open question if such an analysis could be rendered universally composable. The generalization and inclusion of AIM attacks and rational behavior in formal analysis frameworks for Nakamoto consensus based cryptocurrency designs, including approaches such as *Proof-of-Stake*, hence poses an interesting and important open research challenge. On the practical side, our new attack, as well as the existing body of research on AIM, demonstrates that it is not only the hashrate distribution among permissionless PoW based cryptocurrencies that plays a central role in defining their underlying security guarantees. The ratio of *rational* miners and available funds for performing AIM also form a key component, as rational miners can be incentivized to act as accomplices to an attacker. The possibility of trustless out-of-band attacks highlights that being able to cryptographically interlink cryptocurrencies increases this attack surface. Further, smart contract based AIM introduces the possibility to align the interests of multiple attackers who want to perform double-spends during the same time period, making low value double-spends theoretically feasible (as economically analyzed in [5]). Together with the topic of counter bribing, new research directions are opened up that raise fundamental questions on the incentive compatibility of Nakamoto consensus. Real world attacks targeting incentives, such as front-running [7], demonstrate that the existence of incentives cannot be ignored in PoW cryptocurrencies. To accurately reflect the security properties of permissionless PoW cryptocurrencies, some form of rationality has to be taken into account. The problem is, that as soon as rational players are considered, all previously proposed AIM/bribing methods, as well as the attack described in this paper, lead to interesting questions whether or not the incentive structures of prevalent cryptocurrencies actually encourage desirable outcomes. Even more so, in a world where multiple cryptocurrencies coexist it is likely not sufficient to model them individually as closed and independent systems.

Acknowledgements. We would like to thank the participants of the Dagstuhl Seminar 18152 (*Blockchains, Smart Contracts and Future Applications*), especially Samuel Christie and Sebastian Faust, as well as the participants of the Dagstuhl Seminar 18461 (*Blockchain Security at Scale*) for all the frutiful discussions.

This paper is based upon work partially supported by (1) the Christian-Doppler-Laboratory for Security and Quality Improvement in the Production System Lifecycle; The financial support by the Austrian Federal Ministry for Digital and Economic Affairs, the Nation Foundation for Research, Technology and Development and University of Vienna, Faculty of Computer Science, Security & Privacy Group is gratefully acknowledged; (2) SBA Research; the competence center SBA Research (SBA-K1)

funded within the framework of COMET Competence Centers for Excellent Technologies by BMVIT, BMDW, and the federal state of Vienna, managed by the FFG; (3) the FFG Bridge 1 project 864738 PR4DLT. (4) the US-Israel Binational Science Foundation (BSF) (5) the Israel Cyber Bureau (6) the Technion Hiroshi Fujiwara cyber-security research center.

References

1. Badertscher, C., Garay, J., Maurer, U., Tschudi, D., Zikas, V.: But why does it work? A rational protocol design treatment of bitcoin. In: Nielsen, J.B., Rijmen, V. (eds.) EUROCRYPT 2018. LNCS, vol. 10821, pp. 34–65. Springer, Cham (2018). https://doi.org/10.1007/978-3-319-78375-8_2

2. Badertscher, C., Maurer, U., Tschudi, D., Zikas, V.: Bitcoin as a transaction ledger: a composable treatment. In: Katz, J., Shacham, H. (eds.) CRYPTO 2017. LNCS, vol. 10401, pp. 324–356. Springer, Cham (2017). https://doi.org/10.1007/978-3-319-63688-7_11

3. Bonneau, J.: Hostile blockchain takeovers (short paper). In: Zohar, A., et al. (eds.) FC 2018. LNCS, vol. 10958, pp. 92–100. Springer, Heidelberg (2019). https://doi.org/10.1007/978-3-662-58820-8_7

4. Bonneau, J.: Why buy when you can rent? Bribery attacks on bitcoin consensus. In: 3rd Workshop on Bitcoin and Blockchain Research, BITCOIN 2016 (2016)

5. Budish, E.: The economic limits of bitcoin and the blockchain. Technical report, National Bureau of Economic Research (2018)

6. Cunicula: Bribery: The double double spend. https://bitcointalk.org/index.php?topic=122291. Accessed 31 Jan 2021

7. Daian, P., et al.: Flash boys 2.0: frontrunning in decentralized exchanges, miner extractable value, and consensus instability. In: IEEE SP (2020)

8. Eskandari, S., Moosavi, S., Clark, J.: SoK: transparent dishonesty: front-running attacks on blockchain. In: FC 2019 - WTSC Workshop (2019)

9. Garay, J., Kiayias, A., Leonardos, N.: The bitcoin backbone protocol: analysis and applications. In: Oswald, E., Fischlin, M. (eds.) EUROCRYPT 2015. LNCS, vol. 9057, pp. 281–310. Springer, Heidelberg (2015). https://doi.org/10.1007/978-3-662-46803-6_10

10. Garay, J.A., Kiayias, A., Leonardos, N.: The bitcoin backbone protocol with chains of variable difficulty (2016). Accessed 06 Feb 2017

11. Judmayer, A., Stifter, N., Schindler, P., Weippl, E.: Pitchforks in cryptocurrencies: enforcing rule changes through offensive forking- and consensus techniques (short paper). In: Garcia-Alfaro, J., Herrera-Joancomartí, J., Livraga, G., Rios, R. (eds.) DPM/CBT -2018. LNCS, vol. 11025, pp. 197–206. Springer, Cham (2018). https://doi.org/10.1007/978-3-030-00305-0_15

12. Kolluri, A., Nikolic, I., Sergey, I., Hobor, A., Saxena, P.: Exploiting the laws of order in smart contracts (2018)

13. Kroll, J.A., Davey, I.C., Felten, E.W.: The economics of bitcoin mining, or bitcoin in the presence of adversaries. In: Proceedings of WEIS (2013)

14. Li, H.C., et al.: Bar gossip. In: USENIX OSDI (2006)

15. Liao, K., Katz, J.: Incentivizing blockchain forks via whale transactions. In: Brenner, M., et al. (eds.) FC 2017. LNCS, vol. 10323, pp. 264–279. Springer, Cham (2017). https://doi.org/10.1007/978-3-319-70278-0_17

16. Luu, L., Velner, Y., Teutsch, J., Saxena, P.: SmartPool: practical decentralized pooled mining. In: USENIX Security Symposium (2017)

17. McCorry, P., Hicks, A., Meiklejohn, S.: Smart contracts for bribing miners. In: Zohar, A., et al. (eds.) FC 2018. LNCS, vol. 10958, pp. 3–18. Springer, Heidelberg (2019). https://doi.org/10.1007/978-3-662-58820-8_1
18. Meiklejohn, S., Mercer, R.: Möbius: trustless tumbling for transaction privacy. In: Proceedings on Privacy Enhancing Technologies (2018)
19. Nakamoto, S.: Bitcoin: a peer-to-peer electronic cash system, December 2008
20. Pass, R., Seeman, L., Shelat, A.: Analysis of the blockchain protocol in asynchronous networks. In: Coron, J.-S., Nielsen, J.B. (eds.) EUROCRYPT 2017. LNCS, vol. 10211, pp. 643–673. Springer, Cham (2017). https://doi.org/10.1007/978-3-319-56614-6_22
21. Sompolinsky, Y., Zohar, A.: Bitcoin's security model revisited (2016). Accessed 04 July 2016
22. Teutsch, J., Jain, S., Saxena, P.: When cryptocurrencies mine their own business. In: Grossklags, J., Preneel, B. (eds.) FC 2016. LNCS, vol. 9603, pp. 499–514. Springer, Heidelberg (2017). https://doi.org/10.1007/978-3-662-54970-4_29
23. Velner, Y., Teutsch, J., Luu, L.: Smart contracts make bitcoin mining pools vulnerable. In: Brenner, M., et al. (eds.) FC 2017. LNCS, vol. 10323, pp. 298–316. Springer, Cham (2017). https://doi.org/10.1007/978-3-319-70278-0_19
24. Winzer, F., Herd, B., Faust, S.: Temporary censorship attacks in the presence of rational miners. In: IEEE EuroS&P Workshops 2019 (2019)

WTSC – DeFi and Tokens

SoK: Lending Pools in Decentralized Finance

Massimo Bartoletti[1], James Hsin-yu Chiang[2(✉)], and Alberto Lluch Lafuente[2]

[1] Università degli Studi di Cagliari, Cagliari, Italy
[2] Technical University of Denmark, DTU Compute, Copenhagen, Denmark
jchi@dtu.dk

Abstract. Lending pools are decentralized applications which allow mutually untrusted users to lend and borrow crypto-assets. These applications feature complex, highly parametric incentive mechanisms to equilibrate the loan market. This complexity makes the behaviour of lending pools difficult to understand and to predict: indeed, ineffective incentives and attacks could potentially lead to emergent unwanted behaviours. Reasoning about lending pools is made even harder by the lack of executable models of their behaviour: to precisely understand how users interact with lending pools, eventually one has to inspect their implementations, where the incentive mechanisms are intertwined with low-level implementation details. Further, the variety of existing implementations makes it difficult to distill the common aspects of lending pools. We systematize the existing knowledge about lending pools, leveraging a new formal model of interactions with users, which reflects the archetypal features of mainstream implementations. This enables us to prove some general properties of lending pools, and to precisely describe vulnerabilities and attacks. We also discuss the role of lending pools in the broader context of decentralized finance and identify relevant research challenges.

1 Introduction

The emergence of permissionless, public blockchains has given birth to an entire ecosystem of *crypto-tokens* representing digital assets. Facilitated and accelerated by smart contracts and standardized token interfaces [1], these so-called *decentralized finance* (DeFi) applications promise an open alternative to the traditional financial system. One of the main DeFi applications are *lending pools*, which incentivize users to lend some of their crypto-assets to borrowers. Unlike in traditional finance, all the parameters of a loan, like its interests, maturity periods or token prices, are determined by a smart contract, which also defines mechanisms to incentivize honest behaviour (e.g., loans are eventually repaid), economic growth and stability. As of April 2021, the two main lending pool platforms hold $13.5B [26] and $6.4B [24] worth of tokens in their smart contracts.

Lending pools are inherently hard to design. Besides the typical difficulty of implementing secure smart contracts [2–4,38], lending pools feature complex economic incentive mechanisms, which make it difficult to understand when a

© International Financial Cryptography Association 2021
M. Bernhard et al. (Eds.): FC 2021 Workshops, LNCS 12676, pp. 553–578, 2021.
https://doi.org/10.1007/978-3-662-63958-0_40

lending pool actually achieves the economic goals it was designed for. As a matter of fact, a recent failure of the oracle price feed used by the Compound lending pool platform led to $100M of collateral being (incorrectly) liquidated [16]. Indeed, most current literature in DeFi is devoted to study the economic impact of these incentive mechanisms [45, 46, 52, 53, 55, 59].

The problem is made even more complex by the absence of abstract operational descriptions of the behaviour of lending pools. Current descriptions are either high-level economic models [52, 53, 59], or the actual implementations. While, on the one hand, economic models are useful to understand the macroscopic financial aspects of lending pools, on the other hand they do not precisely describe their interactions with users. Still, understanding these interactions is crucial to determine if a lending pool is vulnerable to attacks where some users deviate from the expected behaviour. Implementations, instead, reflect the exact actual behaviour, but at a level of detail that makes high-level understanding and reasoning unfeasible.

Contributions. This paper presents a systematic analysis of the behaviour of lending pools, of their properties, vulnerabilities, and of the related literature. Based on a thorough inspection of the implementations of the two main lending pool platforms, Compound [14] and Aave [7], we synthesise a formal, operational model of the interactions between users and lending pools, encompassing their incentive mechanisms. More specifically, our contributions are:

1. a formal model of lending pools, which precisely describes their interactions as transitions of a state machine. Our model captures all the typical transactions of lending pools, and all the main economic features, like collateralization, exchange rates, token prices, and interest accrual (Sect. 3);
2. the formalization and proof of fundamental behavioural properties of lending pools, which were informally stated in literature, and are expected to be satisfied by any implementation (Sect. 4);
3. the formalization of relevant properties of the incentive mechanisms of lending pools, and a discussion of their vulnerabilities and attacks (Sect. 5);
4. a thorough discussion on the interplay between lending pools and other DeFi archetypes, like stable coins and automated market makers (Sect. 7) and the identification of relevant research challenges (Sect. 8).

Overall, our contributions help address the aforementioned challenges in the design of lending pools. Firstly, our formal model provides a precise understanding of the behaviour of lending pools, abstracting from low-level implementation details. Our model is faithful to mainstream lending pool implementations like Compound [14] and Aave [7]; still, for the sake of clarity, we have introduced high-level abstractions over low-level details: we discuss the differences between our model and the actual lending pool platforms in Sect. 6. Secondly, our formalisation of the properties of the incentive mechanisms of lending pools makes it easier to understand and analyse their vulnerabilities and attacks. In this regard, our model is directly amenable for its interpretation as an *executable specification*, thus paving the way for automatic analysis techniques, which may include

mechanised proofs of contract properties and agent-based simulations of lending pools and other DeFi contracts. Due to space constraints, we provide the proofs of our statements in a separate technical report [56].

2   Background

Lending pools (in short, LPs) are financial applications which create a market of loans of crypto-assets, providing incentive mechanisms to equilibrate the market. We now overview the main features of LPs; a glossary of terms is in Table 1.

Users can lend assets to an LP by transferring *tokens* from their accounts to the LP. In return, they receive a *claim*, represented as tokens *minted* by the LP, which can later be redeemed for an equal or increased amount of tokens, of the same *token type* of the original deposit. Lending is incentivized by interest or fees: the depositor speculates that the claim will be redeemable for a value greater than the value of the original deposit. Users can redeem claims by transferring minted tokens to the LP, which pays back the original tokens (with accrued interest) to the redeemer, simultaneously burning the minted tokens. However, redeeming claims is not always possible, as the LP could not have a sufficient balance of the original tokens, as these may have been lent to other users.

User initiate a *loan* by borrowing tokens deposited to an LP. To incentivise users to eventually repay the loan, borrowing requires to provide a *collateral*. Collaterals can be either tokens deposited to the LP when the loan is initiated,

Table 1. Glossary of financial terms used in Lending Pools.

Token	A digital representation of some asset, transferable between users.
Token type	A set of tokens. Tokens of a given type are interchangeable (or *fungible*), whereas tokens of different token types are not.
Native token	The default token type of a blockchain (e.g., ETH for Ethereum).
Token price	The price of a token type τ is the amount of units of a given native crypto-currency (or fiat currency) needed to buy one unit of τ.
Lender	A user who transfers units of a token type in return for a *claim* on a full repayment in the future, which may include additional fees or interest.
Claim	A right to token units in the future. Claims are represented as tokens, which are *minted* and destroyed as claims are created and redeemed.
Minting	Creation of tokens performed by the LP upon deposits.
Borrower	A user who wishes to obtain a *loan* of token type τ. The borrower is required to hold *collateral* of another token τ' to secure the loan.
Collateral	A user balance of tokens which can be seized if the user does not adequately repay a loan.
Collateralization	The ratio of deposited *collateral* value over the borrower's total loan value.
Liquidation	When the *collateralization* of user A falls below a minimum threshold it is *undercollateralized*: here, a user B can repay a fraction of A's loan, in return for a discounted amount of A's collateral *seized* by B.
Interest rate	The rate of loan growth when accruing interest.

and locked for the whole loan duration, or they can be tokens held by the borrower but *seizable* by the LP when a user fails to repay a loan. An unpaid loan of A can be *liquidated* by B, who pays (part of) A's loan in return for a discounted amount of A's collateral. For this to be possible, the value of the collateral must be greater than that of the loan. To incentivize deposits, loans *accrue* interest, which increase a user's loan amount by the *interest rate*.

3   Lending Pools

In this section we introduce a formal model of lending pools, focussing on the common features implemented by the main LP platforms. We make our model parametric w.r.t. platform-specific features, like e.g. interest rate models, and we abstract from some advanced features, like e.g. governance (we discuss the differences between our model and the main LP platforms in Sect. 6).

3.1   Lending Pools Basics

We assume a set of **users** $\mathbb{A}$, ranged over by $A, A', \ldots$, and a set of **token types** $\mathbb{T}$, ranged over by $\tau, \tau', \ldots$. Units of these token types can be freely transferred between users, deposited into LPs, and borrowed. When a user deposits units of τ into an LP, she receives in return units of a token $\{\tau\}$ *minted* by the LP. We denote with $\mathbb{T}_m = \{\{\tau\} \mid \tau \in \mathbb{T}\}$ the set of **minted token types**. We use v, v', r, r' to range over nonnegative real numbers ($\mathbb{R}_0^+$). We write $r : \tau$ to denote r units of a token type τ (and similarly, we write $r : \{\tau\}$ for minted token types).

Wallets and Lending Pools. We model the **wallet** of a user A as a term $A[\sigma]$, where the partial map $\sigma \in \mathbb{T} \cup \mathbb{T}_m \rightharpoonup \mathbb{R}_0^+$ represents A's token holdings. We model a **lending pool** as a pair of the form $(r : \tau, \delta)$, where r is the amount of tokens of type $\tau \in \mathbb{T}$ deposited in the LP, and the map $\delta \in \mathbb{A} \to \mathbb{R}_0^+$ represents the users' **debts** of tokens of type τ.

Blockchain States and Transactions. We formalise the interaction between users and the blockchain as a labelled transition system. Labels $\mathsf{T}, \mathsf{T}', \ldots$ represent **transactions** (see Table 2), while *states* $\Gamma, \Gamma', \ldots$ are compositions of

Table 2. Transactions.

$A : \mathsf{xfer}(B, v : \tau)$	A transfers v units of τ to B
$A : \mathsf{dep}(v : \tau)$	A deposits v units of τ, receiving units of minted token $\{\tau\}$
$A : \mathsf{mxfer}(B, v : \{\tau\})$	A transfers v units of $\{\tau\}$ to B
$A : \mathsf{bor}(v : \tau)$	A borrows v units of τ
int	All loans accrue interest
$A : \mathsf{rep}(v : \tau)$	A repays v units on A's debt in τ
$A : \mathsf{rdm}(v : \{\tau\})$	A redeems v units of $\{\tau\}$, receiving units of τ
$A : \mathsf{liq}(B, v : \tau, \{\tau'\})$	A repays v units of B's debt in τ, seizing units of $\{\tau'\}$ from B

wallets, LPs, and a single **price oracle** $P \in \mathbb{T} \to \mathbb{R}_0^+ \setminus \{0\}$ which prices tokens. We represent states as terms of the form:

$$\mathsf{A}_1[\sigma_1] \mid \cdots \mid \mathsf{A}_n[\sigma_n] \mid (r_1 : \tau_1, \delta_1) \mid \cdots \mid (r_k : \tau_k, \delta_k) \mid P$$

where all A_i are distinct, and $\tau_i \neq \tau_j$ for all $i \neq j$. A state Γ is *initial* when it only contains a price oracle and a set of wallets, holding only non-minted tokens. We treat states as sets of terms: hence, Γ and Γ' are equivalent when they contain the same terms; for a term Q, we write $Q \in \Gamma$ when $\Gamma = Q \mid \Gamma'$, for some Γ'.

Exchange Rate. The *exchange rate* of a token type τ in a state Γ represents the share of deposited units of τ over the units of the associated minted tokens. Before formalising it, we define the auxiliary notion of *supply* of a token type $t \in \mathbb{T} \cup \mathbb{T}_m$ in a state Γ, i.e. the sum of the balances of t in all the wallets in Γ, and possibly in the LPs. It is defined inductively as:

$$sply_t(\mathsf{A}[\sigma]) = \sigma(t) \qquad sply_t(r : \tau, \delta) = \begin{cases} r & \text{if } t = \tau \\ 0 & \text{otherwise} \end{cases} \tag{1}$$

$$sply_t(P) = 0 \qquad sply_t(\Gamma \mid \Gamma') = sply_t(\Gamma) + sply_t(\Gamma')$$

Then, we define the **exchange rate** $ER_\tau(\Gamma)$ as:

$$ER_\tau(\Gamma) = \begin{cases} \dfrac{r + \sum_\mathsf{A} \delta(\mathsf{A})}{sply_{\{\tau\}}(\Gamma)} & \text{if } \Gamma = (r : \tau, \delta) \mid \Gamma', r > 0 \\ 1 & \text{otherwise} \end{cases} \tag{2}$$

The idea is that, while initially there is a $1/1$ correspondence between minted and deposited tokens, when interest is accrued this relation changes to the benefit of lenders.

Net Worth and Collateralization. The *value* of A's tokens in a state Γ, denoted by $V_\Gamma(\mathsf{A})$, is the sum of the values of all (non-minted) tokens in A's wallet (the value is the product between token amount and price):

$$V_\Gamma(\mathsf{A}) = \sum_{\tau \in \mathbb{T}} \sigma(\tau) \cdot P(\tau) \qquad \text{if } \Gamma = \mathsf{A}[\sigma] \mid P \mid \Gamma' \tag{3}$$

We define similarly the value $V^m(\mathsf{A})$ of minted tokens held by A. To determine the value of $\{\tau\}$, its price is equated to that of the underlying token τ:

$$V_\Gamma^m(\mathsf{A}) = \sum_{\tau \in \mathbb{T}} \sigma(\{\tau\}) \cdot ER_\tau(\Gamma) \cdot P(\tau) \qquad \text{if } \Gamma = \mathsf{A}[\sigma] \mid P \mid \Gamma' \tag{4}$$

The value $V^d(\mathsf{A})$ of A's debt is the sum of the value of tokens borrowed by A:

$$V_\Gamma^d(\mathsf{A}) = \sum_{i \in I} \delta_i(\mathsf{A}) \cdot P(\tau_i) \qquad \text{if } \Gamma = \|_{i \in I}(r_i : \tau_i, \delta_i) \mid \|_j \mathsf{A}_j[\sigma_j] \mid P \tag{5}$$

The **net worth** of a user is the value of the tokens in her wallet (both minted and non-minted), minus the value of her debt:

$$W_\Gamma(\mathsf{A}) = V_\Gamma(\mathsf{A}) + V_\Gamma^m(\mathsf{A}) - V_\Gamma^d(\mathsf{A}) \tag{6}$$

$\Gamma_0 = A[100 : \tau_0] \mid B[50 : \tau_1] \mid P = \{^1/\tau_0, ^1/\tau_1\}$

$\xrightarrow{\text{1. A:dep}(50:\tau_0)}$

$\Gamma_1 = A[50 : \tau_0, 50 : \{\tau_0\}] \mid (50 : \tau_0, \{\}) \mid \cdots$

$\xrightarrow{\text{2. B:dep}(50:\tau_1)}$

$\Gamma_2 = B[50 : \{\tau_1\}] \mid (50 : \tau_1, \{\}) \mid \cdots$

$\xrightarrow{\text{3. B:bor}(30:\tau_0)} \quad \{C_{\Gamma_3}(B) = 1.7\}$

$\Gamma_3 = B[30 : \tau_0, \cdots] \mid (20 : \tau_0, \{^{30}/B\}) \mid \cdots$

$\xrightarrow{\text{4. int}} \quad \{C_{\Gamma_4}(B) = 1.5, ER_{\tau_0}(\Gamma_4) = 1.1\}$

$\Gamma_4 = \cdots \mid (20 : \tau_0, \{^{34}/B\}) \mid \cdots$

$\xrightarrow{\text{5. B:rep}(5:\tau_0)} \quad \{C_{\Gamma_5}(B) = 1.7\}$

$\Gamma_5 = B[25 : \tau_0, \cdots] \mid (25 : \tau_0, \{^{29}/B\}) \mid \cdots$

$\xrightarrow{\text{6. px}} \quad \{C_{\Gamma_6}(B) = 1.3 < C_{min}\}$

$\Gamma_6 = \cdots \mid P' = \{^{1.3}/\tau_0, \cdots\}$

$\xrightarrow{\text{7. A:liq}(B,13:\tau_0,\{\tau_1\})} \quad \{C_{\Gamma_7}(B) = 1.5 = C_{min}\}$

$\Gamma_7 = A[37 : \tau_0, 19 : \{\tau_1\}, \cdots] \mid B[31 : \{\tau_1\}, \cdots] \mid (38 : \tau_0, \{^{16}/B\}) \mid \cdots$

$\xrightarrow{\text{8. A:rdm}(10:\{\tau_0\})} \quad \{ER_{\tau_0}(\Gamma_8) = 1.1\}$

$\Gamma_8 = A[48 : \tau_0, 40 : \{\tau_0\}, \cdots] \mid (27 : \tau_0, \{^{16}/B\}) \mid \cdots$

Fig. 1. Interactions between two users and a lending pool.

The **collateralization** of a user is the ratio of the value of her minted tokens and the value of her debt:

$$C_\Gamma(A) = \begin{cases} V_\Gamma^m(A) / V_\Gamma^d(A) & \text{if } V_\Gamma^d(A) > 0 \\ +\infty & \text{otherwise} \end{cases} \tag{7}$$

State-Update Operators. We use the standard notation $\sigma\{v/x\}$ to update a partial map σ at point x: namely, $\sigma\{v/x\}(x) = v$, while $\sigma\{v/x\}(y) = \sigma(y)$ for $y \neq x$. Given a partial map $\sigma \in \mathbb{T} \cup \mathbb{T}_m \rightharpoonup \mathbb{R}_0^+$, a partial operation $\circ \in \mathbb{R}_0^+ \times \mathbb{R}_0^+ \rightharpoonup \mathbb{R}_0^+$, $t \in \mathbb{T} \cup \mathbb{T}_m$ and $v \in \mathbb{R}_0^+$, we define the partial map $\sigma \circ v : t$ as follows:

$$\sigma \circ v : t = \begin{cases} \sigma\{v'/t\} & \text{if } t \in \text{dom}\,\sigma \text{ and } v' = \sigma(t) \circ v \in \mathbb{R}_0^+ \\ \sigma\{v/t\} & \text{if } t \notin \text{dom}\,\sigma \end{cases}$$

3.2   An Overview of Lending Pools Behaviour

Before formalizing the behaviour of lending pools, we give some intuition through an example involving users A and B. We display their interactions in Fig. 1.

In the initial state, A has 100 units of τ_0, B has 50 units of τ_1, and the price of both token types is 1. In the first two steps, A and B deposit 50 units of τ_0

and τ_1, for which they receive equal amounts of minted tokens $\{\tau_0\}$ and $\{\tau_1\}$. We denote with $\{\}$ the function $\lambda A.0$ (i.e., no user has debts).

Next, B borrows $30 : \tau_0$. The 50 minted tokens of type $\{\tau_1\}$ in B's wallet serve as *collateral* for the loan. The *collateralization* of B is the ratio between the *value* of B's balance of $\{\tau_1\}$ and the value of B's debt of τ_0, according to (7). Assuming a minimum collateralization threshold of $C_{min} = 1.5$, B could borrow up to 33 units of τ_1, given the collateral of $50 : \{\tau_1\}$. Nonetheless, B decides to leave some margin to manage future price volatility and the accrual of interest, which can both negatively affect collateralization. In the state Γ_3, the map $\{30/\mathsf{B}\}$ in the LP for τ_0 represents that B's debt of τ_0 is 30, while the other users have no debt.

In step 4, interest accrues on B's debt. Here, we assume that the interest rate is 12%, so B's debt grows from 30 to 34 units of τ_1. In step 5, B repays 5 units of τ_0 to reduce the risk of becoming *liquidated*, which can occur when B's collateralization falls below the threshold $C_{min} = 1.5$.

Despite this effort, the price is updated in step 6, increasing $P(\tau_0)$ by 30% relative to $P(\tau_1)$, thereby decreasing the relative value of B's collateral to B's loan. As a result, the collateralization of B drops below the threshold C_{min}.

In step 7, A liquidates $13 : \tau_0$ of B's debt, restoring B's collateralization to C_{min}, and simultaneously seizing $19 : \{\tau_1\}$ from B's balance. The exchange of $13 : \tau_0$ for $19 : \{\tau_1\}$ implies a liquidation discount, which ensures that the liquidation is profitable for any user performing it.

In step 8, A redeems $10 : \{\tau_0\}$, receiving $11 : \tau_0$ in exchange. Here, each unit of $\{\tau_0\}$ is now exchanged for more than 1 unit of τ_0, due to accrued interest.

3.3 Lending Pools Semantics

We now present the rules which define the transitions between lending pool states. In all the rules, denote with Γ_0 the state *before* the transition, and with Γ_1 the state *after* the transition. An extended running example (Figs. 2, 4, 5, 6 and 7) illustrates all the peculiar aspects of these rules.

Token Transfer. The transaction $A : \mathsf{xfer}(B, v : \tau)$, represents the transfer of $v : \tau$ from A to B. Its effect on the state is specified by the following rule:

$$\frac{\sigma_A(\tau) \geq v}{A[\sigma_A] \mid B[\sigma_B] \mid \Gamma \xrightarrow{A:\mathsf{xfer}(B,v:\tau)} A[\sigma_A - v : \tau] \mid B[\sigma_B + v : \tau] \mid \Gamma} \text{[XFER]}$$

Rule [XFER] states that the transfer is permitted whenever the sender has a sufficient balance. Note that the rule only allows transfers of *non*-minted tokens; transfers of minted tokens is specified by rule [MXFER] below.

Deposit. A user A can deposit v units of a (non-minted) token τ by performing the transaction $A : \mathsf{dep}(v : \tau)$. Upon the first deposit of τ, A receives exactly v units of the minted token $\{\tau\}$:

$$\frac{\sigma(\tau) \geq v \qquad (\_ : \tau, \_) \notin \Gamma}{A[\sigma] \mid \Gamma \xrightarrow{A:\mathsf{dep}(v:\tau)} A[\sigma - v : \tau + v : \{\tau\}] \mid (v : \tau, \lambda A.0) \mid \Gamma} \text{[DEP0]}$$

$$\Gamma_0 = \mathsf{A}[100 : \tau_0, 300 : \tau_1] \mid \mathsf{B}[50 : \tau_0, 50 : \tau_2] \mid \mathsf{C}[100 : \tau_2] \mid P$$

$$\xrightarrow{\text{1. A:dep}(100:\tau_0)}$$

$$\Gamma_1 = \mathsf{A}[0 : \tau_0, 100 : \{\tau_0\}, \cdots] \mid (100 : \tau_0, \{\}) \mid \cdots$$

$$\xrightarrow{\text{2. A:dep}(150:\tau_1)}$$

$$\Gamma_2 = \mathsf{A}[150 : \{\tau_1\}, \cdots] \mid (150 : \tau_1, \{\}) \mid \cdots$$

$$\xrightarrow{\text{3. B:dep}(50:\tau_0)}$$

$$\Gamma_3 = \mathsf{B}[0 : \tau_0, 50 : \{\tau_0\}, \cdots] \mid (150 : \tau_0, \{\}) \mid \cdots$$

$$\xrightarrow{\text{4. B:dep}(50:\tau_2)}$$

$$\Gamma_4 = \mathsf{B}[0 : \tau_2, 50 : \{\tau_2\}, \cdots] \mid (50 : \tau_2, \{\}) \mid \cdots$$

$$\xrightarrow{\text{5. C:dep}(100:\tau_2)}$$

$$\Gamma_5 = \mathsf{C}[0 : \tau_2, 100 : \{\tau_2\}] \mid (150 : \tau_2, \{\}) \mid \cdots$$

Fig. 2. Running example: deposit actions

The first rule premise ensures that A's balance is sufficient. The second premise checks that no LP for τ is already present in the state. The map $\lambda \mathsf{A}.0$ represents the fact that, in the newly created LP, the debt of each user is 0.

For further deposits of τ, the LP mints new units of $\{\tau\}$. Their amount v' is the ratio between the deposited amount v and the exchange rate $ER_\tau(\Gamma_0)$ between τ and $\{\tau\}$, defined in (2).

$$\frac{\sigma(\tau) \geq v \qquad v' = v/ER_\tau(\Gamma_0)}{\mathsf{A}[\sigma] \mid (r : \tau, \delta) \mid \Gamma \xrightarrow{\text{A:dep}(v:\tau)} \mathsf{A}[\sigma - v : \tau + v' : \{\tau\}] \mid (r + v : \tau, \delta) \mid \Gamma} \text{ [Dep]}$$

Figure 2 exemplifies users depositing funds to the LP. In step 1, A deposits 100 units of τ_0. Since this is the first deposit of τ_0, the LP mints exactly 100 units of $\{\tau_0\}$, and transfers these units to A. In step 2, A deposits 150 units of τ_1; similarly to the previous case, A receives 150 units of $\{\tau_1\}$. In step 3, B deposits 50 units of τ_0. Since τ_0 was already deposited, the LP mints 50 units of $\{\tau_0\}$, and transfers them to B. Finally, in steps 4 and 5, B and C deposit units of τ_2; after that, the balances of tokens τ_0, τ_1, τ_2 in the LP total 150 units.

Transfer of Minted Tokens. Minted tokens can be transferred between users, provided that, after the transfer, the sender has enough minted tokens to use as collateral. More specifically, we require that the *collateralization* of the sender in the target state is above a constant threshold $C_{min} > 1$.

$$\frac{\sigma_\mathsf{A}(\{\tau\}) \geq v \qquad C_{\Gamma_1}(\mathsf{A}) \geq C_{min}}{\mathsf{A}[\sigma_\mathsf{A}] \mid \mathsf{B}[\sigma_\mathsf{B}] \mid \Gamma \xrightarrow{\text{A:mxfer}(\mathsf{B},v:\{\tau\})} \mathsf{A}[\sigma_\mathsf{A} - v : \{\tau\}] \mid \mathsf{B}[\sigma_\mathsf{B} + v : \{\tau\}] \mid \Gamma} \text{ [MxFer]}$$

Borrow. Any user can borrow units of a token type τ from an LP, provided that the LP has a sufficient balance of τ, and that the user has enough minted

tokens to use as collateral.

$$\frac{r \geq v > 0 \qquad C_{\Gamma_1}(\mathsf{A}) \geq C_{min}}{\mathsf{A}[\sigma] \mid (r : \tau, \delta) \mid \Gamma \xrightarrow{\mathsf{A}:\mathsf{bor}(v:\tau)} \mathsf{A}[\sigma + v : \tau] \mid (r - v : \tau, \delta\{\delta(\mathsf{A})+v/\mathsf{A}\}) \mid \Gamma} \text{[Bor]}$$

We exemplify bor transactions in Fig. 3. Users B and C borrow amounts of τ_0 and τ_1 at steps 6–8, keeping their collateralization above C_{min}, which is assumed to be 1.5. C's collateralization decreases from 3.3 to 1.7 upon step 8: this is due to the increase in $V^d(\mathsf{C})$, whilst $V^m(\mathsf{C})$ remains constant at 100.

$$\Gamma_5 = \mathsf{B}[50 : \{\tau_0\}, 50 : \{\tau_2\}] \mid \mathsf{C}[100 : \{\tau_2\}] \mid (150 : \tau_0, \{\}) \mid (150 : \tau_1, \{\}) \mid \cdots$$

$$\xrightarrow{6.\ \mathsf{B}:\mathsf{bor}(50:\tau_1)} \{C_{\Gamma_6}(\mathsf{B}) = 2.0\}$$

$$\Gamma_6 = \mathsf{B}[50 : \tau_1, \cdots] \mid (100 : \tau_1, \{50/\mathsf{B}\}) \mid \cdots$$

$$\xrightarrow{7.\ \mathsf{C}:\mathsf{bor}(30:\tau_0)} \{C_{\Gamma_7}(\mathsf{C}) = 3.3\}$$

$$\Gamma_7 = \mathsf{C}[30 : \tau_0, \cdots] \mid (120 : \tau_0, \{30/\mathsf{C}\}) \mid \cdots$$

$$\xrightarrow{8.\ \mathsf{C}:\mathsf{bor}(30:\tau_1)} \{C_{\Gamma_8}(\mathsf{C}) = 1.7\}$$

$$\Gamma_8 = \mathsf{C}[30 : \tau_1, \cdots] \mid (70 : \tau_1, \{50/\mathsf{B}, 30/\mathsf{C}\}) \mid \cdots$$

Fig. 3. Running example: borrow actions

The collateralization of a user depends on the amount of minted tokens she possesses, the amount of tokens she has borrowed, and the price of all tokens involved. Hence, collateralization is potentially sensitive to all actions that can affect those values. This includes both interest accrual and changes in token prices (which are unpredictable), as we shall see. Borrowers must therefore maintain a safety margin in order to protect against potential liquidations.

Interest Accrual. Interest accrual models the periodic application of interest to loan amounts and can be executed in any state. The action applies a token-specific interest to each loan, updating the debt mapping for *all* users.

$$\frac{\forall i \in I : \forall \mathsf{A} : \delta_i'(\mathsf{A}) = \delta_i(\mathsf{A}) \cdot (1 + I_{\Gamma_0}(\tau_i)) \qquad (\_ : \_, \_) \notin \Gamma}{\|_{i \in I} (r_i : \tau_i, \delta_i) \mid \Gamma \xrightarrow{\text{int}} \|_{i \in I} (r_i : \tau_i, \delta_i') \mid \Gamma} \text{[Int]}$$

Existing lending pool platforms deploy different algorithmic interest rate models [53]. We leave our model parametric w.r.t. interest rates, and only require that the interest rate is positive, a property that all models in [53] satisfy:

$$I_{\Gamma}(\tau) > 0 \tag{8}$$

We extend our running example with three interest updates in Fig. 4, resulting in the increase of all loan amounts. Each subsequent execution of int *decreases* the collateralization of users B and C, since the V^d of both borrowers *increases* as interest is applied (7).

$$\Gamma_8 = (120 : \tau_0, \{^{30}/\text{C}\}) \mid (70 : \tau_1, \{^{50}/\text{B}, {}^{30}/\text{C}\}) \mid \cdots$$

$$\xrightarrow{9.\ \text{int}} \quad \{C_{\Gamma_9}(\text{B}) = 1.9, C_{\Gamma_9}(\text{C}) = 1.6\}$$

$$\Gamma_9 = (120 : \tau_0, \{^{31}/\text{C}\}) \mid (70 : \tau_1, \{^{53}/\text{B}, {}^{32}/\text{C}\}) \mid \cdots$$

$$\xrightarrow{10.\ \text{int}} \quad \{C_{\Gamma_{10}}(\text{B}) = 1.8, C_{\Gamma_{10}}(\text{C}) = 1.5\}$$

$$\Gamma_{10} = (120 : \tau_0, \{^{32}/\text{C}\}) \mid (70 : \tau_1, \{^{56}/\text{B}, {}^{34}/\text{C}\}) \mid \cdots$$

$$\xrightarrow{11.\ \text{int}} \quad \{C_{\Gamma_{11}}(\text{B}) = 1.7, C_{\Gamma_{11}}(\text{C}) = 1.4\}$$

$$\Gamma_{11} = (120 : \tau_0, \{^{33}/\text{C}\}) \mid (70 : \tau_1, \{^{59}/\text{B}, {}^{36}/\text{C}\}) \mid \cdots$$

Fig. 4. Running example: interest accrual

Repay. A user with a loan can repay part of it by executing a rep transaction:

$$\frac{\sigma(\tau) \geq v > 0 \qquad \delta(\text{A}) \geq v}{\text{A}[\sigma] \mid (r : \tau, \delta) \mid \Gamma \xrightarrow{\text{A:rep}(v:\tau)} \text{A}[\sigma - v : \tau] \mid (r + v : \tau, \delta\{^{\delta(\text{A})-v}/\text{A}\}) \mid \Gamma} \quad [\text{Rep}]$$

This increases the collateralization of the repaying user, as V^d is reduced (7). Users must always maintain a sufficient collateralization, to cope with adverse effects of interest accruals and price updates.

In Fig. 5, C is suffering from low collateralization after the last interest accrual in step 11. Here, $C_\Gamma(\text{C})$ is equal to $C_{min} = 1.5$. The subsequent repayment of 15 units of τ_0 increases C's collateralization back to 1.9.

$$\Gamma_{11} = \text{C}[30 : \tau_0, 30 : \tau_1, 100 : \{\tau_2\}] \mid (120 : \tau_0, \{^{33}/\text{C}\}) \mid (70 : \tau_1, \{^{59}/\text{B}, {}^{36}/\text{C}\}) \mid \cdots$$

$$\xrightarrow{12.\ \text{C:rep}(15:\tau_0)} \quad \{C_{\Gamma_{12}}(\text{C}) = 1.9\}$$

$$\Gamma_{12} = \text{C}[15 : \tau_0, \cdots] \mid (135 : \tau_0, \{^{18}/\text{C}\}) \mid \cdots$$

Fig. 5. Running example: repay actions

Redeem. A user without any loans can redeem minted tokens $\{\tau\}$ for the underlying tokens if enough units of τ remain in the LP. A user with a *non-zero* loan amount of any token can only redeem minted tokens such that the resulting collateralization is not below C_{min}. This constraint does not apply to users without loans, as minted tokens are not used as collateral.

$$\frac{\sigma(\{\tau\}) \geq v \qquad v' = v \cdot ER_\tau(\Gamma_0) \qquad r \geq v' \qquad C_{\Gamma_1}(\text{A}) \geq C_{min}}{\text{A}[\sigma] \mid (r : \tau, \delta) \mid \Gamma \xrightarrow{\text{A:rdm}(v:\{\tau\})} \text{A}[\sigma - v : \{\tau\} + v' : \tau] \mid (r - v' : \tau, \delta) \mid \Gamma} \quad [\text{Rdm}]$$

We exemplify rdm transactions in Fig. 6. From Fig. 5, B has a non-zero loan amount, hence she can only redeem $11 : \{\tau_2\}$ before her collateralization decreases to $C_{min} = 1.5$, at which B cannot further redeem. Since A has no loans, she can redeem as many tokens $\{\tau_0\}$ as the LP balance permits. For A's

redeeming of $50 : \{\tau_0\}$ for $51 : \tau_0$ the exchange rate is > 1, because of the accrued interest during the prior execution of int. By contrast, the exchange rate for B is 1, as no loan exists on τ_2, and thus no interest was accrued. The minted tokens $\{\tau_2\}$ and $\{\tau_0\}$ returned to the LP by B and A are burnt.

$$
\begin{aligned}
\Gamma_{12} = \ &A[0 : \{\tau_0\}, 100 : \{\tau_0\}, \cdots] \mid (135 : \tau_0, \{^{18}/c\}) \mid \\
&B[50 : \{\tau_0\}, 50 : \tau_1, 50 : \{\tau_2\}] \mid (150 : \tau_2, \{\}) \mid \cdots \\
&\xrightarrow{\;13.\; B:rdm(11:\{\tau_2\})\;} \quad \{C\Gamma_{13}(B) = 1.5,\; ER_{\tau_2}(\Gamma_{12}) = 1\} \\
\Gamma_{13} = \ &B[11 : \tau_2, 39 : \{\tau_2\}, \cdots] \mid (139 : \tau_2, \{\}) \mid \cdots \\
&\xrightarrow{\;14.\; A:rdm(50:\{\tau_0\})\;} \quad \{ER_{\tau_0}(\Gamma_{13}) = 1.02\} \\
\Gamma_{14} = \ &A[51 : \tau_0, 50 : \{\tau_0\}, \cdots] \mid (84 : \tau_0, \{^{18}/c\}) \mid \cdots
\end{aligned}
$$

Fig. 6. Running example: redeem actions

Liquidation. When the collateralization of a user B is below the threshold C_{min}, another user A can *liquidate* part of B's loan, in return for a discounted amount of minted tokens *seized* from B. A can execute liq if she has enough balance to repay a fraction of the lent token, and if B has a sufficient balance of seizable, minted tokens. The maximum seizable amount is bounded by B's balance of the minted token and by the resulting collateralization of B, which cannot exceed C_{min}. After this threshold, B's collateralization is restored, and B is no longer liquidatable.

$$
\frac{\begin{array}{ccc} \sigma_A(\tau) \geq v & \sigma_B(\{\tau'\}) \geq v' & \delta(B) \geq v \\ C_{\Gamma_0}(B) < C_{min} & C_{\Gamma_1}(B) \leq C_{min} \end{array} \qquad v' = v \cdot \frac{P(\tau)}{P(\tau')} \cdot \eta_{liq}}{\begin{array}{l} A[\sigma_A] \mid B[\sigma_B] \mid (r : \tau, \delta) \mid \Gamma \xrightarrow{\;A:liq(B,v:\tau,\{\tau'\})\;} \\ A[\sigma_A - v : \tau + v' : \{\tau'\}] \mid B[\sigma_B - v' : \{\tau'\}] \mid (r + v : \tau, \delta\{^{\delta(B)-v}/_B\}) \mid \Gamma \end{array}} \; \text{[Liq]}
$$

where we require that:

$$
C_{min} > \eta_{liq} > 1 \tag{9}
$$

The constraint $\eta_{liq} > 1$ implies a discount applied to the seized amount received by the liquidator, as more value is received than repaid. In [Liq], there are no constraints on the collateralization of the liquidator A: its balance of minted tokens increases whilst its lent token amounts remain unchanged, thus always increasing its collateralization (7).

For the liquidations in Fig. 7, we set $\eta_{liq} = 1.1$. After the price update in step 15, both B and C are undercollateralized. C is liquidated by A in step 16, which restores $C_\Gamma(C)$ to 1.5. By contrast, $C_\Gamma(B)$ is 0.9 after the price update. Subsequent liquidations by A seize *all* units $\{\tau_0\}$ and $\{\tau_2\}$ from B's wallet. However, B still has a debt of $11 : \tau_1$. This debt is *unrecoverable*, since there is no incentive to repay or liquidate it, given the lack of collateral.

$$\Gamma_{14} = \mathsf{A}[150:\tau_1, 50:\{\tau_0\}, \cdots] \mid \mathsf{B}[50:\{\tau_0\}, 39:\{\tau_2\}, \cdots] \mid \mathsf{C}[100:\{\tau_2\}, \cdots] \mid$$
$$(70:\tau_1, \{^{59}/\mathsf{B}, {}^{36}/\mathsf{c}\}) \mid \cdots \mid P = \{^1/\tau_0, {}^1/\tau_1, {}^1/\tau_2\}$$

$$\xrightarrow{\text{15. px}} \quad \{C_{\Gamma_{15}}(\mathsf{B}) = 0.9, C_{\Gamma_{15}}(\mathsf{C}) = 1.3\}$$

$$\Gamma_{15} = \cdots \mid P' = \{^{1.7}/\tau_1, \cdots\}$$

$$\xrightarrow{\text{16. A:liq(C,27:}\tau_1,\{\tau_2\})} \quad \{C_{\Gamma_{16}}(\mathsf{C}) = 1.5\}$$

$$\Gamma_{16} = \mathsf{A}[123:\tau_1, 50:\{\tau_2\}, \cdots] \mid \mathsf{C}[50:\{\tau_2\}, \cdots] \mid (97:\tau_1, \{^{59}/\mathsf{c}, {}^9/\mathsf{c}\}) \mid \cdots$$

$$\xrightarrow{\text{17. A:liq(B,27:}\tau_1,\{\tau_0\})} \quad \{C_{\Gamma_{17}}(\mathsf{B}) = 0.7\}$$

$$\Gamma_{17} = \mathsf{A}[96:\tau_1, 100:\{\tau_0\}, \cdots] \mid \mathsf{B}[0:\{\tau_0\}, \cdots] \mid (124:\tau_1, \{^{32}/\mathsf{B}, {}^9/\mathsf{c}\}) \mid \cdots$$

$$\xrightarrow{\text{18. A:liq(B,21:}\tau_1,\{\tau_2\})} \quad \{C_{\Gamma_{18}}(\mathsf{B}) = 0\}$$

$$\Gamma_{18} = \mathsf{A}[75:\tau_1, 89:\{\tau_2\}, \cdots] \mid \mathsf{B}[0:\{\tau_2\}, \cdots] \mid (145:\tau_1, \{^{11}/\mathsf{B}, {}^9/\mathsf{c}\}) \mid \cdots$$

Fig. 7. Running example: liquidation actions

Price Updates. Finally, the price oracle can be updated non-deterministically:

$$P \mid \Gamma \xrightarrow{\text{px}} P' \mid \Gamma \quad \text{[Px]}$$

4   Fundamental Properties of Lending Pools

We now establish some fundamental properties of lending pools. A crucial property is that the exchange rate of any token τ either strictly increases, when users have loans on τ, or remain stable otherwise. This guarantees that stocks of the minted token $\{\tau\}$ will gain value.

Lemma 1. *Let* $\Gamma = (r:\tau,\delta) \mid \cdots$, *and let* $\Gamma \xrightarrow{\mathsf{T}} \Gamma'$. *Then:*

(a) if $\mathsf{T} = \mathsf{int}$ *and* $\delta(\mathsf{A}) > 0$ *for some* A, *then* $ER_\tau(\Gamma) < ER_\tau(\Gamma')$;
(b) otherwise, $ER_\tau(\Gamma) = ER_\tau(\Gamma')$.

Lemma 2 establishes that the supply of any (non-minted) token is constant.

Lemma 2. *Let* $\Gamma \xrightarrow{\mathsf{T}} \Gamma'$. *Then, for all* $\tau \in \mathbb{T}$: $sply_\tau(\Gamma) = sply_\tau(\Gamma')$.

The net worth of a user can be increased in short or long sequences of transitions. In general, there is no winning strategy (in the game-theoretic sense) for a single user who wants to increase her net worth, unless she can control price updates. However, under certain conditions, winning strategies exist. We consider first a simple 1-player game where a user can choose her next action to improve her net worth in the next state. Lemma 3 shows that liquidation is the only action that allows the user to increases her net worth in a single step.

Lemma 3. *Let* $\Gamma \xrightarrow{\mathsf{A}:\ell(\cdots)} \Gamma'$. *Then:*

(a) if $\ell = \mathsf{liq}$, *then* $W_\Gamma(\mathsf{A}) < W_{\Gamma'}(\mathsf{A})$;
(b) otherwise, $W_\Gamma(\mathsf{A}) = W_{\Gamma'}(\mathsf{A})$.

Since this is the winning strategy for all users, but liquidations may be limited by the amount of debts and collaterals, an adversary with the power to drop or reorder transactions could potentially monopolize liquidations for itself. We refer to Sect. 7 for additional discussion of such attacks.

We now consider a slightly extended game, where A guesses that the adversary is going to fire int, resulting in $\Gamma_0 \xrightarrow{\text{int}} \Gamma_1$, but can still perform an action $A : \ell(\cdots)$ before int, resulting in $\Gamma_0 \xrightarrow{A:\ell(\cdots)} \Gamma_0' \xrightarrow{\text{int}} \Gamma_1'$. Here, A's goal is to choose her action ℓ such that $W_{\Gamma_1'}(A) \geq W_{\Gamma_1}(A)$. Lemma 4 shows that A can achieve this goal by performing deposit, repay, or liquidation actions.

Lemma 4. Let $\Gamma_0 \xrightarrow{\text{int}} \Gamma_1$ and $\Gamma_0 \xrightarrow{A:\ell(\cdots)} \Gamma_0' \xrightarrow{\text{int}} \Gamma_1'$. Then:

(a) if $\ell \in \{\text{liq}, \text{dep}, \text{rep}\}$, then $W_{\Gamma_1'}(A) \geq W_{\Gamma_1}(A)$;
(b) otherwise, $W_{\Gamma_1'}(A) \leq W_{\Gamma_1}(A)$.

Overall, Lemmas 3 and 4 determine the set of actions to consider (together with their parameters) to maximize short-term improvements in net worth.

5 Lending Pool Safety, Vulnerabilities and Attacks

In this section we discuss further properties of lending pools, focusing on risks which could lead to unsecured loans or exploitation by malicious actors. In particular, we consider user collateralization and availability of token funds in LPs (utilization): if these can be targeted by an attacker, the motivation is to limit the LP functionality (denial-of-service) or to make the victim incur losses, which in some cases may imply a gain for the attacker. We consider attackers with the ability to perform some of the actions of the LP model, or even update the price oracle. More powerful attackers that can drop or reorder transactions are discussed in Sect. 7.

5.1 Collateralization Bounds and Risks

The lending pool design assumes that loans are *secured* by collateral: liquidations thereof are incentivised in order to recover loans if the borrowing users fail to repay. However, collateral liquidation is exposed to risks. First, the incentive to liquidate is only effective if the liquidator values the seized collateral higher than the value of the repaid loan amount, implying a profit. Second, large fluctuations in token price may reduce the relative value of the collateral, eventually making loans partially unrecoverable. Further, an attacker with the ability to update token prices can force users to become undercollateralized, and then seize the collateral of victims without repaying any loans.

LP-Minted Token Risk. Lending pools must determine the appropriate levels of collateralization based on token prices given by the oracle. However, the value of minted tokens is unpredictable, since they are not determined by price oracles (recall that the domain of P does not include minted tokens). The definition of

collateralization in (7) values minted tokens at the same price as their underlying token, just like LP implementations [9,18] do. However, since minted tokens are only redeemable if the LP has sufficient funds (see rule [RDM]), it may happen that users value minted tokens at a lower price than their underlying tokens. This happens e.g. when the funds in LPs are low, which may prevent users from performing redeem actions. Lending pool designs do not account for this, running the risk of incorrectly pricing minted tokens.

Safe Collateralization. Assuming a correct valuation of minted tokens, under-collateralized loans should be swiftly liquidated, given the incentivization provided by the liquidation discount. Furthermore, the user collateral value should be high enough, such that the user's loan amount is sufficiently repaid by liquidations to recover the user collateralization back to C_{min}. Therefore, we introduce two notions of safe collateralization.

Inspired by [55], we say that a state is ε-*collateralization safe* when the ratio between the value of the debt of undercollateralized users and the total value of the debt is less than ε:

$$\frac{\sum_{C_\Gamma(A) < C_{min}} V_\Gamma^d(A)}{\sum_A V_\Gamma^d(A)} \leq \varepsilon \tag{10}$$

If the liquidation incentive is effective, a value below ε should not persist, as users promptly execute liquidations. The efficiency of liquidations has been studied in [59]. Note that large volumes of seized collaterals which are immediately sold on external markets may delay further liquidations, as investigated in [52], due to the external market's finite capacity to absorb such a sell-off.

The notion of ε-collateralization safety does not account for undercollateralized loans which are *non-recoverable*, as previously illustrated in Fig. 7. The set of non-recoverable, undercollateralized users are those with a collateralization below r_{liq}. The non-recoverable value of a user's debt is defined as V_Γ^{nrd} in (11). It represents the remaining value of the debt of a user A should it be fully liquidated, such that no further collateral can be seized.

$$V_\Gamma^{nrd}(A) = \begin{cases} V_\Gamma^d(A) - \frac{V_\Gamma^m(A)}{r_{liq}} & \textit{iff } C_\Gamma(A) < r_{liq} \\ 0 & \textit{otherwise} \end{cases} \tag{11}$$

From (9) and (11) it follows that when A's collateralization is below r_{liq}, the discounted value of the collateral can no longer reach the remaining debt value.

We say that a state is *strongly ε-collateralization safe* when the fraction of the non-recoverable debt value over the total debt value is below ε:

$$\frac{\sum_A V_\Gamma^{nrd}(A)}{\sum_A V_\Gamma^d(A)} \leq \varepsilon \tag{12}$$

Condition (12) is stronger than (10): if a state is strongly ε-collateralization safe, then it is also ε-collateralization safe. Given equal denominators of (10) and (12), this is a consequence of comparing numerators: note that the numerator

of (10) is greater than that of (12), as $V_\Gamma^d(A)$ is greater than $V_\Gamma^{nrd}(A)$, and the set $\{A \mid C_\Gamma(A) < C_{min}\}$ is a superset of $\{A \mid C_\Gamma(A) < r_{liq}\}$ by (11).

Strong price volatility is a risk for ε-collateralization safety, as a sharp drop in price can immediately reduce a previously over-collateralized user to become under-collateralized: such an immediate drop leaves the user with no opportunity to maintain its collateralization with repayments.

Attacks on Safe Collateralization. Malicious agents which can perform price updates can therefore influence the evolution of the LP to lead it to a state that is not ε-collateralization safe or not strongly ε-collateralization safe.

For example, an attacker controlling the price oracle could act as follows. First, she would perform price updates to make any user undercollateralized. The attacker can then perform liquidations on these users and benefit from the discount resulting from both the price update and r_{liq}. The attacker has maximized her profits by updating P such that $V_\Gamma^m(B)$ in (7) is close to zero, where B is a user under attack. In this case, $A : \mathsf{liq}(B, v : \tau, \{\tau'\})$ can be performed with a small v, and repeated liquidations for different minted tokens can be executed to seize the full balance of B's collateral.

As a matter of fact, a recent failure of the oracle price feed used by the Compound LP implementation led to \$100M of collateral being (incorrectly) liquidated [16]. Though it is unclear whether this was an intentional exploit or not, it illustrates the feasibility of such a price oracle attack.

5.2 Utilization Bounds and Risks

The **utilization** of a token τ is the fraction of units of τ currently lent to users:

$$U_\tau(\Gamma) = \frac{\sum_A \delta(A)}{r + \sum_A \delta(A)} \qquad \text{if } \Gamma = (r : \tau, \delta) \mid \Gamma' \qquad (13)$$

This notion plays a crucial role in the incentive mechanism of LPs, as explained in [53]: as a matter of fact, it is often used as a key parameter of interest rate models in implementations [25,27] and literature [53].

Over- and Under-Utilization. Note that $U_\tau(\Gamma)$ ranges between 0 and 1. We say that τ is *under-utilized* if its utilization is 0 and *over-utilized* when it is 1. A state is under-utilized (resp. over-utilized) if it contains under-utilized (resp. over-utilized) tokens.

Under-utilization occurs when some units of τ have been deposited, but not borrowed by any user. This implies that interest accrual does not increase the debt of any user, as so the exchange rate of τ in (2) remains constant, thereby not resulting in any gain for lenders.

On the other hand, over-utilization occurs when some users have borrowed units of τ, but the LP has no deposited funds of τ. In this case, users can neither borrow nor redeem.

Under- and over-utilization should be avoided. An optimal utilization of a token type τ strikes a balance between the competing objectives of interest maximization and the ability for users to borrow τ tokens or to redeem $\{\tau\}$ tokens. The interest rate models described in [53] intend to incentivize actions of both borrowers and lenders to discover an equilibrium between under- and over-utilization. Informally, this is achieved with interest rate models which rise and fall with utilization: increasing utilization and interest rates incentivize deposits and repayment of loans. Decreasing utilization and interest rates incentivize redeems and additional loan borrowing.

We now discuss under- and over-utilization attacks: note that the first kind of attacks is weaker than the second kind, as funds can still be safely recovered in case of under-utilization.

Under-Utilization Attacks. Under-utilization attacks can be achieved by an attacker interested in reducing the interest accrual for depositors, or in discouraging the borrowing of a token τ. The attacker can temporarily reduce utilization by repaying large amounts of loans. The effectiveness of this approach will depend on the amounts of τ repaid by the attacker, as a lowered utilization can also reduce the interest rate (in certain models [53]), thereby incentivizing additional borrowing. An attacker which can update the price oracle can lower the collateralization of borrowers arbitrarily, thereby incentivizing repayments and liquidations to target lower utilization of specific tokens.

Over-Utilization Attacks. Over-utilization attack could be achieved by an attacker interested in preventing redeem or borrow actions on τ. The attacker can do this by redeeming all units of τ, while avoiding loans to be repaid or liquidated. We illustrate an over-utilization attack in Fig. 8. Users A and C initially hold the entire supply of τ_0 in their wallets. User A colludes with B to steal C's balance of τ_0: in steps 1–2, A and B deposit $100 : \tau_0$ and $100 : \tau_1$, respectively. User B uese her balance of $100 : \{\tau_1\}$ as a collateral to borrow $50 : \tau_0$ from the LP in step 3. At this point, A and B are acting as lender and borrower of τ_0, for which the utilization is 0.5. User C, having observed an opportunity to earn interest on τ_0 decides to deposit $50 : \tau_0$ in step 4. However, user A still has a balance of $100 : \{\tau_0\}$, which she redeems in step 5. Now, users A and B have removed all units of τ_0 from the LP, pushing the utilization of τ_0 to 1, and preventing C from redeeming the minted tokens in his wallet. Of course, B cannot redeem her minted tokens of type $\{\tau_1\}$, since her loan has not been repaid, but this can be considered the cost of the attack.

$$\Gamma_0 = A[100 : \tau_0] \mid B[100 : \tau_1] \mid C[50 : \tau_0] \mid P = \{1/\tau_0, 1/\tau_1\}$$

$$\xrightarrow{\text{1.A:dep}(100:\tau_0)}$$

$$\Gamma_1 = A[100 : \{\tau_0\}] \mid (100 : \tau_0, \{\}) \mid \cdots$$

$$\xrightarrow{\text{2. B:dep}(100:\tau_1)}$$

$$\Gamma_2 = B[100 : \{\tau_1\}] \mid (100 : \tau_1, \{\}) \mid \cdots$$

$$\xrightarrow{\text{3. B:bor}(50:\tau_0)} \quad \{C_{\Gamma_3}(B) = 2\}$$

$$\Gamma_3 = B[50 : \tau_0, 100 : \{\tau_1\}] \mid (50 : \tau_0, \{50/B\}) \mid \cdots$$

$$\xrightarrow{\text{4. C:dep}(50:\tau_0)}$$

$$\Gamma_4 = C[50 : \{\tau_0\}] \mid (100 : \tau_0, \{50/B\}) \mid \cdots$$

$$\xrightarrow{\text{5. A:rdm}(100:\{\tau_0\})}$$

$$\Gamma_5 = A[100 : \tau_0] \mid C[50 : \{\tau_0\}] \mid (0 : \tau_0, \{50/B\}) \mid \cdots$$

Fig. 8. An over-utilization attack.

6 Differences Between Our Model and LP Platforms

We have synthesised our model from informal descriptions in the literature and the implementation and documentation of lending pools Compound [27] and Aave [25]. To distill a usable, succinct model we have abstracted some implementation details, that could be incorporated in the model at the cost of a more complex presentation. We discuss here some of the main abstractions we made.

The original implementations of Compound and Aave gave administrators control over the economic parameters of the LP, i.e. C_{min}, r_{liq}, and the interest rate function. This made administrators of such early versions privileged users, who could in principle prevent honest depositors, borrowers and liquidators from withdrawing funds. A Compound administrator, for example, can replace application logic which computes collateralization and authorizes supported tokens [13]. Later versions of these platforms have introduced *governance tokens* (respectively, COMP and AAVE), which are allocated to initial investors or to LP users, who earn units of such tokens upon each interaction. Governance tokens allow holders to propose, vote for, and apply changes in economic parameters, including interest rate functions. By contrast, our model assumes that economic parameters are fixed, and omits governance tokens.

In implementations, adding a new token type to the LP must be authorized by the governance mechanisms. By contrast, in our model any user can add a new token type to the LP by just performing the first deposit of tokens of that type. Implementations also allow administrators or governance to assign weights to each token type. This is intended to adjust collateralization and liquidation thresholds C_{min} and r_{liq} for the predicted price volatility of token types present in a user's loan and collateral. Further, implementations require users to pay *fees* upon actions. These fees are accumulated in a reserve controlled by the

governance mechanisms of the LP, and intended to act as a buffer in case of unforeseen events. Our model does not feature token-specific weights and fees.

User liquidations in implementations are limited to repay a maximum fraction of the loan amount [5,15]. However, this implementation constraint can be bypassed by a user employing multiple accounts, so we omit it in our model.

Lending pool platforms implement the update of interest accrual in a *lazy* fashion: since smart contracts cannot trigger transactions, periodic interest accrual would rely on a trusted user to reliably perform such actions, introducing a source of corruption. Therefore, interest accrual is performed whenever a user performs an action which requires up-to-date loan amounts. Here, the interest rate in implementations is not recomputed for each time period. Instead, a single interest rate is applied to the period since the last interest accrual [8,17] in order to reduce the cost of execution, leading to inaccuracies in loan interest.

Comparison with Other LP Models. Besides the actual LP platforms, we compare our model with other models of LPs in the literature.

The *liquidation model* of [52] is meant to simulate interactions between lending pool liquidations and token exchange markets in times of high price volatility. Unlike in our model, [52] performs liquidations in aggregate, and it omits individual user actions. The interest rate functions of [53] formalize various interest rate strategies used by LP implementations, and can be seen as complementary to our work. Indeed, even if we did not incorporate such functions directly in our model (for brevity), they could be easily included as instances of $I_\Gamma(\tau)$ in rule [INT]. The work [59] introduces an LP state model, which is instantiated with historical user transactions observable in the Compound implementation deployed on Ethereum. The model abstraction facilitates the observation of state effects of each interaction, and investigates the (historical) latency of user liquidations following the undercollateralization of borrowing accounts. Aforementioned work prioritizes high-level analysis over model fidelity: indeed, the lending pool properties and attacks we present are a direct consequence of the precision in our lending pool semantics.

The emergent behaviour of lending pools in times of high price volatility is examined in [52] by simulation of a lending pool liquidation model. Here, a large price drop can cause many accounts to become undercollateralized: assuming liquidators sell off collateral at an external market for units of the repaid token type, the authors suggest that limited market demand for collateral tokens may prevent liquidations from being executed, thereby posing a risk to ε-*collateralization safety* as we have defined in Eqs. (10 and (12).

Lending pool behaviour at the user level is modelled in [55], which simulates agents interacting with the Compound implementation to examine the evolution of *liquidatable* and *undercollateralized* debt, notions similar to *(strong)* ε-*collateralization safety* (10) (12). [45,46] examine the competition for user deposits between staking in proof-of-stake systems and lending pools: in the case where lending pools are believed to be more profitable, users may shift deposits away from the staking contract of the underlying consensus protocol towards lending pools, thereby endangering the security of the system.

Lending pool interest rate behaviour is examined in [53], where empirical behaviour of interest rate models in Compound [27], Aave [25] and dYdX [19] are analyzed. In particular, the authors observe a statistically significant coupling in interest rates between deployed lending pools, suggesting that the dynamic interest models are effective in discovering a global interest rate equilibrium for a given token. Our formal model is parameterized by the interest rate, that must always be positive (8): since this property holds for all interest rate functions in [53], our model can be instantiated with them.

7 DeFi Archetypes Beyond Lending Pools

We now discuss the interplay between lending pools and other DeFi archetypes, like algorithmic stable coins, automated market makers, margin trading and flash loans, which are all predominantly deployed on the Ethereum blockchain [41]. We refer to [66] for an overview of these DeFi archetypes.

Algorithmic Stable Coins. MakerDAO [21] is the leading algorithmic stable coin and is credited with being one of the earliest DeFi projects. It incorporates several features found in lending pools, such as deposits, minting, and collateralization. As of April 2021, $7.5B [30] worth of crypto-tokens are locked in the MakerDAO implementation. Users are incentivized to interact with the smart contract to mint or redeem DAI tokens. This, in turn, adjusts the supply of DAI such that a stable value against the reference price (e.g., USD) is maintained. Synthetic tokens are similar to algorithmic stable coins but may track an asset price such as gold or other real-world assets. Reference asset prices are determined by price oracles.

The work [58] introduces a taxonomy for various price stabilization mechanisms, providing insight into the functionality of such contracts. The work [52] uncovers a vulnerability in the governance design of MakerDAO, allowing an attacker to utilize flash loans to steal funds from the contract. The empirical performance of MakerDAO's oracles is studied in [51], which also proposes alternate price feed aggregation models to improve oracle accuracy. Finally, [48] investigates the optimal bidding strategy for collateral liquidators in MakerDAO, which is executed by through user auctions.

Stable coins which track prices of real-world currencies (e.g. USD) exhibit a price stability useful for lending pools: users with stable collateral or loan values have a lower likelihood of suddenly becoming undercollateralized.

Automated Market Makers. Automated market makers (AMMs) allow users to exchange units of a token τ for units of another token τ' and vice-versa. AMMs do not match opposing actions of buyers and sellers: users simply exchange tokens with an AMM, where the exchange rate is determined algorithmically as a function of the AMM state. Hence, the dynamic exchange rate of an AMM is affected with each user interaction. As of April 2021, leading AMMs Uniswap [34] and Curve Finance [28] hold $5.3B [33] and $4.6B [28] worth of tokens and feature an estimated $1.3B [33] and $180M [28] worth of token exchanges every day.

The work [36] investigates algorithmic exchange rate models and defines the user arbitrage problem, where a profit-seeking agent must determine the optimal set of AMMs (with differing exchange rates) to interact with: given such arbitrage opportunities will be exploited by rational users, it is expected that exchange rates across AMMs remain consistent. AMM price models can fail: the *constant product* exchange rate model implemented by Uniswap [34] and Curve [29] is simple, but can theoretically reach a state where the the exchange rate is arbitrarily high. The work [65] proposes bounded exchange rate models to address this issue.

A theory of AMMs is proposed in [57], formally specifying their possible interactions and their economic mechanisms. This allows [57] to develop a concurrency theory of AMMs: in particular, it shows that sequences of deposit and redeem actions can be ordered interchangeably, resulting in observationally equivalent states. Be leveraging the formal model, [57] establishes fundamental properties of AMMs, like e.g. the preservation of deposited token supplies, and *token liquidity*, which ensures that deposited tokens cannot be frozen in an AMM. Further, it devises a general solution to the arbitrage problem, the main game-theoretic foundation behind the economic mechanisms of AMMs.

The work [35] suggests that AMMs track global average token prices effectively. As such, AMMs can inform price oracles: such oracles, however, only update price information with each new block [23] computed from time-weighted price averages of AMMs over the past block interval. This increases the cost of manipulating prices of the oracle, as the manipulated price must be sustained over a period of time. We note that lending pool implementations do not rely on oracles which derive prices from AMM states.

AMMs suffer from *front-running attacks*, where an attacking user observes the victim's unconfirmed token exchange transaction, and sequences its own transaction prior to that of the victim. A front-running attack on an AMM user takes advantage of the update in exchange rate resulting from the victim's token exchange, who ends up paying a higher price, as illustrated in [67]. Front-running of smart contracts is investigated more generally in [50]: mitigations such as commit-and-reveal schemes are proposed, which come with an increased cost for user-contract interactions. In the context of AMMs, [47] introduces the notion of gas auctions, where adversarial users compete to front-run a given AMM exchange transaction by outbidding each others transaction fee.

We note that similar attacks can be modeled with an attacker that can drop or reorder transactions in our lending pool model. Such an attacker can trivially defer attempts of a borrower to repay a loan: subsequent interest accrual will eventually cause the user to become undercollateralized, so that the attacker can liquidate the victim. Such an attacker can also monopolize all liquidations for herself, preventing other users from executing such an action. The work [47] suggests that miners may be incentivized to perform such attacks due to gain resulting from liquidation discounts.

Margin Trading. An important use case of lending pools are *leveraged* long or short positions initiated by users, also referred to as margin trading. In a

leveraged long position of τ against τ', the user speculates that the price of the former will increase against the price of the latter: a user borrows τ' at a lending pool against collateral deposited in τ, and then exchanges the borrowed units of τ' back to τ at a token exchange or an AMM. The user will now earn an amplified profit if the price of τ appreciates relative to τ', since both the borrowed balance and redeemable collateral in τ appreciates in value whilst only the loan repayable with τ' decreases in value. A leveraged short position simply reverses the token types. Margin trading contracts such as bZx Fulcrum [11] combine lending and AMM functionalities to offer margin trades through a single smart contract. However, since such margin trading contracts perform large token exchanges at external AMMs, attackers can use such actions to manipulate AMM prices, as shown in [60]. Furthermore, the scope of such attacks is magnified when performed with flash loans.

Flash Loans. Any smart contract holding tokens can expose flash loan functionality, allowing users to borrow and return a loan within a single atomic transaction group. Atomic transaction groups are sequences of actions from a single user, which must execute to completion or not execute at all. They can be implemented in Ethereum by user-defined smart contracts [6], and they are natively supported by Algorand [39]. As such, flash loans are guaranteed to be repaid or not executed at all. The work [64] introduces a framework to identify flash loan transactions on the Ethereum blockchain for an analysis of their intended use-cases, which include arbitrage transactions, account liquidations (in lending pools or stable coins) and attacks on smart contracts. Our model can be easily extended to encompass flash loan semantics.

Flash loans have been used in several recent attacks [10,12,20,22,60]. The flash loan attack on bZx Fulcrum described in [60] involves sending the borrowed tokens to a margin trading contract, which, in turn, initiates a large token exchange at an external AMM: here, the large amount of exchanged tokens causes a significant shift in dynamic AMM exchange rate, which represents an arbitrage opportunity exploited by the attacker in several execution steps involving other contracts. Flash loans provide attackers with access to very large token values to initiate attacks.

8   Research Challenges

Our model already allows us to formally establish properties of LPs (Sect. 4), and to precisely describe potential attacks to LPs as sequences of user actions (Sect. 5). However, lending pools operate within a wider DeFi ecosystem, composed by a set of collaborating or competing agents, interacting through possibly separate contract execution environments enabled by miners, who may have transaction ordering privileges and their own goals [47]. We highlight some open research challenges for the compositional security in DeFi systems, where we expect lending pool applications to play a central role.

Agent Strategies. As shown in Lemma 3, there exist *rational* lending pool actions which always increase the net worth of the agent. We contrast such

risk-free rational strategies against those which are *speculative*, driven by an agent's expectation of a future system state which is not guaranteed: depositing or borrowing from an LP are speculative strategies, as they are motivated by an expectation of future interest, which are, in turn, regulated by future actions of borrowers and depositors.

Whereas there appears to be a clear path towards formal specification of rational strategies in DeFi systems, the specification of speculative agent behaviour in DeFi remains an open question. For individual DeFi archetypes, agent-based models have been proposed [23,55] with a focus on rational behaviour, yet the specification of economically speculative strategies in a richer composition of DeFi application remains an open research challenge.

Classical agent-based models from economic disciplines feature specification techniques of economically (speculative) agent behaviour: here, we also observe that stochastic model checking tools from formal methods are increasingly deployed [63] in the economic research community and suggest that stochastic model checking of agent-based models of DeFi systems may provide a path forward towards the automatic analysis of agent strategies.

A Model of Transaction Concurrency. As exemplified by Lemma 4, actions in lending pools and DeFi are generally not concurrent. In particular, the exploitation of non-concurrency in AMM's has received much attention, where an actor with transaction ordering privileges can benefit from ordering its own transaction before and after that of the victim [54,67] for financial benefit. More generally, the ability of miners to extract value beyond transaction fees from specific sequences of DeFi interactions has been denoted miner-extractable-value (MEV) [47]. For LP applications, a rational miner is incentivized to perform liquidation actions itself, thereby invalidating liquidation attempts by other users: this may support the security of LP's, as loans are quickly liquidated. However, it also highlights the challenges in developing a formal model of a DeFi system composed of different DeFi applications. Such a model must feature a notion of *incentive-consistent* action sequences in the presence of rational agents with transaction ordering privileges, such that any miner interaction with DeFi applications are intended and beneficial the security of the DeFi system.

Towards the goal of exploring action concurrency in a composed DeFi system, [62] models user functionality enabled by composing an AMM and LP: here, an AMM pair offers swaps between two stable coin types, which are provided by depositors. The deposited stable coins, however, are forwarded to lending pools, thus enabling AMM depositors to also earn interest in addition to swap fees. The resulting agent model is implemented as communicating sequential processes, allowing the exploration of different action sequences.

We note, however, that such analysis is further complicated by atomic chains of transactions, such as those obtained by nested contract calls in Ethereum. Here, the sequencing of individual actions within the call-chain is determined by the authorizing user: this can result in DeFi exploits amplified by flash loans [42,60,64]. As transactions, call-chains must also exhibit consistency with miner transaction ordering incentives: here, a lack of formal models to integrate call-chain semantics with formal models of MEV remains apparent.

A model of transaction ordering may ultimately facilitate the automated analysis of a DeFi system specification which includes lending pools, given that it narrows the set of *valid* interaction sequences. Given sufficiently specified agent strategies, such a theory may pave the way towards novel model checking techniques in DeFi.

Cryptographic Protocol Composition. Cryptographic protocols play an increasingly central role in DeFi systems, as they allow DeFi applications to keep private selected parts of the application state: public execution introduces incentives (MEV) which challenge DeFi security, but the public execution of user actions also compromises privacy. The popularity of crypto-asset mixers [32] powered by ZK-SNARK proofs on the Ethereum blockchain foreshadows the emergence of privacy-focused DeFi applications, which in turn, may open new approaches to mitigate MEV. Private order-matching has been proposed with multi-party-computation techniques [43], and we foresee similar techniques for DeFi applications. Furthermore, advanced cryptographic protocols improve scalability: many DeFi applications have migrated to ZK-rollups [31] in order to absorb the increased user demand on the Ethereum blockchain.

For the secure composition of cryptographic protocols deployed for both privacy and scalability, the formal methods community may contribute both classical information flow [44] analysis techniques and cryptographic protocol composition analysis [49]: as a multitude of privacy-focused and scalable applications are composed in a single system, we highlight the formal analysis of safe cryptographic protocol composition in DeFi as an new research frontier.

Domain-Specific Languages. Since the analysis of security aspects of DeFi applications will invariably involve specifications of agents and miners, higher abstractions of DeFi specification will arguably be of interest to the DeFi and formal methods communities. Domain-specific languages with formal semantics (e.g. [37,40,61]) provide suitable specification means for such abstractions. Moreover, they fulfill two purposes: firstly, they enable formal reasoning and security proofs. Secondly, DeFi-specific languages can provide built-in security guarantees, given a foundational theory of the underlying DeFi system.

Acknowledgements. Massimo Bartoletti is partially supported by Conv. Fondazione di Sardegna & Atenei Sardi project F74I19000900007 *ADAM*. James Hsin-yu Chiang is supported by the PhD School of DTU Compute. Alberto Lluch Lafuente is partially supported by the EU H2020-SU-ICT-03-2018 Project No. 830929 CyberSec4Europe (cybersec4europe.eu).

References

1. ERC-20 token standard (2015). https://github.com/ethereum/EIPs/blob/master/EIPS/eip-20.md
2. Understanding the DAO attack, June 2016. http://www.coindesk.com/understanding-dao-hack-journalists/
3. Parity Wallet security alert, July 2017. https://paritytech.io/blog/security-alert.html

4. A Postmortem on the Parity Multi-Sig library self-destruct, November 2017. https://goo.gl/Kw3gXi
5. Aave maximum liquidation amount (2020). https://github.com/aave/aave-protocol/blob/efaeed363da70c64b5272bd4b8f468063ca5c361/contracts/lendingpool/LendingPoolLiquidationManager.sol#L181
6. Aave v1 flashloan receiver interface (2020). https://github.com/aave/aave-protocol/blob/efaeed363da70c64b5272bd4b8f468063ca5c361/contracts/flashloan/interfaces/IFlashLoanReceiver.sol#L11
7. Aave v1 implementation (2020). https://github.com/aave/aave-protocol/tree/efaeed363da70c64b5272bd4b8f468063ca5c361
8. Aave v1 simplified interest (2020). https://github.com/aave/aave-protocol/blob/efaeed363da70c64b5272bd4b8f468063ca5c361/contracts/libraries/CoreLibrary.sol#L423
9. Aave valuation of atokens (2020). https://github.com/aave/aave-protocol/blob/efaeed363da70c64b5272bd4b8f468063ca5c361/contracts/lendingpool/LendingPoolDataProvider.sol#L114
10. Akropolis Defi attack (2020). https://cryptonews.com/news/defi-akropolis-drops-20-following-a-usd-2m-heavy-hack-8299.htm
11. bzx fulcrum website (2020). https://fulcrum.trade
12. Coindesk: Value DeFi attack (2020). https://www.coindesk.com/value-defi-suffers-6m-flash-loan-attack
13. Compound comptroller setter (2020). https://github.com/compound-finance/compound-protocol/blob/a5591d5f9a7f6f7ad3601ec89b126a8c2af159f6/contracts/CToken.sol#L1152
14. Compound implementation (2020). https://github.com/compound-finance/compound-protocol/tree/a5591d5f9a7f6f7ad3601ec89b126a8c2af159f6
15. Compound maximum liquidation amount (2020). https://github.com/compound-finance/compound-protocol/blob/a5591d5f9a7f6f7ad3601ec89b126a8c2af159f6/contracts/ComptrollerG5.sol#L510
16. Compound oracle attack (2020). https://news.bitcoin.com/100-million-liquidated-on-defi-protocol-compound-following-oracle-exploit
17. Compound simplified interest (2020). https://github.com/compound-finance/compound-protocol/blob/a5591d5f9a7f6f7ad3601ec89b126a8c2af159f6/contracts/CToken.sol#L423
18. Compound valuation of ctokens (2020). https://github.com/compound-finance/compound-protocol/blob/a5591d5f9a7f6f7ad3601ec89b126a8c2af159f6/contracts/ComptrollerG5.sol#L753
19. dydx website (2020). https://dydx.exchange
20. Harvest Finance flashloan attack post-mortem (2020). https://medium.com/harvest-finance/harvest-flashloan-economic-attack-post-mortem-3cf900d65217
21. Makerdao website (2020). https://makerdao.com
22. Origin Dollar attack (2020). https://cryptonews.com/news/4th-major-defi-hack-in-a-month-origin-dollar-loses-usd-7m-8331.htm
23. Uniswap oracle template (2020). https://github.com/Uniswap/uniswap-v2-periphery/blob/dda62473e2da448bc9cb8f4514dadda4aeede5f4/contracts/examples/ExampleOracleSimple.sol
24. Aave markets website (2021). https://app.aave.com/markets
25. Aave website (2021). https://www.aave.com
26. Compound markets website (2021). https://compound.finance/markets
27. Compound website (2021). https://www.compound.finance

28. Curve statistics (2021). https://www.curve.fi/dailystats
29. Curve website (2021). https://www.curve.fi
30. Defi pulse website (2021). https://defipulse.com
31. Starkware (2021). https://starkware.co/
32. Tornado (2021). https://tornado.cash/
33. Uniswap statistics (2021). https://info.uniswap.org
34. Uniswap website (2021). https://www.uniswap.org
35. Angeris, G., Chitra, T.: Improved price oracles: Constant function market makers. arXiv preprint arXiv:2003.10001 (2020), https://arxiv.org/abs/2003.10001
36. Angeris, G., Kao, H.T., Chiang, R., Noyes, C., Chitra, T.: An analysis of uniswap markets. Cryptoeconomic Systems Journal (2019). https://ssrn.com/abstract=3602203
37. Arusoaie, A.: Certifying Findel derivatives for blockchain. J. Logical Algebraic Methods Programm. **121**, 100665 (2021). https://doi.org/10.1016/j.jlamp.2021.100665
38. Atzei, N., Bartoletti, M., Cimoli, T.: A survey of attacks on ethereum smart contracts (SoK). In: Maffei, M., Ryan, M. (eds.) POST 2017. LNCS, vol. 10204, pp. 164–186. Springer, Heidelberg (2017). https://doi.org/10.1007/978-3-662-54455-6_8
39. Bartoletti, M., Bracciali, A., Lepore, C., Scalas, A., Zunino, R.: A formal model of Algorand smart contracts. In: Financial Cryptography (2021). (to appear) https://arxiv.org/abs/2009.12140
40. Bartoletti, M., Zunino, R.: BitML: a calculus for Bitcoin smart contracts. ACM CCS (2018). https://doi.org/10.1145/3243734.3243795
41. Buterin, V.: Ethereum: a next generation smart contract and decentralized application platform (2013). https://github.com/ethereum/wiki/wiki/White-Paper
42. Cao, Y., Zou, C., Cheng, X.: Flashot: a snapshot of flash loan attack on DeFi ecosystem. arXiv preprint arXiv:2102.00626 (2021). https://arxiv.org/abs/2102.00626
43. Baum, C., David, B., Frederiksen, T.: P2DEX: Privacy-Preserving Decentralized Cryptocurrency Exchange. Cryptology ePrint Archive, Report 2021/283 (2021). https://eprint.iacr.org/2021/283
44. Cecchetti, E., Yao, S., Ni, H., Myers, A.C.: Compositional Security for Reentrant Applications. arXiv preprint arXiv:2103.08577 (2021). http://arxiv.org/abs/2103.08577
45. Chitra, T.: Competitive equilibria between staking and on-chain lending. arXiv preprint arXiv:2001.00919 (2019). https://arxiv.org/abs/2001.00919
46. Chitra, T., Evans, A.: Why stake when you can borrow? Available at SSRN 3629988 (2020). http://dx.doi.org/10.2139/ssrn.3629988
47. Daian, P., et al.: Flash boys 2.0: Frontrunning in decentralized exchanges, miner extractable value, and consensus instability. In: IEEE Symposium on Security and Privacy, pp. 910–927. IEEE (2020). https://doi.org/10.1109/SP40000.2020.00040
48. Darlin, M., Papadis, N., Tassiulas, L.: Optimal bidding strategy for maker auctions. arXiv preprint arXiv:2009.07086 (2020). https://arxiv.org/abs/2009.07086
49. Dolev, D., Yao, A.: On the security of public key protocols. IEEE Trans. Inf. Theory **29**(2), 198–208 (1983)
50. Eskandari, S., Moosavi, S., Clark, J.: SoK: transparent dishonesty: front-running attacks on blockchain. In: Bracciali, A., Clark, J., Pintore, F., Rønne, P.B., Sala, M. (eds.) FC 2019. LNCS, vol. 11599, pp. 170–189. Springer, Cham (2020). https://doi.org/10.1007/978-3-030-43725-1_13

51. Gu, W.C., Raghuvanshi, A., Boneh, D.: Empirical measurements on pricing Oracles and decentralized governance for stablecoins. Available at SSRN 3611231 (2020). http://dx.doi.org/10.2139/ssrn.3611231

52. Gudgeon, L., Pérez, D., Harz, D., Livshits, B., Gervais, A.: The decentralized financial crisis. In: Crypto Valley Conference on Blockchain Technology (CVCBT), pp. 1–15. IEEE (2020). https://doi.org/10.1109/CVCBT50464.2020.00005

53. Gudgeon, L., Werner, S., Perez, D., Knottenbelt, W.J.: Defi protocols for loanable funds: Interest rates, liquidity and market efficiency. In: ACM Conference on Advances in Financial Technologies, pp. 92–112 (2020). https://doi.org/10.1145/3419614.3423254

54. Qin, K., Zhou, L., Gervais, A.: Quantifying Blockchain Extractable Value: How dark is the forest? arXiv preprint arXiv:2101.05511 (2021). https://arxiv.org/abs/2101.05511

55. Kao, H.T., Chitra, T., Chiang, R., Morrow, J.: An Analysis of the Market Risk to Participants in the Compound Protocol https://scfab.github.io/2020/FAB2020_p5.pdf

56. Bartoletti, M., Chiang, J.H., Lluch-Lafuente, A.: SoK: Lending Pools in Decentralized Finance. arXiv preprint arXiv:2012.13230 (2020). https://arxiv.org/abs/2012.13230

57. Bartoletti, M., Chiang, J.H., Lluch-Lafuente, A.: A theory of Automated Market Makers in DeFi. arXiv preprint arXiv:2102.11350 (2021). https://arxiv.org/abs/2102.11350

58. Moin, A., Sekniqi, K., Sirer, E.G.: SoK: a classification framework for stablecoin designs. In: Bonneau, J., Heninger, N. (eds.) FC 2020. LNCS, vol. 12059, pp. 174–197. Springer, Cham (2020). https://doi.org/10.1007/978-3-030-51280-4_11

59. Perez, D., Werner, S.M., Xu, J., Livshits, B.: Liquidations: Defi on a knife-edge. In: Financial Cryptography (2021). (to appear) https://arxiv.org/abs/2009.13235

60. Qin, K., Zhou, L., Livshits, B., Gervais: Attacking the DeFi Ecosystem with Flash Loans for Fun and Profit. In: Financial Cryptography (2021). (to appear) https://arxiv.org/abs/2003.03810

61. Lamela Seijas, P., Thompson, S.: Marlowe: financial contracts on blockchain. In: Margaria, T., Steffen, B. (eds.) ISoLA 2018. LNCS, vol. 11247, pp. 356–375. Springer, Cham (2018). https://doi.org/10.1007/978-3-030-03427-6_27

62. Tolmach, P., Li, Y., Lin, S.W., Liu, Y.: Formal analysis of composable DeFi protocols. In: 1st Workshop on Decentralized Finance (2021), (to appear) https://arxiv.org/abs/2103.00540

63. Vandin, A., Giachini, D., Lamperti, F., Chiaromonte, F.: Automated and Distributed Statistical Analysis of Economic Agent-Based Models. arXiv preprint arXiv:2102.05405 (2021) https://arxiv.org/abs/2102.05405

64. Wang, D., et al.: Towards understanding flash loan and its applications in defi ecosystem. arXiv preprint arXiv:2010.12252 (2020). https://arxiv.org/abs/2010.12252

65. Wang, Y.: Automated market makers for decentralized finance (defi). arXiv preprint arXiv:2009.01676 (2020). https://arxiv.org/abs/2009.01676

66. Werner, S.M., Perez, D., Gudgeon, L., Klages-Mundt, A., Harz, D., Knottenbelt, W.J.: Sok: Decentralized Finance (DeFi). arXiv preprint arXiv:2101.08778 (2021), https://arxiv.org/abs/2101.08778

67. Zhou, L., Qin, K., Torres, C.F., Le, D.V., Gervais, A.: High-Frequency Trading on Decentralized On-Chain Exchanges. arXiv preprint arXiv:2009.14021 (2020). https://arxiv.org/abs/2009.14021

Standardized Crypto-Loans
on the Cardano Blockchain

Dmytro Kondratiuk[1], Pablo Lamela Seijas[1] , Alexander Nemish[1],
and Simon Thompson[1,2]

[1] IOHK, Wan Chai, Hong Kong
{dmytro.kondratiuk,pablo.lamela,alexander.nemish,simon.thompson}@iohk.io
[2] School of Computing, University of Kent, Kent, UK
s.j.thompson@kent.ac.uk

Abstract. Crypto-loans are innovative financial instruments that allow
trustless peer-to-peer lending, and potentially providing a safe and convenient source of liquidity for cryptocurrency holders. In this paper we
explore a smart contract framework for building standardised crypto-
loans using the Marlowe domain-specific language and the ACTUS standard for financial contracts.

Keywords: ACTUS · Blockchain · Cardano · Finance · Haskell ·
Marlowe · Smart contract · Static analysis

1 Introduction

Smart contracts – programs that run in a blockchain environment – can be
defined in a variety of ways [9]. Many such approaches are general purpose,
and can be used to program any kind of contract it makes sense to run on a
blockchain; moreover, they tend to be expressive enough to be Turing complete
(in some cases with restrictions on the runtime environment). For example, Plutus, the general-purpose language running on the Cardano blockchain [3], is a
dialect of Haskell. Another approach is to develop special-purpose or *domain-
specific languages* (DSLs) which embody a particular application domain: Marlowe [8] is a high-level DSL for writing financial contracts on the Cardano
blockchain.

In this paper we explore ways in which contracts described in ACTUS (Algorithmic Contract Types Unified Standards) can be defined in the contract languages Marlowe, Plutus and Haskell. Of course, Plutus or Haskell are able to
express these contracts, but rendering them in Marlowe brings extra advantages. Marlowe is defined to provide a range of guarantees by design: a Marlowe
contract will only make a finite number of interactions with its environment,
and its lifetime can be read off from the code for a contract; moreover, when
the contract terminates, any assets held by the contract will automatically be
returned to the participants. None of these guarantees can be provided by a
general purpose language.

© International Financial Cryptography Association 2021
M. Bernhard et al. (Eds.): FC 2021 Workshops, LNCS 12676, pp. 579–594, 2021.
https://doi.org/10.1007/978-3-662-63958-0_41

Each Marlowe contract has a finite set of possible execution paths, and so it is possible to analyse the complete behaviour of a contract without running it. Such *static analysis*, based on SMT solving [12], can be used to check properties of a contract; for example, it is possible to check whether a contract will honour all the Pay constructs that it contains, however it is executed. In the case that a Pay can fail, the analysis gives an example trace showing how that failure happens. The language design and static analysis provide assurance that Marlowe contracts are much less likely to "misbehave" than contracts written in a general-purpose language like Solidity.

In implementing ACTUS on Cardano we are able to provide further assurance in three other ways. First, we are able to use the declarative nature of Haskell to transliterate ACTUS formulas term-by-term into an executable form in Haskell. Secondly, we are able to use random, property-based testing to validate the Haskell implementation against another written in Java. Finally, we are able to automatically *generate* ACTUS contracts in Marlowe from the terms – i.e. parameters – of the contracts; for a simple loan these would include the start and end dates of the loan and the amount loaned (the 'principal'). The generated contracts use the executable spec in calculating values of cash flows in the Marlowe contracts.

The contribution of our work is to show that that it is possible to implement financial contracts on blockchain in a way that *multiple forms of assurance* are provided: from the language itself, from the static analysis, and from custom property verification. The development environment for Marlowe, the Marlowe Playground[1], also provides a simulation environment for contracts for stepping forwards (and backwards) through contract execution, and thus allowing users to validate that contracts perform as they should. In addition, implementing ACTUS provides a suitable benchmark against which to assess the design of Marlowe; we illustrate how the implementation has led to the addition of a conditional expression construct to the language.

In the remainder of the paper, Sect. 2 covers the relevant financial background, including the ACTUS financial standard. Section 3 builds an executable specification of ACTUS in Haskell, and this is used in Sect. 4 to generate the Marlowe code for an ACTUS contract from the contract terms. Section 5 explains how tokens are used to represent ownership of roles in a running contract, and Sect. 6 describes how we provide assurance that contracts behave as they should. Section 7 examines related work and Sect. 8 concludes.

A note on notation: typewriter font will be used for Marlowe constructs while *math* font will be used for mathematical formulas and pseudo-code.

2   Financial Contracts

In this section we give a brief introduction to financial contracts, and to loans in particular, and then describe the ACTUS financial standard.

---

[1] https://alpha.marlowe.iohkdev.io/#/.

2.1 Crypto-Loans

A loan is a form of debt incurred by an individual or other entity. The lender advances a sum of money to the borrower. In return, the borrower agrees to a certain set of terms including any finance charges, interest, repayment date, and other conditions [10].

Cryptocurrency-backed loans must have *collateral* when there is no trust between party and counterparty. While a loan is usually settled in a stable-coin currency (e.g. USDT/USDC), collateral is typically denominated in a cryptocurrency (e.g. BTC). The purpose of such a loan is to give the borrower access to the fiat value of their crypto-funds without actually selling them for fiat. The borrower pays interest in exchange for gaining liquidity.

Every loan has a positive net payoff (return minus investment) that is either rendered as a one-time payment – often called a *zero-coupon bond* (ZCB) – or by scheduling payment of the interest. The rate of interest could be fixed throughout the lifetime of a contract: for example, zero-risk bonds have a fixed interest proportional to the inflation rate. However, in the generic case the interest rate is variable and depends on an external factor agreed in advance, and the rate is periodically updated by observing the state of that factor.

Such loans often represent an investment in a particular venture or industry. As a somewhat fictional example, one could imagine a cryptocurrency miner who decided to scale their crypto-farm: a loan (in USD) with variable interest that directly depends on cryptocurrency prices would be more attractive for a miner because it would directly correlate with miner's profits. For example, if the price of the cryptocurrency goes down in a particular month the borrower would have to pay lower interest, and so would always pay a fixed share of the profits. In a more traditional setup the interest rate could depend on prices of other commodities: a canonical example would be a power plant taking a loan with interest depending on electricity prices.

In both cases, the prices of cryptocurrency or electricity become a driver for the interest rate. However, one cannot simply take the bare price of the asset and turn it into a rate. In order to make *units of measurement* compatible with each other adjustments should be made. Fluctuations of the interest rate driver are embedded thus:

$$\Delta_r = capfloor(driver * multiplier + spread - interestRate_{t-1})$$

$$interestRate_t = capfloor(interestRate_{t-1} + \Delta_r),$$

where *capfloor* is a function that limits the range of fluctuation, and so limiting the lender's exposure to risk:

$$capfloor(x) = max(min(x, floor), cap)$$

The *spread* parameter here loosely represents the difference between the average prime rate that the lender expects – the *benchmark yield* – and the rate imposed

by the driver: the higher the spread, the higher the resulting interest rate. The multiplier rescales the interest rate curve in order to represent the changes to be made converting between different units of measurement: how many rate percentage points you would get for a USD-to-kilowatt conversion and so on.

In the context of ACTUS and similar frameworks, there is one more factor influencing interest rates thorough scaling:

$$interestPayment = interestScalingFactor * interestRate * notional$$

This scaling is dynamic and loosely adjusts for variance (volatility) of the asset that the interest rate driver represents.

Interest Accrual and Capitalisation. A counterparty might decide to reinvest profit received as interest from the loan. In the simplest case, this renders as compound interest. This can be modelled through interest accrual and capitalisation (conversion of income or assets into capital); for instance, contracts from the ACTUS specification accrue interest between interest payments and can transfer interest to a notional during interest capitalisation event (IPCL).

Overall, variable interest rates introduce a certain risk for a lender, thus they can be subject to hedging. While any instrument that depends on the same risk factor (interest rate driver) would suffice, the most popular way to hedge a variable interest rate loan is an interest rate swap. This instrument allows two (or more) parties to exchange their incomes - one from a fixed interest rate loan, the other from a variable-rate loan.

Counterparty Risk. Trustless setups, especially ones in the cryptocurrency world, including decentralised smart-contracts and exchanges, require no trust between party and counterparty involved in a contract. In case of a loan, this literally means that counterparty has zero obligation to pay the money back, thus rendering the loan useless for a party. Such risks are usually addressed by introducing collaterals, as in the following scenario.

1. Alice would like to borrow 1000 USD
2. She has Bitcoin assets cost around 1500 USD, which she intends to hold throughout a year, so Alice has high confidence in the market (she expects prices to double or triple)
3. Bob would like to lend 1000 USD and get an interest higher than traditional interest rate offered by banks (let's say 15% instead of 10%). He is either bearish or neutral towards Bitcoin.
4. Alice transfers her BTC as collateral to a contract, and Bob transfers his USD to Alice
5. If Alice pays the interest and notional on time, and the BTC price does not render collateral worthless, she can get her collateral back; otherwise the loan gets liquidated and the collateral is transferred to Bob.

2.2 ACTUS

The Algorithmic Contract Types Unified Standards (ACTUS) [1] define the logic embedded in legal agreements that eventually turn the contract terms into actual cash flows, or more generally business events. Most of its basic contract types represent different variations of lending contracts. ACTUS provides additional benefit of being regulatory friendly, and the ACTUS foundation provides a set of tools allowing Monte-Carlo simulations of ACTUS contracts.

ACTUS relies on a state machine formalism in order to describe the behaviour of a given contract. Every *payoff* – i.e. transfer of funds in or out of a contract – can be inferred for any given state. Every state can be derived from previous events and observed risk factors:

$$payoff_i = POF(state_i)$$

$$path_i = STF(ct, ev_1) \circ STF(ct, ev_2) \circ \ldots \circ STF(ct, ev_i)$$

$$state_i = path_i(INIT(ct)),$$

where *ct* stands for contract terms, *INIT* returns initial state, *sched* returns scheduled events, *STF* takes *contract terms*, *event*, and *state* and returns the next state, and *POF* returns the *payoff* in a state.

2.3 Oracles

In order to support variable interest rates and scaling, ACTUS requires a smart contract to be able to observe the value of a given risk factor, such as an interest rate, at a particular point in time t. This is due to the state of the risk factor not being known at instantiation time.

$$riskfactor_{it} = O_{rf}(i, t)$$

In the case of the Cardano blockchain, these values are usually provided through an *oracle* mechanism [11]. An oracle could be a trusted party providing necessary data or network of parties under consensus [2].

From a Marlowe DSL perspective, the exact mechanism that provides external data is less important, as Marlowe abstracts over IO by requiring a particular type of input – a Choice – that is protected with a cryptographic signature by the source of the choice. As a result, the event of receiving data from an oracle is treated the same as receiving numeric input in other languages.

3   Building an Executable Specification of ACTUS

ACTUS is defined in a textual specification[2] which, while expressed in mathematical notation, is essentially informal. In this section we describe how this

---

[2] Available from https://www.actusfrf.org/techspecs.

specification is turned into an *executable* version by rendering it in Haskell. This translation is in fact a transliteration, since notation, variable names and so forth are respected.

3.1 Rendering the Specification in Haskell

The ACTUS standard is specified in terms of scheduling, payoff and state transition functions that are polymorphic on event and contract type, as noted in Sect. 2.2 above. The specification also follows quite specific naming conventions that are incompatible with Haskell's conventions. The executable specification follows original ACTUS conventions as closely as possible in order to ease code base maintenance when faced with updates of the ACTUS spec repository[3].

Using Haskell itself as a DSL for explicitly encoding formulas without using advanced language idioms also simplifies code generation. In case of ACTUS this comes at a cost reduced type-safety, handling nullable values explicitly introduces risk of exceptions. However this risk is addressed using property-based testing, and in particular QuickCheck generators. This is discussed in more detail in Sect. 6 below.

3.2 Utilising Polymorphism to Abstract over Basic Operations

In order to keep our executable specification independent of the carrier – whether it is a smart-contract engine, proof assistant, analytical framework or even machine learning model – we abstract over the underlying representation of state variables,

```haskell
-- Definitions/ContractState.hs
data ContractStatePoly a b = ContractStatePoly
{
  tmd    :: b
, nt     :: a
, ipnr   :: a
, ipac   :: a
, feac   :: a
, fac    :: a
, nsc    :: a
, isc    :: a
, prf    :: ContractStatus
, sd     :: b
, prnxt  :: a
, ipcb   :: a
} deriving (Show)
```

and arithmetic operations,

---

[3] https://github.com/actusfrf/actus-techspecs.

```
-- Ops.hs
class ActusOps a where
    _min :: a -> a -> a
    _max :: a -> a -> a
    _zero :: a
    _one :: a

class ActusNum a where
    (+) :: a -> a -> a
    (-) :: a -> a -> a
    (*) :: a -> a -> a
    (/) :: a -> a -> a

class YearFractionOps a b where
    _y :: DCC -> a -> a -> a -> b

class DateOps a b where
    _lt :: a -> a -> b --returns pseudo-boolean

class RoleSignOps a where
    _r :: ContractRole -> a
```

Thus, every formula in the executable spec could be instantiated to:

- a formula on some *atomic* type, like `Double` or `Day`, which could be used to directly compute cash-flows for analytical purposes or precompute payoffs for smart contracts that do not depend on oracles; or
- a formula representing a piece of *abstract syntax*, e.g. a Marlowe `Value` or `Observation`, that could be used to generate smart contracts that depend on oracles or to generate code in another language, such as Agda.

This approach of abstracting formulas has a limitation of not allowing conditionals to be expressed in an abstract way: in other words, there is no `ActusIf` typeclass. Luckily most of conditional expressions in ACTUS specification don't depend on variable state of a contract, they depend on `ContractTerms` that are known in advance during contract generation. This allows us to dispatch appropriate formulas during generation rather than execution.

The only exception to this are the rare situations where we need to compare 2 state variables and choose either $formula'$ or 0 depending on the result of the comparison result:

$$formula(st) = \begin{cases} formula'(st) & var_1(st) < var_2(st) \\ 0 & otherwise \end{cases}$$

We rely on a pseudo-Boolean *less than* function in order to address that:

$$formula(st) = pseudoLt(var_1(st), var_2(st))) * formula'$$

$$pseudoLt(a, b) = Cond(a > b, 1, 0)$$

3.3   Contract Term Representation and Explicit Applicability

In order to simplify serialisation and deserialisation of contract terms across ACTUS related services maintained by Cardano we rely on "superposed" representation of contract terms: all ACTUS contract types are represented with the same type.

While such a representation allows both encoder and decoder to express any ACTUS contract terms, it also allows for invalid combinations of terms (for example $PRNXT$ cannot be applied to a PAM contract), which means contracts require specific validation that is implemented by means of the applicability function:

$$Applicability : ContractTerms \rightarrow Bool$$

ACTUS standard defines a family of applicability functions polymorphic on contract type:

$$Applicability : ContractType \times ContractTerms \rightarrow ApplicabilityType$$

where applicability could be: *none*, *always*, *nullable*, or *multiple*.

In order to build superposed contract terms type for such functions, we have to resolve conflicting applicability types for merged contract terms using the following resolution rules:

$$weaken(a_1, a_2) = \begin{cases} nullable & a_1 = none \land a_2 = always \\ nullable & a_1 = always \land a_2 = none \\ a_1 & priority(a_1) > priority(a_2) \\ a_2 & otherwise \end{cases}$$

$$priority(x) = \begin{cases} 0 & x = always \\ 1 & x = none \\ 2 & x = nullable \\ 3 & x = multiple \end{cases}$$

where a_i is the applicability of a given term of the ith contract type to be merged.

4   Generating Marlowe Contracts from Standardised ACTUS Contract Terms

In this section we describe how concrete Marlowe contracts are generated from the terms – i.e. parameters – of standard ACTUS contracts. We also describe the implementation of the system, and reflect on the limitations of generating contracts in Marlowe, where contracts have a predefined lifetime and a predefined collection of interactions with contract participants.

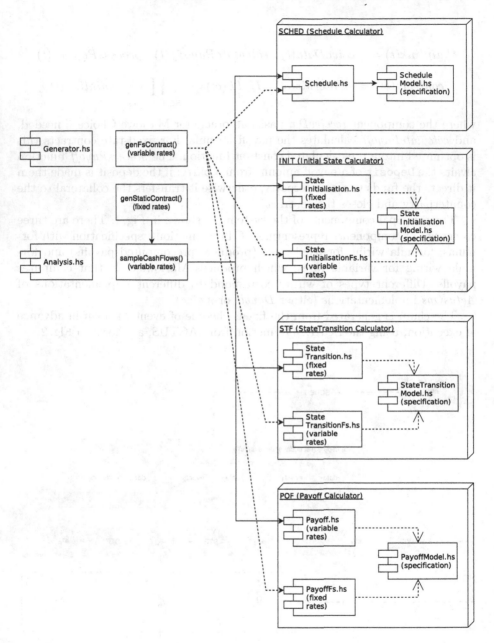

Fig. 1. Modules responsible for contract generation.

4.1   Overall Architecture

A generated contract is essentially a continuation chain of smaller contracts:

$$chainlink(t) = receiveData(t) \circ calculatePayoff(t) \circ processPayoff(t)$$

$$contract(ct) = collaterals(ct) \circ INIT(ct) \circ \prod_{t \in SCHED(ct)} chainlink(t)$$

where the component *receiveData* asks an oracle for Marlowe Choice if needed, and *calculatePayoff* calculates the payoff formula. For fixed-rate contracts this is optimised into a pre-calculated constant function. The *processPayoff* function awaits the `Deposit` of a payoff amount from a party: if the deposit is made them it directs the funds to a counterparty, otherwise it transfers the collateral to the counterparty and closes the contract.

The principal components of the system are shown in Fig. 1. There are three categories of components representing ACTUS functions: specification with formulas, formula wiring for fixed rates (produces precomputed payoffs) and formula wiring for variable rates, which produces Marlowe-code that computes payoffs. Different types of wiring correspond to different implementations of *ActusOps* implementations (either *Double* or *Value*).

The chain is generated from the fixed schedule of events, known in advance of execution, using the SCHED() function from ACTUS, as shown in Fig. 2.

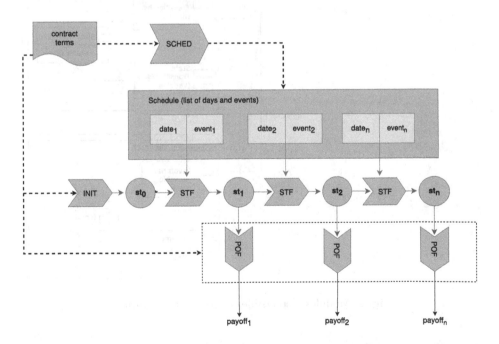

Fig. 2. Chain of sub-contracts representing ACTUS logic

4.2   Avoiding Exponential Growth

Marlowe contracts are finite, and in particular Marlowe itself does not have constructs for functions or recursion; these *are* available in the Haskell and JavaScript embeddings of Marlowe, but they are unrolled on translation to pure Marlowe. Naive usage of the If operator in Marlowe could lead to exponential growth of a contract, as in this pesudocode example:

```
if condition
  then
    perform_something1()
    continue()
  else
    perform_something2()
    continue()
```

Translating this to Marlowe would inline contents of `continue()` twice and, given that ACTUS contracts are essentially generated using continuation as an accumulator, this would lead to exponential explosion of the size of any ACTUS contract that has conditionals in their state transition logic.

An example of such logic would be cap/floor limitations on interest rates:

$$adjusted = max(min(original, floor), cap)$$

We addressed this issue by introducing the Cond expression construct in order to represent conditional expressions, rather than only conditional contracts as was the case before. Instead of using the If contract to decide the value of some variable, we use a conditional expression instead. Cond is a pure function that returns a value depending on a condition, in contrast to the If contract that chooses between two continuation contracts.

4.3   Limitations Due to Termination

Marlowe doesn't allow contracts that run indefinitely, even if their recursion is productive, as would be the case in a perpetual swap contract, for example. We therefore cannot support certain contract types from ACTUS specification, namely the ones that don't have a defined maturity date (like UMP).

There is a possible workaround: contracts with no maturity date could be represented as actors with a finite number of state transitions. We prototyped this approach, however it does seem more prone to errors comparing to rendering predefined schedules. More importantly, it greatly affects static analysis because the number of reduction steps in the contract grows from

$$N_{scheduled} = count(event)$$

to

$$N_{stateTransitions} = \frac{max(date(event)) - min(date(event))}{precision}$$

4.4   Fixed-Point Precision

For numeric types Marlowe supports Integers, while ACTUS is expressed in terms of real numbers. In order to model a real number in such a setup we rely on fixed-point precision. The algebra looks like this:

```
(+)   = AddValue -- x/n + y/n = (x + y)/n
(-)   = SubValue -- x/n - y/n = (x - y)/n
a * b = Scale (1 % marloweFixedPoint) $ MulValue a b
                        -- x/n * y/n = (x * y)/n^2
```

We scale all numbers with *marloweFixedPoint* factor - which only requires modification of *MulValue*. We plan to move Marlowe to fixed-precision numbers for the on-blockchain implementation available later in 2021.

4.5   Representing Actus State in a Marlowe Contract

The Marlowe DSL does not support any notion of records, the variables can only be of type "Integer". In order to map contract state – *ContractStatePoly* – we pack a set of Marlowe variables of type `Value` and `Observation`, representing the previous $(t-1)$ state, passing them into *ContractStatePoly*, to apply the polymorphic state transition, and finally unpack *ContractStatePoly* into a set of Marlowe variables representing the state at t:

$$st_t = unpack(stateTransition(pack(st_{t-1}))),$$

where *pack* is a chain of `UseValue` constructs and unpack is a chain of `Lets`.

Representing State Transitions in Marlowe. Marlowe is a declarative language, and so in particular it does not support mutable variables. We therefore represent the state at stage t (st_t) literally through this naming convention:

```
variableName(name, t) =
  concat(name, '_', t)
generateAccessor(name, t) =
  UseValue variableName(name, t)
generateSetter(name, t, formula) =
  Let variableName(name, t) formula
```

4.6   Actus Labs

In order to demonstrate and test the capabilities of Actus generators, a visual online Blockly-based tool was developed for the Marlowe Playground. The Actus Labs tool, shown in Fig. 3, allows users to construct contract terms visually to generate a corresponding Marlowe contract and then to try it out in a simulation environment.

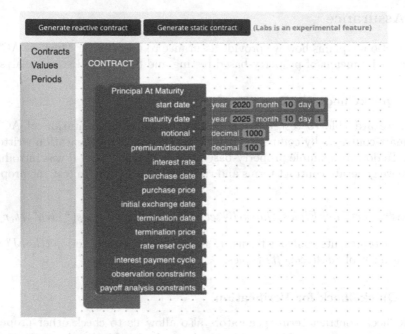

Fig. 3. Actus Labs - an online tool for generating Actus contracts for Marlowe.

5   Tokenization

Every participant of a Marlowe contact is described by a `Role` which is in its turn represented through a unique non-fungible token, created at the time that the contract instantiated on the blockchain. This makes every ACTUS contract a tradable security, allowing a participant to sell its *share* in a contract by selling a corresponding role token.

Role tokens can potentially allow more complex manipulation over such shares, especially when the share represents an incoming cash flow; in that case, participants send funds to a party represented by a given token. Such a token would represent a positive cashflow, which in turn could not only become tradable but could also allow the derivation of tokens representing *fractional parts* of a particular cash flow in a contract.

Moreover, this process turns ACTUS loans into derivatives. For example, contracts like *Interest Rate Swap* (and *Swaps* in general) could be approximated by an *Atomic Swap* of tokens representing incoming cash flows from loans. For example, if Alice has fixed income from a loan, or some other investment, and Bob has comparable but variable (fluctuating) income, Bob can hedge by swapping cash flows with Alice. If Alice's income is locked with `token1` and Bob's income is locked with `token2` then an atomic swap of those tokens is equivalent to a swap of cash flows.

6 Assurance

This section explains how we provide assurance to users of our Marlowe ACTUS contracts by means of property-based testing and SMT-based static analysis.

6.1 QuickCheck for Cross-Testing

First, we are able to test the executable Haskell implementation of ACTUS for smart-contracts by comparing it with an existing implementation written in Java. To do this a simple property-based test in QuickCheck [4] was introduced, where we generate contract terms and risk factors randomly to test the property.

$$\forall ct. \forall rf.\ getCashFlows(\text{``}haskell\text{''}, ct, rf) \equiv getCashFlows(\text{``}java\text{''}, ct, rf)$$

where ct represents contract terms, rf is risk factor model, and $getCashFlows()$ returns a set of *(date,payoff)* tuples.

6.2 QuickCheck for Verification

QuickCheck contract terms generators also allow us to check other properties of a Marlowe contract. This could be enhanced with Marlowe's static analysis feature by utilising the `Assert` operator:

```
do
let contractTerms = sample(qcgenerator)
    contract = generateMarloweContract(contractTerms)
    contractWithAssert = appendAssertion(contract, assertion)
runStaticAnalysis(contractWithAssert)
```

While this scenario does not cover all possible contracts, it could guarantee that property holds for a statistically significant fraction of a contract.

6.3 Static Analysis for Verification

Using static analysis for Marlowe [7] it is possible fully to check a particular contract instead of using random sampling. That would allow some refinements:

– More balanced sampling: the space of contract terms depends linearly on the coverage, e.g. if we cover 10% of all possible contract terms - we'll cover 10% of all contracts. Different contracts have different sets of risk factors, and so different search spaces. For instance, a contract with 3 risk factors (e.g. 3 observations of an interest rate) would span over n^3 values while a contract with 10 risk factors would span over n^{10}. Meanwhile, the space of all risk factors in all contracts doesn't linearly depend on the coverage - some contracts might have more risk factor observations and some - less. So covering 10% of all contracts doesn't necessary mean covering 10% of all possible observations.

- Dependency tracking: an SMT-solver is potentially likely to be more aware of execution paths that lead to the failure of a test, a feature that could significantly reduce search space.
- Completeness: if not timed out, SMT-solving is decidable while sampling is semi-decidable.

6.4 Securing Collateral Logic with Auto-Refund Warnings

By design, the Marlowe interpreter always refunds any assets held in a contract when it terminates. This is done in order to ensure that no funds are lost forever. The funds are returned to whoever's internal account holds them.

However, this presents as a problem in certain cases where ownership of the funds could not be determined automatically. For example, if Alice puts collateral in a crypto-loan contract she would formally maintain ownership, which means she would get automatically refunded when a `Close` construct is reached.

This implicit refund is easy to overlook by smart-contract developers as plain `Close` is often used as a default action in case of unexpected behaviour like timeouts, and especially a `Choice` timeout. This can easily lead to costly mistakes if Alice maliciously decides to exploit an auto-refund feature in order to get her collateral without paying back, as in the following scenario:

1. Alice creates a contract where a `Deposit` timeout would lead to `Close`
2. Bob doesn't know or test the timeout path. Even if Bob is a programmer, he might not be aware that `Close`ing the contract would cause the collateral to be refunded to Alice.
3. Alice puts her collateral in the contract and gets the notional from Bob.
4. Alice doesn't pay for the loan: i.e. there is a `Deposit` timeout.
5. Alice gets her collateral back.
6. Bob loses his notional.

In order to prevent this from happening, an additional `Auto-Refund` security check was introduced as part of static analysis tooling which notifies users about all `Close` constructs that can lead to automatic refunds and encourages users to write explicit logic for edge cases.

7 Related Work

Unlike current mainstream DeFi lending approaches, our Marlowe ACTUS implementation relies on trade-matching instead of pooling. While asset pooling is proven to be superior to order-book based approaches when it comes to automated market makers, using it for lending has been shown to be more susceptible to attacks [5]. Moving trade matching off-chain also improves scalability of the protocol - every loan is a separate contract, thus there is no global state.

There are frameworks, like ISDA Common Domain Model [6], providing a more precise representation of the business processes involved in institutional trading as well as code-generation capabilities. However, due to its structural simplicity, ACTUS is more suited for generating code in a financial domain specific language like Marlowe rather than general-purpose one like Java or Haskell.

8 Conclusion

The Marlowe language was explicitly designed as a set of building blocks for financial contracts that could be combined by anyone familiar with basic programming. The Marlowe ACTUS generators improve on that by providing a way to automatically combine blocks based on standardised requirements specified by the user. Marlowe ACTUS also provides a toolkit for cross-testing against the original ACTUS spec, a framework for adding new contract types, cash-flow visualisations and verification tooling.

We are very grateful to colleagues at Quanterall and Finley and Kegan McIlwaine of the University of Wyoming for their contributions to this project.

References

1. ACTUS. https://www.actusfrf.org/. Accessed 02 Feb 2020
2. Beniiche, A.: A study of blockchain oracles (2020). https://arxiv.org/abs/2004.07140. Accessed 04 Feb 2020
3. Brünjes, L., Gabbay, M.J.: UTxO- vs account-based smart contract blockchain programming paradigms. In: Margaria, T., Steffen, B. (eds.) ISoLA 2020. LNCS, vol. 12478, pp. 73–88. Springer, Cham (2020). https://doi.org/10.1007/978-3-030-61467-6_6
4. Claessen, K., Hughes, J.: QuickCheck: a lightweight tool for random testing of haskell programs. In: ICFP 2000. ACM, New York (2000). https://doi.org/10.1145/351240.351266
5. Flash-loan attack definition. https://www.coindesk.com/harvest-finance-24m-attack-triggers-570m-bank-run-in-latest-defi-exploit. Accessed 02 Feb 2020
6. ISDA Common Domain Model. https://www.isda.org/2019/10/14/isda-common-domain-model/. Accessed 02 Feb 2020
7. Lamela Seijas, P., Smith, D., Thompson, S.: Efficient static analysis of Marlowe contracts. In: Margaria, T., Steffen, B. (eds.) ISoLA 2020. LNCS, vol. 12478, pp. 161–177. Springer, Cham (2020). https://doi.org/10.1007/978-3-030-61467-6_11
8. Lamela Seijas, P., Thompson, S.: Marlowe: financial contracts on blockchain. In: Margaria, T., Steffen, B. (eds.) ISoLA 2018. LNCS, vol. 11247, pp. 356–375. Springer, Cham (2018). https://doi.org/10.1007/978-3-030-03427-6_27
9. Lamela Seijas, P., Thompson, S., McAdams, D.: Scripting smart contracts for distributed ledger technology. Cryptology ePrint Archive, Report 2016/1156 (2016). https://eprint.iacr.org/2016/1156
10. Loan definition. https://www.investopedia.com/terms/l/loan.asp. Accessed 02 Feb 2020
11. Mammadzada, K., Iqbal, M., Milani, F., García-Bañuelos, L., Matulevičius, R.: Blockchain oracles: a framework for blockchain-based applications. In: Asatiani, A., et al. (eds.) BPM 2020. LNBIP, vol. 393, pp. 19–34. Springer, Cham (2020). https://doi.org/10.1007/978-3-030-58779-6_2
12. Vanegue, J., Heelan, S., Rolles, R.: SMT solvers for software security. In: Proceedings of the 6th USENIX Conference on Offensive Technologies, WOOT 2012, p. 9. USENIX Association, USA (2012)

Fairness in ERC Token Markets: A Case Study of CryptoKitties

Kentaro Sako[1,2], Shin'ichiro Matsuo[1,3]([✉]), and Sachin Meier[1]

[1] Georgetown University, Washington, USA
Shinichiro.Matsuo@georgetown.edu
[2] Waseda University, Shinjuku City, Japan
[3] NTT Research Inc., Tokyo, Japan

Abstract. Fairness is an important trait of open, free markets. Ethereum is a platform meant to enable digital, decentralized markets. Though many researchers debate the market's fairness, there are few discussions around the fairness of automated markets, such as those hosted on Ethereum. In this paper, using pilot studies, we consider unfair factors caused by adding the program. Because CryptoKitties is one of the major blockchain-based games and has been in operation for an extended period of time, we focus on its market to examine fairness. As a result, we concluded that a gene determination algorithm in this game has little randomness, and a significant advantage to gain profit is given to players who know its bias over those who do not. We state incompleteness and impact of the algorithm and other factors. Besides, we suppose countermeasures to reduce CryptoKitties' unfairness as a market.

Keywords: CryptoKitties · Smart contracts · Financial market fairness

1 Introduction

1.1 Background

After Bitcoin was proposed, many challenges are conducted to make economic activities performed autonomously without any trusted party. Bitcoin tries to realize such a space for a simple application like payment. On the other hand, with Solidity and another language, Ethereum tries to realize "smart contract" beyond the payment process. The movement is recently expanding decentralized finance. When we deal with the simple payment process, requirements on application-level security are a bit simple, preventing double-spending in the case of bitcoin. The amount of payment is assumed to be correctly agreed among the payer and payee. KYC/AML is the other regulatory requirement under debate.

On the other hand, in the case of a smart contract, such requirements become complicated. Throughout our experience regarding Initial Coin Offering, there is a potential to scam due to asymmetric knowledge and some unfair situation for

© International Financial Cryptography Association 2021
M. Bernhard et al. (Eds.): FC 2021 Workshops, LNCS 12676, pp. 595–610, 2021.
https://doi.org/10.1007/978-3-662-63958-0_42

participants. One of the significant expectations of permissionless block-chain mechanism and smart contract as its application is to provide transparency and fairness of an economic system. It may be true for payment applications like Bitcoin and cryptocurrency, but it is unknown if we can expect the same fruits for smart contracts.

Although there is a lot of research about the ordinary financial system's fairness, they do not discuss the fairness of markets run autonomously by programming code. When we try to discuss the fairness of smart contracts, we need to consider two aspects, at minimum, addition to the concept of the fairness of the ordinary financial system; (1) effect by autonomous execution, and (2) trust of the programming code. Autonomous execution may make users difficult to manage their assets and strategy and understand their financial transactions are executed over a fair setting. The user should trust the programming code of the smart contract platform. Though the developers claim that the programming code is disclosed at GitHub for transparency, average users do not have enough capability to understand the code. As an example of supply chain risks, it is hard to prove the execution code is the same as the source code at GitHub repository.

At the time of writing of this paper, we do not have good criteria to evaluate if a specific smart contract platform/application is fair or not. Though it is big research to discuss the fairness of smart contracts, it is worth conducting research on the source of the unfairness of smart contracts. This direction will be the basis of such evaluation criteria.

1.2   Our Contribution

This research discusses the potential unfairness of a market created by the smart contract. For example, we analyze CryptoKitties, a blockchain-based game, and make the most use of smart contracts on Ethereum to evaluate if it is fair or not as a market. According to [1], its economic effect is more than forty million dollars. Thus, the existence of potential unfairness may lead to a question regarding legitimacy as a place to exchange cryptocurrency.

We investigated the internal algorithm of CryptoKitties to determine the price of each kitty potentially. In particular, we focus on how an ERC-721 token is created in CryptoKitties. In this game, a kitty is produced as an ERC-721 token. We assume that each player's goal at this game is the player earns ETHs by exchanging tokens and enjoying the kitty. The characteristics of a newly born kitty, an ERC-721 token, are determined by this game's gene algorithm. If this algorithm is not fair, there is a risk that users will unfairly lose ETHs. We also research trading tokens among the owner of kitties. It gives all users a chance to get kitties.

As a result of the research, we found that CryptoKitties does not satisfy some fair market conditions. The gene determination algorithm does not have qualified randomness, and it has a huge influence on the determination of characters of a newborn kitty. It is a source of asymmetry of knowledge. Only a person who knows the nature of the random function can predict a potential new kitty's characteristics. Therefore, it is possible to guess which kitty produces the most

valuable kitty. We found that only users, who know this bias and can buy kitties that give birth to valuable kitties, can earn more ETHs. When a player tries to sell a kitty cheaper than the breeding fee, his revenue will be smaller than his cost. Thus, the game has an unrealistic assumption on players' literacy; all people must have the ability to understand the algorithm. This fact does not mean that all players have opportunities to gain profit. Moreover, currently, the auction format in CryptoKitties has information asymmetry by conspiring with a seller and a bidder. When the seller tells the bidder when his auction starts, it is difficult for other players to participate because the bidder makes a successful bid as soon as it starts. We indicate that CryptoKitties may be providing an unfair environment for many users.

In this paper, we argue conditions that a fair market should be kept in Sect. 2. In practice, we compare CryptoKitties and fair market conditions in Sect. 3. Section 4 mentions countermeasures that may make CryptoKitties fairer. We consider other vulnerabilities of CryptoKitties in Sect. 5. Finally, we conclude our research in Sect. 6.

1.3   Related Work

Alesja Serada et al. He studied CryptoKitties as a subject to see how blockchain will shape future game design. He examined the relationship between token ownership and the value construction of CryptoKitties. In addition, he showed how the breeding and market aspects of kitty work concerning maintaining the game economy. As a result, the authors showed that the kitty's value is decreasing because there is no upper limit to the number of kitties that can be bred. The value of Gen 0 kitties, which cannot be created by breeding, decreased as well. We also showed that the existence of a transaction fee GAS could hinder the intervention of new users. He concludes that these are the points that make the game economy unsustainable [8].

Charlotte et al. Based on that "trust without trust," blockchain has emerged as a disruptive technology that is considered an alternative to law. The authors doubt that whether participants can transact with each other without the need for legally sanctioned trust. The authors specifically highlight the need for users to verify that a Dapps (short for decentralized applications) really does fall under it. He focuses on some Dapps, including CryptoKitties. Since it is possible that CryptoKitties is not decentralized, it is marketing as a Dapp may be misleading to users. The reason why kitty is considered immutable and cannot be taken away from others is the blockchain's immutability. However, only the market uses its properties. Charlotte points out that it is vulnerability, and some can cheat others to execute a dishonest contract [9].

2   Considering the Fairness of Financial Services Based on Smart Contract

2.1   Preliminary

Blockchain. Blockchain is a database commonly used as a ledger for cryptocurrencies. Satoshi Nakamoto proposed it as a bitcoin ledger in 2008 [2]. Blockchain has some special characteristics; it enables decentralized systems, immutable data, transparency, and anonymity.

A blockchain consists of many blocks. Each block contains transactions, a timestamp, a previous hash, and a nonce. A transaction has a sender address, a recipient address, and a value.

There is no administrator in the blockchain. Instead, every member of the network manages blockchain data. A Peer-to-Peer network connects the participants as nodes. Each node has blockchain data. If someone creates a new block, he sends all nodes connecting him to the block. These nodes will send other nodes when they receive the block. Soon, everyone will have that information.

So, how do users make a new block? First, block creators called miners to determine a block which they want to connect their block. If they determine the transactions in that block are legitimate, they will make the previous hash of their block the hash value of that block. It is called a blockchain because the blocks are connected like a chain by hashing. If there are six or more blocks connected behind a block, it is considered correct. Next, miners select transactions which they thought right ones and each transaction fee is high. If a miner creates a new block, he is rewarded with new coins. In the Ethereum blockchain, the transaction fee is called "Gas". Then, miners calculate "nonce" so that a block hash value is less than the threshold. As a block has nonce, if they change its value, they will also alter the block hash value. This threshold is set so that miners can find a nonce in 10 min, making it difficult for multiple blocks to be created simultaneously. If a block is easily created, it is immediately assumed to be the correct one, and they will approve suspicious transactions. The threshold prevents this. Finding the nonce is called Proof-of-Work (PoW), and the process of making the block is called mining. In this way, blockchain is a decentralized system.

Other properties also meet. A block created once will be saved in every node's server. an attacker must attack all nodes in order to successfully alter the blockchain. As it is too difficult, no participants can alter blocks. Besides, they can see all blocks. So, blockchain has transparency, such as seeing which address a cryptocurrency originated from. By using this, we know an address' balance. Of course, other users can not steal its cryptocurrency, thanks to the UTXO system [12]. However, bitcoin and Ethereum addresses have anonymity so that people can not figure out a real person who has the address.

Smart Contract. Nick Szabo proposed smart contracts in the 1990s [3]. We define a contract in advance. Nobody can change it once defined. When a person

agrees with its definition, the contract is executed. His and the contractor's settlement will be run automatically. We need no third party to run this contract. Take a vending machine, for example; the pre-definition is the product's price and pictures displayed. By selecting a juice, it is correctly executed until settlement. If the input is the same, smart contracts must have the same output. For a given contract, if the input is the same, the result must be the same regardless of who performs it. If this is not the case, then different people can buy the same juice at different prices. This would make smart contracts unreliable and different from the concept.

Ethereum is the first cryptocurrency to be able to operate smart contracts. With writing pre-definition on blockchain, all users do not re-write it. After pre-definition, programming code Solidity runs contracts. Thanks to this system, we can exchange ETHs safely.

CryptoKitties. CryptoKitties is one of the most famous blockchain-based games [5]. Axiom Zen created this game in 2017 [4]. We show CryptoKitties' overview in Fig. 1. In this game, users exchange ERC-721 tokens for ETHs. ERC-721 is a non-fungible token(NFT) transferred on the Ethereum blockchain. Unlike cryptocurrency tokens, NFTs are unique tokens, with specific parameters. Each ERC-721 token differs in its value. In CryptoKitties, an ERC-721 token is treated as a kitty. Again, players exchange kitties for ETHs. These kitties have an ID, gene, and generation. A kitty's ID is assigned in the order of birth, and the algorithm written in Solidity determines the gene that determines the appearance of the kitty. The generation of a child kitty is one greater than the generation of the parent kitties. By birthing and trading kitties, users aim to earn ETHs. This game's source code is written in Solidity. So, all trades and

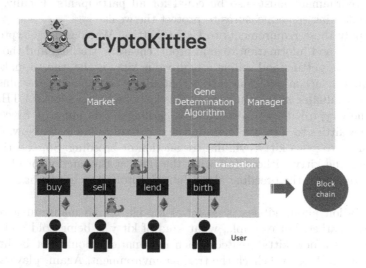

Fig. 1. CryptoKitties' overview

breeding in this game are executed by smart contracts. Not only transactions but kitties' data are on the blockchain.

There are two ways to obtain a kitty. One is winning an auction. The auction of CryptoKitties is the dutch system that the exhibited kitty's price goes down as time passes. There are two types of auctions; standard and rental. When a user wins a standard auction, he can get a kitty. In case of rental, a winner has to return the kitty after breeding, another way to get a kitty.

A user can get a new kitty created by the gene determine algorithm. To make a kitty, he needs to choose two parent kitties that are inputs of the algorithm. Parent kitties can be chosen from those that they already own, or one of them can be a kitty he won at a rental auction. After selecting parent kitties, he can get a baby kitty. Thus, this process is called breeding. A player has to pay 0.008 Eth when he lets two kitties breed. Since a kitty has no gender, it can be either a matron or a sire. The breeding defines a baby kitty's gene and generation. A kitty's generation settles its breeding period [13,14]. It has 14 kinds; the longest is two weeks, and the shortest is one minute. When a kitty is created, the cooldown period is determined by its generation, and from that point on, each time it is bred, the period increases by one kind. After breeding, until this period of time has passed, a matron kitty cannot reproduce. Through this kind of trading and breeding, players can get expensive kitties and sell them for a profit.

2.2   Fairness in CryptoKitties Market

In terms of economics, a fair market should keep the following criteria [10]. According to [10], every player has opportunities to profit and take risks equally. A market must also prevent cheating. Information asymmetry, where some people know information about making a profit, must remain relatively small. The trading environment must also be equal for all participants. Finally, fairness includes adopting some measures to protect the weak.

We apply these requirements to CryptoKitties. We consider CryptoKitties' opportunities and information to gain profit, cheating, trading, and the inferior. To begin with, getting and selling a high-value kitty is the way to make ETHs. So, an equal chance in this game means that all players can get high-value kitties. If you get a valuable kitty, you can sell this kitty and earn a lot of ETHs. Then, we assume CryptoKitties' cheating. It is earning ETHs unfairly. Since players need to get kitties to earn, we focus on the way to obtain a kitty. Now, there are two methods to get a kitty: winning an auction or breeding. So, cheating could be winning the kitty without following the rules at the auction or obtaining it by tampering with the breeding algorithm. CryptoKitties must not allow either action.

Information about kitties is essential for players to maintain a relatively symmetric market. For example, what kind of kitty is being sold and at what price? How is a new kitty created? Such information should not be limited to certain parties. Next, we check the trading environment. Again, players can get a kitty by auction or breeding. It is the only auction that a player trades his kitty with other players. Therefore, there should be a rule of auction so that no

one has a disadvantage. Finally, we define the weak. According to [10], it takes players without sufficient information, or poor negotiation skills or judgement as examples. We have already defined information as about a kitty. Then, a player's bargaining power has to do with how valuable kitties he can get to gain profit. If he has a lot of ETHs, he will get many kitties and valuable ones. Thus, this ability is related to financial resources. Besides, as mentioned in Sect. 2.1, the gene determination algorithm expresses how it creates a kitty. If a player understands this game's algorithm, he knows and can judge which kitty he should get and how kitties he should select as parents to gain profit. With understanding this game's algorithm and making the right decisions, he can make money, so judgment is affected by how well he understands this game. He will try to understand the algorithm written in Solidity to obtain an expensive kitty. Hence, the ability to judge is related to the ability to read Solidity. Therefore, we define the weak as players who have little ETHs and can not read Solidity.

We found that CryptoKitties does not meet some of condition that we show below. Condition No.1 and No.2 are rules for protecting the socially vulnerable. Other states provide opportunities to gain profit for all users. We point out that this game does not meet all conditions except for No.4 in Sect. 3. In Sect. 5, we mention that CryptoKitties may not satisfy with Condition No.1, 3, and 4 as a future work.

1. Non-rich players should not be at a disadvantage.
2. Players who cannot read Solidity should not be at a disadvantage.
3. All users should be notified of all trading opportunities.
4. The player base should be large enough such that the supply and demand behavior of a few players do not affect the entire market.
5. All users should have an equal chance for profit.

3   Analysis on CryptoKitties and Its Impact to Fairness

3.1   Analysis on Gene Determination Algorithm

Gene, one of a kitty's parameters, is a 240-bits number and depends on Gene Determine Algorithm, as shown in Algorithm 1. Again, gene defines a kitty's appearance. In detail, each of the five bits determines an element of appearance. For example, the ninth five bits correspond to the kitty's eyes' color. This algorithm is a smart contract and determined by the mixGenes function defined in GeneScience.sol [13]. In this algorithm, a gene array is used. Its length is 48, and each cell corresponds to an element of the kitty's appearance. The first cell is the last 5 bits of gene value. The second cell is second last 5 bits of gene value. In the same way, determine all of the cells of the gene array. We elucidate whether the gene determination algorithm that builds the value of tokens to be traded in the market makes the market unfair.

The first step in this algorithm between lines 6 and 21 is a swap for parents' genes. Before the swap, the gene value is divided into twelve groups. Each group has four cells; the first four cells are group 0, the second four cells are group 1,

Algorithm 1. Gene determination algorithm

1: $matron$:= matron gene array
2: $sire$:= sire gene array
3: $child$:= child gene array
4: $hash$:= SHA-256(target block), $hash[i]$ means i-th bit of $hash$
5: $k = 0$, k uses for $hash$
6: **for** $i = 0 \ldots 11$ **do**
7:     **for** $j = 2 \ldots 0$ **do**
8:         **if** $hash[k : k + 2] == 0$ **then**
9:             $swap(matron[i * 4 + j], matron[i * 4 + j + 1])$
10:         **end if**
11:         $k+ = 2$
12:     **end for**
13: **end for**
14: **for** $i = 0 \ldots 11$ **do**
15:     **for** $j = 2 \ldots 0$ **do**
16:         **if** $hash[k : k + 2] == 0$ **then**
17:             $swap(sire[i * 4 + j], sire[i * 4 + j + 1])$
18:         **end if**
19:         $k+ = 2$
20:     **end for**
21: **end for**
22: **for** $i = 0 \ldots 47$ **do**
23:     $mutated = false$
24:     **if** $i\%4 == 0$ **then**
25:         **if** $abs(matron[i] - sire[i]) == 1$ and $min(matron[i], sire[i])\%2 == 0$ **then**
26:             **if** $hash[k : k + 3] <= 1$ **then**
27:                 $child[i] = smallT/2 + 16$
28:                 $k+ = 3$
29:                 $mutated = true$
30:             **end if**
31:         **end if**
32:     **end if**
33:     **if** $!mutated$ **then**
34:         **if** $hash[k] == 1$ **then**
35:             $child[i] = matron[i]$
36:         **else**
37:             $child[i] = sire[i]$
38:         **end if**
39:         $k+ = 1$
40:     **end if**
41: **end for**
42: **return** child

---

and so on. Within each group, the swap operation is to change the order of the gene array. Next, the SHA-256 hash value of a block on the Ethereum blockchain, called the "target block", is involved in the algorithm. Six bits of the hash value of the target block are used for the swap in one group. Group 0 uses the last

six bits, and group 1 uses the second last six bits and same as below. Let us say the four cells in the group are a_0, a_1, a_2, and, a_3 starting with the one with the smallest index. If the last two bits among six bits are both zero, a_2 and a_3 are swapped. Then, if the next two bits are both zero, a_1 and a_2 are swapped. Finally, if the remaining two bits are both zero, a_0 and a_1 are swapped.

The next operation between lines 22 and 41 is to fill the cells of the gene array of the kitty to be born one by one (for $i = 0$ to 47). There are two methods for genetic determination in a new kitty; inheritance and mutation. The first action executed in this operation is checking to see if cell i of the parents meets the requirements for mutation. If two cells are satisfied with 1, 2, and either 3-a or 3-b in below, the formula (1) determines the child gene's cell i. In that formula, smallT means the smaller of the two parents' cell i.

1 i is multiple of 4
2 the absolute value of matron cell i and sire cell i is 1, and smaller one is even
3-a the smaller cell value is less than 22, and the lower three bits of the unused bits in the hash value of the target block are 001 or 000
3-b the smaller cell value is not less than 22, and the three bits of the unused bits in the hash value of the target block are 000.

$$cell = smallT/2 + 16 \tag{1}$$

When any of the conditions are not met, a baby inherits either parent's cell. If the lowest bit of the unused bits in the hash value of the target block is 1, cell i of a child gene will inherit from the matron's gene. If not, it inherits from the sire.

In short, a baby gene is dependent on its parents' gene and target block hash. All kitties' genes are on the blockchain. By using Etherscan, a block explorer, we can check any kitty's genes. Then, what is the target block? We can predict the attributes of the newborn kitty if we know the hash value. If it means randomly choosing one of all the blocks on the blockchain, we can not infer the outcome. However, the target block is somewhat limited and predictable. The target block is a block that will be issued when a matron kitty becomes fertile again. Specifically, there is a variable that stores the frequency of block creation. The product of that value and the matron kitty's breeding period corresponds to the blocks issued when she can breed again. We can find out the breeding period of the kitty from CryptoKitties' official page. Therefore, we know when the breeding period ends, and the block issued at that time or thereabouts becomes the target block. By calculating the hash value of the target block, we can predict a baby gene.

The fact that the results are predictable means that CryptoKitties has not satisfied conditions 1 and 5 of fairness, as defined in Sect. 2. First, the blockchain's transparency allows us to see what kitties are traded at a high price. Since we can expect breeding results, it is also possible to predict parent kitties to produce ones that match these trends. If we can make a successful bid to them, we will get a good kitty and earn ETHs.

However, in this kind of competition, well-financed users will have an immense advantage. We prove this by setting up a simple environment for CryptoKitties. First of all, we use "Cattribute", which means the attribute of kitty. In this market, there are 324 kinds of Cattribute. Let us take "driver" one of the Cattribute as an example. As of Jan. 13, there were 24 "driver" kitties in the market, and when we checked the gene sequences of all of them, we found that No.0 is 15 and No.36 is 23 in common. Therefore, we expect that the kitty that satisfies these two points has a "driver". In case of other Cattribute, "dominator", we found that No.0 is 28 and No.28 is 23 in common. Also, for a Cattribute, the first kitty to belong to that it gets Diamond. The first 10 kitties to belong to it will get Gilded, and the first 100 kitties will get Amethyst. From then on, these jewels are called "Family Jewels", as shown in Fig. 2 [6]. According to [7], the minimum price for each Jewel 5, 0.5, 0.07, 0.009 Eth, in order of rarity. Since these prices are positively correlated with rarity, the environment we set up should be like that. So, our environment will adopt these prices. In other words, all kitties with Diamonds are assumed to be 5 Eth. For the other kitties, we take those kitties that can produce X Eth will be sold for $X/2$ Eth. For example, a kitty that can reliably give birth to a Diamond kitty would be priced at 2.5 Eth. Also, the minimum price of kitty to be sold in the market shall be 0.004 Eth. This amount is the minimum in the market as of Jan. 13.

Fig. 2. Family jewels

The key is a kitty who has Diamond, which represents the pioneer of that Cattribute. Assuming that the value remains the same as more kitties of the same Cattribute, if a player gets a Diamond kitty, his earnings will be greater. Specifically, he gives birth to 499 kitties so that all of them inherit Diamond kitty's Cattribute and gets family jewels. He will get 9 Gilded, 90 Amethyst, and 400 Lapis kitties. Selling all kitties, his earnings will be 14.4 Eth. $(0.5 * 9 + 0.07 * 90 + 0.009 * 400)$ Thus, the player who has a Diamond kitty will gain a lot of profit. Then, in order for this to happen, you need to win the auction or mutate and give birth to the Diamond kitty.

This algorithm will make it easy to know which kitty to get to make money. Such kitty is in high demand, and users with financial resources are more likely to win its competition. Alice, who has little ETHs, can not buy Diamond kitty because it is expensive. So, to get these kitties, she has to birth them from other kitties. She can expect which kitties will give birth to a Diamond kitty. However, it is difficult for her to get these parent kitties because they are also in high demand. Since we can predict what kind of kitty will be born, some players

may repeatedly give birth to make a parent kitty. But, Alice can not do a lot of births because a fee of 0.008 Eth is charged for breeding, and again she has little ETHs. For Alice to make a profit, she needs to do things differently. It is a way of making a lot of small gains. However, since the birth fee is 0.008 Eth, selling a normal kitty will negatively affect revenue. So, she ought to sell kitties that have family jewels. As with the Diamond kitty, they are in high demand, and Alice will be hard to get them. Therefore, since the profit opportunity for players who do not have a lot of ETHs is small, CryptoKitties violates Condition No.1 and No.5.

3.2   The Readability of Solidity Source Code

Based on the previous section, rich players can make money, but is it really possible for everyone? Next is focusing on the actions they will take. The most important thing is understanding the gene determination algorithm. When they try to give birth to a Diamond kitty, they must find a Cattribute that has not been found yet. Besides, they have to choose parent kitties so that a Diamond kitty is created. In other cases, when trying to produce other kitties with family jewels from a Diamond kitty, they also have to select parent kitties so that the child inherits a parent's Cattribute. So, to make money, they need to understand the algorithm's behavior.

However, not everyone will be able to understand this algorithm. The main reason is that it is only written in Solidity. People who lack this Solidity knowledge will find it almost impossible to understand the algorithm. There is a considerable gap between those who can understand Solidity and those who can not, which significantly affects profit opportunities. Therefore, the results are contrary to Condition No.2 and No.5.

3.3   Auction in CryptoKitties

In this game, trading kitty is always done through auctions. So, users can only get kitties from the auction market. In this auction, a seller decides the starting and ending price of a kitty and auction. He can then start the auction with his intentions. According to condition No.3, a fair market should provide information about a trade for all players. It needs an environment where they know what kind of transactions exist. In this sense, information on what type of kitty is on sale and what price must be shared with all users.

However, when an exhibitor, Alice, and a user, Bob, are colluding, other players, Charlie, have little chance to get Alice's kitty. Alice tells when her auction tries to start Bob. As soon as her auction starts, Bob bids for her kitty, and Charlie can not see its trading. For example, Alice gives Bob a Diamond kitty in this way. Other players can not get her kitty, and Bob can get it and its child kitties. They have the same Cattribute as their parent's one and with family jewels. As a result, only Bob can gain profits. It contradicts condition No.3 and also enhances to expand information asymmetry.

4   Enhancement of Fairness

4.1   Gene Determination Algorithm

As mentioned in Sect. 3.1, we state that it is difficult for not rich players to gain profit in CryptoKitties. To earn ETHs, players need to sell a valuable kitty. Users with financial resources have the advantage of getting that kitty at auction. The method of giving birth and getting a kitty is also easier for wealthy users than ordinary ones. Because the outcome of the gene determination algorithm is predictable, so the demand for parent cats is high, and only rich people can afford them.

If the gene determination algorithm's output is unpredictable, everyone may have a chance to gain profit. With that algorithm, no one would know which parent kitties would produce a valuable kitty. All players do not know which kitties as parent kitties will make a profitable kitty. As a result, the value of the parent kitties that produced a valuable kitty is unknown. Therefore, there is a possibility that even a cheap kitty can give birth. If kitties are reasonable, many users can bid on them, so we think the game will be fairer than it is now.

However, there is a problem with introducing an unpredictable system in CryptoKitties. Again, this algorithm is written in Solidity and a smart contract. Smart contracts should output the same result if the input is the same. So, smart contracts cannot use a random number generator. Therefore, we have to consider the system without a random number.

To add an external randomness source in deciding the gene, we suggest that the users jointly create a random number as an input to a hash function (e.g., SHA-256) instead of creating value from a block on the blockchain. For instance, when someone does breeding, everyone chooses a random number. The sum of them which users submit will be the input to a secure hash function. As all users cannot predict the input, its output is also unpredictable. This game's usability will not be compromised if a tool automatically generates and sends a random number every time requested. It provides fairness to the market since every user can have a chance to get a high-value ERC-721 token and trade it to ETHs. We show this system's overview in Fig. 3.

4.2   The Readability of Solidity Source Code

We explained that if a player does not understand the gene determination algorithm, he may not get a valuable kitty from kitties he has in Sect. 3.2. For example, though he has a Diamond kitty, he could not inherit its Cattribute to its kitties. He will not make ETHs because the kitty without Family Jewels is expected not popular. On the other hand, if another player understands the algorithm, he will create expensive kitties. We expect that there are significant gaps to opportunities to gain profit between knowledgable players and not. In order to make CryptoKitties fair, this game should fix so that all users can understand its gene algorithm. However, we insist that CryptoKitties does not achieve it.

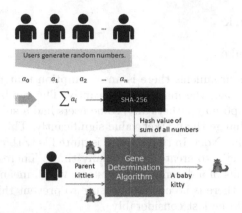

Fig. 3. Countermeasure's overview

The readability of a programming language is not high, and it is estimated that more people cannot read it than those who can.

We assert that CryptoKitties needs a system so that more people can figure it out. These game managers should explain the gene determination algorithm through multiple representations, such as diagrams, flowcharts, language descriptions, and so on. It is expected that a variety of explanatory methods will reduce the information asymmetry about the gene determination algorithm, and this game will satisfy with Condition No.2.

4.3 Auction in CryptoKitties

We show that when a seller and a bidder are colluding, it is difficult for other players to get the seller's kitty in Sect. 3.3. Though other players will not create this kitty's children, the bidder can. The seller gains profit by selling the kitty, and the bidder also earns ETHs by selling child kitties of his kitty. But, other players have no opportunities to gain profit from the kitty. We argue that CryptoKitties does not provide an environment that every user has a chance to gain profit equally. This problem stems from the fact that the time between the auction starts and the winning bid is too short.

CryptoKitties should create an auction where everyone has an opportunity to trade. One countermeasure is that supposing it allows a certain amount of time, one hour, for example, between when a player puts up an item and when it becomes available for bidding. During this period, players not a part of the collusion will have time to react to the new bid. We hope that it enhances to give all users a chance to get the kitty and make money.

5   Future Work

5.1   Proof of Stake

In the Ethereum environment, there is an assumption that no user has a large enough ETHs to impact the market significantly. The symbol of this is Proof-of-Stake (PoS). Suppose hypothetically; some users had a substantial amount of ETHs that could change the market value significantly. This situation does not satisfy with Condition No.4. In that case, it is quite likely that those users would be able to get the right to create blocks many times. There is no bias towards users' chance in terms of a decentralized system, which means to have the right to create. However, there is no countermeasure to prevent this from happening, and playability could be lost considerably.

5.2   Secret Trading

We state that transactions can only be conducted through the auction market. However, is this really the case? We find some suspicious transactions, as shown in Fig. 4 [11]. They traded kitties for nothing and without going through an auction possibly. We can not figure out how they traded. If they really sold using other methods, this game does not meet condition 3 and will be reversed. This game should also take measures for such cases.

Fig. 4. One example of suspicious transactions

5.3   Blockchain Anonymity

Ethereum's blockchain is guaranteed to be anonymous. The blockchain address cannot be tied to who the actual user. However, there is one problem that arises from this. That is insider trading. Insider trading is that a person who has

inside information about a company buys or sells shares before the information about the material fact is made public. Or, this company's employees trade them. In order to discover these transactions, it is necessary to know who made them. Unfortunately, it is so difficult to find them in the blockchain environment because of blockchain anonymity. If its developer takes part in this game and earns a large ETHs, we will not see his illegal activity. Thus, CryptoKitties should prove that they are not doing this.

5.4   Fee of Transaction and Breeding

We argue that Gas and breeding fee are an obstacle to the motivation to trade for the average user. Transaction fee, GAS, is a miner's motivation because if he has a right to create a block, all GAS in the block will be his income. Miners tend to select transactions whose GAS are high. Therefore, wealthy users' transactions are apt to be approved since they pay high GAS. Besides, the Breed fee, currently 0.008 ETH, is CryptoKitties managers' income. When a player tries to breed his kitties, he pays ETHs in exchange for a new kitty. Users who can afford ETHs can breed or exchange kitty without hesitation. In contrast, for other players, their action count is limited. So, they have fewer opportunities to get good kitties and gain profit than rich people. In other words, these fees do not help conditions No.1 and No.5. One countermeasure we think is decreasing the breeding fee to prevent limiting trading for non-rich players.

6   Conclusion

We found that CryptoKitties does not satisfy with four conditions that fair markets should keep. The gene determination algorithm which affects the ERC-721 token value is not satisfied with two conditions. We show that we can predict we will produce a token. In the environment we set, a player can not gain profit when he creates and sells a cheap token because of the breeding fee. So, he needs to sell a token whose value is more than the breeding fee. Since we can know which token we will make, we know tokens that can produce high-value ones. Thus, rich players are more easily to get lucrative limited tokens. Besides, people who can understand this algorithm are limited because the only Solidity expresses this algorithm. Users who know this bias have a significant advantage in playing CryptoKitties. Also, there is a problem with its trading market. When two players are colluding, it will be possible to attack them by not allowing others to trade with them.

We also mention the countermeasures of these problems. In the case of the gene algorithm, it needs an unpredictable system. Since this problem stems from seeing the input of the hash value it is using; we propose that all users determine its input. Moreover, describing these mechanisms in other than a programming language will reduce the information asymmetry. Finally, as one cause is that from the auction starts to the winning bid is too short, one method would be noticing what kitties have been put up for auction and allow a certain amount

of time before the bidding starts. We propose these improvements, but there are other problems in CryptoKitties.

References

1. CryptoKitties' users and economic effect. https://corporate.coincheck.com/2020/10/06/117.html. Accessed 9 Jan 2021
2. Satoshi Nakamoto: Bitcoin: A Peer-to-Peer Electronic Cash System (2008). https://bitcoin.org/bitcoin.pdf
3. Nick Szabo: Smart Contracts: Building Blocks for Digital Markets (2018)
4. Axiom Zen: CryptoKitties: Collectible and Breedable Cats Empowered by Blockchain Technology (2017). https://drive.google.com/open?id=1soo-eAaJHzhw_XhFGMJp3VNcQoM43byS
5. CryptoKitties. https://www.cryptokitties.co/. Accessed 9 Jan 2021
6. CryptoKitties official guide for family jewels. https://guide.cryptokitties.co/guide/cat-features/family-jewels. Accessed 9 Jan 2021
7. Family jewels' minimum amount. https://kittyhelper.co/. Accessed 9 Jan 2021
8. Serada, A., Sihvonen, T., Tuomas Harviainen, J.: CryptoKitties and the new ludic economy: how blockchain introduces value, ownership, and scarcity in digital gaming. Games Culture 1555412019898305 (2020). https://doi.org/10.1177/1555412019898305
9. Ducuing, C.: How to make sure my cryptokitties are here forever? The complementary roles of blockchain and the law to bring trust. Eur. J. Risk Regul. **10**(2), 315–329 (2019). https://doi.org/10.1017/err.2019.39
10. The concept of fairness in financial transactions. https://www.imes.boj.or.jp/research/papers/japanese/kk18-5-1.pdf. Accessed 9 Jan 2021
11. One example of suspicious transactions. https://etherscan.io/tx/0x3591b36cfd443fd686ba8015d93f82832a6bf007b438caa6f342679601145e9e. Accessed 20 Jan 2021
12. Bitcoin UTXO. https://river.com/learn/bitcoins-utxo-model/. Accessed 26 Jan 2021
13. CryptoKitties source code. https://github.com/cryptocopycats/awesome-cryptokitties/tree/master/contracts. Accessed 26 Jan 2021
14. CryptoKitties cooldown speed. https://guide.cryptokitties.co/guide/cat-features/cooldown-speed. Accessed 26 Jan 2021

Coins, Covid, Keynes and K-Shaped Recovery

Pepi Martinez[✉], William Huang, and Bud Mishra

New York University & RxCovea, New York, USA
{pepi,wwh237,mishra}@nyu.edu

Abstract. Reckless monetary policy, especially in the wake of a pandemic, amplifies the gap between the extreme ends of the income distribution, thus exacerbating the long term effects of income inequality and loss of human capital. Attempts of federal stimulus bills fall short in timing and size, including directing spending towards those most in need. We propose a general smart contract protocol that distributes funds to targeted individuals with programmatic spending enforceability, alleviating the K-Shape recovery that current monetary policy is creating and turn it into the desired V-Shape. Utilizing incentive structures, our model directs spending to help stimulate the economies of targeted communities and struggling businesses. Smart contracts remove the current inefficiencies in the political trust and permission-based solution and allow for more transparency, verification, and incentives to help one's community in times of need. Such a system allows for a more positive and direct relationship between those with funds and those who need funds.

Keywords: Smart contract · Economic stimulus · Donation fund · Incentive systems · Spend tracking

1 Introduction

Since the middle of March 2020, seeking to reduce the impact of the coronavirus on the economy, the Federal Reserve (FED) has been injecting unprecedented amounts of liquidity into the market with various policies like *Repurchase Agreement Operations (REPO), Quantitative Easing (QE), the purchasing of Corporate Debt Bonds and Mortgage Backed Securities (MBS), direct business lending programs (PMCCF, SMCCF, MSLP, and PPP)*, and "helicopter" cash stimulus to private citizens. Such prescriptions are based on a rigorous framework introduced by John Maynard Keynes, aptly named "Keynesian Economics," which emphasizes the importance of increasing government spending in times of economic crisis to stimulate demand. However, despite the initial objectives of this policy, it does not always stimulate the economy in positive ways according to our analysis of current government spending programs. We introduce the hypothesis

B. Mishra—Mentor.

© International Financial Cryptography Association 2021
M. Bernhard et al. (Eds.): FC 2021 Workshops, LNCS 12676, pp. 611–627, 2021.
https://doi.org/10.1007/978-3-662-63958-0_43

that although Keynesian monetary policy is arguably most responsible for the large income inequality, inflated asset prices, and sub optimal full employment, it could be rescued with the help of a decentralized and transparent fund distribution system based on smart contracts. In this report, we examine this hypothesis in two ways. First, we analyze historical interest rates, Federal Reserve balance sheets, and unemployment data among other data sets to analyze the correlations between these variables and the FED mandate. Second, we propose a design of a decentralized smart contract protocol to generate a V-shaped recovery, thus blunting the sharp inequality that could result from a K-shaped recovery.

Keynes and Modern Stimulus

Keynesian monetary policy rose to prominence as an antidote to the Great Depression, and continued to be deployed during times of crisis, such as the 2008 housing recession and the 2020 COVID-19 crisis. Keynesian monetary policies were aggressively deployed and sustained throughout the Obama administration as an antidote to the 2008 housing recession, leveraging sustained lowered interest rates and increased federal spending to bolster economic growth. They were again installed with the onset of COVID-19 and the ensuing recession as the federal government under the Trump administration plummeted the already deflated interest rate to all-time lows. As shown by both responses to these crises, the three main tools the Federal Reserve utilizes to establish Keynesian monetary policies and stimulate the economy comprise of the following: (1) Interest rates, which determines the spending and borrowing in the economy; (2) Quantitative Easing (QE) and balance sheet expansion, which refers to when the Federal Reserve buys financial assets from the open market; and (3) Repurchase Agreements (Repo), loans the Federal Reserve credits to businesses overnight. With these tools and policies, the Federal Reserve hopes to stimulate recovery during recessions. However, despite the Federal Reserve's initial objectives or intentions, it does not always stimulate the economy in a positive way—one must judge policies by their results, and not by the intentions of its founders. We attempt to prove this in the following sections by analyzing historical macroeconomic data against the Federal Reserve's mandate. We will also further explore possibilities made ubiquitously accessible via decentralized and "trustless" smart contract technologies.

Keynesian Supply Side Shocks

With the onset of the Covid-19 pandemic, governments began shutting down contact-intensive sectors of the economy to protect public health. A shutdown in one sector of the economy, although endogenous, may create a negative spiral in the demand of sectors unaffected by the shutdown [1]. Demand deficiencies can thus spiral multiplicatively, and cause an anemic recovery as opposed to the desired V-shape due to firm exits and heightened levels of unemployment, resulting in permanent losses of human capital.

To illustrate, imagine a market sector comprised of K-12 teachers and daycare firms during the pandemic. Teachers that are sent home no longer require daycare

services as they can take care of their own children. Although the shutdown did not directly affect the daycare sector, its demand is still affected by the shutdown of K-12 schools. Additionally, some of the teachers who are laid off no longer receive income, diminishing their overall spending on other sectors and adding to demand problems of firms originally unaffected by the original shutdown. Another example: restaurants that closed due to pandemic reasons no longer require accounting services, leaving accounting firms with fewer customers and decreased cash flows. This decrease in cash flow results in many firms deciding that they no longer require office spaces, chaining adjacent sector shocks into a series of cross-sector demand losses.

These Keynesian supply shocks can have dire consequences to the overall economy and drag on the recovery process. The exits of firms can also cause a spiral of demand shocks—if sector one shuts down and sector two requires their goods as material inputs, they would also be forced to shut down. These endogenous outcomes feed back into themselves and perpetuate their losses throughout the economy as a whole.

We propose that the fundamental issue with sector imbalances and monetary policy is the lack of complete information acquirable by central planners, as the computation required to understand the holistic needs of individuals and businesses are far greater than any central authority can efficiently manage. Instead, we propose a decentralized smart contract system that allows for intertemporal demand imbalances to level out through programmable monetary policies across sectors, created and governed by trusted agents who decide upon the rules and methods of fund distributions. Decentralization allows for small community-driven structures of distribution that better understand the needs of participants, stemming from locality-driven information completeness.

We will explore the effectiveness of current centralized policies in the following section, and propose a smart contract protocol that would allow anyone to create a fund with enforceable distribution and spending rules.

2   Effectiveness of Current Monetary Policy

Our experiment requires performing linear regression on predictor and response variables. Each of the predictor variables are key levers the Federal Reserve uses to enact monetary policy. Since these variables are ubiquitously observable, suitably parametrized smart-contracts can be implemented to embody Keynesian framework scalably, evolvably and decentrally. In the years prior to 2008, interest rates were the key tool the Federal Reserve used to influence the amount of spending in the economy. However, due to recent economic downturns along with other political and macro economic factors, interest rates have been historically low, hovering between 0–2%. This constraint requires the analysis of the other predictor variable—the Federal Reserve balance sheet. The FED balance sheet shows the outstanding assets and liabilities at the FED, which grows when the Federal Reserve "prints money" to buy assets from financial institutions to inject money into the economy, and contracts when the FED sells these

assets. The combination of these variables shows us a clear picture of Federal Reserve actions. While there are certain desirable qualities to such centralized approaches, it is not necessarily the only plausible approach - especially in the absence of suitable institutions and their governance. To test this hypothesis, we evaluated the "standard" centralized approach as follows.

The Federal Reserve's core mandate states that their mission is to keep unemployment rates low while maintaining a slow and steady growth of inflation, which we measure through Consumer Price Index (CPI). Based on their respective actions through balance sheet expansion and interest rates controls, we are looking to find high correlations between each predictor and response variables that reflect this core mandate. Correlations are measured by adjusted r-squared values and high r-squared values will be interpreted as values above 0.2, which compensates for the lagging indicator effect.

Linear Regression Analysis

For each regression set we provide the regression statistics, coefficient table, and the corresponding interpretation. The full set of predictor vs. response variable analysis can be found in the appendix (Fig.1).

Regression Statistics		Coefficients	Standard Error	t Stat	P-value
Multiple R	0.05777537	Intercept	5.97983406	0.26158063	22.8603852	3.0614E-61
R Square	0.00333799	WALCL	7.0197E-08	7.9807E-08	0.8795789	0.38000156
Adjusted R Square	-0.0009766					
Standard Error	2.04393569					
Observations	233					

Here we see a slight positive relationship between balance sheet and unemployment, which is the opposite of what the Federal Reserve intends. Based on the Federal Reserve mandate, we expected to see balance sheet expansion lower unemployment rates, not increase. This is not what our regression shows. The adjusted r-squared value of ~0.0 shows almost no relationship between the two variables. With a p-value of 0.38, we cannot conclude with statistical significance that the relationship between unemployment and balance sheet even exists. This provides a strong argument against the Federal Reserve employing Keynesian monetary policies for the purposes of lowering unemployment.

Fig. 1. Balance sheet vs unemployment

Statistical Conclusion

The key finding from this study is the statistically significant, negative correlation between interest rates and unemployment and the nonexistent correlation

between the balance sheet expansion and decreased unemployment. This finding is strong evidence against Keynesian monetary policies as it is currently employed, indicating that funds injected through the current monetary system appears to have no effect on unemployment—and where there is an effect, the effect is counterproductive. There are a couple of reasons we conjectured to explain this phenomenon. First, it is possible that decreased interest rates make it easier for businesses to accrue bad debt as the threshold for borrowing is substantially lowered. This disparity increases the rate of loan delinquencies and decreases the rate of overall productivity in the economy as businesses with low productivity are propped up in the system. Secondly, our data hints that the majority of monetary stimulus does not trickle down to people in the bottom rings of the socioeconomic ladder, and is instead captured by large corporations and the wealthy in the form of equity markets. The current rise of stock market prices is strong evidence of the latter, as we see the stock market grow to unprecedented highs as an effect of FED monetary policies.

Through careful analysis and examination of our hypothesis, we reach the conclusion that Keynesian monetary policy as enacted today, whose initial objective is to stimulate the economy in times of crisis, is responsible for inflated asset prices and sub-optimal full employment. Therefore, we propose a smart contract system that would alleviate the fundamental issues of fund distribution and accountability, as discussed in our analysis in modern US monetary policy, by removing existing bottlenecks between those with funds and those in need of funds.

In the full research paper, we will perform realistic simulations to gauge the effect of monetary policies deployed on smart contracts.

3   Maneki Protocol Fundamentals

From our analysis of US monetary policies, we identified several glaring problems with money distribution today. These problems plague both public and private fund distribution, eroding trust between fund providers and recipients due to the lack of built-in accountability and transparency at the fund distribution layer. We discuss each of the problems below.

Distribution Fraud

Perhaps the most prominent problem with current government spending programs is the lax verification (fraud detection and credible threat) systems associated with them, leading to billions of taxpayer money lost annually. A couple of figures illustrate the enormity of the problem:

- A report from Association of Certified Fraud Examiners (ACFE) in 2012 finds that non-profit fraud amounts to roughly 5% of organizational revenue each year. Using this estimation, public charities alone are losing around $100B to fraud annually [3].

- The US Department of Labor estimates a loss of around \$3.5B in unemployment insurance fraud annually [4].
- In September 2020, the Department of Justice (DOJ) charged 57 individuals who received more than \$157 million from the Paycheck Protection Program (PPP) with fraud after they revealed the funds collected were not spent on employee wages, and instead on enriching the individuals. The DOJ went on to say that "the total amount of fraud is unclear at this point, and more charges are expected over the coming months and year" [5].
- From the March 2020 \$2.2T government stimulus alone, an estimated \$40B in losses can be attributed to collection fraud [6].

The issues with money distribution cannot only be attributed to a failure of fraud detection systems. Rather, it highlights the inability of current spending programs to prevent recipient and spending fraud at the distribution level. For example, PPP's goal of protecting employee wages falls short because a significant portion of the money received by corporations are not paid out to employees. This narrative signals a need for change in the way we distribute money today.

Loss of Trust

The inability to prevent monetary fraud in existing distribution systems directly results in a loss of public trust in money distribution systems. This trustlessness is especially prevalent in young adults today, who according to Pew Research Center, have become increasingly less confident in key institutions (e.g., police officers and business leaders) compared to older generations [10]. The loss of trust creates a negative cycle of people becoming less willing to make donations compared to before. Indeed, donation levels are at historic lows, only 73% of the US population donated to charities in 2020, down 9% nationwide since 2017 [9].

Large and Fragmented Network

A large part of why distribution fraud exists can be attributed to the scale, complexity, and fragmented nature of many spending programs. A popular form of collection fraud (Sybil fraud) occurs when an individual creates multiple aliases to double or triple one's collection from the same or different programs. Because many programs do not have robust Know-Your-Customer (KYC) infrastructure in place, and programs generally do not share recipient information with other organizations, it becomes impossible to track fund distribution across many spending programs.

Today's spending programs are very limited in their ways to control recipient spending once the funds are distributed. Government programs rely on financial audits, which are costly and non-exhaustive; nonprofits and charities rely on accumulating funds to distribute capital and goods to the right people, using distribution processes that often lack transparency and public trust. This issue deflates the overall amount people would donate, perpetuating the negative cycle.

Centralized Brokers

A centralized broker provides a single point of failure for any fund. If the broker is compromised, all donations are instantly compromised as well. Donors looking for a way to ensure their funds are properly distributed to the right people need to trust and verify the entities they donate to, increasing the barrier of entry for donors.

Smart Contract Protocol

To address the many issues surrounding money distribution systems today, we propose the creation of a smart contract protocol that leverages the distributed, synchronized, and immutable nature of blockchain systems to protect donors and recipients, expose spending insights, such as historical spend for a fund, and improve on fund distribution infrastructure through programmatic rules.

Maneki leverages blockchain properties like immutable transaction trails to enable the quick detection of fund abusers, and provide smart contract-enabled guarantees to fund distribution such as fund dispensing from the contracts only if certain conditions are met. These conditions are provided by donors who create the funds at the outset, and enforced at distribution time. Recipients who are eligible may then apply, having fully understood and agreed to the conditions for receiving such funds. More examples of fund rules and their creation process are provided in the next section. By using smart contracts, needed transparency can be provided to both parties.

The next section explores the protocol in more depth, identifying the different stakeholders, smart contract flows, and real-world use cases.

4   Maneki Protocol

The Maneki platform is a smart contract protocol that enables trustless distribution of funds in a transparent manner. The protocol aims to strengthen trust between fund providers and recipients through mechanisms that incentivize people to donate and receive more. Maneki leverages existing benefits of blockchain technology and is designed as a Layer-2 solution built on top of payment tokens. As such, any payment-based blockchain supporting smart contract implementation can deploy the Maneki protocol on top to further secure fund distribution and spend tracking on the original payment system. Such payment tokens may include any form of Central Bank Digital Currencies (CBDC), ERC-20 payment tokens, and other payment-based tokens with smart contract capabilities.

Best conveyed by Hayek, "law and language have been allowed to develop for millennia while the improvement of money has been frozen and restricted from private experimentation." This is no longer the case today. With the advent of Bitcoin and other decentralized payment tokens, governments and citizens alike are rethinking the role and functions of money. Maneki is a smart contract protocol that provides private citizens, philanthropists, and even governments the ability to experiment on an ever-improving system of money distribution

enforced by smart contracts. Maneki utilizes smart contracts to enforce accurate distribution and spending of allocated funds, providing trust through the transparent system of execution. It aims to serve as a general framework to improve upon the money distribution system, much like how language and law have developed in the modern time.

In this paper, we outline several real-world use cases that can greatly benefit from the use of Maneki.

Stakeholders

The Maneki protocol revolves around two types of stakeholders. *Donors* create public or private funds that other donors can also contribute to, and distribute to *recipients* specified by the funds' rules. Fund rules are smart contract enforced logic that is programmed into each fund that specifies how funds are distributed and spent. Public funds can be contributed to by any blockchain-registered entity, and donors can specify and vote on fund rules based on the amount of funds contributed. Private funds are controlled by the entities who create them, and new donors may be added to contribute, determined by the administrators of those funds.

Recipients apply to funds and are approved based on existing fund rules. For example, a fund may only distribute to teenagers in low-income areas, and dictate that the money can only be spent on educational supplies. Once a recipient is approved for a fund and agrees to the fund rules, they receive those funds and can only spend them in accordance to the rules. If recipients try to spend funds against the fund rules, a penalty can be enacted against them in the form of reduced reputation that may influence future fund applications. Repeated offenses may lead to participation withdrawal from the funds they attempt to abuse. Through these smart contracts a crypto coin circulating among the recipients may get connected and valuated against donors' funds, and this in turn may attract or repel additional donors.

Smart Contracts

Maneki's smart contracts leverage blockchain's history of immutable transactions to perform integrity checks and verify distribution before money is deployed. After fund deployment, they serve to enforce and incentivize spending habits dictated by the fund rules. We believe that over time, these fund rules can create creative and complex incentive structures that are utilized to deploy money to the right people and for the right causes. The smart contracts enable distribution and tracking processes that provide security and trust to the fund distribution process for both donors and recipients, ultimately resulting in more efficient money deployment through targeted spending.

Fund Distribution

There are two types of accounts registered on Maneki. Individual accounts, which comprise of donors and recipients who transact in fund distribution, and business

accounts who receive payment from fund recipients. Business accounts will be discussed in more detail in the Recipient Spending section. Account creation is enforced by a KYC/identity solution powered by either the central government, in the case of a CBDC implementation, or existing providers such as Civic. Each account is then remembered by the Maneki protocol, who register each individual with exactly one account to protect from double dipping and reputational Sybil attacks.

Once recipients authorize their accounts, they allow the sharing of certain information with funds, similar to using OAuth solutions like Google login today. Recipients can specify funds to automatically enroll in based on their needs, and also request to join funds they are qualified for. A REQUEST_FUNDS contract allows a business or individual to request funding from a specific program.

Donors create and manage funds using a suite of smart contracts:

- The FUND_CREATE contract allows donors to create a fund. Each fund exists as its own entity and administrators of the fund provide rules that govern it. These rules include requirements for recipient eligibility, fund governance and voting structure, distribution schedules, and spending conditions, among other fund settings. Funds can be donated to by anyone if public, and selected individuals if private.
- The FUND_UPDATE contract updates existing contracts' rules that can change distribution schedule, eligibility criteria, and any other mutable fund rules. Eligibility of a fund can be configured to trigger based on many criteria. For example, a fund may specify that only people who are vaccinated for COVID-19 are eligible to receive funds. This type of interaction requires off-chain validation that may be integrated into the onchain ecosystem via oracle solutions.
- The FUND_DISTRIBUTE contract triggers fund distribution, either manually or programmatically. Prior to distribution, the contract rechecks the rules of the fund, reconciles recipient list against current eligibility criteria, and finally assigns predetermined number of tokens to each member on the recipient list.

Fund rules are flexible by nature, allowing donors to best direct their money to the causes most aligned with their respective values. They can choose and enforce eligibility criteria to a fund based on the geography of recipients, income levels, spending history patterns, among many other parameters as long as the condition is sourced and provided by a trusted oracle. An example of a useful fund rule in a government-backed stimulus funds is a trigger that returns funds if they are not spent after a set period of time. Such a rule encourages individual spending, increasing money velocity and decreasing the risk of hyperinflation. This implementation allows for the enforcement of stimulus packages that are predictably spent, increasing aggregate spending without needing to wait for banks to create credit. Fund rule flexibility ensures that donors feel confident that their funds will truly go where they intended them to, and that funds can adapt based on changing needs for both existing and potential future funds.

Recipient Spending

The protocol allows two types of spend tracking: checking of fund rules in real-time and spend provenance.

Every time a recipient spends from one of their fund allocations, they invoke the SPEND_FUNDS contract in real-time. The SPEND_FUNDS contract is the most widely used contract in the protocol. The contract imposes a set of spending rules specified in the fund rules, then validates the spend transaction based on valid recipient addresses and account balances. Participating businesses on the platform register their products on the Maneki protocol so the contract can deterministically track and enforce spending. The Maneki protocol enforces a protocol-wide product categorization data standard to facilitate spend tracking, similar to a more robust version of Merchant Category Codes (MCC) used by banks and credit card providers today. As such, SPEND_FUNDS contracts confirm that funds are spent as donors intended, while addressing recipient needs.

The second part of spend tracking leverages blockchain's immutable nature to provide spend provenance insights at a fund, individual, and business level. Spend data on the blockchain can be indexed to provide aggregate spending data and money flows to fund creators to improve future fund rule changes, identify distribution errors, and provide protocol-wide reputation for all entities. Each entity's spending data are cryptographically secured and private, accessible only with permission from the entity. In other words, entities have the right to explicitly grant funds to certain spending data or personal information.

5   Use Cases

The general architecture of the Maneki protocol aims to enable the identification of better fund management systems. To this end, we explore several fund use cases that can benefit from the protocol today.

5.1   Government Stimulus Funding

One of the key failures of the Payment Protection Program (PPP) is the vague and non-enforceable language around protecting employee wages due to unforeseen supply shocks, the primary reason for the fund's existence. As a direct result, many corporations receive billions of dollars in taxpayer money in the form of loans and grants, only to lay off their employees shortly after. Through this distribution system alone, millions of working class Americans are defrauded while executives enrich themselves at an unprecedented time of crisis. While these actions of a few are ethically deplorable, it is the money distribution design that is fundamentally flawed, allowing for poorly regulated access to these funds.

A fund built on top of Maneki protocol would ensure that stimulus funds are tracked, businesses are spending the money based on loan and grant stipulations, and individuals are receiving the money from employers. Every interaction would be regulated and enforced by smart contract execution. Throughout the fund lifecycle, the government is able to analyze the efficiency of fund deployment

to change the fund's rules to optimize fund allocation for the future, all based on historical fund performance. Individuals will also be able to receive a larger share of the stimulus spending, decreasing income inequality and encouraging domestic spending as a whole.

5.2 Unemployment Insurance

Similar to PPP, unemployment insurance administered today suffers from distribution errors and collection fraud. In 2015, the US Department of Labor estimated a loss of $3.5B in unemployment insurance at an error rate of 10.7% [7]. A state-administered fund distribution application built on Maneki could provide clear provenance insurance collection and work history of any individual, giving the state department a clear understanding of the fund distribution lifecycle. Smart contract execution serves not only to prevent funds from collection fraud, but can also be used to protect against the common issue of people misunderstanding the rules and terms of unemployment insurance collection. Furthermore, the state may derive spending insights based on an individual's employment history, if permissioned by the individual, to assist in finding employment. For example, if John Doe has not found a job within 2 months, an agent can be dispatched to offer job search assistance. Similar applications can be built to power government funded programs like TANF or SNAP (i.e., food stamp program), where funds can be distributed to a predetermined set of individuals who qualify, and spending rules are attached to each fund allocation.

5.3 Donation and Fundraising Platforms

As Maneki's general structure allows it to be deployed on any type of payment blockchain with smart contract capabilities, donation and fundraising applications powered by Maneki smart contracts pose an interesting alternative for people looking to donate money outside of traditional nonprofits. For example, a donation platform can utilize Maneki smart contracts built on top of Diem (formerly known as Libra) to source capital for user-proposed causes. Donors on the platform can create funds for specific causes and invite others to donate, with the added benefits of recipient spending enforcement and transparent insights on fund usage. These funds provide a way for any individual or group to raise, distribute, and track money spent for any purpose, using a transparent and trustworthy set of smart contracts. As a result, recipients benefit from broader access to capital and a fairer system of fund distribution while donors ensure the security of their funds.

5.4 Managing Keynesian Supply Shocks

As government funds created on smart contracts have access to aggregated information across many different programs, they allow for the creation of intertemporal rules between economic sectors to alleviate the demand imbalances created

by Keynesian supply shocks. Preset conditions triggered by smart contracts can automatically execute the interest rate changes or employer-side tax rates by sector, allowing monetary policies to be deployed in a targeted fashion. In this scenario, agents create contracts on the Maneki platform that correct asymmetries in demand between sectors by either easing or tightening conditions based on the health of a particular sector. For instance, the technology industry has benefited greatly since the onset of pandemic-driven shutdowns, while the restaurant sector is in fallout. By lowering interest rates or payroll taxes for only the affected industries, fund creators can perform macroeconomic policies that are better informed on a per sector basis, resulting in less fund leakage and higher efficiency.

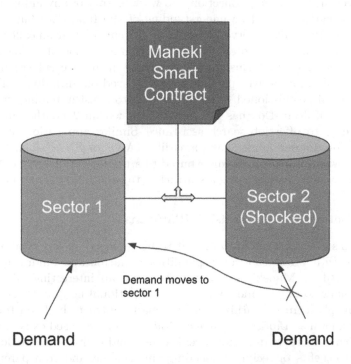

Fig. 2. Maneki contract alleviating sector imbalances. A detailed simulation and analysis of the recovery dynamics will appear in the full paper. Imagine Sector 1 represents college education sector and Sector 2 k-12 education sector: two sectors one thriving under impersonal interaction and the other suffering from a lack of in-person interaction. A pandemic induced supply shock to Sector 2 may lead to a partial demand for the Sector 1 infrastructure to be used by Sector 2, and covered by donations from Sector 1 governed by Maneki Smart contracts.

A realistic simulation will appear in the full paper showing how a Maneki smart contract can help alleviate Keynesian Supply Shocks [1].

Other forms of macroeconomic policies can be deployed and monitored more effectively through the use of smart contracts. For instance, a widely-known correlation between REPO and market volatility can be modeled on the Maneki protocol to create a micro lending structure that automatically eases FED lending conditions based on liquidity conditions between sectors. While we primarily tackle ineffective government spending in this paper, the customizability and adaptability of the protocol allows for the creation and growth of an infinite number of incentive structures (Fig. 2).

6 Conclusion

Modern monetary policies derived from Keynesian economics amplifies the wage gap separating the ends of the income distribution. The attempts of federal stimulus bills fall short in execution and size, including directing spending towards the most in need. A smart contract system that distributes funds directly to communities and individual payments while incentivizing good spending behavior would alleviate the K-Shape recovery that current monetary policies are creating. By utilizing mechanism and incentive designs, our model directs spending to help stimulate the economies of small communities and struggling businesses. The usage of smart contracts removes the many inefficiencies in current bureaucratic solutions and allows for more transparency, trust, and incentives to help one's community in times of need.

Appendix: Linear Regression Charts

(See Figs. 3, 4, 5, 6, 7, 8 and 9)

Type	Name	Description
Predictor	Interest Rate	Interest rate is one of the key levers used by the Federal Reserve to enact monetary policies to stimulate the economy.
Predictor	Balance Sheet	Balance sheet is the size of asset purchases, used by the Federal Reserve to prop up the economy when enacting Quantitative Easing policies (QE)
Response	Consumer Price Index	Consumer Price Index (CPI) is a measure of inflation in the economy. When the Federal Reserve increases its balance sheet or decreases interest rates, we expect to see an increase in CPI.
Response	Delinquency Rates on Commercial and Industrial Loans	Delinquency rates refer to the percentage of loans in the United States that are late or close to default. This is a measure of the health of businesses in the United States that are behind on their payments or close to insolvency.
Response	Unemployment Rate	The percentage of people who are unemployed in the United States. This is the actual unemployment rate as opposed to full employment, which is the target unemployment rate of the US government, normally around 4-5%.

Fig. 3. Predictor and response variables

Balance Sheet vs CPI	We expect a high positive correlation between the Balance Sheet and CPI. We reject the null hypothesis at Significance F-level of 0.05.
Balance Sheet vs Unemployment	We expect a high negative correlation between the Balance Sheet and Unemployment. We reject the null hypothesis at Significance F-level of 0.05.
Balance Sheet vs Delinquency	We expect a high negative correlation between the balance sheet and Delinquency Rates. We reject the null hypothesis at Significance F-level of 0.05.
Interest Rate vs CPI	We expect a high negative correlation between Interest Rate and CPI. We reject the null hypothesis at Significance F-level of 0.05.
Interest Rate vs Unemployment	We expect a high positive correlation between Interest Rate and Unemployment. We reject the null hypothesis at Significance F-level of 0.05.
Interest Rate vs Delinquency	We expect a high positive correlation between Interest Rate and Delinquency Rates. We reject the null hypothesis at Significance F-level of 0.05.

Fig. 4. Response expectations

Regression Statistics		Coefficients	Standard Error	t Stat	P-value
Multiple R	0.59135009	Intercept 7.04869811	0.13682247	51.5171096	8.411E-135
R Square	0.34969493	FEDFUNDS -0.6179169	0.05340038	-11.571396	4.5865E-25
Adjusted R Square	0.34708326				
Standard Error	1.60390145				
Observations	251				

This regression analysis shows a strong negative relationship between interest rates and unemployment, with an adjusted r-squared value of ~0.35 and a coefficient p-value of ~0.0. This is similar to the balance sheet vs unemployment regression as it indicates that increased interest rates actually lowers unemployment, contrary to Federal Reserve monetary policies. This analysis shows that the negative correlation is statistically significant, and provides a strong argument against Keynesian monetary policies of increased subdued interests rates.

Fig. 5. Interest rate vs unemployment

Regression Statistics		Coefficients	Standard Error	t Stat	P-value
Multiple R	0.91682028	Intercept 190.039543	1.13180502	167.908376	1.408E-242
R Square	0.84055942	WALCL 1.2125E-05	3.4819E-07	34.8215979	1.1398E-93
Adjusted R Square	0.8398662				
Standard Error	8.78536508				
Observations	232				

Our regression shows a very strong correlation between balance sheet and CPI with an adjusted r-squared value of ~.84. The p-value of ~0.0 is extremely low, providing strong evidence that balance sheet movement has a strong positive correlation with inflation at any reasonable confidence level, including a=0.05. This behavior is expected.

Fig. 6. Balance sheet vs CPI

Regression Statistics		Coefficients	Standard Error	t Stat	P-value
Multiple R	0.43537084	Intercept 2.41887675	0.17685809	13.676936	2.1603E-23
R Square	0.18954777	WALCL -2.47E-07	5.4758E-08	-4.5108175	2.0085E-05
Adjusted R Square	0.18023223				
Standard Error	0.84550997				
Observations	89				

This regression shows a semi-strong negative relationship between balance sheet expansion and delinquency rates, with an adjusted r-squared value of ~0.18. This is close to our benchmark for a large correlation of 0.2, and the coefficient's p-value of ~0.0 indicates the relationship to be statistically significant. This observed relationship reflects our initial hypothesis for the regression pair.

Fig. 7. Balance sheet vs delinquencies

Regression Statistics			Coefficients	Standard Error	t Stat	P-value
Multiple R	0.54376937	Intercept	230.114297	1.896123	121.360427	7.405E-223
R Square	0.29568513	FEDFUNDS	-7.5360675	0.73856313	-10.203688	1.2244E-20

Adjusted R Square   0.29284515

Standard Error   22.1500697

Observations   250

This regression shows a strong negative relationship between interest rates and inflation, with an adjusted r-squared value of ~0.3 and a coefficient p-value of ~0.0. This indicates that the regression is statistically significant. The strong negative correlation between the two variables is expected from our initial hypothesis as it is logical that lowered interest rates will lead to lower inflation with lowered velocity in the monetary system.

Fig. 8. Interest rate vs CPI

Regression Statistics			Coefficients	Standard Error	t Stat	P-value
Multiple R	0.0298674	Intercept	1.95242207	0.1342553	14.5426072	1.0308E-26
R Square	0.00089206	FEDFUNDS	-0.0159864	0.0527157	-0.3032563	0.76230622

Adjusted R Square   -0.008808

Standard Error   1.01587351

Observations   105

This regression shows a very slight negative correlation between interest rates and delinquencies, with an adjusted r-squared value of ~-0.0 and a coefficient p-value of ~0.76. The correlation coefficient indicates that there is almost no relationship between the two variables.

Fig. 9. Interest rate vs delinquencies

References

1. Guerrieri, V., Lorenzoni, G., Straub, L., Werning, I.: Macroeconomic Implications of COVID-19: Can Negative Supply Shocks Cause Demand Shortages? National Bureau of Economic Research (2020)
2. Deloitte. https://www2.deloitte.com/us/en/insights/industry/public-sector/fraud-waste-and-abuse-in-entitlement-programs-benefits-fraud.html
3. Association of Certified Fraud Examiners. https://www.acfe.com/press-release.aspx?id=4294973129
4. Semiannual Report to Congress. https://www.oig.dol.gov/public/semiannuals/84.pdf
5. Washington Post. https://www.washingtonpost.com/business/2020/09/10/ppp-fraud-charges/
6. Los Angeles Times. https://www.latimes.com/politics/story/2021-03-02/criminals-stole-billions-in-covid-19-unemployment-benefits-new-relief-bill-wont-stop-a-repeat
7. Gov Tech. https://www.govtech.com/data/Aiming-Analytics-at-Our-35-Billion-Unemployment-Insurance-Problem.html

8. Small Business Administration. https://www.sba.gov/funding-programs/loans/coronavirus-relief-options/paycheck-protection-program
9. Gallup.   https://news.gallup.com/poll/310880/percentage-americans-donating-charity-new-low.aspx
10. Pew   Research   Center.   https://www.pewresearch.org/fact-tank/2019/08/06/young-americans-are-less-trusting-of-other-people-and-key-institutions-than-their-elders

Author Index

Printed in the United States
by Baker & Taylor Publisher Services

Printed in the United States
By Bookmasters